Vacuum in Particle Accelerators

Vacuum in Particle Accelerators

Modelling, Design and Operation of Beam Vacuum Systems

Oleg B. Malyshev

Author

Dr. Oleg B. Malyshev
ASTeC
STFC Daresbury Laboratory
Keckwick Lane
WA4 4AD Daresbury, Cheshire
United Kingdom

Cover Image: Courtesy of Mr. Clive Hill, STFC Daresbury Laboratory

All books published by **Wiley-VCH** are carefully produced. Nevertheless, authors, editors, and publisher do not warrant the information contained in these books, including this book, to be free of errors. Readers are advised to keep in mind that statements, data, illustrations, procedural details or other items may inadvertently be inaccurate.

Library of Congress Card No.:
applied for

British Library Cataloguing-in-Publication Data
A catalogue record for this book is available from the British Library.

Bibliographic information published by the Deutsche Nationalbibliothek The Deutsche Nationalbibliothek lists this publication in the Deutsche Nationalbibliografie; detailed bibliographic data are available on the Internet at <http://dnb.d-nb.de>.

© 2020 Wiley-VCH Verlag GmbH & Co. KGaA, Boschstr. 12, 69469 Weinheim, Germany

All rights reserved (including those of translation into other languages). No part of this book may be reproduced in any form – by photoprinting, microfilm, or any other means – nor transmitted or translated into a machine language without written permission from the publishers. Registered names, trademarks, etc. used in this book, even when not specifically marked as such, are not to be considered unprotected by law.

Print ISBN: 978-3-527-34302-7
ePDF ISBN: 978-3-527-80916-5
ePub ISBN: 978-3-527-80914-1
oBook ISBN: 978-3-527-80913-4

Typesetting   SPi Global, Chennai, India
Printing and Binding   Markono Print Media Pte Ltd, Singapore

Printed on acid-free paper

10 9 8 7 6 5 4 3 2 1

Contents

Acknowledgements *xv*
Nomenclature *xvii*
Introduction *1*
Oleg B. Malyshev
References *3*

1 Vacuum Requirements *5*
Oleg B. Malyshev
1.1 Definition of Vacuum *5*
1.2 Vacuum Specification for Particle Accelerators *6*
1.2.1 Why Particle Accelerators Need Vacuum? *6*
1.2.2 Problems Associated with Beam–Gas Interaction *8*
1.2.2.1 Beam Particle Loss *8*
1.2.2.2 Background Noise in Detectors *8*
1.2.2.3 Residual Gas Ionisation and Related Problems *9*
1.2.2.4 Contamination of Sensitive Surfaces *9*
1.2.2.5 Safety and Radiation Damage of Instruments *10*
1.2.3 Vacuum Specifications *11*
1.2.4 How Vacuum Chamber Affects the Beam Properties *12*
1.3 First Considerations Before Starting Vacuum System Design *13*
1.3.1 What Is the Task? *13*
1.3.2 Beam Lattice *14*
1.3.3 Beam Aperture and Vacuum Chamber Cross Section *15*
1.3.3.1 Required Mechanical Aperture *15*
1.3.3.2 Magnet Design *17*
1.3.3.3 Mechanical Engineering *17*
1.3.3.4 Other Factors Limiting a Maximum Size of Beam Vacuum Chamber *17*
1.3.4 Vacuum Chamber Cross Sections and Preliminary Mechanical Layout *18*
1.3.5 Possible Pumping Layouts *19*
1.4 First and Very Rough Estimations *20*
1.5 First Run of an Accurate Vacuum Modelling *22*
1.6 Towards the Final Design *22*

1.7	Final Remarks 25
	References 25

2	**Synchrotron Radiation in Particle Accelerators** 29
	Olivier Marcouillé
2.1	Emission of a Charged Particle in a Magnetic Field 29
2.1.1	Radiated Energy Density and Power Density 31
2.1.2	Angular Flux 32
2.2	SR from Dipoles 32
2.2.1	Emission Duration and Critical Energy 33
2.2.2	Photon Flux 34
2.2.3	Vertical Angular Distribution of Photon Flux 37
2.2.4	Photon Power 39
2.2.5	Vertical Angular Distribution of Power 41
2.3	SR from Quadrupoles 42
2.4	SR from Insertion Devices 43
2.4.1	Motion of Charged Particles Inside a Planar Insertion Device 44
2.4.2	Resonance Wavelength 45
2.4.3	Radiation from Undulators and Wigglers 46
2.4.4	Angular Aperture of ID at Resonant Wavelength 51
2.4.5	Estimation of Power Distribution Radiated in a Wiggler 52
2.4.6	Estimation of the Power Collected by Simple Geometry Aperture 54
2.4.7	Method for Estimation Absorbed Power on the Complex Shapes 54
2.5	Software Dedicated to Evaluation of the Photon Flux and Power Distribution from the Insertion Devices 55
2.5.1	XOP 56
2.5.2	Synchrotron Radiation Workshop (SRW) 56
2.5.3	SPECTRA 57
2.5.4	SYNRAD 58
2.5.5	OSCARS 59
	Acknowledgements 59
	References 60
	Further Reading 60

3	**Interaction Between SR and Vacuum Chamber Walls** 61
	Vincent Baglin and Oleg B. Malyshev
3.1	Photon Reflectivity 61
3.2	Photoelectron Production 69
3.2.1	Total Photoelectron Yield 69
3.2.2	Effect of the Photon Energy 72
3.2.3	Effect of the Incidence Angle 76
	References 76

4	**Sources of Gas in an Accelerator Vacuum Chamber** 79
	Oleg B. Malyshev and Junichiro Kamiya
4.1	Residual Gases in Vacuum Chamber 79

4.2	Materials Used for and in Vacuum Chambers and Built-In Elements	*81*
4.2.1	Stainless Steel	*82*
4.2.2	Aluminium Alloys	*83*
4.2.3	Copper and Its Alloys	*84*
4.2.4	Titanium and Its Alloys	*85*
4.2.5	Ceramics	*85*
4.2.6	Other Vacuum Materials	*86*
4.3	Thermal Outgassing	*87*
4.3.1	Thermal Outgassing Mechanism During Pumping	*88*
4.3.2	Equilibrium Pressure	*89*
4.3.3	Vapour Pressure	*91*
4.3.4	Thermal Outgassing Rate of Materials	*93*
4.3.5	Outgassing Rate Measurements	*97*
4.3.5.1	Throughput Method	*97*
4.3.5.2	Conductance Modulation Method	*98*
4.3.5.3	Two-Path Method	*98*
4.3.5.4	Gas Accumulation Method	*99*
4.3.6	Thermal Desorption Spectroscopy	*100*
4.4	Surface Treatments to Reduce Outgassing	*102*
4.4.1	Cleaning	*102*
4.4.2	Bakeout	*105*
4.4.3	Air Bake	*106*
4.4.4	Vacuum Firing	*106*
4.4.5	Surface Coatings	*108*
4.4.5.1	Coating the Surface by Thin Films of Material with Low Hydrogen Permeability and Low Outgassing	*108*
4.4.5.2	Coating the Surface by Thin Film of Getter Materials	*108*
4.5	Electron-Stimulated Desorption	*109*
4.5.1	ESD Definition and ESD Facilities	*109*
4.5.2	ESD for Different Materials as a Function of Dose	*112*
4.5.3	ESD as a Function of Amount of Desorbed Gas	*113*
4.5.4	Effect of Pumping Duration	*114*
4.5.5	ESD as a Function of Electron Energy	*119*
4.5.6	Effect of Bakeout on ESD	*122*
4.5.7	Effectiveness of Surface Polishing and Vacuum Firing on ESD	*123*
4.5.8	A Role of Oxide Layer on Copper	*125*
4.5.9	Effect of Surface Treatment	*125*
4.5.10	Effect of Vacuum Chamber Temperature	*125*
4.6	Photon-Stimulated Desorption	*128*
4.6.1	PSD Definition and PSD Facilities	*128*
4.6.2	PSD as a Function of Dose	*131*
4.6.3	PSD for Different Materials	*131*
4.6.4	PSD as a Function of Amount of Desorbed Gas	*135*
4.6.5	PSD as a Function of Critical Energy of SR	*136*
4.6.6	Effect of Bakeout	*137*
4.6.7	Effect of Vacuum Chamber Temperature	*140*

4.6.8	Effect of Incident Angle	*142*
4.6.9	PSD versus ESD	*144*
4.6.10	How to Use the PSD Yield Data	*145*
4.6.10.1	Scaling the Photon Dose	*145*
4.6.10.2	Synchrotron Radiation from Dipole Magnets	*145*
4.6.10.3	PSD Yield and Flux as a Function of Distance from a Dipole Magnet	*148*
4.6.10.4	PSD from a Lump SR Absorber	*151*
4.6.10.5	Combining PSD from Distributed and Lump SR Absorbers	*153*
4.7	Ion-Stimulated Desorption	*155*
4.7.1	ISD Definition and ISD Facilities	*155*
4.7.2	ISD as a Function of Dose	*156*
4.7.3	ISD Yield as a Function of Ion Energy	*158*
4.7.4	ISD Yield as a Function of Ion Mass	*159*
4.7.5	ISD for Different Materials	*160*
4.7.6	Effect of Bakeout and Argon Discharge Cleaning	*161*
4.7.7	ISD versus ESD	*161*
4.7.8	ISD Yield as a Function of Temperature	*161*
4.7.9	ISD Yields for Condensed Gases	*163*
	Acknowledgements	*166*
	References	*166*
5	**Non-evaporable Getter (NEG)-Coated Vacuum Chamber**	*175*
	Oleg B. Malyshev	
5.1	Two Concepts of the Ideal Vacuum Chamber	*175*
5.2	What Is NEG Coating?	*177*
5.3	Deposition Methods	*179*
5.4	NEG Film Characterisation	*181*
5.5	NEG Coating Activation Procedure	*182*
5.6	NEG Coating Pumping Properties	*188*
5.6.1	NEG Coating Pumping Optimisation at CERN	*188*
5.6.2	NEG Coating Pumping Optimisation at ASTeC	*190*
5.7	NEG Coating Lifetime	*193*
5.8	Ultimate Pressure in NEG-Coated Vacuum Chambers	*195*
5.9	NEG-Coated Vacuum Chamber Under SR	*196*
5.10	Reducing PSD/ESD from NEG Coating	*200*
5.10.1	Initial Considerations	*200*
5.10.2	ESD from Vacuum Chamber Coated with Columnar and Dense NEG Films	*201*
5.10.3	Dual Layer	*202*
5.10.4	Vacuum Firing Before NEG Deposition	*204*
5.11	ESD as a Function of Electron Energy	*204*
5.12	PEY and SEY from NEG Coating	*204*
5.13	NEG Coating Surface Resistance	*206*
5.14	NEG at Low Temperature	*207*
5.15	Main NEG Coating Benefits	*207*

5.16	Use of NEG-Coated Vacuum Chambers *208*	
	References *209*	
6	**Vacuum System Modelling** *215*	
	Oleg B. Malyshev	
6.1	A Few Highlights from Vacuum Gas Dynamics *215*	
6.1.1	Gas in a Closed Volume *216*	
6.1.1.1	Gas Density and Pressure *216*	
6.1.1.2	Amount of Gas and Gas Flow *217*	
6.1.2	Total Pressure and Partial Pressure *218*	
6.1.3	Velocity of Gas Molecules *218*	
6.1.4	Gas Flow Rate Regimes *220*	
6.1.5	Pumping Characteristics *221*	
6.1.6	Vacuum System with a Pump *223*	
6.1.7	Vacuum Conductance *223*	
6.1.7.1	Orifice *224*	
6.1.7.2	Vacuum Conductance of Long Tubes *224*	
6.1.7.3	Vacuum Conductance of Short Tubes *225*	
6.1.7.4	Serial and Parallel Connections of Vacuum Tubes *226*	
6.1.8	Effective Pumping Speed *226*	
6.2	One-Dimensional Approach in Modelling Accelerator Vacuum Systems *228*	
6.2.1	A Gas Diffusion Model *229*	
6.2.2	A Section of Accelerator Vacuum Chamber in a Gas Diffusion Model *231*	
6.2.3	Boundary Conditions *232*	
6.2.4	Global and Local Coordinates for Each Element *238*	
6.2.5	Using the Results *240*	
6.2.6	A Few Practical Formulas *241*	
6.2.6.1	Gas Injection into a Tubular Vacuum Chamber *241*	
6.2.6.2	Vacuum Chamber with Known Pumping Speed at the Ends *241*	
6.2.6.3	Vacuum Chamber with Known Pressures at the Ends *244*	
6.3	Three-Dimensional Modelling: Test Particle Monte Carlo *245*	
6.3.1	Introduction *245*	
6.3.2	A Vacuum Chamber in the TPMC Model *246*	
6.3.3	TPMC Code Input *246*	
6.3.4	TPMC Code Output *248*	
6.3.4.1	Gas Flow Rate *248*	
6.3.4.2	Gas Density and Pressure *250*	
6.3.4.3	Transmission Probability and Vacuum Conductance *250*	
6.3.4.4	Pump-Effective Capture Coefficient *251*	
6.3.4.5	Effect of Temperature and Mass of Molecules *251*	
6.3.5	What Can Be Done with TPMC Results? *251*	
6.3.5.1	A Direct Model with a Defined Set of Parameters *252*	
6.3.5.2	Models with Variable Parameters *253*	
6.3.6	TPMC Result Accuracy *256*	

6.4	Combining One-Dimensional and Three-Dimensional Approaches in Optimising the UHV Pumping System 257
6.4.1	Comparison of Two Methods 257
6.4.2	Combining of Two Methods 258
6.5	Molecular Beaming Effect 260
6.6	Concluding Remarks 265
6.A	Differential Pumping 265
6.B	Modelling a Turbo-Molecular Pump 266
	Acknowledgements 267
	References 267

7 Vacuum Chamber at Cryogenic Temperatures *269*
Oleg Malyshev, Vincent Baglin, and Erik Wallén

7.1	Pressure and Gas Density 269
7.2	Equilibrium Pressure: Isotherms 272
7.2.1	Isotherms 273
7.2.2	Cryotrapping 279
7.2.3	Physisorption on Gas Condensates 281
7.2.4	Temperature Dependence of the H_2 Isotherms 282
7.2.5	Choice of Operating Temperature for Cryogenic Vacuum Systems 286
7.3	Gas Dynamics Model of Cryogenic Vacuum Chamber Irradiated by SR 289
7.3.1	Infinitely Long Vacuum Chamber Solution 291
7.3.1.1	Vacuum Chamber Without a Beam Screen 292
7.3.1.2	Vacuum Chamber with Holes in the Beam Screen 292
7.3.2	Short Vacuum Chamber Solution 294
7.3.2.1	Solution for a Short Vacuum Chamber with a Given Pressure at the Ends 296
7.3.2.2	Solution for a Short Vacuum Chamber with a Given Pumping Speed at the Ends 298
7.4	Experimental Data on PSD from Cryogenic Surface 300
7.4.1	Experimental Facility for Studying PSD at Cryogenic Temperatures 301
7.4.2	Discovery of Secondary PSD 301
7.4.3	Calculation of the Desorption Yields from Experimental Data 306
7.4.4	Primary PSD Yields 308
7.4.5	Secondary PSD Yields 310
7.4.6	Photon-Induced Molecular Cracking of Cryosorbed Gas 312
7.4.6.1	Experimental Measurements 312
7.4.6.2	How to Include Cracking into the Model 315
7.4.6.3	Example 316
7.4.7	Temperature of Desorbed Gas 318
7.5	In-Depth Studies with COLDEX 321
7.5.1	COLDEX Experimental Facility 321
7.5.2	PSD of Cu as a Function of Temperature 324
7.5.3	Secondary PSD Yields 325

7.5.4	PSD of a BS with Sawtooth for Lowering Photon Reflectivity and PEY	*326*
7.5.5	Vacuum Transient	*328*
7.5.6	Temperature Oscillations	*329*
7.6	Cryosorbers for the Beam Screen at 4.5 K	*331*
7.6.1	Carbon-Based Adsorbers	*333*
7.6.1.1	Activated Charcoal	*333*
7.6.1.2	Carbon Fibre	*334*
7.6.2	Amorphous Carbon Coating Absorption Properties	*337*
7.6.3	Metal-Based Absorbers	*338*
7.6.3.1	Aluminium-Based Absorbers	*338*
7.6.3.2	Copper-Based Absorbers	*340*
7.6.3.3	LASE for Providing Cryosorbing Surface	*341*
7.6.4	Using Cryosorbers in a Beam Chamber	*341*
7.7	Beam Screen with Distributed Cryosorber	*342*
7.8	Final Remarks	*343*
	References	*344*

8 Beam-Induced Electron Multipacting, Electron Cloud, and Vacuum Design *349*
Vincent Baglin and Oleg B. Malyshev

8.1	BIEM and E-Cloud	*349*
8.1.1	Introduction	*349*
8.1.2	E-Cloud Models	*351*
8.2	Mitigation Techniques and Their Impact on Vacuum Design	*356*
8.2.1	Passive Methods	*357*
8.2.2	Active Methods	*363*
8.2.3	What Techniques Suit the Best	*365*
8.3	Secondary Electron Emission (Laboratory Studies)	*365*
8.3.1	SEY Measurement Method	*365*
8.3.2	SEY as a Function of the Incident Electron Energy	*367*
8.3.3	Effect of Surface Treatments by Bakeout and Photon, Electron, and Ion Bombardment	*367*
8.3.4	Effect of Surface Material	*368*
8.3.5	Effect of Surface Roughness	*369*
8.3.6	'True' Secondary Electrons, Re-Diffused Electrons, and Reflected Electrons	*371*
8.3.7	Effect of Incidence Angle	*374*
8.3.8	Insulating Materials	*374*
8.4	How the BIEM and E-Cloud Affect Vacuum	*376*
8.4.1	Estimation of Electron Energy and Incident Electron Flux	*376*
8.4.2	Estimation of Initial ESD	*378*
8.5	BIEM and E-Cloud Observation in Machines	*379*
8.5.1	Measurements in Machines	*379*
8.5.1.1	Vacuum Pressure	*381*
8.5.1.2	Vacuum Chamber Wall Properties	*382*
8.5.1.3	Specific Tools for BIEM and Electron Cloud Observation	*386*

8.5.2	Machines Operating at Cryogenic Temperature *390*
8.5.2.1	Surface Properties at Cryogenic Temperature *391*
8.5.2.2	Observations with Beams *394*
8.5.2.3	The CERN Large Hadron Collider Cryogenic Vacuum System *401*
8.6	Contribution of BIEM to Vacuum Stability *405*
8.7	Past, Present, and Future Machines *407*
	Acknowledgements *409*
	References *409*

9   Ion-Induced Pressure Instability *421*
Oleg B. Malyshev and Adriana Rossi

9.1	Introduction *421*
9.2	Theoretical *422*
9.2.1	Basic Equations *422*
9.2.2	Solutions for an Infinitely Long Vacuum Chamber *425*
9.2.2.1	Room Temperature Vacuum Chamber *425*
9.2.2.2	Cryogenic Vacuum Chamber *426*
9.2.2.3	Summary for an Infinitely Long Vacuum Chamber *427*
9.2.3	Short Vacuum Chamber *428*
9.2.3.1	Solution for a Short Vacuum Chamber with a Given Gas Density at the Ends *428*
9.2.3.2	Solution for a Short Vacuum Chamber with a Given Pumping Speed at the Ends *431*
9.2.3.3	Solution for a Short Vacuum Chamber Without a Beam Screen Between Two Chambers With a Beam Screen *434*
9.2.3.4	Some Remarks to Solutions for Short Tubes *437*
9.2.4	Multi-Gas System *437*
9.2.5	Two-Gas System *438*
9.2.5.1	Solutions for an Infinitely Long Vacuum Chamber *439*
9.2.5.2	Solution for a Short Vacuum Chamber in the Equilibrium State *439*
9.2.6	Some Comments to the Analytical Solutions *440*
9.2.7	Effect of the Ion-Stimulated Desorption on the Gas Density *441*
9.2.7.1	Infinitely Long Vacuum Chamber (One Gas) *441*
9.2.7.2	Vacuum Chamber with a Given Pumping Speed at the Ends (One Gas) *441*
9.2.7.3	Two-Gas System *443*
9.2.8	Some Numeric Examples from the LHC Design *443*
9.2.8.1	The Critical Current for an Infinitely Long Vacuum Chamber *444*
9.2.8.2	Short Vacuum Chambers *445*
9.2.8.3	Effect of the Ion-Stimulated Desorption on the Gas Density *445*
9.3	VASCO as Multi-Gas Code for Studying the Ion-Induced Pressure Instability *447*
9.3.1	Basic Equations and Assumptions *447*
9.3.2	Multi-Gas Model in Matrix Form and Fragmentation in Several Vacuum Chamber Elements *448*
9.3.2.1	Boundary Conditions *449*
9.3.3	Transformation of the Second-Order Differential Linear Equation into a System of First-Order Equations *450*

9.3.3.1	Boundary Conditions	*451*
9.3.4	Set of Equations to be Solved	*451*
9.3.5	'Single Gas Model' Against 'Multi-Gas Model'	*452*
9.4	Energy of Ions Hitting Vacuum Chamber	*455*
9.4.1	Ion Energy in the Vacuum Chamber Without a Magnetic Field	*455*
9.4.1.1	Circular Beams	*455*
9.4.1.2	Flat Beams	*458*
9.4.2	Ion Energy in a Vacuum Chamber with a Magnetic Field	*460*
9.4.2.1	Vacuum Chamber in a Dipole Magnetic Field	*461*
9.4.2.2	Vacuum Chamber in a Quadrupole Magnetic Field	*461*
9.4.2.3	Vacuum Chamber in a Solenoid Magnetic Field	*462*
9.5	Errors in Estimating the Critical Currents I_c	*464*
9.5.1	Beam–Gas Ionisation	*465*
9.5.2	Ion Impact Energy	*465*
9.5.3	Ion-Stimulated Desorption Yields	*465*
9.5.4	Pumping	*466*
9.5.5	Total Error in Critical Current	*466*
9.6	Summary	*467*
	References	*467*

10 Pressure Instabilities in Heavy Ion Accelerators *471*
Markus Bender

10.1	Introduction	*471*
10.2	Pressure Instabilities	*472*
10.2.1	Model Calculations of the Dynamic Pressure and Beam Lifetime	*476*
10.2.1.1	Closed System (Vessel)	*476*
10.2.1.2	Vessel Including Collimation	*478*
10.2.1.3	Longitudinal Profile	*478*
10.2.2	Consequences	*479*
10.3	Investigations on Heavy Ion-Induced Desorption	*480*
10.3.1	Desorption Yield Measurements	*481*
10.3.2	Materials Analysis	*483*
10.3.3	Dedicated Set-ups to Measure Ion-Induced Desorption Yields	*485*
10.3.4	Results	*489*
10.3.4.1	Materials	*490*
10.3.4.2	Surface Coatings	*493*
10.3.4.3	Cleaning Methods	*494*
10.3.4.4	Energy Loss Scaling	*495*
10.3.4.5	Angle Dependence	*496*
10.3.4.6	Conditioning	*497*
10.3.4.7	Cryogenic Targets	*498*
10.3.5	Theoretic	*499*
10.3.5.1	Interaction of Ions with Matter	*499*
10.3.5.2	Inelastic Thermal Spike Model	*501*
10.4	Conclusion: Mitigation of Dynamic Vacuum Instabilities	*505*
	Acknowledgement	*507*
	References	*507*

Index *515*

Acknowledgements

The editor hereby acknowledges with deepest appreciation and thanks the support of many people who helped at different stages of writing and editing this book from its original idea to final polishing.

First of all, I would like to thank Dr. Vincent Baglin (CERN, Switzerland) for providing me the greatest support in discussing a structure and a content of the book, for helping to build a team of co-authors, and for being my co-author in three chapters.

Great thanks to my co-authors who worked hard aiming to write a book that would be useful to particle accelerator vacuum community and keeping consistency from chapter to chapter. Dr. Olivier Marcouillé (SOLEIL, France) worked on describing synchrotron radiation in terms required for a vacuum system designer (this information is usually hard to find in other books), Dr. Junichiro Kamiya (J-PARC, Japan) made a detailed overview of thermal outgassing of materials used in particle accelerators and methods of measurements, Dr. Erik Wallén (LBNL, USA) kindly agreed to review theoretical cryosorption models and results of cryosorption experiments, Dr. Adriana Rossi (CERN, Switzerland) worked with me on the ion-induced pressure instability analysis, and Dr. Markus Bender (GSI, Germany) summarised a present state of problem solving for heavy ion-induced pressure instability.

I would also like to thank many colleagues from different research centres around the world for useful suggestions, corrections, reviewing different parts of the book, and following feedback. I would also acknowledge my colleagues from the STFC Daresbury Laboratory for providing necessary information, images, and graphs, for useful suggestions, and for support. My Great thanks are to Mr. Clive Hill for the cross-sectional view of EMMA accelerator shown on a cover of this book.

I would also wish to give a special thanks to my wife Larisa and my children Dmitry and Daria for the daily support, interest to a progress, reading and making corrections, and great patience when I was spending my free time with a computer instead of a family.

And finally, many thanks to Wiley teams for the great help with the book production: to Dr. Martin Preuss for setting up the process of publishing and for answering my numerous questions, to Ms Shirly Samuel for assistance with the submission process, to the production team lead by Mr. Ramprasad Jayakumar for careful reading and corrections and for being responsive to author's concerns and suggestions.

Dr. Oleg B. Malyshev
Editor

Nomenclature

A [m^2]	Vacuum chamber cross section area or volume per unit of axial length
$A(r, t)$ [W/m^3]	Source term of energy input on the electrons
a [m]	Vacuum chamber or channel height or width
a	A constant in equations
B [T]	Magnetic field
$B(r, t)$ [W/m^3]	Source term of energy input on the lattice
b [m]	Vacuum chamber or channel width or height
C [m^2/s]	Distributed pumping speed of pumping holes or slots per unit axial length
$C_{e/a}$ [J/(kg·K)]	Specific heat of the electronic/lattice system
c [m^2/s]	Distributed pumping speed per unit axial length
D [m^2/s]	Knudsen diffusion coefficient
D	Accumulated dose of particle bombarding a surface
– D or D_γ	Photon dose
○ D or D_γ [photons]	Total photon dose
○ D or D_γ or D_L [photons/m]	Photon dose per unit of axial length
○ D or D_γ or D_A [photons/m^2]	Photon dose per unit of area
– D or D_e	Electron dose
○ D or D_e [electrons]	Total electron dose
○ D or D_e or D_L [electrons/m]	Electron dose per unit of axial length
○ D or D_e or D_A [electrons/m^2]	Electron dose per unit of area
– D or D_i	Ion dose
○ D or D_i [ions]	Total ion dose
○ D or D_i or D_L [ions/m]	Ion dose per unit of axial length
○ D or D_i or D_A [ion/m^2]	Ion dose per unit of area

d [m]	Tube or orifice diameter
E	Energy of charged particles
– E [MeV, GeV, TeV]	Energy of particles in the beam
– E_0	Rest energy, e.g. $E_0 = 0.511$ MeV for electron and $E_0 = 938.27$ MeV for proton
– E_e or E [eV, keV]	Energy of test electron in ESD and SEY measurements
– E_i or E [eV, keV]	Energy of test ion in ISD
– E_{des} [eV]	Desorption energy
$\mathcal{E}$ [V/m]	Electric field
F [m]	Vacuum chamber cross section circumference or surface area per unit axial length
f	Fraction of beam ions ($0 < f < 1$)
g [W/(m³·K)]	Electron–phonon coupling
$H_{(index)}$ [ions/s] or [ions/(s·m)]	Ion flux
I [A]	Charged particle beam current
– I_e [mA]	(Photo)electron current
– I_i [mA]	Ion current
I [J]	Mean ionization potential
J [molecules/(s·m²)]	An impingement rate
$K_{e/a}$ [W/(m·K)]	Thermal conductivity of the electronic/lattice system
Kn	Knudsen number
K_q	Charge state of ions
L [m]	Length of vacuum chamber
M [kg/mol] or [amu]	Molecular molar mass
Mh_i	A number of hits on facet i' in TPMC model
Mp_i	A number of particles pumped by facet i in TPMC model
m [kg]	(molecular) mass
N [molecules]	A number of molecules in a volume
N	A number of generated molecules in TPMC model
n [molecules/m³]	Number density of gas
n_e [molecules/m³]	Thermal equilibrium gas density (in Chapters 7 and 9)
n_e [electrons/m³]	Electron density (in Chapters 8 and 10)
P [Pa]	Pressure
$\mathbf{P}$ [W/m]	Power dissipation per unit axial length
R [m]	Bending radius of dipole magnet
R or ρ	Photon reflectance (reflectivity coefficient)
R_z [μm]	Mean surface roughness
r [m]	Radius

Q [molecules/s] or Q^* [Pa·m³/s]	Local gas flux
q [molecules/(s·m²)] or q^* [Pa·m/s]	Specific outgassing rate
q [molecules/(s·m)] or q^* [Pa·m²/s]	Gas desorption flux per unit axial length
S [m²/s]	Distributed pumping speed per unit axial length
S_{eff} [m³/s]	Effective pumping speed
$S_{\text{id}} = A\bar{v}/4$ [m³/s]	Ideal pumping speed
S_p [m³/s]	Pumping speed of a lumped pump
$S = FL\bar{v}/4$ [m³/s]	Ideal wall pumping speed of accelerator vacuum chamber of length L
S_A [m/s]	Specific pumping speed (pumping speed per unit of surface area)
s [molecules/m²]	Surface molecular density, a number of adsorbed molecules
s_0 [molecules/m²]	A number of adsorption sites
T [K]	Temperature of gas or walls of vacuum chamber
t [s]	Time
$U = u/L$ [m³/s]	The vacuum chamber conductance
$u = AD$ [m⁴/s]	Specific vacuum chamber conductance per unit axial length
V [m³]	Vacuum chamber volume
v [m/s]	Bulk velocity
$\bar{v}$ [m/s]	Average molecular velocity
v_{rms} [m/s]	Root-mean-square molecular velocity
$\mathbf{W}$	Transmission probability matrix
w	Transmission probability
x and y [m]	Transversal coordinate
Z	Atomic number
Z_{eff}	Effective charge of projectile ion, screened by electrons
z [m]	Longitudinal coordinate along the beam vacuum chamber

α	Sticking probability of molecules on vacuum chamber walls
α	Exponent in Eqs. (4.29), (4.34), and (4.35) for $\eta(D)$
β	Capture coefficient
Γ	Photon flux
– Γ [photons/s]	Total photon flux
– Γ or Γ_L [photon/(s·m)]	Linear photon flux (photon flux per unit of axial length)
– Γ or Γ_A [photon/(s·m²)]	Photon flux per unit surface area

– Γ_{mrad} [photon/(s·mrad)]	Photon flux from the beam in dipole magnetic field into 1 mrad bend
γ	The Lorentz factor: $\gamma = E/E_0$
δ	Secondary electron yield
ε	Photon energy
ε_c	Critical energy of SR
η or η_e or ξ [molecules/electron]	ESD yield
η or η_γ [molecules/photon]	PSD yield
η_t [molecules/(s·m²)] or [Pa·m]	Specific thermal outgassing rate
η' or η_e' or ξ' [molecules/electron]	ESD yield from cryosorbed gas (secondary ESD)
η' or η_γ' [molecules/photon]	PSD yield from cryosorbed gas (secondary PSD)
Θ	Electron flux (surface bombardment intensity)
– Θ [electron/s]	Total electron flux
– Θ or Θ_L [electron/(s·m)]	Electron flux per unit axial length
– Θ or Θ_A [electron/(s·m²)]	Electron flux per unit surface area
Θ [mrad or °]	Incidence angle of bombarding particles
$\theta = s/s_0$	Normalised surface coverage
ν_0 [s⁻¹]	Oscillation frequency of bound atom/molecule
ρ	A pump capture efficiency (or a capture coefficient), pump mesh or beam screen transparency
$\rho(x, y)$ [C/m³]	Beam charge density
τ [s]	Beam lifetime, an average residence time of sorbed molecule on a surface
σ [m²]	An ionisation cross section of the residual gas molecules by beam particles, an interaction cross section (in Chapter 1)
σ_x and σ_y [m]	Transverse r.m.s. beam sizes
χ [molecules/ion]	ISD yield
χ' [molecules/ion]	ISD yield from cryosorbed gas (secondary ISD)

Physical Constants

c	Speed of light in vacuum	$c = 299\,792\,458$ m/s
k_B	Boltzmann constant	$k_B = 1.380\,650\,4(24) \times 10^{-23}$ J/K
		$= 1.380\,650\,4(24) \times 10^{-23}$ Pa·m³/K
h	Plank's constant	$h = 6.626\,069\,57 \times 10^{-34}$ m²·kg/s
q_e	Elementary charge	$q_e = 1.602\,176\,46 \times 10^{-19}$ C
N_A	Avogadro constant	$N_A = 6.022\,140\,76 \times 10^{23}$ mol⁻¹
R	Ideal gas (Regnault) constant	$R = 8.314\,459\,8(48)$ J/(mol·K) or Pa·m³/(mol·K) or kg·m²/(mol·K·s²)

List of Abbreviations

AC	angular coefficient method
ESD	electron-stimulated desorption
ISD	ion-stimulated desorption
NEG	non-evaporable getter
PEE	photoelectron emission
PEY	photoelectron yield
PSD	photon-stimulated desorption
RGA	residual gas analyser
SEE	secondary electron emission
SEY	secondary electron yield
SIP	sputter ion pump
SR	synchrotron radiation
TD	thermal desorption
TPMC	test particle Monte Carlo method
TMP	turbo-molecular pump
TSP	titanium sublimation pump
UHV	ultra-high vacuum
XHV	extreme high vacuum

Frequently Used Vacuum Units and Their Conversion

Vacuum Units

	Pa	mbar	Torr	bar	Atmosphere at sea level
Pa	1	10^{-2}	$7.500\,62 \times 10^{-3}$	10^{-5}	$9.869\,2 \times 10^{-6}$
mbar	100	1	0.750 062	10^{-3}	$9.869\,2 \times 10^{-4}$
Torr	133.322	1.333 22	1	$1.333\,22 \times 10^{-3}$	$1.315\,8 \times 10^{-3}$
bar	10^5	10^3	750.062	1	0.986 92
atm	$1.013\,25 \times 10^5$	$1.013\,25 \times 10^3$	760	1.013 25	1

Conversion of Frequently Used Units

Amount of gas	PV	$N = \dfrac{PV}{k_B T}$	$n_{\text{mol}} = \dfrac{PV}{RT}$	$m = M\dfrac{PV}{RT}$
Units	$\text{Pa·m}^3 = 10\,\text{mbar·l}$	molecules	mol	kg
Gas flow	$\dfrac{d(PV)}{dt}$	$\dfrac{dN}{dt}$	$\dfrac{dn_{\text{mol}}}{dt}$	$\dfrac{dm}{dt}$
Units	$\text{Pa·m}^3/\text{s} = 10\,\text{mbar·l/s}$	molecules/s	mol/s	kg/s
Specific outgassing rate	$\dfrac{1}{A}\dfrac{d(PV)}{dt}$	$\dfrac{1}{A}\dfrac{dN}{dt}$	$\dfrac{1}{A}\dfrac{dn_{\text{mol}}}{dt}$	$\dfrac{1}{A}\dfrac{dm}{dt}$
Units	$\text{Pa m/s} = 10^5\,\text{mbar·l}/(\text{s·cm}^2)$	molecules/(s·cm^2)	mol/(s·cm^2)	kg/(s·m^2)

Monolayer (ML)

A monolayer (ML) is a one-molecule thick layer of closely packed molecules of gas on a geometrically flat surface.

In practical estimations for the gases present on rough surface of accelerator vacuum chamber, an approximate value of $1\,\text{ML} \approx 10^{15}$ molecules/cm^2 can be used.

Introduction

Oleg B. Malyshev

ASTeC, STFC Daresbury Laboratory, Keckwick Lane, Daresbury, Warrington WA4 4AD Cheshire, UK

A large number of good books related to vacuum science and technology have already been written. Thus, the International Union for Vacuum Science, Technique and Applications (IUVSTA) and American Vacuum Society (AVS) have published on their websites a list of 'Textbooks on vacuum science and technology published, 1922–2003', prepared by Kendall B.R. [1] which a list of textbooks on vacuum science and technology published in 1922–2003, prepared by Kendall B.R., which has 136 book titles, including [2–6]. A few more books were published in recent years to represent a modern level of knowledge in the rarefied gas dynamics and modelling, design of vacuum system and vacuum technology, and vacuum instrumentation and materials [7–10]. However, these books do not cover a number of specific problems related to vacuum systems of charged particle accelerators and other large vacuum systems. The lack of this specialist education materials was covered by CERN Accelerator Schools in 1999, 2007, and 2017 (published in their proceedings [11–13]) and in vacuum-related articles in the *Handbook of Accelerator Physics and Engineering* [14], related to a number of different aspects of vacuum science, technology, and engineering for particle accelerators. The proceedings of two workshops on vacuum design of synchrotron radiation (SR) sources were also published by AIP [15, 16]. However, there are a very small number of publications related to accelerator vacuum chamber modelling and optimisation, including selecting and manipulating the input data to the model [17–19], although there were a few presentations at conferences, workshops, schools, and short courses on this subject.

This book aims to help vacuum scientists and engineers in the gas dynamics modelling of accelerator vacuum systems. It brings together the main considerations, which have to be discussed and investigated during modelling and optimisation in a design of particle accelerator vacuum system, as well as to give some analytical solutions that could be useful in vacuum system design optimisation. This includes, first of all, an analysis of experimental data that should be used as inputs to analytical models; secondly, an understanding of what physical and chemical processes are happening in the vacuum chamber with and without a

Vacuum in Particle Accelerators: Modelling, Design and Operation of Beam Vacuum Systems,
First Edition. Oleg B. Malyshev.
© 2020 Wiley-VCH Verlag GmbH & Co. KGaA. Published 2020 by Wiley-VCH Verlag GmbH & Co. KGaA.

beam; and thirdly, choosing and applying a model (or available software) and interpreting the results. It is expected that readers have theoretical knowledge and practical experience in vacuum science and technology, thermodynamics, gas dynamics and some basic knowledge in particle accelerators.

The structure of the book corresponds to a workflow in the design of accelerator vacuum chamber:

(1) *Chapter 1* describes *first considerations* at the beginning of work on a new machine such as what type of machine and what vacuum specifications, rough vacuum estimations, etc.
(2) *Chapters 2–5* provide an *input data* for gas dynamics models:
    – Synchrotron radiation(SR) is one of the main characteristics required in modelling of vacuum systems of many particle accelerators. Chapter 2 describes photon flux, critical energy, power, and angular distribution from dipoles, quadrupoles, wigglers, and undulators. The authors were writing the formulas in the format that could be useful for the vacuum designers.
    – Chapter 3 is focused on two important effects in the interaction between SR and vacuum chamber walls: photon reflectivity and photoelectron production. These two effects play a significant role in the photon-stimulated desorption processes in room temperature and cryogenic beam chambers, and the beam-induced electron multipacting and should also be considered in the ion induced pressure instability.
    – Chapter 4 describes the main materials used in accelerator vacuum chambers, their cleaning procedure, thermal outgassing, and electron-, photon-, and ion-stimulated desorption.
    – Chapter 5 is devoted to a very special vacuum technology – non-evaporable getter coating.
(3) *Chapters 6–10* describe the *gas dynamics models*:
    – Chapter 6 describes vacuum system modelling using two main approached: a one-dimensional diffusion model and a three-dimensional test particle Monte Carlo method. We recommend reading this chapter before the following Chapters 7–10.
    – Chapter 7 describes specific problems of particle accelerators at cryogenic temperature.
    – Chapter 8 demonstrates how vacuum chamber design of positively charged machines can be affected by mitigation of beam-induced electron multipacting and e-cloud.
    – Chapter 9 describes the ion-induced pressure instability, another potential problem of positively charged machines, gas dynamics model, a number of analytical solutions, and stability criteria.
    – Chapter 10 is fully devoted to the heavy ion machine vacuum problems and solutions. We recommend reading Chapters 6–9 before this chapter.

The authors believe that vacuum scientists and engineers, postdocs and PhD students will find the book very helpful in their work related to the gas dynamics modelling and vacuum design of charged particle accelerator vacuum systems. The authors would be happy to receive a feedback or comments to any part of this book. This includes questions related to clarity, consistency, typos, missing points, wish-to-see, etc.

References

1 Kendall, B.R. (2003). Textbooks on vacuum science and technology. IUVSTA, http://iuvsta.org/iuvsta-publications/#textbooks (accessed 5 June 2019), American Vacuum Society: http://www2.avs.org/historybook/links/vactextbook.htm (accessed 5 June 2019).
2 Lafferty, J.M. (ed.) (1998). *Foundations of Vacuum Science and Technology*. New York, NY: Wiley.
3 Hoffman D.M., Singh B., and Thomas J.H. III. (eds). Handbook of Vacuum Science and Technology. *Academic Press*, 1998.
4 Dushman, S. and Lafferty, J.M. (1962). *Scientific Foundations of Vacuum Technique*, 2e. New York, NY: Wiley.
5 Berman, A. (1992). *Vacuum Engineering Calculations, Formulas and Solved Exercises*. New York, NY: Academic Press.
6 Saksaganskii, G.L. (1988). *Molecular Flow in Complex Vacuum Systems*. New York, NY: Gordon and Breach.
7 Jousten, K. (ed.) (2008). Handbook of Vacuum Technology. In: *Wiley*. ISBN: 3527407235.
8 O'Hanlon, J.F. (2003). *A User's Guide to Vacuum Technology*. Wiley. ISBN: 9780471270522.
9 Yoshimura, N. (2007). *Vacuum Technology: Practice for Scientific Instruments*. Springer Science & Business Media. ISBN: 9783540744320.
10 Sharipov, F. (2016). *Rarefied Gas Dynamics. Fundamentals for Research and Practice*. Wiley-VCH.
11 Turner, S. (ed.) (1999). *CERN Accelerator School: Vacuum Technology*, 28 May to 3 June 1999, Snekersten, Denmark, CERN 99-05, CERN, Geneva, 19 August 1999.
12 Brandt, D. (ed.) (2007) *CERN Accelerator School: Vacuum in Accelerators*, 16–24 May 2007, Platja d'Aro, Spain, CERN-2007-003, CERN, Geneva, 11 June 2007.
13 Strasser, B., Raynova, I. (2017). *CERN Accelerator School: Vacuum for Particle Accelerators*, 6–16 June 2017, Lund, Sweden.
14 Chao, A.W. and Tigner, M. (eds.) (2002). *Handbook of Accelerator Physics and Engineering*. Singapore: World Scientific publishing Co. Pte. Ltd.
15 Halama, H.J., Schuchmann, J.C., and Stefan, P.M. (ed.) (1988). Vacuum design of advanced and compact synchrotron light sources. *AIP Conference Proceedings 171*, AVS Series 5, Upton, NY.
16 Krauss, A.R., Amer, Y.G., and Bader, S.D. (ed.) (1998). Vacuum design of synchrotron light sources. *AIP Conference Proceedings 236*, AVS Series 12, Upton, NY.
17 Gröbner, O. (1991). General considerations in the design of accelerator vacuum systems. *J. Vac. Sci. Technol., A* 9: 2074–2080.

18 Mathewson, A.G. (1993). Vacuum problems in particle accelerators due to interactions of synchrotron radiation, electrons and ions with surfaces. *Vacuum* 44: 479–483.

19 Malyshev, O.B. (2012). Gas dynamics modelling for particle accelerators. *Vacuum* 86: 1669–1681.

1

Vacuum Requirements
Oleg B. Malyshev

ASTeC, STFC Daresbury Laboratory, Keckwick lane, Daresbury, Warrington WA4 4AD Cheshire, UK

1.1 Definition of Vacuum

The content of this book is fully related to vacuum, so it is reasonable to begin with its definition. It appears that the 'common sense' definition is very different from the scientific one. For example, Oxford Dictionaries [1] defines vacuum as 'a space entirely devoid of matter'. A space or container from which the air has been completely or partly removed, while Cambridge Dictionaries Online [2] gives a more accurate definition: 'a space from which most or all of the matter has been removed, or where there is little or no matter'. However, the scientific community refers to the ISO standards, ISO 3529-1:1981 [3], where the definition of vacuum is given as follows:

> "1.1.1
>
> **vacuum**
> A commonly used term to describe the state of a rarefied gas or the environment corresponding to such a state, associated with a pressure or a mass density below the prevailing atmospheric level."

In other words, in rarefied gas dynamics, a gas is in vacuum conditions as soon as its pressure per standard reference conditions is below 100 kPa. In practice, vacuum conditions apply when a vacuum pump connected to a closed vacuum vessel is switched on.

Theoretically, there is no limit for rarefication. However, in practice, there is a limit of what can be achieved and what can be measured. Nowadays, some modern vacuum systems may cover up to 15–16 orders of magnitude of gas rarefication, whereas the total pressure measurements are technologically limited to $\sim 10^{-11}$ Pa.

For convenience, 'to distinguish between various ranges or degrees of vacuum according to certain pressure intervals', ISO 3529-1:1981 also defines the ranges of vacuum:

Vacuum in Particle Accelerators: Modelling, Design and Operation of Beam Vacuum Systems,
First Edition. Oleg B. Malyshev.
© 2020 Wiley-VCH Verlag GmbH & Co. KGaA. Published 2020 by Wiley-VCH Verlag GmbH & Co. KGaA.

Low (rough) vacuum:	100 kPa to 100 Pa
Medium vacuum:	100 to 0.1 Pa
High vacuum (HV):	0.1 Pa to 10 µPa
Ultra-high vacuum (UHV):	below 10 µPa

A vacuum system designer should be aware that regardless the definition of the vacuum ranges given by ISO 3529-1:1981, a few alternative ranges with different boundaries and two more ranges (very high vacuum [VHV] and extremely high vacuum [XHV]) are used in vacuum community, for example, when each range covers exactly 3 orders of magnitude:

Low (rough) vacuum:	10^5 to 10^2 Pa
Medium vacuum:	10^2 to 10^{-1} Pa
High vacuum (HV):	10^{-1} to 10^{-4} Pa
Very high vacuum (VHV):	10^{-4} to 10^{-7} Pa
Ultra high vacuum (UHV):	10^{-7} to 10^{-10} Pa
Extremely high vacuum (XHV):	below 10^{-10} Pa

1.2 Vacuum Specification for Particle Accelerators

1.2.1 Why Particle Accelerators Need Vacuum?

All particle accelerators are built to meet certain user's specifications (e.g. certain luminosity in colliders; defined photon beam parameters in synchrotron radiation (SR) sources; specified ion or electron beam intensity, timing and a spot size on a target; etc.). The user's specifications are then translated to the specification to the charged particle beam parameters, which, in their turn, are translated the specifications to all accelerator systems where the specifications to vacuum system are one of the most important for all types of particle accelerators.

Ideally, charged particles should be generated, accelerated, transported, and manipulated without any residual gas molecules. However, residual gas molecules are always present in a real vacuum chamber. The energetic charged particles can interact with gas molecules and these interactions cause many unwanted effects such as loss of the accelerated particle, change of a charge state, residual gas ionisation, and many others [4, 5].

In practice, vacuum specifications for particle accelerators or other large vacuum system are set to minimise these effects of beam–gas interaction *to a tolerable level* when their impact on beam parameters is much lower than one from other physical phenomena. Thus, the particle accelerator vacuum system should provide the required (or specified) vacuum in the presence of the charged particle beam.

Not only the residual gas affects the beam, but the beam can also cause an increase of gas density by a beam-induced gas desorption in its vacuum chamber.

1.2 Vacuum Specification for Particle Accelerators

There are a number of such effects such as photon-, electron-, and ion-stimulated desorption, inductive heat, radiation damage of vacuum chamber material, etc.

There are a number of different types of charged particle accelerators with various specifications to vacuum. Common in all these specifications is that the unwanted effect due to the presence of residual gas in a vacuum system should be negligible. Generally speaking, the particle accelerators are designed to generate, accelerate, form, and transport the charged particle beams with some required beam characteristics [6, 7] to an area of application such as an interaction point in colliders [8–10] and solid, liquid, or gaseous targets [11, 12], or to the device(s) where the beam used for generating photons in SR sources [13, 14] (in dipoles, wigglers undulators or free electron lasers [FELs]), etc. In all these cases the loss rate of charged particles due to unwanted *beam–gas interactions* should be below a tolerable level, defined by a process or a phenomenon for which the beam is generated. The beam–gas interactions of different natures are well described in literature (for example: see Ref. [7], p. 155 in Ref. [15]) and are summarised in Table 1.1.

The interactions of high energy particles with gas atoms, molecules, or any other type of a target (other particles in gaseous, liquid, or solid state) are determined in terms of an interaction cross section, σ, a probability the beam particles to interact with the atoms of target. When a charged particle beam of intensity, I, crosses a target of thickness dx with a density of atoms n, the change in beam current is

$$dI = -I\sigma n\, dx. \tag{1.1}$$

The charged particle beam moves with a velocity, v, passing through the thickness dx with time dt: $dx = v\, dt$. Thus, Eq. (1.1) can be rewritten as follows:

$$\frac{dI}{dt} = -I\sigma n v. \tag{1.2}$$

The cross section is a constant having the dimension of an area, i.e. m^2 in SI. However, a widely practical unit is also a barn: 1 barn = 10^{-28} m^2. The interaction

Table 1.1 The beam–gas interactions.

	Beam–gas interactions	Type of affected beam particles
Inelastic	Bremsstrahlung	e^+, e^-
	Ionisation energy loss	All particles
	Electron capture	Low energy A^+, A^{Z+}
	Electron loss	A^+, A^-, A^{Z+}
	Nuclear reactions	All particles
Elastic	Single Coulomb scattering	All particles
	Multiple Coulomb scattering	$A^{Z+}, \bar{p}$
	Gas ionisation	
Space charge	Ion cloud space charge	Negatively charged beams
	Electron cloud space charge	Positively charged beams

cross sections depend on the nature and energy of colliding particles. These cross sections can be found in specialised literature, for example, in the booklets, provided by the Particle Data Group (PDG) [16], in Refs. [17, 18] and on p. 213 in Ref. [7].

1.2.2 Problems Associated with Beam–Gas Interaction

The potential problems for particle accelerators associated with beam–gas interaction were shortly described in the following.

1.2.2.1 Beam Particle Loss

In the case of storage rings, the beam current, I, decays with time t as

$$I = I_0 \exp\left(-\frac{t}{\tau}\right), \tag{1.3}$$

where τ is the total beam lifetime. There are numerous effects that define the intrinsic beam lifetime τ_{beam} such as quantum effect, Touschek effect, and particle lifetime, etc., and a beam–gas interaction lifetime τ_{gas} is defined as

$$\frac{1}{\tau_{gas}} = v \sum_i \sigma_i n_i, \tag{1.4}$$

where n is the residual gas density for a gas species i, σ is the beam–gas interaction cross section, and v [m/s] is a velocity of beam-charged particles [19, 20].

Then the total beam lifetime is defined as

$$\frac{1}{\tau} = \frac{1}{\tau_{beam}} + \frac{1}{\tau_{gas}}. \tag{1.5}$$

Shorter lifetime requires more often interruption of the user's operation of particle accelerator to top up the beam; therefore the longer the total beam lifetime, the better.

Thus, the criteria for a **good vacuum** in the storage rings can be defined as

$$\tau_{gas} > \tau_{beam}. \tag{1.6}$$

In linacs, the beam lifetime is not an issue, so the criterion for a 'good vacuum' would be a tolerable beam loss rate due to a beam–gas interaction.

1.2.2.2 Background Noise in Detectors

The beam–gas interaction debris and Bremsstrahlung radiation may increase background noise in a detector at interaction points in colliders and in other sensitive instruments in a machine. In this case the criterion for a 'good vacuum' is a tolerable noise in detectors or instruments due to the beam–gas interactions in the interaction region. One should consider that the source of the debris or radiation could be quite far away from a detector or an instrument or within a line of sight for radiation and upstream/downstream of nearest dipoles for charged debris/particles.

1.2.2.3 Residual Gas Ionisation and Related Problems

The beam–gas interaction causes not only the beam losses but also gas molecule ionisation. Therefore, gas species ions and electrons with low energies are generated along the beam pass. These ions can create an ion cloud with a *space charge* that affects the negatively charged beam quality such as an emittance grow, a tune shifts, tune spreads, coherent collective multi-bunch instabilities, and a reduced beam lifetime due to increased local pressure (see pp. 129 and 165 in [15]). These effect are called the *fast ion instability* and the *ion trapping instability*.

In the case of positively charged beams, the electrons generated from the beam–gas interaction, together with photoelectrons and secondary electrons, are added in the *electron cloud* (see p. 133 in [15] and Chapter 8), also causing the beam emittance to grow.

The ions generated with positively charged beams can cause an *ion-induced pressure instability* (see Chapter 9), a quick pressure increase in the beam vacuum chamber.

The ionisation cross section of the residual gas molecules by beam particles is one of the key parameters for the ion-induced pressure instability.

The ionisation cross sections of the residual gas molecules for positrons and protons were reported in the literature, for example, see [7], Refs. [7, 21–26], and references within. Following the Bethe theory, the ionisation cross sections can be calculated with the following equation:

$$\sigma = 4\pi \left(\frac{\hbar}{mc}\right)^2 (M^2 x_1 + C x_2) = 1.874 \times 10^{-20} \text{ cm}^2 (M^2 x_1 + C x_2), \quad (1.7)$$

where $x_1 = \frac{1}{\beta^2} \ln\left[\frac{\beta^2}{1-\beta^2}\right] - 1$, $x_2 = \beta^{-2}$, $\beta = \frac{v}{c} = \sqrt{1 - \left(\frac{E_0}{E}\right)^2}$, and E and E_0 are the total and rest energy of a particle, respectively. The coefficients M^2 and C for various gases were reported in Ref. [21]. Table 1.2 reproduces the reported data for the gases that are usually present in vacuum chamber of particle accelerators. The ionisation cross sections calculated with Eq. (1.7) are shown in Figure 1.1 for positron and electron (or proton) beams as a function of their particle energy. It should be noted that the ionisation cross sections depend only on the velocity of the ionising particle, but neither on its charge nor on its mass. However, the energy of the ionising particles depend on their mass; thus the graph for the electron and positron ionisation cross sections and a function of energy are the same, while the proton energy for the same velocity is larger by a proton/electron mass ratio, thus shifting the proton energy axis by this ratio.

1.2.2.4 Contamination of Sensitive Surfaces

In some specific areas of the accelerator, the vacuum requirement might be specified by a *surface–gas interaction*. For example, the Ga–As photocathode lifetime is very sensitive to oxygen-containing gases such as CO, CO_2, H_2O, O_2, etc. [27, 28]; the FEL mirrors are sensitive to hydrocarbon gases [29]. In such cases, the specification for vacuum can include the maximum *total* pressure and the maximum *partial* pressure for particular gas species.

Table 1.2 Values and standard deviation (s.d.) of M^2 and C in Eq. (1.7).

Gas	M^2 Value	M^2 s.d.	C Value	C s.d.
H_2	0.695	0.015	8.115	0.021
He	0.774	0.030	7.653	0.037
He[a]	0.7525	—	8.068	—
CH_4	4.23	0.13	41.85	0.20
H_2O	3.24	0.15	32.26	0.47
CO	3.70	0.15	35.17	0.19
N_2	3.74	0.14	34.84	0.20
O2	4.20	0.18	38.84	0.47
Ar	4.22	0.15	37.93	0.19
CO_2	5.75	0.073	57.91	0.27

a) Theoretical value.
Source: From Ref. [31].

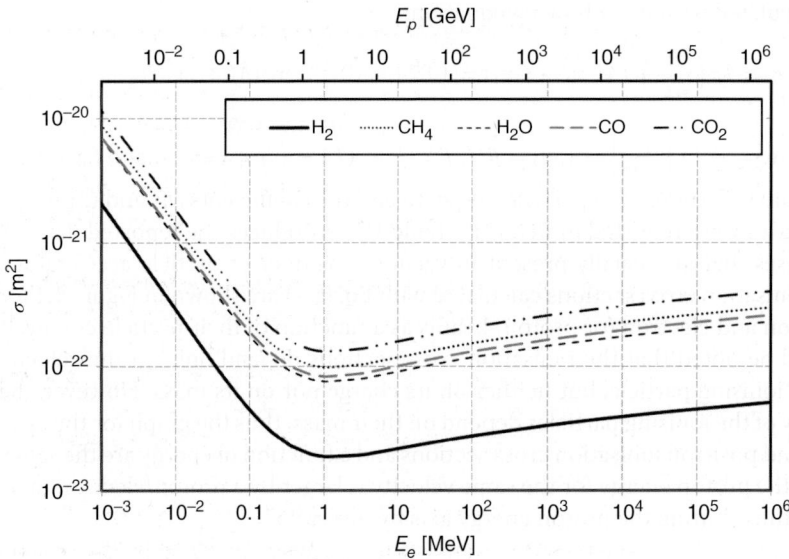

Figure 1.1 The ionisation cross sections for H_2, CH_4, H_2O, CO, and CO_2 as a function of beam energy for protons (top horizontal axis E_p) and electrons and positrons (bottom horizontal axis E_e).

1.2.2.5 Safety and Radiation Damage of Instruments

The beam–gas interaction may affect not only the beam. The Bremsstrahlung radiation due to beam–gas interaction should be seriously considered in the design phase because it may be a source of the following:

- A risk to personnel safety:
    - A residual radioactivity of the vacuum chamber and equipment in an accelerator tunnel.
    - It can also be important for radiation safety during accelerator operation, for example, Bremsstrahlung radiation on an SR beamline could be so significant that it is unsafe for a human to operate the beamline.
- Radiation damage of instruments:
    - An induced material damage and corrosion, damage of accelerator instrumentation, cables, and comptrollers inside and outside of vacuum chamber or even in an accelerator tunnel.
    - An increased risk of quench in superconducting magnets and radio frequency (RF) cavities.

1.2.3 Vacuum Specifications

Thus, vacuum specifications are defined by *a tolerable level* of direct or indirect disturbance, primarily to the quality of charged particle beam, as well as to all accelerator components due to the presence of residual gas species in an accelerator vacuum chamber. The vacuum specifications could be defined for each part, section, or sector of the machine in relation to a *location and time*:

- Location (where a specified vacuum is required):
    - Local for a component, for examples:
        - 10^{-10} Pa between the electron gun and first bending magnet.
        - Average pressure along an incretion device should be less than 10^{-9} Pa.
    - Average for a large section of the machine, for examples:
        - Average pressure along the storage ring should be less than 10^{-8} Pa.
        - Average pressure along transfer line should be less than 3×10^{-7} Pa.
    - Combination of both, for examples:
        - Average pressure along the storage ring should be less than 10^{-8} Pa and local pressure bump should not be greater than 2×10^{-7} Pa.
- Time (when this specification should be reached):
    - 100 hours vacuum lifetime at $I = 560$ mA after 100 A h conditioning (for Diamond Light Source, DLS)
    - $P(N_2$ eqv$) = 10^{-6}$ Pa after bakeout and a week of pumping (for a booster)
    - $n(H_2$ eqv$) = 10^{15}$ m^{-3} after two years conditioning, corresponding to 100 hours vacuum lifetime for the Large Hadron Collider (LHC)

The residual gas composition may include a number of different gas species, and their relative concentration may vary along the beam path due to the fact that vacuum conductivity and pumping speed of the pumps are different for each gas. The best way to specify the required pressure or gas density is to express it in nitrogen equivalent (for room temperature machines) or hydrogen equivalent (for cryogenic machines). To calculate the equivalent pressure, all what is needed is the beam–gas interaction cross sections for each gas, σ_i. Then the N_2 *equivalent pressure* can be calculated as

$$P_{N_2 \text{ eqv}} = \sum_i \frac{\sigma_i}{\sigma_{N_2}} P_i, \tag{1.8}$$

where P_i is a partial pressure for gas i. Similarly, the H_2 *equivalent gas density* can be calculated as

$$n_{H_2 \text{ eqv}} = \sum_i \frac{\sigma_i}{\sigma_{H_2}} n_i. \tag{1.9}$$

It is worth mentioning that, in many cases, accelerator scientists consider that residual gas can be somehow 'completely eliminated' from a beam chamber. So, it may take some time and effort on working together with accelerator scientists to calculate the effect of beam–gas interactions, to set up a tolerable level of vacuum and define vacuum specifications for a whole machine and all of its components. It is very important that at early stages of accelerator design, a vacuum scientist is involved in feasibility studies to check whether and how these vacuum specifications can be met.

Based on these, the accelerator vacuum design objectives are as follows:

– Defining all sources of residual gas in vacuum chamber with and without a beam;
– Calculating required pumping, type of pumps, and their locations;
– Defining the means of pressure measurements, their types, and locations;
– Defining the necessary procedure for material selection, cleaning, and treatments (polishing, coating, firing baking, etc.);
– Providing the results of modelling for the gas density (or pressure) profile along the beam path and at any specific location.

1.2.4 How Vacuum Chamber Affects the Beam Properties

A vacuum chamber is required to provide vacuum to meet vacuum specification for the beam particles.

The walls of a vacuum chamber set the boundary conditions for the beam electromagnetic field and, therefore, can interact directly with a beam. The electric conductivity and a shape of vacuum chamber walls can affect the longitudinal and transversal wakefields limiting a maximum current, increasing beam emittance and energy spread. In general, an electric conductivity of vacuum chamber walls could be specified in a wide range from high conductivity to insulating.

The surface of vacuum chamber or the components may be additionally specified for low photoelectron yield (PEY) and secondary electron yield (SEY), low or high photon reflectivity.

The walls of the vacuum chamber should be transparent for the magnetic field of magnetic components of accelerator such as dipoles, quadrupoles, sextupoles, and kickers: i.e. vacuum chamber walls should be non-magnetic.

The pressure (or gas density) inside vacuum chamber could be up to 15 orders of magnitude lower than outside; thus the material should be suitable for vacuum chamber, i.e. sufficiently dense for providing an efficient barrier for gas molecule and atom penetration and diffusion through vacuum chamber walls. Desorption of molecules from vacuum chamber inner walls and in-vacuum components are the main source of gas in a vacuum system. Therefore the material should be UHV (or even XHV) compatible.

Finally, a vacuum chamber should be produced. Thus it should be mechanically strong and stable, the cost and availability of material should be accounted for, and it is important to consider how easy to manufacture, store, make joints, etc.

All the required properties limit a list of materials that could be used for an accelerator vacuum chamber. The most common materials are 316LN and 304L stainless steel, copper, aluminium, titanium, and their alloys, ceramics, and glass (see Chapter 4 for details). Other materials can also be used, such as carbon or beryllium tubes in detector vacuum chamber.

Various surface treatments can be applied to change some surface properties. Surface polishing reduces the RF surface resistance and increases the photon reflectivity, while the surface roughing increases the RF surface resistance and photon absorption and also reduces PEY and SEY.

Thin and thick film coatings are often applied to provide the required properties. For example,

- Stainless steel chamber can be coated with copper to provide better electric conductivity.
- A low SEY coating can be applied to suppress electron cloud (see Chapter 8).

1.3 First Considerations Before Starting Vacuum System Design

1.3.1 What Is the Task?

When the design phase of a new machine starts, a vacuum system designer needs a lot of information that should be included or considered.

(1) What type of machine is going to be designed and built?
  - Collider (circular or linear).
  - SR (or photon) source.
    - Which generation of the SR source (defined by a key parameter – beam emittance, ε)?
      - First generation uses 'parasitic' SR from dipoles in storage rings and synchrotrons ($\varepsilon \sim 300\text{--}1000$ nm·rad, incoherent radiation).
      - Second generation is a specialised SR source with SR from dipoles and wigglers ($\varepsilon \sim 100$ nm·rad, incoherent radiation).
      - Third generation is a specialised SR source with SR from incretion devices (wigglers and undulators: $\varepsilon_x \sim 3\text{--}20$ nm·rad, $\varepsilon_y \sim 0.01\varepsilon_x$, incoherent radiation) and can also use SR from dipoles.
      - Fourth generation is a specialised photon source (such as FELs) with coherent beam and small emittance $\varepsilon_x \sim 10\text{--}300$ pm·rad.
  - Charged particle beam acceleration, transport, and delivery to the users.
  - Is your task designing the whole machine or only a part? Which part?
(2) What are the main beam parameters?
  - The type of accelerated charged particles.
    - Electrons, positrons, protons, (heavy) ions, and other particles.
  - A charge of particles.
  - Beam energies.

- Beam intensity (current and peak current, number of particles per bunch, bunch length and spacing).
- Beam transversal sizes
  - Often it is sufficient to know minimum and maximum values for whole machine or for different sectors of machine.
(3) Is there SR and what are the main parameters?
  - Sources of SR: dipoles, quadrupoles, wigglers, undulators, FEL, etc.
  - Critical photon energy ε_c for dipole and wigglers, or photon energy/-ies for undulators and FELs.
  - Photon flux, Γ, onto vacuum chambers, SR absorbers, beam collimators, etc..
  - Photon reflectivity.
(4) Are there any of the following problems in consideration and what mitigation techniques can be applied?
  - Electron machines
    - Fast ion instability.
    - Ion trapping instability.
  - Positron, proton, and other machines with a positive charge
    - Electron cloud
    - Ion-induced pressure instability.
  - Heavy ion machines
    - Heavy ion induced pressure instability.
    - Ion induced pressure instability.
    - Electron cloud.
(5) Are there specific components?
  - Electron, proton or ion guns.
  - Solid, liquid or gaseous targets.
  - Antimatter sources (for positrons, antiprotons).
  - Interaction regions.
  - Incretion devices (wigglers, undulators, FELs).
  - Mirrors.
  - Beam windows.
(6) Beam pipe temperature.
  - Room temperature.
  - Mainly room temperature with short cryogenic sections.
  - Mainly cryogenic.
  - Fully cryogenic.
(7) Are there specific problems affecting vacuum design?
  - High power loss, high radiation damage, etc.

This list is of required information is certainly not complete and may significantly vary per task; however, it could be a good starting checklist for a beam vacuum system design.

1.3.2 Beam Lattice

A beam lattice is a magnet structure for the accelerator to drive and focus the charged particle beam. The beam lattice defines the location, orientation, length, and magnetic field strength of dipole, quadrupole, and sextupole magnets and insertion devices. It also defines the best location for beam instrumentation, such as the beam position monitors, the beam scrappers, and SR power absorbers and tapers.

The choice of a vacuum system philosophy, pumping system methods, location and size of vacuum pumps, and the choice and location of other vacuum instrumentations is dictated by the beam lattice components and beam instrumentation. Ideally, vacuum instrumentation should fit within an available space. However, if the provided space is insufficient to meet the required vacuum specification, possible solutions should be discussed with accelerator lattice, magnet, and RF scientists and engineers and mechanical designers.

1.3.3 Beam Aperture and Vacuum Chamber Cross Section

1.3.3.1 Required Mechanical Aperture

The size of the beam vacuum chamber should be sufficient to accommodate the beam. The beam can be round, elliptical, or even 'flat' (when one transversal dimension is much greater than the other one). The charge density $\rho(x, y)$ of a beam for a Gaussian distribution of particles can be described as [30]

$$\rho(x, y) = \frac{Nq_e}{2\pi\sigma_x\sigma_y} \exp\left(-\frac{x^2}{2\sigma_x^2} - \frac{y^2}{2\sigma_y^2}\right), \tag{1.10}$$

where N is a number of particles of charge q_e in the beam, x and y are the horizontal and vertical distances from the centre of the beam, and σ_x and σ_y are the horizontal and vertical transverse r.m.s. beam sizes. That means that the charge density is lower by a factor of $e^{-1/2}$ for particles located at coordinates $(x, y) = (\pm\sigma_x, 0)$ or $(x, y) = (0, \pm\sigma_y)$:

$$\rho(\sigma_x, 0) = \rho(0, \sigma_y) = \rho(0, 0)e^{-1/2} = 0.607\rho(0, 0). \tag{1.11}$$

To avoid the loss of beam particles due to collision with vacuum chamber walls, the vacuum chamber's horizontal and vertical dimensions, a and b, should be much larger than σ_x and σ_y, respectively. The relative charge density of the beam for various x/σ_x and y/σ_y ratios is shown in Table 1.3.

The beam size can be calculated from beta functions $\beta_{x,y}$, emittance $\varepsilon_{x,y}$, dispersion function $D(z)$, and momentum p of the beam provided by a beam lattice design as follows [31]:

$$\sigma_x = \sqrt{\beta_x \varepsilon_x}, \quad \sigma_y = \sqrt{\beta_y \varepsilon_y} + \left|D(z)\frac{\Delta p}{p}\right| \tag{1.12}$$

These values should normally be provided by accelerator scientists from the results of their lattice design of the machine. However, this gives just some ideas about the required size of a vacuum chamber. In the beam lattice design, there is an ideal beam orbit and also, ideally, the centre of the beam should travel along this ideal orbit, and the longitudinal axis of the beam vacuum chamber should coincide with the ideal beam orbit (see Figure 1.2a). However, in practice, the beam can fluctuate with time around the *ideal orbit* occupying a greater space

Table 1.3 Reduction of relative beam charge density as a function of distance from the beam centre.

x/σ_x or y/σ_y	$\rho(a)/\rho(0)$
0	1
1	0.607
2	0.368
3	0.223
4	0.135
5	8.21×10^{-2}
6	4.98×10^{-2}
7	3.02×10^{-2}
8	1.83×10^{-2}
10	6.74×10^{-3}
15	5.53×10^{-4}
20	4.54×10^{-5}

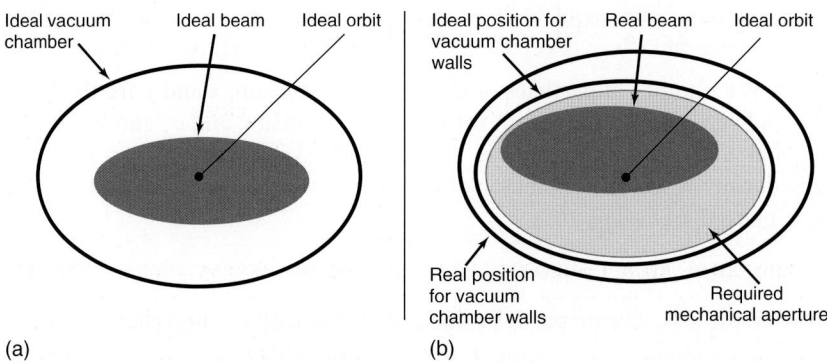

Figure 1.2 Ideal (a) and real (b) positions of a beam and vacuum chamber in respect to the ideal orbit.

called a 'close orbit'. A complicated numerical analysis with nonlinear beam optics performed by accelerator scientists will come with a *required mechanical aperture* (also known as a 'beam stay clear') as a function of the longitudinal coordinate (see Figure 1.2b).

The beam size may significantly vary along the beam trajectory. A real mechanical vacuum chamber should not be exactly the same as a required mechanical aperture but should fully accommodate it.

The mechanical design should also consider mechanical tolerances and misalignments in the shape of vacuum chamber, axial twist, longitudinal bends (after manufacturing or after placing of vacuum chamber supports due to gravity), etc., so the real vacuum chamber cross section is usually slightly larger than the required mechanical aperture.

A beam vacuum chamber could be of the same cross section for the entire machine, making this the most economical solution for vacuum chamber; however, it could be not optimal and cost effective for an accelerator. In practice, in many machines there are a few typical cross sections of the vacuum chamber for different components or sections of the machine.

1.3.3.2 Magnet Design

A large cross section would provide more space for the beam and better vacuum conductance of its vacuum chamber. However, an upper limit for the dimensions of the beam vacuum chamber is dictated by the magnet design. The cost of magnets (dipoles, quadrupoles, etc.) increases approximately quadratically with a gap between the magnet poles where a beam vacuum chamber is placed. The smaller the gap, the smaller the size of the magnet coils and the lower the cost of the magnet. The larger the gap, the harder (or even impossible) it is to reach the required magnetic field strengths. Thus, there must be balanced considerations in choosing a vacuum chamber cross section between the required mechanical aperture to accommodate the beam and the magnet design.

1.3.3.3 Mechanical Engineering

A vacuum chamber should meet a number of specifications related to mechanical engineering. The vacuum chamber should be mechanically stable: i.e. it should not deform due to gravity, the atmospheric pressure, and the Laurence force during magnet quenches; it should not crack due to temperature expansion and cooling; it shouldn't vibrate; etc.

The material choice may also be dictated by required electrical conductivity, thermal conductivity, the cost of material and the space, and cost restrictions. The choice of material and mechanical stability will define the vacuum chamber wall thickness, which is an additional element in a trade-off between the required mechanical aperture and the magnet design.

A real vacuum chamber is not ideal, the vacuum chamber dimensions have certain accuracy; there could be some imperfections at the welds and joints, mechanical and thermal deformations, twists and bends of vacuum chamber; and the shape of vacuum chamber may also deform under vacuum. The position of the vacuum chamber is not ideal: there are misalignments and non-linearity of straight components, especially elastic and plastic deformation of long vacuum chambers between two supports due to the gravity, so the axis of the beam vacuum chamber may be offset from the ideal orbit (see Figure 1.2b). All these considerations must also be included in the specification of the vacuum chamber minimum aperture and require either larger dimension than an ideal vacuum chamber and/or small tolerances for these dimensions.

1.3.3.4 Other Factors Limiting a Maximum Size of Beam Vacuum Chamber

As it was described above, the beam chamber should accommodate the beam, so the beam chamber size's *lower limit* is defined by the beam size (a required mechanical aperture) and mechanical imperfection of vacuum vessel. The beam chamber size's *upper limit* is defined by available gap(s) in magnetic components and vacuum chamber wall thickness. A few other factors limiting the maximum size of a beam chamber cross section should be also considered:

- The cost of the vacuum chamber and components increases with its size.
- There are components in particle accelerators (for example, undulators) where the vacuum chamber apertures and shape are defined by these components.
- A vacuum chamber could be also an integral part of other components, for example, a cryogenic vacuum chamber inside a superconducting magnet.
- A beam screen could be placed inside a bigger vacuum chamber, for example, as a part of a bellows assembly or as an SR screen in cryogenic vacuum chamber.

1.3.4 Vacuum Chamber Cross Sections and Preliminary Mechanical Layout

The shape of a beam is either round or elliptic. So-called flat beams are the elliptic ones with $\sigma_x \gg \sigma_y$). Therefore, the shape of the beam vacuum chamber is also often either round or elliptic. These shapes are easy to manufacture and quite convenient in mechanical design for placing inside many magnetic components, dipoles, quadrupoles, sextupoles, etc., without changing the vacuum chamber shape. However, other shapes are also widely used: square, rectangular, hexagonal, and octagonal (see Figure 1.3), as well as a variety of other shapes.

The vacuum conductance of a beam vacuum chamber could be insufficient (too low) to meet required vacuum specification. In this case another chamber (usually called an antechamber) is placed parallel to the beam chamber and is connected to it with a slot over the entire length of antechamber. The antechamber can be used either over the entire length of the machine, on some sections only, or just for specific components.

There are three main types of antechamber:

- Large antechamber with a cross section much larger than that of beam chamber for increasing vacuum conductance of a narrow beam vacuum chamber (see Figure 1.4a).

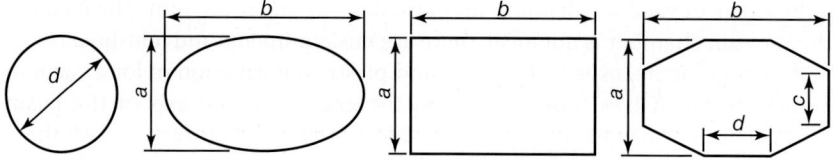

Figure 1.3 Examples of a beam chamber cross sections: round, elliptic, rectangular, and octagonal with inner dimensions (wall thickness and outer dimensions are not shown).

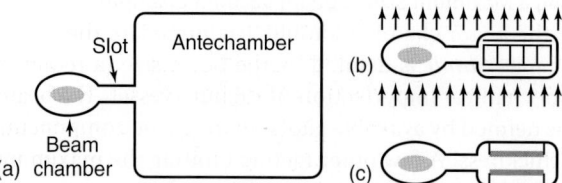

Figure 1.4 Examples of a beam chamber with an antechamber: (a) a large antechamber for increasing a vacuum conductance, (b) an antechamber with a distributed SIP in a dipole magnetic field, and (c) an antechamber with a distributed NEG strip pumps.

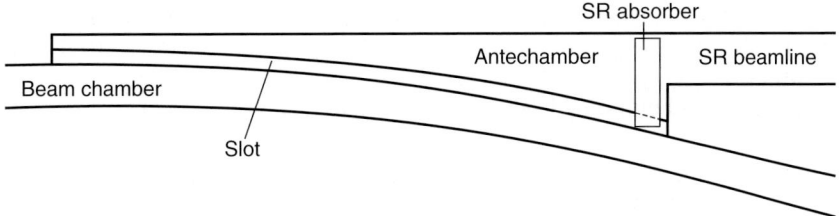

Figure 1.5 An example of a beam chamber with an SR antechamber, a slot between them, SR absorber, and beginning of SR beamline.

- Relatively small antechamber for placing distributed vacuum pumps, which usually are sputter ion pumps (SIPs) operating in a dipole and quadrupole magnetic field or the non-evaporable getter (NEG) strips (see Figure 1.4b,c).
- SR beam antechamber (see Figure 1.5)
  - for separating SR and charged particle beam at the beginning of SR beamlines where SR and charged particle beam coexist in the same vacuum chamber,
  - for more efficient absorption of SR with specially designed SR absorbers,
  - for reducing photoelectron production in the beam chamber,
  - for reducing SR background in detector,
  - for protecting cryogenics or sensitive equipment.

In general, the vacuum chamber cross sections for each machine are optimised in a multi-iteration process considering all limits, wishes, and acceptable and unacceptable solutions for each special field involved in the design: beam lattice, magnets, cryogenics, vacuum, mechanical design, radiation protection, health and safety, etc.

1.3.5 Possible Pumping Layouts

Pumping technology provides a wide range of possible vacuum solution based on different pumping layouts.

1. The lumped pumping layout shown in Figure 1.6a consists of
   - a beam vacuum chamber
     - simple, without an antechamber,

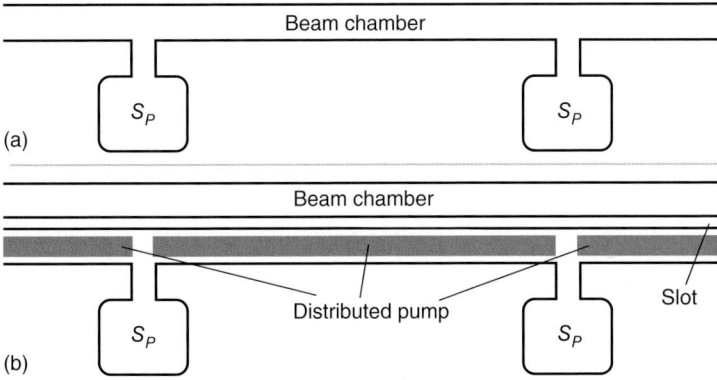

Figure 1.6 A beam chamber between two lumped pumps (a) without an antechamber and (b) with an antechamber containing a distributed pump.

     ○ or with a large antechamber for increasing vacuum conductance,
   - and the pumps with pumping units located at a certain distance from each other.
2. The distributed pumping layout shown in Figure 1.6b consists of
   - a beam vacuum chamber,
   - with an antechamber containing the distributed pump along entire length of the antechamber,
   - and the pumps or pumping units located at a certain distance from each other.
3. The NEG coated vacuum chamber is used for UHV/XHV conditions. In this case an entire vacuum chamber is coated with NEG film – no antechamber required. NEG coated chambers are discussed in detail in Chapter 3.

The lumped pumps can be any of the following pumps (but not limited to) or combination of two or even three of them in one same unit:

- Sputter ion pump (SIP)
- Titanium sublimation pump (TSP)
- NEG pumps
- Turbo-molecular pump (TMP) backed up by a roughing pump or pumping station
- Cryopumps

1.4 First and Very Rough Estimations

Before choosing a possible pumping layout and starting mechanical design of vacuum system, we have to *roughly estimate* the number and size of pumps required to meet vacuum specifications. First of all it is necessary to define all sources of residual gas in a vacuum chamber with and without a beam. Most commonly these are thermal outgassing of vacuum chamber and in-vacuum components, beam-stimulated gas desorption (photon-stimulated desorption [PSD], electron-stimulated desorption [ESD], ion-stimulated desorption [ISD], heavy ion-stimulated desorption [HISD]) and gas injection.

Knowing what is available and from previous experience, we have to work out what thermal outgassing can be expected from the material of a vacuum chamber and likely applied cleaning and preparation procedures. For example, the specific thermal outgassing rate of $\eta_t = 10^{-12}$ [mbar·l/(s·cm^2)] (or 10^{-9} [Pa·m/s] in SI units) is routinely obtained on baked stainless steel chambers, in the following analysis, we will use this value as a constant average value.

A total internal surface area of a vacuum chamber, A_{tot} [cm^2], can be roughly estimated (or guessed). Then the total outgassing rate is

$$Q_{tot} = \eta_t A_{tot}. \tag{1.13}$$

To reach the required pressure of P_{spec}, the total pumping speed S_{tot} should be greater than

$$S_{tot} > \frac{Q_{tot}}{P_{spec}} = \frac{\eta_t A_{tot}}{P_{spec}}. \tag{1.14}$$

Since a vacuum conductance of the beam chamber was not considered here, *this is a lower limit estimate of total required pumping speed.*

A minimum required number of pumps N_p and the average distance between them $\langle L_p \rangle$ can be estimated for a machine or its section with a total length of accelerator, L_{tot} [m], considering that the pumping speed of lumped pumps used in accelerators usually varies in the range $100\, l/s < S_p < 1000\, l/s$, as follows:

$$N_p = \frac{S_{tot}}{S_p} = \frac{\eta_t A_{tot}}{P_{spec} S_p}; \tag{1.15}$$

$$\langle L_p \rangle = \frac{L_{tot}}{N_p} = \frac{P_{spec} S_p L_{tot}}{\eta_t A_{tot}}. \tag{1.16}$$

Similarly, rough estimations can be performed for the machines with a beam-stimulated desorption (PSD, ESD, ISD). For example, vacuum modelling of the machines with SR requires the calculated values of the SR critical energy ε_c and the total photon flux Γ_{tot}. The required conditioning time allows estimating the average photon dose D [photons/m], which, in turn, allows estimating the average PSD yields, η_γ [molecules/photon].

In this case, the total outgassing rate is

$$Q_{tot} = \eta_t A_{tot} + \eta_\gamma \Gamma_{tot}. \tag{1.17}$$

The total pumping speed S_{tot} should be greater than

$$S_{tot} > \frac{Q_{tot}}{P_{spec}} = \frac{\eta_t A_{tot} + \eta_\gamma \Gamma_{tot}}{P_{spec}}. \tag{1.18}$$

The minimum required number of pumps N_p and the average distance between them $\langle L_p \rangle$ can be estimated as follows:

$$N_p = \frac{S_{tot}}{S_p} = \frac{\eta_t A_{tot} + \eta_\gamma \Gamma_{tot}}{P_{spec} S_p}; \tag{1.19}$$

$$\langle L_p \rangle = \frac{L_{tot}}{N_p} = \frac{P_{spec} S_p L_{tot}}{\eta_t A_{tot} + \eta_\gamma \Gamma_{tot}}. \tag{1.20}$$

One should not forget that *these estimations are very rough,* as they do not consider the vacuum conductance of beam chamber, they don't consider details of mechanical design, and all the input parameter values are very approximate. Accurate modelling will give the required number of pumps to be greater than N_p and the distance between them to be lower than $\langle L_p \rangle$; but this would be possible only when the mechanical design already exists.

Meanwhile, these simple calculations can quickly provide some initial information on how simple or, on the contrary, how challenging would it be to meet the required vacuum specification. The results and conclusions of these calculations allow estimating a scale of the task and could be sufficient for the first discussion on a vacuum system mechanical design – on how many pumping ports, possible type, and size of the pumps will be required overall.

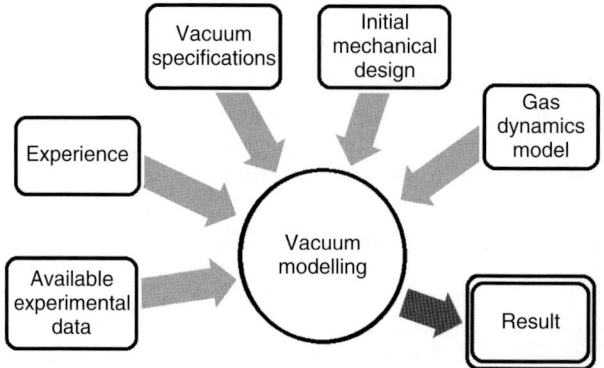

Figure 1.7 Initial considerations for vacuum modelling and design.

1.5 First Run of an Accurate Vacuum Modelling

When the initial mechanical design (or even a layout) of an accelerator and its vacuum chamber is available, a more detailed and accurate vacuum model can be built, taking into account all the vacuum conductances that have been ignored in the preliminary considerations. Chapters 2–5 describe the input data for the modelling (experimental results for TD [thermal desorption], PSD, ESD, and ISD and formulas for SR), the following Chapters 6–9 are devoted to the gas dynamics modelling, and the final Chapter 10 contains both experimental data and models for the heavy ion machine vacuum systems.

Now, using the preliminary mechanical layout, available experimental data, design, and operation experience from the past, one can draw up a rough vacuum design layout and, with the use of in-house or commercial software, perform vacuum modelling and obtain results on pressure (or gas density) profiles along the beam path, average gas density, more accurate results for the number of pumps, their locations, and required pumping speeds (see Figure 1.7). A few options could be considered and investigated to choose then which one is the most suitable and cost effective.

1.6 Towards the Final Design

In practice, a full vacuum system design cycle is more complicated. Schematically it can be shown as in Figure 1.8.

The design of a new machine is always based on experience of designing and operating of previous machines; this also includes the 'it would be nice to have…', 'this was a good (or bad) idea to…', practical 'Do's' and 'Don'ts' (which could be true, or not necessarily true, or folklore).

A significant amount of *experimental data* is available in published journal papers, proceedings, preprints, reports, notes, and personal archives. These are the operation data for the running accelerators, the SR, and particle beamlines, many results were obtained in dedicated experiments in laboratories, on the

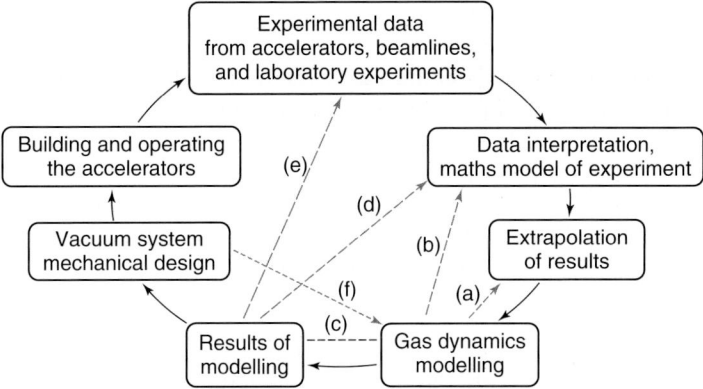

Figure 1.8 Modelling workflow diagram in vacuum system design cycle. (a) – extrapolation revision, (b) and (d) – data interpretation revision, (c) and (f) – vacuum model revision, (e) – experimental data revision or request for a new experimental study.

beamlines, or on the dedicated vacuum chamber inserts in the existing and operating machines. At this point it is extremely important to justify which results are more relevant to a newly designed machine. Ideally, all the experimental conditions should be as close as possible to the operation conditions on the future machine: materials, dimensions of vacuum chamber, surface cleaning and treatments, vacuum firing, bakeout, vacuum chamber temperature; energy, intensity, and dose of bombarding particles (SR, electrons, ions, etc.); and many other relevant parameters.

Furthermore, the following should be carefully considered when applying experiential data for modelling a new machine:

- The experimental data has experimental errors.[1]
- The experimental *data* has undergone an *interpretation* based on a model, which could have an *error*; the model error could be reported or not reported together with the results. Furthermore, one should keep in mind that although the models applied for data interpretation are quite simple, easy to apply, and most likely correctly employed, occasionally, the traditional models could not be applied to new materials and condition of experiment; thus the data interpretation could be incorrect.
- The data could be *insufficient*: for example, there could be no data for some gas species or data were reported for low photon dose, no data for some particles bombarding vacuum chamber walls, no data of desorption yields for certain energies of incident particles, etc. In this case, the *data can be extrapolated* based on existing knowledge, observations, experience, and consideration. The

---

1 Note that presently only the total pressure vacuum gauges can be calibrated down to $\sim 10^{-7}$ Pa. Below this pressure the gauges are not calibrated. Furthermore, there is no ISO standard for the partial pressure measurements and the RGA calibration yet. It is incorrect to apply the gauge correction factor for the RGA data. To address this problem, many vacuum groups in accelerator centres employ some in-house-developed RGA calibration procedures. Thus, the partial pressure data should be used with a great caution.

extrapolations can be used, but a vacuum designer should be careful with a level of confidence in such extrapolations.
- When a vacuum designer starts building a gas dynamics model, he/she might discover that the reported data can't be used for modelling of a new accelerator. For example, occasionally reported 'effective desorption yields' allow a comparison of different samples on the same measurement facility but can't be used for the gas dynamics modelling. However, more careful and detailed analysis of raw data may allow extracting the real desorption yields or indicate the need of new experiment.

A *gas dynamics model* should include all the physical phenomena that affect the gas density along the beam path. On the other hand, the model should not be unnecessary complicated. Room temperature vacuum systems can be modelled with a 1D gas diffusion model and a 3D test particle Monte Carlo (TPMC) model described in Chapter 6. A cryogenic vacuum chamber requires a more complicated model described in Chapter 7. High intensity accelerators with positively charged beam may suffer from two problems: a design of a vacuum system with electron cloud problem is discussed in Chapter 8, and a model for the ion induced pressure instability is described in Chapters 9 and 10. Models in Chapters 7–10 are more complicated than in Chapter 6 because they include more specific phenomena for particular types of accelerators. The applied gas dynamics model should have an error that could be related to the model itself (see Chapters 6–10). One should also consider the difference between a real vacuum chamber and a simplified vacuum chamber shape in the applied gas dynamics model (for example, a bellows unit shown as a tube) could also result in a difference between a predicted (modelled) and real behaviour of vacuum system.

The *results* of gas dynamics modelling may indicate that model is not as accurate as required, and then the gas dynamics model should be modified or changed. The results may also demonstrate that the errors and approximations are too large to take a critical design decision and a new dedicated experimental study is required.

When the results of gas dynamics modelling shows that the vacuum system layout, locations of pumps and their pumping speed allow to meet vacuum specifications; this can be implemented in the *vacuum system mechanical design*. This design is often somehow different from the original model (different location and available space for pumping ports, new components, other material of vacuum chamber or its components, smaller or larger aperture of vacuum chamber adopted for a new version of the beam lattice or a new magnets design). That requires performing the gas dynamic modelling for an updated vacuum system mechanical design. Usually, there could be a number of iterations in the design with various minor and major modifications until the design is finalised (frozen).

When the accelerator is finally built and in operation, one can compare the predicted and actual behaviour of vacuum system (for example, see Figure 1.9 demonstrating a comparison of modelling results and measured data for the DLS [32]). This experience can then be applied for future machines.

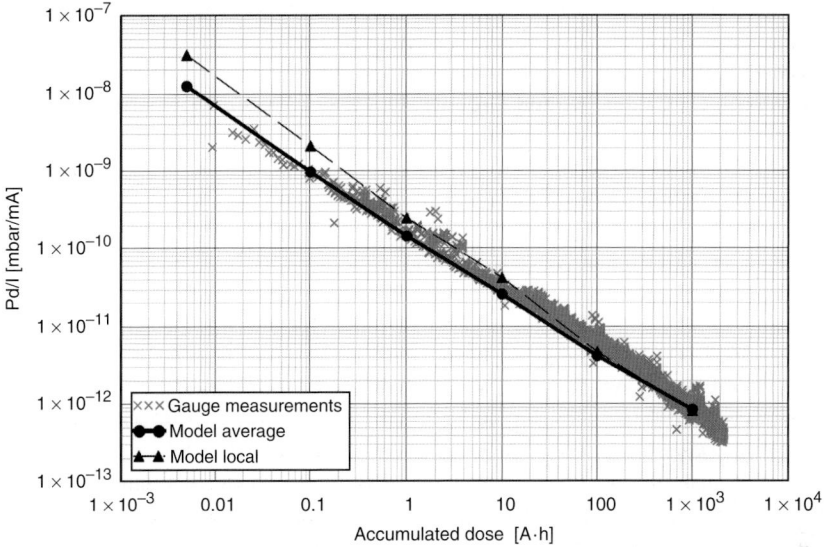

Figure 1.9 Modelling results versus measured data for the Diamond Light Source. Source: Wilson 2001 [26]. Reprinted with permission of Oxford University Press.

1.7 Final Remarks

The level of knowledge and expertise available in accelerator vacuum community is sufficient to design any type of accelerator with a high confidence. We aim to make a good design of accelerator vacuum system, which meets its specifications, allows for later improvements, and is economical, reliable, and maintainable. There could be two, three, or even more possible solutions to build vacuum system to the specification. It is very important to remember that any vacuum system design for an accelerator is a compromise between various system requirements, available technologies, and space restriction as well as considerations of time, cost, and practicality.

Gas dynamics modelling of vacuum system is a key process in verifying that everything is considered correctly in vacuum and mechanical design: input data, their accuracy and relevance, machine parameters, and specific problems.

The authors hope that this book would help with a design work for future charged particle accelerators.

References

1 http://www.oxforddictionaries.com/definition/english/vacuum. Oxford University Press 2016 (accessed 30 May 2019).
2 http://dictionary.cambridge.org/dictionary/british/vacuum. Cambridge University Press 2016 (accessed 30 May 2019).

3 ISO 3529-1:1981(en) Vacuum technology - Vocabulary - Part 1: General terms. https://www.iso.org/obp/ui/#iso:std:iso:3529:-1:ed-1:v1:en (accessed 30 May 2019).
4 Benvenuti, C., Calder, R., and Gröbner, O. (1987). Vacuum for particle accelerators and storage rings. *Vacuum* 37: 699–707.
5 Dabin, Y. (2002). Large vacuum system engineering. *Vacuum* 67: 347–357.
6 Wille, K. (2000). *The Physics of Particle Accelerators*. UK: Oxford University Press.
7 Chao, A.W. and Tigner, M. (eds.) (2002). *Handbook of Accelerator Physics and Engineering*. Singapore: Word Scientific Publishing Co. Pte. Ltd.
8 Brüning, O.S., Collier, P., Lebrun, P. et al. (eds) (2004). LHC Design Report. *CERN-2004-003-V-1*. Geneva: CERN.
9 Brau, J., Okada, Y., Walker, N.J. et al. (eds) (2007). International Linear Collider Reference Design Report: ILC Global Design Effort and World Wide Study. *ILC-Report-2007-001*. Geneva: CERN, 778 p (4.v).
10 SuperB Collaboration: Biagini, M., Raimondi, P., and Seeman, J. SuperB Progress Reports – Accelerator. http://arxiv.org/abs/1009.6178v2.
11 Green Paper. The Modularized Start Version (2009). A stepwise approach to the realisation of the Facility for Antiproton and Ion Research in Europe (FAIR). Darmstadt, Germany: GSI.
12 ISIS Neutron and Muon Source. http://www.isis.stfc.ac.uk/ (accessed 30 May 2019).
13 Diamond Synchrotron Light Source (2003). Report of the design Specification (Green book). Warrington, Cheshire, UK: CCLRC, Daresbury Laboratory.
14 SOLEIL Synchrotron. http://www.synchrotron-soleil.fr/ (accessed 30 May 2019).
15 Turner, S. (ed.) (1999). *CERN Accelerator School - Vacuum Technology*, 28 May – 3 June 1999, Snekersten, Denmark, Proceedings, *CERN 99-05*. Geneva: CERN.
16 The Particle Data Group (PDG). http://pdg.lbl.gov (accessed 30 May 2019).
17 Møller, S.P. (1999). Beam residual gas interactions. In: *Proceedings of CERN Accelerator School - Vacuum Technology*, 28 May – 3 June 1999, Snekersten, Denmark, Proceedings (ed. S. Turner), *CERN 99-05*. Geneva: CERN, p. 155.
18 Grasfström, P. (2007). Lifetime, cross-section and activation. In: *Proceedings of CERN Accelerator School – Vacuum in Accelerators*, 16–24 May 2007, Platja d'Aro, Spain, Proceedings (ed. D. Brandt), *CERN-2007-003*, Geneva: CERN, p. 213.
19 Dobbing, G.S. (1998). A study of the effects of insertion device parameters on gas scattering lifetime for DIAMOND, DPG-98-67. Warrington, DC: CLRC Daresbury Laboratory.
20 Yoshimura, N. (2007). *Vacuum Technology: Practice for Scientific Instruments*. Springer Science & Business Media. ISBN: 9783540744320.
21 Riekej, F. and Prepejchal, W. (1972). Ionization cross sections of gaseous atoms and molecules for high-energy electrons and positrons. *Phys. Rev. A* 6: 1507.
22 Mathewson, A.G. and Grobner, O. (2002). Thermal outgassing and beam induces desorption. In: *Handbook of Accelerator Physics and Engineering*

(eds. A.W. Chao and M. Tigner), 226. Singapore: Word Scientific publishing Co. Pte. Ltd.
23 Hwang, W., Kim, Y.-K., and Rudd, M.E. (1996). New model for electron-impact ionization cross sections of molecules. *J. Chem. Phys.* 104 (8): 2956.
24 Kim, Y.-K., Irikura, K.K., Rudd, M.E. et al. (2004). Electron-Impact Cross Sections for Ionisation and Excitation Database. https://physics.nist.gov/PhysRefData/Ionization/molTable.html, https://dx.doi.org/10.18434/T4KK5C (accessed 30 May 2019).
25 Ishimaru, H., Shinkichi, S., and Inokuti, M. (1995). Ionization cross sections of gases for protons at kinetic energies between 20 MeV and 385 GeV, and applications to vacuum gauges in superconducting accelerators. *Phys. Rev. A* 51: 4631.
26 Mathewson, A.G. and Zhang, S. (1996). The Beam-Gas Ionization Cross Section at 7.0 TeV. Vacuum Technical Note 96-01. CERN.
27 Chanlek, N., Herbert, J.D., Jones, R.M. et al. (2014). The degradation of quantum efficiency in negative electron affinity GaAs photocathodes under gas exposure. *J. Phys. D: Appl. Phys.* 47: 055110.
28 Iijima, H., Shonaka, C., Kuriki, M. et al. (2010). A study of lifetime of nea-GaAs photocathode at various temperatures. *Proceedings of IPAC'10*, Kyoto, Japan (23–28 May 2010), p. 2323.
29 Castagna, J.C., Murphy, B., Bozek, J., and Berrah, N. (2013). X-ray split and delay system for soft X-rays at LCLS. *J. Phys.: Conf. Ser.* 425: 1520210.
30 Wille, K. (2000). *The Physics of Particle Accelerators*, Chapters 2 and 3. Oxford University Press.
31 Wilson, E. (2001). *An Introduction to Particle Accelerators*. Oxford University Press.
32 Malyshev, O.B. and Cox, M.P. (2012). Design modelling and measured performance of the vacuum system of the Diamond Light Source storage ring. *Vacuum* 86: 1692–1696. https://doi.org/10.1016/j.vacuum.2012.03.015.

2

Synchrotron Radiation in Particle Accelerators

Olivier Marcouillé

Synchrotron SOLEIL, L'Orme des Merisiers, Saint-Aubin - BP 48, 91192 Gif-sur-Yvette Cedex, France

Synchrotron radiation (SR) is an electromagnetic radiation emitted by a relativistic charged particle moving in a magnetic field environment. Theory of SR is well developed and described in many reference books [1–3]. In charged particle accelerators, different types of magnets (e.g. dipoles, quadrupoles, sextupoles, etc.) are used to control the charged beam trajectory (or orbit) and the focusing beam parameters and often are the sources of SR. There are also dedicated devices specifically built to generate a high intensity SR such as wigglers and undulators.

Contrary to many books describing SR that could be delivered through the beamlines to SR users, our interest is mainly focused on the photons that did not reach the users and interact with vacuum chamber walls and SR absorbers and that are usually not well described. However few details of SR must be pointed out to anticipate the operation troubles.

2.1 Emission of a Charged Particle in a Magnetic Field

A charged particle of charge q and normalised speed $\vec{\beta}$ radiates light towards the direction $\vec{n}$ of an observer located at a distance r. The light is characterised by the electric field $\vec{E}$ and the associated magnetic field $\vec{B}$ due to the particle motion. Both quantities are derived from the retarded potentials of Lienard–Wiechert and are written as follows [4] (Figure 2.1):

$$\vec{E}(t) = \frac{q}{4\pi\varepsilon_0 c}\left[\frac{c(1-\beta^2)(\vec{n}-\vec{\beta})}{r^2(1-\vec{n}\cdot\vec{\beta})^3} + \frac{(\vec{n}\times\{(\vec{n}-\vec{\beta})\times\dot{\vec{\beta}}\})}{r(1-\vec{n}\cdot\vec{\beta})^3}\right]_{t'=t-\frac{r}{c}} \quad (2.1)$$

$$\vec{B}(t) = \left\{\frac{\vec{n}\times\vec{E}(t)}{c}\right\} \quad (2.2)$$

It is the retarded time due to the delay for the light emitted at the time t to reach the observer located at a distance r from the source and c is the light velocity. It

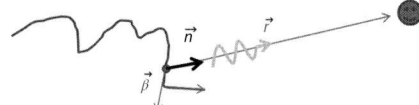

Figure 2.1 Scheme of emitted light of charged particle in an arbitrary path seen by an observer (☺).

means that $t' = t - r(t')/c$. On the other hand, any change of retarded time dt' is linked to the change of time dt:

$$dt = [1 - \vec{n}\vec{\beta}]dt' \tag{2.3}$$

Equations (2.1) as well as (2.2) are composed of two terms named 'far field' component expressed in $1/r$ and 'near field' component expressed in $1/r^2$.

In most applications in the vacuum domain, only the far field term is considered as predominant.

$$\vec{E}(t) = \frac{q}{4\pi\varepsilon_0 c}\left[\frac{(\vec{n} \times \{(\vec{n}-\vec{\beta}) \times \dot{\vec{\beta}}\})}{r(1-\vec{n}\cdot\vec{\beta})^3}\right]_{t'=t-\frac{r}{c}} \tag{2.4}$$

The spectral distribution of photons is calculated via the Fourier Transform $\vec{E}(\omega)$ of the electric field $\vec{E}(t)$:

$$\vec{E}(\omega) = \frac{1}{\sqrt{2\pi}}\int_{-\infty}^{+\infty}\vec{E}(t)e^{-j\omega t}\,dt \tag{2.5}$$

The relation between the retarded time t' at the observer and particle time t enables to make a change of variable: $t' = t - r(t')/c$. By differentiating $dt = [1 - \vec{n}\vec{\beta}]dt'$, it gives

$$\vec{E}(\omega) = \frac{1}{\sqrt{2\pi}}\frac{q}{4\pi\varepsilon_0 c}\int_{-\infty}^{+\infty}\left[\frac{(\vec{n} \times \{(\vec{n}-\vec{\beta}) \times \dot{\vec{\beta}}\})}{r(1-\vec{n}\cdot\vec{\beta})^2}\right]e^{-j\omega\left(t'+\frac{r(t')}{c}\right)}dt.$$

In the far field regime, variations of position of the emitting particles and observation direction $\vec{n}$ are considered to be small: $1/r$ and $\vec{n}$ are supposed to be constant. The Fourier transform expression can be solved using integration by parts:

$$\frac{df}{dt'} = \left[\frac{(\vec{n} \times \{(\vec{n}-\vec{\beta}) \times \dot{\vec{\beta}}\})}{r(1-\vec{n}\cdot\vec{\beta})^2}\right], \quad f = \left[\frac{\vec{n} \times (\vec{n}-\vec{\beta})}{r(1-\vec{n}\cdot\vec{\beta})}\right] \text{ and}$$

$$g = e^{-j\omega(t'+r(t')/c)}, \quad \frac{dg}{dt} = -j\omega(1-\vec{n}\vec{\beta})g$$

$$\vec{E}(\omega) = \frac{1}{\sqrt{2\pi}}\frac{q}{4\pi\varepsilon_0 rc}$$

$$\times \left\{\left[\frac{(\vec{n} \times \{\vec{n} \times \vec{\beta}\})}{1-\vec{n}\cdot\vec{\beta}}\right]_{-\infty}^{\pm\infty} + j\omega\int_{-\infty}^{+\infty}(\vec{n} \times \vec{n} \times \vec{\beta})e^{-j\omega\left(t+\frac{r}{c}\right)}dt\right\}.$$

At $t' = \pm\infty$ the first term tends to zero, leading to

$$\overrightarrow{E(\omega)} = \frac{1}{\sqrt{2\pi}} \frac{jq\omega}{4\pi\varepsilon_0 rc} \int_{-\infty}^{+\infty} \vec{n} \times (\vec{n} \times \vec{\beta}) e^{-j\omega\left(t+\frac{r}{c}\right)} dt. \quad (2.6)$$

As $r = \vec{n} \cdot \vec{r} = \vec{n} \cdot \vec{R} - \vec{n} \cdot \vec{r}_p$, the electric field can be written differently:

$$\overrightarrow{E(\omega)} = \frac{1}{\sqrt{2\pi}} \frac{jq\omega}{4\pi\varepsilon_0 rc} \int_{-\infty}^{+\infty} \vec{n} \times (\vec{n} \times \vec{\beta}) e^{-j\omega\left(t+\frac{\vec{n}\cdot\vec{R}-\vec{n}\cdot\vec{r}_p}{c}\right)} dt. \quad (2.7)$$

The magnetic field of the emitted electromagnetic wave is easily deduced since
$\vec{B}(\omega) = \left\{ \frac{\vec{n} \times \vec{E}(\omega)}{c} \right\}$.

2.1.1 Radiated Energy Density and Power Density

The angular power density is related to the Poynting vector ($\vec{\pi} = \frac{\vec{E} \times \vec{B}}{\mu_0}$) by

$$dP = \vec{\pi} \, d\vec{S} = [\vec{\pi}\,\vec{n}] r^2 d\Omega \quad (2.8)$$

This leads to the analytical expression:

$$\frac{dP}{d\Omega}(\omega) = \frac{r^2 |E(\omega)|^2}{\mu_0 c} \quad (2.9)$$

The total energy radiated by one particle is calculated by integrating the power over the angles and the time:

$$W = r^2 \cdot \int_0^{4\pi} d\Omega \cdot \int_{-\infty}^{+\infty} \frac{|E(t)|^2}{\mu_0 c} dt = r^2 \cdot \int_0^{4\pi} d\Omega \cdot \int_{-\infty}^{+\infty} \frac{E(\omega)^2}{\mu_0 c} d\omega \quad (2.10)$$

It is also written as

$$W = 2r^2 \cdot \int_0^{4\pi} d\Omega \cdot \int_0^{+\infty} \frac{|E(\omega)|^2}{\mu_0 c} d\omega \quad (2.11)$$

As a consequence the energy density is

$$\frac{\partial^2 W}{\partial\Omega\partial\omega} = 2r^2 \frac{|E(\omega)|^2}{\mu_0 c} \quad (2.12)$$

In other terms,

$$\frac{d^2 W}{d\Omega d\omega} = \frac{q^2 \omega^2}{16\pi^3 \varepsilon_0 c} \left| \int_{-\infty}^{+\infty} \vec{n} \times (\vec{n} \times \vec{\beta}) e^{-j\omega\left(t - \frac{\vec{n}\cdot\vec{r}_p}{c}\right)} dt \right|^2 \quad (2.13)$$

The power density $d^2P/d\Omega d\omega$ is obtained by multiplying the energy density $d^2 W/d\Omega d\omega$ by the number of particles per second I/q. It becomes

$$\frac{d^2 P}{d\Omega d\omega} = \frac{q\omega^2 I}{16\pi^3 \varepsilon_0 c} \left| \int_{-\infty}^{+\infty} \vec{n} \times (\vec{n} \times \vec{\beta}) e^{-j\omega\left(t - \frac{\vec{n}\cdot\vec{r}_p}{c}\right)} dt \right|^2 \quad (2.14)$$

2.1.2 Angular Flux

There is a direct relation between the number of photons per second Γ and the radiated power P:

$$\frac{d^2P}{d\Omega} = \frac{\hbar\omega}{2\pi}\frac{d^2\Gamma}{d\Omega}$$

where h is the Planck constant.

Finally expressed in unit of relative bandwidth $d\omega/\omega$ the angular flux is given by

$$\frac{d^2\Gamma}{d\Omega d\omega/\omega} = \frac{q\omega^2}{8h\pi^2\varepsilon_0 c} I \left| \int_{-\infty}^{+\infty} \vec{n}\times(\vec{n}\times\vec{\beta})e^{-j\omega\left(t-\frac{\vec{n}\cdot\vec{r_p}}{c}\right)} dt \right| \quad (2.15)$$

2.2 SR from Dipoles

The motion of a charged particle crossing a constant magnetic field region is governed by the Lorentz equation:

$$\gamma m_0 \frac{d\vec{\beta}}{dt} = q\vec{\beta}\times\vec{B}_0 \quad (2.16)$$

γ is the Lorentz factor and m_0 the mass at rest of the particles. Because of the cross product of expression (2.7) and while the magnetic field $\vec{B}_0$ is spatially constant, the motion remains in the plane $(x, 0, z)$ and the charged particle describes a circle of radius ρ (Figure 2.2) due to the centripetal force $\gamma m_0 [\beta c]^2/\rho$.

ρ is written as follows with E being the particle energy and E_0 energy at rest (Figure 2.3):

$$\rho[m] = \frac{\sqrt{E^2[J] - E_0^2[J]}}{q[C]B_0[T]c[m/s]} = \frac{\sqrt{E^2[eV] - E_0^2[eV]}}{B_0[T]c[m/s]} \quad (2.17)$$

Note: considering the mass of particles at rest,

- for electrons, $E_0 = m_0c^2 = 511\times 10^3$ eV
- for protons, $E_0 = m_0c^2 = 938\times 10^6$ eV

In most storage rings (colliders, synchrotron facilities), the rest energy is much lower than the particle energy. Consequently, by neglecting E_0, Eq. (2.17) can be written as

$$\boxed{\rho[m] = \frac{E[eV]}{B_0[T]c[m/s]}} \quad (2.18)$$

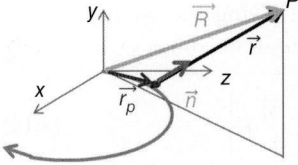

Figure 2.2 Motion of one particle.

Figure 2.3 Motion of a charged particle in a constant field (circle). The radiation seen by an observer (☺) is represented by an undulated line.

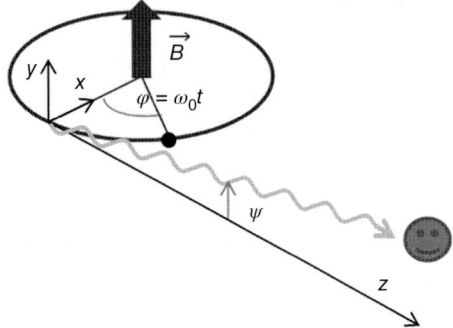

For example,

- Electrons in a SOLEIL dipole: $E = 2.75$ GeV, $B_0 = 1.71$ T, hence $\rho = 5.36$ m
- Electrons in a DLS dipole: $E = 3$ GeV, $B_0 = 1.4$ T, hence $\rho = 7.15$ m
- Protons in the LHC arc: $E = 7$ TeV, $B_0 = 8.33$ T, hence $\rho = 2803$ m

2.2.1 Emission Duration and Critical Energy

Due to the relativistic contraction of angles, the emission coming from the charged particle is reduced to the angular aperture of $\pm 1/\gamma$. In a circular motion of radius ρ, the time Δt_{part} to cross an arc of $2/\gamma$ angle is $\Delta t_{part} = \frac{2\rho}{\gamma c}$ (Figure 2.4). The circular motion is accomplished at the pulsation $\omega_0 = \beta c/\rho$.

The duration of the emission is reduced by the factor $(1 - \beta)$ due to the contraction of time (see Eq. (2.3)), which means that an external observer sees light pulses of duration $\Delta t_{obs} = \frac{\rho}{\gamma^3 c}$.

For example,

- Electrons in SOLEIL dipole: $E = 2.75$ GeV ($\gamma = 5381$), $\rho = 5.36$ m: $\Delta t_{obs} = 1.16 \times 10^{-19}$ seconds
- Electrons in DLS dipole: $E = 3$ GeV ($\gamma = 5870$), $\rho = 7.15$ m: $\Delta t_{obs} = 1.18 \times 10^{-19}$ seconds
- Protons in the LHC arc: $E = 7$ TeV ($\gamma = 7519$), $\rho = 2803$ m: $\Delta t_{obs} = 2.2 \times 10^{-17}$ seconds

The radiation is composed of a train of very short pulses of $\frac{\rho}{\gamma^3 c}$ duration. The photon flux extends in a wide spectral range characterised by the critical energy ε_c expressed as

$$\varepsilon_c = \frac{3 h \gamma^3}{4 \pi \rho} \tag{2.19}$$

Figure 2.4 Path of a charged particle and emission seen by observer (☺).

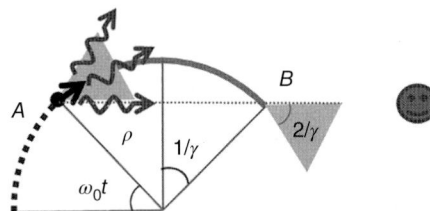

Replacing ρ, ε_c can be also written as

$$\varepsilon_c = \frac{3\hbar c^2}{4\pi} \frac{B_0 E^2}{E_0^3} \quad (2.20)$$

A more practical unit expression of ε_c (in eV) is given by

$$\boxed{\varepsilon_c[\text{eV}] = \frac{3 \times 10^{-9} \hbar c^2 [\text{m/s}]}{4\pi q} \frac{B_0[\text{T}] E^2[\text{GeV}]}{E_0^3[\text{GeV}]}} \quad (2.21)$$

The energy at rest E_0 plays an important role in the spectral domain of the emission. Indeed Eq. (2.16) gives a completely different value of ε_c when considering electrons or protons. Developing Eq. (2.16) ε_c can be written as

- $\varepsilon_c[\text{eV}] = 665.02 E^2[\text{GeV}] B_0[\text{T}]$ for electrons. Operation at SOLEIL ($E = 2.75$ GeV and $B_0 = 1.71$ T) enables to reach radiation at critical energy of 8600 eV (hard X-rays).
- $\varepsilon_c[\text{eV}] = 1.0743 \times 10^{-7} E^2[\text{GeV}] B_0[\text{T}]$ for protons. Operation at LHC ($E = 7000$ GeV and $B_0 = 8.3$ T) enables to reach radiation at critical energy of 43.7 eV (VUV-very soft X-rays).

2.2.2 Photon Flux

The angular flux is calculated with the analytical expression Eq. (2.15). Considering a circular motion described in Figure 2.3, the particle position $\vec{r}_p$ and its normalised velocity $\vec{\beta}$ are written as follows:

- $\vec{r}_p(t) = \rho[(1 - \cos\omega_0 t), 0, \sin\omega_0 t]$
- $\vec{\beta}(t) = \beta[\sin\omega_0 t, 0, -\cos\omega_0 t]$

The observation direction can be also written as

- $\vec{n} = [0, \sin\psi, \cos\psi]$

Then using the properties of triple vector product,

$$\vec{n} \times (\vec{n} \times \vec{\beta}) = \beta[-\sin(\omega_0 t), \cos(\psi)\sin(\psi)\cos(\omega_0 t), -\sin^2(\psi)\cos(\omega_0 t)]$$

Assuming also that ψ is small and the pulses are very short, one can write that

- $\sin\omega_0 t \approx \omega_0 t - \frac{(\omega_0 t)^3}{6}$
- $\cos\omega_0 t \approx 1$
- $\sin\psi \approx \psi$
- $\cos\psi \approx 1 - \frac{\psi^2}{2}$

The triple vector product can be approximated to $\vec{n} \times (\vec{n} \times \vec{\beta}) = \beta[-\omega_0 t, \Psi, 0]$ at first order.

2.2 SR from Dipoles

On the other hand, taking into account that $\rho/c = \beta/\omega_0$ and $\beta \sim 1 - 1/2\gamma^2$, the expansion at third order of the term in the exponential can be written as

$$t - \frac{\vec{n}\vec{r}_p}{c} = t - \frac{\rho}{c}\cos\psi \cdot \sin\omega_0 t \approx t - \frac{\rho}{c}\left[1 - \frac{\psi^2}{2}\right]\left[\omega_0 t - \frac{(\omega_0 t)^3}{6}\right]$$

$$\approx \frac{t}{2\gamma^2}[1 + \gamma^2\psi^2] + \frac{\omega_0^2}{6}t^3$$

Equation (2.15) can be written differently with distinguished terms A_x and A_z:

$$\frac{d^2\Gamma}{d\Omega d\omega/\omega} = \frac{q\omega^2}{8h\pi^3\varepsilon_0 c}I\{A_x^2 + A_z^2\}$$

where

$$A_x = \int_{-\infty}^{+\infty} -\omega_0 t \cdot \exp\left[j\omega\left\{\frac{t}{2\gamma^2}[1 + \gamma^2\psi^2] + \frac{\omega_0^2}{6}t^3\right\}\right]dt$$

$$= -2\omega_0 \int_0^{+\infty} t \cdot \sin\left[j\omega\left\{\frac{t}{2\gamma^2}[1 + \gamma^2\psi^2] + \frac{\omega_0^2}{6}t^3\right\}\right]dt$$

and

$$A_z = \psi \int_{-\infty}^{+\infty} \exp\left[j\omega\left\{\frac{t}{2\gamma^2}[1 + \gamma^2\psi^2] + \frac{\omega_0^2}{6}t^3\right\}\right]dt$$

$$= 2\psi \int_0^{+\infty} \cos\left[j\omega\left\{\frac{t}{2\gamma^2}[1 + \gamma^2\psi^2] + \frac{\omega_0^2}{6}t^3\right\}\right]dt$$

The integrals A_x and A_z can be expressed with modified Bessel functions such as $K_{1/3}$ and $K_{2/3}$:

$$\int_0^{+\infty} u \cdot \sin\left[\frac{3\nu}{2}\left(u + \frac{u^3}{3}\right)\right]du = \frac{1}{\sqrt{3}}K_{2/3}(\nu)$$

$$\int_0^{+\infty} \cos\left[\frac{3\nu}{2}\left(u + \frac{u^3}{3}\right)\right]du = \frac{1}{\sqrt{3}}K_{1/3}(\nu)$$

The expression of the flux density is composed of modified Bessel functions such as $K_{1/3}$ and $K_{2/3}$:

$$\frac{d^2\Gamma}{d\Omega d\omega/\omega} = \frac{3q}{8\pi^2}\frac{I}{\varepsilon_0 hc}\left[\frac{E}{E_0}\right]^2[1 + X^2]^2 y^2 \left\{K_{2/3}^2(\xi) + \frac{X^2}{1 + X^2}K_{1/3}^2(\xi)\right\}$$

(2.22)

With $y = \omega/\omega_c = \varepsilon/\varepsilon_c$, $X = \gamma\psi$, $\xi = y(1 + X^2)^{3/2}/2$.

On axis ($X = 0$) the analytical expression of the flux is reduced to

$$\frac{d^2\Gamma}{d\Omega d\omega/\omega} = \frac{3q}{8\pi^2}\frac{I}{\varepsilon_0 hc}\left[\frac{E}{E_0}\right]^2 y^2 K_{2/3}^2(y/2) = \frac{3q}{8\pi^2}\frac{I}{\varepsilon_0 hc}\left[\frac{E}{E_0}\right]^2 H_2(y) \quad (2.23)$$

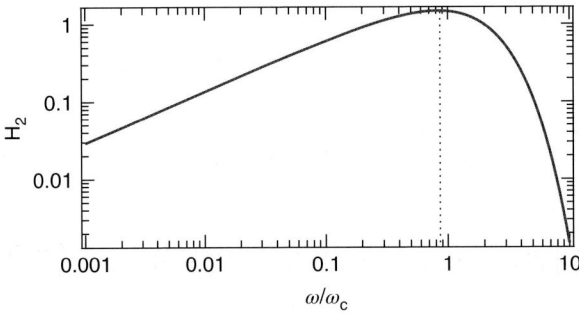

Figure 2.5 Graph of the function H_2.

where $H_2(y) = y^2 K_{2/3}^2(y/2)$. Figure 2.5 shows the function $H_2(y)$ which increases slowly up to a maximum for $\omega/\omega_c = \varepsilon/\varepsilon_c = 0.83$ with a value of 1.47 and drops sharply at high photon energy.

The flux per horizontal angle θ_x and relative bandwidth $d\omega/\omega$ is obtained by integrating Eq. (2.23) over all vertical angles Ψ:

$$\frac{d^2\Gamma}{d\Omega d\omega/\omega} = \frac{3q}{8\pi^2} \frac{I}{\varepsilon_0 hc} \left[\frac{E}{E_0}\right]^2 y^2 K_{2/3}^2(y/2) = \frac{3q}{8\pi^2} \frac{I}{\varepsilon_0 hc} \left[\frac{E}{E_0}\right]^2 H_2(y) \quad (2.24)$$

where $G_1(y) = y \int_y^{+\infty} K_{5/3}^2(u) du$.

The function $G_1(y)$ is plotted in Figure 2.6. $G_1(y)$ is maximum for $\omega/\omega_c = \varepsilon/\varepsilon_c = 0.29$ that is below the critical energy.

Expressed in practical units [photons/s/0.1% bandwidth/mrad horizontal angle] Eq. (2.24) becomes

$$\frac{d^2\Gamma}{d\theta_x d\omega/\omega}[\text{photons}/(s \cdot 0.1\%\text{BW} \cdot \text{mrad})] = 10^{-3} \frac{\sqrt{3}q}{4\pi\varepsilon_0} \frac{I[A]}{hc} \frac{E[\text{GeV}]}{E_0[\text{GeV}]} G_1(y) \quad (2.25)$$

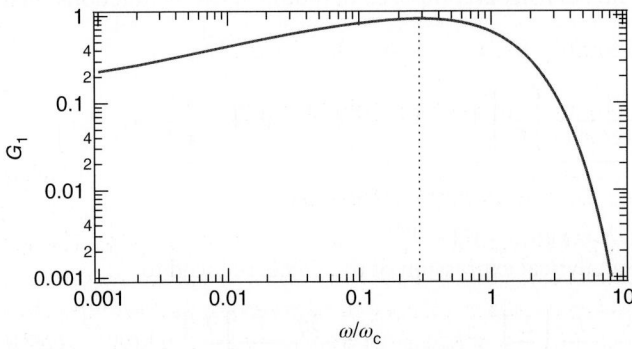

Figure 2.6 Graph of the function G_1.

Note:

- For electrons: $\dfrac{d^2\Gamma}{d\theta_x d\omega/\omega}$ [photons/(s·0.1%BW·mrad)] $= 2.457 \times 10^{13} I[A] E[\text{GeV}] G_1(y)$
- For protons: $\dfrac{d^2\Gamma}{d\theta_x d\omega/\omega}$ [photons/(s·0.1%BW·mrad)] $= 1.3407 \times 10^{10} I[A] E[\text{GeV}] G_1(y)$

At equivalent energy, protons produce less flux than electrons because of their larger rest mass.

The total flux per horizontal angle integrated over all photon energies is written as follows:

$$\frac{d\Gamma}{d\theta_x} = \frac{\sqrt{3}q}{4\pi\varepsilon_0} \frac{I}{hc} \frac{E}{E_0} \int_0^{+\infty} \left[\int_y^{+\infty} K_{5/3}^2(u) du \right] dy \tag{2.26}$$

Note: $d\omega/\omega = dy/y$

$$\frac{d\Gamma}{d\theta_x} [\text{photons}/(\text{s}\cdot\text{mrad})] = \frac{5 \times 10^{-3} q}{4\sqrt{3}} \frac{I[A]}{\varepsilon_0 hc} \frac{E[\text{GeV}]}{E_0[\text{GeV}]} \tag{2.27}$$

Over one storage ring turn ($\Delta\theta_x = 2000\pi$ mrad) the total photon flux is

$$\boxed{\Gamma[\text{photons/s}] = \frac{10\pi q}{4\sqrt{3}} \frac{1}{\varepsilon_0 hc} \frac{I[A]E[\text{GeV}]}{E_0[\text{GeV}]}} \tag{2.28}$$

Note:

- For electrons: $\Gamma[\text{photons/s}] = 8.08 \times 10^{20} I[A] E[\text{GeV}]$
- For protons: $\Gamma[\text{photons/s}] = 4.4 \times 10^{17} I[A] E[\text{GeV}]$

Or also per 1 m of bending magnet:

$$\boxed{\frac{d\Gamma}{ds}[\text{photons}/(\text{s}\cdot\text{m})] = \frac{10\pi q}{4\sqrt{3}(2\pi\rho)} \frac{I[A]}{\varepsilon_0 hc} \frac{E[\text{GeV}]}{E_0[\text{GeV}]} = \frac{10^{-8} q}{8\sqrt{3}} \frac{I[A]}{\varepsilon_0 h} \frac{B_0[T]}{E_0[\text{GeV}]}} \tag{2.29}$$

Note:

- For electrons: $\dfrac{d\Gamma}{ds}[\text{photons}/[\text{s}\cdot\text{m}]] = 3.86 \times 10^{19} I[A] B_0[T]$
- For protons: $\dfrac{d\Gamma}{ds}[\text{photons}/[\text{s}\cdot\text{m}]] = 2.1 \times 10^{16} I[A] B_0[T]$

2.2.3 Vertical Angular Distribution of Photon Flux

The angular distribution of photon flux can be calculated whatever the vertical angle Ψ is, by programming an evoluted calculator or with data manager software such as EXCEL or equivalent. Figure 2.7 presents the vertical profile of the radiation calculated with Eq. (2.22) for several photon energies. The power density has been divided by $\dfrac{d^2\Gamma}{d\Omega d\omega/\omega}(\gamma\Psi = 0)$ for normalisation (given by Eq. (2.24)).

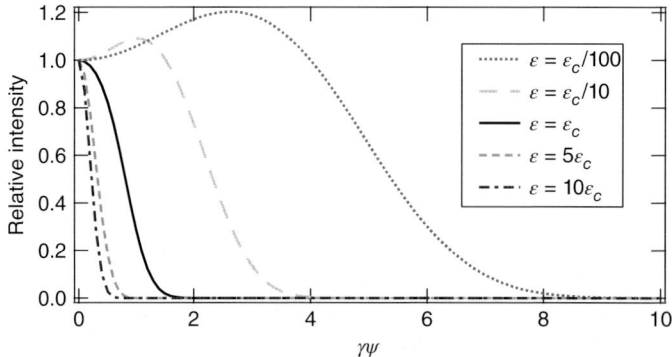

Figure 2.7 Vertical distribution of flux versus vertical angle $\gamma\psi$.

From Figure 2.7 it can be seen that photon of high energies (above ε_c) are concentrated close to the medium plane (small angles). The vertical aperture $\Delta\psi$ of the radiation depends on the photon energy. Three cases can be considered:

- $\varepsilon \ll \varepsilon_c$, $\quad \Delta\Psi \cong \frac{1}{\gamma}\left[\frac{\varepsilon_c}{\varepsilon}\right]^{1/3}$
- $\varepsilon = \varepsilon_c$, $\quad \Delta\Psi \cong \frac{1}{\gamma}$
- $\varepsilon \gg \varepsilon_c$, $\quad \Delta\Psi \cong \frac{1}{\gamma}\left[\frac{\varepsilon}{\varepsilon_c}\right]^{1/2}$

The flux crossing an angular aperture $(\Delta\theta_0, \Delta\psi_0)$ is calculated by numerical integration of Eq. (2.22) for a given ε_c. The angular aperture (Figure 2.8) is defined by the horizontal size ΔX, the vertical size of the slit a inside the vacuum chamber, and the distance from the source L:

$$\Delta\theta_0 \approx \Delta X_0/L \quad \text{and} \quad \Delta\Psi_0 \approx a/L \quad \text{(small angle approximation)}$$

$$\frac{d\Gamma}{d\omega/\omega}(\Delta\theta_0, \Delta\psi_0) = \int \frac{d^2\Gamma}{d\Omega d\omega/\omega} d\theta d\Psi$$

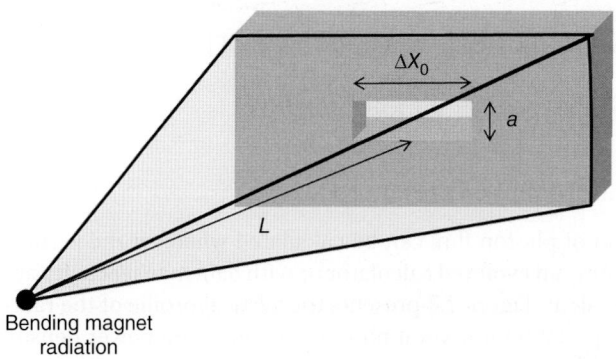

Figure 2.8 Scheme of the bending magnet radiation passing through an aperture of a vacuum chamber.

2.2 SR from Dipoles

which can be written as

$$\frac{d\Gamma}{d\omega/\omega}(\Delta\theta_0, \Delta\psi_0) = \frac{3q}{4\pi^2\varepsilon_0}\frac{\gamma}{hc}\Delta\theta_0 I \int_0^{\frac{\gamma\Delta\psi_0}{2}} T\left(\frac{\varepsilon}{\varepsilon_c}, X\right) dX$$

with

$$T\left(\frac{\varepsilon}{\varepsilon_c}, X\right) = \left(\frac{\varepsilon}{\varepsilon_c}\right)^2 (1+X^2)^2$$

$$\times \left[K_{2/3}^2\left(\frac{\varepsilon}{2\varepsilon_c}[1+X^2]^{3/2}\right) + \frac{X^2}{1+X^2}K_{1/3}^2\left(\frac{\varepsilon}{2\varepsilon_c}[1+X^2]^{3/2}\right)\right]$$

The integral of $T\left(\frac{\varepsilon}{\varepsilon_c}, X\right)$, $\int_0^{\frac{\gamma\Delta\psi_0}{2}} T\left(\frac{\varepsilon}{\varepsilon_c}, X\right) dX$, is presented in Figure 2.9 versus ratio $\varepsilon/\varepsilon_c$ and vertical angular aperture.

By choosing $\varepsilon/\varepsilon_c$ and the vertical angular aperture $\Delta\psi_0$, it is possible from the graph to obtain a good estimation of $\int_0^{\frac{\gamma\Delta\psi_0}{2}} T\left(\frac{\varepsilon}{\varepsilon_c}, X\right) dX$ and thus calculate the flux over the aperture $\Delta\theta_0$ (horizontal) $\times \Delta\psi_0$ (vertical).

Example: A slit of horizontal aperture 10 mm and vertical aperture 330 μm installed at 10 m from the emission corresponds to a normalised vertical angle $\gamma\Delta\psi_0$ of 1.8 at SOLEIL ($\gamma = 5380$). If one chooses the ratio $\varepsilon/\varepsilon_c = 0.02$, the integral of $T\left(\frac{\varepsilon}{\varepsilon_c}, X\right)$ read in figure is $\int_0^{\frac{\gamma\Delta\psi_0}{2}} T\left(\frac{\varepsilon}{\varepsilon_c}, X\right) dX = 0.2$. The flux collected in the angular aperture $\Delta\theta_0$ (0.1 mrad) $\times \Delta\psi_0$ (33 μrad) is obtained by multiplying the result (0.2) by $\frac{3q}{4\pi^2\varepsilon_0}\frac{\gamma}{hc}\Delta\theta_0 I$.

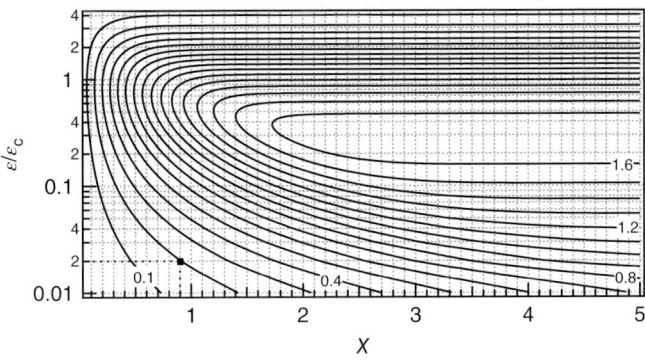

Figure 2.9 $\int_0^X T\left(\frac{\varepsilon}{\varepsilon_c}, u\right) du$ versus X.

2.2.4 Photon Power

Using the relation between the spectral power density and the spectral flux density, $\frac{d^2P}{d\Omega d\omega} = \frac{\hbar\omega}{2\pi}\frac{d^2\Gamma}{d\Omega d\omega}$, and integrating Eq. (2.22), the angular power density can be expressed as

$$\frac{dP}{d\Omega} = \frac{7q}{64\pi\varepsilon_0}\frac{\gamma^5}{\rho}I\frac{1}{(1+\gamma^2\Psi^2)^{5/2}}\left[1 + \frac{5\gamma^2\Psi^2}{7(1+\gamma^2\Psi^2)}\right] \quad (2.30)$$

Or:

$$\frac{dP}{d\Omega}\left[\frac{W}{mrad^2}\right] = \frac{7 \times 10^{-15} qc}{64\pi\varepsilon_0} \frac{E^4[GeV]}{E_0^5[GeV]}$$
$$I[A]\frac{B_0[T]}{(1 + [E/E_0]^2\Psi^2)^{5/2}}\left[1 + \frac{5[E/E_0]^2\Psi^2}{7(1 + [E/E_0]^2\Psi^2)}\right] \quad (2.31)$$

Note:

- For electrons: $\dfrac{dP}{d\Omega}\left[\dfrac{W}{mrad^2}\right] = 5.420 \dfrac{E^4[GeV]B[T]I[A]}{(1 + \gamma^2\Psi^2)^{5/2}}\left[1 + \dfrac{5\gamma^2\Psi^2}{7(1 + \gamma^2\Psi^2)}\right]$

- For protons: $\dfrac{dP}{d\Omega}\left[\dfrac{W}{mrad^2}\right] = 2.952 \times 10^{-3} \dfrac{E^4[GeV]B[T]I[A]}{(1 + \gamma^2\Psi^2)^{5/2}}\left[1 + \dfrac{5\gamma^2\Psi^2}{7(1 + \gamma^2\Psi^2)}\right]$

On axis ($\Psi = 0$):

$$\frac{dP}{d\Omega} = \frac{7 \times 10^{-9} qc}{64\pi\varepsilon_0} \frac{E^4[GeV]}{E_0^5[GeV]} I[A]B_0[T] \quad (2.32)$$

Expressed in W/mrad²:

$$\frac{dP}{d\Omega}\left[\frac{W}{mrad^2}\right] = \frac{7 \times 10^{-15} qc}{64\pi\varepsilon_0} \frac{E^4[GeV]}{E_0^5[GeV]} I[A]B_0[T] \quad (2.33)$$

The integration of Eq. (2.31) over the vertical angle Ψ gives the horizontal angular power density:

$$\frac{dP}{d\theta_x}\left[\frac{W}{mrad}\right] = \frac{10^{-12} qc}{6\pi\varepsilon_0} \frac{E^3[GeV]}{E_0^4[GeV]} I[A]B_0[T] \quad (2.34)$$

After one turn ($\Delta\theta_x = 2000\pi$ mrad), the radiated power P is

$$P[W] = \frac{10^{-9} qc}{3\varepsilon_0} \frac{E^3[GeV]}{E_0^4[GeV]} I[A]B_0[T] \quad (2.35)$$

Note:

- For electrons: $P[W] = 2.652 \times 10^4 E^3[GeV]I[A]B_0[T]$
- For protons: $P[W] = 14.45 E^3[GeV]I[A]B_0[T]$

The power loss per meter is equal to the total power divided by the circumference. $dP/ds = P/2\pi\rho$. In other words,

$$\frac{dP}{ds}\left[\frac{W}{m}\right] = \frac{10^{-18} qc^2}{6\pi\varepsilon_0} \frac{B_0^2[T]E^2[GeV]}{E_0^4[GeV]} I[A] \quad (2.36)$$

Note:

- For electrons: $\dfrac{dP}{ds}\left[\dfrac{W}{m}\right] = 1270.31 B_0^2[T]E^2[GeV]I[A]$

- For protons: $\dfrac{dP}{ds}\left[\dfrac{W}{m}\right] = 1.12 \times 10^{-10} B_0^2[T]E^2[GeV]I[A]$

For example:

- Electrons in SOLEIL dipole: $E = 2.75$ GeV, $B = 1.71$ T, $I = 0.5$ A, hence $\frac{dP}{ds} = 14\,045$ W/m
- Electrons in DLS dipole: $E = 3$ GeV, $B = 1.4$ T, $I = 0.5$ A, hence $\frac{dP}{ds} = 11\,204$ W/m

2.2.5 Vertical Angular Distribution of Power

Equation (2.31) can be also written as

$$\frac{dP}{d\Omega} = \frac{7q}{64\pi\varepsilon_0} \frac{\gamma^5}{\rho} IS(X)$$

where S is a function defined by

$$S(X) = \frac{1}{(1+X^2)^{5/2}} \left[1 + \frac{5X^2}{7(1+X^2)}\right]$$

X represents the vertical normalised angle $\gamma\Psi$. Figure 2.9 presents the shape of the $S(X)$ versus normalised angle X.

The distribution is well approximated by a Gaussian function with standard deviation $\sigma_P \approx \frac{0.608}{\gamma}$. Let's recall that the power distribution is constant over the horizontal angle θ. One can calculate the power transmitted through an angular aperture $(\Delta\theta_0, \Delta\psi_0)$:

$$P(\Delta\theta_0, \Delta\psi_0) = \int \frac{dP}{d\Omega} d\theta d\Psi = \frac{7q}{32\pi\varepsilon_0} \frac{\gamma^4}{\rho} \Delta\theta_0 I \int_0^{+\gamma\Psi_0/2} S(X)dX$$

By integrating $S(X)$, one can find that

$$\int_0^{+\gamma\Psi_0/2} S(X)dX = \frac{0.761905\,(\gamma\Psi_0/2)\left[0.75 + (\gamma\Psi_0/2)^2\right]\left[1.75 + (\gamma\Psi_0/2)^2\right]}{\left[1 + (\gamma\Psi_0/2)^2\right]^{5/2}} \quad (2.37)$$

with $\int_0^{+\infty} S(X)dX = 0.761863$ (see Figure 2.10).

As a conclusion the power passing through the aperture $[\Delta X_0, a]$ is written as

$$\boxed{P(\Delta X_0, a) = \frac{7q}{32\pi\varepsilon_0} \frac{\gamma^4}{\rho} \frac{\Delta X_0}{L} I \cdot \frac{0.761905\left(\frac{\gamma a}{2L}\right)\left[0.75 + \left(\frac{\gamma a}{2L}\right)^2\right]\left[1.75 + \left(\frac{\gamma a}{2L}\right)^2\right]}{\left[1 + \left(\frac{\gamma a}{2L}\right)^2\right]^{5/2}}}$$

(2.38)

Example: SOLEIL ($\gamma = 5360$, $\rho = 5.36$), the power transmitted through a slit of 50 mm (horizontal) × 10 mm (vertical) installed at 10 m from the emission point is 375 W, whereas the total emitted power is 471 kW.

Note: The power absorbed by the vacuum chamber P_{abs} can be evaluated by subtracting $P(\Delta X_0, a)$ from the total radiated power.

$$P_{abs}(\Delta X_0, a) = \frac{7q}{32\pi\varepsilon_0} \frac{\gamma^4}{\rho} \frac{\Delta X_0}{L} I$$

$$\times \left[0.761863 - \frac{0.761905 \left(\frac{\gamma a}{2L}\right) \left[0.75 + \left(\frac{\gamma a}{2L}\right)^2\right] \left[1.75 + \left(\frac{\gamma a}{2L}\right)^2\right]}{\left[1 + \left(\frac{\gamma a}{2L}\right)^2\right]^{5/2}} \right]$$

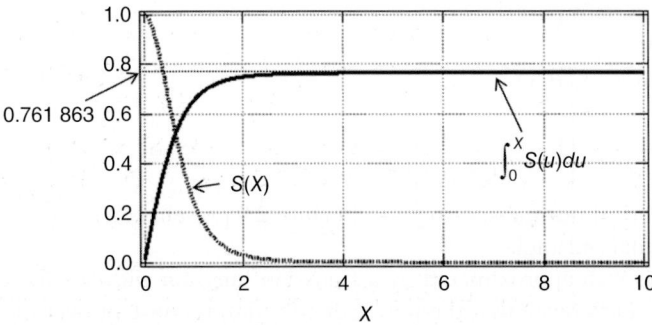

Figure 2.10 Function S and integral of S versus vertical normalised angle $X = \gamma \Psi$.

2.3 SR from Quadrupoles

SR can be generated by other magnets in accelerators. In particle colliders the final focusing quadrupoles have a dual role: to focus the beam to its minimum size and to drive the beams to the interaction point. The charged particles enter the quadrupole at a distance a from the magnetic axis and are submitted to a magnetic field:

$$B[T] = K \left[\frac{T}{m}\right] a[m] \tag{2.39}$$

where K is the quadrupole field gradient. In some cases, such a magnetic field is high enough for SR generations (e.g. in Inner Triplet quadrupoles in the LHC and FCC).

In the analytical expression of the flux and the power, the product $I \cdot B$ should be replaced for a Gaussian beam by $\int_{-\infty}^{+\infty} \frac{t}{\sqrt{2\pi}\sigma} e^{-\frac{(x-a)^2}{2\sigma^2}} |B(x)| dx$

- The particle beam is not necessary centred on the quadrupole axis (the beam could be displaced by the value a). If $a \gg \sigma$, one considers that all electrons of the beam have the same position sign. The magnetic field is replaced by the average magnetic field B_{avg} defined as $B_{avg}[T] = K \left[\frac{T}{m}\right] a[m]$.

Equations (2.29) and (2.36) are independent from the transverse beam size σ and are modified as follows:

$$\frac{d\Gamma}{ds}[\text{photons}/(\text{s}\cdot\text{m})] = \frac{10^{-8} q \, I[A]}{8\sqrt{3} \, \varepsilon_0 h} \frac{B_{\text{avg}}[T]}{E_0[\text{GeV}]} \quad (2.40)$$

$$\frac{dP}{ds}\left[\frac{W}{m}\right] = \frac{10^{-18} q c^2}{6\pi\varepsilon_0} \frac{B_{\text{avg}}^2[T] E^2[\text{GeV}]}{E_0^{\,4}[\text{GeV}]} I[A] \quad (2.41)$$

Example: For proton beam of 14 TeV and 0.5 A entering a quadrupole of 200 T/m with a misalignment a of 1 mm, the total flux and power are respectively 2.11×10^{15} photons/s and 8.8×10^{-4} W/m.

For electron beam of 3 GeV and 0.5 A entering a quadrupole of 100 T/m with a misalignment a of 1 mm, the total flux and power are respectively 1.93×10^{18} photons/s and 57.2 W/m.

- If the particle beam is centred ($a = 0$), particle located off-axis may radiate. The magnetic field is replaced by the mean square magnetic field B_{rms} defined as $B_{\text{rms}}[T] = K\left[\frac{T}{m}\right]\sigma[m]$

$$\frac{d\Gamma}{ds}[\text{photons}/(\text{s}\cdot\text{m})] = \frac{10^{-8} q \, I[A]}{8\sqrt{3} \, \varepsilon_0 h} \frac{B_{\text{rms}}[T]}{E_0[\text{GeV}]} \quad (2.42)$$

$$\frac{dP}{ds}\left[\frac{W}{m}\right] = \frac{10^{-18} q c^2}{6\pi\varepsilon_0} \frac{B_{\text{rms}}^2[T] E^2[\text{GeV}]}{E_0^{\,4}[\text{GeV}]} I[A] \quad (2.43)$$

Example: For proton beam of 14 TeV and 0.5 A entering a quadrupole of 200 T/m with a transverse size of 100 µm, the total flux and power are 2.11×10^{14} photons/s and 8.8×10^{-6} W/m, respectively.

For electron beam of 3 TeV and 0.5 A entering a quadrupole of 100 T/m with a transverse size of 100 µm, the total flux and power are respectively 1.93×10^{17} photons/s and 0.572 W/m.

2.4 SR from Insertion Devices

SR can also be generated in dedicated insertion devices (IDs) such as wigglers and undulators. In this section we will show the ways for simple estimation of the ID power and photon flux loads. Generally, ID consists of a big number of the magnetic poles, which produce an alternating magnetic field along charged particles trajectory.

The proposed semi-analytical methods provide the sufficient accuracy for estimation of the power load and photon flux on the wall of the vacuum chamber and on the radiation absorbers.

For more precise calculation, one can use a dedicated computer code. Some example of such codes will be detailed at the end of this section.

2.4.1 Motion of Charged Particles Inside a Planar Insertion Device

IDs are special organised magnetic structures capable of producing periodic alternating magnetic field and, as a consequence, imposing particles to undulate along the ID axis. A main advantage of IDs compared to bending magnet is the capability to generate much more photon flux, that is, many orders of magnitude higher.

The magnetic field inside an ID is periodic and can be described by the relation: $\vec{B}(z) = B_0 \cos(k_u z)\vec{y}$, where $k_u = \frac{2\pi}{\lambda_u}$. λ_u is called magnetic period. When an electron enters the undulator with a speed $\vec{v} = v_0 \vec{z}$, it is submitted to the Lorentz force already presented (Eq. (2.16)): $\gamma m_0 \frac{d\vec{\beta}}{dt} = q\vec{\beta} \times \vec{B_0}$. This relation is also written as

$$\frac{d\vec{\beta}}{dt} = -\frac{eB_0}{\gamma m_0} \cos(k_0 z)(\vec{\beta} \times \vec{y}) = \frac{eB_0}{\gamma m_0} \cos(k_0 z)\vec{x}$$

In first approximation ($z \sim ct$) and after integration,

$$\beta_x = \frac{K}{\gamma} \sin(k_u z)$$

With

$$\boxed{K = \frac{eB_0 \lambda_u}{2\pi m_0 c} = 0.934 B_0[\text{T}]\lambda_u[\text{cm}]} \tag{2.44}$$

K is called deflection parameter.

It can be seen that $\beta_y = 0$ (the particles undulate in the xz plane). On the other hand,

$$\beta^2 = \beta_x^2 + \beta_y^2 + \beta_z^2 \quad \text{and} \quad \beta = \sqrt{1 - \frac{1}{\gamma^2}}$$

β_z is written as:

$$\beta_z^2 = 1 - \frac{1}{\gamma^2} - \frac{K^2}{\gamma^2} \sin^2(k_u z)$$

However $\gamma \gg 1$, at first order in $1/\gamma$:

$$\beta_z \sim 1 - \frac{1}{2\gamma^2}\left[1 + \frac{K^2}{2}\right] - \frac{K^2}{4\gamma^2}\cos(2k_0 z) = \overline{\beta_z} - \frac{K^2}{4\gamma^2}\cos(2k_0 z)$$

$$\times \text{ with } \overline{\beta_z} = 1 - \frac{1}{2\gamma^2}\left[1 + \frac{K^2}{2}\right]$$

After integration with assumption that $z \sim ct$, the motion of particles is described by

$$x = \frac{K\lambda_u}{2\pi\gamma}[1 - \cos(k_u z)] \cong \frac{K\lambda_u}{2\pi\gamma}[1 - \cos(k_u ct)] \tag{2.45}$$

$$z \cong \overline{\beta_z} ct - \frac{K^2}{8\gamma^2 k_u} \sin(2k_u ct) \tag{2.46}$$

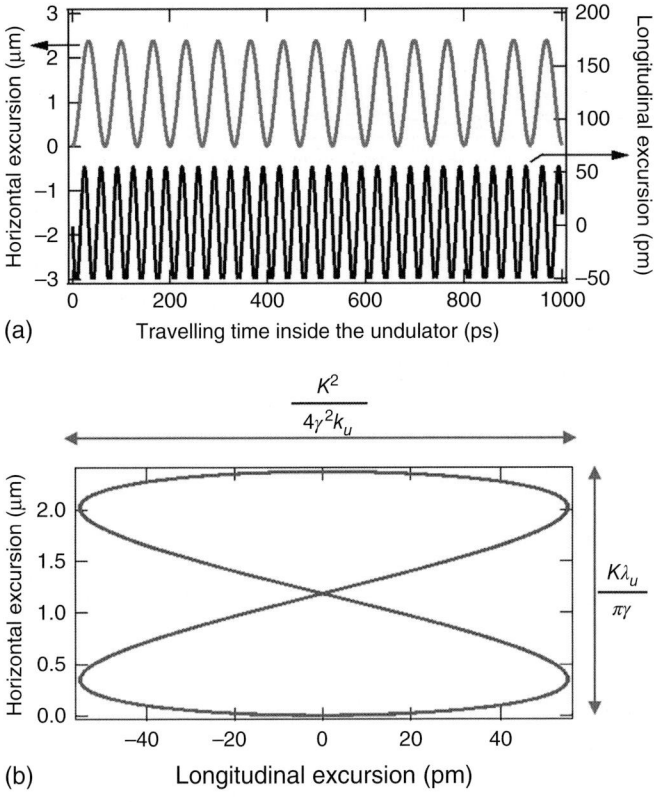

Figure 2.11 Motion of particles in the observer frame (a) and in the particle frame (b). One considers an undulator of 100 periods of 20 mm with a K value of 1.8 at SOLEIL (2.75 GeV).

Particles are submitted to oscillations in the transverse plane ($x0z$) at the spatial frequency k_u and also in the longitudinal axis z at the frequency $2k_u$. In the particle frame, the particle moving at the $\overline{\beta_z}$ speed describes an '8' shape (Figure 2.11). This effect is a characteristic of planar IDs.

Note: The '8' shape motion of particle explains the appearance of the fundamental wavelength as well as higher order harmonics. When $K \ll 1$, the longitudinal excursions are negligible. The particles behave as electrons in a hertzian antenna emitting at the fundamental wavelength. As K grows, the particle motion is periodically disturbed along the longitudinal axis leading to the appearance of higher order harmonic. This phenomenon enables thus to operate the ID at high energy by using high harmonics even for low-intermediate energy storage rings. However in circular polarised undulators, the electrons experiment a helicoidally trajectory with a constant longitudinal speed. The spectrum of the on-axis radiation presents only the resonant energy.

2.4.2 Resonance Wavelength

The resonant wavelength answers to a constructive interference condition imposing that during the travelling of one spatial period the particle is delayed by one electromagnetic temporal period T.

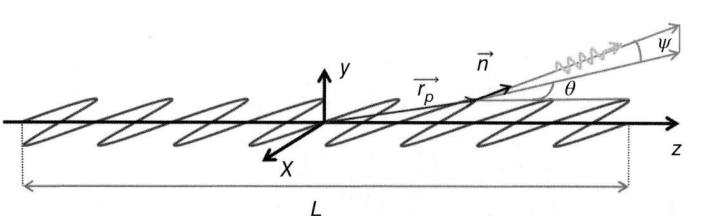

Figure 2.12 Schematic view of the particle motion inside the ID and radiation seen by an observer (☺).

This condition is converted in terms of wavelength as follows: $c\left[\frac{\lambda_0}{\beta_z c} - \frac{\lambda_0}{c}\cos(\theta)\cos(\psi)\right] = n\lambda_r$. Considering small angle observation ($\theta, \psi \ll 1$), the resonant wavelength can be written as

$$\lambda_r(\theta, \psi) = \frac{\lambda_u}{2\gamma^2}\left[1 + \frac{K^2}{2} + \gamma^2\theta^2 + \gamma^2\psi^2\right] \tag{2.47}$$

θ and ψ are the observation angles. It is also usual to talk about resonant energy defined as: $\varepsilon_r = \hbar\omega_r/2\pi$ (Figure 2.12).

2.4.3 Radiation from Undulators and Wigglers

Depending on the field amplitude and the photon energy, there are two distinguished modes of operation of IDs: wiggler mode and undulator mode.

- In the wiggler mode, photons emitted from different parts of the trajectory are absolutely independent of each other. Radiation coming from the magnetic poles interferes *with random phase* leading to a continuous photon spectrum similar to the spectrum radiated by bending magnets. The emission is said to be in*coherent*.
- In the undulator mode, radiations coming from each part of the trajectory interfere *coherently*. As a consequence, the spectrum presents a series of lines (harmonics) peaked at the resonant wavelength (Eq. (2.47)) and its harmonics.

To determine in which mode an ID is operated, one must find if the duration of the photon pulse Δt_{rad} is negligible (Figure 2.13a) or not (Figure 2.13b) with respect to the temporal periodicity of the radiation T.

In Section 2.2.1, we have seen that the time duration can be written as $\Delta t_{rad} = \frac{\rho}{\gamma^3 c}$ (FWHM), which can be also expressed as $\Delta t_{rad} = \frac{\lambda_0}{2\pi\gamma^2 cK}$. On the other hand, the temporal periodicity of the radiation T is defined by the delay between photons and electrons to traverse through one spatial period λ_0:

$T = \frac{\lambda_0}{2\gamma^2 c}\left(1 + \frac{K^2}{2}\right)$. If $\Delta t_{rad} > T/4$, photon pulses overlap each other resulting in *coherent* interferences. This situation defines the undulator regime. In the opposite situation (wiggler regime or wiggler mode), the radiation is composed of a train of individual short pulses, which do *not* interfere *coherently*. From the above requirement, this situation is encountered when $K > 1$.

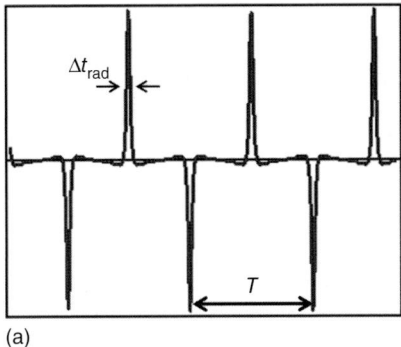 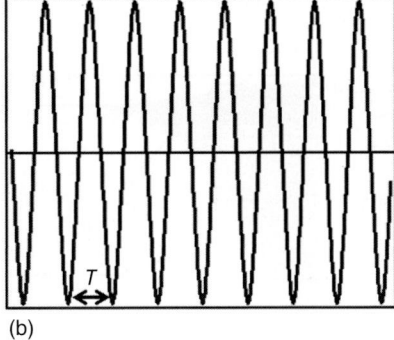

Figure 2.13 Temporal distribution of the radiation intensity for large K value (a) and low K value (b).

The K-factor is an important quantitative parameter that defines what type of radiation will be produced by an ID. K could own various physical meanings. On the one hand, it is the ratio of the maximal deflection angle of the electron trajectory from the longitudinal axis to the typical angle of the SR divergence, i.e. $(1/\gamma)$. From another point of view, K parameter can be considered as the ratio of the period length to length of radiation formation. If K is a smaller than 1 (undulator mode), the radiation formation length is larger than the spatial period: i.e. the wave field leave the charge particle where particles pass a few periods. This means that photons emitted from different points of the particle trajectory have a consistent phase and the constructive or destructive interference is possible between them. In the other case (wiggler mode), the radiation formation length is essentially smaller than the spatial period, and photons emitted in the same direction from different periods are absolutely independent. There is no interference between them in this case, and the observer can see radiation like a number of the short separated photon flashes. Every flash is similar to flash that an observer can see in the more simple case of the bending magnet with constant field along particle trajectory. The total flux in the wiggler case is higher than the irradiated from bending magnet flux by a factor of $2N$, N corresponding to the number of the wiggler periods.

In the undulator mode, because the angular deflection of the particle is lower than SR divergence fan, the observer can see a small variation of the SR intensity during particle passing through the periodic field. The time profile of the intensity, which the observer can see, has a near-sinusoidal shape. The Fourier transform of such profile defines a spectrum with discrete line structures.

Because harmonics are the result of coherent sum of the electric fields, the spectral width of the harmonics is proportional to $1/nN$, and peak intensities are proportional to N^2.

Note: Even if $K \gg 1$, one can distinguish the two modes of operation of the ID. In this case the spectrum presents a series of lines at low photon energy and a continuous evolution at high energy. The frontier between the 2 regimes is reached when the undulator structure disappear in the spectrum (Figure 2.14).

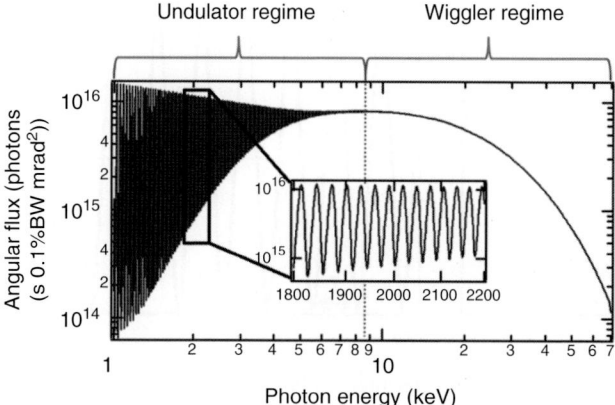

Figure 2.14 Spectrum radiated by a 2-m-long wiggler ($K = 9.8$, $\lambda_u = 50$ mm) at SOLEIL (2.75 GeV).

To calculate the photon flux emitted by an ID, let us recall the spectral intensity Eq. (2.13):

$$\frac{d^2W}{d\Omega d\omega} = \frac{q^2\omega^2}{16\pi^3\varepsilon_0 c}\left|\int_{-\infty}^{+\infty} \vec{n}\times(\vec{n}\times\vec{\beta})e^{-j\omega\left(t-\frac{\vec{n}\cdot\vec{r_p}}{c}\right)}dt\right|^2$$

- $\vec{n} = (\cos(\theta)\cos(\psi), \cos(\theta)\sin(\psi), \sin(\psi))$
- $\vec{\beta} = \frac{K}{\gamma}\cos(k_u z), 0, 1 - \frac{1}{2\gamma^2}\left[1+\frac{K^2}{2}\right] - \frac{K^2}{4\gamma^2}\cos(2k_u z)$
- $\vec{r_p} = \left(\frac{K\lambda_u}{2\pi\gamma}[1-\cos(k_u ct)], 0, \overline{\beta}_z ct - \frac{K^2}{8\gamma^2 k_u}\sin(2k_u ct)\right)$ with $\overline{\beta}_z = 1 - \frac{1}{2\gamma^2}\left[1+\frac{K^2}{2}\right]$

To calculate the spectral intensity, one uses the development in Bessel functions: $e^{ja\sin\theta} = \sum_{n=-\infty}^{n=+\infty} J_n(a)e^{jn\theta}$.

The spectral intensity is superimposition of spectral lines of magnitude I_n:

$$\frac{\partial^2 W}{\partial\Omega\partial\omega} = \sum_1^n I_n \text{ where } I_n = \frac{q^2\omega^2 K^2 N^2 \lambda_0^2}{16\pi^3\varepsilon_0 c(4\gamma^2 c^2)}\cdot$$

$$F_n(\alpha,\xi)\cdot\text{sinc}^2\left[\pi N\frac{\omega-n\omega_r}{\omega_r}\right] \quad (2.48)$$

$$F_n(\alpha,\xi) = \left|\sum_{k,l} J_k(n\alpha)J_j(n\xi)\cdot\{\delta(n+1,-k-2l)+\delta(n-1,-k-2l)\}\right|^2$$

δ is the Kroeneker symbol defined as $\delta(i,j) = 1$ if $i = j$ and $\delta(i,j) = 0$ otherwise. $\omega_r = \frac{2\pi c}{\lambda_r}$ is the resonance pulsation; $\alpha = \frac{2K\theta\gamma\cos\psi}{1+\frac{K^2}{2}+\gamma^2\theta^2}$ and

$\xi = \frac{K^2}{4\left(1+\frac{K^2}{2}+\gamma^2\theta^2\right)}$; N is the number of periods.

The spectral intensity differs from zero for pulsations close to $n\omega_r$. With the approximation of $\omega \sim n\omega_r$, I_n is written as

$$I_n = \frac{q^2\gamma^2 K^2 N^2 n^2}{4\pi\varepsilon_0 c\left(1 + \frac{K^2}{2} + \gamma^2\theta^2\right)^2} \cdot F_n(\alpha, \xi) \cdot \text{sinc}^2\left[\pi N \frac{\omega - n\omega_r}{\omega_r}\right]$$

which is also expressed as

$$I_n = \frac{q^2\gamma^2 N^2}{4\pi\varepsilon_0 c} \cdot G_n(\alpha, \xi) \cdot \text{sinc}^2\left[\pi N \frac{\omega - n\omega_r}{\omega_r}\right]$$

$\times$ with $G_n(\alpha, \xi) = \dfrac{K^2 n^2}{\left(1 + \frac{K^2}{2} + \gamma^2\theta^2\right)^2} \cdot F_n(\alpha, \xi)$

or

$$I_n = \frac{q^2\gamma^2 N^2}{4\pi\varepsilon_0 c} \cdot G_n(\alpha, \xi) \cdot \text{sinc}^2\left[\pi N \frac{\varepsilon - n\varepsilon_r}{\varepsilon_r}\right] \qquad (2.49)$$

with $E_r = \frac{h}{2\pi}\omega_r$

On axis ($\theta = 0$, $\psi = 0$) G_n can be simplified:

$$G_n(K) = \frac{K^2 n^2}{\left(1 + \frac{K^2}{2}\right)^2} \left|\sum_k J_k\left(\frac{nK^2}{4\left(1 + \frac{K^2}{2}\right)}\right) \cdot \{\delta(n+1, -2l) + \delta(n-1, -2l)\}\right|^2$$

Two cases appear:

- n is even: $G_n(K) = 0$
- n is odd: $G_n(K) = \dfrac{K^2 n^2}{\left(1 + \frac{K^2}{2}\right)^2} \cdot \left[J_{\frac{n+1}{2}}\left(\frac{nK^2}{4\left(1 + \frac{K^2}{2}\right)}\right) - J_{\frac{n-1}{2}}\left(\frac{nK^2}{4\left(1 + \frac{K^2}{2}\right)}\right)\right]^2$

$$\frac{\partial^2 W}{\partial\Omega\partial\omega} = \sum_n^\infty \frac{q^2\gamma^2 N^2}{4\pi\varepsilon_0 c} \cdot G_n(K) \cdot \text{sinc}^2\left[\pi N \frac{\varepsilon - n\varepsilon_r}{\varepsilon_r}\right] \qquad (2.50)$$

On axis no even harmonics appear. The amplitude scales as N^2 and the relative FWHM width is given by the sinc part of the expression. $\left.\frac{\Delta E}{E}\right|_{\text{Nat}} \cong \frac{0.9}{nN}$

For example:

- 2-m-long undulator composed of 40 periods operating at SOLEIL on the first harmonic: $\left.\frac{\Delta E}{E}\right|_{\text{Nat}} = 0.018$
- 2-m-long undulator composed of 100 periods operating at SOLEIL on the fifteenth harmonic: $\left.\frac{\Delta E}{E}\right|_{\text{Nat}} = 6 \times 10^{-4}$
- 90-m-long undulator composed of 4500 periods operating at SACLA FEL on the first harmonic: $\left.\frac{\Delta E}{E}\right|_{\text{Nat}} = 2 \times 10^{-4}$

The G_n function determines the amplitude of the spectral intensity. Figure 2.15 shows the evolution of G_n versus K for several harmonics n.

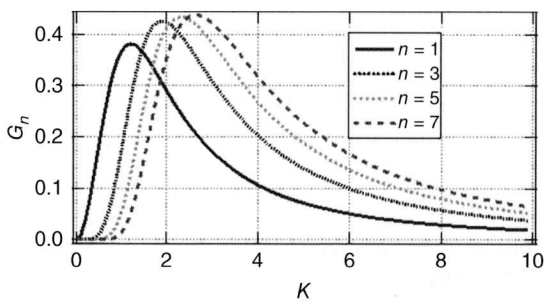

Figure 2.15 G_n versus K value.

To calculate the angular flux $\frac{\partial^2 \Gamma}{\partial \Omega \partial \omega}$, one takes into account that $\frac{I}{q}\frac{\partial^2 W}{\partial \Omega \partial \omega} = \frac{\hbar \omega}{2\pi}\frac{\partial^2 \Gamma}{\partial \Omega \partial \omega}$:

$$\frac{\partial^2 \Gamma}{\partial \Omega \partial \omega / \omega} = \frac{q}{2\varepsilon_0 hc} IN^2 \gamma^2 \sum_n^\infty G_n(K) = \frac{q}{2\varepsilon_0 hc} IN^2 \left[\frac{E}{E_0}\right]^2 \sum_n^\infty G_n(K) \quad (2.51)$$

Considering electrons (or positrons) Eq. (2.51) can be written as

$$\frac{\partial^2 \Gamma}{\partial \Omega \partial \omega / \omega}[\text{photons}/(\text{s} \cdot 0.1\%\text{BW} \cdot \text{mrad}^2)]$$

$$= 1.744 \times 10^{14} I[\text{A}] N^2 E^2[\text{GeV}] \sum_n^\infty G_n(K) \quad (2.52)$$

Integrated over all angles the total flux is written:

$$\frac{\partial \Gamma}{\partial \omega / \omega} = \sum_n^\infty \frac{q \pi N}{2\varepsilon_0 hc} IQ_n(K) \quad (2.53)$$

with

$$Q_n(K) = \frac{K^2 n}{1 + \frac{K^2}{2}} \cdot \left[J_{\frac{n+1}{2}}\left(\frac{nK^2}{4\left(1 + \frac{K^2}{2}\right)}\right) - J_{\frac{n-1}{2}}\left(\frac{nK^2}{4\left(1 + \frac{K^2}{2}\right)}\right) \right]^2$$

Figure 2.16 presents the evolution of Q_n versus K value.

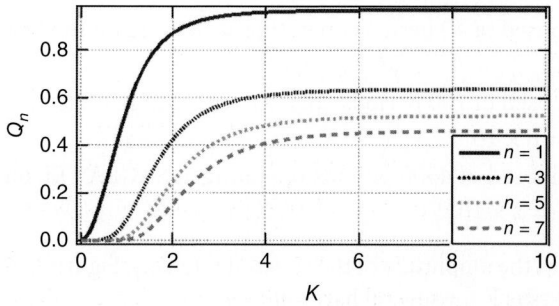

Figure 2.16 Q_n versus K value.

Considering electrons (or positrons) Eq. (2.53) can be written as

$$\frac{\partial \Gamma}{\partial \omega/\omega}[\text{photons}/(s \cdot 0.1\%\text{BW})] = 1.431 \times 10^{14} NI[\text{A}] \sum_{n}^{\infty} Q_n(K). \quad (2.54)$$

2.4.4 Angular Aperture of ID at Resonant Wavelength

To calculate the angular aperture of the radiation in the undulator regime, one can approximate the distribution as Gaussian [5]. A simplest method is to start from the analytical expression of the resonant wavelength λ_r (see Eq. (2.47)) and to find the angles θ_1 and θ_2 corresponding to a shift of $-1/2nN$ in regard to λ_r on axis. The difference $\Delta\theta = \theta_2 - \theta_1$ is the angular of the emission coming from the undulator:

$$\Delta\theta = \frac{1}{2\gamma}\sqrt{\frac{1+K^2/2}{nN}} = \sqrt{\frac{\lambda_r}{2L}} \quad (2.55)$$

From Eq. (2.55) it can be seen that high energy photons are concentrated in small angles. On the contrary low energy photons tend to span with large angles. The length of the ID also impacts the angular aperture of the radiation.

Example:

- IR-visible: $\varepsilon \sim 1.55$ eV ($\lambda \sim 800$ nm), $\Delta\theta = 35$ μrad with 3.28 m long ID of K value of 28
- VUV: $\varepsilon \sim 5$ eV ($\lambda \sim 250$ nm), $\Delta\theta = 11.2$ μrad with 10 m long ID of K value of 6.55 on first harmonic
- Soft X-rays: $\varepsilon \sim 1$ keV ($\lambda \sim 1$ nm), $\Delta\theta = 15.8$ μrad with 2 m long ID of K value of 2 on first harmonic
- Hard X-rays: $\varepsilon \sim 21$ keV ($\lambda \sim 0.0476$ nm), $\Delta\theta = 3.45$ μrad with 2 m long ID of K value of 2 on twenty-first harmonic

Note 1: In fact the angular aperture is not only defined by the resonant wavelength close to the axis. The expression of $\Delta\theta$ should be modified by the fact that radiation could be observed not only at the wavelength λ_r ($\theta = 0, \psi = 0$) on harmonic m but also for angles different from 0 coming from harmonics n higher than m. One starts from the equality:

$$\lambda_{rm}(\theta = 0, \psi = 0) = \lambda_{rn}(\theta \neq 0, \psi \neq 0)$$

$$\frac{\lambda_u}{2m\gamma^2}\left[1 + \frac{K^2}{2}\right] = \frac{\lambda_u}{2n\gamma^2}\left[1 + \frac{K^2}{2} + \gamma^2\theta^2 + \gamma^2\psi^2\right]$$

By defining $\phi^2 = \theta^2 + \psi^2$, one establishes that $\gamma\phi = \sqrt{(n-m)\left[1+\frac{K^2}{2}\right]}$. Around the main angular aperture $\Delta\theta$, it appears several radiation rings of diameter: $\frac{2}{\gamma}\sqrt{(n-m)\left[1+\frac{K^2}{2}\right]}$. Figure 2.17 shows schematically the angular distribution for a K value of 2 considering the fundamental harmonic ($m = 1$) of an undulator

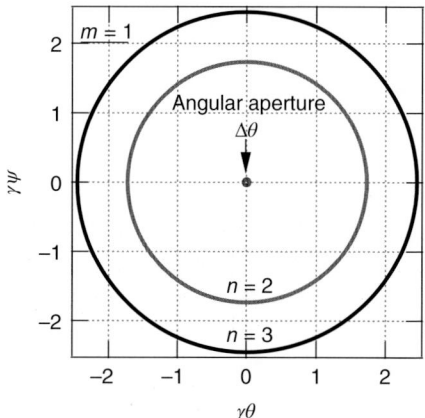

Figure 2.17 Angular aperture on first harmonic ($m = 1$) and rings of higher harmonics ($n = 2$ and $n = 3$). K value of 2 and number of period of 100.

composed of 100 periods of 20 mm. The diameter of the rings may exceed the natural angular aperture by many orders of magnitude.

Note 2: The particle bunch owns its own angular divergence in both horizontal and vertical planes σ'_x et σ'_z (assuming a Gaussian distribution). The total angular divergence should be modified by summing quadratically the contributions:

$$\Sigma'_x \approx \sqrt{\sigma'^2_x + \Delta\theta^2}$$

$$\Sigma'_y \approx \sqrt{\sigma'^2_y + \Delta\theta^2}$$

2.4.5 Estimation of Power Distribution Radiated in a Wiggler

The total power P_T and power density $\frac{dP}{d\Omega}$ emitted by particles crossing a wiggler can be estimated by using analytical expressions established by Kwang-Je Kim [6]. In other words,

$$P_T[\text{kW}] = \frac{10^{-21} qc^2}{6\pi\varepsilon_0} \frac{E_e^2[\text{GeV}] \cdot I[\text{A}]}{E_0^4[\text{GeV}]} \int_0^L B^2(z)dz. \tag{2.56}$$

$$\frac{dP}{d\Omega} = \frac{d^2P}{d\theta\,d\psi} = P_T \frac{21\gamma^2}{16\pi K} G(K) f\left(\gamma\theta/K, \gamma\psi\right), \tag{2.57}$$

where

$$G(K) = K \frac{K^6 + \frac{24}{7}K^4 + 4K^2 + \frac{16}{7}}{(1+K^2)^{7/2}}. \tag{2.58}$$

For large K ($K > 10$), the function f can be estimated with a good accuracy by the following formula:

$$f(a,b) = \sqrt{1-a^2} \left\{ \frac{1}{(1+b^2)^{5/2}} + \frac{5b^2}{7(1+b^2)^{7/2}} \right\} \quad a \leq 1.$$

$$f(a,b) = 0, \quad a > 1 \tag{2.59}$$

Because the radiation is concentrated in the region of the small cones around trajectory axis, the angles can be connected with observer coordinates with paraxial approximation:

$$\theta \approx x/L,$$
$$\psi \approx y/L; \qquad (2.60)$$

where L is the distance between wiggler centre and the observer and x and y are the observer coordinates in the plane perpendicular of the wiggler axis. Function $G(K)$ and $f(\gamma\theta/K, \gamma\psi)$ are plotted in Figures 2.18 and 2.19 respectively. Figure 2.18 represents only one quarter of whole curve since the function $f(a, b)$ is symmetrical for both arguments.

From Figure 2.19, one can see that most of the radiated power is concentrated in an angular aperture of $\pm K/\gamma$ in horizontal and $\pm 1/\gamma$ in vertical.

Formula (2.57) can be slightly modified in the case of a close orbit distortions presence:

$$\frac{dP}{d\Omega} = \frac{d^2P}{d\theta\,d\psi} = P_T \frac{21\gamma^2}{16\pi K} G(K) f\left(\gamma(\theta-\theta_c)/K, \gamma(\psi-\psi_c)\right) \qquad (2.61)$$

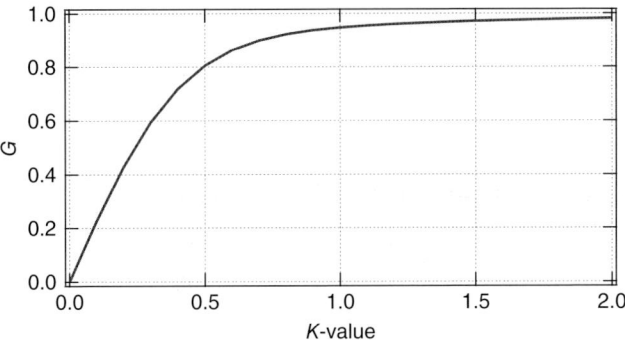

Figure 2.18 Function $G(K)$.

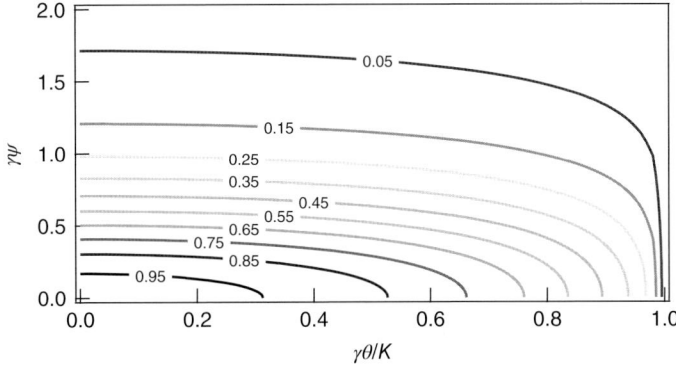

Figure 2.19 Contour plot of the function $f(\gamma\theta/K, \gamma\psi)$.

where θ_c, ψ_c are the horizontal and vertical emission angles resulting from an angular distortions of the orbit.

2.4.6 Estimation of the Power Collected by Simple Geometry Aperture

Expressions (2.57) and (2.61) give the local distribution of the power density. For calculation of the total SR power collected through the surface element, it is possible to integrate these formulas over the projection surface on the plane normal to the beam axis. For rectangular region, if one considers the surface limited by angles θ_1 and θ_2 in horizontal direction and by ψ_1 and ψ_2 in vertical, the integration can be performed analytically:

$$P(\theta_1, \theta_2, \psi_1, \psi_2) = P_T \frac{21}{16\pi} G(K) \left[F\left(\gamma\theta_2/K, \gamma\psi_2\right) - F\left(\gamma\theta_1/K, \gamma\psi_1\right) \right], \quad (2.62)$$

where

$$F(a,b) = \int_0^b dA \int_0^a dB\, f(A,B) = \frac{1}{2}(a\sqrt{1-a^2} + a\sin a)$$
$$\times \frac{b}{21} \left(\frac{16b^4 + 40b^2 + 21}{(1+b^2)^{5/2}} \right). \quad (2.63)$$

This formula is valid when horizontal angles do not exceed maximal angular divergence for wiggler radiation in horizontal plane, i.e. should be $|\theta_i| \leq K/\gamma$. In another case the horizontal angles in the formula (2.62) should be replaced by the limit value $\theta_{max} = K/\gamma$.

2.4.7 Method for Estimation Absorbed Power on the Complex Shapes

Considering more complex geometry, formulas (2.57) and (2.61) can be integrated numerically by dividing the various regions into big number of small parts. Sometimes when it is necessary to take into account the shadows of the chamber elements before calculated surface, the shape of regions can be very complex with a number of long irregular branches. In these cases the element of division should be smaller than width of such fragments, and the number of elements becomes big. The integration result becomes dependent on the way of division and integration accuracy is not good.

There is one way, proposed by Dr. K. Zolotarev, to reduce the dimension of integration and replace two dimensions surface integration by one dimension integration over the surface boundary contour with using Stokes theorem:

$$\iint_S \overrightarrow{\text{rot}(\vec{A})} ds = \int_L \vec{A}\vec{dl}. \quad (2.64)$$

Thanks to this property, if we can express integrated function f defined in Eq. (2.59) as a component of the rotor vector, we can define of the vector **A**. Function f can be expressed in factor form:

$$f(x,y) = u(x)v(y)$$
$$u(x) = \sqrt{1-x^2}$$

$$v = \frac{1}{(y^2+1)^{5/2}} + \frac{5}{7}\frac{y^2}{(y^2+1)^{7/2}} = \frac{1}{7}\frac{12y^2+7}{(y^2+1)^{7/2}} \quad (2.65)$$

If we define a new function $V(y)$, like

$$V(y) = \int v(y)\,dy = \frac{1}{21}\frac{y(16y^4+40y^2+21)}{(y^2+1)^{5/2}} \quad (2.66)$$

We can express vector **A** as follows:

$$\begin{aligned} A_y &= u(x)\,V(y) \\ A_x &= A_z = 0 \end{aligned} \quad (2.67)$$

In this case we obtain,

$$\overrightarrow{\text{rot}(\mathbf{A})} = (0,0,u(x)v(y)) = (0,0,f(x,y)) \quad (2.68)$$

Thus the integration can be expressed by contour integral:

$$P(S) = \iint_S \frac{dP}{d\Omega}\,d\Omega$$
$$= P_T \frac{21\gamma^2}{16\pi K}\iint_S f(\gamma\theta/K,\gamma\psi)\,d\theta\,d\psi = P_T \frac{21}{16\pi}\iint_{S'} f(x,y)\,dx\,dy \quad (2.69)$$

$$P(S) = P_T \frac{21}{16\pi}\int_{L(x,y)} u(x)V(y)\,dy \quad (2.70)$$

This integration can be performed by using standard integration routines from popular mathematic software, which can make automatic division of the contour for providing necessary accuracy (for example, function *quad* from MatLAB package).

2.5 Software Dedicated to Evaluation of the Photon Flux and Power Distribution from the Insertion Devices

As previously seen in various sections, analytical expressions enables to calculate quickly the radiated power, the power density and photon flux collected inside an aperture of simple geometry. However, formulas are not well adapted to complex geometry and a minimum of programming is needed. In addition, the equations do not take into account the real particle beam characteristics (transverse size, angular divergence, and energy spread) and also the magnetic defaults of the IDs, which strongly impact the optical and spectral performances of the photon source in terms of peak intensity and photon divergence. For this purpose several codes available in freeware version have been developed for 20 years for ID designer, vacuum scientists, SR beamline scientists, and free electron laser users and are presented in the following subsections.

2.5.1 XOP

X-ray Oriented Programs (XOP) is a widget-based driver program that is used as a common front end interface for computer codes of interest to the SR community. It provides codes for modelling of X-ray sources (e.g. SR sources, such as undulators and wigglers), calculate the characteristics of optical devices (mirror, filters, crystals, multilayers, etc.), and also contains tools for multipurpose data visualisations and analyses. The current version of XOP (v2.3) runs on most Unix and Windows platforms and is available free of charge to the scientific community. The point-and-click interface is used as a driver for a variety of codes from different authors written in different computer languages. XOP runs under most UNIX machines and Windows. XOP includes a flexible DAta BAse for X-ray applications (DABAX), which is a compilation tables for X-ray applications with a collection of codes to access, visualise, and process these tables.

OrAnge SYnchrotron Suite (OASYS) is a new generation simulation toolbox. The OASYS suite drives several simulation tools providing new mechanisms of interoperability and communication within the same software environment. OASYS includes most of the XOP tools for simulating spectra, power density, and radiation of most SR sources. It has also been designed to perform efficient beamline simulations using the most powerful software available, such as Shadow3 for ray tracing calculations and Synchrotron Radiation Workshop (SRW) for wave optics simulations.

Authors: Manuel Sanchez del Rio (ESRF) and Roger J. Dejus (APS).

Website: http://www.esrf.eu/Instrumentation/software/data-analysis/xop2.4.

2.5.2 Synchrotron Radiation Workshop (SRW)

SRW is a software dedicated to evaluate the spectral, spatial and polarisation characteristics of the radiation in the near field and/or far field approximation produced by a relativistic electron beam travelling through an arbitrary magnetic field. The magnetic field data is either created by a macro or can be read from a file coming from a design or magnetic measurements. The computed wavefront of the radiation can be propagated through drift spaces, lenses (Mirror, Refractive, Fresnel Zone, rectangular slit or cylindrical hole, etc.), apertures or arbitrary 2D phase shifting elements.

- Fast numerical algorithms are used for bending magnets radiation and periodic field IDs such as linear, ellipsoidal, figure-8, tapered, and optical klystron type of undulators or wigglers. The computation includes electron beam emittance and energy spread.
- The pre and post processing is made in the Igor Pro (low cost) graphing and analysing software. All computations are driven by dialogue box and/or command lines. Command lines can easily be grouped into a user defined macro to parameterise and/or automate a sequence of computation. The result can be visualised in linear, contour, image, and surface plots. Publication quality transparencies can be produced from within Igor. Extension and/or customisation is easily done using either the powerful Igor macro language or using C, C++, or Fortran.

A few areas of use of SRW:

- Beamline design and optimisation in the IR, UV, and X-ray range.
- Processing of the magnetic field measured on an Undulator/Wiggler and simulation of field errors on the spectral performance.
- Electron beam diagnostic in a storage ring or free electron laser.
- Estimating the brilliance (brightness) from undulators, wigglers and bending magnet sources.
- Edge radiation.

Recently, a new version of SRW has been proposed under Python environment.
First Release: December 1997, current Version: 3.76
Authors: O. Chubar, P. Elleaume
Main Publication: Chubar, O. and Elleaume, P. (1998). Accurate and efficient computation of synchrotron radiation in the near field region. *Proceedings of the EPAC98 Conference*, 22–26 June 1998, pp. 1177–1179.
Website: http://www.esrf.eu/Accelerators/Groups/InsertionDevices/Software/SRW.

2.5.3 SPECTRA

SPECTRA is an application software to calculate optical properties of SR emitted from bending magnets, wigglers (conventional and elliptical), and undulators (conventional, helical, elliptical, and figure-8). Calculations of radiation from an arbitrary magnetic field distribution are also available. Parameters on the electron beam and the source can be edited completely on graphical user interfaces (GUIs) and it is possible to show the calculation result graphically. The energy spectrum and radiation power after transmitting various filters and convolution of detector's resolution are also available.

The graphical part of SPECTRA is written in the C++ language with wxWidgets GUI tool kit and OpenGL graphic library. Thanks to portability of these libraries, SPECTRA will run on most available operating systems such as Microsoft Windows, Mac OS X, Linux, and most unix-like operating systems.

At present, the main functions supported in SPECTRA are as follows:

- Radiation power
  - Power density distribution observed at a certain longitudinal position.
  - Partial power passing through a finite aperture (circular and rectangular).
  - Filtered power and power density. The filter can be composed of any material and several typical filters are already built in. If not in the built-in list, the user can set up their original material. The user can also import a custom data to specify the transmission rate (transmission versus photon energy).
  - Surface power density, i.e. radiation power per unit area incident on the surface of a target object, whose normal vector is not necessarily in parallel to the optical axis.
- Photon flux
  - Spectrum of flux density, partial flux, and total flux.
  - Spatial profile of the flux density at a given photon energy.

Authors: Takashi Tanaka, RIKEN SPring-8 Center
Reference: Tanaka, T. and Kitamura, H. (2001). SPECTRA: a synchrotron radiation calculation code. *J. Synchrotron Radiat*. 8, 1221–1228.
Website: http://radiant.harima.riken.go.jp/spectra.

2.5.4 SYNRAD

'SYNRAD+' is the evolution of the SYNRAD code, which has been used by R. Kersevan, since the early 1990s. The original code was written in TurboPascal, compiled under DOS, and due to severe memory limitations could handle only rather simple geometries. In spite of this it has been used by the author to analyse and design the vacuum system upgrade of the CESR $e+/e-$ B-factory collider (Cornell University, Ithaca, USA), and many new chambers at the European Synchrotron Radiation Facility (ESRF, Grenoble, France).

The new version of the code, SYNRAD+, is the companion of Molflow+, and shares with the latter most of the ray-tracing engine algorithm. It takes full advantage of modern ray-tracing techniques and has been optimised for speed and accuracy and benchmarked against many analytical, numerical and experimental data and results. SYNRAD+ has some built-in features, which allow its user to load a geometry generated by CAD programs, of arbitrary complexity, in STL file format, i.e. a sequence of triangular facets. The user then assigns to each facet its 'optical' properties, such as specification of the material (Cu, stainless steel, aluminium, etc.), its surface finish value (in terms of surface finish or rugosity), and allows also the user to choose among a predefined set of energy- and angle-dependent reflectivity curves, which are usually found in literature or interpolated from existing databases. SYNRAD+ sets the position and direction of the beam and then computes the transverse source particle distribution as per relevant formula, which depend on the optics of the beam, i.e. the beta functions, emittance, dispersion, horizontal-to-vertical coupling, etc. It then computes the trajectory of the beam centroid using the dipole approximation: i.e. it bends the beam according to its energy and local magnetic field vector. The user can define dipolar and quadrupolar magnetic fields, in addition to periodic ones like those of undulators and wigglers, either defined analytically or numerically (e.g. obtained from measurements on magnetic benches). Global spectra are also calculated for the single facets, and photon flux and power densities can be obtained (in units of photons/s/cm^2, or W/cm^2, respectively). SYNRAD+ can export these data to finite element codes, like ANSYS, which then carry out the thermo-mechanical analysis of the absorber. SYNRAD+ shares with Molflow+ a set of simple 'geometry editing' routines, which allow the user to create or modify the geometry of the model within the code itself, without the need to use additional codes, or go back to the CAD program. A typical use is to create a cross or 'tee' from two intersecting tubes, of arbitrary cross-sections, which is a rather common feature of particle accelerators. Benchmarking has been also performed. A modern multi-core CPU workstation can run in a reasonable time (i.e. few to tens of hours) a SYNRAD+ simulation with 100 000 facets, each of them with tens to hundreds of texture elements. The actual limit is given by the graphic card memory installed on the computer, although there is a way to send the core of the Monte Carlo ray-tracing simulation to a faster host server and use the graphic card of a separate computer only to visualise the results.

Presently SYNRAD+ is available to run only under MS Windows, although some users run it under a simulated environment in Linux.

Authors: R. Kersevan, M. Ady; (CERN)

Website: https://molflow.web.cern.ch/content/about-molflow

2.5.5 OSCARS

OSCARS is a modern code for the computation of radiative properties of charged particles in electromagnetic fields. OSCARS is capable of calculating spectra, flux, and power densities for arbitrary field configurations, multi-particle and mixed-particle beams, with user configurable precision. Notably, OSCARS is capable of calculating flux and power density distributions on arbitrary shaped surfaces in 3D.

Characteristics of OSCARS:

- Accurate calculation with user defined precision.
- Multi-threaded and capable of using your graphical processing unit (GPU).
- Designed with very large scale computing in mind.
- Simple and very powerful python API (application program interface).
- 100% open source.
- We welcome feedback and contributions.

The core of OSCARS is written in modern C++ for speed with a simple python user interface. No additional packages are required to run the core of OSCARS. One can easily run OSCARS on their desktop or laptop computer. It also comes with utilities to use message passing interface (MPI) for your local machine and cluster usage. Significant gains are achieved through the use of GPUs and OSCARS makes this very easy for compatible NVIDIA GPUs. OSCARS was also designed with very large scale computing in mind and easily runs on 'the cloud' and such facilities as the Open Science Grid.

Author: Dean Hidas, NSLSII Synchrotron Center

Website: oscars@bnl.gov.

Acknowledgements

This chapter has been especially dedicated to estimate the flux and the power produced by charged particles crossing a bending magnet or an ID. The radiation can hit the wall of vacuum chamber resulting in outgasing due to local heating. The main issue to evaluate flux and power comes from the potential complex geometry of vacuum chamber or obstacle even if this subject has been treated by many scientists. This why the authors would like to thank particularly Dr. K. Zolotarev who proposed analytical and semi-numerical solutions as well as all the codes authors who propose efficient tools in freeware enabling to design beamlines from emission point to sample environment.

References

1 Landau, L.D. and Lifshitz, E.M. (1971). *The Classical Theory of Fields*. Pergamon Press.
2 Jackson, J.D. (1962). *Classical Electrodynamics*. Wiley.
3 Sokolov, A.A. and Ternov, I.M. (1968). *Synchrotron Radiation*. Oxford: Pergamon Press.
4 Lienard, A. (1898). Champ électrique et Magnétique. *L'Eclairage Electr.* 16: 5, 53, 106.
5 Elleaume, P. (2003). *Undulators, Wigglers and Their Applications*. Taylor & Francis. ISBN: 0-415-28040-0.
6 Kim, K.-J. (1986). Angular distribution of undulator power for an arbitrary deflection parameter K. *Nucl. Instrum. Methods Phys. Res., Sect. A* 246: 67–70.

Further Reading

Hofman, A. (2004). *The Physics of Synchrotron radiation*. Cambridge University Press. ISBN: 9780511534973.
Kim, K.-J. (1989). Characteristics of synchrotron radiation. In: *AIP Conference Proceeding 184*, New York, vol. 1, 567.
Walker, R.P. (1996). Synchrotron radiation and free electron lasers. In: *Proceedings of CAS-CERN Accelerator School*, CERN 98-04, Grenoble-France, 129.
Clarke, J.A. (2004). *The Science and Technology of Undulators and Wigglers*. Oxford Science publication. ISBN: 9780198508557.

3

Interaction Between SR and Vacuum Chamber Walls

Vincent Baglin[1] and Oleg B. Malyshev[2]

[1] *CERN, Organisation européenne pour la recherche nucléaire, Espl. des Particules 1, 1211 Meyrin, Switzerland*
[2] *ASTeC, STFC Daresbury Laboratory, Keckwick lane, Daresbury Warrington WA4 4AD Cheshire, UK*

The synchrotron radiation (SR) generated by the beam particle from the magnetic components of particle accelerator can irradiate inner surfaces of vacuum chamber and various component inside the accelerator vacuum chamber. In relation to the vacuum design of particle accelerators, there are three main effects that critically affect it and, therefore, are under consideration in this book:

- Photon reflectivity,
- Photoelectron production,
- Photon-stimulated gas desorption.

These effects depend on both SR parameters (energy of photon, intensity, and incident angle) and surface parameters (material, roughness, and treatments). In this chapter we will summarise the experimental results for the photon reflectivity and the photoelectron production, while the photon-stimulated gas desorption data are reported in Chapter 4.

3.1 Photon Reflectivity

When a photon interacts with an atom of the surface material, it might be absorbed (at the condition of equal energies of the atom resonance energy and the interacting photons), putting the atom into an excited state. After some time the atom relaxes back to the ground state by emitting a photon with energy equal to that of the original photon but in any direction. This emitted photon may interact with other atoms of the surface and be eventually absorbed, thereby heating the vacuum chamber wall, or transmitted through the outer surface, or emitted backwards by diffused reflection from the inner surface. High energy photons above ~100 keV can be transmitted through the vacuum chamber walls without interacting with the vacuum chamber material and be deleterious for the surrounding material. In addition, there are photons that are not absorbed and actually 'bounce' off of the surface, forming specular reflection.

Vacuum in Particle Accelerators: Modelling, Design and Operation of Beam Vacuum Systems,
First Edition. Oleg B. Malyshev.
© 2020 Wiley-VCH Verlag GmbH & Co. KGaA. Published 2020 by Wiley-VCH Verlag GmbH & Co. KGaA.

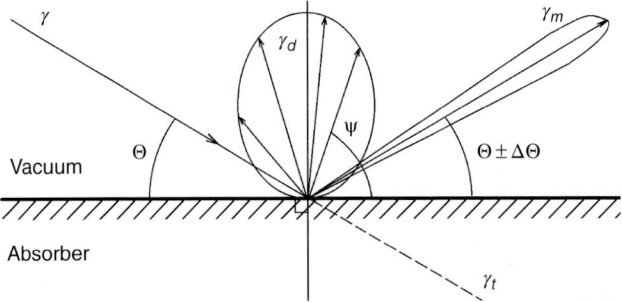

Figure 3.1 Incident photons, γ, hit an absorber surface at an incident angle Θ, which can be absorbed, transmitted (γ_t), diffused reflected (γ_d) with an angle Ψ, or mirror reflected (γ_m) with angle $\Theta \pm \Delta\Theta$.

In application to the accelerator vacuum system design, the main interest is on where a photon with an energy ε (or wavelength λ) and an incident angle Θ could be absorbed or diffusely or specularly reflected (see Figure 3.1). The incident angle Θ may vary from grazing incidence on vacuum chamber walls to normal incident at beam collimators and SR absorbers.

The X-ray interaction with matter can be computed from photo absorption cross sections and scattering models [1–3]. Among other parameters, the X-ray reflectivity of surfaces under several conditions is available at the Berkeley Laboratory [1]. For example, the reflectivity of a copper mirror was obtained with this online calculations [1] and is shown in Figure 3.2 in two graphs (a) as a function of incident photon energy in a range from 30 eV to 30 keV for a few incidence angles on a range from grazing 0.1° to normal and (b) as a function of incidence angle (Θ) for different photon energies in a range from 30 eV to 3 keV. One can see that

- for a grazing incident of $\Theta = 0.1°$ nearly all photons are reflected in the photon energy range from 30 eV to 30 keV;
- starting from $\Theta = 0.2°$, the higher incident angle, the more the higher energy photons are absorbed by the surface;
- the lowest reflectivity values are shown for normal incident; however, the reflectivity of photons is practically independent on incidence for $60° \leq \Theta \leq 90°$.

The surfaces have a reflectivity spectrum, which is, on one hand, a function of the angle of incidence and the energy of the incoming photons and, on another hand, depends on the surface material composition, surface roughness, and other surface characteristics. Dedicated and specific measurements on real samples are of great importance for a proper design.

In application to SR reflection in accelerator vacuum chamber, an incoming SR flux Γ_{in} after interacting with a wall surface can be divided in a few parts:

- Absorbed Γ_a
- Transmitted
- Diffusional reflected, Γ_d

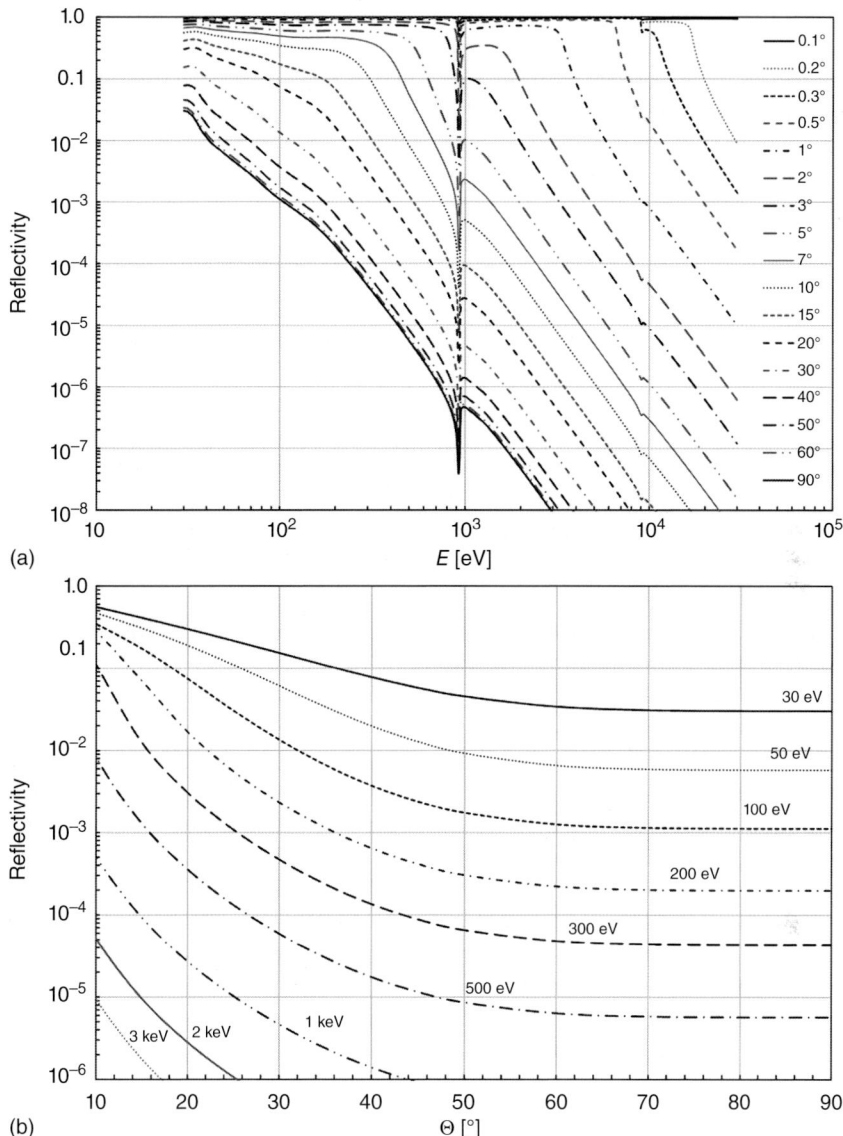

Figure 3.2 Reflectivity of a copper mirror (a) as a function of photon energy (E) for a few incidence angles and (b) as a function of incidence angle (θ) for different photon energies.

- Specular reflected, Γ_s
- Backscattered, Γ_b

These characteristics were studied in application to accelerators and reported in a few articles, which are in reasonable agreement and complementary to each other [4–7].

A diffuse and forward scattered reflectivity at 20 mrad grazing incidence was studied at BINP (Novosibirsk, Russia) on as-received rolled 316L stainless steel and 50 μm co-laminated copper (as-received and oxidised by heating the samples under vacuum to 300 °C and exposing then to air at atmospheric pressure for five minutes) [6]. The experimental set-up shown in Figure 3.3 consisted of a 34 cm long and a 55 mm internal diameter beam pipe. Inside this tube, six strips, each 2 cm wide and 30 cm long, were configured such that they formed a hexagonal shape covering the complete perimeter of the beam pipe. When the experimental system is aligned with respect to the axis of the SR beam, all photons were incident on the end calorimeter (CAL in Figure 3.3), allowing to measure full photon flux and power. In the inclined position, the photon beam was incident at ~20 mrad along the electrode strip 1; in this case, only forward scattered photons were able to reach the CAL. The diffused reflectivity was measured with electrode strips 1–6. The results of the study are shown in Table 3.1. One can see that reflectivity for measured SR power is significantly lower than photon reflectivity, which indicates that a spectrum of forward scattered photons have less number of high energy photons, or, in other words, the reflectivity for high energy photons is lower than for low energy photons. It is also shown that a shiny co-laminated copper has shown the highest forward scattered reflectivity and the air baking procedure allows its reduction; however, it is still high compared to as-received stainless steel sample.

Forward scattering photon reflectivity R was studied at CERN (Geneva, Switzerland) at critical photon energies ε_c of 45 and 194 eV at 11 mrad grazing incidence [4]. Four copper samples were prepared: (a) 50 mm Cu co-laminated onto a high-Mn-content stainless steel and annealed under H_2 atmosphere at 920 °C for 7.5 minutes; (b) Ex situ air baked at 350 °C for 5 minutes; (c) Cu electrodeposited from a Cu-sulphate bath onto 316LN stainless steel; (d) A Cu sawtooth structure, 0.5 mm step height and 10 mm periodicity, mounted such that the photons were incident quasi-normal to the vertical face of the sawtooth. This surface was studied in three different surface conditions: as-received, baked at 150 °C for nine hours, and baked at 150 °C for 24 hours. The sample surface roughness, R_a, was measured in this study for samples (a)–(c). A summary of the results for the forward scattering photon reflectivity R is given in Table 3.2. As expected, the higher the roughness, the lower the forward scattering photon reflectivity, and the sawtooth sample provides the lowest forward scattering photon reflectivity values. The reflectivity for photons with $\varepsilon_c = 45$ eV is higher than for photons with $\varepsilon_c = 194$ eV (except for the Cu electrodeposited sample).

A detailed study of the photon reflectivity for energies between 8 and 200 eV was performed at ELETTRA (Trieste, Italy) from industrial materials that could be used in the construction of the beam screen for the Large Hadron Collider (LHC) arcs [7]. Figure 3.4 shows the cross-sectional view of the sawtooth structure performed on the Cu-co-laminated beam screen surface adapted for the construction of the LHC arcs to intercept the SR at quasi-perpendicular incidence (N. Kos, CERN, private communication). The experimental set-up shown in Figure 3.4 allows to determine the space distribution of the scattered light by computer-controlled movements of the detector over the entire space above the sample with the exception of the small region (close to $\Theta_A = 180°$ and $\Phi_A = 0°$) (Figure 3.5).

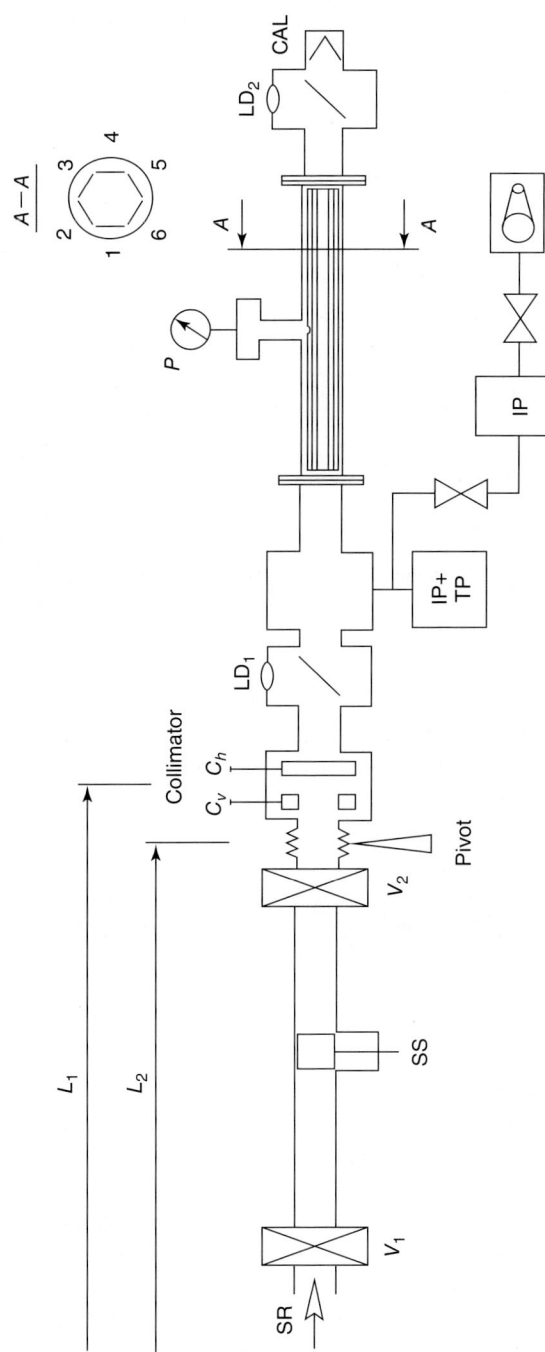

Figure 3.3 The experimental set-up for reflectivity measurements. Source: Reprinted with permission from Anashin et al. [6], Fig. 1. Copyright 2000, Elsevier.

Table 3.1 Diffuse and forward scattered reflectivity at 20 mrad grazing incidence on strip 1.

Sample	ε_c [eV]	Absorption at incidence	Diffuse reflectivity			Forward scattered reflectivity	
			Absorbed photons [%]				Absorbed power [%]
		Strip 1	Strips 2 and 6	Strips 3 and 5	Strip 4	CAL	CAL
Stainless steel as-received	243	60	1.5	3.5	8.5	22	<2
Cu co-laminated as-received	245	4.5	0.1	0.1	0.1	95	50
Cu co-laminated air baked	205 to 113	30	0.3	0.3	0.5	65	20

Source: Adapted from Anashin et al. 2000 [6], Tables 1 and 3.

Table 3.2 Forward scattering photon reflectivity R for SR with critical photon energies of 45 and 194 eV.

Surface	R_a or sawtooth height	Status	R (%) at 45 eV	R (%) at 194 eV
Cu co-laminated	$R_a = 12$ nm	As-received	81	77
Cu co-laminated	$R_a = 64$ nm	Air baked	22	18
Cu electrodeposited	$R_a = 1.6$ nm	As-received	5	7
Cu sawtooth	0.5 mm step height and 10 mm periodicity	As-received	2	—
		150 °C, 9 h	1.3	1.2
		150 °C, 24 h	1.3	1.2

Source: Adapted from Baglin et al. 1998 [4], Table 1.

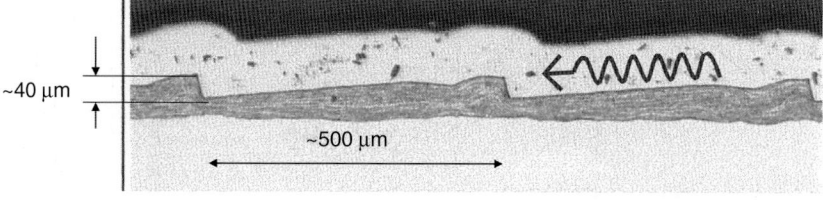

Figure 3.4 A cross-sectional view of the sawtooth structure produced for the construction of the LHC beam screens. Source: Courtesy to Nicolas Kos (CERN, Geneva, Switzerland).

Photons impinged onto the sample with 26 mrad incidence angle with $\varepsilon_c = 44$ eV. The photon reflectivity as a function of the azimuthal position Θ_A of the detector on the scattering plane is shown in Figure 3.6. In the case of the flat Cu surface, most (80%) of the reflected light is collected by the detector when placed around the geometrically defined specular (i.e. forward) direction, a very small part of the incident light is back reflected or diffused (<2%), and 18% of the incoming light is absorbed on the sample. In the case of the sawtooth sample,

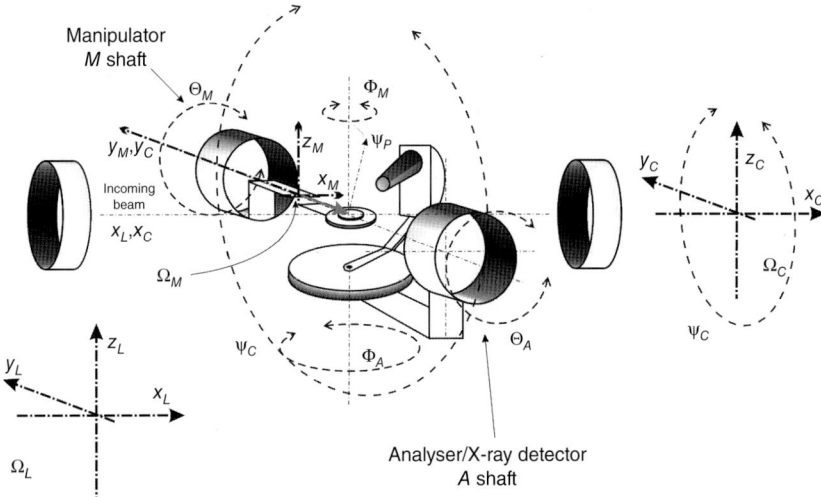

Figure 3.5 Schematic view of the experimental set-up showing the degrees of freedom for the positions of the sample and of the photon detector. Source: Reprinted with permission from Mahne et al. [7], Fig. 1. Copyright 2004, Elsevier.

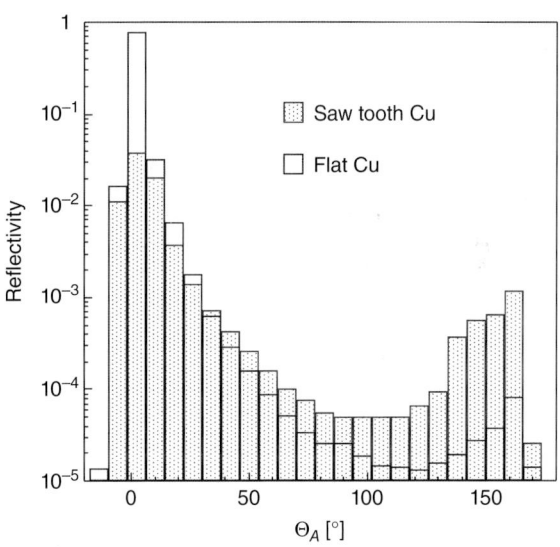

Figure 3.6 Measured reflectivity, on the scattering plane, from a flat Cu sample (empty bars) and from the sawtooth sample (filled bars). Source: Reprinted with permission from Mahne et al. [7], Fig. 2. Copyright 2004, Elsevier.

the forward scattering is reduced to only about 4%, the total reflectivity over the entire space is around 10%, and thus 90% of the incoming light was absorbed on the sample (see Table 3.3).

In order to determine the spectral composition of the reflected light, measurements have been performed with monochromatic light in the range of energies 8–200 eV. Figure 3.7 shows the results for forward, backward, and diffuse scattering for the flat Cu and sawtooth samples. For the flat Cu sample, the forward scattered reflectivity reduces from ~98% at 10 eV to ~78% at 27 eV and remains

Table 3.3 Measured values of the forward scattering, back scattering, and diffused light expressed in percentage of the incoming light.

	Flat sample	Sawtooth sample
Forward scattering (%)	80	4
Back scattering (%)	0	2
Diffused (%)	2	4
Total reflected (%)	82	10

Source: Reprinted with permission from Mahne et al. [7], Table 1. Copyright 2004, Elsevier.

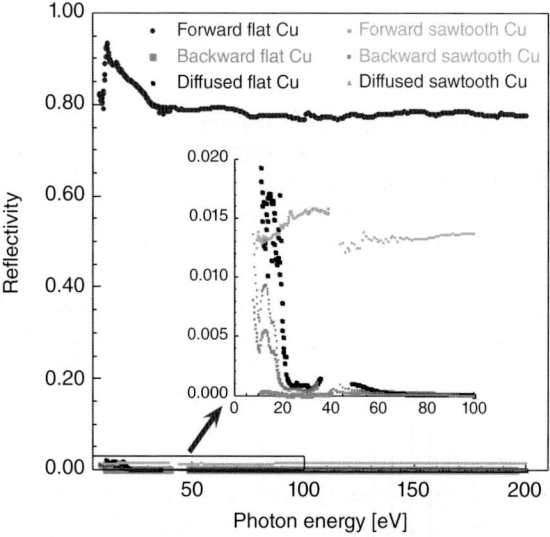

Figure 3.7 Reflectivity of the measured flat and sawtooth Cu surface versus impinging monochromatic photon energy between 8 and 200 eV. Source: Reprinted with permission from Mahne et al. [7], Fig. 3. Copyright 2004, Elsevier.

approximately the same up to a maximum measured energy of 200 eV, most of the diffused scattered photons were detected for energies 8–25 eV, and backscattered photons are negligible. For the sawtooth sample, the forward scattered reflectivity increases between 10 and 30 eV and remains approximately the same as at 10eV in the range of energies 40–200 eV; most of the diffused and backscattered photons were detected for energies 8–25 eV.

The presented results highlight the importance to consider the reflectivity of the SR irradiated surface when studying photon stimulated desorption (PSD):

- Up to 95% of incident photons can be forward scattered; thus calculated PSD yields from the experimental measurements may be underestimated by up to a of factor 20 if forward scattered reflectivity is not considered.
- Up to 18% of incident photons can be diffuse scattered. These photons irradiate the parts of vacuum chamber that are not irradiated by direct SR and cause PSD and PEY (photoelectron yield) in the locations that are in a shadow from direct SR.

Figure 3.8 Reflectivity of flat Cu surface versus impinging monochromatic photon energy between 130 and 1600 eV for various incidence angle, Θ, and emission angle, Θ_{det}. Source: Reprinted with permission from Schäfers and Cimino [8], Fig. 10. Copyright 2013, CERN, Geneva, Switzerland.

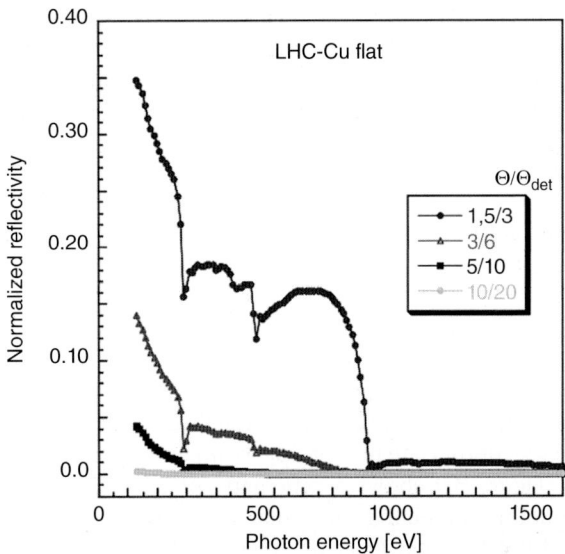

When increasing the photon energies and the incidence angle, the reflectivity decreases due to longer penetration depth. Figure 3.8 shows the reflectivity measured for a flat LHC Cu sample for monochromatic photon energy up to 1600 eV for various incidence angle, Θ, and emission angle Θ_{det} [8, 9]. At grazing angle (1.5°), the reflectivity is the largest. It decreases while decreasing the incidence angle. Absorption k-edges of C and O surface contaminants at 284.4 and 543.1 eV are clearly visible. Also observable is the Cu L_3-edge at 932.7 eV from the bulk of the sample. The reflectivity of a material is therefore sensitive to the bulk, the surface contaminants, and its roughness.

The effect of the incidence angle of the SR on the vacuum chamber wall is therefore extremely important and deeply influences the design of vacuum systems. SR machines have an antechamber type design with an absorber located at the pipe end to cope for large heat loads, e.g. ESRF, SOLEIL, and ALBA. The LHC has a sawtooth structure located in the equatorial plane to adsorb the photons. Future colliders, under study, are also based on these proven absorbers and sawtooth technologies and could also exploit, for the first time, the high reflectivity property of some materials when irradiated at grazing angle [10].

3.2 Photoelectron Production

Under SR irradiation, free electrons can be emitted from a metal surface due to photon–electron emission (PEE) effect. Photoelectrons are emitted when the photon energy is larger than the energy separation between the top of the valence band and the vacuum level.

3.2.1 Total Photoelectron Yield

The production of photoelectrons is characterised by the (total) PEY, which is the amount of photoelectrons emitted by incident absorbed photons.

In circular machines, since the SR is irradiating the vacuum chamber surface at grazing angle (from one to a few tens of mrad), a significant part of the incident photon reflects in the specular direction. Thus, photon reflectivity measurements of tubes are usually associated with photoelectron measurements. In a very simple formalism, the PEY per absorbed photon, PEY*, is derived from the measured PEY, and forward reflectivity, R, by Eq. (3.1):

$$\text{PEY}^* = \frac{\text{PEY}}{(1-R)} \tag{3.1}$$

Obviously, this approach is a very simplistic view of the underlying mechanism of interaction between a photon and a surface. Indeed, the reflectivity of photons on materials is a strong function of the incidence angle and the photon energy [5]. However, this simple approach has the following advantage that it can easily be implemented in any machine and does not require access to specific SR beamlines with appropriate photon spectrum and allocated dedicated beam time.

Figure 3.9 shows a schematic of an experimental set-up used to measure photon-stimulated molecular desorption, PEY, and photon reflectivity in the CERN Electron Positron Accumulator (EPA) [4]. First, the system is put in a straight through position, and a photoelectron current, directly proportional to the incoming photon flux, is measured at the end collector, which is negatively biased to −60 V to repel the photoelectrons. Second, the experimental chamber is tilted in a position in which the SR photons irradiate the chamber at a grazing angle (11 mrad in this case). A photoelectron current is then also recorded on the end collector. Since the specular reflected photons irradiates the end collector at a quasi-perpendicular angle, the ratio of the current measured in the second situation to the current measured in the first situation is a measure of the forward reflectivity, R. In order to collect the photoelectrons produced on the side of the vacuum chamber, a 200 mm long wire is stretched inside the vacuum chamber and polarised up to 1 kV. The measured current is a function of the applied voltage onto the wire: the larger the voltage, the larger the photoelectron current due to larger volume of collection. The collection volume can be computed with a software solving electromagnetic equations. For the set-up shown in Figure 3.9, the collection length was estimated to be ~500 mm at 1 kV. This value was estimated from a measurement done with a specific vacuum chamber equipped with sawtooth intercepting the SR light in quasi-perpendicular incidence. The measurement result, normalised to the end collector photoelectron current, allowed computing the collection length.

Table 3.4 gives a compilation of forward reflectivity and PEY per absorbed photons obtained with the above set-up. The values are measured with a SR spectrum, which is mainly in the UV range, namely, with 45 and 194 eV critical energies (it is recalled that ~90% of the emitted photons by the SR mechanism have an energy below the critical energy). In the UV range, typical values of PEY* at grazing angle is rather constant in the range of 0.1 e^-/photon. Conversely, the forward reflectivity is a strong function of the nature and the geometry of the surface.

In the X-ray range, for smooth copper irradiated with SR of 4 keV critical energy, the forward reflectivity is 33% and the measured photon electron yield per absorbed photon equals 0.43 e^-/photon [11]. For baked aluminium, the

3.2 Photoelectron Production

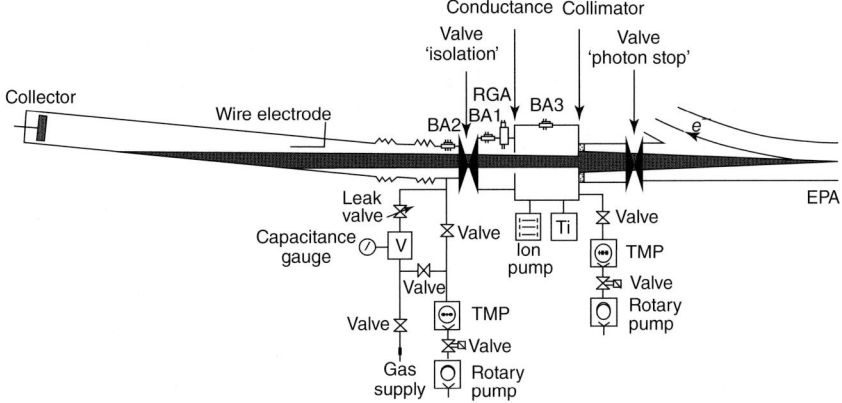

Figure 3.9 Schematic of a synchrotron radiation beamline for the measurement of photon-stimulated molecular desorption, photon reflectivity, and photoelectron yield in the CERN Electron Positron Accumulator (EPA). Source: Baglin et al. 1998 [4], Fig. 1. Reprinted with permission of CERN.

Table 3.4 Forward scattering photon reflection and photoelectron yield for materials subjected to SR at 11 mrad grazing angle with 45 and 194 eV critical energies.

		45 eV		194 eV	
Material	Status	R (%)	PEY* (e/ph)	R (%)	PEY* e^-/photon
Al	Unbaked	—	0.11	—	0.32
Cu smooth	Unbaked	81	0.11	77	0.32
	Air baked	22	0.10	18	0.18
Cu electrodeposited	Unbaked	5	0.08	7	0.08
Cu sawtooth	Unbaked	8	0.03	7	0.04
Ti–Zr	Unbaked	20	0.06	17	0.08
Ti–Zr	Activated at 350 °C	20	0.02	17	0.03

Source: Baglin et al. 1998 [4]. Reproduced with permission of CERN.

forward reflectivity is 20% and the measured photon electron yield per absorbed photon equals $\sim 2 \times 10^{-3}$ e^-/photon [12].

When bombarding the technical surfaces with photons, a conditioning is usually observed. However, as opposed to molecular desorption yields, which can decrease by several orders of magnitude, the PEY* is much less reduced. This is illustrated in Figure 3.10 where the PEY per absorbed photon as a function of photon dose of SR with 194 eV critical energy is shown [13]. The sample is an LHC beam screen prototype made of Cu co-laminated onto stainless steel with a sawtooth structure. As shown in Figure 3.4, the sawteeth have steps of the order of 40 μm height with 500 μm pitch in such a way that the photon irradiates the sample in a quasi-perpendicular incidence. During the conditioning process, the PEY* is reduced by about a factor of 2. However, in the meantime, the forward reflectivity remained unchanged. A similar observation was made under the same

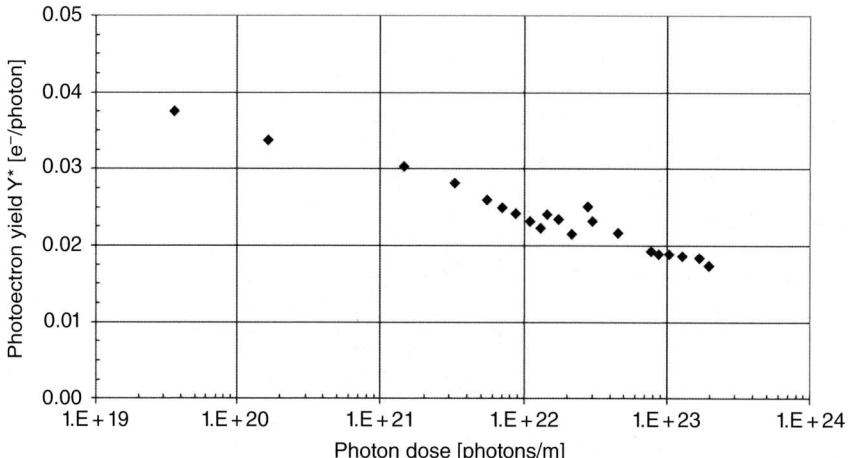

Figure 3.10 Photoelectron yield per absorbed photon of a Cu co-laminated with sawtooth surface as a function of SR photon dosed with 194 eV critical energy. Source: Reprinted with permission from Baglin et al. [13], Fig. 2. Copyright 2001, CERN, Geneva, Switzerland.

experimental circumstances when irradiating a Cu-co-laminated surface without a sawtooth structure.

3.2.2 Effect of the Photon Energy

The SR spectrum extends from infrared to UV, X-rays, or even γ-rays for large lepton colliders. Thus, the measured PEY* value presented above is the sum of the contributions of photons bombarding the vacuum chamber wall with energies ranging from meV to keV or even MeV. Increasing the photon energies causes several effects:

- A larger photon penetration depth means that electron excitation happened at larger distances from the surface.
- More photoelectrons are produced per impact photon.
- The Compton electron scattering for MeV range photons.

A closer look to the impact of photons on technical surfaces can be done with top-class instrumented beamlines installed at SR facilities. This is briefly introduced as follows.

Figure 3.11 shows an example of possible information, which can be obtained for such systems [5]. The figure shows the photoelectron energy distribution curves of evaporated gold when irradiated at 45° by photon with energies from 20 to 110 eV covering the UV spectrum range. The curves have been vertically shifted for clarity. In this photon range, most of the photoelectrons have energies below 10 eV. Only a few parts of the emitted photoelectron, ~0.1–1%, have larger energy.

The photoemission process is surface sensitive and probes the metallic surface within the nanometer range. Photons can directly interact with the solid and eject core electrons or valence band electrons from it. Therefore this process

Figure 3.11 Photoelectron energy distribution curves of Au for photon energies in the range 20–110 eV. For clarity, the curves have been vertically shifted. Source: Cimino et al. 1999 [5], Fig. 3. https://journals.aps.org/prab/abstract/10.1103/PhysRevSTAB.2.063201. Licensed under CC BY 3.0.

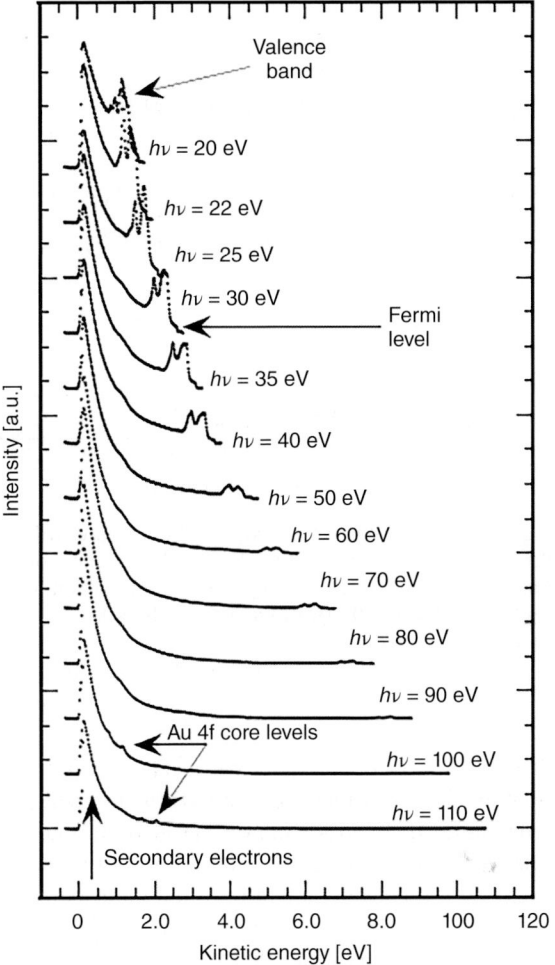

is widely used in surface science to evaluate the surface contamination of any material by X-ray photoelectron spectroscopy (XPS). The kinectic energy of the photoemitted electrons is simply given by the incoming photon energy minus the binding energies of the electrons minus the work function energy. In our example, the work function energy for clean polycrystalline gold is well known and equals 5.1 eV. Thus, only photons with energies above this value can initiate PEE from gold. Typical values of work function for technical materials range from 4 to 6 eV. Besides these electrons emitted from valences bands, core levels, etc., which are of paramount importance for the surface scientist, there is a class of low energy electrons present in any photoemission spectra: the secondary electrons. These electrons are due to the absorption of the photon within the solid creating electrons that can diffuse into it, while losing their energy by inelastic collision producing a cascade of secondaries. Secondary electrons, which are produced within 3–5 nm from the surface, are emitted from the material. As shown, they exhibit a characteristic energy distribution, which can be fitted by a Lorentzian.

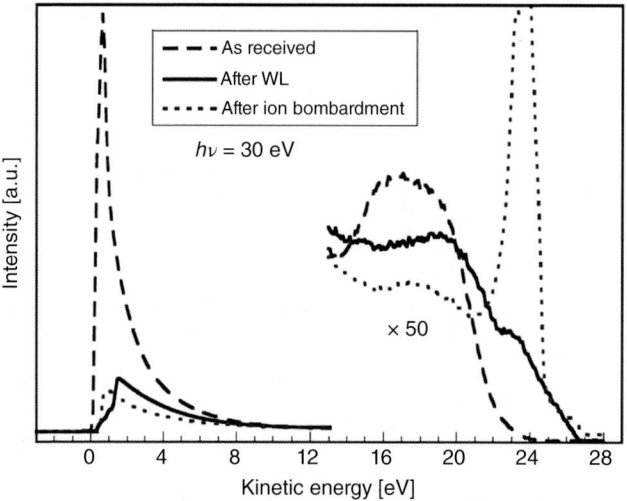

Figure 3.12 Photoelectron energy distribution curves of an LHC-type beam screen material for different surface treatments when irradiated with monochromatic photons of 30 eV. Source: Cimino et al. 1999 [5], Fig. 5. https://journals.aps.org/prab/abstract/10.1103/PhysRevSTAB.2.063201. Licensed under CC BY 3.0.

Apart from the incoming particle, these secondary electrons have the same origin from the material due to electron bombardment (see Chapter 8).

Due to the nature of the surface, evaporated gold, the spectra shown above are stable in time. But technical surfaces, which are building parts of accelerator machine, behave differently. This is illustrated in Figure 3.12 where the photoelectron energy distribution curves of Cu co-laminated on stainless steel when irradiated with monochromatic (30 eV) photons are shown for different surface treatments. As-received Cu exhibit a high and narrow peak of secondary electrons with less than 2 eV. SR irradiation with 'white light' (WL) reduces and broaden the secondary electron peak. Ion bombardment (sputtering) remove the contaminants from the surface (in the 16–20 eV region) resulting in the appearance of the so-called Fermi edge at ∼24 eV, signature of the cleanliness of the sputtered sample [5].

In a synchrotron machine, the vacuum chamber wall is irradiated by SR. The emitted photons have energies that cover the UV and X-ray (and even γ-ray for a Large Electron–Positron Collider [LEP] type machine) range, referred here as WL. The resulting photoemission spectra is therefore the sum of monochromatic spectra similar to the one in Figure 3.11. Figure 3.13 shows the modification of the photoelectron energy distribution curves of a technical material when subjected to WL photon irradiation in the UV range [5]. As shown, apart from stable surfaces such as evaporated gold (Au), the photon irradiation strongly modifies the photoelectron energy distribution. In a general manner, the secondary electron peak (below ∼5 eV) is reduced and broadened. However, the peak shape differs from one sample to another, which might, consequently, strongly impact phenomena that depend on the vacuum chamber wall properties, for instance, the build-up mechanism of the electron cloud (see Chapter 8).

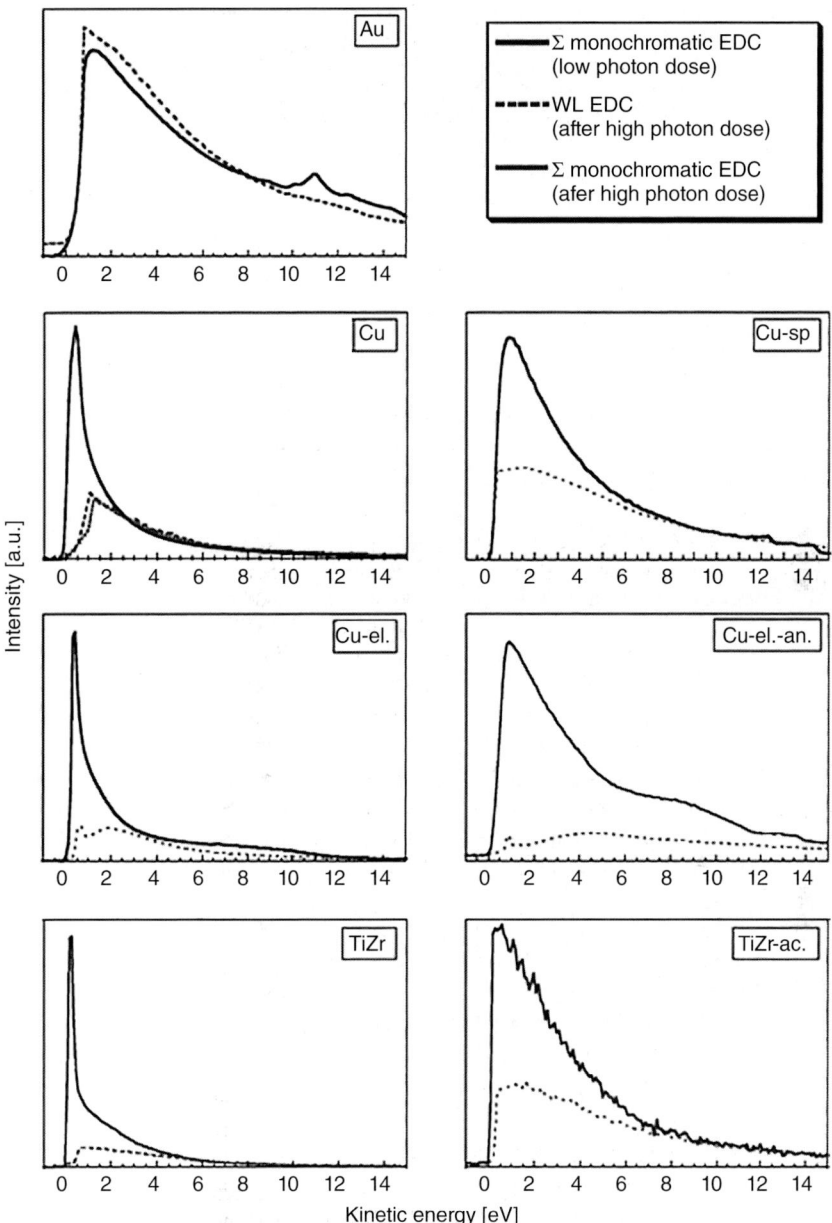

Figure 3.13 Modification of the photoelectron energy distribution curves of some technical materials under 'white light' photon irradiation in the UV range. Cu sputtered (Cu-sp), Cu electropolished (Cu-el), Annealed electropolished Cu at ~330 °C for two hours (Cu-el.-an), Activated TiZr at ~300 °C for four hours (TiZr-ac). Source: Cimino et al. 1999 [5], Fig. 17. https://journals.aps.org/prab/abstract/10.1103/PhysRevSTAB.2.063201. Licensed under CC BY 3.0.

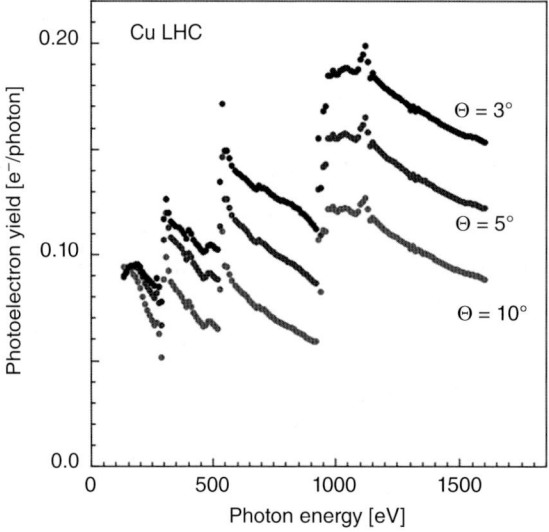

Figure 3.14 Total photoelectron yield of flat Cu surface versus impinging monochromatic photon energy between 130 and 1600 eV for various incidence angles, Θ_t. Source: Reprinted with permission from Schäfers and Cimino [8], Fig. 15. Copyright 2013, CERN, Geneva, Switzerland.

3.2.3 Effect of the Incidence Angle

When reducing the incidence angle, the photoelectrons are produced very close to the surface, thus increasing the PEY per absorbed photons. This is illustrated in Figure 3.14 where the total PEY for several incidence angles, Θ, is plotted versus monochromatic photon energy in the range 130–1600 eV.

The total PEY increases by about a factor of 2 when reducing the incidence angle from 10° to 3°. Similarly to Figure 3.8, the photo-absorption edges, characteristic of the surface, are visible: C at 284.4 eV, O at 543.1 eV, and Cu at 932.7 eV. The production of photoelectron is therefore a strong function of the surface state and cleanliness.

References

1 Henke, B.L., Gullikson, E.M., and Davis, J.C. (1993). X-ray interactions: photoabsorption, scattering, transmission, and reflection at E = 50-30,000 eV, Z = 1-92. *At. Data Nucl. Data Tables* 54 (2): 181–342. http://henke.lbl.gov/optical_constants.

2 Hubbell, J.H., Veigele, W.J., Briggs, E.A. et al. (1975). Atomic form factors, incoherent scattering functions, and photon scattering cross sections. *J. Phys. Chem. Ref. Data* 4: 471–538; erratum in 6, 615–616 (1977).

3 Cook, R.L. and Torrance, K.E. (1981). A reflectance model for computer graphics. In: *Proceedings of SIGGRAPH 81, Computer Graphics*, vol. 15, 307–316.

4 Baglin, V., Collins, I.R., and Gröbner, O. (1998). Photoelectron yield and photon reflectivity from candidate LHC vacuum chamber with implications to the vacuum chamber design. In: *Proceedings of 6th European Particle Accelerator Conference (EPAC'98)*, Stockholm, Sweden, 22–26 June 1998, 2169–2171.

5 Cimino, R., Collins, I.R., and Baglin, V. (1999). VUV photoemission studies of candidate Large Hadron Collider vacuum chamber materials. *Phys. Rev. Spec. Top. Accel. Beams* 2: 063201.
6 Anashin, V.V., Malyshev, O.B., Fedorov, N.V. et al. (2000). Reflection of photons and azimuthal distribution of photoelectrons in a cylindrical beam pipe. *Nucl. Instrum. Methods Phys. Res., Sect. A* 448: 76–80. https://doi.org/10.1016/S0168-9002(99)00747-0.
7 Mahne, N., Baglin, V., Collins, I.R. et al. (2004). Photon reflectivity distributions from the LHC beam screen and their implications on the arc beam vacuum system. *Appl. Surf. Sci.* 235: 221–226.
8 Schäfers, F. and Cimino, R. (2013). Soft X-ray reflectivity: from quasi perfect mirrors to accelerator walls. In: *Proceedings of ECloud-12, Isola Elba, 2012, CERN 2013-002*, 105–114. https://doi.org/10.5170/CERN-2013-002.
9 Cimino, R. and Schäfers, F. (2014). Soft X-ray reflectivity and photoelectron yield of technical materials: experimental input for instability simulations in high intensity accelerators. In: *Proceedings of IPAC14*, Dresden, Germany, 15–20 June 2014, wepme034, 2335.
10 Cimino, R., Baglin, V., and Schäfers, F. (2015). Potential remedies for the high synchrotron-radiation-induced heat load for future highest-energy-proton circular colliders. *Phys. Rev. Let.* 115 (26): 264804.
11 Suetsugu, Y., Tsuchiya, M., Nishidono, T. et al. (2003). Application of a sawtooth surface to accelerator beam chambers with low electron emission rate. *J. Vac. Sci. Technol., A* 21 (186).
12 Gröbner, O., Mathewson, A.G., Strubin, P. et al. (1989). Neutral gas desorption and photoelectric emission from aluminum alloy vacuum chambers exposed to synchrotron radiation. *J. Vac. Sci. Technol., A* 7: 223.
13 Baglin, V., Collins, I.R., Gröbner, O. et al. (2001). Measurement at EPA of vacuum and electron cloud related effect. CERN-SL-2001-003, Geneva 2001, *Proceedings of the LHC Performance Workshop 2001*, Chamonix, France.

4

Sources of Gas in an Accelerator Vacuum Chamber

Oleg B. Malyshev[1] and Junichiro Kamiya[2]

[1] ASTeC, STFC Daresbury Laboratory, Keckwick lane, Daresbury, Warrington WA4 4AD Cheshire, UK
[2] Japan Atomic Energy Agency (JAEA), J-PARC Center, Accelerator Division, Shirakata 2-4, Tokai, Naka, Ibaraki 319-1195, Japan

4.1 Residual Gases in Vacuum Chamber

Residual pressure and composition of gas in a closed vacuum volume connected to a working vacuum pump depend on initial conditions, flow rate of gas coming into a vacuum chamber, and layout of vacuum systems. In this chapter we will focus on the origins of gas molecules in vacuum chamber and how they can be inhibited or reduced.

The following are origins of residual gas in a vacuum chamber (see Figure 4.1):

- Gas from outside atmosphere
    ○ Atmospheric gases that remains in a vacuum chamber during or after pumping down.
    ○ Vacuum leaks at flanges, welds, cracks, valves, and other joints and seals.
    ○ Trapped volume and virtual leaks.
- Gas injection
- Evaporation of liquids left in a vacuum system
- Back-streaming from the vacuum pumps
    Outgassing and induced desorption from vacuum chamber walls and in-vacuum components
    ○ Thermal outgassing that includes the following processes:
        ■ Gas permeation from outside atmosphere through the vacuum chamber walls.
        ■ Gas diffusion from the bulk of the vacuum chamber walls and in-vacuum components.
        ■ Atomic diffusion on the surface and recombination into molecules.
        ■ Desorption of gas molecules from the surface.
    ○ Gas desorption induced by bombardment (irradiation)
        ■ Photon-stimulated desorption (PSD)
        ■ Electron-stimulated desorption (ESD)

Vacuum in Particle Accelerators: Modelling, Design and Operation of Beam Vacuum Systems,
First Edition. Oleg B. Malyshev.
© 2020 Wiley-VCH Verlag GmbH & Co. KGaA. Published 2020 by Wiley-VCH Verlag GmbH & Co. KGaA.

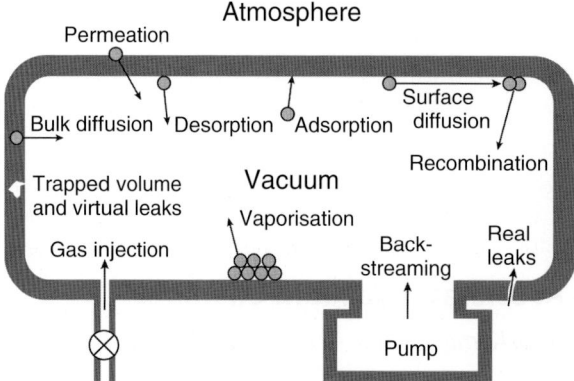

Figure 4.1 Mechanisms contributing to residual gas in a vacuum chamber.

- Ion-stimulated desorption (ISD) and heavy ion-stimulated desorption (HISD).
- Other particle-induced gas desorption.
○ Outgassing due to mechanical movements, stress, and deformation.
○ Products of chemical reactions.

New vacuum chambers are usually manufactured, stored, exposed, and assembled in atmospheric air. Major constituents of dry air are nitrogen (78% in volume), oxygen (21%), and argon (0.9%); see Table 4.1. The actual air also includes water vapour. The amount of water vapour depends on geographical region and current weather, varying from nearly zero to about 5%.

Initial pump down removes most of this gas out of an interior of a sealed vacuum chamber. Presence of atmospheric gases after sufficient pumping time is an indication of a likely atmospheric leak(s). Other sources of atmospheric residual gases are as follows:

– Trapped volume, i.e. *a small volume inside a larger vacuum chamber that has insufficient vacuum conductance to pumping*; for example, a volume between two contacting parallel flat surfaces, a bolt and a tapped hole, etc.

Table 4.1 Gas composition (highest at the top) in atmosphere and in different vacuum chambers.

Atmosphere (at sea level)	Unbaked vacuum chamber	Baked vacuum chamber	NEG-coated vacuum chamber	At cryogenic temperatures (1–80 K)
N_2 (78%)	H_2O	H_2	H_2	H_2
O_2 (21%)	H_2	CO	CH_4	CO
Ar (0.93%)	CO	CO_2	C_xH_y	CH_4
CO_2 (0.04%)	CO_2	CH_4	CO	CO_2
H_2O (0.1–5%)[a]	CH_4	C_xH_y		

a) H_2O concentration is not included in the above dry atmosphere.

- Porous material, i.e. when a very porous material is presented in a vacuum chamber, the pressure is determined by its outgassing from its large practical surface area and insufficient vacuum conductance (like trapped volume).
- Valves between atmosphere and vacuum that do not fully close or leak.

It is self-evident that air leaks above the detection limit of helium leak detectors need to be eliminated before further steps are applied. In the absence of vacuum leak, trapped volumes or gas injection into the vessel, the residual gas composition in a ultra-high vacuum (UHV) system is very different from atmospheric gas composition. In unbaked vacuum systems, partial pressure is dominated by water. After bakeout the water is usually eliminated and residual gas composition consists mainly of hydrogen, as demonstrated in Table 4.1, then carbon monoxide, and carbon dioxide. No concentration values or ratios between these values can be shown for vacuum chambers in general case. The gas composition is varied depending on many factors: choice of material, initial condition, cleaning procedure, bakeout, pumping system design, type of pumps, vacuum chamber temperature, the intensity and accumulated dose of photon, electron or ion bombardment on the surface, and many others.

In the following parts of this book, we will focus on the main trends in behaviour of a vacuum chamber of particle accelerators under different conditions in the absence of vacuum leak.

4.2 Materials Used for and in Vacuum Chambers and Built-In Elements

The main purpose of any vacuum chamber wall is to separate the air atmosphere from the inner pumped volume and at the same time to give it strength and rigidity to prevent its own collapse or significant deformation due to atmospheric pressure. Depending on the demands for the ultimate achievable pressure range, the wall can be made of any nonporous material that enables machining of parts, their joining, and tight sealing.

One of the most basic demands is low vapour pressure in the whole temperature range of operation. For most inorganic materials with a high melting point, a vapour pressure is very low and does usually not represent a relevant contribution to the ultimate achievable pressure. This is not true for many good engineering materials where in the manufacturing process many volatile additives remained in the bulk. Consequently, the ultimate pressure is often governed by outgassing of these gases for the most time of vacuum system operation.

The main requirement for accelerators to operate most of the time in the low UHV range greatly reduces the selection of suitable materials that can fulfil the main demand: very low specific outgassing rate q, expressed in Pa·m/s in SI units (or widely used units of mbar l/(cm^2·s)). Requirement for low q in accelerators is even more stringent than in large UHV vessels because in long beam pipes it is difficult to realise adequate pumping speed per unit area due to limited vacuum conductance. Moreover, even when applied materials are well selected, there is another stage of processing and cleaning, which must be realised correctly; otherwise machined parts may still exhibit unacceptably high q. This main requirement for low q is obvious for all elements assembled into a complex UHV system. Most of engineering techniques to achieve the low q in accelerator field overlap with general techniques in other UHV and XHV (extreme high vacuum) fields.

The most common materials used for vacuum chamber are stainless steel, aluminium alloys, and oxygen free copper. Some other materials are less common or used for special purpose: ceramic elements for electrical isolation and feedthroughs, ceramic vacuum chambers where it should be electrically non-conductive, titanium-based alloys for vacuum chambers and components as a low residual radioactivation material, and some other materials. Glass is used to be the main material in vacuum studies and applications for quite long. It has very good vacuum properties but it has a serious disadvantage: it is fragile. Presently, glass is mainly used for so-called view ports and for some components where its transparency and electrical non-conductivity are essential.

New or uncommon material can also be used for accelerator vacuum chambers or their components because they may be irreplaceable in analytical beam-end work chambers and preparation chambers due to their specific properties that are essential for the devices. They are used for special purposes and need to be assembled before application into a specific component. It is critically important to check that those materials meet UHV/XHV specification. There are materials that are not compatible with UHV/XHV because of high outgassing rate and/or desorbing of gas species with high molecular mass, such as hydrocarbons, organic molecules, some chemically active molecules, etc. Such materials cannot be used for and in the particle accelerator vacuum chambers. There are good references of materials used in vacuum [1–3]. Below some most common materials for using in accelerator vacuum systems are described.

4.2.1 Stainless Steel

Special grades of stainless steel are the most often applied for construction of UHV systems, i.e. chambers, components, and built-in elements, as holders, manipulators, etc. As there are many metallurgical grades of steels specified mainly by chemical composition, various designations exist in different parts of the world, related to various national or international designations. As most relevant scientific and technical papers deal with North American convention system introduced by American Iron and Steel Institute (AISI), it is applied in this text. Conversion of specific designation into any of national designation system is readily found in conversion tables. Two designation types that are most widely applied in relation to UHV are austenitic stainless steel: AISI 304 and AISI 316. Their chemical composition is determined by three main constituents, chromium, nickel, and iron, while impurities must be kept within prescribed limits. The composition of AISI 304 (in wt%) is as follows: Cr 18–20, Mn < 2, Ni 8–10.5, C < 0.08, P < 0.045, S < 0.03, Si < 1, Fe balance. The composition of AISI 316 is as follows: Cr 16–18, Ni 10–14, Mo 2–3, Mn < 2, Si < 1, C < 0.08, P < 0.04, S < 0.03, Fe balance.

Adding a suffix letter 'L' means low carbon content to prevent chromium carbide causing poor corrosion resistance, and 'N' means nitrogen addition for higher mechanical strength and corrosion resistance. Both elements play an important role in mechanical and thermal stability. AISI 316LN – whose composition is Cr 16–18, Ni 10–14, N 0.1–0.16, Mo 2–3, Mn < 2, Si < 1, C < 0.03, P < 0.045, S < 0.03, Fe balance – is the most often used material in UHV systems even if low carbon and high nitrogen do not represent a noticeable role in relation to UHV.

The main attribute, which makes it so useful in several other applications, is its corrosion resistance in the air up to 400 °C. The barrier, which blocks further oxidation and chemical reactions, is a dense and stable Cr oxide layer. Other excellent properties that are advantageous compared to other metals are simple welding by standard techniques and being non-magnetic, chemically inert, and relatively cheap. It should be noted that even the authentic stainless steel can become magnetised by processes such as cutting, bending, and welding. AISI 304 is easier to be magnetised than AISI 316. And nitrogen-doped stainless steel with suffix letter 'N' is more difficult to be magnetised. The perfect non-magnetism is sometimes required for the beam pipes because even a little magnetisation can distort the magnetic field to control the beam orbit. In such case, annealing for demagnetisation, which is a heat treatment at suitable temperature, is performed after all processes. To further improve the mechanical properties during the cold work and to expel the excess of hydrogen, a vacuum re-melting phase is a common procedure in modern metallurgy to improve these two steels further. After this phase, the content of hydrogen is low enough for safe rolling and mechanical reshaping, which could otherwise lead to hydrogen embrittlement. Unfortunately, even if hydrogen concentration for metallurgical application is acceptably low, it is still high for UHV as it generates a stable and virtually perpetual q.

Apart from these two grades of austenitic stainless steel, recent report on ferritic stainless steel such as AISI 430 and other metals with higher permeability opens an interesting new approach for vacuum material due to their preferable magnetism characteristics [4, 5]. Namely, in accelerators, beam pipes made of metals with high magnetic permeability are sometimes very useful to shield the unnecessary stray magnetic field from near magnet. Depending on the magnetic field to be shielded and the thickness of the metals, AISI 430, carbon steel, or nickel–iron alloy like permalloy (Ni80Fe20) are the candidates. Vacuum firing is effective both to reduce the source of gas in the bulk and to demagnetise.

4.2.2 Aluminium Alloys

Aluminium alloys are often used for vacuum system design, such as a vacuum chamber, a gasket, a rotor of turbo-molecular pump, electrode, and so on. Attractive characteristics of aluminium alloys are the low outgassing, high electrical thermal conductivity, non-magnetism, good material workability, and low density or lightness. A very low q is expected, which follows from two facts: low hydrogen solubility and very low alumina permeability. It is known that alumina is one of the best hydrogen diffusion barriers and it is instantly formed on aluminium surface. High thermal conductivity of aluminium alloys enables easier and more uniform heating even when the heat is not delivered uniformly. Especially in synchrotrons, thermal conductivity has very important characteristics. With high thermal conductivity, heat generated in the beam pipes by synchrotron radiation (SR) rapidly diffuses. Therefore, local temperature rise can be prevented. Aluminium alloys are also easily formed to a beam pipes with a complex cross section by extrusion process. However, tungsten inert gas (TIG) welding of aluminium alloys is not as easy as stainless steel, because local heating is difficult due to the high thermal conductivity. Therefore, electron beam welding is often used. The low residual radioactive characteristics, which means the dose rate after irradiation by the high energy particle reduces more rapidly than other metals, is another very attractive characteristic of an accelerator vacuum material.

A number of aluminium bases alloys are widely used presently: Al–Cu (2000 series), Al–Mg (5000 series), Al–Mg–Si (6000 series), and Al–Zn–Mg (7000 series). Most of these alloys consist of more than 90% Al and less than 10% other materials. For example, widely used Al–Mg alloy 5052 is based on 97% Al, 2.5% Mg, and 0.25% Cr. In the TRISTAN vacuum system, 6063 and 2219 alloys are used for the beam pipes and the flanges, respectively [6].

Similar outgassing rate to stainless steel is obtained for aluminium alloys. The outgassing rate of 10^{-6} to 10^{-7} Pa·m/s after 10 hours pumping without baking and 10^{-10} to 10^{-11} Pa·m/s with bakeout at 150 °C for about 24 hours are reported [7, 8]. Bakeout temperature for the aluminium alloy is limited to 150 °C at a maximum due to the depression of the mechanical strength.

Special attention is necessary in the extrusion process for beam pipes because the active surface at high temperature is covered with a porous aluminium oxide–hydride film, which traps machining oil components. In the TRISTAN, special extrusion in oxygen and argon atmosphere was performed to form the stable clean oxide layer in the aluminium surface [6].

4.2.3 Copper and Its Alloys

Copper is often used for vacuum material in an accelerator as, for example, beam pipe, electrode, current feedthrough, RF contact, cooling water pipe, beam pipe, and gasket. Advantages of the copper are high electric and/or thermal conductivity, non-magnetism, low q, and effective radiation shielding property due to the high density. One of the disadvantages is the difficulty for welding due to high thermal conductivity; thus electron beam welding is usually necessary. Other disadvantages are the necessity of joint between copper beam pipe and stainless steel flange, the heaviness, etc. Oxygen-free high conductivity copper (OFHC), especially C10100 with minimum 99.99% copper and maximum 0.0005% oxygen and C10200 with minimum 99.95% copper and maximum 0.001% oxygen are widely used. There are several types of oxygen-free copper with different contained amounts of oxygen and hydrogen owing to the process of manufacture.

Many types of copper alloys exist to compensate for copper's weak point, low strength. Beryllium copper, which is a copper alloy with 0.5–3% beryllium, has high mechanical strength. Thus, it is widely used, for example, as RF fingers in bellows in an accelerator to make a smooth pass for beam image current.

Other strengthened copper alloy is oxide dispersion strengthened (ODS) copper, commercially known as Glidcop®. Dispersed 0.3–1.1 wt% aluminium oxide ceramic particles play a role to increase the mechanical strength more than twice of pure copper. Large Hadron Collider (LHC) collimator is a recent example, which used Glidcop in an accelerator.

4.2.4 Titanium and Its Alloys

Titanium and its alloys for specific vacuum chambers and components have become attractive materials for vacuum chamber along with stainless steel, aluminium alloys, and copper alloys, because of the corrosion-resistant and low q characteristics. These characteristics owe the stable surface oxide layer. Titanium and its alloys have good material workability such as cutting, bending, and welding between titanium. They also demonstrate non-magnetism. Thermal conductivity and electrical resistivity of titanium are the same order as those of stainless steel, while the thermal expansion coefficient of titanium is almost half of the stainless steel. Therefore, the titanium is also used for the joint metal for alumina ceramics, as described in the next section. Because the titanium is also a low radioactive material, the pure titanium is used as a standard material for beam pipes, bellows, and inner RF shields in a high-power proton accelerator [9]. Titanium alloys such as Ti–6Al–4V is sometimes used for flanges to compensate the lower mechanical strength of the pure titanium than stainless steel.

There is another application of pure titanium in UHV practice, applying it for sublimation pumps. As hydrogen solubility in titanium is high, freshly evaporated films exhibit high pumping speed until the concentration approached the equilibrium. In a real case, the capacity depends on other gas species simultaneously pumped by the film.

4.2.5 Ceramics

Ceramics is a collective term for sintered compacts of inorganic compounds such as oxide, carbide, nitride, and boride. Crystalline ceramics fall roughly into two categories, oxide and non-oxide ceramics. Oxide ceramics includes alumina: Al_2O_3, sapphire, which is a single-crystal alumina, steatite ($MgO \cdot SiO_2$), zirconia (ZrO_2), magnesia (MgO), and ferrite, whose main component is ferric oxide. Non-oxide ceramics includes silicon nitride (Si_3N_4), silicon carbide (SiC), boron nitride (BN), and aluminium nitride (AlN). Machinable ceramics, such as MACOR® or Photoveel®, is another category of ceramics. Although a glass has several same manufacturing steps as crystalline ceramics, it is not often classified in the ceramics due to being non-crystalline and amorphous.

Alumina is the most widely used ceramics for a vacuum material in accelerators due to better electrical insulating property, hardness, low q characteristics, and cost. Alumina is used as feedthrough, support for heaters, insulation spacer, beam pipe, and RF window. The higher the purity of alumina, the better the insulating properties and the mechanical properties, and, furthermore, the lower the helium permeability. Metallisation brazing is a widely applied method for metal–ceramics tight bonding. Metallisation of ceramics requires several sophisticated manufacturing steps at high temperature and in pure gases or in a high vacuum. Mo–Mn metalising method is widely used for alumina. The metal, which has similar thermal expansion coefficient to alumina is used for the bounding metal to reduce the stress during brazing. Kovar®, whose composition (in wt%) is Ni 29, Co 17, Si 0.2, Mn 0.3, Fe balance, is widely used. Attention is needed when Kovar is used in an accelerator because it has magnetism. Recently in a high-power proton accelerator, titanium is used as the jointed metal between alumina beam pipe and metal flange because of its non-magnetism, similar thermal expansion coefficient to alumina, and the low radioactive characteristics [10]. Brazing using active metal is also applied for direct bounding of alumina and metal. The advantage of the active metal brazing is that the metalising process can be omitted.

In many accelerators, alumina beam pipes are used to prevent the induced current, which would be generated in the case of metals by rapid change of magnetic flux and causes the temperature rise of the chamber and distortion of the magnetic field [11, 12]. Alumina is also used for the RF window, which is transparent for the high-power RF voltage for particle acceleration with separating the atmosphere and UHV. Because alumina generally has high secondary electron yield, high voltage or electron impact in vacuum would cause an electrification and induce a creeping discharge. Especially in the high-frequency electric field, electron avalanche due to multipactor effect would break the RF window. In such case, coating on the alumina surface to reduce the secondary electron yield such as titanium nitride (TiN) and attention that emitted electron does not inject to the alumina surface are necessary.

4.2.6 Other Vacuum Materials

There are still a lot of other materials used in an accelerator vacuum, which were not covered above, such as graphite (used, for example, as beam stopper, beam collimator, or heater in vacuum), ferrite (as core of fast pulsed magnets like kicker), silicon steel (as many magnetic cores), and so on.

Glass and *elastomer* should be mentioned because they are used in many accelerators even though seldom nowadays.

Glass used to be the main material in vacuum studies and applications for quite a long time from late nineteenth century until mid of twentieth century. There are several grades of glass and many of them have acceptable properties related to UHV. Unfortunately, all of them have a serious engineering disadvantage: fragility. Besides this, shaping of glass by heat requires special engineering skills. Glass can be joined only to a few selected metals, which must be matched to the thermal expansion coefficient of a particular glass type. Nowadays, glass is mainly used for so-called view ports and for some components where its transparency or electrical insulation is essential. By sophisticated thermal procedures, specific glass types are tightly joined first to a thin oxidised specific metal element with matched thermal expansion coefficient. The other side of the thin-walled element is welded to standardised austenitic stainless steel flange. By using a similar sealing principle to glass view ports, elements for electromagnetic radiation transmission into UHV can be designed. There are requests for a wide band or some specific narrow spectral range, which can be fulfilled by other non-glass materials such as sapphire, zinc oxide, germanium, diamond, beryllium, and some other exotic materials. Sealing of these materials to metal flange is made by gold, silver, or any soft metal. As all these elements are designed for UHV, they withstand elevated temperature treatments and the achieved final outgassing rate is low.

Among the polymers, *elastomer* is the most popular vacuum material because it is used as demountable gasket and O-ring. Fluoroelastomer and perfluoroelastomer as represented by Viton® and Karretts®, respectively, are widely used in HV and UHV regions. They can be baked out at about 150 °C to obtain low q. Special care is needed for storage after bakeout because elastomer has water absorption characteristics. In addition, elastomer has much larger permeability than metals. From these reasons, in modern accelerators, where XHV is required to achieve high beam power or low beam emittance, metal seals is preferred to elastomer seals.

4.3 Thermal Outgassing

All materials, which are used to build vacuum chamber and vacuum components, desorb gas into the vacuum system. Thermal desorption is a spontaneous process of releasing of gas molecules from the materials into vacuum. Thermal outgassing means that a number of thermally desorbed molecules is greater than a number of reabsorbed ones or, in other words, that a number of molecules leaving a surface of a material in vacuum is greater than a number of molecules arriving at and adsorbing on the surface. Thermally desorbed molecules mainly consist of two things [13]:

1. Molecules diffusing through the bulk material of a vacuum chamber, entering the surface and desorbing from it.
2. Molecules, which have been adsorbed previously, that desorb again, when the chamber is pumped to vacuum.

As listed first in this chapter, the thermal outgassing can be divided to the following processes:

- Permeation of gas species from outside atmosphere through the bulk material of the vacuum chamber towards the vacuum sided surface.
- Diffusion of gas species contained in the bulk of the vacuum chamber material towards the vacuum sided surface.
- Diffusion and recombination of molecules on the surface.
- Desorption of gas molecules from the surface.
- Desorption of gas molecules that were (re-)adsorbed on the surface (initially or after the air venting).
- Desorption of products of chemical reaction on the surface (for example, generation of hydrocarbons on metal surfaces).

In this section, mechanism of thermal outgassing and thermal outgassing rate of vacuum materials are described.

4.3.1 Thermal Outgassing Mechanism During Pumping

In the absence of vacuum leaks, trapped volumes, or gas injection into the vessel, the pumping process of the unbaked system from atmospheric pressure is divided into two main processes. The first process is pumping of gases, with which the vacuum chamber was initially fulfilled with, for example, atmospheric gases or pure nitrogen. The other process is pumping of gases, which are outgassing from the vacuum chamber inner walls. This process includes the surface desorption, bulk diffusion, and permeation. The pressure $P(t)$ in the vacuum chamber at the time t is written as

$$P(t) = P_0 \exp\left(-\frac{S}{V}t\right) + \frac{Q(t)}{S}, \qquad (4.1)$$

where P_0 is the initial pressure (for example, an atmospheric pressure), S is effective pumping speed, V is a chamber volume, and $Q(t)$ is a thermal outgassing rate from the vacuum chamber inner walls at a time t. In the real case, the effective pumping speed is time dependent just after the start of the pumping, because a vacuum conductance in viscous flow range depends on the pressure. Anyhow the first term becomes negligibly small in a relatively short time after the start of the pumping at $t = 0$. For example, as just a thought experiment, in the case of the vacuum chamber, which has the volume of $1\,\text{m}^3$ with the pumping speed of $0.5\,\text{m}^3/\text{s}$, the first term in Eq. (4.1) is estimated to be 9.5×10^{-9} Pa for $t = 1$ minute and 8.9×10^{-22} Pa for $t = 2$ minutes. Of course, the real pressure does not act like that. First, it should also be noted that pumping speed of real pumps depends on pressure range and has a back streaming flow. But more importantly, the second term in Eq. (4.1) becomes dominant after removing of the volume gas, which is around 10^{-3} Pa. Thermal outgassing rate, which is the second term in Eq. (4.1), is further divided into several terms corresponding to different phenomenon in outgassing process as follows:

$$Q(t) = Q_s(t) + Q_d(t) + Q_p, \qquad (4.2)$$

where $Q_s(t)$ is the outgassing rate defined by surface desorption, $Q_d(t)$ by bulk diffusion, and Q_p by permeation. $Q_s(t)$ and $Q_d(t)$ are the time dependent, while Q_p is a time independent value. Figure 4.2 shows the typical pumping curve from atmospheric pressure to XHV region with a contribution of each outgassing term [2]. In a typical unbaked system after realistic pumping time, $Q_s(t)$ is the dominant outgassing term. Most of gas load in this region is water, although hydrogen, nitrogen, oxygen, carbon oxides, and hydrocarbons are also present. If a system is exactly in adsorption equilibrium state, $Q_s(t)$ decays in proportion as $e^{(-t/\tau)}$, where τ is average residence time of the adsorbed molecule on a surface. However, such simple model cannot be applied to the water. There are several adsorption states for the water on a surface, resulting in a range of activation energy of desorption, about 92–100 kJ/mol on a metal [2]. Therefore, several τ values exist for the water adsorption. Delay of $Q_s(t)$ proportional to t^{-1} is explained as superposition of several $e^{(-t/\tau)}$ [14].

After $Q_s(t)$ becomes negligible, thanks to the long pumping or bakeout, $Q_d(t)$ will be dominant term, which decays as $t^{-1/2}$. In the usual UHV system, permeation Q_p is negligibly small comparing to other two terms. However, because

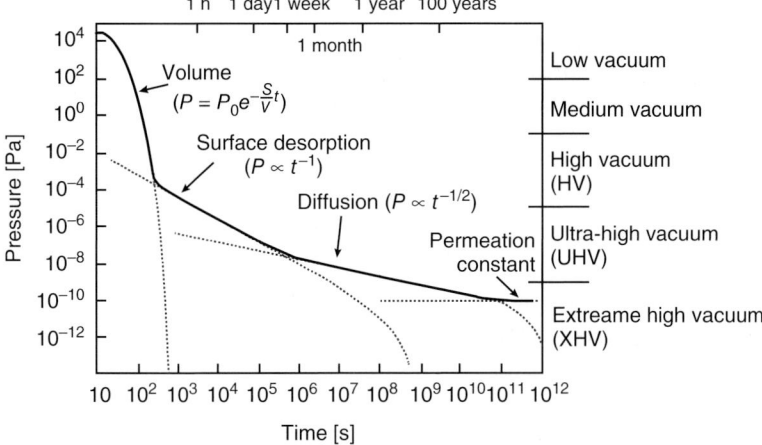

Figure 4.2 Typical pumping curve and contribution of each phenomenon. The values on vertical and horizontal axes were added as a rough indication. Source: O'Hanlon 2003 [2]. Reproduced with permission of John Wiley & Sons.

the permeation of the atmospheric gases through the elastomer is not negligible, elastomer gaskets are recommended not be used in the UHV system. This could be a case for thin wall vacuum chambers as well. Reduction of the outgassing, originated from surface desorption $Q_s(t)$ and bulk diffusion $Q_d(t)$, is discussed in Sections 4.4.2–4.4.4.

4.3.2 Equilibrium Pressure

Understanding of equilibrium pressure is important because pressure in a vacuum chamber becomes stable at the adsorption equilibrium. Adsorption equilibrium is a state where the desorption and adsorption rate of the gas molecules are equal. When temperature is constant in adsorption equilibrium, the amount of adsorption is a function of pressure. This relation between the amount of adsorption and the pressure is called adsorption isotherm. Several equations were experimentally and theoretically derived to express adsorption isotherm, such as Henry, Langmuir, Freundlich, Temkin, BET (Brunauer, Emmett, Teller), and so on. There are reviews, which introduce each adsorption isotherm and the pump-down behaviour with adsorbed layer obeying the adsorption isotherm [14, 15]. For example, Langmuir isotherm is based on assumption that when an adsorption site is occupied by a molecule, another molecule cannot adsorb on that site. The adsorption energy is assumed to be independent of surface coverage. The Langmuir isotherm is induced by considering the equilibrium between the desorption and adsorption rate:

$$\alpha \frac{s_0 - s}{s_0} \frac{P_{eq}}{\sqrt{2\pi m k_B T}} = \frac{s}{\tau}, \qquad (4.3)$$

where s_0 is the number of adsorption sites, s is the number of adsorbed molecules or occupied sites, α is an adsorption (or sticking) probability, P_{eq} is an equilibrium pressure, m is the mass of a molecule, k_B is the Boltzmann constant, T is temperature, and τ is an average residence time.

The term $(s_0 - s)/s_0$ represents the rate of unoccupied sites. The term $P_{eq}/\sqrt{2\pi m k_B T}$ represents the induced molecules to the unit surface area per unit time. The coverage $\theta = s/s_0$ becomes

$$\theta = \frac{aP_{eq}}{1 + aP_{eq}}, \tag{4.4}$$

where $a = \alpha\tau/(s_0\sqrt{2\pi m k_B T})$ is a constant number if the temperature is fixed. Figure 4.3a shows the Langmuir isotherm. From the assumption the maximum coverage in the Langmuir isotherm is a monolayer.

In many cases, the multilayer surface coverage $\theta > 1$ is observed along with the equilibrium pressure increase. This indicates the additional layers grow on top of the monolayer. BET isotherm is derived by considering the adsorbed multilayer, which is formed after the monolayer coverage on the surface. The $(n+1)$th layer is assumed to start to grow after completion of nth layer. The adsorption energy of the first layer is assumed to be different from that of the multilayer and independent of surface coverage. The coverage is written as

$$\theta = \frac{a\frac{P_{eq}}{P_s}}{\left(1 - \frac{P_{eq}}{P_s}\right)\left[1 + (a-1)\frac{P_{eq}}{P_s}\right]}, \tag{4.5}$$

where P_s is vapour pressure. Constant number a is determined by the difference between adsorption energy of the first layer E_1 and over second layers E_2, which is written as

$$a = \exp\left(\frac{E_1 - E_2}{k_B T}\right). \tag{4.6}$$

Right panel of Figure 4.3 shows the BET isotherms. In the BET isotherm, pressure in adsorption equilibrium increases with the amount of adsorption on the surface along an adsorption isotherm until it reaches the vapour pressure P_s. Equilibrium pressure is the pressure on the adsorption isotherm including vapour pressure and a function of adsorption surface concentration s.

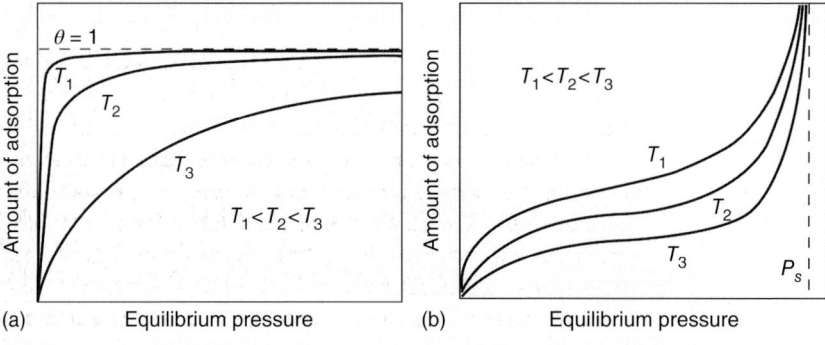

Figure 4.3 Typical adsorption isotherm according to Langmuir (a) and BET (b) formula.

4.3 Thermal Outgassing

For practical use we need to consider that there are two gas flows from and to adsorption phase:

$$q_{out} = n_{eq}(T,s)\frac{\bar{v}(T)}{4} = \frac{P_{eq}(T,s)\bar{v}(T)}{4k_B T}; \quad q_{in} = n\frac{\bar{v}(T)}{4} = \frac{P\bar{v}(T)}{4k_B T}; \quad (4.7)$$

where P_{eq} and n_{eq} are equilibrium pressure and gas density, P and n are the actual pressure and gas density in vacuum chamber, $\bar{v}$ is the average molecular velocity, and s is the amount of adsorption surface concentration. In the equilibrium when $P = P_{eq}$, two gas flows from and to the adsorption phase or liquid phase are equal: $q_{out} = q_{in}$. However, as soon as pressure in gas phase has been changed (for example, by opening the valve to a pump or changing a pumping speed), these flows are no longer equal and amount of surface concentration s will either reduce, if $P < P_{eq}$, for example, during pumping, or increase, if $P > P_{eq}$, for example, gas is inlet, according to $ds/dt = q_{in} - q_{out}$. Change of s will proceed until the system reaches adsorption isothermal, and finally the pressure becomes equilibrium pressure: $P = P_{eq}$.

This also explains why water can be pumped away only during a bakeout. At higher temperature of vacuum chamber, an equilibrium pressure increases, while both a wall adsorption probability α and, therefore, an average residence time decrease. If there is no pumping connected, it will lead to a significantly higher pressure (much higher than thermal expansion), and there will be still $P = P_{eq}$ (corresponding higher temperature) and $q_{out} = q_{in}$, but as soon as a valve to an external pump is open, the balance will be broken and, therefore, $q_{out} > q_{in}$.

4.3.3 Vapour Pressure

Understanding of vapour pressure is important because the pressure in a vacuum chamber is determined by the vapour pressure when liquid or condensed gas is present in vacuum. In addition, it is also important because when a system is pumped by a cryopump, the pressure is very much tied to the vapour pressure at low temperature. At first, it would be helpful to understand vapour pressure by looking at the phase diagram in the pressure–volume plane for the gaseous and liquid phase (Figure 4.4). Gas can be compressed to liquid at certain suitable temperatures. However, it becomes more difficult to liquefy the gas at higher temperature because the kinetic energy of the gas particle increases. Finally, the substance cannot be liquefied at and above a certain temperature, no matter how much pressure is applied. Such temperature is defined as critical temperature, T_c, of a substance. For example, for water $T_c = 374\,°C$ and for nitrogen $T_c = -147\,°C$. Considering the system filled with gas at the temperature bellow T_c, the pressure increases from point A by compressing the volume along the line with the isotherm until it reaches point B, a dew point; here liquid droplets begin to form in the system. As the system is further compressed, it moves along the horizontal line BC, at constant pressure, until the whole amount of gas has condensed into liquid at C, a bubble point. From there the system follows the same isotherm with increasing values of the pressure. At all points between B and C, the system is a mixture of gas and liquid. The constant pressure BC is called equilibrium vapour pressure, saturated vapour pressure, or simply vapour pressure.

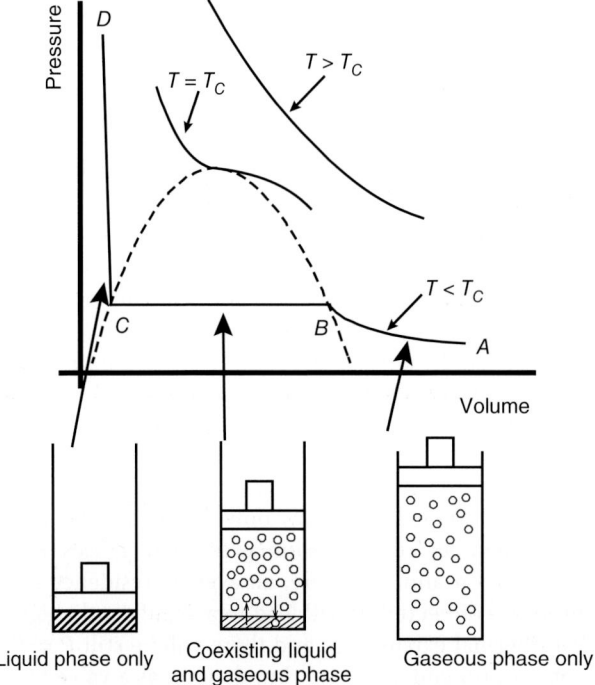

Figure 4.4 Schematic of the pressure–volume phase diagram for different temperature. T_c represents the critical temperature. Dashed curve is the envelope of dew points and bubble points for different temperature bellow T_c. Line AB: gaseous phase only. Line BC: coexisting liquid and gaseous phase. Line CD: liquid phase only.

The process of evaporation in a closed system at a given temperature proceeds until the system reaches thermodynamic equilibrium state, where there are the same number of molecules returning to the liquid as there are escaping. Vapour pressure is the pressure in such system, which is exerted by a vapour in thermodynamic equilibrium with its condensed phases (solid or liquid) at a given temperature in a closed system. Vapour pressure strongly depends on a thermally activated process, so the higher the temperature, the greater the thermal agitation that causes the escape of molecules from the surface. As an example, Figure 4.5 shows the temperature dependence of vapour pressure about water. Water boils at 100 °C with vapour pressure 1.013×10^5 Pa in the standard atmosphere. It is worth remembering that the vapour pressure of water at room temperature (22 °C) is 2.6×10^3 Pa.

As described earlier in this section, when liquid or condensed gas is present in vacuum, the vacuum is filled with the vapour and the pressure is limited by the vapour pressure. Therefore, oil free pumps must be used in modern accelerators. Oil vapour in accelerator vacuum not only merely makes ultimate pressure high but also contaminates the surfaces, which sometimes causes the serious damage to high voltage devices such as accelerator tubes. For general vacuum systems, oil with low vapour pressure is used for the pumps such as diffusion pumps and rotary pumps or vacuum greases. For example, Fomblin® is widely used for rotary

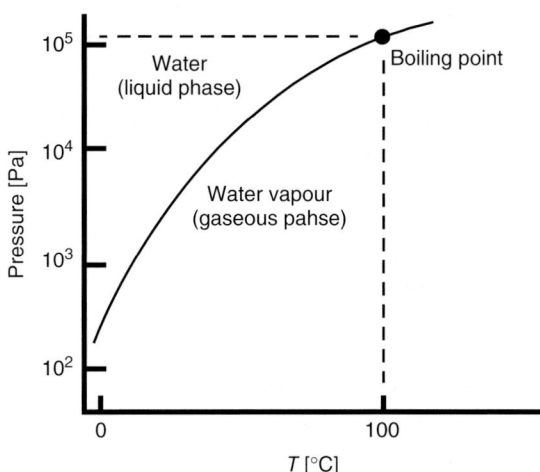

Figure 4.5 Temperature dependence of vapour pressure for water. Boiling point of water is 100 °C with vapour pressure 1.013 × 10⁵ Pa in the standard atmosphere.

pump, whose vapour pressure is about 3×10^{-4} Pa at 20 °C and 3×10^{-1} Pa at 100 °C. For diffusion pumps, oil with much lower vapour pressure with less than 10^{-5} Pa is used.

Temperature dependence of the vapour pressure, $P_s(t)$, is provided by the Clausius–Clapeyron equation:

$$\frac{dP_s}{dT} = \frac{1}{T}\frac{\Delta H}{\Delta V}, \tag{4.8}$$

where T is the absolute temperature, ΔH is an enthalpy change of evaporation per mole or molar heat of vaporisation, and ΔV is an amount of volume change per mole when liquid phase changes to saturation vapour, and it can be described as $\Delta V \cong RT/P_s$ because a vapour volume is generally much larger than a liquid volume. The integration of Eq. (4.8) then gives the dependence of vapour pressure on the temperature as

$$\ln P_s = -\frac{1}{T}\frac{\Delta H}{R} + \text{const.} \tag{4.9}$$

The constant in the equation is obtained by P_s, ΔH and T. The vapour pressure increases with temperature. Vapour pressures for common gases are shown in Figure 4.6 as temperature dependence [16]. Table 4.2 summarises temperature at some typical vapour pressure for selected gases. This table is written in terms of temperature for a given vapour pressure. The behaviour of vapour pressure at cryogenic temperatures will be discussed in Chapter 7.

Figure 4.7 shows vapour pressure curves of selected metals. Metals with high vapour pressure at low temperature, such as zinc or lead, are not are not suitable for the UHV and XHV vacuum components, which needs bakeout.

4.3.4 Thermal Outgassing Rate of Materials

Thermal outgassing determines the base pressure in the accelerator vacuum chamber without a charged particle beam, SR, or charged particles bombarding the vacuum chamber wall. The amount of thermal desorption is described per

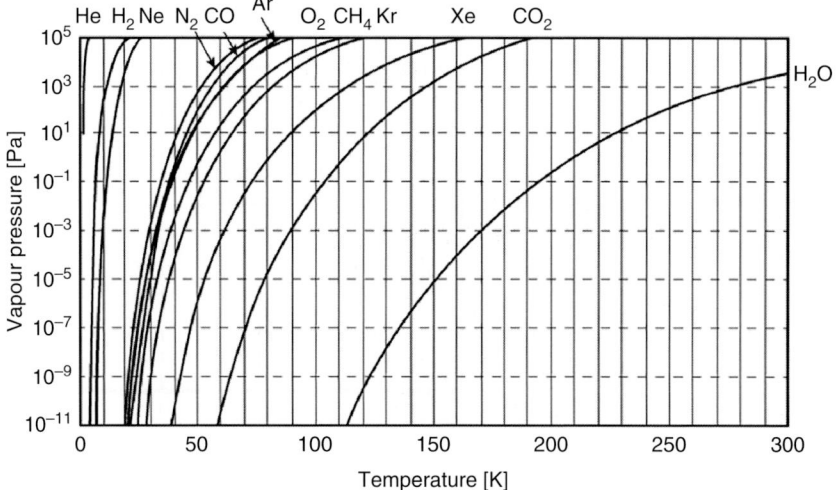

Figure 4.6 Vapour pressure curves on common gases. Source: Adapted from Honing and Hook 1960 [16], Fig. 1(b) on p. 366.

Table 4.2 Temperatures for some vapour pressures.

		Vapour pressure [Pa]				
		10^{-8}	10^{-4}	1	10^2	10^5
Symbol	Compound	Corresponding temperatures [K] for vapour pressures				
H_2O	Water	129	161	213	253	373
H_2	Hydrogen	3.14	4.35	6.8	9.4	20.3
N_2	Nitrogen	20.9	26.8	37.0	46.1	77.4
O_2	Oxygen	25.1	31.7	42.8	53.2	90.0
CO_2	Carbon dioxide	68.0	85.0	114	136	194

unit area by an outgassing rate q_{th} [Pa·m/s] or a thermal desorption yield η_t [molecules/(s·m^2)]. Thermal outgassing rates may vary in orders of magnitude depending on many factors such as material, cleaning procedure, history of material, pumping time, temperature, etc. For example, a value of about 10^{-8} Pa·m/s (or 2.5×10^{13} molecules/(s·m^2) at room temperature) will be easily obtained for carefully chosen and well-prepared vacuum materials (e.g. stainless steel) after a few hundred hours of pumping [17]; for the baked metals, a value of outgassing rate could be reduced by approximately an order of magnitude.

Many research laboratories have the outgassing rate measurement facilities to test all new materials before their use. A number of measurements of thermal outgassing rates of various materials were done in the past. There are a few good overviews of such studies, e.g. [18–21]. Unfortunately, several data were not published and available only as scientific workshop presentations.

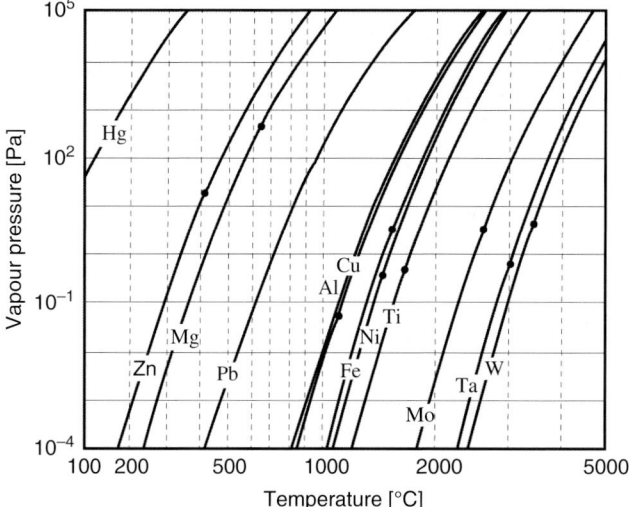

Figure 4.7 Vapour pressure curves on selected metals. Dots represent the melting points. Source: Jousten 2016 [1], Fig. 3.26 on p. 78. Reproduced with permission of John Wiley & Sons.

Table 4.3 summarises examples of thermal outgassing rates (in nitrogen equivalent) for the materials used in vacuum systems; more data can be found at references [30–33]. However, these numbers should be used for a design with a great care as the materials used in each case may have properties different from shown here.

A vacuum system designer has to consider that the outgassing rate of vacuum materials is not an intrinsic value rather than that it changes with time and depends on a 'history' of this material:

- Manufacturing conditions:
  o Manufacturers, sites, batches of material, date of manufacturing, and storage conditions.
  o Differences and modifications in production process between different manufacturers or between two items from the same manufacturer.
- Cleaning procedures:
  o A variety of procedures, chemicals, duration of each process, and quality control.
- Surface treatments:
  o Surface roughness
    ■ Smother surface the less outgassing.
  o Bakeout duration and temperature
    ■ Higher temperature allows a faster degassing, but there is a maximum bakeout temperature for each material, e.g. 250–300 °C for stainless steel, 220–250 °C for copper, and 150–180 °C for aluminium alloys.
    ■ Duration of bakeout may vary from a few hours to a few weeks.
  o Vacuum firing duration, temperature, and pressure during firing.
  o Duration of exposure to air after *ex situ* bakeout or vacuum firing

Table 4.3 Examples of thermal outgassing rates.

Material	Conditions	q_{th} [Pa·m/s]	References
Stainless steel (fresh)	After 10 h pumping	3×10^{-5}	[18]
Stainless steel (304L, electropolished)	30 h at 450 °C, 24 h at 150 °C, 1 h air exposure, after 10 h pumping	2×10^{-7}	[22]
Stainless steel (316L, vacuum remelted, electropolished)	24 h at 150 °C, 1 h air exposure, after 10 h pumping	2×10^{-7}	[22]
Stainless steel (304L, glass beads blasted)	23 h at 150 °C (*in situ*)	1×10^{-9}	[23]
Stainless steel (304L, electropolished)	48 h at 450 °C, air exposure, 23 h at 150 °C (*in situ*)	$<1 \times 10^{-11}$	[23]
Stainless steel (316L, air baked at 100 °C for 2 h)	20 h at 100 °C (*in situ*)	1×10^{-10}	[24]
Stainless steel (316LN, vacuum fired)	60 h at 100 °C (*in situ*)	1×10^{-11}	[13]
Aluminium (fresh)	After 10 h pumping	8×10^{-7}	[18]
Aluminium	20 h at 100 °C (*in situ*)	5×10^{-11}	[18]
Aluminium alloy (A6063, extrusion)	48 h at 150 °C, 1 h air exposure, after 10 h pumping	8×10^{-7}	[25]
Aluminium (A6063, extrusion)	20 h at 140 °C (*in situ*)	4×10^{-11}	[26]
OFHC copper (fresh)	After 10 h pumping	$1-7 \times 10^{-6}$	[18]
OFHC copper (mechanical polished)	After 10 h pumping	2×10^{-7}	[18]
Copper (fresh)	After 10 h pumping	6×10^{-6}	[18]
Copper (mechanical polished)	After 10 h pumping	5×10^{-7}	[18]
Copper	20 h at 100 °C (*in situ*)	1×10^{-9}	[18]
Titan (chemical polished)	After 5 h pumping	7×10^{-9}	[27]
Titan (chemical polished)	Baked (*in situ*)	7×10^{-13}	[27]
Alumina (96%)	After 20 h pumping	1×10^{-4}	[28]
Zirconia	After 20 h pumping	7×10^{-5}	[28]
MACOR®	After 20 h pumping	7×10^{-6}	[28]
Pyrex® (fresh)	After 10 h pumping	7×10^{-7}	[18]
Viton® A	Fresh, after 1 h pumping	2×10^{-3}	[18]
Viton® A	4 h at 150 °C in air, after 1 h pumping	1×10^{-6}	[29]
Viton® A	4 h at 150 °C in air, after 10 h pumping	3×10^{-7}	[29]
Teflon®	After 10 h pumping	3×10^{-5}	[18]

- Coatings:
  - Protective layer
  - Gas diffusion barrier layer
  - Pumping layer
- Pumping 'history':
  - Duration of pumping.
    - Unbaked materials depend mainly on pumping time.
  - The number and duration of air vents.
    - Short (a few minutes) exposure to air has little effect, while a few months of exposure eliminates an effect of earlier pumping.
  - The number, duration, and pressure of exposure to gases.

4.3.5 Outgassing Rate Measurements

There are several methods for measuring the outgassing rate, such as throughput method, gas accumulation method (build-up method, pressure rise method), mass loss measurements, and so on. Details of those methods are described in the reference [34]. Here outlines of the typical outgassing measurement methods are introduced.

4.3.5.1 Throughput Method

A throughput method is the most common method for outgassing measurements; a schematic diagram of this method is shown in Figure 4.8. In this method, the chamber under test or the chamber including the samples is pumped through an orifice with known conductance C. When the pressure in the test chamber P_1 is large relative to the pressure in the pumping system P_2, the pumping speed for the test chamber is only defined by the conductance of the orifice. The outgassing rate q are derived as

$$q = (P_1 - P_2)\frac{C}{A}, \tag{4.10}$$

where A is the surface area of the samples or the chamber under test. When the outgassing of the samples in the chamber is measured, the background outgassing of the chamber must be considered and measured without samples in another

Figure 4.8 Schematic diagram of the throughput method, G_1 and G_2 are pressure gauges, C is known vacuum conductance.

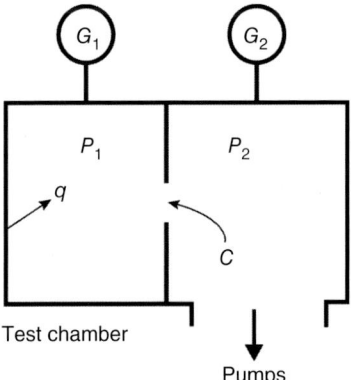

experiment with exactly same experimental conditions. If the sample has a low outgassing rate, the sample area should be increased (by using a larger sample or by increasing a number of samples) to obtain the total outgassing sufficiently high for its measurements.

4.3.5.2 Conductance Modulation Method

The conductance modulation method is a type of throughput method, which uses the orifice with the variable conductance [35]. In a schematic diagram shown in Figure 4.9, the conductance is modulated by changing the separation of a plunger from an opening hole in an annular disc. The pressures in the test chamber P_1 and P_2 are measured for conductance C_1 and C_2 defined by moving the plunger in positions 1 to 2, respectively. The outgassing rate q can be expressed for two plunger positions as follows:

$$q = \frac{P_1 - P_p}{A} S_1 = \frac{P_2 - P_p}{A} S_2. \tag{4.11}$$

where S_1 and S_2 are the effective pumping speed for the test chamber with the plunger positions 1 and 2, respectively, and P_p is the pressure in the pump chamber (which does not depend on a plunger position). The effective pumping speeds are written as

$$\frac{1}{S_1} = \frac{1}{C_1} + \frac{1}{S_p} \quad \text{and} \quad \frac{1}{S_2} = \frac{1}{C_2} + \frac{1}{S_p}, \tag{4.12}$$

where S_p is the pumping speed of the pumps. Then, the outgassing rate is derived as

$$q = \frac{P_1 - P_2}{A} \left(\frac{1}{C_1} - \frac{1}{C_2} \right)^{-1}. \tag{4.13}$$

Hence, in this method, the outgassing rate q is calculated from the pressures P_1 and P_2 (measured in the test chamber with the same gauge) corresponding to the defined conductances C_1 and C_2.

4.3.5.3 Two-Path Method

Two-path method is another type of throughput method [23] as shown in a schematic diagram in Figure 4.10. The advantage of this method is that the X-ray limit and outgassing rate of the gauges can be cancelled out. Furthermore,

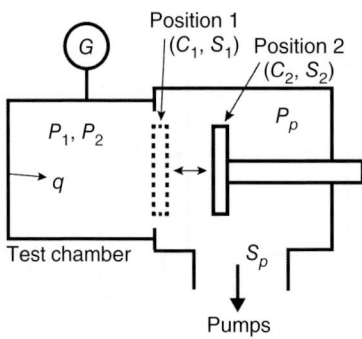

Figure 4.9 Schematic diagram of the conductance modulation method. G is a pressure gauge, C_1 and C_2 are a vacuum conductance in plunger positions 1 and 2, correspondingly.

Figure 4.10 Schematic diagram of the two-path method.

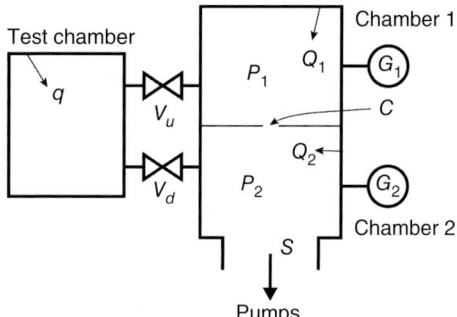

the outgassing from the chambers in the system, which are Chambers 1 and 2 in Figure 4.10, are also cancelled out. Thus, the two-path method is suitable for measuring a very low outgassing rate. When the upstream path is selected by opening the valve V_u and closing V_d, the pressure P_{1u}, which is the corresponding pressure in Chamber 1, is written as

$$P_{1u} = P_{2u} + \frac{qA + Q_1}{C}, \quad (4.14)$$

where the Q_1 [Pa m³/s] is the outgassing rate from the whole inner wall of Chamber 1. When the downstream path was selected by closing the valve V_u and opening V_d, the corresponding pressure in Chamber 1, P_{1d}, is written as

$$P_{1d} = P_{2d} + \frac{Q_1}{C}. \quad (4.15)$$

The pressure in the Chamber 2 remains constant for each path measurement because it is determined only by the total outgassing rate, namely, $P_{2u} = P_{2d} = (qA + Q_1 + Q_2)/S$, where S is the pumping speed of the pumps. Therefore, the outgassing rate q can be estimated as

$$q = \frac{C}{A}((P_{1u} - P_{2u}) - (P_{1d} - P_{2d})) = \frac{C}{A}(P_{1u} - P_{1d}). \quad (4.16)$$

Thus, the outgassing rate of the test chamber can be obtained by measuring pressure difference $P_{1u} - P_{1d}$ without the effect of the outgassing of Chambers 1 and 2. Because the pressure P_{1u} and P_{1d} are measured by the same gauge G_1, the X-ray limit of the gauge was cancelled by subtraction in Eq. (4.16).

4.3.5.4 Gas Accumulation Method

When a test chamber during evacuation is isolated from the pump by closing the valve between them, the pressure begins to rise. An average outgassing rate can be defined by using initial and final pressure measurements, P_i and P_f, taken correspondingly at time t_i and t_f:

$$q_{av} = \frac{P_f - P_i}{t_f - t_i} \frac{V}{A}, \quad (4.17)$$

where A and V are the area and volume of the test chamber, respectively.

Using the rate of pressure rise, the outgassing rate q can be measured as a function of time:

$$q(t) = \frac{dP}{dt}\frac{V}{A}. \tag{4.18}$$

A spinning rotor gauge or capacitance diaphragm gauge are preferred in this method, while ionisation gauges will cause errors in the measurement due to their pumping action and outgassing from the hot filaments.

4.3.6 Thermal Desorption Spectroscopy

Thermal desorption spectroscopy (TDS) is an effective method for evaluating the thermal properties of outgassing of vacuum material by measuring the desorbed gas while heating a sample. Analysis of the TDS results will provide important knowledge about adsorption, absorption, and desorption behaviour of atoms and molecules on surfaces because the TDS spectra include the information about (i) the number of the desorbing molecules and the population of the individual molecules, (ii) the activation energy of desorption, and (iii) the order of the desorption reaction [36, 37]. The gas species can be easily quantified by a residual gas analyser (RGA), usually a quadrupole mass spectrometer. The schematic diagram of a typical TDS measurement system is shown in Figure 4.11. In this example, a sample is mounted on a quartz stage, and the stage temperature is elevated by the infrared light, which is guided through the quartz rod. Usually, the temperature is linearly elevated as $T = T_0 + \beta T$. Figure 4.12 shows an example of the measured TDS spectra for typical gas species about austenitic stainless steel sample.

One of the most important information obtained by the TDS is derivation of the adsorption energy of desorbed species on the vacuum materials. The desorption rate is generally written as

$$\frac{ds}{dt} = s^n v_0 \exp\left(-\frac{E_d}{k_B T}\right), \tag{4.19}$$

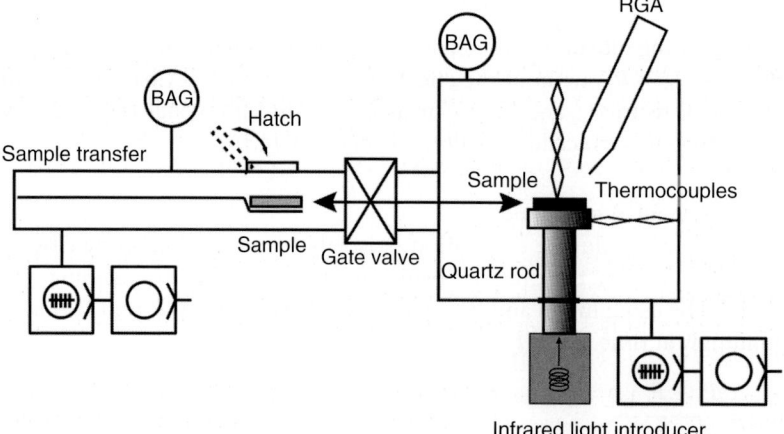

Figure 4.11 Typical measurement set-up of the TDS.

Figure 4.12 Examples of the thermal desorption spectra for austenitic stainless steel sample.

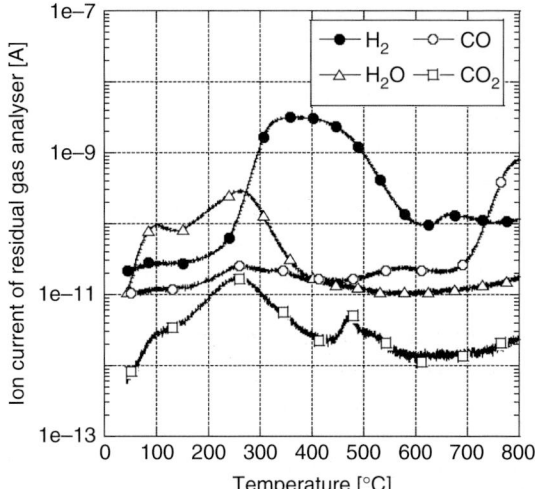

where s is a number of adsorbed molecules (molecules/m²), n is an order of the desorption reaction, v_0 is a rate constant, E_d is an adsorption energy or an activation energy of desorption, and T is the temperature: $T = T_0 + \beta T$. The temperature of the desorption peak T_p is easily of obtained from the equation

$$\frac{d}{dT}\left(\frac{ds}{dt}\right) = 0. \tag{4.20}$$

The most basic desorption process is the case that the order of the desorption reaction $n = 1$, where the adsorbed molecules on the surface are thermally desorbed from the surface without any surface diffusion. In that case, if n, v_0, and E_d do not depend on s and the order of the desorption reaction $n = 1$, then the solution of Eq. (4.19) can be written in its approximate form as

$$E_d = k_B T \left(\ln\left(\frac{T_p v_0}{\beta}\right) - 3.64 \right). \tag{4.21}$$

Thus, E_d can be immediately obtained from the peak temperature T_p in the TDS spectrum assuming the rate constant to be $v_n = 10^{13}$ s^{-1}, which is the typical value when the surface coverage is low [38]. This useful relation is called Redhead's equation [36]. It should be noted that this equation is adequate only for the case where the adsorption amount is low.

There is another method for estimating E_d and also v_0. The surface coverage is considered to be constant in the start of the thermal desorption. Thus, from the Arrhenius plot for the desorption rate in the rising part of the TDS spectrum, E_d and v_0 for the initial surface coverage can be obtained as the gradient and intercept of the plot, respectively [39].

TDS also gives the information about the specific gas content in the material by integrating the TDS spectrum about the object gas. Bacher discussed the hydrogen content [wt ppm] in the austenitic stainless steels with various treatments [40].

4.4 Surface Treatments to Reduce Outgassing

This section describes the surface treatments to fulfil demands to accelerator vacuum such as cleaning, bakeout, heat treatments like vacuum firing, or surface coating. The numerous treatments are applied to remove surface contaminants, reduce outgassing, and suppress harmful effects of particle hit to the surface. To choose the most suitable treatment processes for specific vacuum systems, vessels, or components, one should take into account the following: materials from which the items are made, the level of required vacuum, a particular performance requirement (e.g. low desorption), a particular contaminant (e.g. hydrocarbons) whose partial pressure must be minimised, how the items will be assembled or installed, etc. [41].

4.4.1 Cleaning

The vacuum chambers and components used in accelerators are manufactured through grinding, cutting, bending, welding, and so on. The surface of the substances after such process covered by a natural oxide or hydro-oxide layer. Beyond them, there will be particles such as dust, fibres, metallic powder, swarf, and organic substances such as cutting oil, grease, or solvent. These contaminations are attached on the surface during the machining, packing, storage, and carriage. Therefore, the first process to eliminate potential unwanted sources of gas originated from machining is proper cleaning. Many of organic solvents and strong cleaning agents recommended in old texts related to UHV were omitted due to harmful effects on the environment. They have been replaced by hot water solutions of detergents followed by rinsing in demineralised water.

There are some kinds of adherence patterns of contamination onto the surface: (i) van der Waals's force, (ii) electrical force due to the charge, (iii) chemical reaction such as in oxide layer, and (iv) diffusion into the surface. Rinsing will be effective to remove the contamination (i) and (ii), while the removal of the surface layer will be needed for elimination of (iii) and (iv).

For the modern accelerators, which requires UHV or XHV, leaving a surface free from foreign materials by degreasing, washing and drying is indispensable, but not sufficient. It is necessary depending on the required vacuum quality to eliminate or reduce chemical layer and gas diffused within the bulk material and adsorbed on its surface by the additional operations, such as electropolishing, vacuum firing, surface coating, and so on. The cleaning procedure depends on the state of the material to be cleaned and the required pressure. Figure 4.13 shows an example of workflow diagram for general metal cleaning procedures and their relation to required vacuum qualities; this diagram is an updated and extended version of the one shown in Fig. 2 on p. 590 in Ref. [42]. It should be noted that the process must be a little different from each metal.

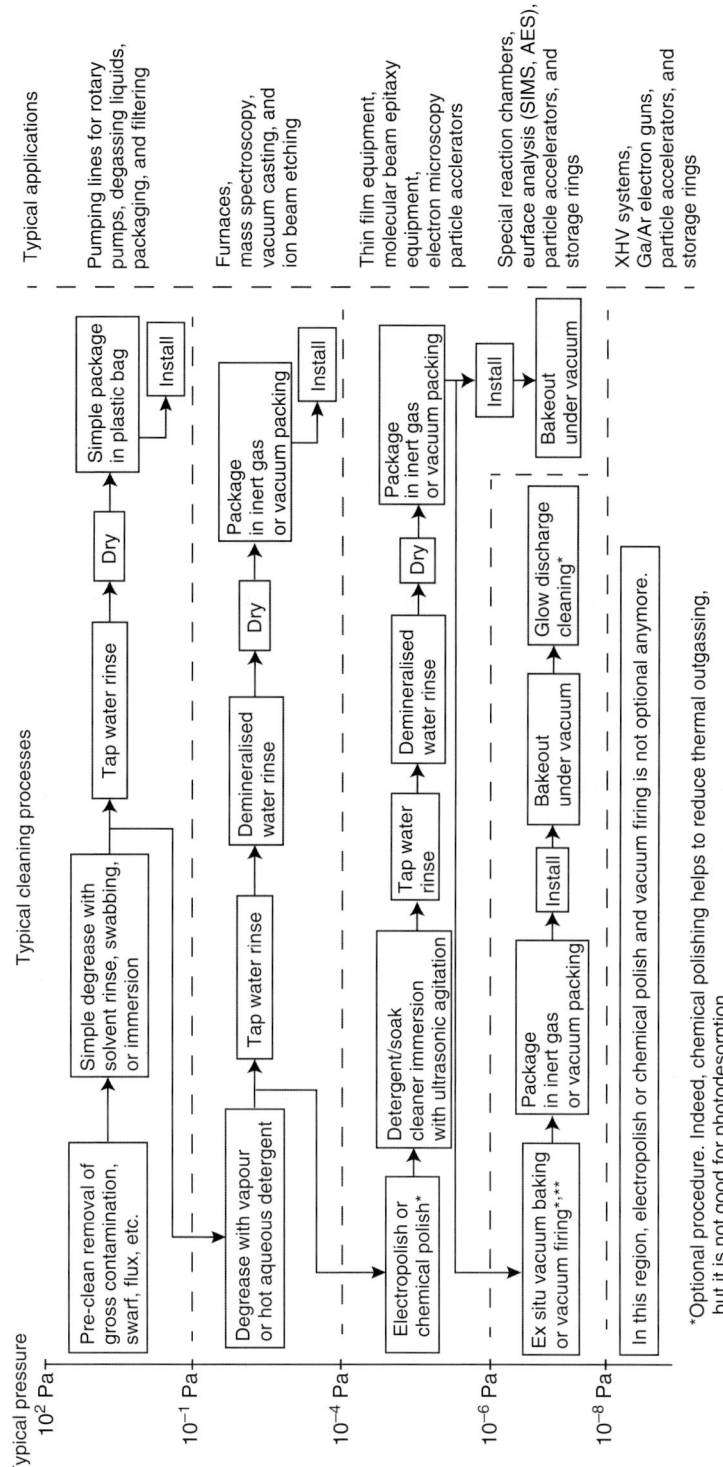

Figure 4.13 Examples of cleaning procedures including bakeout and other processes related to required vacuum qualities. These general cleaning procedures apply particularly to metals, although each metal must be treated a little differently. Nowadays, required pressure for typical applications becomes lower than the figure.

Austenitic stainless steels, AISI304 and AISI316, represent the main choice as constructional material. The stainless steel cleaning procedure usually includes (but may vary) the following operations [1, 43]:

- Degreasing
  - Initial mechanical removing of grease, wrapping, dirt, glue, etc.
- Initial washing in a hot (~80 °C) bath or with a jet
  - With various cleaning detergents (usually mild alkaline or acetone).
  - Rinsing with tap water.
- Washing in a hot (~60 °C) ultrasonic bath or with a jet
  - With aqueous alkaline cleaning solution.
  - Rinsing with a hot (~80 °C) demineralised and/or deionised water.
- Drying using an air blower with clean dry air, hot if possible.

The cleaning procedure may also include the following additional operations:

- Mechanical and/or electrical polishing
  - To reduce surface roughness to Ra = 0.2 mm, which reduces thermal outgassing.
- Chemical etching
  - To remove an outside layer, usually a natural oxide layer followed by surface passivation.
- Cleaning with ozonised water (for aluminium alloy chambers) [44]
- Argon discharge cleaning [45, 46]
  - To reduce all types of gas desorption.
- Pre-baking (ex situ baking) [47]
  - To reduce all types of gas desorption.
  - To avoid *in situ* bakeout.
- Vacuum firing [48]
  - For example, for stainless steel to 950 °C for one to two hours, depending on wall thickness.
  - To reduce all types of gas desorption.
  - For deep H_2 depleting.
- Air baking [49, 50]
  - Deep H_2 depleting,
  - Creating an oxide layer protecting from contaminating;
- Surface coating with other materials.
  - To provide desired surface properties such as electrical conductivity, reduced photon and secondary electron emission, gas diffusion barrier, etc.
- *In situ* baking
  - Temperature and duration can vary:
    - For example, 24 hour bakeout at up to 300 °C for stainless steel, up to 250 °C for copper, and up to 180 °C for aluminium alloys.
- This is the most common procedure for UHV systems.

For aluminium alloy, natural oxide layer is formed after machining such as cutting and extrusion. Because such oxide layer is thick and porous, it would be the source of gas absorption. The alkaline detergent or alkaline etching is effective to remove such natural oxide layer.

The cleaning of other materials such as copper and its alloy, ceramics, glass, and molybdenum is reported by the Tito [41]. The exact procedure depends upon the material, its history, required vacuum, contaminants, cost, availability of certain facilities, and cleaning agents.

4.4.2 Bakeout

When a brand-new vacuum chamber or a vacuum chamber, which was exposed to the atmosphere, is pumped down, the residual outgassing is dominated by surface desorption, which is $Q_s(t)$ in Eq. (4.2). The main outgassing component is water, which is desorbed from the chamber wall. Before the water plunges into the pump port, desorbed water sticks on another side of the wall, spends for average residence time, and repeats that process. Average residence time of water on the chamber wall at room temperature typically varies between 10 seconds and one hour. This is why the water continues to remain in vacuum chamber even after long pumping time. Bakeout is a well-known method to effectively obtain UHV region. It plays a role to shorten the average residence time resulting in the shorter time for the water to plunge in to the pump. The average residence time can be reduced to 10^{-2} to 10^{-3} seconds by bakeout at 150–200 °C. Figure 4.14 shows the typical pressure behaviour during the bakeout. After pumping at moderate temperatures up to 150 °C, water is mostly removed and the prevalent gas representing q becomes hydrogen. The residence time on the surface τ exponentially becomes shorter with higher temperature as

$$\tau = \tau_0 \exp\left(\frac{E_d}{k_B T}\right) \tag{4.22}$$

where τ_0 has the similar value to inverse of frequency of the molecule on the surface due to thermal vibration, usually 10^{-13} seconds, and E_d is activation energy of desorption, about 92–100 kJ/mol for water on a metal [2]. Therefore, bakeout

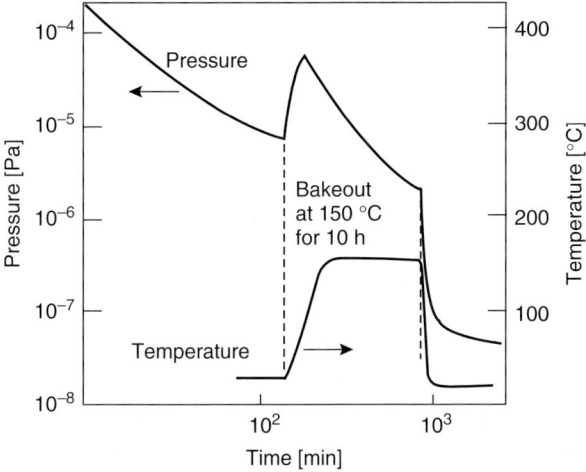

Figure 4.14 Typical pressure behaviour during bakeout effective for the outgassing reduction.

at higher temperature is effective to reduce outgassing from such standpoint. However, it should be cared that the higher temperature causes other problems such as leak due to the thermal expansion of the chamber and gasket. The adequate temperature should be selected from a material point of view.

Bakeout up to a temperature of 500 °C would not cause any problem for the stainless steel (excluding sealed flanges; for example, carbon fibre (CF) with cooper gasket is limited to 300 °C) case. Meanwhile, maximum bakeout temperature for aluminium (and aluminium alloys – *it should be different for different alloys*) is usually below or equal to 150 °C.

An important technique of bakeout is to uniformly heat the vacuum chamber. Popular method is to attach the sheathed heater on the outer surface of the chamber and cover them by heat insulator such as glass fibres. Proportional–integral–derivative (PID) controller is widely used to control the temperature with minimum labour.

4.4.3 Air Bake

An oxide layer formed on the stainless steel surface is expected to serve as a diffusion barrier for hydrogen diffusing from the bulk and to reduce the outgassing rate [51]. Bakeout in the controlled oxygen atmosphere or just in the air to positively form the oxide layer has been practiced for nuclear fusion devices and the gravitational wave detectors [52, 53]. By several days of air bake at about 400 °C and following vacuum bakeout, the low outgassing rates in the order of 10^{-11} Pa·m/s were obtained.

Practical merit of air bake is that it does not need a vacuum furnace, which is essential for the vacuum firing. Thus, air bake is applied for the large vacuum chambers as referred above.

4.4.4 Vacuum Firing

After eliminating the outgassing originated from surface desorption, main outgassing is dominated by bulk diffusion, which is described by Q_d in Eq. (4.2). In that stage, main outgassing component is hydrogen, whose atoms diffuses from the bulk of materials and recombine to the molecules at the surface. Hydrogen content in untreated stainless steel is in the order of 2×10^{19} atoms/cm^3, or about 3 wt ppm [54]. To deplete the bulk from diffusing atoms, it is recommended to perform the high temperature treatment in high vacuum, namely, 'vacuum firing' [55, 56]. In the vacuum firing procedure, cleaned components are put in a high vacuum furnace, where reaction with residual atmosphere is indeed very weak. For stainless steel, vacuum firing was performed at 800–1000 °C, while there are little examples for aluminium alloys and copper alloys due to their mechanical strength deterioration at such high temperature. Titanium is one of a few examples to which the vacuum firing was applied other than stainless steel as described later in this section.

The simple expression about the effect of the vacuum firing on outgassing reduction is based on the assumption that the outgassing rate is determined by only diffusion process, which obeys the diffusion equation, without any regard for the reaction processes at the surface such as recombination and desorption. In such simple case, the outgassing rate at processing time t is given as

$$q \cong \frac{4C_0 D_R}{d} \exp\left(-\pi^2 \frac{D_T t}{d^2}\right), \tag{4.23}$$

where main parameters are plane thickness d [cm]; diffusivity at room temperature and vacuum firing temperature D_R and D_T [cm²/s], respectively; and initial hydrogen concentration C_0 [atoms/cm³] [57]. This is the case where both sides of a plane with infinite area are revealed to vacuum and hydrogen concentration is zero at the surface. Diffusivity D is thermally activated process described by

$$D(T) = D_0 \exp\left(-\frac{E_d}{k_B T}\right), \tag{4.24}$$

where $D_0 \cong 0.012$ cm²/s is a characteristic pre-exponential term related to specific metal and $E_d \cong 55$ kJ/mol is activation energy of diffusion process between interstitial sites. These equation means that the higher the temperature, the lower the concentration, and consequently the outgassing rate decreases. Thus, the above model qualitatively explains the effect of the vacuum firing on outgassing reduction. However, expected outgassing rate from this model is much smaller than the experimental data. For example, the outgassing rate calculated by above equations for a vacuum firing at 950 °C for two hours is in the order of 10^{-18} Pa·m/s, while the typical measured data is about 10^{-11} Pa·m/s. One of the explanations for this discrepancy was attempted by considering the hydrogen recombination at the surface [58]. The atomic hydrogen concentration at the surface is assumed to be zero at the above expression; the hydrogen desorbs as rapidly as it diffuses to the surface. The physical fact is that hydrogen diffuses as individual atoms to the bulk surface but escapes from the metal in molecular form. A hydrogen atom at the surface must wait a partner hydrogen to form a molecule. By including this fact, the diffusion equation cannot be solved by analytical approach but only by numerical algorithms. There is a review of hydrogen outgassing suppression, which mainly discusses about the abovementioned diffusion model, the discrepancy to outgassing data, and approach to explain the discrepancy [59].

From the practical point of view, duration and temperature of the treatment must be limited not just because of costs but also because of impurities precipitation or grain size growth that change mechanical properties. The effect of vacuum firing on stainless steel morphology versus outgassing rates was studied at CERN, and the outgassing as a function of temperature shows a few peaks. It was found that vacuum firing should not exceed 950 °C and its duration depends on thickness. J-PARC has investigated the effect of vacuum firing on the hydrogen concentration in titanium. Hydrogen concentration in untreated titanium is more than 10 wt ppm. They measured the relation between the vacuum firing temperature and hydrogen concentration; resulting optimal temperature is 650–700 °C to obtain less than 1 wt ppm hydrogen concentration [9].

4.4.5 Surface Coatings

Coatings on the inner surface of vacuum chambers by thin films are performed not only to reduce thermal outgassing but also to suppress non-thermal outgassing such as particle-induced molecule desorption yield. Furthermore, recently in accelerators, coatings of getter materials are intensively performed to solute and diffuse gas molecules acting as getter pump. A kind of coating may have several roles, for example, TiN film is coated to suppress outgassing of water vapour due to the shorter average residence time, reduce electron emission yield, and barrier the diffusion from bulk due to its low permeability. Because details of coatings for each purpose are discussed in other sections, only brief introduction is mentioned here.

4.4.5.1 Coating the Surface by Thin Films of Material with Low Hydrogen Permeability and Low Outgassing

So far, coating of such as titanium nitride, boron nitride, silicon, or silicon oxide has been investigated to reduce outgassing by acting as diffusion barrier or low activation energy of desorption [60–63]. From the aspect of applying to vacuum chambers, TiN could be one of the best coatings because the uniform coating and control of film thickness have been established.

There is another benefit of applying them when such coating exhibit low secondary electron emission. Here, TiN is one of the most widely applied coatings to the accelerator vacuum chambers [64, 65]. Although the secondary electron yields (SEYs) depend on coating condition, bakeout condition, particle irradiation dose, and so on, TiN coating usually has SEY of about 1 or less. TiN coating has been also applied to the RF window as described in Section 4.2.5 to suppress the electron avalanche caused by the high-frequency electric field. Recently CERN has developed carbon coatings to obtain much lower SEY for applying the beam pipes in Super Proton Synchrotron (SPS) [66]. Low SEY materials, coatings and treatments are discussed in detail in Chapter 8.

In many technical areas hydrogen causes severe problems by embrittlement of steel and the problem is solved by cladding impermeable coatings. Tritium accumulation in the steel wall of the next generation of nuclear fusion reactor (DEMO) is another issue where coating by highly efficient permeation barrier effectively reduces the risk high accumulated dose [67]. Many dielectric materials can be deposited on metal samples when the permeation reduction factor can be easily confirmed. Unfortunately, uniform coating of inner side of UHV chamber is not always an easy task and the benefits of thin films obtained on testing small samples are not always realised when applied to large vacuum chambers.

4.4.5.2 Coating the Surface by Thin Film of Getter Materials

The idea is known in accelerator community for several years. The whole inner surface is coated by a carefully designed film of non-evaporable getter (NEG) materials, which are capable of capturing a noticeable amount of hydrogen and other gases. Distributed pumping speed is achieved because beam pipes become getter pumps. Transition metals such as Ti, V, Zr, Nb, and Hf were already known for their high solubility and diffusivity for hydrogen and other gases and used as the constituent materials of getter pumps. Before as-deposited film can be applied as a pump, a thermal activation process in high vacuum is needed. Mixing of getter elements has been developed to decrease the activation temperature and became suitable for the coating of accelerator beam pipes [68, 69]. Ti–Zr–V coating, which is activated at 180–200 °C, was applied to the warm section of the LHC in CERN, and the technique has expanded to worldwide accelerator community.

The main drawback of this approach is high reactivity of the film for other active gas species, which needs a very careful maintenance of low pressure. Otherwise the accumulated gases cause irreversible deterioration of pumping capability as the full pumping cannot be restored.

The NEG coating production, characterisation, activation process, and vacuum properties are discussed in detail in Chapter 5.

4.5 Electron-Stimulated Desorption

4.5.1 ESD Definition and ESD Facilities

To study the gas desorption from various surfaces and materials induced by energetic particles (e.g. SR, charged or neutral particles), one needs a source of these particles. The electrons are the most easily available particles for the laboratory study compared with the others: a hot filament and a bias up to a few kilovolts are sufficient to have a few milliampere electron current with electron energy proportional to the bias (see Figure 4.15a). Alternatively, a sample can be bombarded with the use of electron gun (see Figure 4.15b); in this case no bias is required. Combining such a simple source of electrons with UHV gauges, RGAs, and known (calculated or measured) vacuum conductance, U, or effective pumping speed prompts to building an ESD research facility for measuring ESD yields, defined as *a number of gas molecules desorbed from the surface per incident electron*, η_e [molecules/e$^-$]:

$$\eta_e \left[\frac{\text{molecules}}{\text{e}^-}\right] = \frac{\dot{N}_{\text{molecules}}}{\dot{N}_{\text{electrons}}} = \frac{Q[\text{Pa·m}^3/\text{s}]q_e[\text{C}]}{k_B T[\text{K}]I[\text{A}]}, \quad (4.25)$$

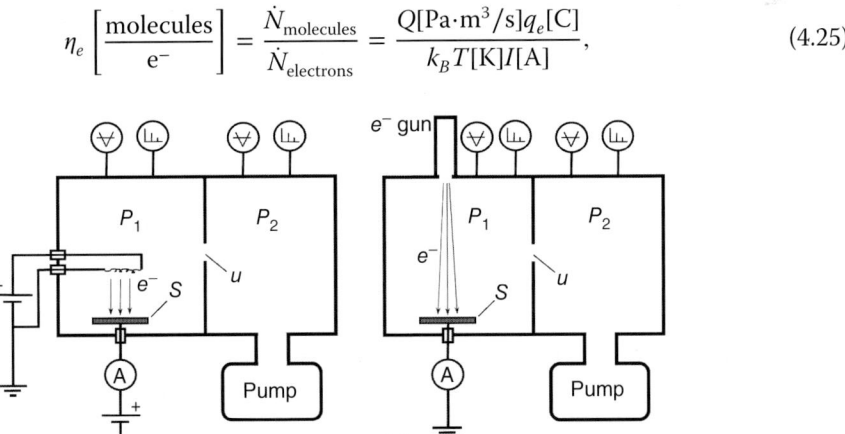

Figure 4.15 Typical layouts for ESD measurements based on a known vacuum conductance U (between the test chamber with a sample S and a pump chamber) and an electron source: (a) a hot filament and a biased sample S or (b) an electron gun.

where Q is a flux of molecules desorbed due to electron bombardment with electron current I and q_e is the electron charge. The desorption flux Q can be calculated for each gas i as follows:

$$Q_i[\text{Pa·m}^3/\text{s}] = (\Delta P_{1,i} - \Delta P_{2,i})[\text{Pa}]U_i[\text{m}^3/\text{s}], \quad (4.26)$$

where $\Delta P_{1,i} = P_{1,i,e} - P_{1,i,\text{bg}}$ and $\Delta P_{2,i} = P_{2,i,e} - P_{2,i,\text{bg}}$ have a partial pressure difference in measurements during electron bombardment (shown with an index 'e') and without electron bombardment (shown with index 'bg' meaning background). The vacuum conductance between two chambers can be realised in a form of orifice (as shown in Figure 4.15) or in a form of a tube. In case of using an effective pumping speed S_{eff}, the desorption flux Q can be calculated with a pressure measured in a test chamber only:

$$Q_i[\text{Pa·m}^3/\text{s}] = \Delta P_{1,i}[\text{Pa}]S_{\text{eff},i}[\text{m}^3/\text{s}]. \quad (4.27)$$

Note: ESD discovered almost a century ago has also a wider definition as the process of desorption of ions, both positive and negative, neutrals, and excited neutrals from surfaces as a direct result of electron bombardment and intensively used for analysis of surface, its adsorbate, bonding energies, and the mechanism of desorption [70, 71]. In this chapter, we are interested in the net effect only: how many gas molecules entered into a vacuum volume after electron–surface interaction (after ion neutralisation, atom recombination, etc.)?

The ESD can be a significant source of gas in a vacuum system when electrons bombard a surface in vacuum. This may happen in the case of electron mulipacting in beam vacuum chambers, RF cavities, and waveguides; in vacuum chambers where the electron beam is used for surface investigation, analysis, or modification; and where electron field emission from the surfaces at high electric field intensity or in the electron-based vacuum instrument the energetic electrons bombard the vacuum chamber or in-vacuum component. So, the experimental data obtained from such an ESD facility are directly applicable to the vacuum system design.

From another side, the ESD results are also applicable for a cost-effective evaluation of different materials and the efficiency of their treatments for other particle bombardment: for example, if a cleaning procedure A allows obtaining ESD yields lower than a cleaning procedure B, then it is very likely that a cleaning procedure A will also result in desorption yields for bombardment with other particles lower than a cleaning procedure B.

The ESD study in application to large vacuum system design was intensively performed just in a few research centres. The most of published work has been done at LNL (USA) [72], CERN (Geneva, Switzerland) [73–76], KEK (Japan) [77], and ASTeC (STFC Daresbury Laboratory, UK) [78–83]. The method of ESD measurements described earlier and shown in Figure 4.15 allows to study ESD from small samples with an advantage of low cost and fast turnover of samples, so it was employed for most of ESD studies. Acknowledging all advantages of this method and great usefulness of the results, there is a question that researchers are asking themselves: how are these measurements applicable to an accelerator vacuum chamber?

4.5 Electron-Stimulated Desorption

- The sample are very small in comparison to a large area vacuum chamber. Thus, the conditions of cleaning, polishing, film deposition, or other treatment are not exactly the same as for the vacuum chamber: small sample surface is easily accessible (due to so-called 'open geometry'), while there is a limited access to the inner surface of a tubular vacuum chamber (because of 'close geometry').
- Experimental conditions are different from those in the accelerator vacuum chamber: a test vacuum chamber in these ESD measurements if much larger than a sample, i.e. desorption or sorption of a test vacuum chamber, may potentially affect the measurements results.

To answer this question, an ESD facility for tubular samples was developed in ASTeC (see Figure 4.16). This layout allows to mimic the 'close geometry' conditions for a tubular vacuum chambers of particle accelerators. A tubular sample undergo the same cleaning and treatment procedures as applied to a real accelerator vacuum chamber. During the ESD measurements, the surface area of the sample is comparable with the surface area of other components: i.e. a possible influence of a test walls is much less than with small samples.

The facility shown in Figure 4.16 allows to measure the desorption flux Q for each gas i by two methods – (i) using a sample tube vacuum conductance or (ii) using an effective pumping speed as follows:

$$Q_{a,i}[\text{Pa}\cdot\text{m}^3/\text{s}] = (\Delta P_{1,i} - \Delta P_{2,i})[\text{Pa}]U_{t,i}[\text{m}^3/\text{s}],$$
$$Q_{b,i}[\text{Pa}\cdot\text{m}^3/\text{s}] = \Delta P_{2,i}[\text{Pa}]S_{\text{eff},i}[\text{m}^3/\text{s}]. \tag{4.28}$$

It is expected that two results $Q_{a,i}$ and $Q_{b,i}$ should be the same: i.e. measuring with two RGAs bring more confidence in the results. However, when only one RGA is available, P_2 measurement is sufficient to obtain the results for $Q_{b,i}$.

When ESD results from different papers are compared, it is important to note what layout of the experiment was used to study as the conditions of the experiments could be very different.

The ESD results are shown and the main functional tendencies of ESD yields under different conditions are summarised in the following subsection. Similarly

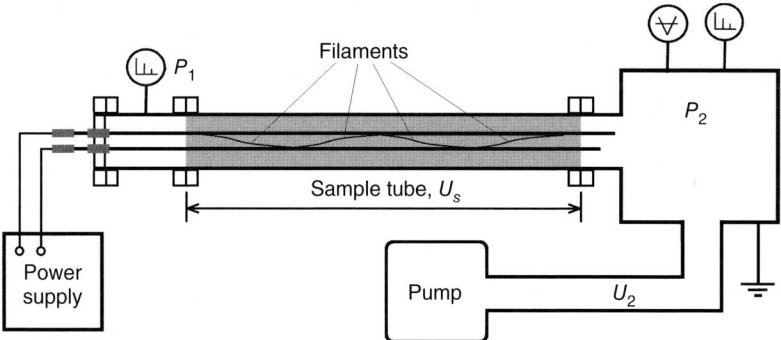

Figure 4.16 A layout for ESD measurements from a tubular sample with vacuum conductance U_s, pressure P_1 (measured with an RGA) and pressure P_2 (measured with an UHV total pressure gauge and an RGA), known vacuum conductance U_2, and pumping speed S (or effective pumping speed S_{eff}).

to thermal desorption, the ESD depends on the choice of material, cleaning procedure, surface processing (etching, polishing, coatings, etc.), history of material, bakeout procedure, and pumping time. Additionally it also depends on the energy of electrons, their current, integrated electron dose, photon dose (if exposed to SR), temperature, etc. These concussions are based on the experimental data from various research groups. Before the measurements the samples were cleaned following the CERN (or similar) standard cleaning procedure, the ESD yields from the samples were measured with electrons bombarding the sample surface at normal incidence.

4.5.2 ESD for Different Materials as a Function of Dose

Figures 4.17 and 4.18 demonstrate the dependence of ESD over a wide range of electron dose (six to seven orders of magnitude). For large doses this dependence is exponential:

$$\eta(D) = \eta(D^*)\left(\frac{D^*}{D}\right)^\alpha; \tag{4.29}$$

where an accumulated electron dose D^* and a corresponding ESD yield $\eta(D^*)$ can be taken at any point on this slope; the exponent α lies between $0.5 \leq \alpha \leq 1$ for vacuum chambers at room temperature. This ESD that yields behaviour as a function of dose has been demonstrated for copper, Glidcop, gold-coated copper, aluminium, stainless steel, and aluminium-coated stainless (see Figures 4.17–4.20).

The comparative ESD study was performed CERN (Switzerland) with 300 eV electrons [76]. The planar samples were cut sheets with dimensions 47 mm × 50 mm from 316LN stainless steel, OFHC copper, and AA6082 aluminium samples, cleaned, and installed in vacuum chamber, baked to 150 °C for 24 hours and then to 300 °C for two hours. The ESD yields results are shown in Figure 4.19. One can see that initial ESD yields from baked samples are lower than from unbaked ones; however the ESD yields are also reduced with an electron dose with a dependence described with Eq. (4.29).

ESD was measured in ASTeC/STFC (UK) with 500 eV electrons from the tubular samples made of three different materials. Before the ESD measurements the samples were baked: 316L stainless steel [78] and copper [80] baked to 250 °C for 24 hours and AA6082 aluminium baked to 220 °C for 24 hours [79]. ESD yields are shown in Figure 4.20.

One can see that in a well-cleaned vacuum chamber, the main desorbed gas species are H_2, CO, CO_2, CH_4, and H_2O, but other species such as Ar and O_2 can also be reported.

It was shown in Ref. [72] and confirmed in later studies that there is a little difference in the ESD yields of three main materials used for manufacturing of vacuum chambers such as copper, aluminium, and stainless steel. Gold plating does not help to reduce ESD. Argon discharge cleaning allows to reduce initial ESD, but the ESD at large doses are higher due to a likely surface area increase during discharge treatment. The ESD yields of stainless steel were reduced by a factor of two after bakeout to 250 °C and by a factor of 5–10 after vacuum firing at 900–950 °C.

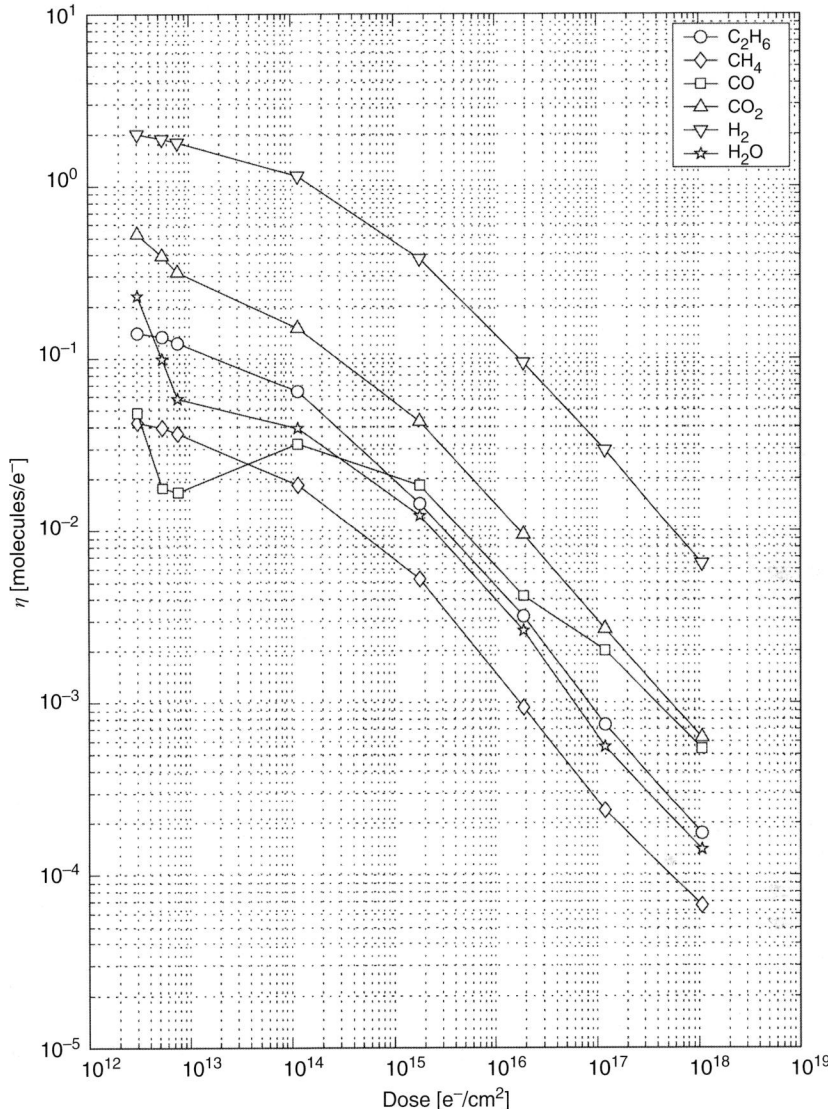

Figure 4.17 ESD yields of unbaked OFHC copper after 24 hour pumping as a function of electron dose at $E_e = 300$ eV. Source: Reprinted with permission from Billard et al. [75], Fig. 3. Copyright 2000, CERN.

4.5.3 ESD as a Function of Amount of Desorbed Gas

The total amount of each gas desorbed in ESD experiments as a function of electron dose can be calculated as follows:

$$N(D)[\text{molecules}] = \int_0^t \eta_e(t_a) \frac{I_e(t_a)}{q_e} dt_a = \int_0^D \eta_e(D_a) dD_a \quad \text{or}$$

$$\Theta(D)[\text{Pa·m}^3] = \int_0^D Q_{\text{ESD}}(D_a) dD_a = k_B T \int_0^D \eta_e(D_a) dD_a. \quad (4.30)$$

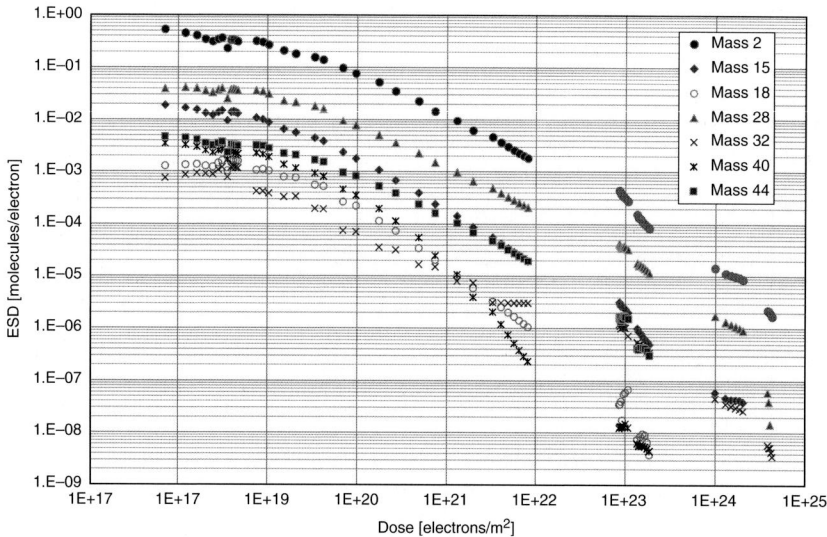

Figure 4.18 ESD yields of 316LN stainless steel baked to 250 °C for 24 hours as a function of electron dose at electron energy E_e = 500 eV. Source: Reprinted with permission from Malyshev and Naran [81], Fig. 1. Copyright 2012, Elsevier.

The total amount of desorbed gas is usually normalised to the total irradiated area of vacuum chamber, A:

$$N_A(D) = \frac{N(D)}{A}; \quad \Theta_A(D) = \frac{\Theta(D)}{A}. \tag{4.31}$$

These amounts obtained in the ESD experiments can be plotted in two ways:

- ESD as a function of amount of desorbed gas (see Figure 4.21 for the same data as shown in Figure 4.20).
- Amount of desorbed gas as a function of electron dose (see Figure 4.22 for the same data as shown in Figure 4.20 and Figure 4.21).

4.5.4 Effect of Pumping Duration

It is well known that pressure in vacuum chamber slowly decreases with pumping time. In application to particle accelerator, we need to know how the ESD yields depend on the duration of pumping. The comparative study on the 316LN stainless steel and aluminium chambers measured after bakeout following short and long pumping times was reported in Refs. [78, 79]. The 316LN stainless steel samples were baked to 250 °C for 24 hours, and the ESD yields from a sample pumped for 24 days are ~10 lower than from a similar sample pumped for 24 hours (1 day). The aluminium samples were baked to 220 °C for 24 hours, and the ESD yields from an aluminium sample pumped for 26 days are practically the same for H_2 and factor 1.5–3 lower for other gas species than from a similar sample pumped for 1 day.

Figure 4.19 ESD yields from different materials baked to 150 °C for 24 hours and then to 300 °C for 2 hours as a function of electron dose. Source: Reprinted with permission from Gómez-Goñi and Mathewson [76], Figs. 3–5. Copyright 1997, American Vacuum Society.

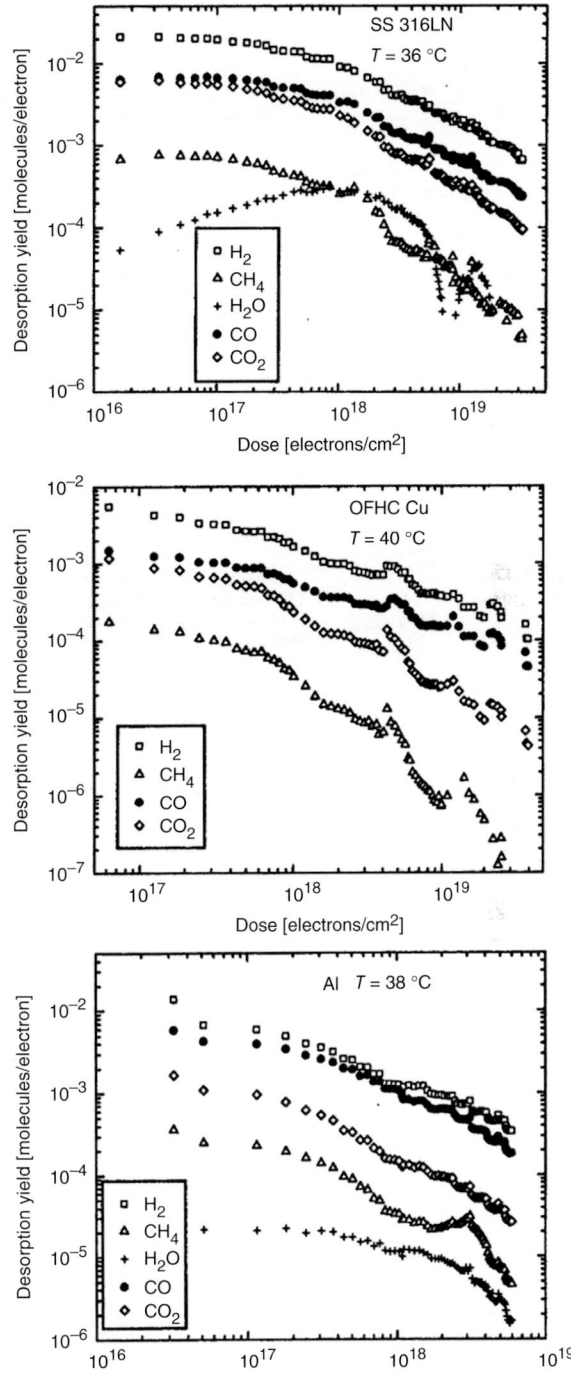

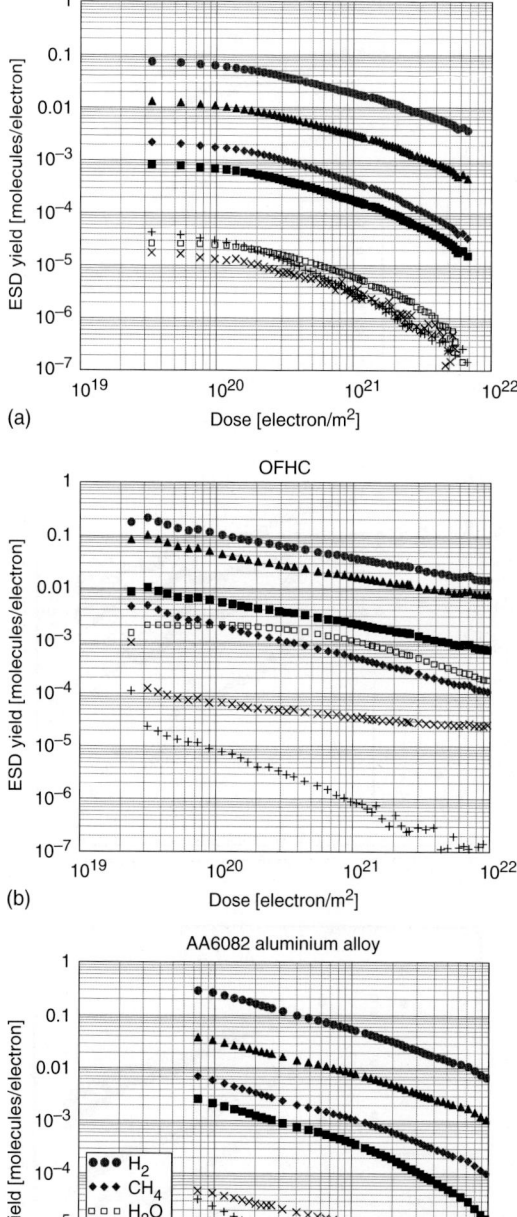

Figure 4.20 ESD yields from different baked materials (to 250 °C for 316L stainless steel and copper and to 220 °C or AA6082 aluminium, for 24 hours) as a function of electron dose at electron energy $E_e = 500$ eV. Source: Original data reported in Refs. [78–80].

Figure 4.21 ESD yields as a function of amount of desorbed gas for the same data as shown in Figure 4.20.

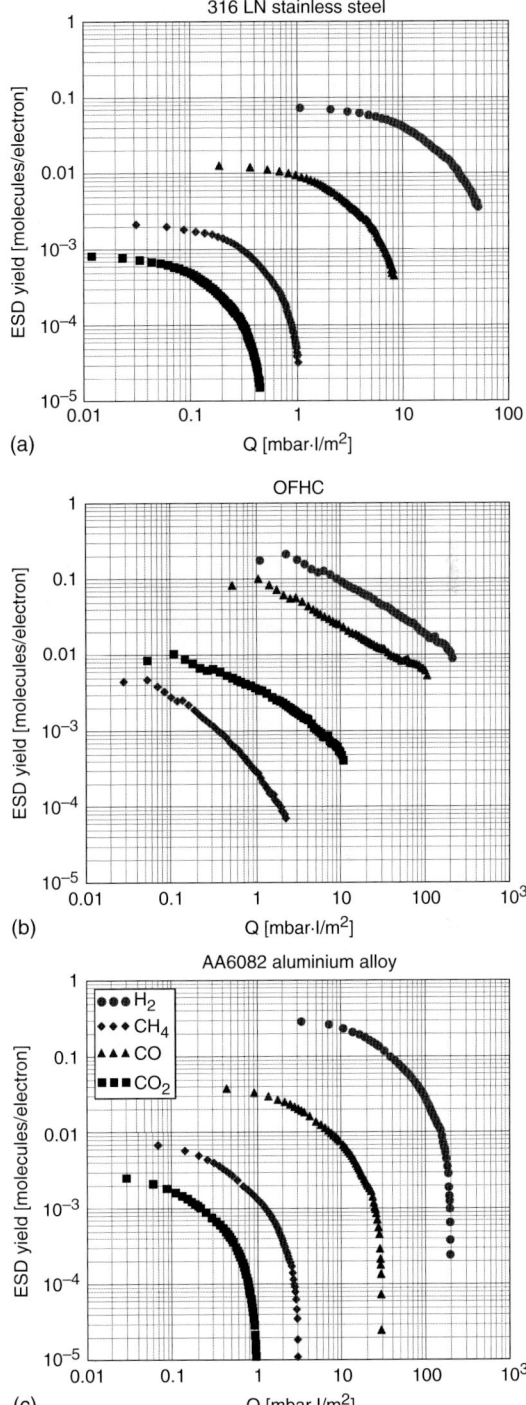

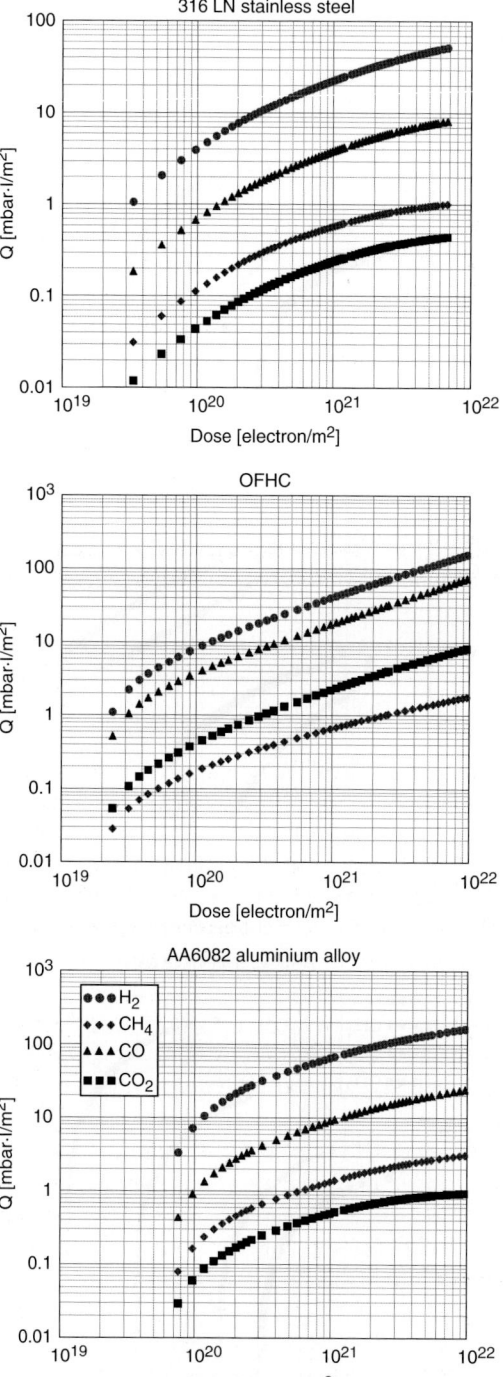

Figure 4.22 Amount of desorbed gas as a function of electron dose for the same data as shown in Figures 4.20 and 4.21.

The main conclusion is that long pumping after bakeout indeed helps to reduce ESD yields from a metal vacuum chamber, but it is too early to speculate how strong this effect could be in accelerator vacuum chamber. More data should be collected in the future.

4.5.5 ESD as a Function of Electron Energy

Different electron energies were used for ESD studies in different laboratories, for example, 1.5 keV in LBL [72], 3 keV and KEK [77], 300 eV [76], and 1.4 keV [74] at CERN and 500 eV in ASTeC [79]; therefore the ESD yield dependence on the electron energy should be taken into account in comparing different results.

The ESD yields depend on the energy of incident electrons, as demonstrated in a few studies [73, 75, 82]. The challenge in such a study is that the measurements should be done on a surface under the same condition, while during the ESD measurements the surface is bombarded, outgasses, and therefore is not the same as initially for the next measurement. Therefore, two different methods were applied.

First method was using a few samples made of the same materials and prepared and treated together; then the measurements were performed following exactly the same procedure [75, 82]. The measurements of four unbaked OFHC copper samples, shown in Figure 4.23, were performed at 20, 50, 100, and 300 eV after the same electron dose of $D = 1.4 \times 10^{17}$ e$^-$/cm^2. The slope of the ESD yield as a function of the electron energy E can be described as $\eta_e(E) \propto E^\beta$ where $\beta \approx 0.85$ in these experiments.

Another method was to measure the ESD yields as a function of electron energy after a very large dose. In two examples, $\eta_e(E)$ for a stainless steel sample baked to 250 °C for 24 hours and for aluminium alloy AA6082 baked to 250 °C for 24 hours, shown in Figure 4.24a,b, were measured in the energy range between 40 eV and 5 keV after reaching an electron dose of $D \approx 10^{21}$ e$^-$/cm^2 [78, 79]. The ESD yield at 500 eV measured before and after the energy scan were the same, verifying that there were no conditioning effects during the energy scans.

A different method was employed in Ref. [82]: three identical 316L stainless steel samples S1, S2, and S3 went through the same cleaning, installation, and bakeout procedure (at 250 °C for 24 hours), when the samples underwent the same two-stage experimental procedure.

In Stage 1, the samples were bombarded with electrons with different electron energies: 50, 500, and 5000 eV, correspondingly, reaching an accumulated electron dose of between 1×10^{23} and 2×10^{23} e$^-$/m^2. The measured ESD yields for H_2 are shown in Figure 4.25a, where the behaviour of ESD yields for other species is very similar. The initial ESD yields are higher for higher energy of impact electrons. However, the ESD yields decreases with electron dose slightly quicker for greater electron energy; therefore, similar ESD yields for the same gas species were measured at an approximate dose of 10^{23} e$^-$/m^2. The same ESD results plotted as a function of amount of desorbed gas (see Figure 4.25b) demonstrate that the ESD yields at 50 eV are 10–100 times lower than those measured at 5 keV for the same amount of desorbed gas.

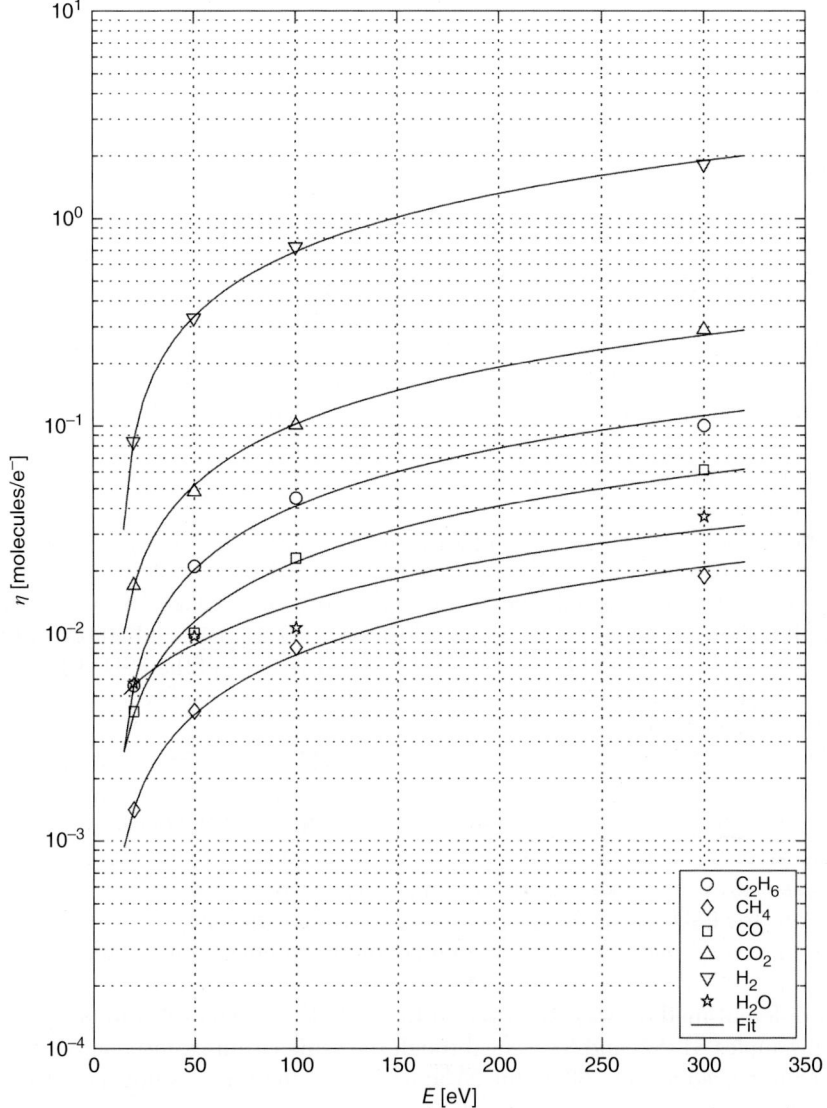

Figure 4.23 ESD yields of unbaked OFHC copper as a function of electron energy at the dose $D = 1.4 \times 10^{14}$ e$^-$/cm^2. Source: Reprinted with permission from Billard et al. [75], Fig. 5. Copyright 2000, CERN.

In Stage 2, an energy scan was performed with samples S1 and S2 by increasing electron energy from 10 eV to 6.5 keV and then decreasing it back to 10 eV. The results obtained in Stage 2 are in good agreement with those obtained in Stage 1: i.e. ESD yields at 50, 500, and 5000 eV in Stage 2 (energy scan) experiments for samples S1 and S2 are less than a factor of 2, different from those obtained in Stage 1 for S1 at 50 eV, for S2 at 500 eV, and for S3 at 5000 eV for the same Q. Also,

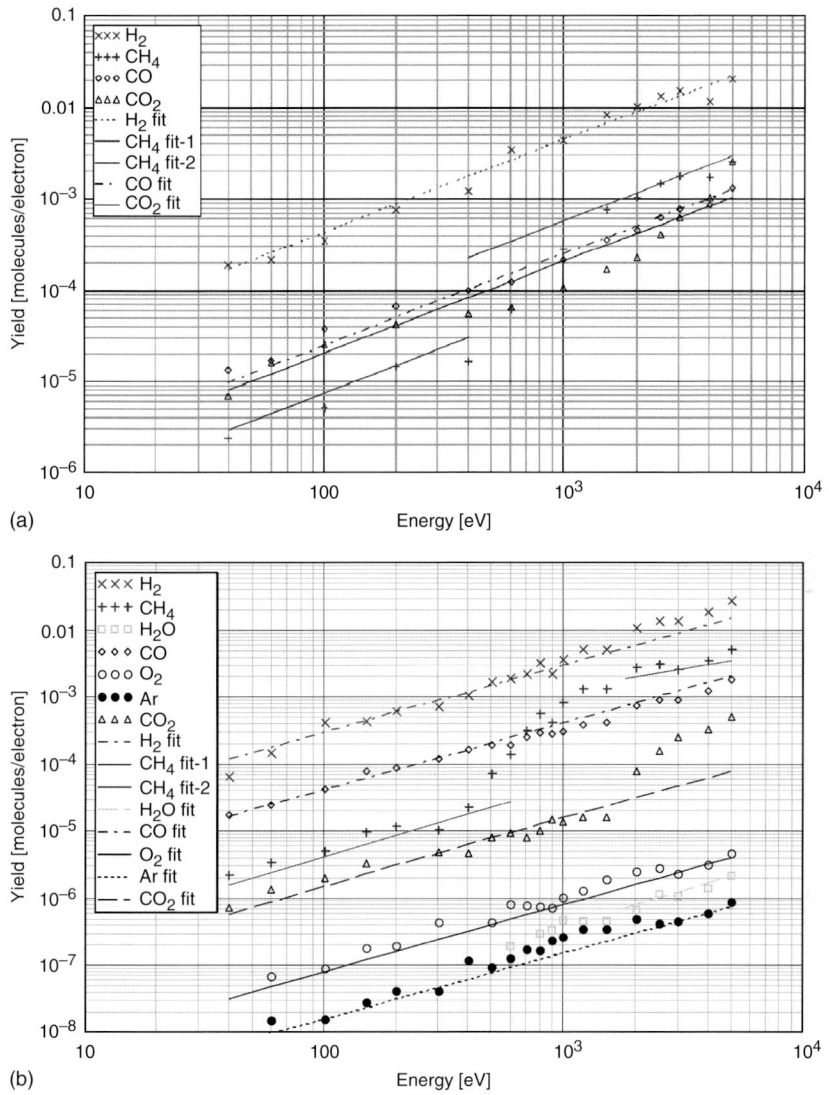

Figure 4.24 ESD yield as a function of electron energy: (a) for stainless steel sample baked at 250 °C for 24 hours at the dose $D = 7 \times 10^{21}$ e$^-$/cm^2 and (b) for aluminium sample baked at 220 °C for 24 hours at the dose $D = 1.3 \times 10^{22}$ e$^-$/cm^2. Source: (a) Reprinted with permission from Malyshev et al. [78], Fig. 7. Copyright 2010, American Vacuum Society. (b) Reprinted with permission from Malyshev et al. [79], Fig. 7. Copyright 2011, Elsevier.

one can see that conditioning during the energy scan is very strong for sample S1, quite significant for sample S2.

All these results are in agreement that for the same state of the surface, the ESD yield increases the electron energy E in the range between 10 eV and 6.5 keV. Further studies are required if the data are needed for higher electron energies.

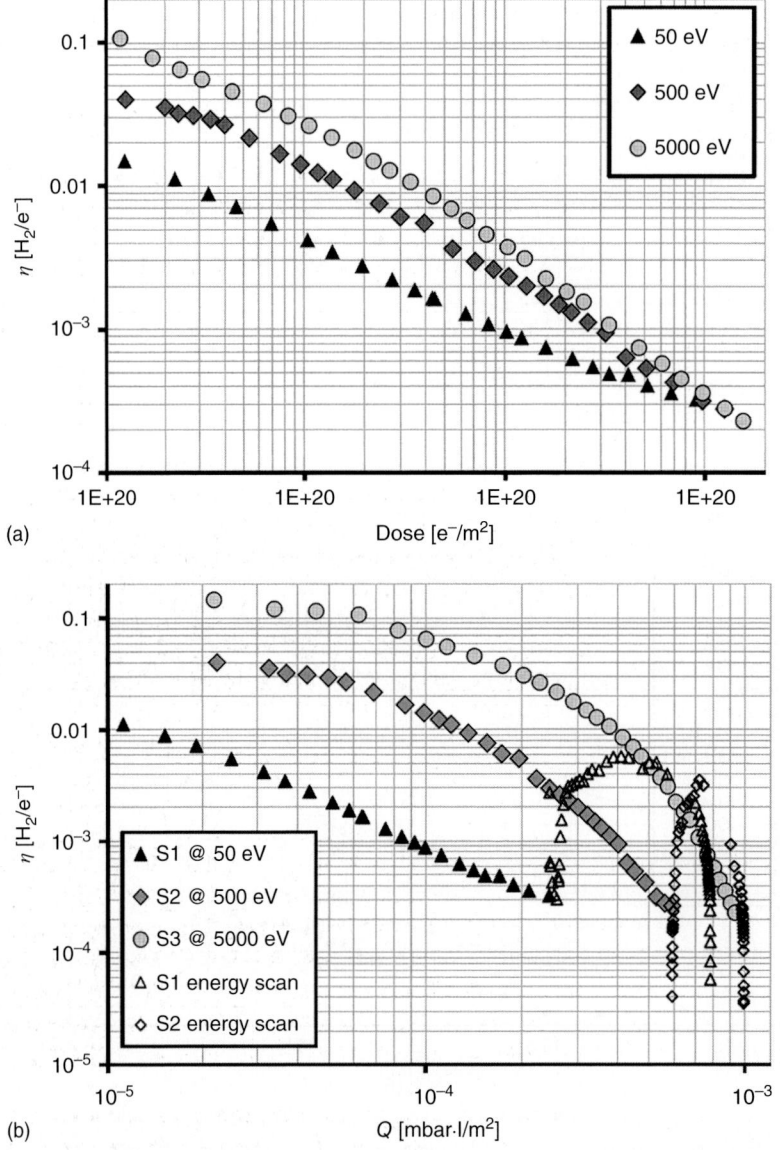

Figure 4.25 The H_2 ESD yields for 316L stainless steel samples baked at 250 °C for 24 hours for four different gas species (a) as a function of electron energy dose and (b) as a function of desorbed gas. Source: Reprinted with permission from Malyshev et al. [82], Figs. 2 and 3. Copyright 2013, American Vacuum Society.

4.5.6 Effect of Bakeout on ESD

The effect of bakeout was demonstrated in a detailed study at CERN in 1978 [74]. The ESD yields were measured for 316LN stainless steel, titanium alloy (TiAl6V4), OFHC copper, and aluminium alloy (5086) before bakeout and after bakeout to temperatures ranging from 150 to 600 °C for stainless steel, titanium alloy, and OFHC copper and ranging from 150 to 300 °C for aluminium alloy. The electron energy was 1.4 keV for all materials except for aluminium alloy bombarded with electron energy of 600 eV.

The results shown in Figure 4.26 demonstrate that the ESD yields for H_2, CH_4, CO, and CO_2 were reduced after bakeout. In general, the higher bakeout temperature, the lower the ESD yields. Thus, in comparison to unbaked samples, the ESD yields after bakeout to 300 °C are lower by a factor of 4–10 for stainless steel, 20–40 for titanium alloy, 10–20 for OFHC copper, and 30–100 for aluminium alloy. Note that a vacuum chamber made of OFHC copper or aluminium alloy should not be baked to 300 °C.

In the reported study only stainless steel and aluminium alloy samples were baked to 150 °C, showing interesting results: the ESD yields obtained after the bakeout to 150 °C (except for H_2 on stainless steel) are lower than or comparable to the ones after the bakeout to 200 °C. This local minimum could be explained as follows. The ESD yield is proportional to gas concentration in the near-surface layers. This concentration is affected by two competing processes that may have a different dependence on temperature: (i) a depletion of near-surface layers from the gases initially contained there by outgassing into vacuum and (ii) increasing due to the gas diffusion from the deeper layers of the samples. As a result, the initial ESD yields for samples baked to 150 °C at least as low as for ones baked to 200 °C. Thus from a practical point of view for an accelerator vacuum chamber, the bakeout to 150 °C could be more cost efficient than to 200 °C considering that bakeout to 150 °C temperature will require a lower electricity consumption and shorter bellows to compensate for vacuum chamber temperature expansion and allows to reach similar or even lower ESD. Unfortunately, only initial ESD were studied in this work; thus one could not generalise this to large electron dose without mere experimental studies.

For more significant reduction of ESD yields, the bakeout temperature of 250–300 °C could be applied to stainless steel vacuum chamber.

Water could be significant or even the main gas in unbaked vacuum chambers; however, the highest ESD yield is always for H_2 (for both unbaked and baked samples) followed by CO, CO_2, and CH_4. In an unbaked vacuum chamber, the ESD yield for H_2O could be comparable to CO or CO_2 (see Figures 4.17 and 4.23), but its significance is much reduced after bakeout (see Figures 4.20 and 4.18).

4.5.7 Effectiveness of Surface Polishing and Vacuum Firing on ESD

In the previous section it was shown that the reduction of thermal outgassing from stainless steel can be achieved by surface polishing and vacuum firing. The aim of this study reported in Ref. [83] was to identify the effectiveness of surface polishing and vacuum firing for reducing ESD from 316LN stainless steel. The four samples studied were made of 316LN stainless steel. Sample S1 was not polished, while the inner surfaces of samples S2, S3, and S4 were polished by the manufacturer using different techniques. Before the experiments the samples were all treated using the same procedure for installing, pumping, and bakeout to 250 °C for 24 hours. After a seven-day duration bombardment with 500 eV electrons, each sample was removed to be vacuum fired to 950 °C for two hours at

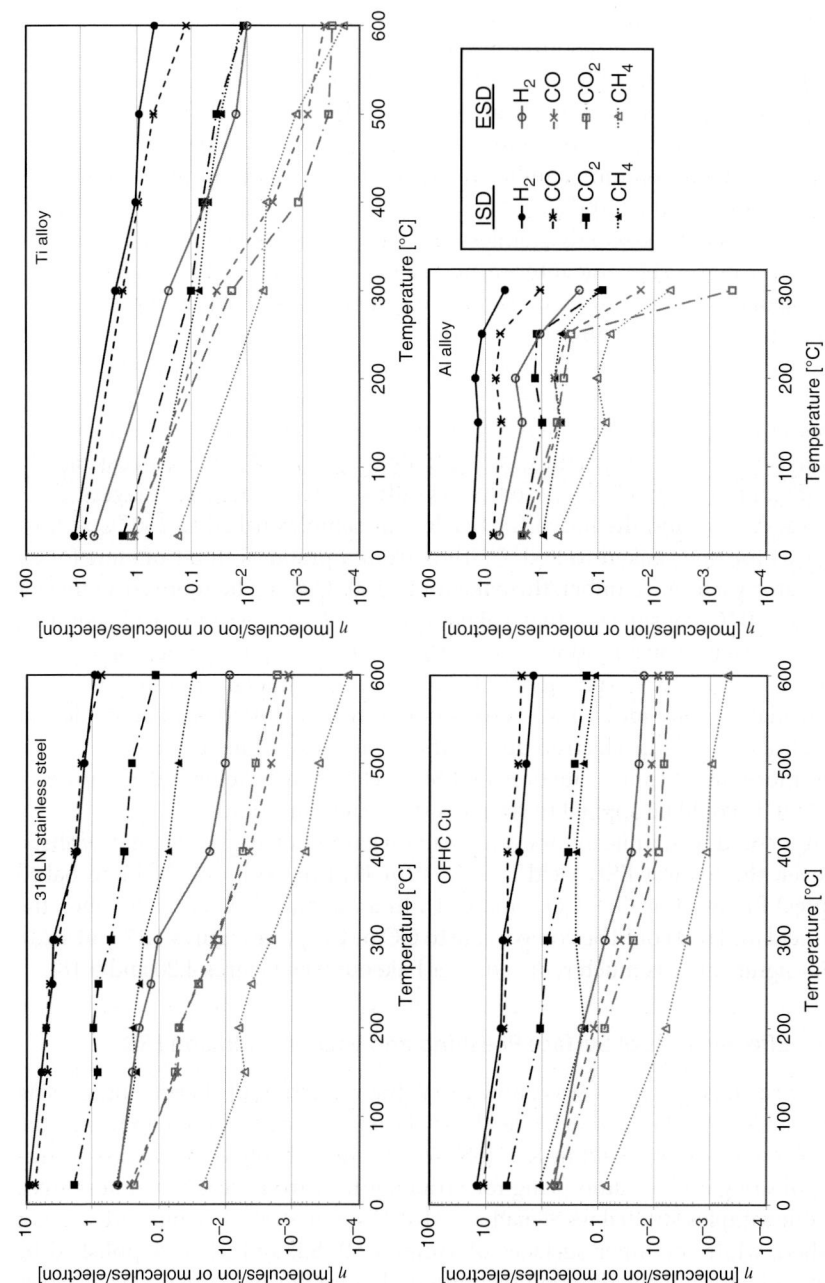

Figure 4.26 The effect of bakeout temperature on the ISD and ESD yields for 316LN stainless steel, titanium alloy (Ti–Al6–N4), OFHC copper, and aluminium alloy (5068) samples bombarded with $^{15}N_2^+$ ions and electrons at 1.4 keV. Source: Achad et al. 1978 [74], Figs. 1, 2, 5, and 6. Reproduced with permission of Elsevier.

a pressure of $\sim 10^{-5}$ mbar. When the firing process was completed, the samples were vented with nitrogen, exposed to air for several hours, and pumped out and baked out to 250 °C for 24 hours. Then the electron bombardment with 500 eV electrons was performed for another week.

All polished samples (S2, S3, and S4) demonstrated very similar result; thus only the results for samples S1 and S3 are shown in Figure 4.27. The results demonstrated that similarly to the thermal outgassing the ESD can be reduced with vacuum firing, but unlike the thermal outgassing, the fine surface polishing of the surface has either no effect to or could even increase the ESD yields.

Thus vacuum firing is a very effective technology for reducing ESD, while the costly and time-consuming surface polishing is an unnecessary treatment for vacuum improvement. However if surface polishing is required for other reasons (such as surface impedance), it should be preferably followed by vacuum firing to avoid degradation of vacuum.

4.5.8 A Role of Oxide Layer on Copper

The efficiency of copper surface preparation was studied at KEK [77]. Copper surface has an oxide layer that affects the ESD. To remove the natural oxide layer, four OFC (ASTM C10100) samples were etched with sulphuric acid and hydrogen peroxide (SH), or with citric acid (CT) or consequently with both (SH + CT). The results reproduced in Figure 4.28 demonstrate that for the total ESD yields as a function, the thickness of the oxide layer is decreasing with it because the total surface density of oxygen and carbon atoms is also increasing with the oxide layer thickness. However, it should be noted that etching with SH was more efficient to reduce initial ESD than CT or SH + CT.

4.5.9 Effect of Surface Treatment

The efficiency of stainless steel (304) surface treatment was studied at BNL (USA) [84]. The treatment procedures were the following:

(1) A degrees treatment (a soap wash, H_2O rinse, acetone and methanol rinse, air dry).
(2) A degrees treatment followed (as in item (1)) by an acid cleaning (five-minute dip in a solution of 1/3 HF, 1/3 HNO_3, 1/3 H_2O).
(3) A degrees treatment followed by an acid cleaning (as in item (2)) with subsequent hydrofluoric acid dip without subsequent H_2O rinse.

Before ESD measurements, the samples were pumped, baked to 200 °C for 60 hours, and cooled down to room temperature. The results for three treatment procedures are shown in Table 4.4.

4.5.10 Effect of Vacuum Chamber Temperature

The effect of vacuum chamber temperature on ESD and PSD will be discussed in Section 4.6.7.

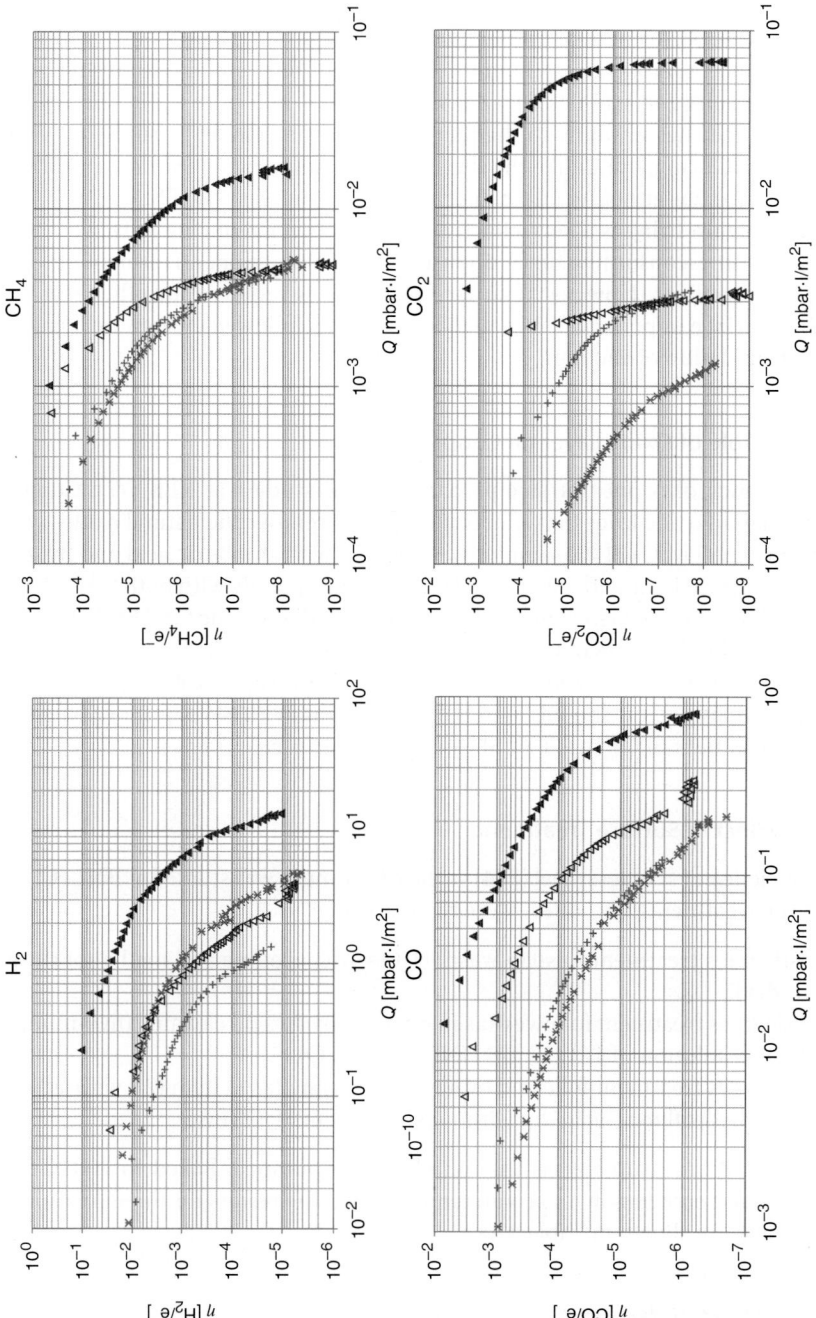

Figure 4.27 ESD yield as a function of the amount of desorbed gas for 316LN stainless steel samples baked at 250 °C for 24 hours: S1, not polished and vacuum fired; S3, polished; and S3VF, polished and vacuum fired. Source: Reprinted with permission from Malyshev et al. [83], Fig. 2. Copyright 2014, American Vacuum Society.

Figure 4.27 (Continued)

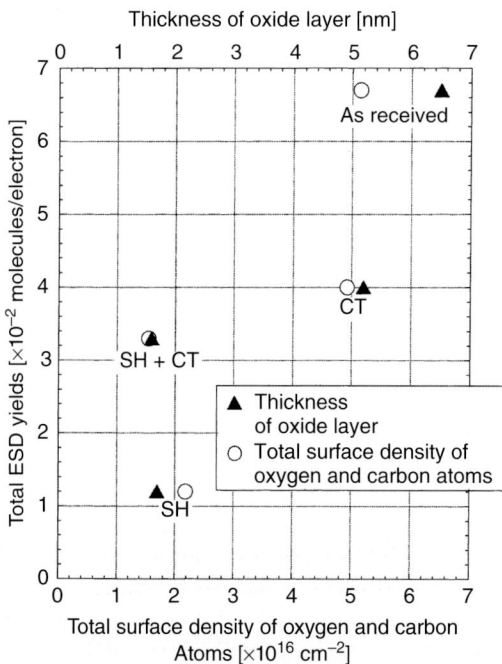

Figure 4.28 ESD yield as a function of the thickness of the oxide layer for OFHC copper samples, $E_e = 1.5$ keV. Source: Reprinted with permission from Nishiwaki and Kato [77], Fig. 4. Copyright 2001, Elsevier.

Table 4.4 ESD from 304 stainless steel with 500 eV electrons after three treatments procedures.

Gas	ESD yield [molecules/electron]		
	(1) Degrease	(2) Acid	(3) HF
H_2	0.18	0.09	0.04
CO	0.05	0.01	0.02
CO_2	0.16	0.02	0.05
CH_4	0.003	0.001	0.001

Source: Edwards 1979 [84]. Reproduced with permission of American Institute of Physics.

4.6 Photon-Stimulated Desorption

4.6.1 PSD Definition and PSD Facilities

Many modern high energy accelerators and storage rings produce SR in their dipoles, quadrupoles, wigglers, and undulators. The characteristics of SR and their relation to beam parameters are described in Chapter 4. In this section we will focus one of the most important sources of gas *in the presence of SR* called photon-stimulated desorption (PSD). PSD can arise from several different mechanisms such as direct excitation, ionisation, electron-hole generation by photon, SR power dissipation in the bulk due to photon or plasmon excitation, etc. [85].

For our purpose, the PSD can be considered as a two-step process. First, photons with energy >10 eV cause the photoelectron emission (PEE), then the photoelectron stimulate gas desorption [86–90]. Direct photon–molecular interactions are negligible. Gas molecules may desorb from a surface when and where *photoelectrons* leave and arrive at the surface. Therefore, to specify where PSD takes place, one should consider not only where SR irradiates a surface in vacuum but also where the photons after initial contact with a wall can be reflected to and where the emitted photoelectrons can arrive to (see Figure 4.29).

The PSD results are reported in terms of PSD yields, which are defined as *a number of gas molecules desorbed from the surface per incident photon*, η_γ [molecules/photon]:

$$\eta_\gamma \left[\frac{\text{molecules}}{\text{photon}} \right] = \frac{Q[\text{Pa·m}^3/\text{s}]}{k_B T[\text{K}]\Gamma[\text{photon/s}]}, \qquad (4.32)$$

where Q is a flux of molecules desorbed due to photon bombardment with the photon flux Γ.

To study the PSD from various surfaces and materials, one needs a source of SR, and this could be generated at charge particle accelerators only. A research facility could either be built on a specialised SR beamlines or be a section of vacuum chamber of an operating machine with SR. A few most commonly used layouts of the facility for PSD measurements are shown in Figure 4.30. In all examples, the SR beamline is equipped with a safety shutter (SS) to stop photon irradiation, insulating vacuum gate valve (V) to separate the experimental facility from the SR beamline, and horizontal and vertical SR beam collimators (C_h and C_v) followed by and luminescent display (LD) to control, observe, and measure the SR beam size. The vacuum conductance method (see Figure 4.30a,b) is the most common for the gas flow measurements. In this case, the PSD yields for each gas i are calculated as

$$\eta_{\gamma i} \left[\frac{\text{molecules}}{\text{photon}} \right] = \frac{(P_{1,i} - P_{2,i})[\text{Pa}]C_i[\text{m}^3/\text{s}]}{k_B T[\text{K}]\Gamma[\text{photon/s}]}. \qquad (4.33)$$

Three gauge method shown in Figure 4.30c is used for tubular samples with a large aspect ratio $R = L/a$, where L is the sample length and a is its

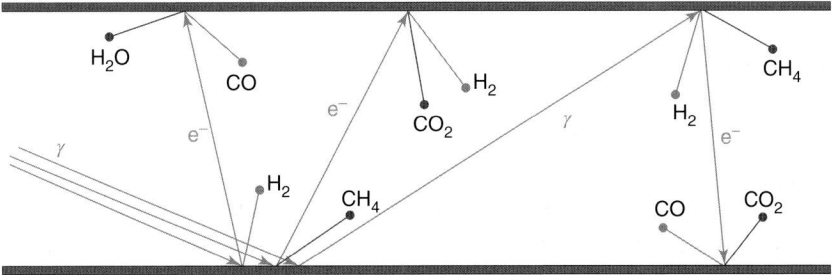

Figure 4.29 SR photons can stimulate PSD or photoelectrons or be reflected; photoelectrons can cause ESD at the interaction with walls; reflected photons can stimulate PSD or photoelectrons or be reflected again.

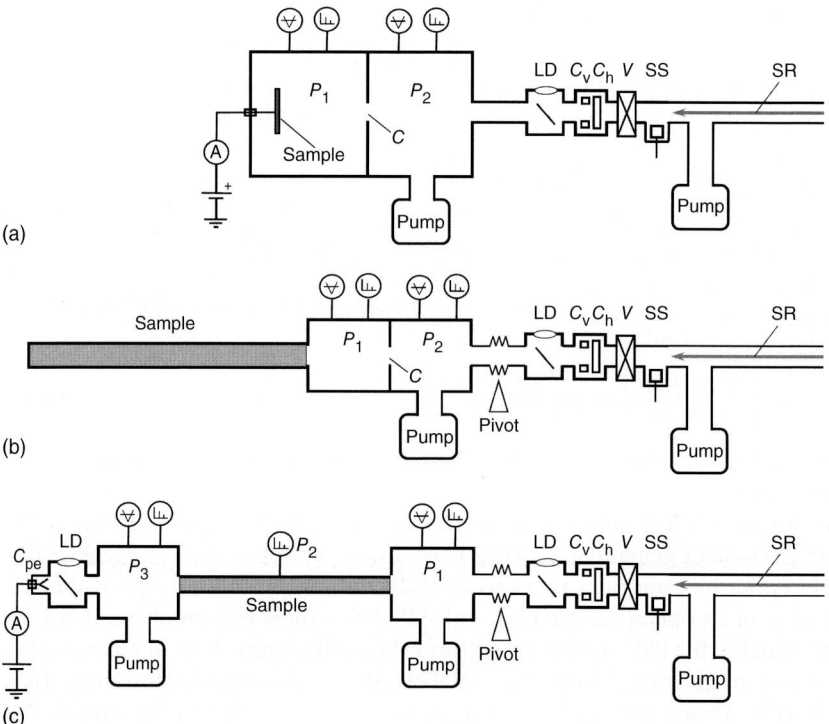

Figure 4.30 Typical layouts for PSD measurements on SR beamline: (a) for small samples and (b) for tubular samples based on a known vacuum conductance C between the test chamber with a sample and a pump chamber and (c) a three-gauge method based on a sample vacuum conductance.

cross-sectional size. The advantage of this method is that this geometry better represents the accelerator vacuum chamber and the pressure P_2 is measured in the middle of the sample. There the pressure has its highest value. A combined collector–calorimeter C_{pe} at the end of the beamline allows measuring an SR photon flux power and photoelectron current to compare full SR (when SR passed through vacuum chamber without irradiating test vacuum chamber) and reflected SR (when sample chamber is irradiated).

Significant progress in studying the PSD for different materials and treatments was made in the end of 1980 to 1990 in NSLS (Brookhaven, USA), CERN (Geneva, Switzerland), KEK (Japan), LURE (Orsay, France), ESRF (Grenoble, France), and BINP (Novosibirsk, Russia). Such studies for new materials, new cleaning procedures, and different experimental conditions continue also in present time. Similarly to thermal desorption and ESD, the PSD depends on the choice of material, cleaning procedure, surface processing (etching, polishing, coatings, etc.), history of material, bakeout procedure, and pumping time. Additionally it depends on the energy of photons (critical energy of SR), photon flux, integrated photon dose, temperature, etc. The main results of these studies are summarised in the following.

4.6.2 PSD as a Function of Dose

The gradual reduction of pressure in an electron and positron storage ring with SR generated by the circulating electron or positron beam is by now well known [91–93]. This phenomenon called 'beam conditioning' or 'beam scrubbing' has originally a phenomenological basis, that is, the experience of operational storage rings around the world as well as the results obtained on dedicated facilities for PSD studies. The results were reported in many articles, for example, [94–108].

PSD yields are studied as a function of *accumulated photon dose, D*. The typical results are shown in Figure 4.31. The PSD yields decrease with accumulated photon dose, the slope of the curve changes with the dose, and it can be described with

$$\eta(D) = \eta_0 \left(\frac{D + D_1}{D_0 + D_1} \right)^\alpha ; \qquad (4.34)$$

where η_0 is an *initial* PSD yield measured at a small accumulated photon dose D_0 (a first measured data point with a low as possible dose), an accumulated photon dose D_1 is a conditional dose when for $D \geq D_1$ the slope of the curve does not change anymore; the exponent α lies between $2/3 \leq \alpha \leq 1$ for vacuum chambers at room temperature. Parameters η_0, D_0, D_1, and α in Eq. (4.34) could be found by fitting experimental data.

For doses $D \geq D_1$ the Eq. (4.34) could be simplifies as

$$\eta(D) = \eta(D^*) \left(\frac{D^*}{D} \right)^\alpha ; \qquad (4.35)$$

where an accumulated photon dose D^* and a corresponding PSD yield $\eta(D^*)$ can be taken at any point on this slope. Note that in some experiments there is an initial grow in PSD at the beginning of SR irradiation and/or after every interruption in SR. This relates to a transition between a quasi-equilibrium balance of desorption–absorption processes on the surfaces of vacuum chamber without SR and with SR.

Let us compare two experimental results obtained by Foerster et al. at NSLS [95] and Herbeaux et al. at LURE [105] on desorption from a stainless steel vacuum chamber pre-baked at 200 °C for 24 hours (but not baked *in situ*) by SR with critical photon energies $\varepsilon_c = 500$ eV and $\varepsilon_c = 3.35$ keV, respectively. Although the initial values of the PSD yields at a dose of 10^{19} photon/(s·m) found by Herbeaux is about four times higher than that measured by Foerster, the difference at doses of 10^{22}–10^{23} photon/m is already negligible and may be related to the different photon critical energies. Herbeaux finds $\alpha = 1$, while Foerster, $\alpha = 2/3$. These results could be compared for CO as it is shown in Figure 4.32. This observation is confirmed by PSD measurements at other experiments concluding that the exponent α is usually measured $\alpha \approx 1$ in the experiments with the critical photon energy in the range $1 \text{ keV} < \varepsilon_c < 8 \text{ keV}$, while $\alpha \approx 2/3$ in experiments with $180 \text{ eV} < \varepsilon_c < 600 \text{ eV}$.

4.6.3 PSD for Different Materials

PSD depends on the choice of material. The main interest in PSD studies was on copper, aluminium, stainless steel, and copper plated stainless steel [91–108, 110];

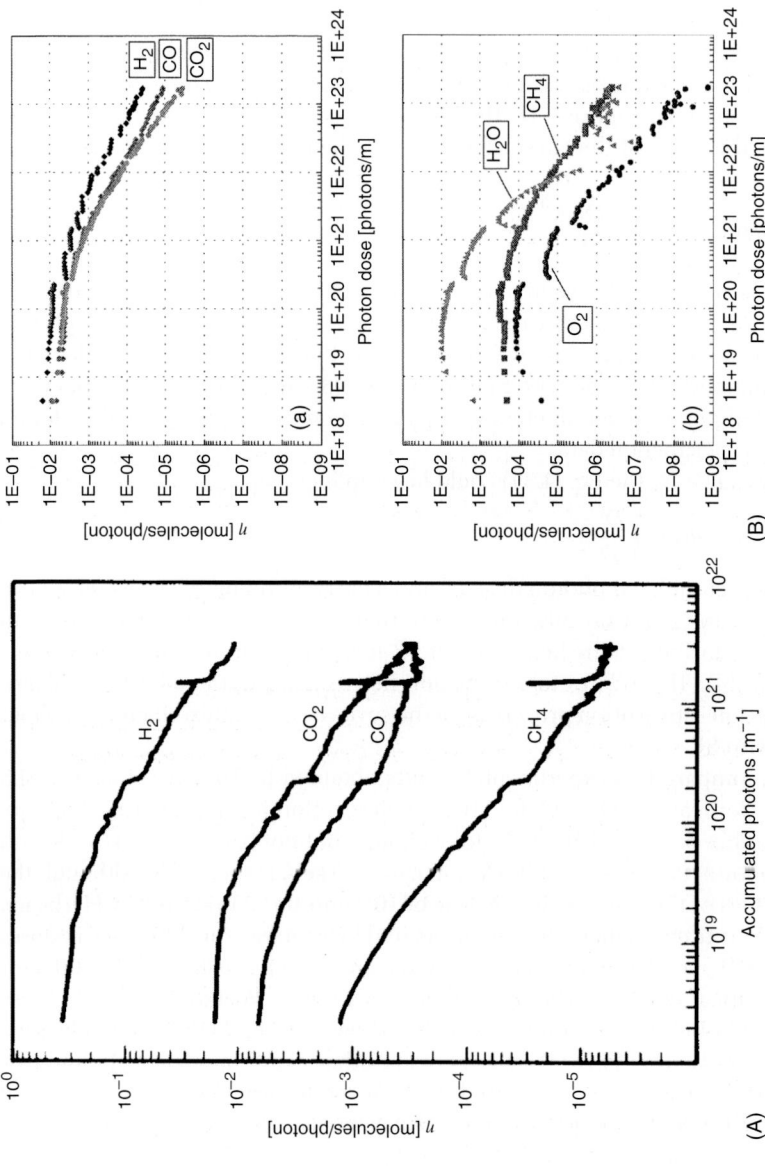

Figure 4.31 The PSD yield η for various gas species as a function of the accumulated photon dose at $\varepsilon_c = 3.75$ keV at DCI from aluminium vacuum chamber on the panel (A) (Fig. 9 in Ref. [91]) and from stainless steel vacuum chamber on panel (B). Source: (B) Reprinted with permission from Herbeaux et al. [102], Fig. 4. Copyright 2001, Elsevier.

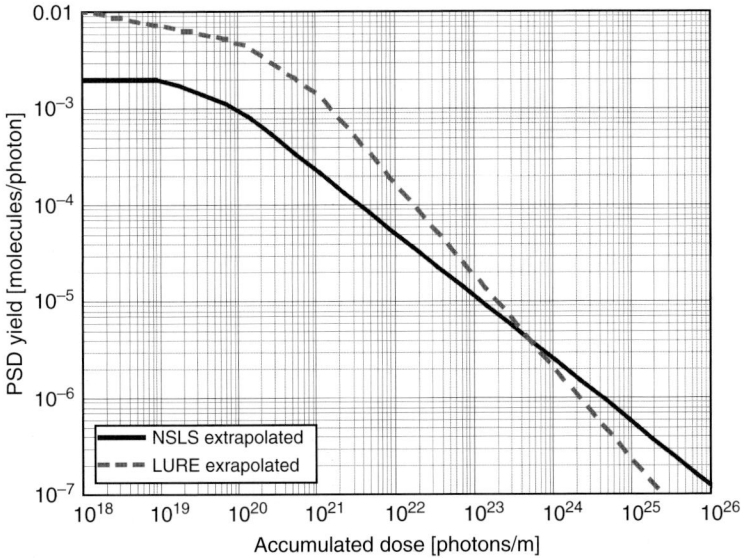

Figure 4.32 Example of CO PSD yields for stainless steel chamber as a function of accumulated photon dose D, based on the results of experiments at NSLS [95] and LURE [105] with critical photon energies $\varepsilon_c = 500$ eV and $\varepsilon_c = 3.35$ keV, respectively. Yields for doses higher than 10^{23} photons/m were extrapolated. Source: Reprinted with permission from Malyshev [109], Fig. 1. Copyright 2012, Elsevier.

however the results for less common materials such as titanium and beryllium [111, 112] have also been reported.

The results of comparative study of PSD from stainless steel baked to 300 °C for 24 hours and aluminium and copper baked to 150 °C for 24 hours are shown in Figure 4.33 [110]. One can see that the main desorbed species in a baked vacuum chamber are H_2, CO, CO_2, and CH_4, shown in order of significance. In unbaked chambers, PSD of water could be significant and even dominant.

One should use the data with great care because there are many effects that could affect the results:

1. Production process may vary from one manufacturer to another and change with time even at the same place affecting composition, structure, and impurities. A little change in production technology may greatly affect vacuum properties of the material. Especial attention should be paid to
    a. what technology and procedures were applied for vacuum chamber production: rolled metal sheet shaped and welded into the vacuum chamber or rolled tubes, extruded, or machined from a single piece;
    b. whether the metal blank was forged before machining;
    c. what type of welding and brazing is applied;
    d. if there were changes in manufacturing process, some modifications in it that do not affect the mechanical properties of the material may have a high impact on desorption.

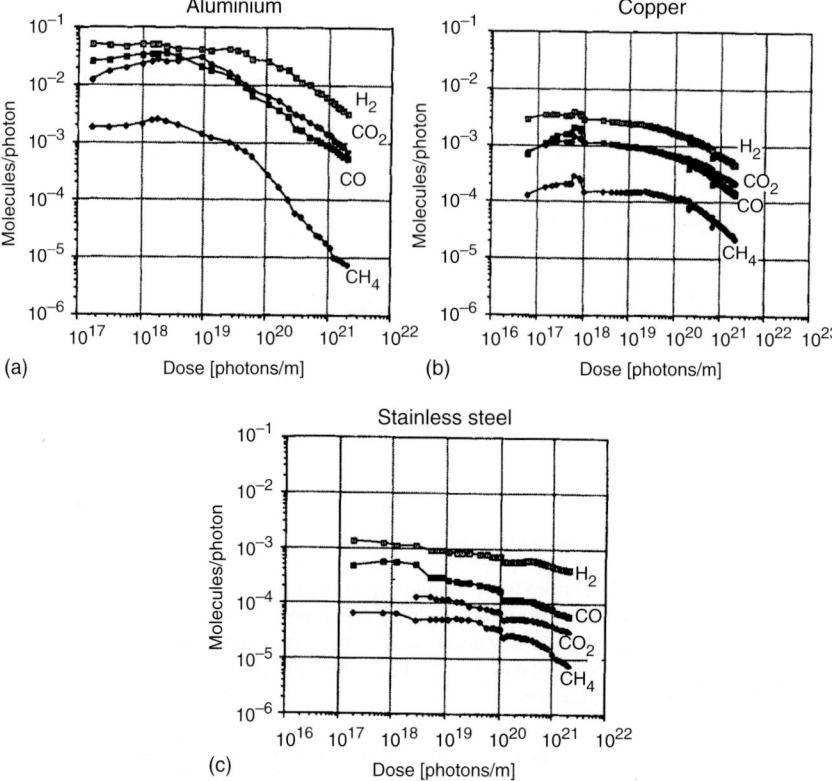

Figure 4.33 The PSD yields as a function of accumulated photon dose for different materials for the vacuum chamber. Source: Reprinted with permission from Mathewson et al. [110], Figs. 1–3. Copyright 1991, American Vacuum Society.

2. Cleaning procedure (discussed in Section 4.3).
3. Additional surface preparation technologies could be applied such as etching, polishing, argon discharge cleaning, coatings, etc.
4. Bakeout: vacuum firing, *ex situ* and *in situ* bakeout.
5. Photon critical energy.
6. Angle of interaction.
7. Temperature of vacuum chamber.

Since technologies are developing and changing, new alloys and materials could appear on a market, different manufacturers may provide the raw materials with different properties; thus vacuum properties of the same type of material may vary significantly. Thus, the literature data is giving just a range of possible PSD values. To minimise such an uncertainty in a design of an accelerator, the most practical approach applied for large project was to perform a set of PSD measurements of a vacuum chamber prototyped from potential manufacturer(s).

4.6.4 PSD as a Function of Amount of Desorbed Gas

The total amount of each gas desorbed in PSD experiments as a function of electron dose can be calculated as follows:

$$N(D)[\text{molecules}] = \int_0^t \eta_\gamma(t_a)\Gamma(t_a)dt_a = \int_0^D \eta_\gamma(D_a)dD_a \quad \text{or}$$

$$\Theta(D)[\text{Pa·m}^3] = \int_0^D Q_{\text{PSD}}(D_a)dD_a = k_BT\int_0^D \eta_\gamma(D_a)dD_a. \quad (4.36)$$

The total amount of desorbed gas is usually normalised to the length of vacuum chamber, L, or to its total area, A:

$$N_L(D) = \frac{N(D)}{L}; \quad N_A(D) = \frac{N(D)}{A};$$

$$\Theta_L(D) = \frac{\Theta(D)}{L}; \quad \Theta_A(D) = \frac{\Theta(D)}{A}. \quad (4.37)$$

These amounts obtained in the ESD experiments can be plotted in two ways:

– PSD as a function of amount of desorbed gas (see Figure 4.34 for the same data as shown in Figure 4.33).

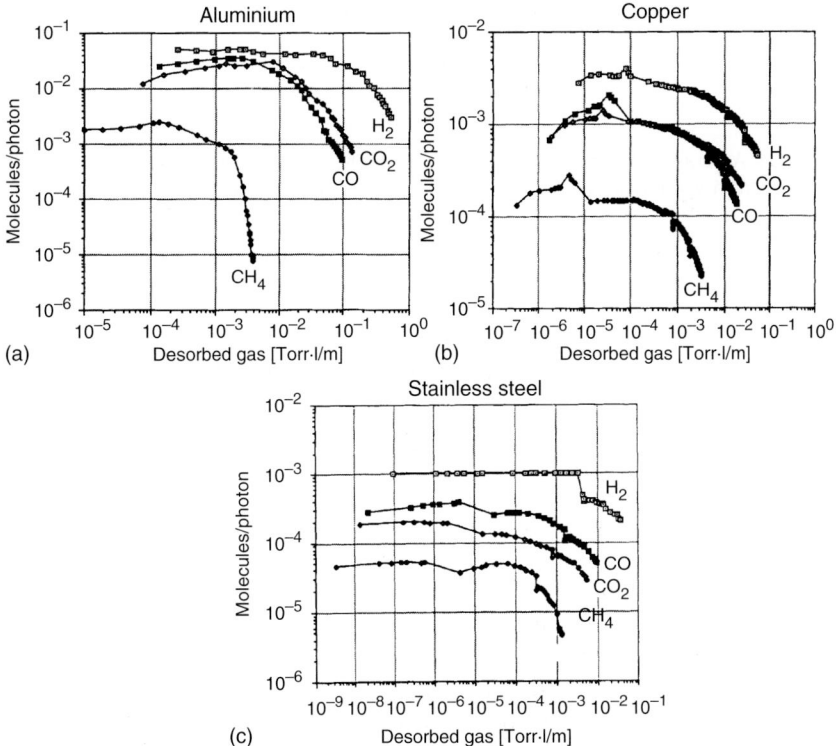

Figure 4.34 PSD as a function of amount of desorbed gas. Source: Reprinted with permission from Mathewson et al. [110], Figs. 8–10. Copyright 1991, American Vacuum Society.

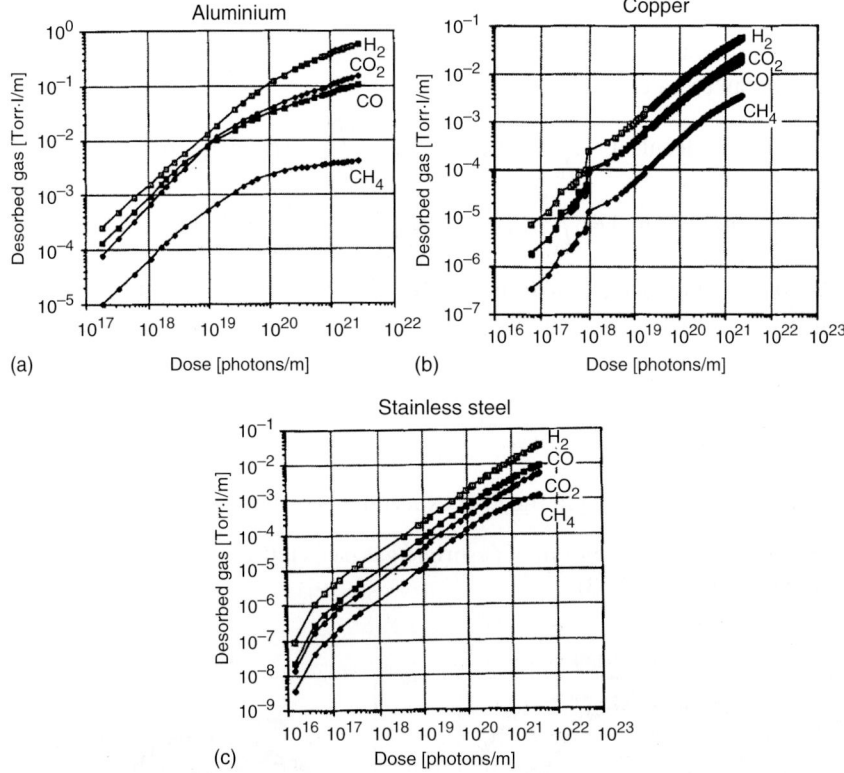

Figure 4.35 Amount of desorbed gas as a function of photon dose. Source: Reprinted with permission from Mathewson et al. [110], Figs. 5–7. Copyright 1991, American Vacuum Society.

- Amount of desorbed gas as a function of electron dose (see Figure 4.35 for the same data as shown in Figures 4.33 and 4.34).

The PSD results can also be presented as shown in Figure 4.36 for an aluminium alloy (6063-T5)-extruded chamber after Ar discharge cleaning at NSLS [92].

4.6.5 PSD as a Function of Critical Energy of SR

The critical energy is one of the parameters for PSD. Most of the studies on PSD as a function of critical energy were done at CERN. It was shown that PSD yield changes with the photon critical energy: PSD yields are directly proportional to the critical photon energy ε_c when $\varepsilon_c \leq 1$ keV [113] (see Figure 4.37), weakly increases with ε_c when $1\,\text{keV} < \varepsilon_c < 100\,\text{keV}$ [113, 114] and directly proportional to either ε_c ($\propto E^3$) or to the SR power ($\propto E^4$) when $\varepsilon_c \geq 100$ keV [114] (see Figure 4.38). Schematically, this dependence of PSD yields (normalised to PSD yield at $\varepsilon_c = 1$ keV) as a function of critical energy can be plotted in the range $10\,\text{eV} \leq \varepsilon_c \leq 10\,\text{MeV}$ as it is shown in Figure 4.39 [115]. Note that this dependence on critical photon energy was shown *for the surfaces under the same conditions of the surface and material*. That means that it is valid for either initial PSD (no

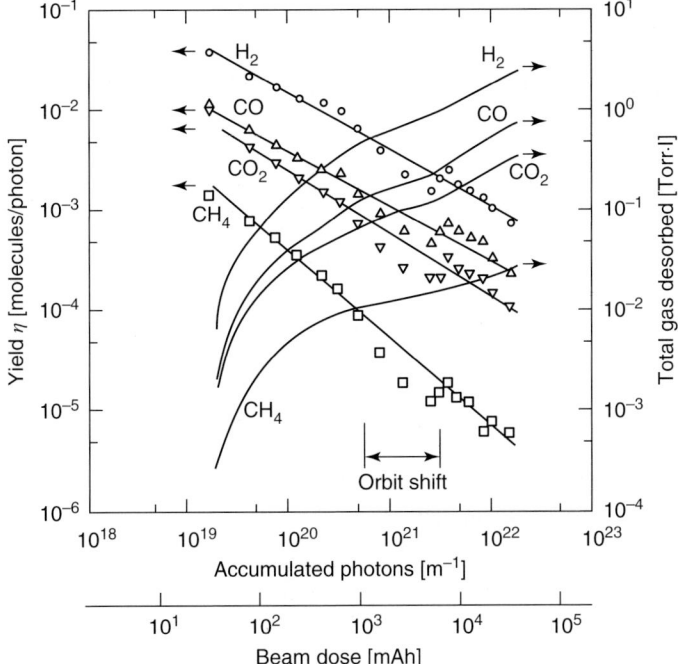

Figure 4.36 PSD yields and amount of desorbed gas as a function of photon dose for $\varepsilon_c = 500$ eV obtained from aluminium alloy (6063-T5) extruded chamber after Ar discharge cleaning at NSLS. Source: Reprinted with permission from Kobari and Halama [92], Fig. 3. Copyright 1987, American Vacuum Society.

pre-conditioning with SR) or an initial difference after changing photon critical energy. Thus two similar vacuum chambers eradicated by SR with different critical energies ε_{c1} and ε_{c2} behave as follows:

- The initial PSD yields can be scaled with ε_c as shown in Figure 4.39.
- The PSD yields of these vacuum chambers after the same photon dose cannot be scaled with ε_c. The total amount of desorbed gas from a vacuum chamber bombarded by photons with higher ε_c is higher: i.e. it is better conditioned, and thus these chambers are different.
- However, if critical energy ε_c was changed at any dose at either of these vacuum chambers, the initial PSD change will be scaled with ε_c as shown in Figure 4.39.

4.6.6 Effect of Bakeout

Proper cleaning procedures, pre-baking, and baking *in situ* will lower the dose required to obtain a required value of the PSD yields. The comparison of baked and unbaked samples was done in many publications. The main concussion of these studies can be summarised as the following (see Table 4.5):

- In the unbaked vacuum system, η_{H_2O} could be the highest or significant yield, a bakeout with $T > 120–150\,°C$ allows to reduce H_2O to negligible level.

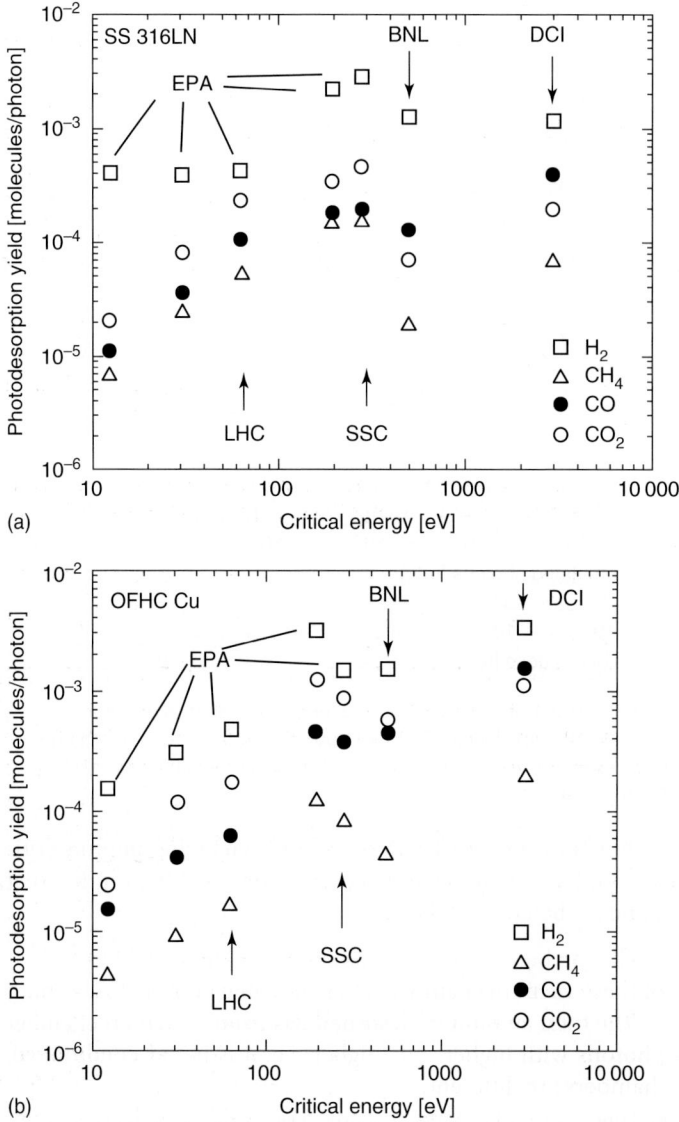

Figure 4.37 The PSD yields as a function of critical photon energy for 316LN stainless steel (a) and copper (b). Source: Reprinted with permission from Gomez-Goni et al. [113], Figs. 2 and 4. Copyright 1994, American Vacuum Society.

- Bakeout to all temperatures allows reducing the PSD yields for all species, and the ratios between these species (except for H_2O) remain practically the same.
- A further reducing of PSD yields can be achieved by increasing the temperature or longer bakeout at the same temperature.

The H_2 PSD yields for different bakeout procedures were intensively studied for ε_c = 284 eV on VEPP-2 in BINP (Novosibirsk, Russia) by a team led by

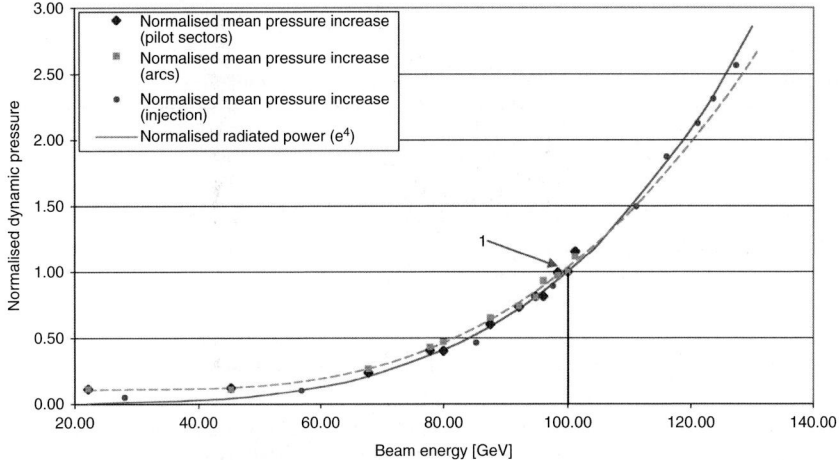

Figure 4.38 Normalised pressure increase in LEP as a function of beam energy. Source: Reprinted with permission from Billy et al. [114], Fig. 4. Copyright 2001, Elsevier.

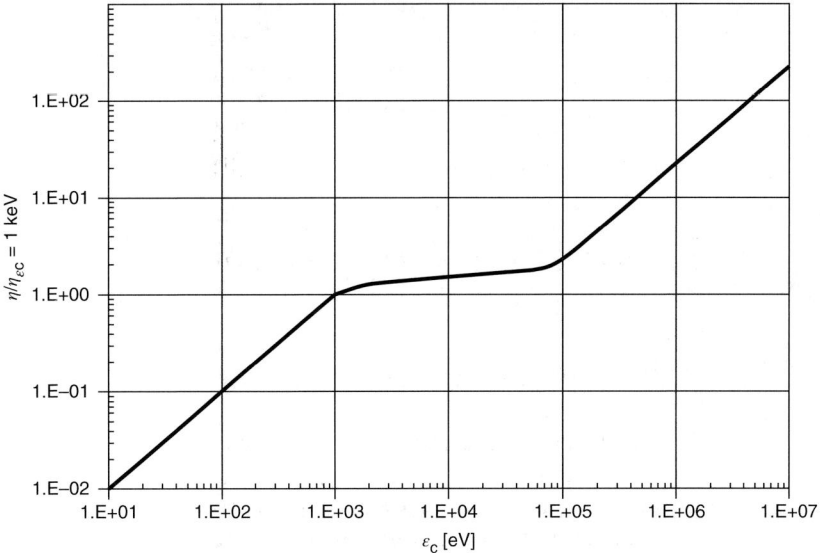

Figure 4.39 PSD yield as a function of critical photon energy. Source: Reprinted with permission from Malyshev [109], Fig. 2. Copyright 2012, Elsevier.

Dr. Anashin. The results were only presented in the author's PhD thesis [97], so they are reproduced here (see Figure 4.40). The samples were 1-m-long tubes with an inner diameter ID = 32 mm. One can see that by applying an *in situ* bakeout (up to 350 °C for 24 hours) and/or vacuum firing (950 °C for two hours), the initial PSD yields can be reduced up to two orders of magnitude. However, the significance of this difference reduces with the photon dose and there is no significant difference between baked and unbaked vacuum chambers at every

Table 4.5 Impact of various bakeout procedures on stainless steel vacuum chamber.

Bakeout	Impact	Comment
In situ at 150 °C for 24 h	Reduction of η_{H_2O} by 5–10 times; reduction of initial PSD yields for other species by 2–4 times	Reducing bakeout temperature to 120 °C requires increasing of bakeout duration to a few days
In situ at 300–350 °C for 24 h	Reduction of initial η_{H_2} by 10–20 times, for other species by 7–15 times	—
Ex situ at 250–300 °C for 24 h	Reduction of initial η_{H_2} by 5–10 times, for other species by 4–8 times	Keep in vacuum; minimise vent to air during installation; purge with dry air, N_2, or noble gases
Vacuum firing at 950 °C for 1–2 h at $P < 10^{-5}$ mbar	Hydrogen depletion in the bulk of vacuum chamber material	Keep in vacuum or fill with N_2 or noble gas
No *in situ* bakeout after vacuum firing	Reduction of η_{H_2} by ~1.5–2 times	—
In situ bakeout after vacuum firing	Reduction of η_{H_2} by ~20–50 times	—

photon dose greater than 3×10^{22} photons/(s·m). This means that the same PSD yields can be achieved either by bakeout or by operating machine for a longer time relying on beam scrubbing (conditioning), whichever is optimum for machine operation based on available resources, funds, and time. The only exception is the result for stainless steel baked *in situ* to 350 °C for 24 hours, where the difference to unbaked chamber remains the same up to the highest dose in the experiment of 3×10^{21} photon/(s·m).

Another conclusion was that using elements made of copper, copper coating, or copper lamination inside the vacuum chambers should not affect the PSD significantly.

4.6.7 Effect of Vacuum Chamber Temperature

Although we are discussing the input data for machines operating at room temperature, the temperature of vacuum chamber and/or its components may vary. The temperature of vacuum chamber may increase during the machine operation due to SR, impedance losses, electron multipacting, heat from warmer components (e.g. hot cathode vacuum gauges or magnets), etc. Even when the vacuum chamber is cooled with cooling water, the vacuum chamber temperature may vary because the temperature of cooling water could increase along its path (as it absorbs the heat from the vacuum chamber), the power dissipation on vacuum chamber could be non-uniform, heat capacity of various components is different, etc. As a result a change in vacuum chamber temperature may affect the pressure inside the vacuum chamber.

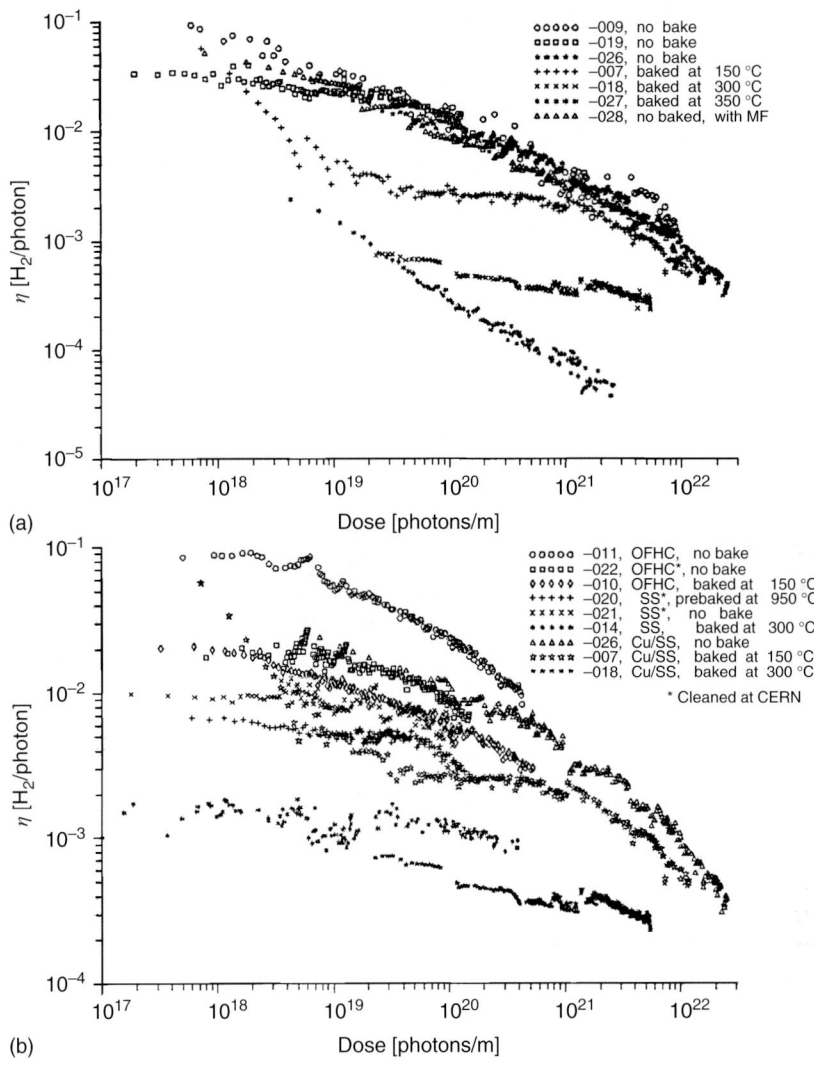

Figure 4.40 The comparison of η_{H_2} from copper-coated stainless steel samples (a) and OFHC, stainless steel, and copper-coated stainless steel samples (b) with and without bakeout. Source: Malyshev (1995) [100].

Unfortunately, little experimental data are available, as only three papers with relevant information were found.

The only study on PSD as a function of vacuum chamber temperature was reported in Ref. [101]. PSD yields were compared at +33 and +70 °C for a well-conditioned copper vacuum chamber, and a ratio of the PSD yields is close to one for many species: $\eta_\gamma(70\,°C)/\eta_\gamma(33\,°C) = 1$ for H_2, 1.01 for CH_4, 1.15 for CO, and 1.67 for CO2. However it is much higher for O_2 and H_2O: $\eta_\gamma(70\,°C)/\eta_\gamma(33\,°C) = 4.5$ and 5.7, correspondingly. It was summarised as follows: 'With the chamber heated to 70 °C, the photodesorption yields for

H$_2$O and O$_2$ increased but the others remained relatively independent of the temperature'.

Two other papers report ESD results. In Ref. [116] the ESD yields for 316LN stainless steel, aluminium, and copper were compared at room temperature and 200 °C. Almost no difference in ESD yields was detected for these samples, with an exception for water that increases with temperature. The results reported in Ref. [80] lie in the range of temperatures between −20 and +70 °C. The ESD yields for 316LN stainless steel samples were measured at three different temperatures: −5, +20, and +70 °C. The ESD yields as a function of electron dose for three samples at these temperatures show very similar results. After long-term electron bombardment at fixed temperature, the ESD temperature dependence was also measured for each sample as a function of temperature between −15 and +70 °C. It was found that the ESD yields increase with temperature (see Figure 4.41) and this dependence is weak for H$_2$ and increases with a molecular mass reaching a maximum difference for CO$_2$: $\eta_e(+70\,°C)/\eta_e(-15\,°C) \approx 3$.

The main conclusion is that the vacuum chamber temperature is not critical in most cases for accelerator vacuum system design. The change of PSD and ESD yields with temperature is quite small compared to desorption yield uncertainties and the significant reduction with an accumulated electron dose.

Note: The PSD at cryogenic temperatures is discussed in Chapter 7.

4.6.8 Effect of Incident Angle

In accelerator vacuum chambers, the SR photons may incident the surface at various angles from grazing incidence on vacuum chamber walls to normal incident at scribers, collimators, and SR absorbers. A few studies were reported to study how the incident angle can affect PSD.

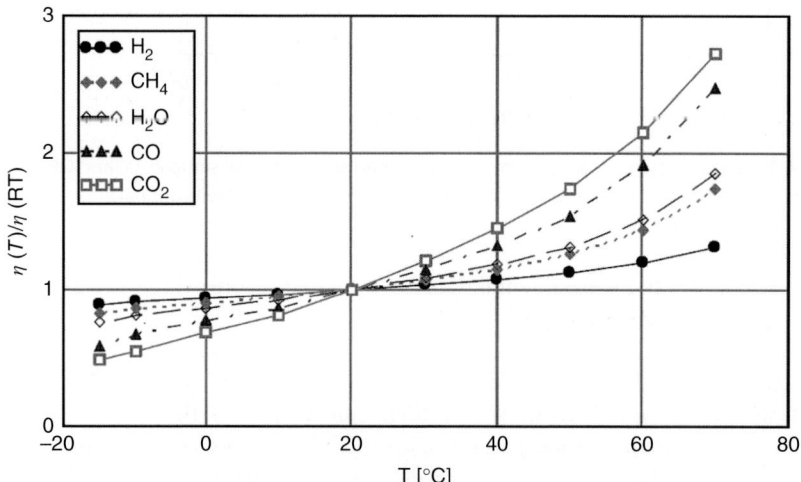

Figure 4.41 The ratio $\eta_e(T)/\eta_e\,(20\,°C)$ as a function of wall temperature. Source: Reprinted with permission from Malyshev and Naran [81], Fig. 2. Copyright 2012, Elsevier.

PSD has the lowest value at normal incident ($\Theta = 90°$, see Figure 3.1) and increases with decreasing incident. This effect was demonstrated on PSD measurements from copper as a function of incident angle at critical photon energy of $\varepsilon_c = 7$ keV [117]. The results are shown in Figure 4.42 as a function of absorber angle measured from normal, i.e. for an angle equal to $(90° - \Theta)$. One can see that there is an insignificant difference in PSD between the results obtained for an incident angles in the range $60 < \Theta < 90°$; however for smaller angles the difference in PSD is rapidly increasing reaching approximately a factor of 2.6 at $\Theta < 20°$.

PSD yields from an aluminium alloy A6061 were examined both normal and grazing incidence $\Theta = 10°$ (175 mrad) with a few different photon energies on a beamline at the Photon Factory (KEK, Tsukuba, Japan) [90]. It was found that the PSD yield at critical photon energy of $\varepsilon_c = 4$ keV at an incidence $\Theta = 10°$ was about four times higher than those at normal incidence.

These conclusions confirm earlier results published in Refs. [91, 112].

It should be noted that the results discussed previously correspond to *absorbed* photons.

After interacting with a wall surface, photons can be absorbed, transmitted, or reflected: (diffusional, specular, or backscattered). These effects depend on an incident angle Θ, as described in Chapter 3. Thus calculation of the total PSD from accelerator vacuum chamber should include not only the PSD due to *a*bsorbed photons. The reflected photons should carefully be included in the calculation of PSD from different parts of the vacuum chamber. Due to the

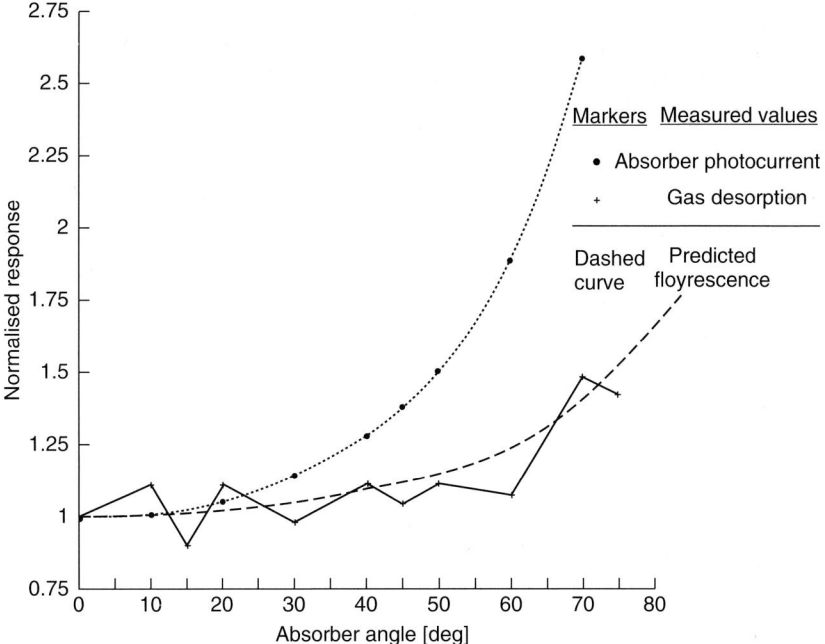

Figure 4.42 PSD and PEY (photoelectron yield) as a function of absorber angle $(90° - \Theta)$ normalised to response at normal incidence for copper surface. Source: Reprinted with permission from Trickett et al. [117], Fig. 4. Copyright 1992, American Vacuum Society.

photon reflectivity, the intensity of absorbed SR could be different from the intensity of direct incident SR. Furthermore, in all machines with SR, there are parts of a vacuum chamber that are not irradiated by direct SR, i.e. considered as being 'in shadow' with no PSD and PEE. However, the reflected photons can irradiate these 'shadowed' locations and cause PSD and PEE.

Absorbed photons can cause PEE, the PSD yield also depends on an incident angle Θ. The electrons may cause ESD where they hit the vacuum chamber wall; thus the photoelectron trajectories have to be considered in the gas desorption pattern from different parts of the vacuum chamber.

Transmitted photons do not play any role in vacuum if they leave vacuum chamber, but may have to be considered if they interact with in-vacuum components.

4.6.9 PSD versus ESD

At the beginning of this chapter, we have already mentioned that the PSD can be considered as a two-step process: PEE and ESD. Thus, how can the ESD results be used when no PSD data are available?

First of all, as ESD is a part of PSD process, then the data obtained with ESD allow comparing the effectiveness of different cleanings, treatments, coatings, and fictional dependence of ESD on dose and temperature. For example, if cleaning or treatment Procedure A is reducing the ESD yields in comparison to Procedure B, the one should expect that it will similarly reduce the PSD yields as well.

A comparative study of PSD and ESD yields on identically prepared samples measured at the DCI storage ring at LURE (Orsay, France) was reported for two different alloys (aluminium type IS0 AlMgSi and a high temperature steel type 'Nimonic') in Ref. [88]. The PSD yields obtained with a critical energy of the photon spectrum was $\varepsilon_c = 713\,\text{eV}$ and the ESD yields were obtained with a 500 eV electrons. It was reported that

> in view of the different primary incident particles, a direct comparison of the desorption yield expressed as molecules per photon and as molecules per electron cannot be made. However, it has been found that satisfactory agreement may be obtained if the data are expressed in terms of the desorbed molecules per photoelectron for PSD and converted to desorbed molecules per secondary electron for ESD.

PSD and ESD yields were also studied from an aluminium alloy A6061 on a beam line at the Photon Factory (KEK, Tsukuba, Japan) [90] and from an aluminium alloy A6082 on a beamline on Electron Positron Accumulator (EPA) at CERN [118]. A similar conclusion was reported:

> The photoelectron emission is the most dominant process determining the PSD by synchrotron radiation The PSD yield is approximately proportional to the calculated photoelectron yield, though it is slightly different from the photoelectron yield obtained in this experiment.

Therefore, in the absence of PSD data, the combined PEE and ESD data can be used to calculate (to estimate) the PSD yields.

4.6.10 How to Use the PSD Yield Data

4.6.10.1 Scaling the Photon Dose

Most of the PSD data are reported in respect to linear photon flux Γ [photon/(s·m)] and linear dose D [photons/m]. This is convenient when the cross-sectional dimensions in the experiment and in the designed future machine are the same. Thus most of the data were obtained with circular or elliptic tubes with a cross section similar to a machine designed by the research team at the time. However, if the vacuum chamber cross sections in the experiment and design are significantly different, the data should be used with care.

As it was mentioned previously, gas molecules are desorbed from a surface when and where the photoelectrons leave and arrive at a surface and the reflected photons and their photoelectrons can reach, so in a circular or elliptic tubes whole surface is desorbing. Therefore, the dose could be normalised to the area rather than length of the sample. For example, if experimental data were for a tube with $d_1 = 130$ mm [102], and we are designing a vacuum chamber with $d_2 = 50$ mm, then the dose for using in our design should be scaled as a ratio of sample areas:

$$D_{d2} = D_{d1} \frac{\text{Area}(d_2)}{\text{Area}(d_1)} = D_{d1} \frac{d_2}{d_1}. \tag{4.38}$$

Thus the original PSD yields data and a new scale for a photon dose calculated as $D_2 = 0.38 D_1$ should be used for a design of a vacuum chamber with $d_2 = 50$ mm as shown in Figure 4.43.

4.6.10.2 Synchrotron Radiation from Dipole Magnets

The photons can be emitted by the charged particle in a magnetic field, and the source of magnetic field can by any magnet in the machine lattice (dipoles, quadrupoles, wigglers, undulators). Dipole magnets are the most common source of SR and can be characterised by bending radius R_d, bending angle φ_d, and the length $L_d = \varphi_d R_d$. Let's consider a typical arc design of the storage ring where straight vacuum chambers with a radius a and a length L_s are placed between dipoles with vacuum chambers with the same radius and the length L_d bent with the dipole bending radius R_d, as shown in Figure 4.44 [109].

In the following calculations, the ideal conditions were applied:

- An ideal circular vacuum chamber, where its axis coincides with an ideal beam orbit.
- The passing beam axis coincides with an ideal beam orbit.

SR generated in a dipole with a bending angle φ can only hit a vacuum chamber inside this dipole as shown in Figure 4.44 when $\varphi_d < \Theta_0$, where Θ_0 is defined as

$$\Theta_0 = a \cos\left(\frac{R_d}{R_d + a}\right). \tag{4.39}$$

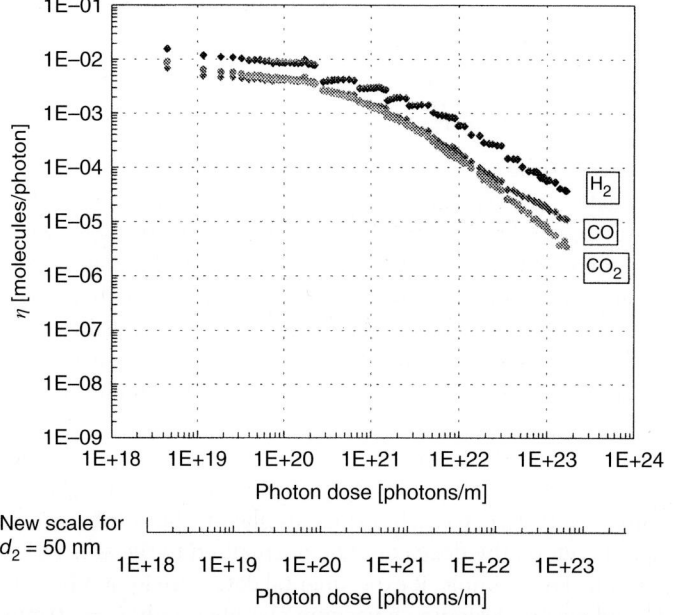

Figure 4.43 The photon dose scaling using the experimental data for a tube with $d_1 = 130$ mm. With a new axis for modelling of vacuum chamber with $d_2 = 50$ mm. Source: Reprinted with permission from Herbeaux et al. [102], Fig. 4. Copyright 2001, Elsevier.

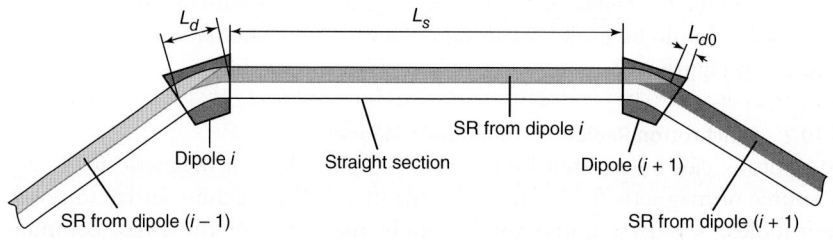

Figure 4.44 Straight vacuum chamber between two dipoles. Source: Reprinted with permission from Malyshev [109], Fig. 3. Copyright 2012, Elsevier.

SR generated at the beginning of a dipole collides with vacuum chamber walls on distance L_{d0} from the end of the dipole:

$$L_{d0} = |L_d - \sqrt{(R_d + a)^2 - R_d^2}|. \tag{4.40}$$

Thus, then the condition $\varphi_d < \Theta_0$ is met, SR generated in the dipole irradiates a part of its vacuum chamber of the length L_{d0} starting from a first collision point and the following downstream straight and dipole chambers.

4.6 Photon-Stimulated Desorption

Outside the dipole the SR irradiates the straight vacuum chamber and part of the dipole $(i+1)$ vacuum chamber from its beginning to distance $L_d - L_{d0}$. The SR incident angle is described as

$$\Theta(z) = \begin{cases} \Theta_0 & \text{for } -L_{d0} \leq z \leq 0 \\ a \sin\left[\dfrac{(R_d+a)\sqrt{z^2 + 2R_d a + a^2} - zR_d}{(R_d+a)^2 + z^2}\right] & \text{for } 0 < z \leq L_s \\ a \sin\left[\dfrac{(R_d+a)\sqrt{z^2 + 2R_d a + a^2} - zR_d}{(R_d+a)^2 + z^2}\right] \\ \quad + \dfrac{z - L_s}{R_d} & \text{for } L_s < z < L_s + L_d - L_{d0}. \end{cases} \quad (4.41)$$

The SR flux intensity on the vacuum chamber wall from a dipole varies along the vacuum chamber with a longitudinal coordinate z as follows:

$$\Gamma(z) = \begin{cases} \dfrac{\Gamma_{\text{1rad}}}{R_d + a} & \text{for } -L_{d2} \leq z \leq 0, \\ \Gamma_{\text{1rad}} \dfrac{(R_d+a)\sqrt{z^2 + 2R_d a + a^2} - zR_d}{[(R_d+a)^2 + z^2]\sqrt{z^2 + 2R_d a + a^2}} & \text{for } 0 < z \leq L_s, \quad (4.42) \\ \Gamma_{\text{1rad}} \dfrac{\sin(\Theta(z))}{\sqrt{z + 2R_d a + a^2}} & \text{for } L_s < z < L_s + L_{d0}; \end{cases}$$

where Γ_{1rad} is a photon flux into 1 rad for electron rings can be calculated as follows:

$$\Gamma_{\text{1rad}}[\text{photons}/(\text{s·rad})] = 1.28 \times 10^{20} E[\text{GeV}] I[\text{A}]. \quad (4.43)$$

Figure 4.45 shows an example of the SR photon flux as a function of distance from dipoles $(i-1)$, i, and $(i+1)$ for the following parameters: $E = 3.0$ GeV, $I = 0.5$ A, $B = 1.4$ T, $R_d = 7.151$ m, $a = 41$ mm, $L_d = 0.936$ m, and $L_s = 15$ m. One can see that in this example the SR incident angle reduces more than 30 times and the SR photon flux reduces three orders of magnitude on a distance of 15 m from the end of dipole.

If $\varphi_d > \Theta_0$, then SR passes the dipole vacuum chamber without hitting it. Similar calculations can be done to write equations for the SR incident angle and the flux intensity in this case when the SR does not irradiate its source dipole vacuum chamber and it bombards the following downstream straight and dipole chambers (excluding a part of the straight camber of length L_{d0} from the end of the next dipole).

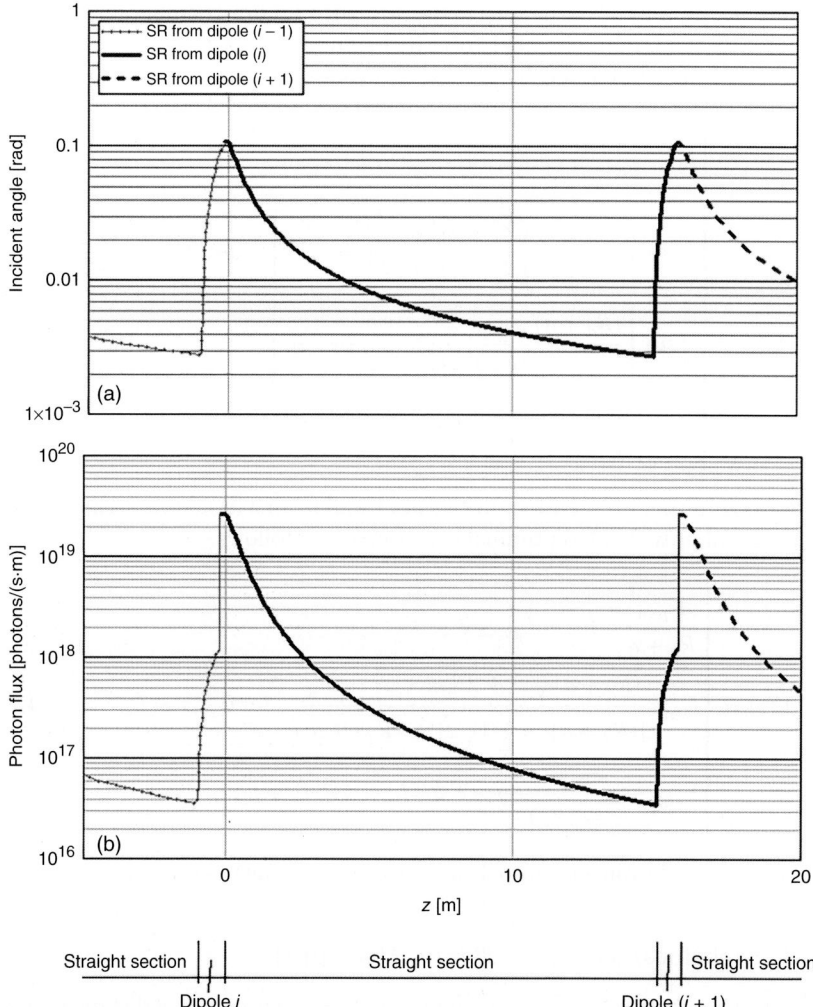

Figure 4.45 SR incident angle (a) and photon flux (b) as a function of distance from a dipole magnet. Source: Reprinted with permission from Malyshev [109], Fig. 4. Copyright 2012, Elsevier.

4.6.10.3 PSD Yield and Flux as a Function of Distance from a Dipole Magnet

Initial PSD should be quite similar for all parts of the vacuum chamber made of the same material and treated the same way. However, since the photon intensity varies significantly along the vacuum chamber, the parts where the SR photon flux is higher will accumulate a larger photon dose than the parts where the SR photon flux is lower; therefore, the PSD yield reduces quicker where the SR photon flux is higher. This is demonstrated in Figure 4.46 for the input data used in

Figure 4.46 The PSD yield as a function of distance from a dipole magnet after different operation time of the accelerator. Source: Reprinted with permission from Malyshev [109], Fig. 5. Copyright 2012, Elsevier.

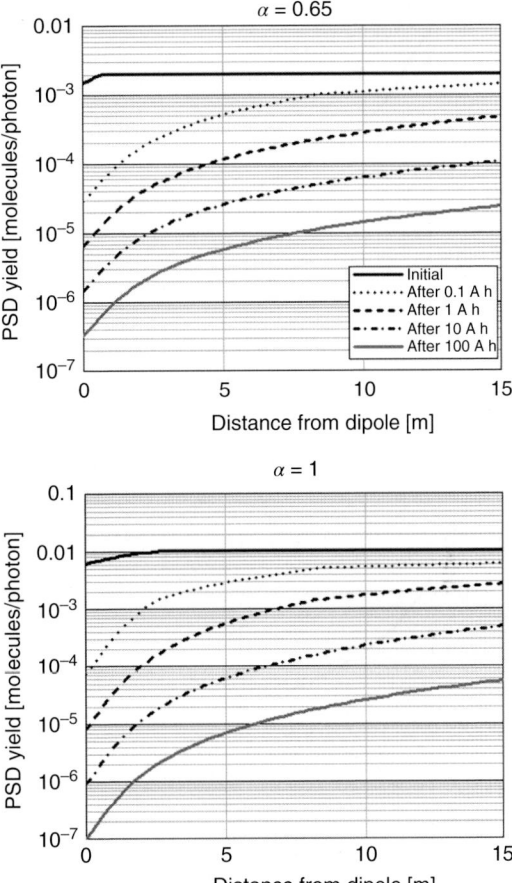

the preceding text for Figures 4.32 and 4.45. The result for the PSD gas flow shown in Figure 4.47 was calculated as

$$q\left[\frac{\text{molecule}}{\text{s}\cdot\text{m}}\right] = \eta\left[\frac{\text{molecule}}{\text{photon}}\right]\Gamma\left[\frac{\text{photon}}{\text{s}\cdot\text{m}}\right]. \quad (4.44)$$

Such calculations can be done for each gas and used in the gas dynamics models discussed in the following chapters.

The main conclusion from this model is that only the initial PSD gas flow is proportional to the incident photon flux and varies in orders of magnitude, while after the beam conditioning the difference is much smaller. In our example, the initial difference was three orders of magnitude and it reduces to a factor of 8 after 100 A h conditioning for $\alpha = 0.65$ and to a factor of 1.5 for $\alpha = 1$. This is quite an expectable result, indeed, for large doses the Eq. (1.2) can be used to

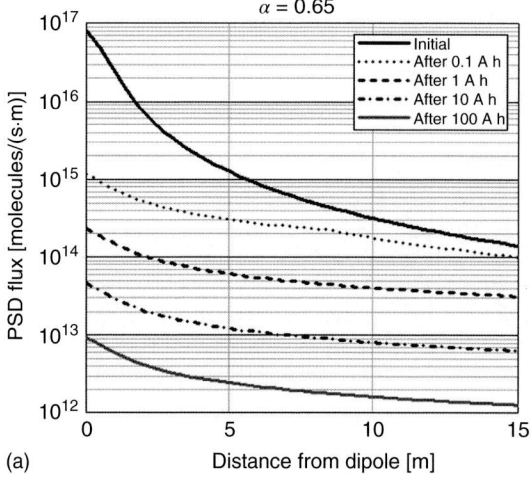

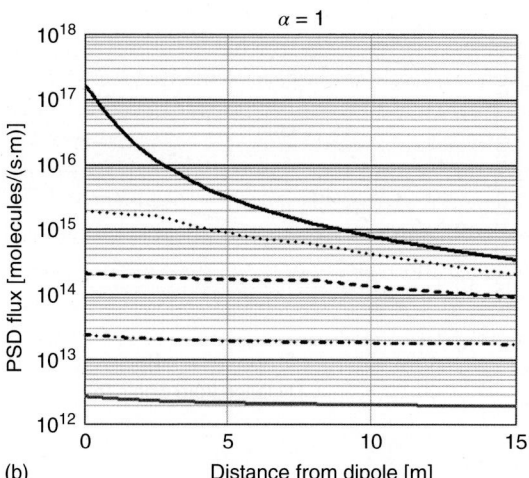

Figure 4.47 The PSD flux as a function of distance from a dipole magnet after different operation time of the accelerator. Source: Reprinted with permission from Malyshev [109], Fig. 6. Copyright 2012, Elsevier.

write an equation for the distributed gas flux, $q(z, t)$, as a function of longitudinal coordinate and time:

$$\left.\begin{array}{l}\eta(z, t) = \eta^* \left(\dfrac{D(0, t^*)}{D(z, t)} \right)^{\alpha} \\ D(z, t) = \int_0^t \Gamma_t(z, \tau) d\tau \end{array}\right\} \Rightarrow q(z, t) = \eta(z, t)\Gamma_t(z, t). \qquad (4.45)$$

In the case of constant photon flux for $\Gamma_t(z, t) = \Gamma(z)$, the photon dose is $D(z, t) = \Gamma(z)t$; thus $q(z, t)$ can be expressed as

$$q(z, t) = \eta_0 (D(0, t_0))^{\alpha} (\Gamma(z))^{1-\alpha} t^{-\alpha} \qquad (4.46)$$

or $q(z, t)$ can be compared to $q(0, t)$:

$$\dfrac{q(0, t)}{q(z, t)} = \left(\dfrac{\Gamma(0)}{\Gamma(z)} \right)^{1-\alpha}. \qquad (4.47)$$

One can see that for large photon doses the distributed gas flux is independent of coordinate z and photon flux $\Gamma(z)$, then $\alpha = 1$, or weakly dependent on coordinate z and photon flux $\Gamma(z)$, then $\alpha = 0.65$:

$$\frac{q(0,t)}{q(z,t)} = \begin{cases} 1 & \text{for } \alpha = 1 \\ \left(\dfrac{\Gamma(0)}{\Gamma(z)}\right)^{0.35} & \text{for } \alpha = 0.65. \end{cases} \quad (4.48)$$

The real vacuum chamber is usually very different from the ideal geometrical model used in our example: its cross section might have any shape and it might change along the beam path. However, the main result demonstrated in the example above is that even when a photon intensity in different parts is different by two orders of magnitude (or even more), only the initial PSD gas flux is proportional to the photon intensity. This difference quickly reduces due to a difference in beam scrubbing.

Note that in the simple model earlier, the PSD as a function of photon incident angle was not included. The incident angle decreases with a distance from the dipole (see Figure 4.45). Since the PSD yield is higher for smaller incident angle, the initial PSD yield could be increasing with a distance from the dipole; however, since photon flux is decreasing, the PSD flux will vary less than in our example, even at initial SR bombardment. Similarly, the photon reflectivity was not included in the model. If the photon reflectivity was included, the photons will be distributed along the beam pass more uniformly; therefore the initial difference in PSD flux from different locations will be less, i.e. lower PSD flux near the dipole and higher at a distance. So, an inclusion of the photon incident angle and the photon reflectivity into the model will affect initial desorption fluxes, but significance of these two parameters can be reduced with dose. Simple calculations similar to an example previously will allow to check how significant this effect for different machines and different vacuum chamber designs.

4.6.10.4 PSD from a Lump SR Absorber

Lump absorbers and collimators are widely used in accelerators to protect vacuum chamber components and sensitive equipment from an intense SR power. Usually the lump SR absorbers should absorb much larger power and photon flux than walls of vacuum chambers. Another difference is the photon incident angle, which can vary between 10° and 90° to the surface, while SR hit the vacuum chamber at grazing incident (<5°). There is a large variety of lump SR absorber design, and some of them were tested for PSD: for example, the PSD yields for a copper SR crotch absorber for BESSY-II are shown in Figure 4.48, and the irradiated area was 10 mm × 35 mm [104]. One should pay attention that in this case the total photon dose D was measured in photons, while for the tubular sample it was always measured in photons/m.

The analysis of PSD yield and PSD gas flux as a function of time can be done for a lumped SR absorber and collimators similar to how it was done for the tubular vacuum chambers above. An example of a lumped SR absorber inside a vacuum chamber is shown in Figure 4.49. Although most of SR power is absorbed by the absorber, photoelectrons and reflected photons may reach any surface within line

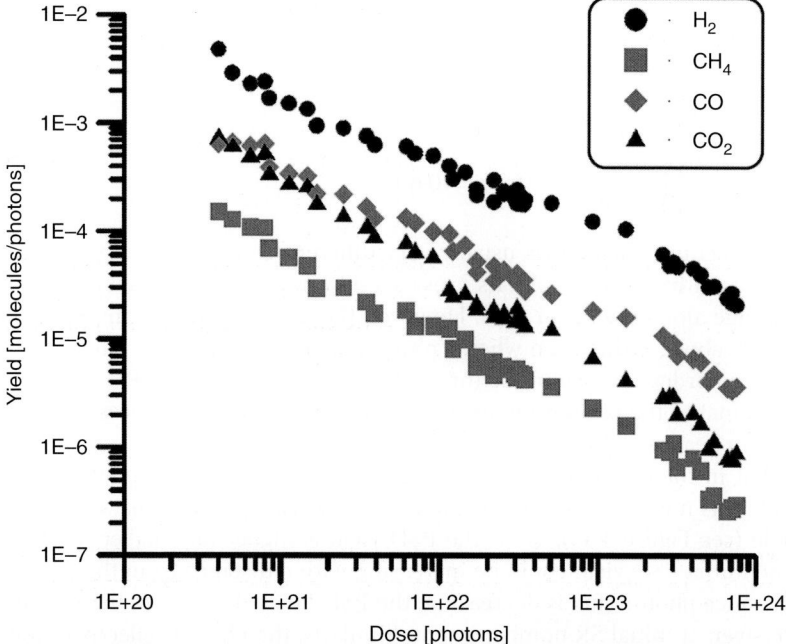

Figure 4.48 The PSD yields of copper SR crotch-absorber as a function of photon dose. Source: Anashin et al. 1998 [104], Fig. B. Reprinted with permission of CERN.

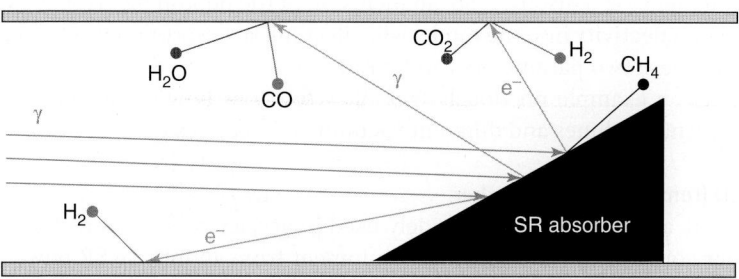

Figure 4.49 The lump SR absorber inside a vacuum chamber.

of sight from a direct incident; therefore, PSD happens not only from the directly irradiated lumped SR absorber but also from a vacuum chamber walls where these absorbers are placed. Since a directly irradiated area of an absorber is much smaller than the surface area of the surrounding vacuum chamber, then initially the absorber will be the main source of gas; however, due to an intensive photon scrubbing, the PSD yield from the absorber will reduce much quicker than for the surrounding vacuum chamber. Thus after some dose, the PSD from the surrounding vacuum chamber due to reflected photons and photoelectrons will be comparable or even higher than one from the absorber. So, the data presented in Figure 4.48 are a superposition of PSD from the absorber and the surrounding vacuum chamber. This introduces an uncertainty for the vacuum system design because the real distribution of reflected photons and photoelectrons is usually not well defined neither in experiment nor in the designed accelerator.

To reduce PSD from the surrounding vacuum chamber, there are many SR absorber designs to trap reflected photons and photoelectrons (see example in Figure 4.50). These designs allow trapping all forward scattered photons, significant part of diffused scattered photons and photoelectrons. Although there is still a remaining part of backscattered photon and photoelectrons, these designs allow creating a complete shadow from SR downstream the entrance to the absorber.

The gas dynamics modelling of such absorbers is also easier. The experimentally obtained PSD yields can be normalised to the total surface area of inner surface of the absorber (irradiated by SR directly and with scattered photons and photoelectrons) and then applied to the designed SR absorbers.

4.6.10.5 Combining PSD from Distributed and Lump SR Absorbers

A typical planar view of vacuum chamber with an antechamber equipped with an SR absorber is shown in Figure 4.51, while a transversal cross section of vacuum chamber with and without an antechamber is shown in Figure 4.52. In the case of a vacuum chamber without an antechamber, all SR is absorbed on beam chamber walls. In the case of a vacuum chamber with an antechamber, most of SR is absorbed on the SR absorber; however a part of SR, κ, is still irradiating the beam chamber walls. The latter should be carefully estimated for each machine; in general it could be anything from a negligible value to $\kappa \approx 20\%$ of total photon flux. Thus there will be the following sources of PSD along a vacuum chamber with an antechamber:

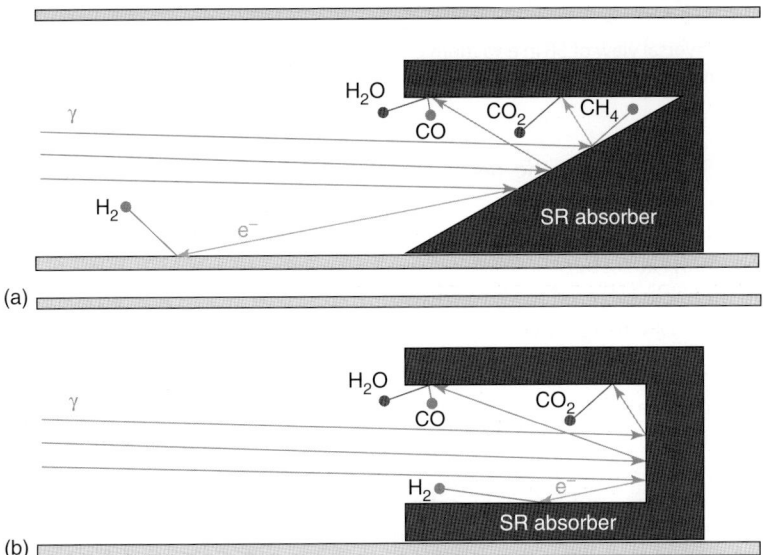

Figure 4.50 The lump SR absorbers inside a vacuum chamber: (a) absorbing SR on a larger area due to a slope, reflected photons, and photoelectrons hitting a cover plate and (b) absorbing SR at normal incident, reflected photons, and photoelectrons intercepted by horizontal plates.

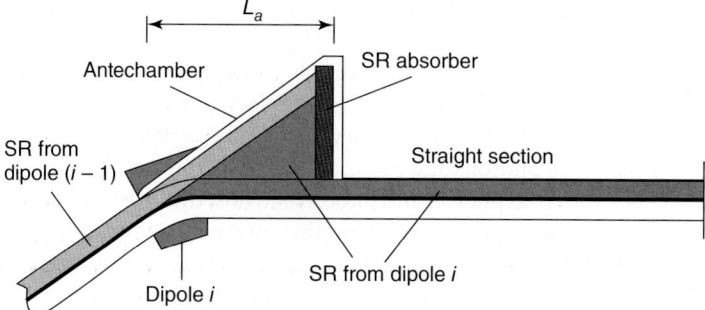

Figure 4.51 Planar view of vacuum chamber with an antechamber equipped with an SR absorber.

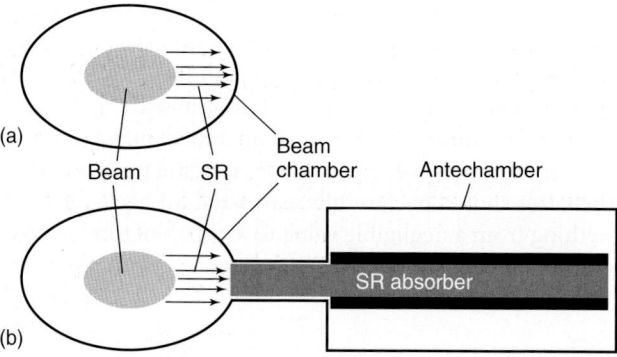

Figure 4.52 Transversal view of SR in a vacuum chamber (a) without an antechamber and (b) with an antechamber.

- q_1 due to SR irradiating the beam chamber walls with a photon flux $\kappa \Gamma$, $0 \leq \kappa \leq 20\%$.
- q_2 due to SR irradiating an SR absorber $\alpha_{abs} \Gamma_{1rad}$, where α_{abs} is a SR planar angle absorbed on SR absorber.
- q_3 due to SR backscattered from the SR absorber (up to $R_{bs} = 2\%$, see Tables 3.1 and 3.2) and irradiating antechamber walls.

These results can be used in gas dynamics modelling described in Chapter 6. If 3D modelling such as test particle Monte Carlo method (TPMC) is used, then these three sources can be modelled accurately, firstly by modelling direct and scattered photon flux into all parts of vacuum chamber, then the photon doses there at different time, and finally by calculating the PSD yields and fluxes from different locations on the beam chamber, the SR absorber, and the antechamber walls.

In the case of 1D modelling where the PSD flux is a function of coordinate z only, this would be a simplification and thus there are a few possible ways to represent the PSD. A simple way is the following:

- PSD due to SR irradiating the beam chamber walls is

$$q_1(z,t) = \eta_b(z,D)\kappa\Gamma_t(z,t), \quad \text{where } D(z,t) = \int_0^t \kappa\Gamma_t(z,\tau)d\tau, \quad (4.49)$$

where η_b is a PSD yield from a beam chamber.
- Molecules desorbed due to SR irradiating an SR absorber will be travelling back to the antechamber; the exact distribution depends on design of both the SR absorber and the antechamber. Thus we can apply an approximation that all these molecules are evenly distributed along the antechamber:

$$q_2(t) = \frac{\eta_{abs}(z,D_{abs})\alpha_{abs}\Gamma_{1rad}(z,t)}{L_a}, \text{ where } D_{abs}(z,t) = \int_0^t \alpha_{abs}\Gamma_{1rad}(z,\tau)d\tau, \quad (4.50)$$

where η_{abs} is a PSD yield from an SR absorber. Another approximation that can be applied for an antechamber design is where the SR absorber length, L_x, is much shorter than an antechamber length, L_a, i.e. for $L_x \ll L_a$. Then we can set that all these molecules are evenly distributed along the antechamber for the length of $3L_x$ from the SR absorber:

$$q_2(t) = \frac{\eta_{abs}(z,D_{abs})\alpha_{abs}\Gamma_{1rad}(z,t)}{3L_x}, \quad \text{where } D_{abs}(z,t) = \int_0^t \alpha_{abs}\Gamma_{1rad}(z,\tau)d\tau. \quad (4.51)$$

- Similarly, PSD due to SR backscattered from the SR absorber can be assumed to be evenly irradiating entire walls of antechamber:

$$q_3(t) = \frac{\eta_a(D_a)R_{bs}\alpha_{abs}\Gamma_{1rad}(z,t)}{L_a}, \quad \text{where } D_a(t) = R_{bs}\alpha_{abs}\int_0^t \Gamma_{1rad}(z,\tau)d\tau, \quad (4.52)$$

where η_a is a PSD yield from an antechamber walls. Effect of dose $\eta_a(D)$ should include a normalisation to the irradiated area as shown in Eq. (4.38). Another approximation that all reflected photons are absorbed within the length of $3L_x$ from the absorber can be applied here as well. In this case we can write:

$$q_3(t) = \frac{\eta_a(D_a)R_{bs}\alpha_{abs}\Gamma_{1rad}(z,t)}{3L_x}, \quad \text{where } D_a(t) = R_{bs}\alpha_{abs}\int_0^t \Gamma_{1rad}(z,\tau)d\tau. \quad (4.53)$$

4.7 Ion-Stimulated Desorption

4.7.1 ISD Definition and ISD Facilities

ISD can be a significant gas source in a vacuum system where the ions or ion beam bombard the surface. For example, in the case of ion-induced pressure instability in positively charged beam machines (described in Chapter 8), the gas species ionised by the beam may collide with vacuum chamber walls with energies up to a few keV. There is very little data published on ISD, as most work has been done at CERN and reported mainly in CERN internal reports or notes [73, 119–122], or not published at all (still stored in personal archives).

The ion guns are widely available on a market for the laboratory use. They are commonly used for finish cleaning or etching of substrate surface with accelerated ions beam with energy up to 2 keV from physi- and chemi-adsorbed gases, water vapours, oxides, and other contaminants before surface characterisation or just before thin film coating deposition. These technologies are focused on a result, which would be an atomically clean surface, not in detail of processing such as ISD yields for different gas species as a function of accumulated ion dose.

In application to particle accelerators, the main interests is shifted to the materials applied for accelerator vacuum chambers and component (such as stainless steel, copper, and aluminium), which undergo a UHV-comparable cleaning procedure, so they are well cleaned and often baked. Thus, an accelerator vacuum design requires the data on the ISD yields of gas species present in the vacuum chamber, such as H_2, CH_4, CO, and CO_2, desorbed under bombardments of their ions (as well as Ar ions due to availability as one of the most common ion guns). Experimental facilities for ISD study were set up similar to ESD ones shown in Figure 4.15b, where electron gun is replaced with an ion gun to bombard samples in the UHV vacuum chamber equipped with UHV gauges and RGAs. A known (calculated or measured) vacuum conductance C between two chambers or effective pumping speed S_{eff} for the test chamber allows measuring ISD yields, defined as *a number of gas molecules desorbed from the surface per incident ion*, χ [molecules/ion]:

$$\chi \left[\frac{\text{molecules}}{\text{ion}} \right] = \frac{N_{\text{molecules}}}{N_{\text{ions}}} = \frac{Q[\text{Pa·m}^3/\text{s}]q_e[\text{C}]n_q}{k_B T[\text{K}]I[\text{A}]}, \qquad (4.54)$$

where Q is a flux of molecules desorbed due to ion bombardment calculated with Eq. (4.26) or Eq. (4.27), I is the ion current, q_e is the elementary charge, and n_q is the ion charge number. Note that symbols η and η_i are often used instead of χ in publications related to ISD yield measurements.

Similar to thermal desorption, PSD, and ESD, the ISD yields were studied for different materials, cleaning, treatment, coatings and bakeout procedures, history of material, and pumping time as a function of accumulated dose and wall temperature. The ISD also depend on mass and energy of ions impacting the surface. The ISD from a cryogenic surface depends not only on a wall temperature but also on a surface density of cryosorbed gases.

Note: All experiments for ISD measurements discussed below were performed with an ion bombardment at the normal incident angle.

4.7.2 ISD as a Function of Dose

The ISD yields as a function of accumulated ion dose from as-received aluminium and copper samples bombarded with Ar^+ ions at 5 keV is shown in Figure 4.53 [119]. One can see that the ISD yields reduce with dose: ISD yields

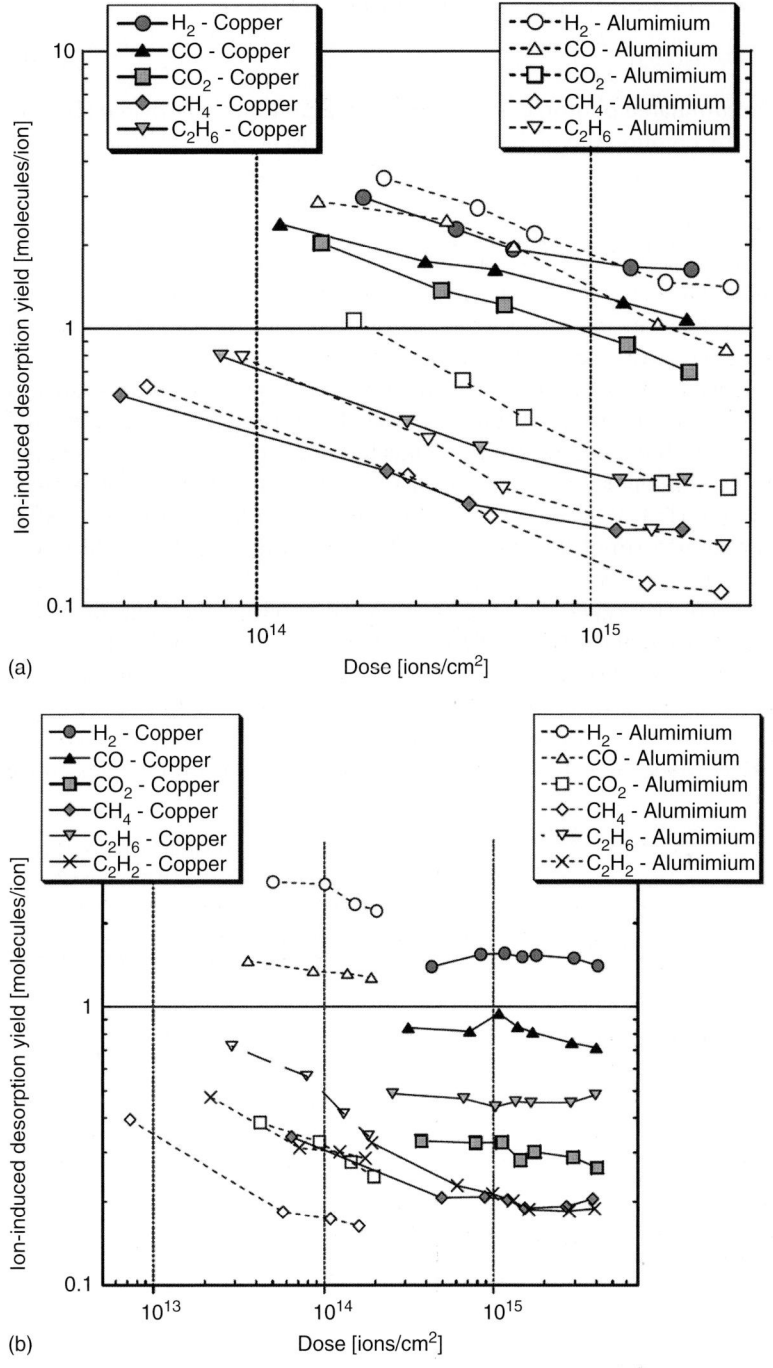

Figure 4.53 The ISD yields as a function of accumulated ion dose from (a) as-received and (b) baked aluminium and copper samples bombarded with Ar^+ ions at 5 keV. Source: Reprinted with permission from Lozano [119], Figs. 2 and 3. Copyright 2002, Elsevier.

can be fitted as a function of *accumulated ion dose, D*, as

$$\chi_i(D) = \chi_i(D^*)\left(\frac{D^*}{D}\right)^\alpha; \quad (4.55)$$

where an accumulated photon dose D^* and a corresponding ISD yield $\eta_i(D^*)$ can be taken at any point on this slope; the exponent α lies between $1/3 \leq \alpha \leq 1/2$ for as-received samples and between $0 \leq \alpha \leq 1/3$ for baked samples.

4.7.3 ISD Yield as a Function of Ion Energy

The ISD as a function of ion impact energy was reported in the papers [73, 119, 120]. The energy of incident ions ($^{15}N_2^+$ and K^+ in [73], Ar^+ in [119], $^{15}N_2^+$ in [120]) was varied in the range from 500 eV to 3 keV.

A typical behaviour of ISD yields as a function of ion energy is shown in Figure 4.54 for unbaked 316LN stainless steel. The ISD yields for all measured gases increase with the ion energy and tends to saturation at the energies larger than 3 keV. It was reported that the results for unbaked 316LN stainless steel, titanium alloy (Ti73–V13–Cr11–Al3), and pure aluminium are similar. After bakeout the same form of the curve was observed with lower ISD yield values [120].

In Ref. [119] the energy of incident Ar^+ ions was varied in the range from 3 to 7 keV. It was shown that the ISD yields for all measured gases either decrease with the ion energy or remains the same for measurements with 5–7 keV ions.

Thus, based on the available data, one can conclude that the ISD yields for all measured gases increase with the ion energy up to approximately 5 keV, remain approximately the same with 5–7 keV ions, and reduce for ion energies above 7 keV.

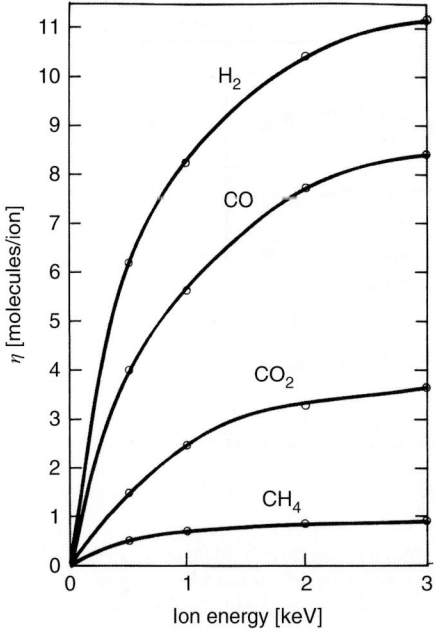

Figure 4.54 ISD yield from 316LN stainless steel sample bombarded with 1.4-keV $^{15}N_2^+$ ions as a function of ion energy. Source: Reprinted with permission from Mathewson [120], Fig. 21. Copyright 1976, CERN.

4.7.4 ISD Yield as a Function of Ion Mass

The ISD yields increase with a mass of ions bombarding the surface. The dependence of the ISD yields on the ion mass for different gas species was studied by N. Hilleret on stainless steel samples with ion impact energy of 5 keV [121, 122]. The H_2, CH_4, CO, and CO_2 ISD yields from unbaked and baked (to 300 °C for 24 hours) stainless steel sample are shown in Figures 4.55 and 4.58 as a function of ion mass (H_2^+, CH_4^+, CO^+, and CO_2^+). Unfortunately, the data on the ISD yields for CO_2 is shown only for baked sample in these papers.

The ISD yields for all desorbed gases can be normalised to ISD yields obtained with the CO^+ ions and compared as ratios. These ratios for unbaked stainless steel samples at room temperature, as found in paper [121], are given below:

$$\chi_{H_2,H_2^+} : \chi_{H_2,He^+} : \chi_{H_2,H_2O^+} : \chi_{H_2,CO^+} : \chi_{H_2,CO_2^+}$$
$$= 0.11 : 0.12 : 0.67 : 1 : 1.2,$$

$$\chi_{CH_4,H_2^+} : \chi_{CH_4,He^+} : \chi_{CH_4,H_2O^+} : \chi_{CH_4,CO^+} : \chi_{CH_4,CO_2^+}$$
$$= 0.09 : 0.11 : 0.64 : 1 : 1.3,$$

$$\chi_{CO,H_2^+} : \chi_{CO,He^+} : \chi_{CO,H_2O^+} : \chi_{CO,CO^+} : \chi_{CO,CO_2^+}$$
$$= 0.06 : 0.09 : 0.56 : 1 : 1.4; \tag{4.56}$$

with a random error of ±30%. This random error is a spread from sample-to-sample reproducibility. The absolute uncertainty is not known, but it is expected

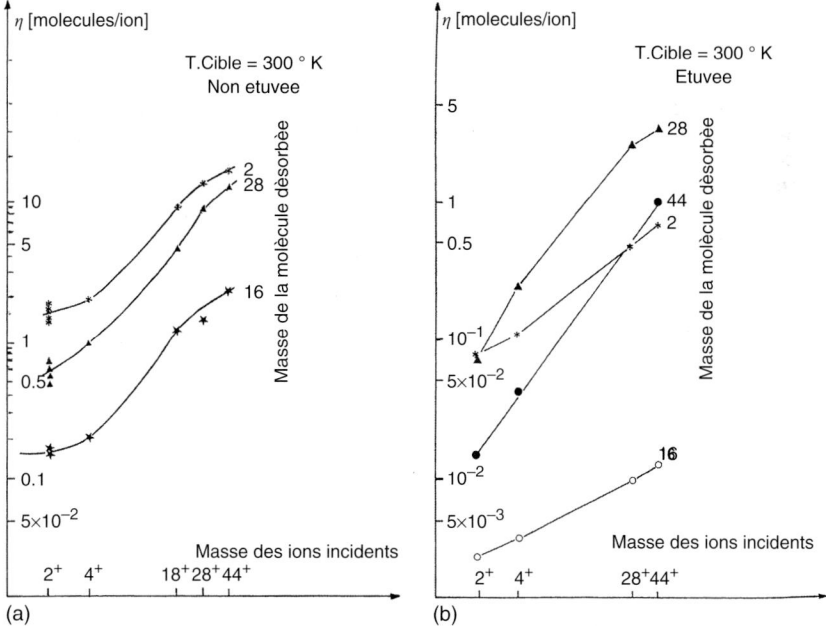

Figure 4.55 ISD yield from (a) unbaked and (b) baked stainless steel sample as a function of incident ion mass. Source: Reprinted with permission from Hilleret [121], Figs. 1 and 4. Copyright 1978, CERN.

to be a factor between 0.5 and 2. From the ratios shown in Eq. (4.56) and knowing the value of the CO^+ yield, one can calculate the values of the desorption yield for different ions and gas species. These ratios can be compared to the ratio of the ion masses:

$$M(H_2) : M(He) : M(H_2O) : M(CO) : M(CO_2)$$
$$= 0.07 : 0.14 : 0.64 : 1 : 1.57 \qquad (4.57)$$

One can see that scaling of the ISD yield with the mass of the incident ions (linear approximation) is reasonable (considering the experimental errors) and can be applied for ISD yield estimations from unbaked samples, if no experimental data is available.

Similarly, the ratios for baked stainless steel samples at room temperature [121] were calculates as

$$\chi_{H_2,H_2^+} : \chi_{H_2,He^+} : \chi_{H_2,CO^+} : \chi_{H_2,CO_2^+} = 0.16 : 0.25 : 1 : 1.4,$$
$$\chi_{CH_4,H_2^+} : \chi_{CH_4,He^+} : \chi_{CH_4,CO^+} : \chi_{CH_4,CO_2^+} = 0.25 : 0.35 : 1 : 1.4,$$
$$\chi_{CO,H_2^+} : \chi_{CO,He^+} : \chi_{CO,CO^+} : \chi_{CO,CO_2^+} = 0.03 : 0.1 : 1 : 1.2,$$
$$\chi_{CO_2,H_2^+} : \chi_{CO_2,He^+} : \chi_{CO_2,CO^+} : \chi_{CO_2,CO_2^+} = 0.03 : 0.08 : 1 : 2.0. \qquad (4.58)$$

In this case there is much greater difference from linear approximation and therefore such no simple scaling of the ISD yield with the mass of the incident ions should not be applied.

The results of M.H. Achard for 2 keV ions of $^{15}N_2^+$ and K^+ confirm that the ISD yields increase with the ion mass [73]. The ion desorption yield for K^+ is measured to be higher than that for $^{15}N_2^+$:

$$\chi_{H_2,^{15}N_2^+} : \chi_{H_2,K^+} = 1 : 1.3;\ \chi_{CH_4,^{15}N_2^+} : \chi_{CH_4,K^+} = 1 : 1.5;$$
$$\chi_{CO,^{15}N_2^+} : \chi_{CO,K^+} = 1 : 1.1;\ \chi_{CO_2,^{15}N_2^+} : \chi_{CO_2,K^+} = 1 : 1.4;$$
$$M(^{15}N_2^+) : M(K^+) = 1 : 1.3. \qquad (4.59)$$

4.7.5 ISD for Different Materials

In a number of experiments, the ISD yields were compared based on different materials used for vacuum chambers and components, such as stainless steel, titanium alloys, pure aluminium, and stainless steel [73, 74, 119, 120]. Thus, Figure 4.53 demonstrates insignificant difference between aluminium and copper samples bombarded with Ar^+ ions at 5 keV [119]. Similarly, the measurements with 1.4 keV $^{15}N_2^+$ ions demonstrate that the ISD yields as-received samples are quite similar for 316LN stainless steel, titanium alloy (Ti–Al6–N4), OFHC copper, and aluminium alloy (5068) (see Figure 4.26) and can be considered the same within a factor of 2 accuracy [74]. The results presented in Ref. [120] also demonstrate insignificant difference between the studied materials and similar ISD yield reduction after applied treatments.

4.7.6 Effect of Bakeout and Argon Discharge Cleaning

The ISD yields from baked and argon discharge cleaned samples were analysed in Refs. [73, 74, 120].

Figure 4.26 demonstrates the effect of 24 hour bakeout temperature (up to 600 °C) for 316LN stainless steel, titanium alloy (Ti–Al6–N4), OFHC copper, and aluminium alloy (5068) samples bombarded with $^{15}N_2^+$ ions. The general observed tendency is the higher bakeout temperature, the lower ISD yields after bakeout. It was found that the effect of 24 hour baking at 300 °C is to reduce the ISD yield by a factor from 3 to 6 for stainless steel, OFHC, and titanium alloy samples [74].

The ISD yields for unbaked 316LN stainless steel, titanium alloy (Ti73–V13–Cr11–Al3), and pure aluminium were compared for as-received sample and after various treatments such as bakeout (to 200 and 300 °C for a duration of 6, 12, 18, and 24 hours), argon discharge cleaning (for argon ion doses of $\sim 8 \times 10^{16}$ and $\sim 8 \times 10^{17}$ Ar^+/cm^2), air vent, and argon discharge cleaning followed by a bakeout [120]. Figure 4.56 shows the that argon discharge cleaning with a dose of $\sim 8 \times 10^{16}$ Ar^+/cm^2 reduces the ISD yields by a factor of ~ 3 for H_2, a factor of ~ 2–6 for CH_4, and a factor 20–70 for CO and CO_2; while the prolonged argon discharge cleaning with a dose of $\sim 8 \times 10^{17}$ Ar^+/cm^2 reduces the ISD yields below 0.01 molecule/ion, i.e. approximately up to three orders of magnitude.

The largest reduction in the ISD yields can be achieved with an *ex situ* glow discharge cleaning followed by *in situ* bakeout to 300 °C for 24 hours, which resulted in reducing the ISD yields by a factor between 6 and 10, see Figure 4.57 [120, 123].

4.7.7 ISD versus ESD

A quite unique comparative study of ISD and ESD from various materials for vacuum chambers was reported in Refs. [74, 124]. The measurements were performed with K^+ ions and electrons at the same energy of 1.4 keV on six different materials: 316LN stainless steel, titanium alloy (Ti–Al6–N4), Inconel 600, Inconel 718, OFHC copper, and aluminium alloy (5068) (see Figure 4.26). The main observations are that (i) the ISD yields are practically insensitive to the tested metals treated at the same conditions, while ESD yields vary by an order of magnitude, and (ii) the ESD yields reduce with a bakeout temperature more efficiently than the ISD yield. Hence, the ESD yield measurement may only indicate a possible tendency in reduction of ISD yields after different cleaning/treatment procedure and, therefore, the ESD yield measurements are not directly applicable for comparing and scaling the results for ISD yields.

4.7.8 ISD Yield as a Function of Temperature

The ISD yields from unbaked and baked stainless steel surface were studies at various temperatures (300, 77, and 4.2 K) [122]. The research was focused on

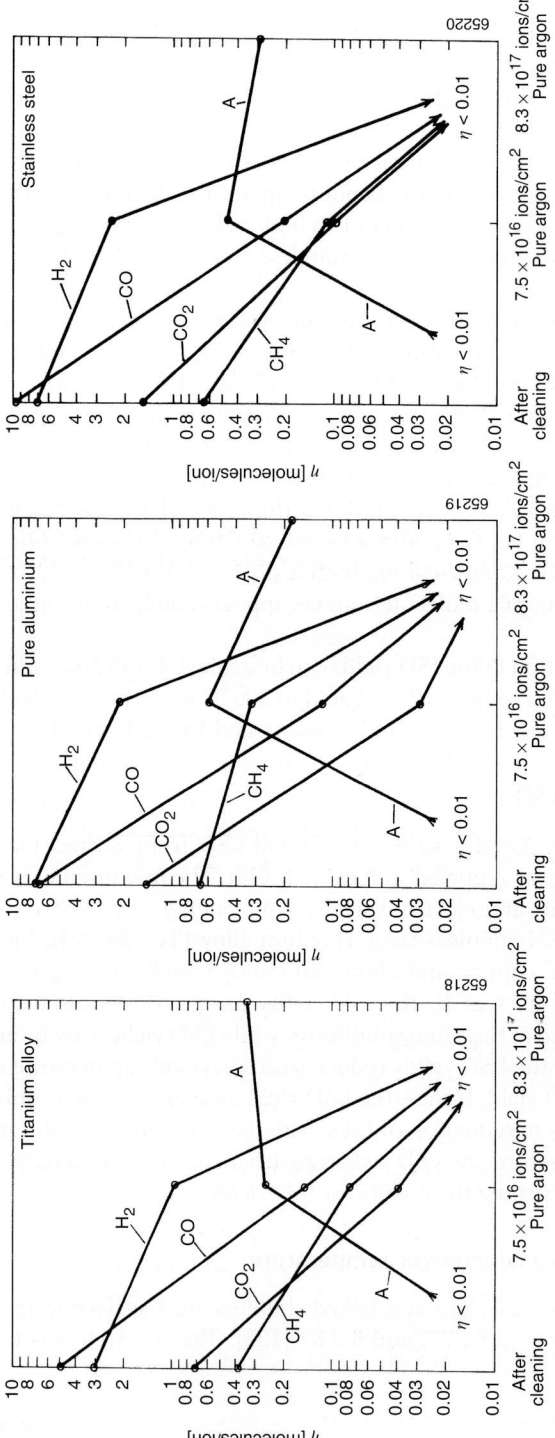

Figure 4.56 Effect of argon discharge cleanings on the ISD yields from titanium alloy (Ti73–V13–Cr11–Al3), pure aluminium, and 316LN stainless steel samples bombarded with 1.4-keV $^{15}N_2^+$ ions. Source: Reprinted with permission from Mathewson [120], Figs. 10–12. Copyright 1976, CERN.

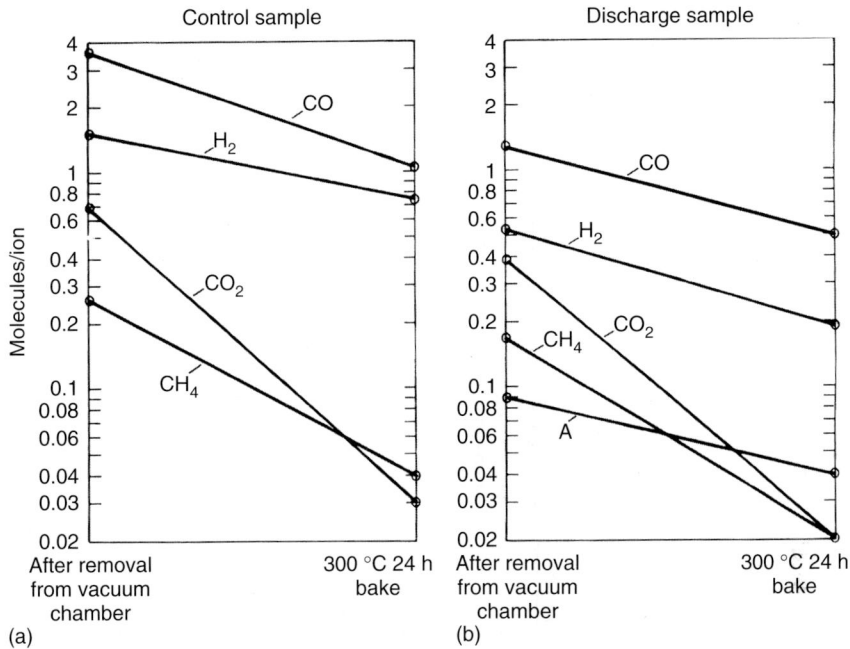

Figure 4.57 Effect of bakeout and argon (+10% oxygen) discharge cleaning on 316LN samples bombarded with 2-keV $^{15}N_2^+$ ions. Source: Reprinted with permission from Mathewson [120], Fig. 20. Copyright 1976, CERN.

studying of ISD of H_2, CH_4, CO, and CO_2 performed using different ions of these gas species: H_2^+, CH_4^+, CO^+, and CO_2^+. The results of this study are shown in Figure 4.58, where ISD yields are plotted as a function of the incident ion mass. The bombarded surface temperature does not affect the ISD yields with an exception of ISD for H_2 at 4.2 K, which are a factor of ~8 lower.

4.7.9 ISD Yields for Condensed Gases

The ISD yields from a layer of condensed gases were studied in the papers [125–127] at low temperature and different ion impact energies. The results of ISD yield measurements from condensed layers of H_2, He, N_2, or CO_2 bombarded by H_2^+, He, and Ar^+ ions were reported for ion energies from 0.5 to 10 keV. Unfortunately there are no experimental data on the CO^+ ISD yields from layers of different condensed gases nor from the layer of condensed CO by different bombarding ions.

Among the studied gases, the typical dependence is the following: the ISD yield increases linearly with the molecular surface coverage up to about 10^{20} molecules/m^2 and then it increases more slowly, reaches its maximum, and starts to decrease from about 10^{21} molecules/m^2; for example, see Figure 4.59. The results also show the ISD yield linear dependence on the ion energy from 0.5 to 10 keV [125, 126].

The ISD yields from cryosorbed gases increase with the ion energy (see Figure 4.60). In the absence of more data, one can consider that the ISD yield linearly increases with the ion mass up to 44 amu.

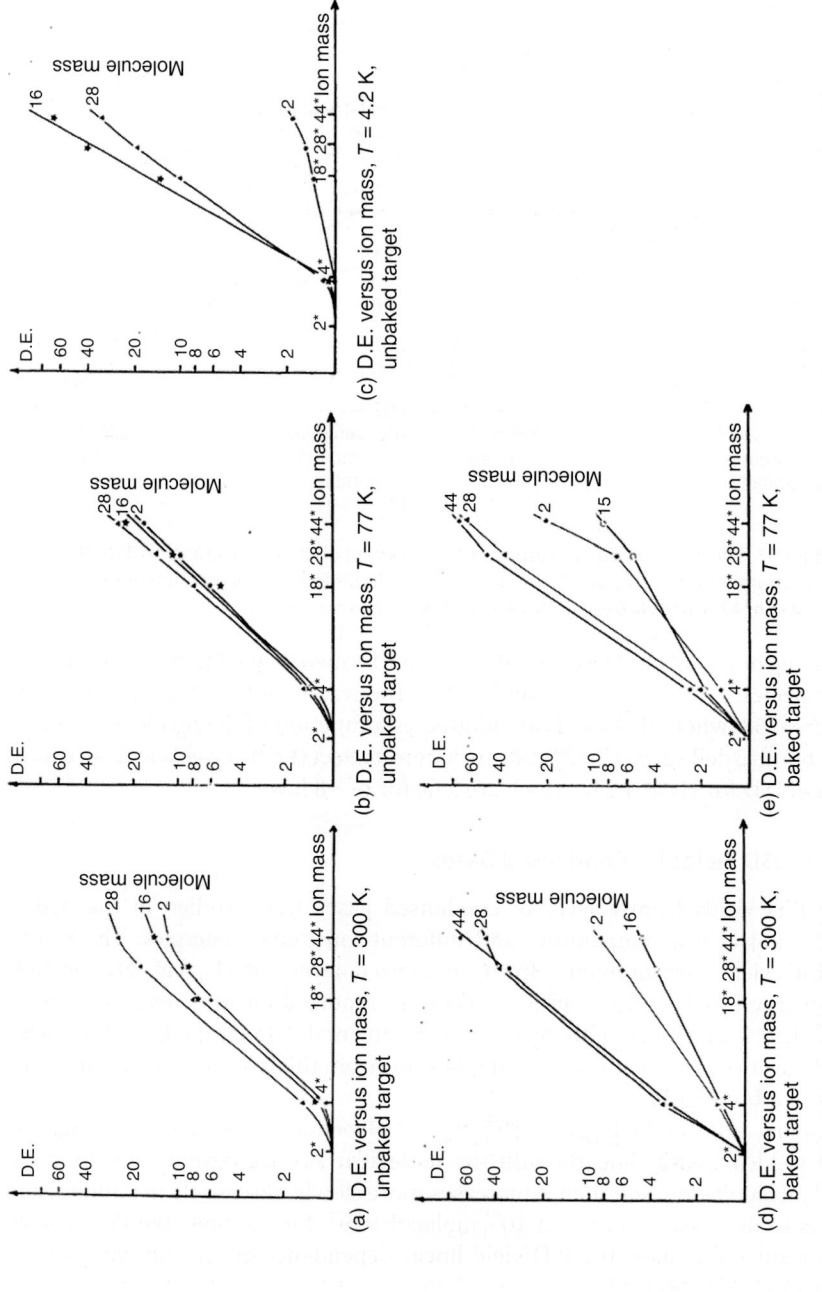

Figure 4.58 ISD yields as a function of ion mass for unbaked (a–c) and baked (d–e) stainless steel samples at 300 K (a and d), 77 K (b and e) and 4.2 K (c) bombarded with 5-keV ions. Source: Reprinted with permission from Hilleret [122], Figs. 1–5. Copyright 1980, CERN.

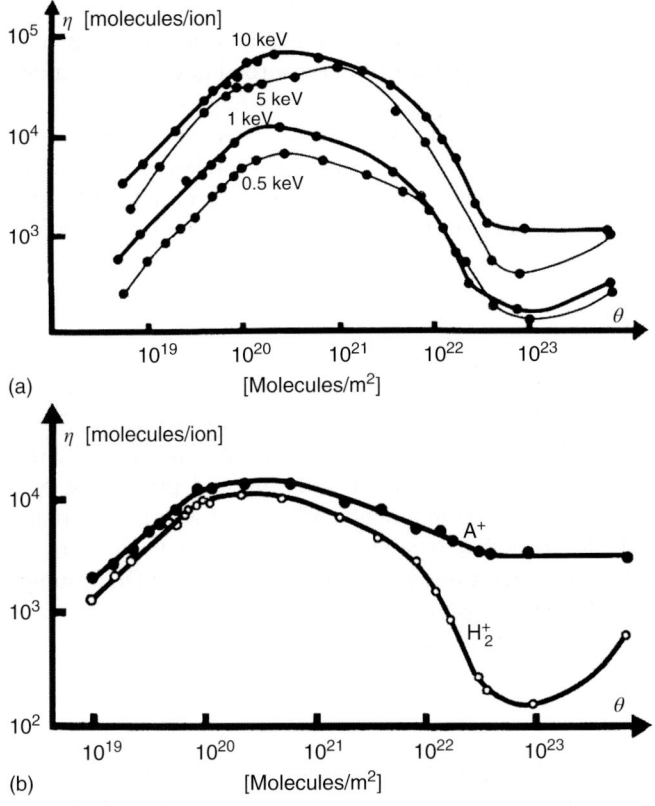

Figure 4.59 ISD yields as a function of condensed H_2 surface coverage at $T = 3.18$ K: (a) bombarded by H_2^+ ions at four different energies and (b) bombarded by 1-keV H_2^+ and Ar^+ ions. Source: Reprinted with permission from Hillert and Calder [125], Figs. 2 and 3. Copyright 1977, CERN.

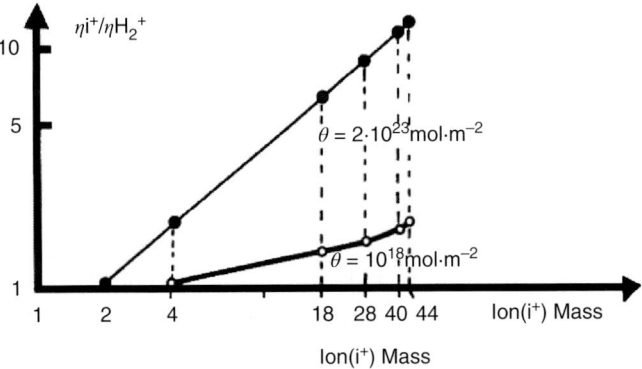

Figure 4.60 ISD yields from condensed H_2 for two different surface coverage as a function of ion mass, the ion energy -5 keV, and $T = 3.18$ K. Source: Reprinted with permission from Hillert and Calder [125], Fig. 4. Copyright 1977, CERN.

There is too little data available for an accurate vacuum system design, so ideally more experiments should be performed. However, in the lack of experimental data, one can roughly estimate ISD yields from cryosorbed gas using obtained patterns.

It should be noted that the ISD yields from a mixture of condensed gases could be quite different from the one measured for a single gas condensate [125]. This effect should be studied in the future for possible mixtures of condensed gas, for example, a very likely mixtures of H_2 and CO with various concentration ratios.

Acknowledgements

The authors want to thank Dr. Vincenc Nemanič from Jožef Stefan Institute (Ljubljana, Slovenia) for an initial thought and suggestions for Sections 4.2–4.4. Many thanks to Dr. Keith Middleman and Mr. Thomas Weston from ASTeC (Warrington, UK) for providing the details on cleaning and processing procedures for vacuum chambers and components. We would also like to thank our colleagues from KEK (Tsukuba, Japan): Dr. Kyo Shibata from KEK for useful discussions and many suggestions for Sections 4.1–4.4 and Dr. Yasunori Tanimoto for many useful suggestions for Sections 4.5–4.7.

References

1 Jousten, K. (2016). *Handbook of Vacuum Technology*, seconde. Wiley-VCH.
2 O'Hanlon, J.F. (ed.) (2003). *A User's Guide to Vacuum Technology, third edition*. Wiley-interscience.
3 Weston, G.F. (1975). Materials for ultrahigh vacuum. *Vacuum* 25: 469.
4 Kamiya, J., Ogiwara, N., Hotchi, H. et al. (2014). Beam loss reduction by magnetic shielding using beam pipes and bellows of soft magnetic materials. *Nucl. Instrum. Methods Phys. Res. A* 763: 329.
5 Park, C., Ha, T., and Cho, B. (2016). Thermal outgassing rates of low-carbon steels. *J. Vac. Sci. Technol. A* 34: 021601.
6 Ishimaru, H. (1984). All-aluminum-alloy ultrahigh vacuum system for a large-scale electron–positron collider. *J. Vac. Sci. Technol. A* 2: 1170.
7 Chen, J.R., Narushima, K., and Ishimaru, H. (1985). Thermal outgassing from aluminum alloy vacuum chambers. *J. Vac. Sci. Technol. A* 3: 2188.
8 Tajiri, K., Saito, Y., Yamanaka, Y., and Kabeya, Z. (1998). Pit-free electropolishing of aluminum and its application for process chamber. *J. Vac. Sci. Technol. A* 16: 1196.
9 Ogiwara, N., Suganuma, K., Hikichi, Y. et al. (2008). Reduction of hydrogen content in pure Ti. *Journal of Physics: Conference Series* 100: 092024.
10 Kinsho, M., Saito, Y., Kabeya, Z., and Ogiwara, N. (2007). Titanium flanged alumina ceramics vacuum duct with low impedance. *Vacuum* 81: 808.
11 Mimashi, T., Iida, N., Kikuchi, M. et al. (2017). Ceramic chamber used in SuperKEKB high energy ring beam abort system. In: *Proc. of IPAC2017*, 2936.

12 Suetsugu, Y., Shibata, K., Ishibashi, T. et al. (2016). First commissioning of the SuperKEKB vacuum system. *Phys. Rev. Accel. Beams* 19: 121001.
13 K. Jousten. Thermal outgassing. Proc. of CERN Accelerator School: Vacuum Technology, Snekersten, Denmark, 28 May to 3 June 1999, p. 111–125; doi: 10.5170/CERN-1999-005.111.
14 Redhead, P.A. (1995). Modeling the pump-down of a reversibly adsorbed phase. I. Monolayer and submonolayer initial coverage. *J. Vac. Sci. Technol. A* 13: 467.
15 Redhead, P.A. (1995). Modeling the pump-down of a reversibly adsorbed phase. II. Multilayer coverage. *J. Vac. Sci. Technol. A* 13: 2791.
16 Honing, R.E. and Hook, H.O. (1960). Vapor pressure data for some common gases. *RCA Review XXI*: 360–368.
17 Akaishi, K., Kubota, Y., Motojima, O. et al. (1996). On the effect of bakeout for the pumping speed dependence of outgassing rate in a stainless steel chamber. *Vacuum* 47: 741.
18 Elsey, R.J. (1975). Outgassing of vacuum materials-II. *Vacuum* 25: 347.
19 P.A. Redhead. Extreme high vacuum. Proc. of CERN Accelerator School: Vacuum Technology, Snekersten, Denmark, 28 May to 3 June 1999, pp. 213–226. doi: 10.5170/CERN-1999-005.213.
20 Peacock, R.N. (1980). Practical selection of elastomer materials for vacuum seals. *J. Vac. Sci. Technol.* 17: 330.
21 Chen, J.R., Lee, C.H., and Liu, Y.C. (1988). A comparison of thermal outgassing rates of aluminum alloy and stainless steel vacuum chambers. *AIP Conf. Proc.* 171: 244.
22 Odaka, K., Ishikawa, Y., and Furuse, M. (1987). Effect of baking temperature and air exposure on the outgassing rate of type 316L stainless steel. *J. Vac. Sci. Technol. A* 5: 2902.
23 Saito, K., Sato, Y., lnayoshi, S., and Tsukahara, S. (1996). Measurement system for low outgassing materials by switching between two pumping paths. *Vacuum* 47: 749.
24 Ishikawa, Y. and Odaka, K. (1990). Reduction of outgassing from stainless surfaces by surface oxidation. *Vacuum* 41: 1995.
25 Dylla, H.F., Manos, D.M., and LaMarche, P.H. (1993). Correlation of outgassing of stainless steel and aluminum with various surface treatments. *J. Vac. Sci. Technol. A* 11: 2623.
26 Saitoh, M., Shimura, K., Iwata, T. et al. (1993). Influence of vacuum gauges on outgassing rate measurements. *J. Vac. Sci. Technol. A* 11: 2816.
27 Kurisu, H., Ishizawa, K., Yamamoto, S. et al. (2008). Application of Titanium Materials to Vacuum Chambers and Components. *Journal of Physics: Conference Series* 100: 092002.
28 Colwell, B.H. (1970). Outgassing rates of refractory and electrical insulating materials used in high vacuum furnaces. *Vacuum* 20: 481.
29 Holkeboer, D.H., Jones, D.W., Pagano, F., and Santeler, D.J. (1993). *Vacuum Technology and Space Simulation*. American Institute of Physics.
30 NASA Outgassing Data for Selecting Spacecraft Materials: https://outgassing.nasa.gov/.

31 ESA Outgassing Data: http://esmat.esa.int/Services/outgassing_data/outgassing_data.html.
32 LIGO Vacuum Compatible Material List: https://dcc.ligo.org/E960050/public.
33 Foundations of Vacuum Science and Technology, J. M. Lafferty, Wiley-Interscience, 1998.
34 Redhead, P.A. (2002). Recommended practices for measuring and reporting outgassing data. *J. Vac. Sci. Technol. A* 20: 1667.
35 Minato, M. and Itoh, Y. (1993). Measurement of outgassing rate by conductance modulation method. *J. Vac. Soc. Jpn.* 36: 175.
36 Redhead, P.A. (1962). Thermal desorption of gases. *Vacuum* 12: 203.
37 Carter, G. (1962). Thermal resolution of desorption energy spectra. *Vacuum* 12: 245.
38 Habenshaden, E. and Küppers, J. (1984). Evaluation of flash desorption spectra. *Surface Science* 138: L147.
39 Miller, J.B., Siddiqui, H.R., Gates, S.M. et al. (1987). Extraction of kinetic parameters in temperature programmed desorption: A comparison of methods. *J. Cham. Phys.* 87: 6725.
40 Bacher, J.P., Benvenuti, C., Chiggiato, P. et al. (2002). Thermal desorption study of selected austenitic stainless steels. *J. Vac. Sci. Technol. A* 21: 167.
41 Tito Sasak, Y. (1990). Survey of vacuum material cleaning procedure: A subcommittee report of the American Vacuum Society Recommended Practices Committee. *J. Vac. Sci. Technol. A* 9: 2025.
42 Halliday, B.S. (1987). Cleaning materials and components for vacuum use. *Vacuum* 37: 587.
43 R. Reid. Cleaning for vacuum service. Proc. of CAS - CERN Accelerator School: Vacuum Technology, Snekersten, Denmark, 28 May to 3 June 1999, p. 139–153. doi: 10.5170/CERN-1999-005.139.
44 Chan, C.K., Hsiung, G.Y., Chang, C.C. et al. (2008). Cleaning of aluminium alloy chambers with ozonized water. *Journal of Physics: Conference Series* 100: 092025.
45 Mathewson, A.G., Alge, E., Gröbner, O. et al. (1987). Comparison of the synchrotron induced gas desorption in aluminium vacuum chamber after chemical and argon discharges cleaning. *J. Vac. Sci. Technol. A* 5: 2512.
46 Dylla, H.F. (1988). Glow discharge technique for conditioning high vacuum systems. *J. Vac. Sci. Technol. A* 6: 1276.
47 Diamond synchrotron light source: report of the design specification. Warrington, Cheshire, UK: CCLRC, Daresbury Laboratory; June 2003.
48 Westerberg, L., Hjörvarsson, B., Wallén, E., and Mathewson, A. (1997). Hydrogen content and outgassing of air-baked and vacuum-fired stainless steel. *Vacuum* 48: 771.
49 Bennett, J.R.J. and Elsey, R.J. (1992). The design of the vacuum system for the joint German-British interferometric gravitational wave detector, GEO. *Vacuum* 43: 35.
50 Tito Sasaki, Y. (2007). Reducing SS 304/316 hydrogen outgassing to 2×10^{-15} torr l/cm^2 s. *J. Vac. Sci. Technol. A* 25: 1309.

51 Ishikawa, Y., Yoshimura, T., and Arai, M. (1996). Effect of surface oxide layers on deuterium permeation through stainless steels with reference to outgassing reduction in ultra- to extremely high vacuum. *Vacuum* 47: 357.
52 Nuvolone, R. (1977). Technology of low-pressure systems—establishment of optimum conditions to obtain low degassing rates on 316 L stainless steel by heat treatments. *J. Vac. Sci. Technol.* 14: 1210.
53 Bennett, J.R.J., Elsey, R.J., and Malton, R.W. (1996). Convoluted vacuum tubes for long baseline interferometric gravitational wave detectors. *Vacuum* 47: 357.
54 Caskey, G.R. and Sisson, R.D. (1981). Hydrogen solubility in austenitic stainless steels. *Scr. Metall.* 15: 1187.
55 Calder, R. and Lewin, G. (1967). Reduction of stainless-steel outgassing in ultra-high vacuum. *J. Appl. Phys.* 18: 1459.
56 Jousten, K. (1998). Dependence of the outgassing rate of a "vacuum fired" 316LN stainless steel chamber on bake-out temperature. *Vacuum* 49: 359.
57 Nemanic, V. and Bogataj, T. (1998). Outgassing of thin wall stainless steel chamber. *Vacuum* 50: 431.
58 Moore, B.C. (1995). Recombination limited outgassing of stainless steel. *J. Vac. Sci. Technol. A* 13: 545.
59 Ishikawa, Y. and Nemanic, V. (2003). An overview of methods to suppress hydrogen outgassing rate from authentic stainless steel with reference to UHV and XHV. *Vacuum* 69: 501.
60 Saito, K., Inayoshi, S., Ikeda, Y. et al. (1995). TiN thin film on stainless steel for extremely high vacuum material. *J. Vac. Sci. Technol.* A13: 556.
61 Chung, K.H., Lee, S.K., Shin, Y.H. et al. (1999). The outgassing from TiN and BN films grown on stainless steel by IBAD. *Vacuum* 53: 303.
62 Dong, C., Mehrotra, P., and Myneni, G.R. (2003). Methods for reducing hydrogen outgassing. *AIP Conference Proceedings* 671: 307.
63 Mamun, M.A.A., Elmustafa, A.A., Stutzman, M.L. et al. (2014). Effect of heat treatments and coatings on the outgassing rate of stainless steel chambers. *J. Vac. Sci. Technol. A* 32: 021604.
64 Shibata, K., Hisamatsu, H., Kanazawa, K. et al. (2008). Development of TiN coating system for beam ducts of KEK B-factory. *Proc. of EPAC* 2008: 1700.
65 He, P., Hseuh, H.C., Todd, R. et al. (2004). Secondary electron emission measurements for TiN coating on the stainless steel of SNS accumulator ring vacuum chamber. *Proc. of EPAC* 2004: 1804.
66 Pinto, P.C., Calatroni, S., Neupert, H. et al. (2013). Carbon coatings with low secondary electron yield. *Vacuum* 98: 29.
67 Federici, G., Biel, W., Gilbert, M.R. et al. (2017). European DEMO design strategy and consequences for materials. *Nucl. Fusion* 57: 092002.
68 Benvenutti, C., Chiggiato, P., Pinto, P.C. et al. (2001). Vacuum properties of TiZrV non-evaporable getter films. *Vacuum* 60: 57.
69 Chiggiato, P. and Pinto, P.C. (2006). Ti-Zr-V non-evaporable getter films: From development to large scale production for the Large Hadron Collider. *Thin Solid Films* 515: 382.

70 Redhead, P.A. (1997). The first 50 years of electron stimulated desorption. *Vacuum* 48: 585–596.

71 de Segovia, J.L. (1996). A review of electron stimulated desorption processes influencing the measurement of pressure or gas composition in ultrahigh vacuum systems. *Vacuum* 47: 333–340.

72 Kennedy, K. Electron stimulated desorption rates from candidate vacuum chamber surfaces. Engineering Note SL0104-M6534, 2 July 1986, LBL. *Berkley*.

73 Achad, M.-H. (Aug. 1976). Desorpion des gaz induite par des electrons et des ions de l'actier inoxydable, du cuivre OFHC, du titane et de l'aluminium purs. CERN-ISR-VA/76-34. *Geneva*: 11.

74 Achad, M.-H., Calder, R., and Mathewson, A. (1978). The effect of bakeout temperature on the electron and ion induced gas desorption coefficient of some technological materials. *Vacuum* 29: 53–65.

75 Billard, F., Hilleret, N., and Vorlaufer, G. Some Results on the Electron Induced Desorption Yield of OFHC Copper. Vacuum Technical Note 00-32, December 2000, CERN. *Geneva*.

76 Gómez-Goñi, J. and Mathewson, A.G. (1997). Temperature dependence of the electron induced gas desorption yields on stainless steel, copper, and aluminium. *J. Vac. Sci. Technol. A* 15: 3093–3103. https://doi.org/10.1116/1.580852.

77 Nishiwaki, M. and Kato, S. (2001). Electron stimulated gas desorption from copper material and its surface analysis. *Appl. Sur. Sci.* 169–170: 700–705. https://doi.org/10.1016/S0169-4332(00)00764-9.

78 Malyshev, O.B., Smith, A.P., Valizadeh, R., and Hannah, A. (2010). Electron stimulated desorption from bare and NEG coated stainless steel. *J. Vac. Sci. Technol. A* 28: 1215–1225. https://doi.org/10.1116/1.3478672.

79 Malyshev, O.B., Smith, A.P., Valizadeh, R., and Hannah, A. (2011). Electron stimulated desorption from aluminium. *Vacuum* 85: 1063–1066. https://doi.org/10.1016/j.vacuum.2011.01.028.

80 O.B. Malyshev, A.P. Smith, R. Valizadeh and A. Hannah. Electron stimulated desorption from copper. Presented at the 11th European Vacuum Conference, 20–24 September 2010, Salamanca, Spain.

81 Malyshev, O.B. and Naran, C. (2012). Electron stimulated desorption from stainless steel at temperatures between −15 and +70 °C. *Vacuum* 86: 1363–1366. https://doi.org/10.1016/j.vacuum.2011.01.028.

82 Malyshev, O.B., Jones, R.M.A., Hogan, B.T., and Hannah, A. (2013). Electron stimulated desorption from the 316L stainless steel as a function of impact electron energy. *J. Vac. Sci. Technol. A* 31: 031601.

83 Malyshev, O.B., Hogan, B.T., and Pendleton, M. (2014). Effect of surface polishing and vacuum firing on electron stimulated desorption from 316LN stainless steel. *J. Vac. Sci. Technol. A* 32: 051601. https://doi.org/10.1116/1.4887035.

84 Edwards, D. Jr., (1979). Ion and electron desorption of neutral molecules from stainless steel (304). *J. Vac. Sci. Technol.* 16: 758.

85 J.B. Hudson, Surface Science: an Introduction. Butterworth-Heinemann Limited, 1992, Chapter 16.

86 Gröbner, O., Mathewson, A.G., Strubin, P., and Alge, E. (1989). Neutral gas desorption and photoelectric emission from aluminium alloy vacuum chamber exposed to synchrotron radiation. *J. Vac. Sci. Technol. A* 7: 223.

87 Trickett, B.A., Schmied, D., and Williams, E.W. (1992). Hard synchrotron radiation and gas desorption process at copper absorbers. *J. Vac. Sci. Technol. A* 10: 217.

88 Andritschky, M., Gröbner, O., Mathewson, A.G. et al. (1988). Synchrotron radiation induced neutral gas desorption from samples of vacuum chamber. *Vacuum* 38: 933.

89 Maslennikov, I., Anashin, V., Foerster, C. et al. (1993). Photodesorption Experiments on SSC Collider Beam Tube Configurations. In: *Proc. of the Particle Accelerator Conference (PAC 93), 17–20 May 1993, Washington D.C.*, p. 3876.

90 Hori, Y., Kobayashi, M., Matsumoto, M., and Kobari, T. (1993). Photodesorption and photoelectron yields at normal and grazing incidence. *Vacuum* 44: 531.

91 Gröbner, O., Mathewson, A.G., Stori, H., and Strubin, P. (1983). Studies of photon induced gas desorption using synchrotron radiation. *Vacuum* 33: 397.

92 Kobari, T. and Halama, H. (1987). Photon stimulated desorption from a vacuum chamber at the National Synchrotron Light Source. *J. Vac. Sci. Technol. A* 5: 2355. https://doi.org/10.1116/1.574451.

93 Reid, R.J. (1993). The attainment of UHV in synchrotron light sources - a review. *Vacuum* 44: 473.

94 Ueda, S., Matsumoto, M., Kobari, T. et al. (1990). Photodesorption from stainless steel, aluminum alloy and oxygen free copper test chambers. *Vacuum* 41: 1928.

95 Foerster, C.L., Halama, H., and Lanni, C. (1990). Photon-stimulated desorption yields from stainless steel and copper-plated beam tubes with various pretreatments. *J. Vac. Sci. Technol. A* 8: 2856.

96 Halama, H.J. and Foester, C.L. (1991). Comparison of photodesorption yields from aluminum, stainless and Cu-plated beam tubes. *Vacuum* 42: 185.

97 Foester, C.L., Halama, H.J., Korn, G. et al. (1993). Desorption measurements of copper and copper alloys for PEP-II. *Vacuum* 44: 489.

98 Foerster, C.L., Lanni, C., Perkins, C. et al. (1995). Photon stimulated desorption measurements of extruded copper of welded copper beam chambers for PEP=II asymmetric B factory. *J. Vac. Sci. Technol. A* 13: 581.

99 Foerster, C.L., Lanni, C., Noonan, N.J., and Rosenberg, R.A. (1996). Photon stimulated desorption measurements of an extruded aluminium beam chambers for the Advance Photon Source. *J. Vac. Sci. Technol. A* 14: 1273.

100 O.B. Malyshev Study of photodesorption processes in the prototypes of superconducting collider vacuum chambers. PhD Thesis, Budker Institute of Nuclear Physics, Novosibirsk,1995.

101 Gröbner, O., Mathewson, A.G., and Marlin, P.C. (1994). Gas desorption from an oxygen free high conductivity copper vacuum chamber by synchrotron radiation photons. *J. Vac. Sci. Technol. A* 12: 846.

102 Herbeaux, C., Marin, P., Rommeluère, P. et al. (2001). Photon stimulated desorption of unbaked stainless-steel vacuum chambers. *Vacuum* 60: 113. https://doi.org/10.1016/S0042-207X(00)00363-8.

103 Hory, Y., Kobayashi, M., and Takiyama, Y. (1994). Vacuum characteristics of an oxygen-free high-conductivity copper duct at the KEK Photon Factory ring. *J. Vac. Sci. Technol. A* 12: 1644.

104 Anashin, V., Bulygin, A., Kraemer, D. et al. (1998). Photodesorption and power testing of the SR crotch-absorber for BESSY-II. Proc. of EPAC-98, Stokholm, 22–26 June, 1998, v. 3: 2163.

105 Herbeaux, C., Marin, P., Baglin, V., and Gröbner, O. (1999). Photon stimulated desorption of an unbaked stainless steel chamber by 3.75 keV critical energy photons. *J. Vac. Sci. Technol. A* 17: 635.

106 J. Gómez-Goñi, et al, *Vacuum Technical Note 92-06*, CERN, 1992.

107 J. Gómez-Goñi, O. Gröbner and A.G. Mathewson. Continued exposure of an OFHC Cu-lined chamber to photons. CERN-AT Vacuum Tech. Note 92-12, CERN, Oct. 1992.

108 Foerster, C.L., Lanni, C., Maslennikov, I., and Turner, W. (1994). Photon desorption measurements of copper and copper plated beam tubes for the SSCL 20 TeV proton collider. *J. Vac. Sci. Technol. A* 12: 1673.

109 Malyshev, O.B. (2012). Gas dynamics modelling for particle accelerators. *Vacuum* 86: 1669. https://doi.org/10.1016/j.vacuum.2012.03.047.

110 A.G. Mathewson, O. Gröbner, P. Strubin1, P. Marin and R. Souchet. Comparison of synchrotron radiation induced gas desorption from Al, stainless steel and Cu chambers. AIP Conf. Proc. **236 (1)**, 313 (1991). doi.org/10.1063/1.41124. (also in CERN/AT-VA/90-21, CERN, Geneva, 1990).

111 Miyauchi, E., Okamura, S., Taguchi, T. et al. (1991). Photon Stimulated Desorption from Finely Polished Metals Irradiated with Synchrotron Radiation. *Jap. J. Appl. Phys.* 30 (2): 2B.

112 Foerster, C.L., Halama, H., and Korn, G. (1992). Photodesorption from copper, beryllium and thin films. *J. Vac. Sci. Technol. A* 10: 2077.

113 Gomez-Goni, J., Gröbner, O., and Mathewson, A.G. (1994). Comparison of photodesorption yields using synchrotron radiation of low critical energies for stainless steel, copper, and electrodeposited copper surfaces. *J. Vac. Sci. Technol.* **A12**: 1714. https://doi.org/10.1116/1.579042.

114 J. C. Billy, J. P. Bojon, B. Henrist, N. Hilleret, M. J. Jimenez, I. Laugier and P. Stubin. The pressure and gas composition evolution during the operation of the LEP accelerator at 100 GeV Vacuum **60**, 183 (2001). doi.org/10.1016/S0042-207X(00)00383-3

115 Malyshev, O.B., Scott, D.J., Bailey, I.R. et al. (2007). Vacuum Systems for the ILC Helical Undulator. *J. Vac. Sci. Technol.* **A25**: 791. https://doi.org/10.1116/1.2746876.

116 Gómez-Goñi, J. and Mathewson, A.G. (1997). Temperature dependence of the electron induced gas desorption yields on stainless steel, copper, and aluminium. *J. Vac. Sci. Technol.* A15: 3093.

117 Trickett, B.A., Schmied, D., and Williams, E.M. (1992). Hard synchrotron radiation and gas desorption processes at a copper absorber. *J. Vac. Sci. Technol.* A10: 217.

118 Gómez-Goñi, J. (2007). Photon stimulated desorption from copper and aluminium chambers. *J. Vac. Sci. Technol. A* 25: 1251.

119 Lozano, M.P. (2002). Ion-induced desorption yield measurements from copper and aluminium. *Vacuum* 67: 339–345. https://doi.org/10.1016/S0042-207X(02)00223-3.

120 Mathewson, A.G. (1976). Ion induced desorption coefficients for titanium alloy, pure aluminum and stainless steel. In: *CERN-ISR-VA/76-5*.

121 Hilleret, N. (1978). Influence de la nature des ions incidents sur les taux de desorption par bombardement ionique de molécules adsorbées sur une surface d'acier inoxydable. In: *CERN-ISR-VA/78-10*.

122 N. Hilleret. Variation of the ion induced desorption yield with temperature and the nature of incident ions. Proc. of 4th ICSS and 3d ECSS, Paris (1980). France, September 1980, v. 2: 1221–1224.

123 I.R. Collins, O. Gröbner, P. Lepeule, R. Veness. Mechanical and vacuum stability criteria for the LHC experimental vacuum chamber. LHC Project Report 205. CERN, July 1998.

124 Mathewson, A.G. (1993). Vacuum problems in particle accelerators due to interaction of synchrotron radiation, electrons and ions with surfaces. *Vacuum* 44: 479–483.

125 N. Hillert and R. Calder. Ion desorption of condensed gases. Proc. of 7th IVC and 3d ICSS, Vienna, Austria, September 1977, pp. 227–230 (1977).

126 J.-C. Barnard, I. Bojko and N. Hillert. Desorption of H_2 and CO_2 from Cu by H_2^+ and Ar^+ ion bombardment. LHC Project Note 44. CERN, April 1996.

127 E. Dietrichs. Desorption of condensed gases from cryogenic surfaces by energetic ions bombardment. CERN, Geneva, Switzerland, 30 July 1993, 63 pages.

5

Non-evaporable Getter (NEG)-Coated Vacuum Chamber

Oleg B. Malyshev

[1] ASTeC, STFC Daresbury Laboratory, Keckwick Lane, Daresbury, Warrington WA4 4AD Cheshire, UK

5.1 Two Concepts of the Ideal Vacuum Chamber

A residual gas pressure in a vacuum chamber increases with gas load (thermal desorption or photon-, electron-, or ion-stimulated gas desorption) and decreases with available pumping speed.

In the traditional approach the emphasis in the ultra-high vacuum/extreme high vacuum (UHV/XHV) system design was focused on looking for various cleaning, surface treatments, and outgassing procedures to prepare a surface of vacuum chamber and components

- that *outgasses as little as possible* ('nil' ideally)
- that *does not pump*; otherwise that surface will contaminated over time with pumped gases.

That means that a vacuum designer has to specify surface cleaning, conditioning, coatings, vacuum firing, ex situ and in situ bakeout temperature (usually as high as possible for each material, e.g. up to 300 °C for stainless steel), and duration (between a few hours to a few weeks), then based on available experimental data for thermal and particle-induced gas desorption to calculate the necessary pumping speed and optimise locations and pumping speeds of the vacuum pumps. An example of commonly used accelerator vacuum chamber is shown in Figure 5.1a. It consists of a beam chamber to accommodate the charged particle beam and an antechamber. There are a few types of antechamber for different purposes: it can used to increase vacuum conductance of vacuum chamber placed between two lumped pumps or to contain in-built distributed pumps. In-built sputter ion pumps (SIP) operating in a magnetic field of dipoles or quadrupoles are successfully used in many circular accelerators; the only problem with these SIP is that they do not operate when accelerator magnets are switched off during shutdowns. To avoid this problem, a design of Large Electron Positron–Collider (LEP) at CERN (Geneva, Switzerland) has implemented an alternative solution by replacing distributed SIP with the St707 non-evaporable getter (NEG) strips, which do not require magnetic field and high voltage. After activation the NEG strips pump H_2, CO, CO_2, and H_2O without any power supply; the only disadvantage

Vacuum in Particle Accelerators: Modelling, Design and Operation of Beam Vacuum Systems,
First Edition. Oleg B. Malyshev.
© 2020 Wiley-VCH Verlag GmbH & Co. KGaA. Published 2020 by Wiley-VCH Verlag GmbH & Co. KGaA.

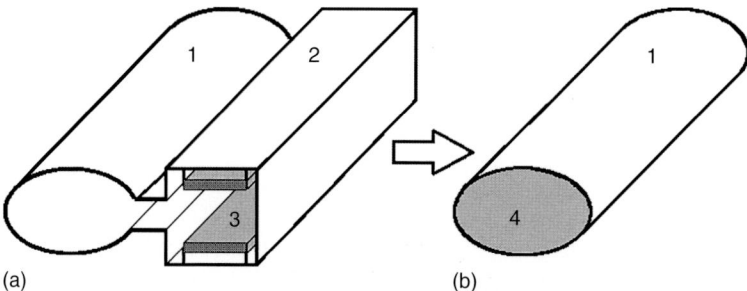

Figure 5.1 From (a) a traditional vacuum chamber design with an antechamber to (b) an NEG-coated vacuum chamber. 1 - beam chamber; 2 - antechamber; 3 - distributed pumps (in-built SIP or NEG strips), and 4 - thin film NEG coating.

is that they don't pump hydrocarbons and noble gases. The NEG strips were activated by heating with a 100 A current passing along these strips to 400 °C for 45 minutes or to 350 °C for about 24 hours, and this requires to provide the current feedthroughs and the strips to be electrically isolated from the vacuum chamber.

In 1990s a very new approach in vacuum systems for accelerators was invented: the NEG coatings [1–8]. The new approach suggested by C. Benvenuti (CERN) was a vacuum chamber, which is fully coated with a thin layer of NEG (as shown in Figure 5.1b), therefore creating a surface

- that *outgasses as little as possible* ('nil' ideally)
- that *does pump*, which, however, will not be contaminated due to a very low outgassing rate from the coated surfaces.

In this approach, an entire surface of vacuum chamber and vacuum components should be coated with NEG film (0.5–3 μm thick). Ideally, there should be no uncoated parts; in practice, the uncoated surface area should be reduced to a minimum (less than 1–2%) with a great caution of its low outgassing. To reach an operation conditions, the NEG coating should be activated by baking in situ to an activation temperature, which could be as low as 150–180 °C for 24 hours. The NEG coating *activation temperature* is much lower than usual bakeout temperatures for vacuum chambers without NEG coating. This is an additional benefit for the mechanical design of vacuum system because it not only reduces the cost of bakeout but also simplifies the mechanical design of the vacuum chamber: the longitudinal expansion of the vacuum chamber is proportional to the temperature increase and should be compensated by bellows – the lower temperature, the lower number of bellows or/and the length of bellows required. Furthermore, the cost of a vacuum chamber manufacturing is much lower: a vacuum chamber cross section is simple (usually round or elliptic), which is not complicated by an antechamber with various designs for different locations on a machine, and it requires flanges of smaller size.

The nature of the getter materials used for NEG coating (Ti, Zr, Hf, V, etc.) is that they pump H_2, CO, CO_2, and H_2O but don't pump hydrocarbons and noble gases. Therefore, small lumped pumps for pumping C_xH_y and noble gases are still

5.2 What Is NEG Coating?

required, however with much smaller size of the pumps and much larger distance between them.

Getters are a subgroup of sorption pumps made of reactive materials. A family tree of sorption pumps is shown in Figure 5.2. Residual gas molecules interacting with a getter surface will be either physically or chemically sorbed. All these types of pumps are used in application to particle accelerators:

- An SIP is an example of a getter pump with gas ionisation, where a glowing discharge of ionised gas molecules allows to sputter titanium atoms from bulk titanium cathode and continuously create a fresh film of pure Ti on neighbouring surfaces.
- A titanium sublimation pump (TSP) is an example of an evaporation getter pump. A fresh Ti film is created periodically by heating a titanium filament with a high current up to a Ti sublimation temperature, and thus the surrounding surfaces are coated with a Ti film.
- Bulk NEG pumps are widely used in many forms: lumped pumps, cartridges, strips, granules, etc. The pure surface can be activated or regenerated by heating the NEG to an activation temperature T_a for a defined duration t_a. The main advantages of NEG pumps are their large pumping speed and sorption capacity; after activation they don't need comptrollers to operate.

NEG coatings have a few differences from the NEG pumps:

- The films deposited directly on vacuum chamber walls.
- The NEG coatings use less getter material, and the film thickness is usually less than a few microns.

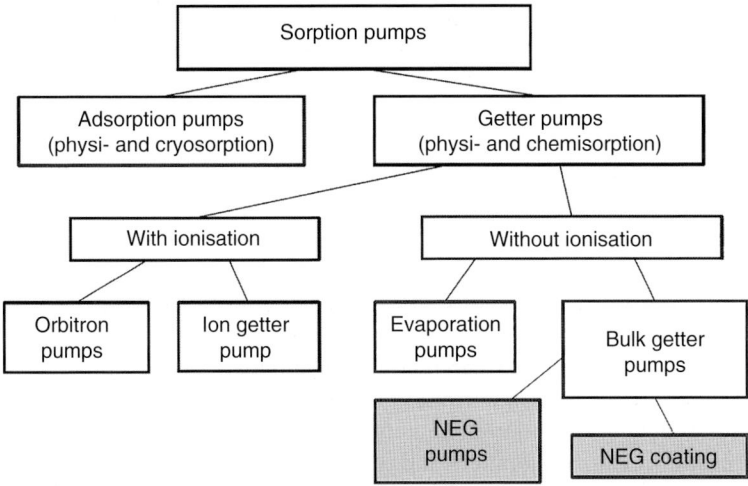

Figure 5.2 NEG coating per classification of sorption pumps.

- The activation temperature should be as low as possible because a bakeout temperature of vacuum chambers is limited to 250–300 °C for stainless steel, 200–250 °C for copper and its alloys, and 150–180 °C for aluminium and its alloys.
- *The main function of the NEG film is a barrier for gas diffusion* from the bulk of vacuum chamber (or its component) material to the surface and following desorption from vacuum chamber walls.
- NEG coating provides a high pumping speed but has a limited sorption capacity; thus *NEG coating should not be used as a pump* for pumping a gas desorbed from uncoated parts of vacuum chamber.
- Combination of low thermal and particle-stimulated desorption (photon-stimulated desorption [PSD], electron-stimulated desorption [ESD], and ion-stimulated desorption [ISD]) and large pumping speed allows to reach the XHV conditions in various layouts of vacuum chamber and, in some cases, is the only solution to meet vacuum specification, for example, in a narrow vacuum chamber with apertures less than 20 mm.

The NEG coatings on accelerator vacuum chamber walls are usually 0.5–3 µm thick films of transitional metals with high oxygen solubility and diffusivity (Ti, Zr, Hf, V, Nb, Ta) or their alloys deposited on the inner walls of vacuum chambers, which are usually made of copper, aluminium alloys, or stainless steel. NEG coating could also be the alloys of Cu, Al, or Fe with these transitional metals.

Note: two terms, 'NEG coating' and 'NEG film', are used in practical application and, in most of cases, can be considered as equal.

After depositing a vacuum chamber with an NEG film under appropriate conditions (the NEG coating is a pure metal alloy), it also has a pure metal surface, which can capture gas molecules such as CO, CO_2, O_2, H_2O, and N_2 by chemical reaction on its active surface, while H_2 does not react chemically but dissolved atomically into the bulk of the film. After venting to air or to gases such as CO, CO_2, O_2, H_2O, and N_2, a layer of carbides, oxides, and/or nitrides is formed on the NEG surface. Formation of this layer defines a pumping capacity of the film, but from another side, it plays a role of a protection layer for deeper parts of the NEG film. Thus, NEG-coated chambers can be safely stored or transferred in N_2 atmosphere; to minimise an exposure to air, they are usually vented to air just before the installation on their locations in an accelerator.

To make NEG coating pumping again, it should be activated by heating under UHV conditions to the NEG coating *activation temperature* for the necessary *duration of activation*. During the activation process, the earlier formed carbides, oxides, and/or nitrides are diffused into deeper layers of the NEG film, while hydrogen atoms desorbed from the NEG film and should be pumped away. A 24-hour-long duration for the NEG coating activation is commonly applied; however, other durations can be used when required: a higher activation temperature allows reaching the same level of activation in a shorter time, while longer activation helps in reducing activation temperature.

Since its invention, the development and further optimisation of NEG coating for various applications have been continued. Initial effort was made to increase pumping speed and sorption capacity and to reduce an activation temperature.

The later effort was to reduce PSD, ESD, and ISD, to suppress photoelectron yield (PEY) and secondary electron yield (SEY), and to investigate surface resistance of various NEG coatings.

It is worth mentioning that there is a special interest to Pd and Pd–Ag thin films used as overlayers for protecting a getter film from oxidation [7, 9, 10].

Most of the work for NEG coating optimisation was performed at CERN and ASTeC. However, a lot of depositions and characterisations of NEG films were performed in many research laboratories around the world; see the references at the end of this chapter.

5.3 Deposition Methods

Among many existing film deposition methods and techniques, NEG coatings are deposited with physical vapour deposition (PVD) methods. Magnetron sputtering is the most common deposition method for the NEG films. It is a plasma coating process whereby sputtering material is ejected due to bombardment of ions to the target surface.

The typical schematic layout of the planar magnetron deposition is shown in Figure 5.3. The vacuum chamber of the PVD coating machine is filled with an inert gas. Plasma is created by an injected noble gas (usually Ar, Kr, or Xe) in a magnet field of permanent magnet and electric field between a biased target and a sample. A glow discharge results in ionising the injected gas and in acceleration of these ions to the target surface. The ions from plasma will kick off atoms from the target surface. These sputtered (or evaporated) atoms will then travel away from the target and grow a film of spattered material on all the surfaces along the line of sight from the target.

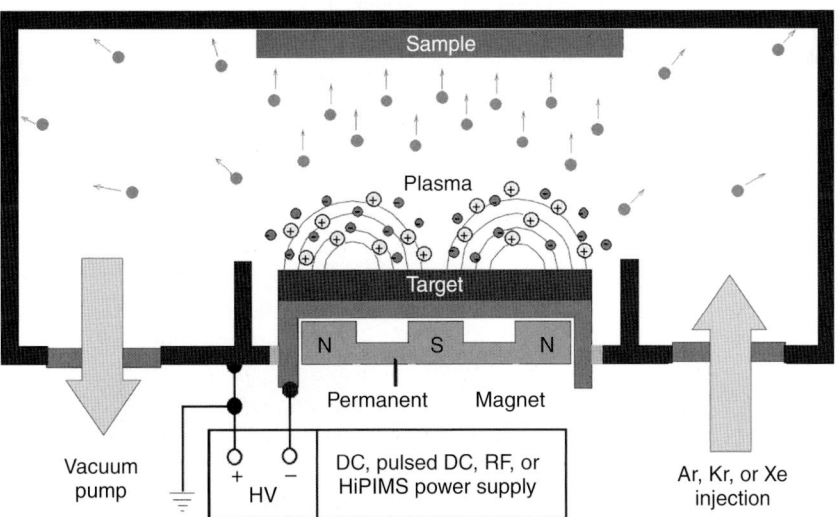

Figure 5.3 Typical schematics of planar magnetron deposition.

Various deposition parameters can be varied during the preposition: working gas and its pressure, temperature of vacuum chamber walls, power supply parameters and a distance from a target to a sample, and size and orientation of sample and target in respect to each other. Additional instruments can be easily added for plasma and ion characterisation instruments, for deposition from multiple targets, and for ion beam, gas clusters, or laser-assisted deposition. Thus, planar deposition is a great tool with a lot of flexibility allowing depositing films from required materials and with desired structure morphology. However, it is difficult or impossible to apply the planar magnetron method to inner surfaces of accelerator vacuum chamber such as a whole tube or extruded vacuum chamber. In application to accelerator vacuum chambers, a planar magnetron can only be applied when a vacuum chamber is made of two halves. From one side, it allows to employ an advantage of developed industrial deposition machines for continuous deposition of thin films on a substrate of practically any length. From another side, these two halves must be welded after the NEG coating; thus the vacuum chamber will have an uncoated weld, and therefore such vessel would not be fully NEG coated.

To perform an NEG coating inside a tubular vacuum chamber, a cylindrical magnetron deposition configuration shown in Figure 5.4 is more suitable. In this case, a target is made of a single metal wire, twisted wires of few different materials, or an alloy rod, which is placed in the middle of the vacuum chamber; an axial solenoid magnetic field is created by a coil located outside of the vacuum chamber. The parameter to control the film morphology are working gas (e.g. Ke or Ar) and its pressure, temperature of vacuum chamber walls, and power supply parameters; the latter allows to choose between DC, pulsed DC, RF, and HiPIMS deposition.

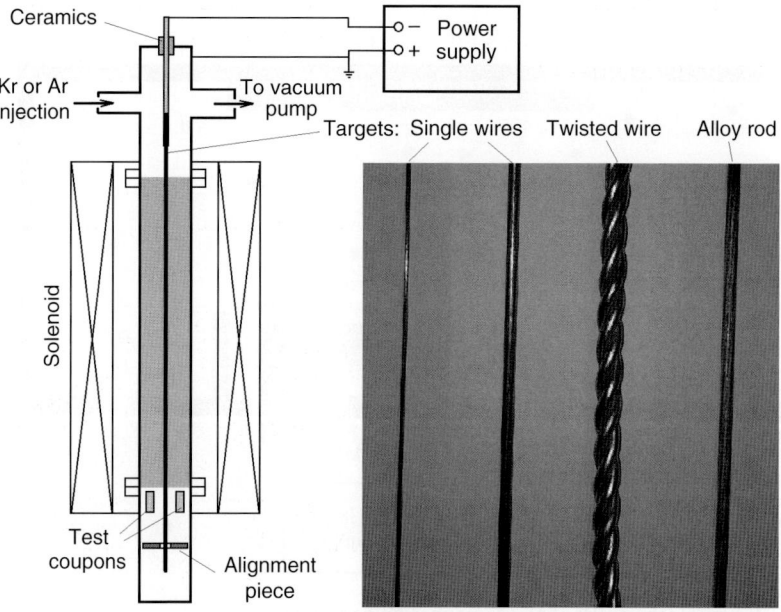

Figure 5.4 Schematic layout of cylindrical magnetron deposition.

The main limitation of this technology is the smallest size of vacuum chamber, presently limited to 5 mm inner diameter, defined by a target size and distance between the target and the vacuum chamber walls required to create plasma.

There exist a few more possible layouts for depositing on vacuum chamber of non-circular cross sections. Elliptic and narrow vacuum chambers (vacuum chamber with ports and/or an antechamber) can be deposited with a few targets in the form of wires. Large cross-sectional vacuum chambers can be deposited with specially designed targets, for example, an alloy tube containing permanent magnets inside (thus there is no need in large solenoid) and cooling channel (to avoid target melting and/or magnet heating above the Curie temperature and, therefore, their demagnetisation).

5.4 NEG Film Characterisation

Many different target materials can be used for depositing NEG films. NEG coating composition could be a single element or dual, ternary, and quaternary alloys of various combinations of transitional metals (Ti, Zr, Hf, V) and their mixtures with Cu, Al, or Fe. Moreover, each material can be deposited under various conditions resulting in different film structures and morphologies. The NEG film could be dense or columnar as shown in Figure 5.5. The grain size of the film can vary in a wide range from a few to a hundreds nanometers. The NEG film thickness can be anything between 0 and 3 μm.

Various surface characterisation techniques are used to characterise the coatings:

- Surface composition and chemical bounding can be determined with an X-ray photoelectron spectroscopy (XPS) and Auger electron spectroscopy (AES).
- Film compositions can be examined with Rutherford backscattering (RBS), secondary ion mass spectrometry (SIMS) and energy dispersion X-ray spectroscopy (EDAX).
- Film morphology – by scanning electron microscopy (SEM) and atomic force microscopy (AFM).
- Grain size, crystal structure, and phase – by transmission electron microscopy (TEM), glancing angle X-ray diffraction (XRD), and electron backscattered diffraction (EBSD).

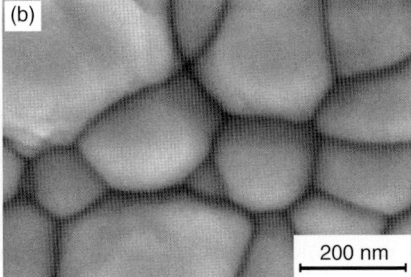

Figure 5.5 SEM image of (a) columnar and (b) dense Ti–Zr–Hf–V NEG coating. Source: Courtesy to Dr. Reza Valizadeh, STFC Daresbury Laboratory, Warrington, UK.

- Film adhesion – by a scratch test.
- Surface roughness and nucleation – with AFM.

All these techniques were employed in various laboratories around the world to find optimum film composition and optimum deposition parameters, as well as to better understand the processes of NEG activation and saturation with gases [2–6, 11–18].

In this chapter, we are not going to overview and explore all these finding, as it is outside of the main scope of the book. However, the main thing that we would like to bring attention to is that the *NEG coating* is not a single well-defined product; it *is a large family of various coatings* that could have the following:

- Different chemical composition, structure, morphology, grain size, crystal structure and phase.
- Different thickness.
- Different vacuum properties (such as PSD, ESD, ISD, sticking probability, and pumping capacity) as a function of bakeout or activation temperature.
- Different surface resistance.
- Different PEY and SEY.

Therefore, the resulting pressure, impact on beam impedance, efficiency of e-cloud suppression, and other properties may vary with what kind of NEG coating is applied.

5.5 NEG Coating Activation Procedure

After NEG deposition, the coated vacuum chamber is usually vented to air to create a protected oxide layer, removed from a deposition facility, pumped again, and filled with N_2 to avoid further contamination and poisoning of NEG coating with active gases from atmosphere. The NEG-coated vacuum chamber will be vented to air again just before the installation on its place in a vacuum system and pumped again as soon as possible. Thus, the installation procedure should be planned in such a way that the exposure to air is minimised to (preferably) minutes or hours rather than days or weeks.

As it was already mentioned above, after installing and pumping the NEG-coated vacuum chamber, the NEG coating should be *activated* (or regenerated) by heating the NEG-coated vacuum chamber to the *activation temperature* for the necessary *duration of activation*. A 24-hour-long bakeout is quite common in vacuum technology; thus, in the same duration it is often applied to NEG-coated chambers as this allows reaching a sufficient level of NEG activation (see details in the following sections). Although the '24-hour-long baking for NEG activation' sound very simple, it is extremely important to know in detail how to activate the NEG-coated chamber connected to non-coated components and chambers.

Since NEG coating capacity is limited to approximately a monolayer of CO or a few monolayers of CO_2 (see the following sections), the NEG coating should not be considered as a pump, various types of lumped NEG pumps, NEG strips and tablets, or NEG-based pumping units. Although the abbreviation 'NEG' is a first word in all these names, the *NEG pumps* were optimised for providing a large pumping speed and large pumping capacity, while the *NEG coatings* were optimised to reduce desorption to such a low level that even a small sorption capacity is sufficient for a lifetime or long-term operation of vacuum system based on NEG coating.

Ideally, an entire surface of vacuum chamber should be coated with NEG. In practice, even in the best effort, this is not always possible and uncoated parts of vacuum chamber or in-vacuum components will be present. There could be various reasons for this, such as required high electric conductivity or, opposite, insulation. It could be that the full coating of some components with a complex shape is very difficult and expensive. These uncoated parts could be pumping and gauges ports, beam instrumentation, windows, bellows, valves, feedthroughs, etc.

Thus, a vacuum system designer need to know how to combine the NEG-coated and -uncoated parts without compromising the properties of NEG coatings.

If the NEG-coated and -uncoated parts of the same vacuum chamber are baked simultaneously, when the gas molecules desorbed from uncoated parts can be pumped not only by the external pump but also by freshly activated NEG coating; thus the NEG coating could be saturated by gas desorption from the non-coated parts and will be either not activated at all or activated only partially. Furthermore, a lifetime sorption capacity of NEG coating is limited by a few hundreds of CO/CO_2 monolayers, so in case of large outgassing for uncoated parts, the NEG coating could be completely saturated during its first bakeout.

To address this problem, the bakeout should have two main stages:

(I) Baking and outgassing of all the uncoated parts while the NEG-coated parts are at the temperature below the NEG activation temperature
(II) The NEG coating activation.

This philosophy was realised, for example, in the procedure shown in Figure 5.6, which was developed in 2000 in the CERN–BINP collaboration [19] as follows:

(1) *Installation and roughing*:
    - The NEG-coated chamber will be mounted and pumped out to $P \leq 10^{-6}$ mbar.
(2) *Baking of the installation, removing water from the NEG-coated chamber*:
    - All the elements of the installation are baked at their maximum acceptable temperature for 24 hours. This is about 300 °C for an SIP, stainless steel pumping port, pumping station, gauges, and residual gas analysers (RGAs), but this should be less for a collimator and luminescent screens.
    - At the same time the NEG-coated chamber shall be kept at ~80 °C to remove the water from the NEG surface and avoid absorption of gases desorbed from uncoated parts. Note: the temperature of the NEG-coated chamber is very important since at the temperature higher than 100 °C, the NEG coating will adsorb the water.
(3) *Outgassing of pumps and gauges*:
    - The NEG cartridge of the NEG lumped pumps is activated at the end of bakeout following its manual.
    - The SIPs should be outgassed by switching on for short time (flashing).

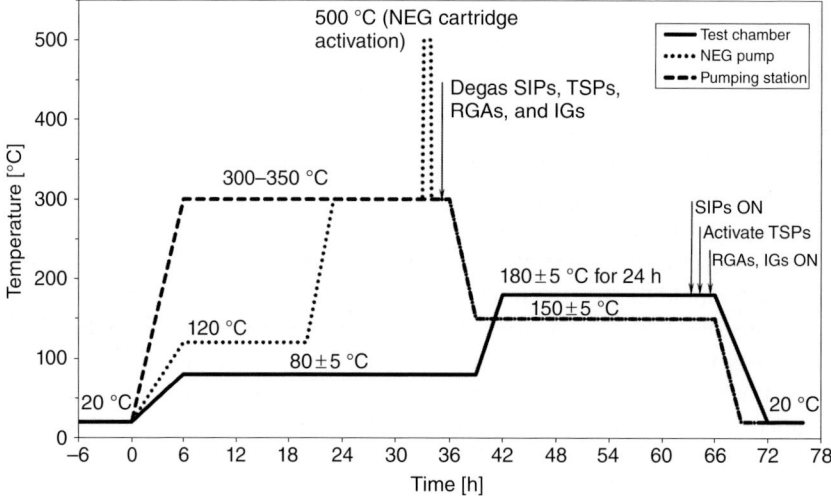

Figure 5.6 CERN–BINP activation procedure for NEG-coated vacuum chambers for activation temperature of 180 °C.

- The Ti evaporator of the TSPs should be outgassed.
- The RGA's and IG's filaments should be degassed with a low current.
- After the degassing all the elements of the installation are cooled down to 150 °C.
(4) *Activation of the NEG coating*:
- All the elements of the installation and the NEG cartridges remain at 150 °C.
- Then the NEG-coated chamber is baked at activation temperature $T_{act} = 180\,°C$ for 24 hours.
(5) *Starting the pumps and gauges*:
- The TSPs should be outgassed and activated before cooling down the system.
- The SIP should be switched on.
- The NEG cartridges of the NEG lumped pumps and the NEG-coated chamber are cooled down to 180 °C.
- The RGAs and IGs are switched on and degassed.
(6) *Cooling down to room temperature*:
- The whole ensemble is cooled down to room temperature ensuring that the NEG-coated chamber remains at a higher temperature than all other parts.

Following this detailed procedure was important for a successful demonstration of performance of NEG-coated chamber under synchrotron radiation (SR) [19].

Consider that the capacity of the NEG film is approximately only a few monolayers for CO, CO_2, H_2O, O_2, and N_2, i.e. the activated NEG could be saturated within seconds with the partial pressure of these gases in the 10^{-6} mbar range

5.5 NEG Coating Activation Procedure

(or within a day for 10^{-11} mbar), leading to a few conditions for a successful NEG coating activation and operation [20]:

Condition 1: The background pressure due to thermal desorption from uncoated part should be better than 10^{-11} mbar for CO, CO_2, H_2O, O_2, and N_2, i.e. no atmospheric leaks can be tolerated, and bakeout and good UHV pumping of the test vacuum chamber (i.e. sufficient pumping speed and ultimate pressure below 10^{-11} mbar) are essential.

Condition 2: NEG film activation (i.e. baking it to activation temperature) should be performed only after the test chamber bakeout, when desorption from uncoated parts of the test system is low. This means that the temperature of the test chamber and the NEG-coated sample should be maintained independently.

Condition 3: No 'short pressure increase' can be tolerated after NEG coating activation. Such pressure increase could happen due to switching on the gauge and the RGA, by opening or closing a valve, etc.

Later studies in ASTeC demonstrated that the NEG coating sorption capacity is larger when the uncoated parts cooled to room temperature after approximately two hours after the beginning of the NEG-coated chamber activation; therefore the activation procedure was modified as follows [20] (see also Figure 5.7):

(1) Ramping up the temperature and baking the uncoated parts at 250–300 °C for 24 hours (or longer if necessary) while the NEG-coated parts are kept at 80 ± 5 °C (to avoid re-condensation but below the temperature that the activation process starts).
(2) Cooling the uncoated parts to 150 °C.

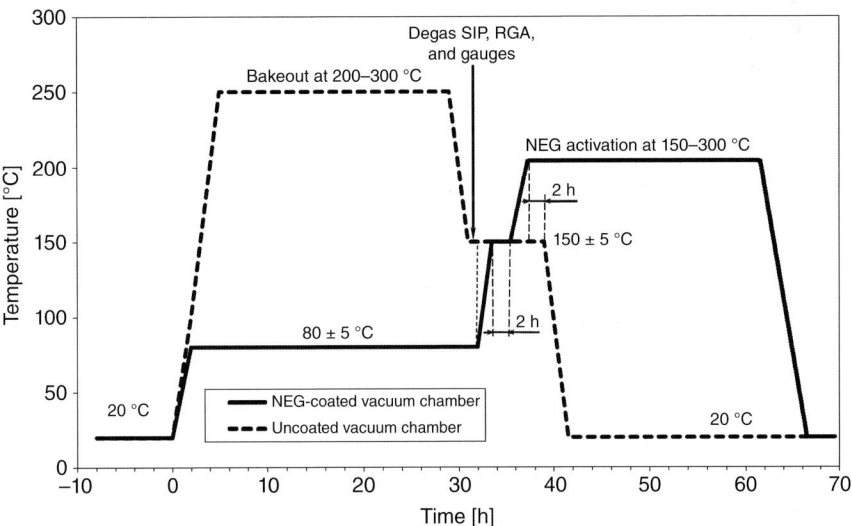

Figure 5.7 ASTeC activation procedure for NEG-coated vacuum chambers.

(3) Switching on and degassing the gauges and the RGAs. If possible, this is also the best time to condition movable vacuum component, for example, by full or partial opening and closing the valves, moving the collimator plates, etc.
(4) Increasing the NEG-coated part temperature to 150 °C and keep all parts of the installation for two hours to outgas at a temperature near transition.
(5) Increasing the temperature of the NEG-coated part to the chosen activation temperature (150–300 °C). The gauges and the RGAs remain switched on if they operate at the pressure below $\sim 10^{-6}$ mbar; otherwise they should be switched off.
(6) One to two hours after the NEG-coated part has reached the activation temperature (usually this is sufficient to reduce the H_2O level by a few orders of magnitude), cool and maintain the uncoated parts as close as possible to room temperature.
   a. If the gauges and the RGAs were switched off, it is a good time to switch them on. Check whether the pressure has reduced to below $\sim 10^{-6}$ mbar (most likely), and if not, repeat in one hour. After that it is recommended to avoid switching off and on.
   b. If there are cooling channels installed, their use commences from this moment.
(7) Activate NEG coating at a chosen temperature for a selected duration (usually 24 hours but may vary) while the uncoated parts are maintained as close as possible to room temperature.
(8) Cool down the NEG-coated part to room temperature.
(9) During cooling down it is preferable to change the pump to a 'clean' UHV pump, which was not used previously during bakeout and activation. In our system it means opening a valve to an SIP and closing the valve to a turbo molecular pump.

Meeting the conditions for a successful NEG coating activation and operation is easy with some layouts of vacuum and could be complicate or even impossible with others.

Cross sections of vacuum chamber with possible NEG coating layouts are shown in Figure 5.8:

(a) A circular or elliptic tube is an ideal vacuum chamber for NEG coating.
(b) Partial NEG coating would be useless because the NEG coating could not be activated due to poisoning from uncoated parts.
(c) Even a small uncoated part, such as an uncoated weld or a distributed SR absorber, could be sufficient to significantly reduce the efficiency of NEG coating.
(d) In the case of a vacuum chamber with an antechamber, both a beam chamber and an antechamber must be coated.
(e) A distributed pump could be used if degassed and activated before NEG activation.
(f) Using the NEG coating in antechamber as a distributed pump would not work due to a small pumping capacity and poisoning from an uncoated beam chamber during NEG activation and following operation.

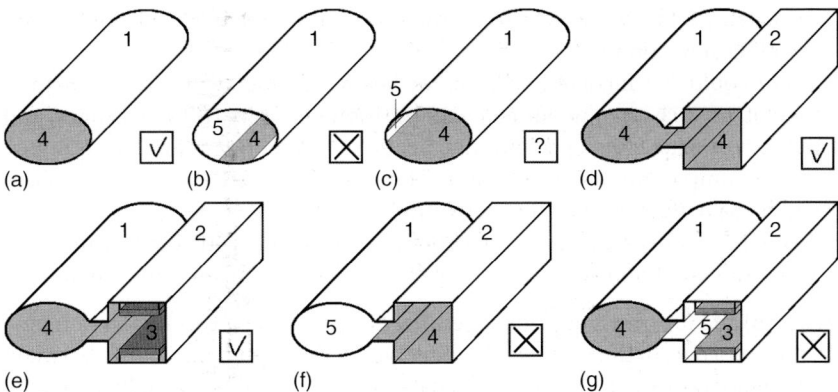

Figure 5.8 Cross sections of vacuum chamber with possible NEG coating layouts: (a) an NEG-coated vacuum chamber, (b) a partially NEG-coated vacuum chamber, (c) even a small uncoated part, such as an uncoated weld or a distributed SR absorber, could be a sufficient to significantly reduce the efficiency of NEG coating, (d) a fully NEG-coated vacuum chamber with an antechamber, (e) a fully NEG-coated vacuum chamber with an antechamber containing a distributed pump, (f) an uncoated beam chamber with NEG-coated antechamber, and (g) an NEG-coated beam chamber with an uncoated antechamber with a distributed pump. 1 - beam chamber; 2 - antechamber; 3 - distributed pumps (in-built SIP or NEG strips); 4 - thin film NEG coating; and 5 - uncoated inner surface.

(g) The NEG coating of beam chamber only is also inefficient: there will be poisoning from an uncoated antechamber during NEG activation and the following operation.

There is a number of possible layouts of using the NEG-coated vacuum chambers along the beam path from a fully NEG-coated chamber to a NEG-coated chamber in one or just a few locations along the beam path. A few possible layouts are shown in Figure 5.9:

(a) Ideally, the whole vacuum chamber should be NEG coated; in this case the NEG surface is fully activated – there is no or very limited poisoning from

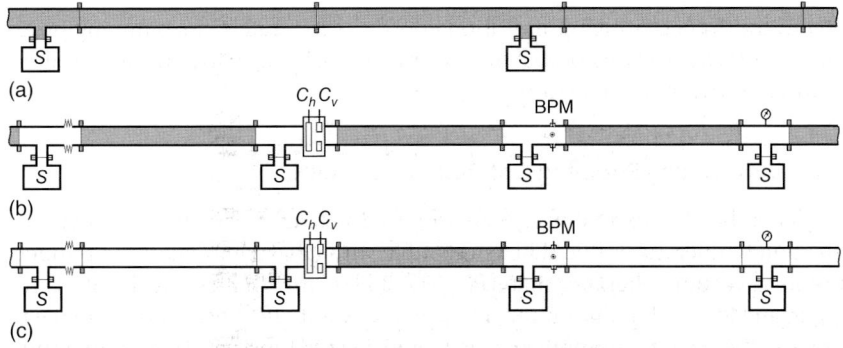

Figure 5.9 An NEG-coated chamber along the beam path: (a) a fully NEG-coated beam chamber, (b) a mostly NEG-coated beam chamber, and (c) a mostly uncoated beam chamber.

uncoated parts. In this case, only the UHV pumps should be baked before the NEG activation.
(b) There could be uncoated sections of beam chamber with various instrumentation such as collimators, beam position monitors (BPMs), gauges, and pumping ports, etc. Such uncoated sections should be preferably equipped with a pump to minimise poisoning of neighbouring NEG-coated chambers and be baked before the NEG activation.
(c) A single NEG-coated chamber should be equipped with UHV pumps on either side to minimise its poisoning from neighbouring uncoated chambers.

In cases (b) and (c), one should consider that during a bakeout and activation process, there is a part of the vacuum chamber with a transient temperature between two set temperatures. Thus the NEG-coated chamber will be partially poisoned at the ends for a length of one to a few cross sections with characteristic size a (e.g. diameter). Since this effect is impossible to avoid, then the larger the length L_{NEG} of NEG-coated chamber, the better vacuum performance; or more accurately, the best results with an NEG-coated vacuum chamber will be obtained for a large aspect ratio L_{NEG}/a.

It is important to mention here that for the vacuum system design, the NEG coating should not be considered as a conventional pump to absorb desorbed gases. Its pumping capacity for CO and CO_2 varies between 0.3 and 5 ML. Total lifetime pumping capacity of NEG coating is limited by an amount of the getter material. In practice, the number of NEG film activations does not exceed 100 times. Moreover, the pumping properties (such as sticking probability and pumping capacity) slowly degrade with a number of performed activations [6]. Therefore, NEG coating is a solution for UHV/XHV vacuum systems, which should not be often vented to air (less than 30–50 times in a lifetime) and where no injection of active gases will be employed. The NEG coating is an ideal solution for narrow undulators and wiggler vacuum chambers, storage rings, dumping rings, particle detectors, etc.

5.6 NEG Coating Pumping Properties

Initially, the NEG coating optimisation effort was focused on its pumping properties: to increase sticking probability, α, and pumping capacity, ϖ, and to reduce the lowest activation temperature, T_a.

5.6.1 NEG Coating Pumping Optimisation at CERN

In order to be able to apply NEG coating to accelerator vacuum chambers, its activation temperature should be compatible with a bakeout maximum temperature for a vacuum chamber material: 300–350 °C for stainless steel, 250 °C for copper, and 200 °C for aluminium alloy. A significant effort was initially made to increase NEG coating pumping capacity and reduce the activation temperature (temperature required to activate the NEG by baking to it for 24 hours), which results in a reasonable sticking probability and pumping capacity. A ternary alloy of titanium, zirconium, and vanadium was found to display full activation after four hours heating at 200 °C at the end of 1997 [5].

The succeeding studies were aiming to obtain a better understanding and further improvements of NEG coating vacuum properties specifically focusing on the following aspects [6]:

- Influence of the elemental composition on activation temperature.
- Sticking factors for H_2 and CO.
- Room temperature saturation capacity for CO.
- Dependence of the vacuum performance on the activation – air venting cycles.
- Trapping of discharge gas atoms and their release during the various stages of the vacuum cycle.
- Testing of a coated chamber in a real accelerator environment.

The influence of the elemental composition of the Ti–Zr–V alloy on its activation temperature was studied in a dedicated system equipped with AES [17, 18]. The surfaces of the NEG alloys were analysed on the as-received state (after air exposure) and then after in situ heating for one hour at a given temperatures: 120, 160, 200, 250, 300, and 350 °C. The criterion for the degree of activation of the NEG surface, R, was defined as a ratio between the intensity of the metallic Zr peak at 147 eV and the Zr peak at 141 eV: i.e. a high R value indicates a high degree of activation. The disadvantage of this method is that R can be used only for samples containing zirconium. The results of the study are shown in quality–composition map of Ti–Zr–V films based on the 'R criterion' shown in Figure 5.10. Two groups of samples are shown, with $R > 0.5$ (white dots) and

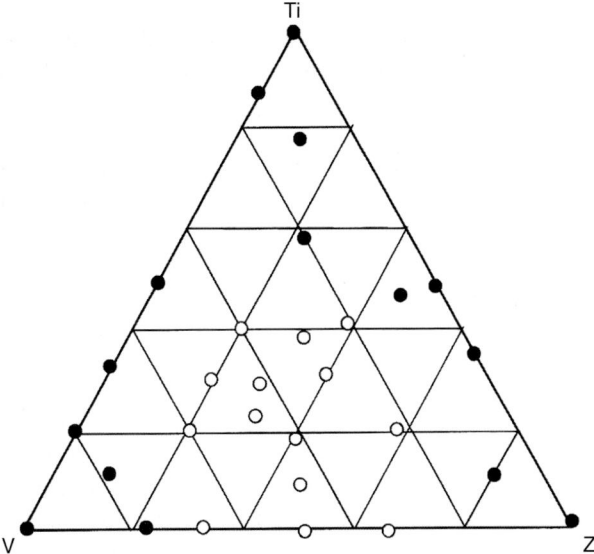

Figure 5.10 Quality–composition map of Ti–Zr–V films based on the 'R criterion' as a function of the in-depth elemental composition. The samples with $R > 0.5$ after 200 °C heating for one hour are indicated by white dots and the others by black dots. Source: Reprinted with permission from Prodromides et al. [17], Fig. 2. Copyright 2001, Elsevier.

$R < 0.5$ (black dots). The following XRD and EDX analysis has shown that the samples with larger degree of activation (with $R > 0.5$) and a lower activation temperature have an amorphous or nanocrystal structure, while the sample with $R < 0.5$ demonstrates the grain size larger than 100 nm.

The Ti–Zr–V films deposited on a tubular sample from three singe metal wires twisted together demonstrated that the NEG coating can be activated by baking to 180 °C for 24 hours. After such activation the typical pumping properties are sticking probabilities $\alpha_{H_2} = 6 \times 10^{-3}$ to 1×10^{-2}, $\alpha_{CO} = 0.2–0.4$, and $\alpha_{CO_2} = 0.4–0.6$ and sorption capacity $\alpha_{CO} = 1–3$ ML and $\alpha_{CO_2} = 2–5$ ML. Sorption capacity of hydrogen is usually not measured because it is much higher than for CO and CO_2, but a monolayer of CO or CO_2 blocks the H_2 pumping, so the sorption capacity of hydrogen is not a performance-limiting factor.

A fully NEG-coated chamber allows to reach XHV conditions after NEG activation: pressures in the order of 10^{-13} mbar were measured in Ref. [8]. The results has demonstrated that the ultimate pressure is likely limited by the degassing of the measuring instrument (which is however among the best available). It was concluded that if an ideal non-degassing gauge was available, an ultimate pressure of approximately 10^{-14} mbar range would be reachable.

5.6.2 NEG Coating Pumping Optimisation at ASTeC

Previous studies have determined that the greatest sticking probability and capacity are provided when the NEG film was deposited with the columnar structure, the lowest NEG activation temperature was reached at 180 °C for 24 hours bakeout [6, 21, 22].

The aim of the following studies was to determine the role of different components and pumping properties of single, binary, ternary, and quaternary NEG films of transitional metals such as Ti, Zr, Hf, and V deposited under the same conditions.

The conditions for NEG coating activation and the best activation have been discussed in Section 5.5. These conditions can be met with tubular samples with a large length-to-aperture ratio. Furthermore, these experimental conditions represent closely the accelerator vacuum chamber. Thus, the pumping properties of NEG coating studied in ASTeC were with tubular samples with ID = 38 mm and $L = 0.5$ m. The experiment layout and typical results are shown in Figure 5.11. NEG coating pumping capacity was defined with an amount of desorbed gas corresponding to the arbitrary chosen pressure ratio $P_2/P_1 = 10$. For example, pumping capacity for the Ti–Zr–Hf–V film activated at $T_4 = 250$ °C corresponds to $x = 3.1$ ML in Figure 5.11. A pressure ratio P_2/P_1 at the beginning of gas injection is used to obtain an initial sticking probability (α_0) in each experiment by using the results of test particle Monte Carlo method (TPMC) modelling. The TPMC results for a pressure ratio P_2/P_1 as a function of sticking probability α for a facility with standard ASTeC tubular sample [23] are shown in Figure 5.12. For example, dashed lines show how an initial sticking probability (α_0) is obtained from the pressure ratio $y = P_2/P_1$ measured for the Ti–Zr–Hf–V film activated at $T_4 = 250$ °C.

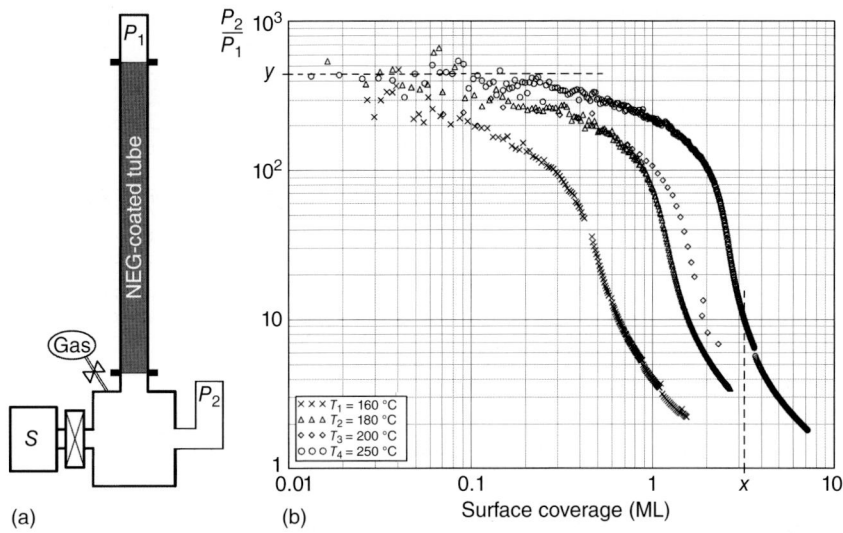

Figure 5.11 (a) A layout for pumping properties evaluation and (b) an example of the results for a pressure ratio P_2/P_1 during CO injection with Ti–Zr–Hf–V film, where y corresponds to an initial pressure ratio P_2/P_1 used for obtaining an initial sticking probability and x is a pumping capacity for the NEG film activated at $T_4 = 250\,°C$.

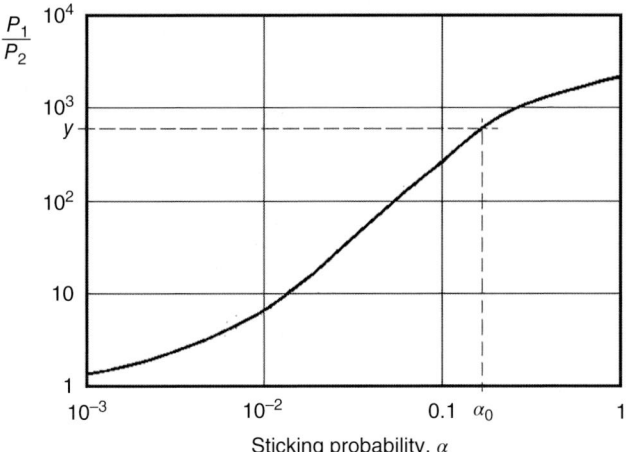

Figure 5.12 The TPMC results for a pressure ratio P_2/P_1 as a function of sticking probability α for the experiment shown in Figure 5.11. Dash lines show how an initial sticking probability α_0 can be obtained from the initial pressure ratio y for Ti–Zr–Hf–V film activated at $T_4 = 250\,°C$.

The following has been demonstrated (see Figure 5.13) [24]:

- The more the elements, the smaller the film grain size and, therefore, the higher sticking probability and the pumping capacity, and the lower the activation temperature.

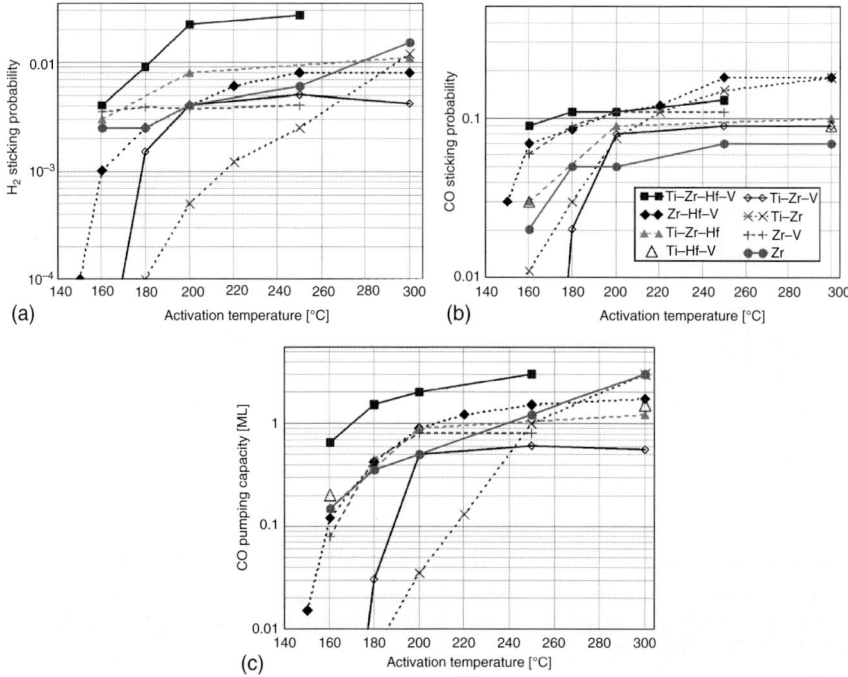

Figure 5.13 (a) H_2 and (b) CO sticking probability and (c) CO pumping capacity for the most promising films [24].

  o The ternary alloy films provide better properties than single or binary alloy films.
  o Among all studied materials, a quaternary Ti–Zr–Hf–V alloy film with a columnar structure has demonstrated the lowest minimum activation temperature of ~160 °C and the higher pumping properties [25].
  o Presence of Zr in all alloys is critical for highest sticking probability and the pumping capacity and lowers the lowest activation temperature; even a pure Zr film was partially activated at ~160 °C.
– Between ternary alloys, Hf–Zr–V, Ti–Zr–Hf, and Ti–Hf–V are quite comparable to each other and better than Ti–Zr–V; however the latter is cheaper and thus the Ti–Zr–V film is nowadays the most widely used NEG coating composition.

All results described above were obtained by using a twisted wire target. That means that the NEG film has the same non-uniformity in the film composition. To check how this non-uniformity may affect NEG film pumping properties, an allowed rod target can be used (see Figure 5.4). Further development of the NEG coating has demonstrated that deposition from the ternary Ti–Zr–V alloy rod allows reduction of the lowest activation temperature to ~160 °C [26], and the use of quaternary Ti–Zr–Hf–V alloy rod allows reduction to even lower temperature: ~140–150 °C [25]. See Figure 5.14.

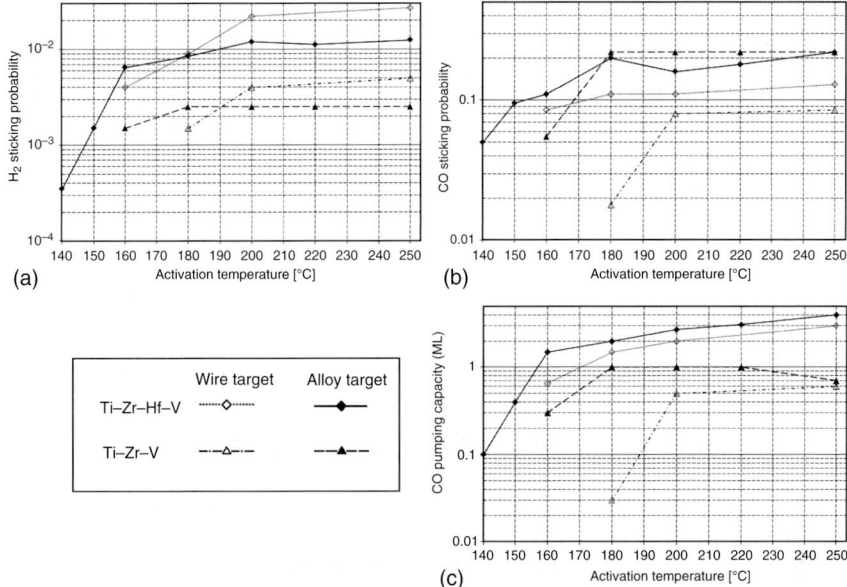

Figure 5.14 (a) H_2 and (b) CO sticking probability and (c) CO pumping capacity for Ti–Zr–V and Ti–Zr–Hf–V films deposited with twisted wire and alloy targets. Source: Malyshev et al. 2014 [25]. Reproduced with permission of Elsevier.

It is worth mentioning here that the activation temperature can be lowered even further. Longer activation time will be required with lower activation temperature to obtain the same pumping properties of NEG coating. To obtain more detailed data, more experiments should be performed to measure the required duration of activation at lower activation temperature.

5.7 NEG Coating Lifetime

NEG-coated film after deposition is a high purity metal or metal alloy. During an exposure to atmosphere or to active gases such as O_2, CO, CO_2, H_2O, etc., an oxide layer is formed on the surface. The oxides on the surface reduce sticking probability not only for oxygen containing gases but also for hydrogen. During the NEG activation process, the oxygen adsorbed at the surface diffuses to dipper layers of NEG film. Due to the nature of getter films, the total lifetime pumping capacity is defined by the amount of NEG material, and due to NEG film thickness it is quite limited. For practical application we need to know how many times NEG coating can be reactivated after full saturation with active gases and how its pumping performance degrades after each saturation–activation cycle.

The effect of repeated activation–air venting cycles on the H_2 sticking probability was studied at CERN; the main result is shown in Figure 5.15 [6]. These results (as well as later unpublished results) demonstrate that NEG coating can be

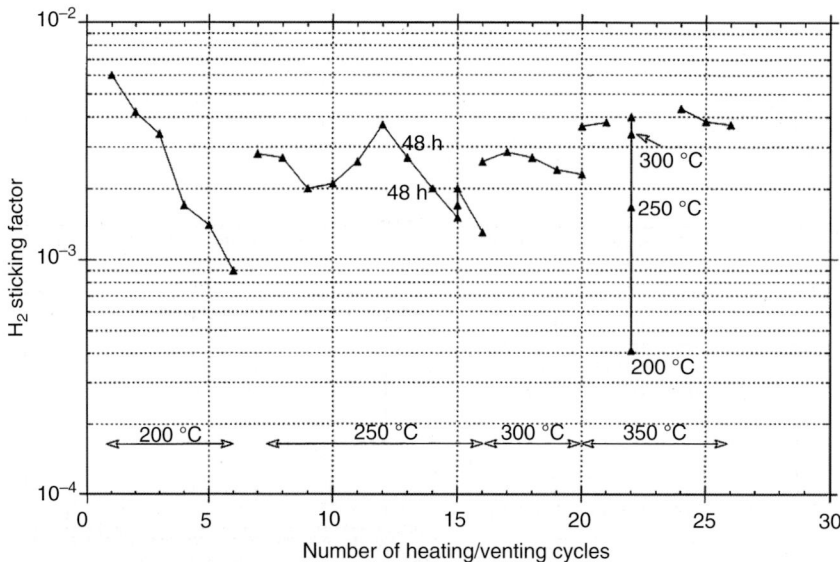

Figure 5.15 Variation of the H_2 sticking probability for a 5-μm-thick Ti–Zr–V film as a function of the number of activation–air venting cycles. The activation temperature was applied for a duration of 24 hours. Source: Reprinted with permission from Benvenuti et al. [6], Fig. 6. Copyright 2001, Elsevier.

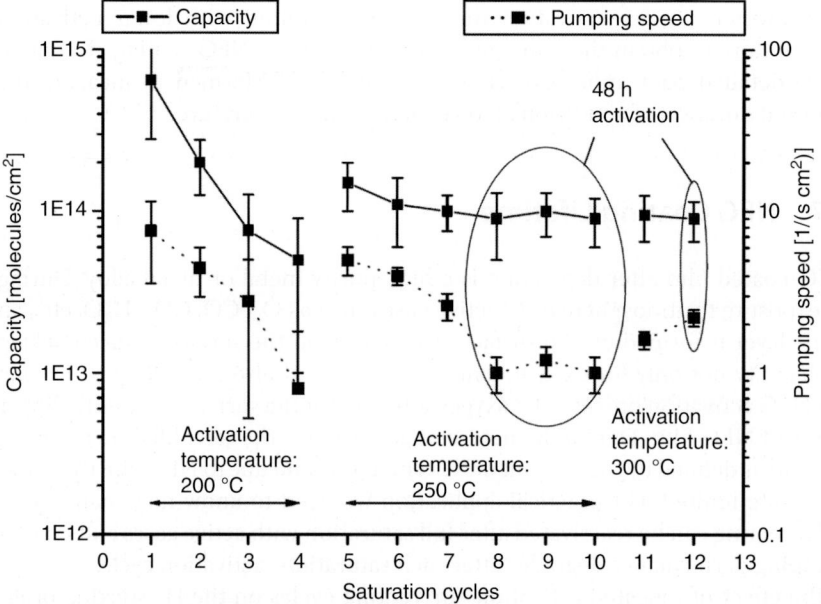

Figure 5.16 CO pumping speed and capacity for a 1-μm thick Ti–Zr–V film as a function of the number of activation–saturation cycles. Source: Reprinted with permission from Bender et al. [27], Fig. 1. Copyright 2010, Elsevier.

vented to air and reactivated for up to more than 55 times; however a degradation of H_2 sticking probability is observed after each cycle. Thus a higher activation temperature is required for providing oxygen diffusion to deeper NEG coating layers.

In another set of repeated activation–venting cycle experiments at GSI (Darmstadt, Germany), the samples were saturated with CO, and the measured CO pumping speed and capacity are shown in Figure 5.16 [27]. Similarly to results at CERN, a degradation of CO pumping speed and capacity was observed after each cycle, and higher activation temperature was required to achieve better pumping speed and capacity. Elastic recoil detection analysis (ERDA) method was employed to study element and depth-resolved evolution of the getter film during gas pumping. It was demonstrated that the oxygen is diffusing into the bulk of the NEG film, whereas the carbon is accumulated at the surface, see (Figure 5.17).

It should be noted that all these studies were performed with NEG coating deposited on CERN's recipe, and although not specified in the papers [6, 27], these coatings have columnar structure. Since an NEG coating lifetime depends on the amount of gas absorbed by NEG and pumping capacity of dense NEG coating is lower than for the columnar one, it is reasonable to expect that lifetime of the dense NEG coating should be greater than of the columnar one. However this should be experimentally checked in the future.

5.8 Ultimate Pressure in NEG-Coated Vacuum Chambers

A special study to measure or estimate ultimate pressures measured in NEG-coated chambers was performed at CERN [8]. A 2-m-long vacuum chamber with inner diameter of 58 mm was fully coated with Ti–Zr–V NEG with krypton as a discharge gas. After installation in a testing facility, the NEG-coated chamber was activated at 250 °C for 24 hours and cooled down to room temperature. The partial pressure measurements were taken while following heating of the activated NEG-coated chamber from room temperature to 250 °C with a ramp up rate of 50 °C/h. No pressure increase was observed up to 120 °C, and then the H_2 peak started to increase followed by Kr at 140 °C and by CH_4 at 170 °C. Figure 5.18 shows only the temperature-dependent measurements (solid line), which were extrapolated down to room temperature to estimate a contribution of the chamber degassing to the ultimate pressure of the order of 10^{-16} Pa (or 10^{-18} Torr as stated in [8]). Similar results were obtained in a few other experiments [8, 28]. Even if the applicability of extrapolation is not well justified, the main conclusion of this study is that after NEG activation, pressures inside an NEG-coated chamber is either close to a low pressure limit of used UHV/XHV gauge or lower.

Thus, after NEG activation, pressures inside a fully NEG-coated chamber without any gas injection or energetic particle bombardment is certainly below 10^{-13} mbar due to a low value of thermal outgassing rate combined with a large pumping speed of NEG coated chamber.

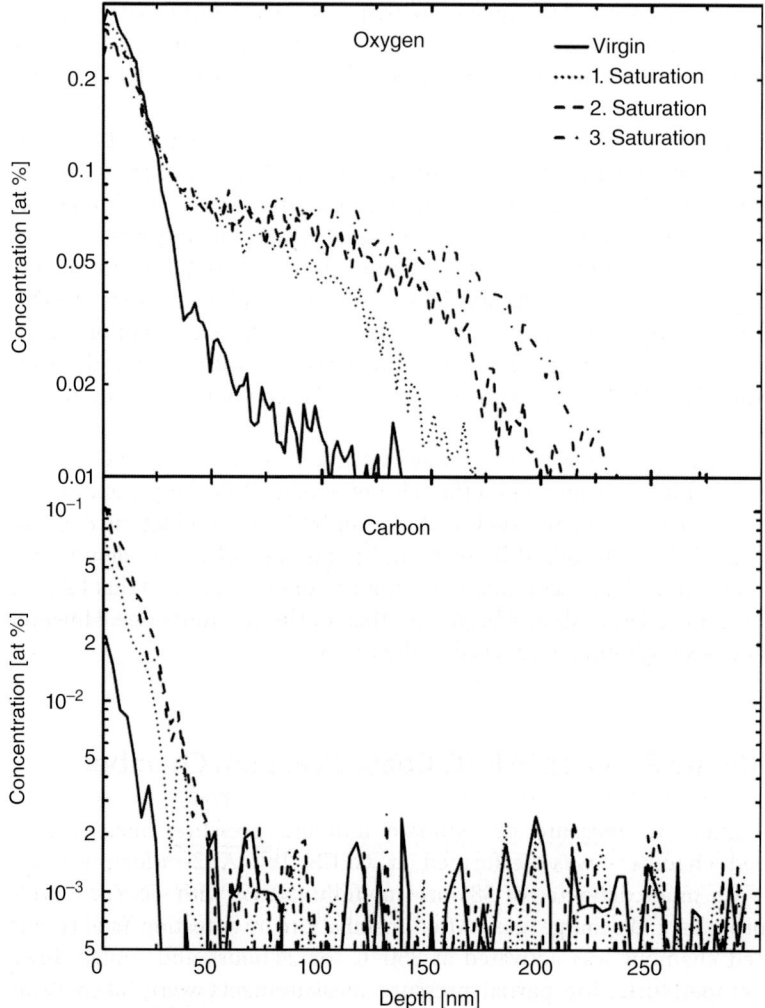

Figure 5.17 Oxygen and carbon depth profiles for four activation cycles of a 1-μm-thick TiZrV film. Source: Reprinted with permission from Bender et al. [27], Fig. 2. Copyright 2010, Elsevier.

If an NEG-coated chamber is equipped with a low thermal outgassing UHV/XHV gauge (for example, an extractor gauge or an RGA), then after the NEG activation this gauge may become the main source of thermal outgassing and a real pressure must be either close or below that gauge reading.

5.9 NEG-Coated Vacuum Chamber Under SR

Since the NEG coating was invented for use in particle accelerators with SR, the behaviour of NEG-coated vacuum chamber was studied under SR performed on SR beamlines. The experiment at the ESRF (Grenoble, France) on a dedicated

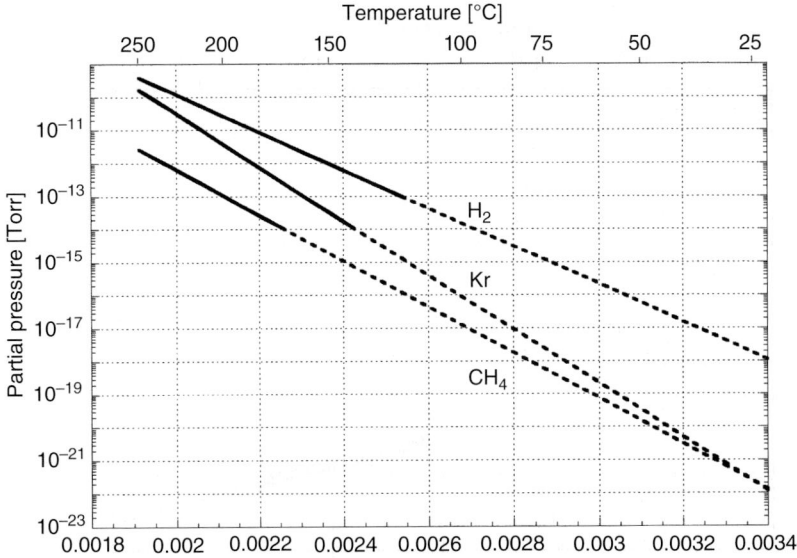

Figure 5.18 Partial pressures evolution measured while heating an activated getter-coated chamber. The solid lines correspond to the measured pressure range, while the dotted ones represent the extrapolation to room temperature. Source: Reprinted with permission from Benvenuti et al. [8], Fig. 2. Copyright 2001, Elsevier.

beamline [29] demonstrated that the NEG coating activation at 250 °C for 20 hours leads to more than the two orders of magnitude reduction of the total desorption yield in N_2 equivalent (see Figure 5.19).

A comparison of two stainless steel samples with and without NEG coating was performed in the experiment at BINP (Novosibirsk, Russia) [19]. Partial pressures were measured in the middle and both extremes of tested sample tubes. In the presence of SR, the H_2, CH_4, CO, and CO_2 pressure increases in the centre of the Ti–Zr–V-coated chamber activated at 190 °C for 24 hours were found to be much lower than those of the stainless steel test chamber baked at 300 °C for 24 hours, see Figure 5.20. Calculated PSD yields for NEG-coated sample were much lower than for bare stainless steel, demonstrating that pressure reduction in the NEG-coated chamber was achieved due to its two main properties: barrier for gas diffusion (reducing PSD) and distributed pumping.

One critical point was considered after NEG coating invention: the CO pumping capacity is in the order of 1 ML. At partial pressure $P_{CO} = 10^{-10}$ mbar and sticking probability $\alpha = 1$, a monolayer of CO is absorbed in approximately 2.5 hours. Therefore many researches worry that NEG coating will be quickly saturated during accelerator operation. However, no saturation was observed neither in NEG-coated vessel in accelerators [30, 31] nor in experiments on SR beamlines [19, 29]. To address this, the CO gas injection into NEG-coated vessel was performed during SR bombardment at BINP [19]. This experiment has demonstrated an important property of NEG exposed to SR: photon-induced pumping of the NEG coating. The rate of pumping was found to be around

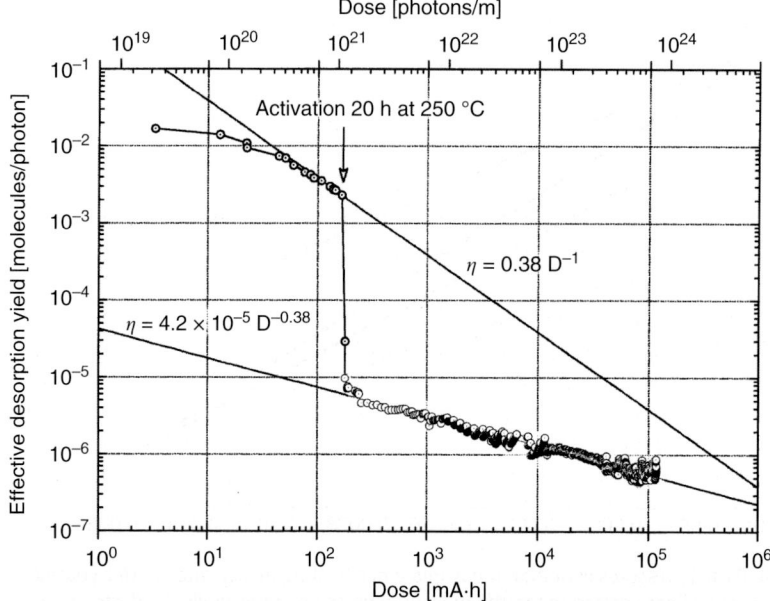

Figure 5.19 Total molecular desorption yield η (N_2 equivalent) of the Ti–Zr–V coated stainless steel chamber as a function of the accumulated dose before and after activation. Source: Reprinted with permission from Chiggiato and Kersevan [29], Fig. 2. Copyright 2001, Elsevier.

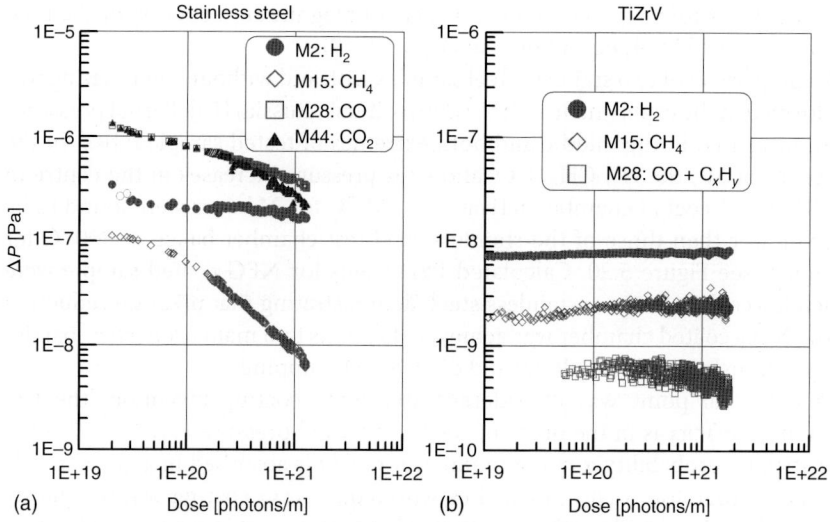

Figure 5.20 Dynamic pressure rise for the (a) stainless steel (baked at 300 °C for 24 hours) and (b) TiZrV-coated vacuum chambers (activated at 190 °C for 24 hours). Source: Reprinted with permission from Anashin et al. [19], Fig. 3. Copyright 2004, Elsevier.

2×10^{-5} molecules/photon. This result explains why no NEG saturation was observed under SR.

Furthermore, in fully NEG-coated accelerator vacuum chamber, the source of CO and CO_2 is mainly the NEG coating itself; thus there are no 'new' molecules in the system, but there is a recycling of the same molecules and atoms (C, H, and O) under SR. The only real source of new CO and CO_2 molecules are uncoated parts and components of vacuum chamber. A partial or even full saturation of NEG coating (or so-called 'poisoning') will take place near these locations where the total amount of pumped molecules should be compared to the total capacity of the NEG film.

A more recent comparative study of PSD was performed in a collaboration between CERN and KEK [32]. The stainless steel samples were cleaned, vacuum fired, and coated with NEG or amorphous carbon at CERN and were installed at KEK's Photon Factory and irradiated with SR at $\varepsilon_c = 4\,\text{keV}$. Figure 5.21 shows the total PSD yield for these samples. One can see that activated NEG-coated samples have significantly lower PSD yield than the uncoated ones.

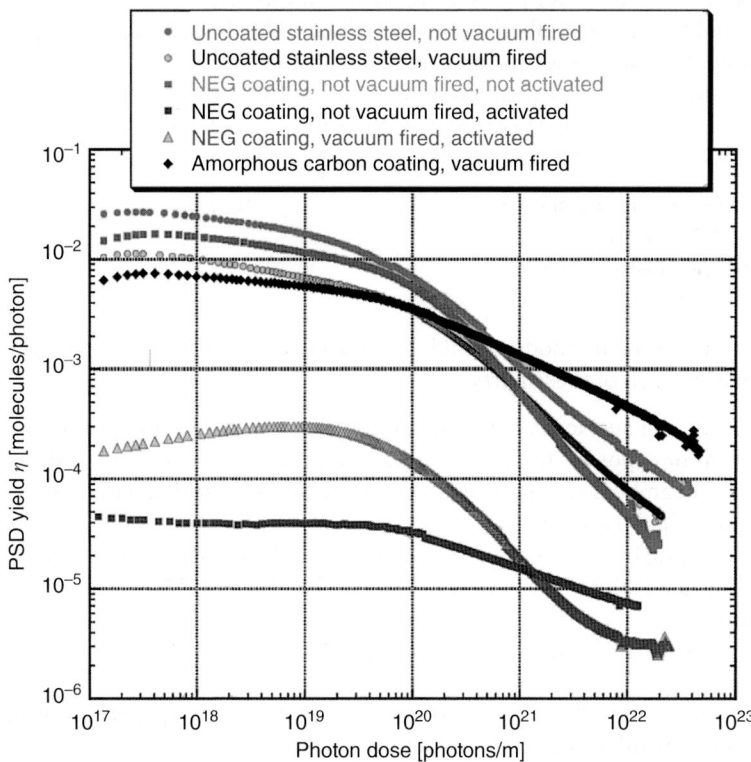

Figure 5.21 Total PSD yields for different samples as a function of photon dose. Source: Ady et al. 2015 [32], Fig. 2. Reprinted with permission of CERN.

5.10 Reducing PSD/ESD from NEG Coating

5.10.1 Initial Considerations

Although the NEG coating allows simplifying a beam vacuum chamber design and solving a problem of vacuum in narrow vessels in particle accelerators, there are new tasks with more demanding specifications. Pressure in the long accelerator vacuum chamber is defined primarily by the desorption yield η and sticking probability α, in many cases as follows:

$$P \propto \frac{\eta}{\alpha}. \tag{5.1}$$

Increasing the NEG coating pumping speed is limited by a simple fact that sticking probability can't be greater than 1: $\alpha \leq 1$. The sticking probabilities of NEG coatings are varied in the following ranges: $0.005 < \alpha_{H_2} < 0.02$, $0.1 < \alpha_{CO} < 0.5$, and $0.4 < \alpha_{CO_2} < 0.6$. That is, it is close to an extreme value for CO and CO_2, and the further increase of α_{H_2} could be possible by increasing surface roughness and porosity. But this will increase a surface RF resistance, which could in turn increase the beam emittance blow up, so it is not good for many accelerators. Therefore, to have significant improvement on vacuum in an NEG-coated vacuum chamber is only by reducing the desorption yields η *in orders of magnitude*, which seems to be quite a realistic task.

Thus, the most important thing is to find out what the sources of gas are and then mitigate each of them. Let's consider an NEG-coated wall in four parts where gas molecules could be contained (see Figure 5.22) [33, 34]

1. on the NEG coating surface
2. inside the NEG coating
3. in subsurface substrate layer
4. in the substrate bulk.

Different mitigation methods should be applied to reduce amount of gas molecules within each of these parts:

1. The NEG coating surface is contaminated during exposure to air; therefore the duration of exposure to air should be minimised. After NEG deposition the vacuum chambers should be vented with N_2, quickly sealed with flanges, pumped and filled back with N_2.
2. The NEG coating bulk can be contaminated by residual gas molecule trapped during film deposition. These residual gas molecules can come from three main sources:
   - Vacuum leaks (mitigated by a proper vacuum design, leak detection and mitigation).
   - Outgassing from walls of deposition vacuum chamber (mitigated by choice of vacuum chamber material, cleaning and bakeout).
   - Impurities from the discharge gas (mitigated by control and purity of discharge gas, clean and leak-free gas injection line).

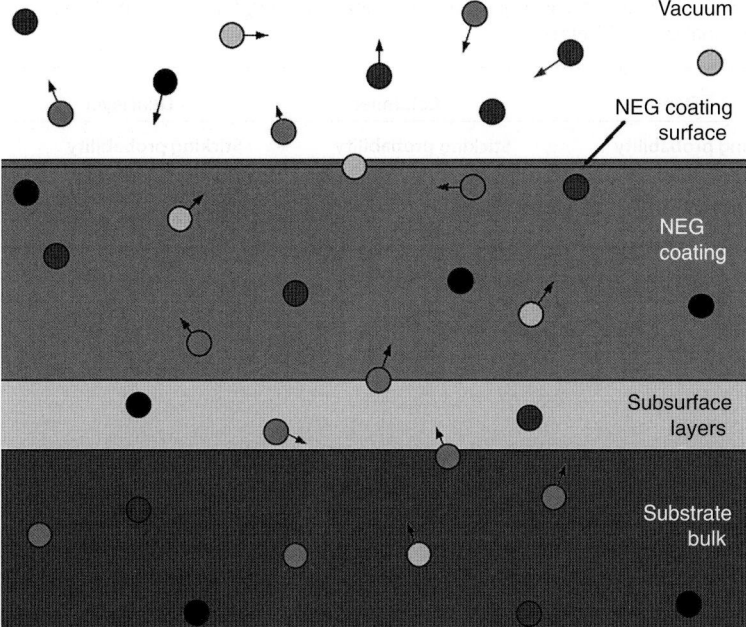

Figure 5.22 NEG-coated wall parts where gas molecules could be contained. Source: Reprinted with permission from Malyshev et al. [33], Fig. 1. Copyright 2014, American Vacuum Society.

3. The subsurface substrate should be properly prepared, cleaned, and baked in situ before NEG deposition to minimise thermal outgassing and gas diffusion from the substrate.
4. The hydrogen depletion from the substrate bulk can be achieved by vacuum firing.

When all these conditions are met, the NEG coating will be a pure metal film without contaminants. After an air vent, a layer of metal oxides will be created on the top of NEG coating as well as a layer of physisorbed gases such as H_2O, CO, CO_2, H_2, and N_2. The oxide layer is protecting the NEG coating from further contamination (poisoning) with gases. It should also be noted here that only hydrogen could diffuse through the NEG film.

5.10.2 ESD from Vacuum Chamber Coated with Columnar and Dense NEG Films

Conditions 1–3 in Section 5.5 can be met in many laboratories equipped with a well-designed and maintained NEG deposition facility. Thus the main source of hydrogen (and methane generated on the NEG surface from diffused hydrogen and carbon on the surface) is substrate bulk [33, 34].

Table 5.1 Pumping properties (sticking probability α and pumping capacity ϖ) of dense, columnar, and dual layer NEG films.

	Dense				Columnar				Dual layer			
	Sticking probability			ϖ_{CO}	Sticking probability			ϖ_{CO}	Sticking probability			ϖ_{CO}
T_a [°C]	H_2	CO	CO_2	[ML]	H_2	CO	CO_2	[ML]	H_2	CO	CO_2	[ML]
150	0.002	0.04	0.075	0.004	0.004	0.2	0.13	3.5	0.009	0.055	0.1	0.5
180	0.0013	0.025	0.012	0.13	0.014	0.2	0.13	3.5	0.018	0.075	0.11	3
250	0.004	0.085	0.02	0.12	0.02	0.2	0.13	—	0.023	0.085	0.2	10

Source: Adapted from Malyshev et al. 2012 [34] and Malyshev et al. 2016 [35].

It was already mentioned above that NEG coating with a columnar structure provides the highest pumping properties and lowest activation temperature. However, due to its nature, the columnar structure is quite transparent for gas diffusion.

NEG coating with a very dense structure is a superior diffusion barrier because it increases the hydrogen diffusion path. In the experiments reported in Ref. [34], two identical 316LN stainless steel tubes of 0.5 m in length and 38 mm in internal diameter were deposited in dense and columnar Ti–Zr–Hf–V NEG films. Table 5.1 shows the comparison of pumping properties of dense and columnar NEG films. One can see that the dense film has much lower pumping properties than the columnar one activated at 150, 180, and 250 °C. In Figure 5.23a,b, the comparison of H_2 ESD yields were obtained under the same experimental conditions. The columnar NEG film demonstrates lower ESD after activation at 150 °C; however the dense film demonstrates lower ESD after activation at 180 and 250 °C. The benefit of dense film as hydrogen diffusion barrier is clearly visible for large doses (greater than 4×10^{21} e$^-$/m^2 activated at 180 °C and 3×10^{22} e$^-$/m^2 activated at 250 °C).

5.10.3 Dual Layer

The benefits of both films – the low ESD yields of the dense film and the high pumping properties of the columnar film – can be combined to achieve an even lower gas density. The results for a dual layer of Ti–Zr–Hf–V NEG consisting of 0.5 μm of dense NEG followed by 1 μm of columnar structure was reported in Ref. [35]. It was reported that the dual layer combines the benefit of both: the ESD yields are like for dense film (see Figure 5.23c) with the pumping properties of columnar film.

The results were reported for 13 consequent ESD experiments (after activation to various temperatures in the range between 80 and 300 °C for 24 hours) followed by full saturation with CO or air; no degradation of this film observed after these 13 cycles.

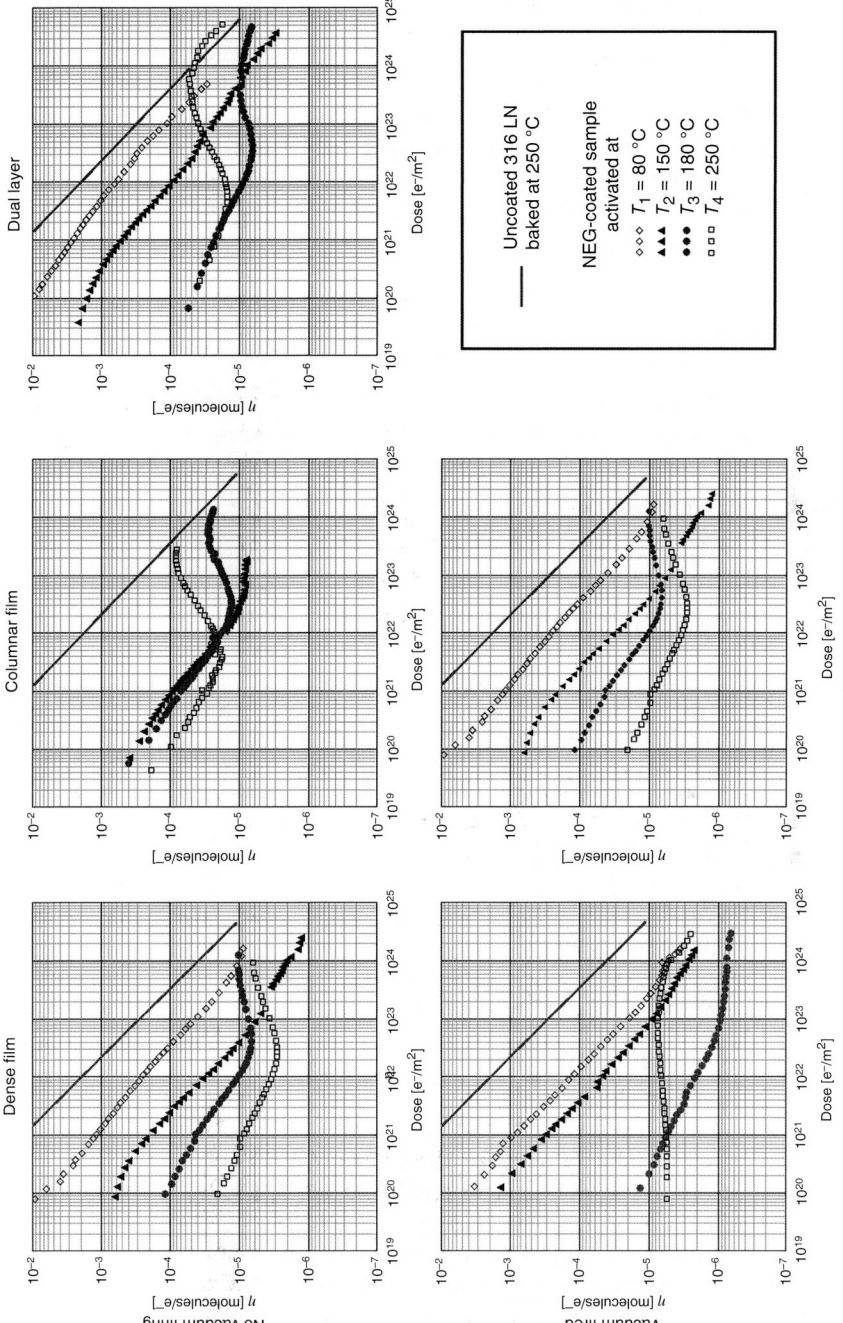

Figure 5.23 Comparison of H_2 ESD yields from Ti–Zr–Hf–V NEG coatings as a function of electron dose for samples baked at 80 °C or activated at 150, 180, and 250 °C and NEG coating [33–35] with a reference to 316LN baked to 250 °C [36].

5.10.4 Vacuum Firing Before NEG Deposition

Mitigation method 4 mentioned in Section 5.10.1 has been met in the experiments reported in Ref. [33]. Two identical 316LN stainless steel tubes of 0.5 m in length and 38 mm in internal diameter were used in this experiment. Vacuum chambers were polished to RA = 0.15–0.2 and then vacuum fired to 950 °C for two hours at a pressure of around 10^{-5} mbar. These chambers were deposited with NEG film the dense and columnar NEG similar to the ones described in Section 5.10.2. The main result is that vacuum firing is an efficient technology to reduce ESD yield by an order of magnitude for both columnar and dense NEG coatings; see Figure 5.23d,e.

5.11 ESD as a Function of Electron Energy

The energy dependence of the ESD yield could be important when no ESD experimental data are available for the electron energy of interest. The ESD yields as a function of electron energy were measured in the energy range between 40 eV and 5 keV [37] (see Figure 5.24). A linear dependence was measured for most of gases except for H_2, which shows $\eta(E) \propto E^{0.5}$.

5.12 PEY and SEY from NEG Coating

NEG coating is not only a solution for achieving a necessary level of vacuum but also an efficient means of mitigation of electron cloud and beam-induced electron multipacting (see Chapter 8). The secondary electron yields (SEY) of Ti–Zr

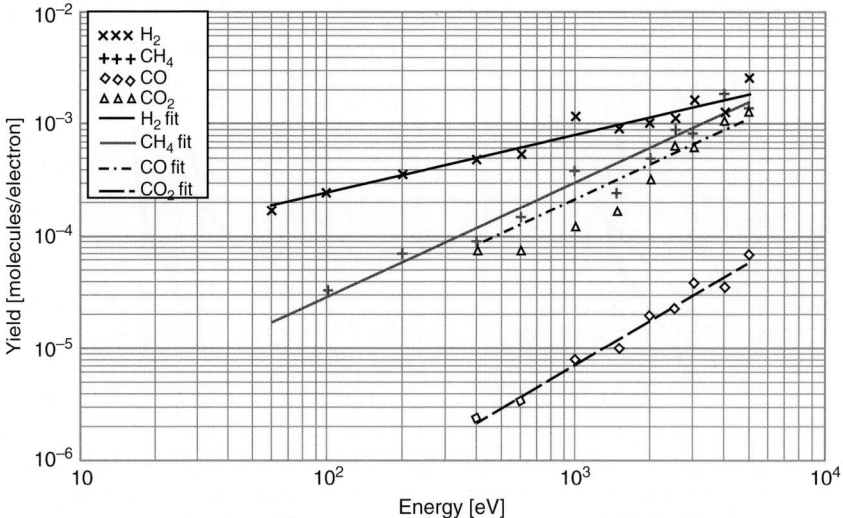

Figure 5.24 η(Ee-) for different gases for NEG coating. Source: Reprinted with permission from Malyshev et al. [37], Fig. 8. Copyright 2010, American Vacuum Society.

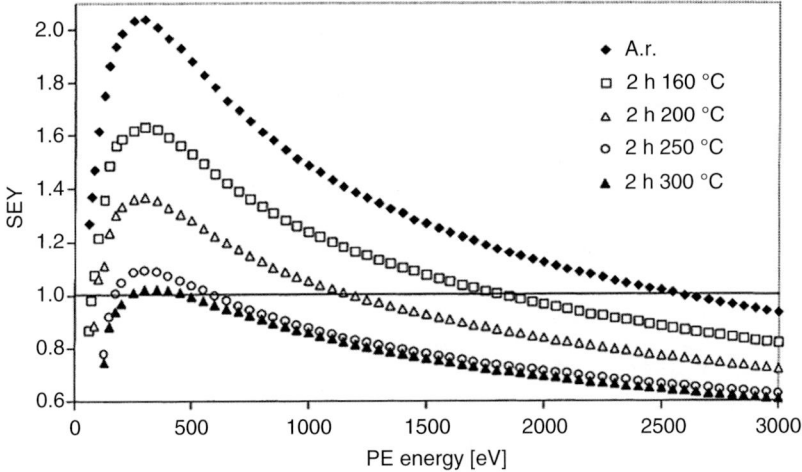

Figure 5.25 SEY as a function of primary electron energy for Ti–Zr–V NEG coating for as-received (A.r.) and after two hours heating to 160, 200, 250, and 300 °C. Source: Reprinted with permission from Henrist et al. [38], Fig. 1. Copyright 2001, Elsevier.

and Ti–Zr–V NEG thin film coatings were reported in Ref. [38, 39]. In a study performed at CERN (Geneva, Switzerland), it was found that δ_{max} reduces with activation temperature from $\delta_{max} \approx 2.0$ on as-received (non-activated) Ti–Zr–V coating to $\delta_{max} \approx 1.0$ measured after activation at 300 °C for two hours [38]; see Figure 5.25.

In the results of experiments in ASTeC reported in Ref. [39], δ_{max} also reduces with activation temperature from ≈ 1.7 on as-received Ti–Zr–V coating to $\delta_{max} \approx 1.35$ measured after activation at 300 °C for one hour. The difference between two results can be explained by differences in NEG coatings (structure, morphology, composition, etc.), a shorter activation in ASTeC (STFC Daresbury Laboratory, Warrington, UK), and differences in other experimental conditions. However, the following electron bombardment demonstrated that a reduction of δ_{max} can also be achieved by electron beam conditioning: $\delta_{max} \approx 1.1$ was measured after achieving a dose of 1.6×10^{-3} C/cm^2, and further electron bombardment to a dose of 5.3×10^{-3} C/cm^2 results in an insignificant reduction of δ_{max} (see Figure 5.26).

The reduction of PEY and SEY under SR was studied at KEK B-factory (Tsukuba, Japan) on Ti–Zr–V samples deposited at BINP (Novosibirsk, Russia) in comparison to uncoated copper with the SR critical energy of 4.1 keV and the incident angle of 51 mrad [40, 41]. The results show that $\delta_{max} = 0.9$–1.1 was obtained on NEG-coated vacuum chamber after NEG activation at 200 °C in comparison to uncoated chamber with $\delta_{max} = 1.1$–1.3. The PEY for the NEG coating and copper were found to be 0.22–0.28 and 0.26–0.34 electrons/photon, respectively. An intensity of electron multipacting (measured with a specially designed monitor) demonstrated a reduction by ~20% due to photon scrubbing after reaching a photon dose of 10^{15} photons/m (and electron dose of $\sim 10^{-2}$ C/mm^2).

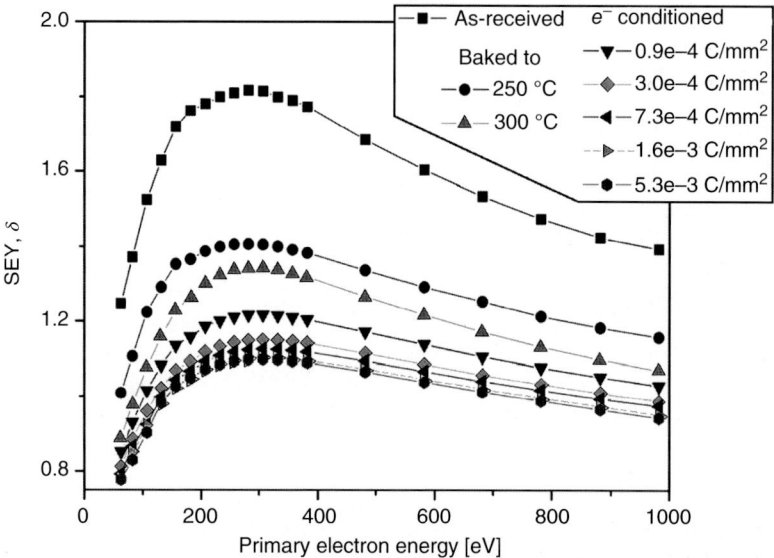

Figure 5.26 SEY as a function of primary electron energy Ti–Zr–V NEG coating for as-received, heated at 250 and 300 °C for one hour and then electron bombarded with a dose of 9×10^{-5}, 3.0×10^{-4}, 7.3×10^{-4}, 1.6×10^{-3}, and 5.3×10^{-3} C/mm². Source: From Wang 2016 [39], Loughborough University, March 2016.

5.13 NEG Coating Surface Resistance

In many particle accelerators the beam parameters could be affected by the beam pipe wakefield impedance. It is vital to understand how the wakefield impedance might vary due to various coatings on the surface of the vacuum chamber; this can be derived from surface resistance measurements. In general, the wakefield impedance of a chamber depends on the material used, its surface characteristics, and the chamber geometry [42, 43].

The bulk conductivity of two types of Ti–Zr–Hf–V NEG films (dense and columnar) was determined using contactless method with the 7.8 GHz three-choked test. This is achieved by calculating the surface resistance of NEG-coated samples from the results obtained with the test cavity measurements and fitting the experimental data to a standard theoretical model [44]. The main findings were as follows:

- The surface resistance of two types of NEG coating (dense and columnar) was investigated at 7.8 GHz. The bulk conductivity was obtained with the analytical model: $\sigma_c = 1.4 \times 10^4$ S/m for the columnar NEG coating and $\sigma_d = 8 \times 10^5$ S/m for the dense NEG coating; the latter is the same value as for the bulk target.
- The standard analytical models for the surface resistance of multilayer structures and the obtained experimental results for the films deposited on copper substrates are in good agreement.
- Based on the NEG conductivity values obtained at 7.8 GHz (and with the assumption that the classical Drude model is still valid), the beam energy loss

Table 5.2 Metals with the highest electric conductivity.

Metal	Ag	Cu	Au	Al	Mo	Zn	Ni	Fe
σ [S/m]	6.2×10^7	5.9×10^7	4.4×10^7	3.7×10^7	1.9×10^7	1.7×10^7	1.4×10^7	1.0×10^7

and the energy spread induced by resistive wall wakefield effects on the beam have been investigated and compared in vessels coated with either of the two types of film.

It must be noted that these results show, from one side, that the bulk conductivity can be varies in a wide range (by a factor of ~60 in these experiments) and depends on deposition conditions; from another side, they correspond to Ti–Zr–Hf–V films only, and other film compositions could result in a different range of values.

The NEG coating surface resistance can be reduced by two ways:

- Reducing the film thickness to 200–400 nm. This requires studying of all vacuum properties (sticking probability, pumping capacity, PSD, ESD, number of activation–saturation cycles), because most of the present evaluation has been done with 1-µ-thick NEG films.
- Developing high conductivity NEG coatings. This can be achieved by including high electrical conductivity metal into the film composition (see Table 5.2). This also requires studying of all vacuum properties for various film compositions and morphologies.

5.14 NEG at Low Temperature

Very little is known about the NEG coating behaviour at low temperature. The only result on PSD at low temperature reported in Ref. [45] showing that the H_2 dynamic pressure is reducing the beam vacuum chamber temperature from 300 to 90 K due to a combination of the two enhancing each other's effects:

- Sticking probability is equal to $\alpha \approx 6.5\times10^{-2}$ and remains constant between 200 and 300 K, but it is increasing when temperature decreases and reaches $\alpha \approx 0.32$ at $T = 90$ K.
- Reduction of PSD yield with temperature by a factor of approximately 4–6.

So, the results are promising, and more experimental data are needed to cover the lower temperature range interesting for the cryogenic vacuum systems: 3–90 K.

5.15 Main NEG Coating Benefits

The main benefits of using NEG coating in accelerator vacuum chamber can be summarised as follows:

- Reduced thermal outgassing.

- Reduced particle stimulated desorption: PSD, ESD, and ISD.
- Provides large distributed pumping speed.
- Can be activated at temperatures in the range of 150–180 °C, which is equal to or lower than the bakeout temperatures applied to uncoated vacuum chambers.
- Allows to use a simple shape of vacuum chamber (round or elliptic) and avoid antechamber for increasing vacuum conductance or for placing a distributed vacuum pump.
- Once activated, it does not require power supply for operation.
- It is a cost-effective solution, due to simplifying vacuum chamber shape, reducing the number and size of pumps required.

Example: In the ILC damping ring design, the vacuum specification in the arcs is required to reach an average dynamic pressure (i.e. a pressure in the presence of the beam) of $\langle P \rangle \leq 10^{-9}$ mbar after 100 A h beam conditioning. This specification can be met with various vacuum designs. Two possible working solutions can be compared [46]:

- A 316LN stainless steel vacuum chamber (without NEG coating) baked at 300 °C for 24 hours requires 200-l/s pumps every 5 m.
- A Ti–Zr–V-coated vacuum chamber baked at 180 °C for 24 hours requires 20-l/s pumps every 30 m. It is important to note that if the 20-l/s pumps are installed every 5 m, the specified pressure of 10^{-9} mbar in the presence of the beam can be met immediately after NEG activation.

Thus, a design based on the NEG-coated chamber requires less number of pumps of smaller size, less number of controllers and cables, less electric power consumption of the pump controller, and less electric power consumption for heating vacuum chamber (because the NEG activation temperature is lower than the required bakeout temperature without NEG coating). Therefore, the NEG coating allows to reach the same vacuum specifications at lower cost, or to meet these specification in a shorter time, or to reach specifications, which can't be met without NEG coating.

5.16 Use of NEG-Coated Vacuum Chambers

NEG coatings are already widely used for the design of particle accelerators. The ESRF located in Grenoble (France) was the first accelerator where the NEG-coated vacuum chambers were installed inside undulators. The ESRF operation conditions occasionally require interventions into some sections of the storage ring vacuum system with air vent followed by vacuum system bakeout and a two-week beam conditioning until the Bremsstrahlung radiation in the user's bunker was reduced to a safe level. Upon the replacement of conventional vacuum chamber with an NEG-coated one, the safe level was reached in three days [30]. After such a remarkable demonstration of benefit of using NEG-coated vacuum chamber, the NEG coating was used on more ESRF beamlines [47] and in many other machines.

SR sources experience similar problems; therefore, the ESRF experience was quickly adopted in other SR machines. Initially, it was carefully used in the most critical sections: in narrow wiggler and undulators vessels at ELETTRA in Italy [31], DLS in the United Kingdom [48], TLS in Taiwan [49], etc. Accumulated positive experience brought more confidence, and NEG coatings of storage ring was implemented for the first time in a Soleil design in France [50], followed by MAX-II and MAX-IV in Sweden [51–53].

Heavy ion accelerators require XHV vacuum conditions and NEG coating is a technology that could create such environment; therefore NEG coating was applied at SIS-18 at GSI in Germany [54], RHIC at BNL in the United States [55], and LEIR at CERN in Switzerland [56, 57].

High energy colliders also require UHV/XHV conditions. In addition the NEG coating provides another benefit: a lower photon and secondary electron yield than bare copper, aluminium, or stainless steel; therefore it is not only vacuum technology but also electron cloud mitigation solution. Thus, large sections of Large Hadron Collider (LHC) at CERN operating at room temperature [58] were coated with NEG. This is the longest NEG-coated vacuum chamber in the world. Studies of NEG coating for high energy colliders are also performed at KEK-B in Japan [40, 41]. The NEG coating was considered for the ILC damping rings [46] and the Interaction Region. It is an option for vacuum systems of the Future Circular Collider Studies [59].

References

1 Benvenuti, C. (1998). Non-evaporable getters: from pumping strips to thin film coatings. In: *Proceedings of EPAC'98*, 22–26 June 1998, Stockholm, Sweden, 200.
2 Benvenuti, C., Chiggiato, P., Cicoira, F., and Ruzinov, V. (1998). Decreasing surface outgassing by thin film getter coatings. *Vacuum* 50: 57.
3 Benvenuti, C. and Francia, F. (1988). Room-temperature pumping characteristics of a Zr–Al nonevaporable getter for individual gases. *J. Vac. Sci. Technol., A* 6 (4): 2528.
4 Benvenuti, C. and Francia, F. (1990). Room temperature pumping characteristics for gas mixtures of a Zr–Al nonevaporable getter. *J. Vac. Sci. Technol., A* 8 (5): 3864.
5 Benvenuti, C., Cazeneuve, J.M., Chiggiato, P. et al. (1999). A novel route to extreme vacua: the non-evaporable getter thin film coatings. *Vacuum* 53: 219–225.
6 Benvenuti, C., Chiggiato, P., Costa Pinto, P. et al. (2001). Vacuum properties of TiZrV non-evaporable getter films. *Vacuum* 60: 57–65.
7 Benvenuti, C., Chiggiato, P., Cicoira, F. et al. (2004). Vacuum properties of palladium thin film coatings. *Vacuum* 73: 139–144.
8 Benvenuti, C., Escudeiro Santana, A., and Ruzinov, V. (2001). Ultimate pressures achieved in TiZrV sputter-coated vacuum chambers. *Vacuum* 60: 279.

9 Johanek, V., Stara, I., and Matolın, V. (2002). Role of Pd–Al bimetallic interaction in CO adsorption and catalytic properties of bulk PdAl alloy: XPS, ISS, TDS, and SIMS study. *Surf. Sci.* 507–510: 92–98.

10 Matolın, V., Johanek, V., Stara, I. et al. (2002). XPS, ISS and TDS study of bimetallic interaction between Pd and Al: CO interaction with supported Pd/alumina catalysts. *Surf. Sci.* 507–510: 803–807.

11 Lozano, M. and Fraxedas, J. (2000). XPS analysis of the activation process in non-evaporable getter thin films. *Surf. Interface Anal.* 30: 623.

12 Matolın, V. and Johanek, V. (2002). Static SIMS study of TiZrV NEG activation. *Vacuum* 67: 177–184.

13 Drbohlav, J. and Matolin, V. (2003). Static SIMS study of Ti, Zr, V and Ti–Zr–V NEG activation. *Vacuum* 71: 323–327.

14 Fabik, S., Chab, V., Dudr, V. et al. (2004). Activation of binary Zr–V non-evaporable getters: a soft X-ray photoemission study of carbide formation. *Surf. Sci.* 566–568: 1246–1249.

15 Matolin, V., Dudr, V., Fabik, S. et al. (2005). Activation of binary Zr–V non-evaporable getters: synchrotron radiation photoemission study. *Appl. Surf. Sci.* 243: 106–112.

16 Scheuerlein, C. and Taborelli, M. (2002). Electron stimulated carbon adsorption in ultrahigh vacuum monitored by Auger electron spectroscopy. *J. Vac. Sci. Technol., A* 20: 93.

17 Prodromides, A.E., Scheuerlein, C., and Taborelli, M. (2001). Lowering the activation temperature of TiZrV non-evaporable getter films. *Vacuum* 60: 35–41.

18 Benvenuti, C., Chiggiato, P., Mongelluzzo, A. et al. (2001). Influence of the elemental composition and crystal structure on the vacuum properties of Ti–Zr–V non-evaporable getter films. *J. Vac. Sci. Technol., A* 19: 2925.

19 Anashin, V.V., Collins, I.R., Dostovalov, R.V. et al. (2004). Comparative study of photodesorption from TiZrV coated and uncoated stainless steel vacuum chambers. *Vacuum* 75: 155.

20 Malyshev, O.B., Middleman, K.J., Colligon, J.S., and Valizadeh, R. (2009). The activation and measurement of non-evaporable getter films. *J. Vac. Sci. Technol., A* 27 (2): 321–327. https://doi.org/10.1116/1.3081969.

21 Benvenuti, C., Chiggiato, P., Costa Pinto, P. et al. (2003). Influence of substrate coating temperature on the vacuum properties of Ti–Zr–V non-evaporable getter film. *Vacuum* 71: 307–315.

22 Malyshev, O.B., Valizadeh, R., Colligon, J.S. et al. (2009). Influence of deposition pressure and pulsed DC sputtering on pumping properties of Ti–Zr–V non-evaporable getter films. *J. Vac. Sci. Technol., A* 27: 521.

23 Malyshev, O.B. and Middleman, K.J. (2009). Test particle Monte-Carlo modelling of installation for NEG film pumping properties evaluation. *Vacuum* 83: 976–979.

24 Malyshev, O.B., Valizadeh, R., A. Hannah, et al. (2012). Optimization of non-evaporable getter coating for accelerator beam pipe 11th European Vacuum Conference, 20–24 September 2010, Salamanca, Spain.

25 Malyshev, O.B., Valizadeh, R., and Hannah, A. (2014). Pumping properties of Ti–Zr–Hf–V non-evaporable getter coating. *Vacuum* 100: 26–28.

26 Valizadeh, R., Malyshev, O.B., Colligon, J.S. et al. (2010). Comparison of Ti–Zr–V non-evaporable getter films deposited using alloy or twisted wire sputter-targets. *J. Vac. Sci. Technol., A* 28: 1404–1412.
27 Bender, M., Kollmus, H., Bellachioma, M.C., and Assmann, W. (2010). UHV-ERDA investigation of NEG coatings. *Nucl. Instrum. Methods Phys. Res., Sect. B* 268: 1986–1990.
28 Rossi, A. (2006). H_2 equilibrium pressure with a NEG-coated vacuum chamber as a function of temperature and H_2 concentration. In: *Proceedings of EPAC 2006*, Edinburgh, Scotland, 1444.
29 Chiggiato, P. and Kersevan, R. (2001). Synchrotron radiation-induced desorption from a NEG-coated vacuum chamber. *Vacuum* 60: 67.
30 Kersevan, R. (2000). Performance of a narrow-gap, NEG-coated, extruded aluminium vacuum chamber at the ESRF. In: *Proceedings of EPAC-2000*, 26–30 June 2000, Vienna, Austria, 2291.
31 Mazzolini, F., Miertusova, J., Pradal, F., and Rumiz, L. (2002). Performance of insertion device vacuum chambers at ELETTRA. In: *Proceedings of EPAC 2002*, 2–7 June 2002, Paris, France, 2577.
32 Ady, M., Chiggiato, P., Kersevan, R. et al. (2015). Photodesorption and electron yield measurements of thin film coatings for future accelerators. In: *Proceedings of IPAC2015*, Richmond, VA, USA, 3123.
33 Malyshev, O.B., Valizadeh, R., Hogan, B.T., and Hannah, A. (2014). Electron-stimulated desorption from polished and vacuum fired 316LN stainless steel coated with Ti–Zr–Hf–V. *J. Vac. Sci. Technol., A* 32: 061601. https://doi.org/10.1116/1.4897932.
34 Malyshev, O.B., Valizadeh, R., Jones, R.M.A., and Hannah, A. (2012). Effect of coating morphology on the electron stimulated desorption from Ti–Zr–Hf–V nonevaporable-getter-coated stainless steel. *Vacuum* 86: 2035.
35 Malyshev, O.B., Valizadeh, R., and Hannah, A. (2016). Pumping and electron-stimulated desorption properties of a dual-layer nonevaporable getter. *J. Vac. Sci. Technol., A* 34: 061602.
36 Malyshev, O.B. and Naran, C. (2012). Electron stimulated desorption from stainless steel at temperatures between −15 and +70 °C. *Vacuum* 86: 1363–1366.
37 Malyshev, O.B., Smith, A.p., Valizadeh, R., and Hannah, A. (2010). Electron stimulated desorption from bare and nonevaporable getter coated stainless steels. *J. Vac. Sci. Technol., A* 28: 1215. https://doi.org/10.1116/1.3478672.
38 Henrist, B., Hilleret, N., Scheuerlein, C., and Taborelli, M. (2001). The secondary electron yield of TiZr and TiZrV non-evaporable getter thin film coatings. *Appl. Surf. Sci.* 172: 95–102.
39 Wang, S. (2016). Secondary electron yield measurements of anti-multipacting surfaces for accelerators. PhD thesis, Chapter 5. Loughborough University, March 2016.
40 Suetsugu, Y., Kanazawa, K., Shibata, K. et al. (2005). First experimental and simulation study on the secondary electron and photoelectron yield of NEG materials (Ti–Zr–V) coating under intense photon irradiation. *Nucl. Instrum. Methods Phys. Res., Sect. A* 554: 92–113. https://doi.org/10.1016/j.nima.2005.08.061.

41 Suetsugu, Y., Kanazawa, K., Shibata, K., and Hisamatsu, H. (2006). Continuing study on the photoelectron and secondary electron yield of TiN coating and NEG (Ti–Zr–V) coating under intense photon irradiation at the KEKB positron ring. *Nucl. Instrum. Methods Phys. Res., Sect. A* 556: 399–409. https://doi.org/10.1016/j.nima.2005.10.113.

42 Zotter, B. and Kheifets, S. (1990). *Impedances and Wakes in High-Energy Particle Accelerators*. World Scientific.

43 Chao, A. (1993). *Physics of Collective Beam Instabilities in High-Energy Accelerators*. Wiley.

44 Malyshev, O.B., Gurran, L., Goudket, P. et al. (2017). RF surface resistance study of non-evaporable getter coatings. *Nucl. Instrum. Methods Phys. Res., Sect. A* 844: 99–107.

45 Anashin, V.V., Dostovalov, R.V., Krasnov, A.A., and Ruzinov, V.L. (2008). Adsorption and desorption properties of TiZrV getter film at different temperatures in the presence of synchrotron radiation. *J. Phys. Conf. Ser.* 100: 092027.

46 Malyshev, O.B. (2006). Vacuum systems for the ILC damping rings. EUROTeV-Report-2006-094, 24 November 2006. http://www.eurotev.org/reports__presentations/eurotev_reports/2006/e1019/EUROTeV-Report-2006-094.pdf.

47 Hahn, M., Kersevan, R., and Parat, I. (2006). Status report on the performance of NEG-coated chambers at the ESRF. In: *Proceedings of EPAC'06*, 26–60 June 2006, Edinburgh, UK, 1420.

48 Herbert, J.D., Malyshev, O.B., Middleman, K.J., and Reid, R.J. (2004). Design of the vacuum system for diamond, the UK 3rd generation light source. *Vacuum* 73: 219.

49 Wang, D.J., Chen, J.R., Hsiung, G.Y. et al. (1996). Vacuum chamber for the wiggler of the Taiwan Light Source at the Synchrotron Radiation Research Center. *J. Vac. Sci. Technol., A* 14: 2624.

50 Herbeaux, C., Béchu, N., and Filhol, J.-M. (2008). Vacuum conditioning of the SOLEIL storage ring with extensive use of NEG coating. In: *Proceedings of EPAC'08*, 23–28 June 2008, Genoa, Italy, 3696.

51 Hansson, A., Wallén, E., Berglund, M. et al. (2010). Experiences from nonevaporable getter-coated vacuum chambers at the MAX II synchrotron light source. *J. Vac. Sci. Technol., A* 28: 220.

52 Calatroni, S., Chiggiato, P., Costa Pinto, P. et al. (2013). NEG thin film coating development for the MAX IV vacuum system. In: *Proceedings of IPAC'13*, Shanghai, China, 3385.

53 Al-Dmour, E. and Grabski, M. (2017). The vacuum system of MAX IV storage rings: installation and conditioning. In: *Proceedings of IPAC2017*, Copenhagen, Denmark, 3468.

54 Bellachioma, M.C., Kurdal, J., Bender, M. et al. (2007). Thin film getter coatings for the GSI heavy-ion synchrotron upgrade. *Vacuum* 82: 435–439.

55 Weiss, D., He, P., Hseuh, H.C., and Todd, R.J. (2005). Development of NEG coating for RHIC experimental beamlines. In: *Proceedings PAC-2005*, Knoxville, Tennessee, USA, 3120.

56 Mahner, E., Hansen, J., Kuchler, D. et al. (2005). Ion-stimulated gas desorption yields of electropolished, chemically etched, and coated (Au, Ag, Pd, TiZrV) stainless steel vacuum chambers and St707 getter strips irradiated with 4:2 MeV/u lead ions. *Phys. Rev. Spec. Top. Accel. Beams* 8: 053201.
57 Mahner, E. (2007). The vacuum system of the Low Energy Ion Ring at CERN: requirements, design, and challenges. *Vacuum* 81: 727–730.
58 Bregliozzi, G., Baglin, V., Blanchard, S. et al. (2008). Achievement and evaluation of the beam vacuum performance of the LHC long straight sections. In: *Proceedings of EPAC'08*, 23–28 June 2008, Genoa, Italy, 3685.
59 Future Circular Collider: Conceptual Design Report 2018. FCC Study Office. CERN. https://fcc-cdr.web.cern.ch/ (Retrieved 15 January 2019)

6

Vacuum System Modelling

Oleg B. Malyshev

ASTeC, STFC Daresbury Laboratory, Keckwick Lane, Daresbury, Warrington, WA4 4AD Cheshire, UK

6.1 A Few Highlights from Vacuum Gas Dynamics

There are a number of books and chapters of vacuum handbooks devoted to the analysis of gas flows in complex vacuum systems, for example, Refs. [1, 2] as well as many other books, papers, lectures, talks, etc., see the references in Chapter 1. However, they are not fully relevant to the vacuum system of particle accelerators. The special things related to accelerators are that (i) the gas load is usually distributed and non-uniform, there could be several sources; (ii) the pumping system may include a number of various ultra high vacuum (UHV) pumps, which are either connected at pumping ports or distributed along a vacuum system; and (iii) the aspect ratio of length to a transverse dimension of vacuum chamber L/d is usually large. Therefore, it is often justified to use a one-dimensional (1D) model to analytically describe the accelerator vacuum systems. In this chapter, we will compare the advantages and disadvantages of an analytical 1D diffusion model (Knudsen–Clausing model) and three-dimensional (3D) test particle Monte Carlo (TPMC) simulations in a particle accelerator vacuum system design and then a possibility to combine the advantages of both in the process of vacuum system design optimisation.

The postulated physical conditions for these models are the following:

- The molecular flow is stationary.
- The particles are uniformly distributed over the vacuum chamber cross section (for a 1D model).
- The angular (or directional) velocity distribution of particles is uniform.
- Free molecular gas flow rate regime, i.e. the mean free path of particles, is much larger than the dimensions of the vacuum chamber.
- The particles are reflected from the walls with a cosine law (diffuse scattering)[1] [3].

---

1 A diffuse scattering assumes a complete accommodation of gas molecules in gas–surface interactions. It must be noted that this condition is incorrect for light species such as H_2, He, Ne, CH_4, and CO [3], deviates from the molecular reflectivity from the diffussed scattering, and affects gas flows at smooth, polished, and single crystal surfaces. However, the deviation from the diffused scattering is insignificant on rough industrial surfaces.

Vacuum in Particle Accelerators: Modelling, Design and Operation of Beam Vacuum Systems,
First Edition. Oleg B. Malyshev.
© 2020 Wiley-VCH Verlag GmbH & Co. KGaA. Published 2020 by Wiley-VCH Verlag GmbH & Co. KGaA.

It is expected that the reader is familiar with the basics of gas dynamics. The following Section 6.1 is not intended to provide a complete cover of vacuum gas dynamics, but it will highlight just a few most important gas dynamics laws and equations that can be used in the accelerator vacuum modelling.

6.1.1 Gas in a Closed Volume

6.1.1.1 Gas Density and Pressure

Particle accelerator vacuum specifications are related to a collision rate between the beam particles and gas molecules; therefore the characteristic that is really important for the beam–gas interaction is a number gas density. Let us consider a closed volume V containing N molecules at temperature T (see Figure 6.1). It is considered here that gas is in equilibrium state and fully thermalised with vacuum chamber walls: i.e. the gas temperature is exactly the same as the wall temperature.[2] Molecules are evenly distributed within this volume with a number gas density (or *gas density*), n, defined as

$$n\left[\frac{\text{molecules}}{\text{m}^3}\right] = \frac{N[\text{molecules}]}{V[\text{m}^3]}. \tag{6.1}$$

However, use of pressure is the most common in vacuum community. Pressure P and a number gas density n for each gas i are bonded with the ideal gas low formulated by Émile Clapeyron in 1834:

$$P_i = n_i k_B T. \tag{6.2}$$

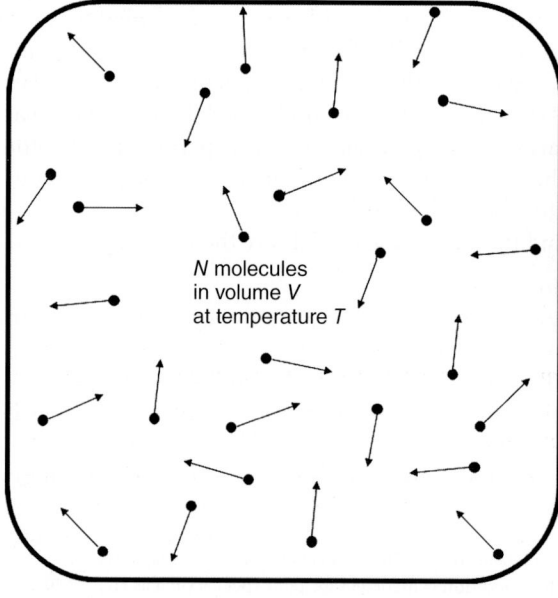

Figure 6.1 Closed volume containing V containing N molecules at temperature T.

---

2 In general, in gas flow dynamics, the gas temperature could be different from the vacuum chamber walls.

This equation is valid for the ideal gas when the rarefied gases are at temperatures far from the liquefying temperature of the species into consideration: i.e. molecules do not stick to each other creating clusters, drops, and a fog.

6.1.1.2 Amount of Gas and Gas Flow

The Boyle–Mariotte law (the seventeenth century empirical discovery) states that for a fixed amount of ideal gas at a fixed temperature, one can write

$$PV = \text{const.} \tag{6.3}$$

After introducing the absolute temperature scale at the end of nineteenth century, this law was generalised to the *ideal gas law* (or the general gas equation), which can be written as

$$PV = Nk_BT. \tag{6.4}$$

This allows to write an equation of the state of ideal gas:

$$\frac{PV}{T} = \text{const.} \tag{6.5}$$

The constants in Eqs. (6.3) and (6.5) are proportional to an amount of gas that can be expressed in SI units either as a mass in kilogrammes [kg] or as an amount of substance in moles [mol]. However, in various applications there are a number of other different units traditionally used for measuring an amount of gas. Thus, in vacuum for particle accelerators, the amount of gas is commonly measured either in a number of molecules, N [molecules], or as a product of pressure and volume, PV [Pa·m³] (or [mbar·l]). It must be noted that a number of molecules N of ideal gas do not depend on its temperature, while the amount of gas measured with pressure units (such as PV [Pa·m³]) must be reported *for a defined gas temperature*. There are a few commonly used defined temperatures:

- The standard temperature is $T = 0\,°C = 273.15\,K$.
- Room temperature: most commonly used are
  - $T = 20\,°C = 293.15\,K$ in SI
  - $T = 300\,K$ for numerical convenience
- An actual temperature of vacuum chamber.

Consequently, a gas flow rate Q can be measured in either [molecules/s] or [Pa·m³/s)] at a defined temperature. Similarly, the specific outgassing rate q_t and gas desorption flux per unit axial length q are defined in Table 6.1.

When the temperature of vacuum chamber (and therefore the temperature of gas) is not changing along the beam path and with time, the use of either pressure or gas density is equally acceptable. However, if the temperature varies, the use of gas density is preferable; see Chapter 7 for more details. Thus, all the following analysis will be mainly written for the gas density. However, when the pressure units have to be used, all the resulting equations and solutions can be easily converted to units by replacing n, Q, q, and q_t with P, Q^*, q^*, and q_t^*, respectively, using formulas shown in Table 6.1 at the defined temperature T.

Table 6.1 Units and their dimensions used in equations for gas density n and pressure P.

Parameter	Units	Parameter	Units	Formula
Amount of gas N	[molecules]	Amount of gas PV	[Pa·m³]	$PV = N k_B T$
Gas density n	[molecules/m³]	Gas pressure P	[Pa]	$P = n k_B T$
Local gas flow rate Q	[molecules/s]	Local gas flow rate Q^*	[Pa·m³/s]	$Q^* = Q k_B T$
Specific outgassing rate q_t	[molecules/(s·m²)]	Specific outgassing rate q_t^*	[Pa·m³/s]	$q_t^* = q_t k_B T$
Gas desorption flux per unit axial length q	[molecules/(s·m)]	Gas desorption flux per unit axial length q^*	[Pa·m²/s]	$q^* = q k_B T$

6.1.2 Total Pressure and Partial Pressure

The gas in a vacuum system is usually composed of several types of molecules (species). The pressure in a vacuum system with several gases is described by the Dalton law: 'The total pressure, P_{tot}, is the sum of all the partial pressure, P_i',

$$P_{tot} = \sum_i P_i. \tag{6.6}$$

6.1.3 Velocity of Gas Molecules

In the kinetic theory of gases, the molecular velocity distribution is determined by the Maxwell–Boltzmann equation (Eq. (3.41) in Ref. [4] or Eq. (2.37) in Ref [5]):

$$F_0\left(\frac{v}{v_{mp}}\right) = \frac{4}{\sqrt{\pi}} \frac{v^2}{v_{mp}^2} \exp\left(-\frac{v^2}{v_{mp}^2}\right), \tag{6.7}$$

where v_{mp} is the most probable velocity (also known as the most probable speed) defined as

$$v_{mp} = \sqrt{\frac{2RT}{M}} = \sqrt{\frac{2k_B T}{m}}. \tag{6.8}$$

In the following analysis the velocity of molecules is applied in many formulas. An average of absolute value of the velocity (also known as mean speed), $\bar{v}$, is represented by

$$\bar{v} = \sqrt{\frac{8RT}{\pi M}} = \sqrt{\frac{8 k_B T}{\pi m}}, \tag{6.9}$$

where M [kg/mol] is the molar mass of the gas, $R = 8.3145$ J/(K·mol) is the molar (or universal) gas constant, and T [K] is the absolute temperature.

The root-mean-square velocity, v_{rms}, is represented by

$$v_{rms} = \sqrt{\frac{3RT}{M}} = \sqrt{\frac{3k_B T}{m}}. \tag{6.10}$$

The ratio between these values are

$$\frac{\bar{v}}{v_{mp}} = \sqrt{\frac{4}{\pi}} = 1.128; \quad \frac{v_{rms}}{v_{mp}} = \sqrt{\frac{3}{2}} = 1.225; \quad \frac{v_{rms}}{\bar{v}} = \sqrt{\frac{8}{3\pi}} = 1.085. \tag{6.11}$$

The normalised molecular velocity distribution is shown in Figure 6.2.

Table 6.2 shows the molar and molecular mass of most common gases present in an accelerator vacuum chamber and their mean speed at LHe, LN$_2$, and room temperatures.

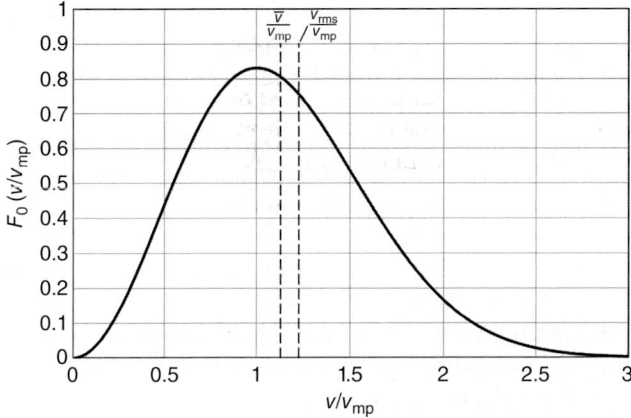

Figure 6.2 The normalised molecular velocity distribution.

Table 6.2 Molar and molecular mass of most common gases in an accelerator vacuum chamber and their mean speed at LHe, LN$_2$, and room temperatures.

Gas	M [g/mol]	m [kg]	$\bar{v}$ [m/s] at $T = 20\,°C$	$\bar{v}$ [m/s] at $T = 77.2\,K$	$\bar{v}$ [m/s] at $T = 4.17\,K$
H_2	2.0159	3.3474×10^{-27}	1755	900.5	209.3
He	4.0026	6.6465×10^{-27}	1245	639.0	148.5
CH_4	16.0425	2.6639×10^{-26}	622.0	319.2	74.19
H_2O	18.0153	2.9915×10^{-26}	587.0	301.2	70.01
CO	28.0101	4.6512×10^{-26}	470.7	241.6	56.14
N_2	28.0134	4.6517×10^{-26}	470.7	241.6	56.14
O_2	31.9988	5.3135×10^{-26}	440.4	226.0	52.53
Ar	39.9480	6.6335×10^{-26}	394.2	202.3	47.01
CO_2	44.0095	7.3079×10^{-26}	375.5	192.7	44.79

6.1.4 Gas Flow Rate Regimes

Gas molecules can interact with vacuum chamber walls and with other gas molecules in a volume. The *equivalent free path* of gas molecules is defined as

$$\lambda = \frac{\mu v_{mp}(T_s)}{P}, \qquad (6.12)$$

where P is the local pressure of the gas, μ is the gas viscosity, and T_s is the surface temperature. The equivalent free path is a more direct and more accurate value than the mean free path used earlier in the rarefied gas dynamics, see Eq. (5.1) in Ref. [4], Eq. (1.32) in Ref. [5] and Eq. (5) in Ref. [6].

A dimensionless parameter known as the Knudsen number, Kn, is used to describe the flows of rarefied gases. The Knudsen number is defined as a ratio between the equivalent free path, λ, and a characteristic vacuum chamber size,[3] d:

$$Kn = \frac{\lambda}{d}. \qquad (6.13)$$

Gas flow rate regimes are defined in respect to the Knudsen number as shown in Table 6.3.

The molecular free regime is most interesting for modelling and design of vacuum systems of particle accelerators, because most of vacuum systems are designed to UHV/XHV vacuum specification (with an exception of initial pumping down of accelerator vacuum chamber and a few specialised devices such as gas targets and ion sources). Thus, the following analysis in this book is related to the free molecular flow regime only.

Since molecules are travelling between the vacuum chamber walls without colliding each other, a behaviour of each molecule does not depend on the others; thus each gas can be modelled independently.

Table 6.3 Gas flow rate regimes.

Gas flow rate regime	Behaviour	λ and d	Kn
Viscous	Molecule–molecule collisions dominate	$\lambda \ll d$	$Kn < 0.01$
Slip	molecule–molecule collisions dominate but molecule–wall collisions are important boundary condition	$\lambda < d$	$0.01 < Kn < 0.1$
Transitional	Molecule–molecule collisions and molecule–wall collisions are equally important for the gas flow	$\lambda \sim d$	$0.1 < Kn < 10$
Free molecular	Molecule–wall collisions dominate behaviour	$\lambda \gg d$	$Kn > 10$

---

3 For circular tubes the characteristic size is its diameter, while for non-circular tubes it can be calculated as $d = 4A_c/F$.

6.1.5 Pumping Characteristics

In a volume with a gas density n, gas molecules are hitting vacuum chamber walls with *an impingement rate*, J, defined as the number of molecular hits per second per unit of surface area, see (Figure 6.3). When gas is in equilibrium state, an impingement rate is calculated for a gas with a molecular mass M and at a defined temperature T with the following formula:

$$J(M,T)\left[\frac{\text{molecules}}{\text{s}\cdot\text{m}^2}\right] = \frac{1}{4}n\left[\frac{\text{molecules}}{\text{m}^3}\right]\bar{v}(M,T)\left[\frac{\text{m}}{\text{s}}\right]. \quad (6.14)$$

Note: In the following text, all equations are written for a single gas with defined M and T; thus the notations (M, T) are omitted.

An impingement rate could be calculated not only for real surfaces but also for any virtual surface. In the latter case, since a virtual surface has two sides, the impingement rates could be calculated in two opposite directions. In static conditions, a number of molecules passing any virtual surface per unit area in each direction is equal to J and the total number of molecules passing any virtual surface per unit area is equal to $2J$, and a net gas flow rate is equal to zero. However, in the presence of gas flow, the impingent rates at two sides of a virtual surface could be different and a net gas flow rate is not equal to zero.

A molecular gas flow rate Q_{in} [molecules/s] towards vacuum chamber walls with surface area A (i.e. the number of molecules hitting this surface every second) is equal to

$$Q_{in}(M,T) = J(M,T)A = nA\frac{\bar{v}(M,T)}{4}. \quad (6.15)$$

In case of using terms of pressure P, a gas flow rate Q^*_{in} [Pa·m/s] is equal to

$$Q^*_{in}(M,T) = PA\frac{\bar{v}(M,T)}{4}. \quad (6.16)$$

A sticking probability α of a sorbing surface can be defined as

$$\alpha = 1 - \frac{Q_{out}}{Q_{in}}, \quad (6.17)$$

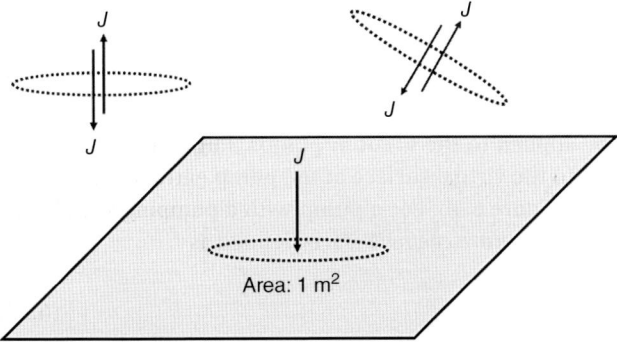

Figure 6.3 Impingement rates J at real and virtual surfaces.

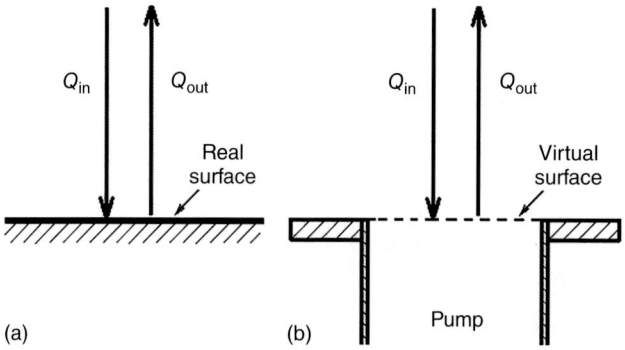

Figure 6.4 Incoming and outcoming gas flows at (a) real and (b) virtual surfaces.

where Q_{out} [molecules/s] is a molecular gas flow rate from vacuum chamber walls; see Figure 6.4a.

Note: In this book we distinguish between sticking coefficient (local parameter) and sticking probability (averaged parameter). A sticking coefficient is defined in the surface science in respect to a well-defined surface (for example, a crystal with certain lattice and its orientation to the surface); thus a sticking coefficient may strongly depend on an incident angle, surface material, and morphology, and reflected molecules may have one or a few discrete preferred directions. In contrast, the word 'probability' is highlighting a statistical nature of sticking probability defined in relation to a real surface of vacuum chamber walls with a surface roughness, various grain sizes, and lattices and orientations. It also implies a diffuse (cosine low) reflectivity of non-absorbed molecules.

An ideal pump is a surface with a sticking probability $\alpha = 1$ (i.e. all interacting molecules stick and remain at the surface for an infinity large time, i.e. sorbed). An equation *for an ideal volumetric pumping speed* of this surface can be written as follows:

$$S_{id} = A\frac{\bar{v}}{4}. \tag{6.18}$$

When sticking probability $\alpha \neq 1$, then the volumetric pumping speed of sorbing wall, S, is

$$S = \alpha A\frac{\bar{v}}{4} = \alpha S_{id}. \tag{6.19}$$

Similar analysis can be applied to the vacuum pumps. Pumping speed of the lumped pumps is defined to the virtual surface at the pump entrance (usually to the connecting flange); see Figure 6.4b. For a pump with a pumping speed S_p, a pump capture efficiency or *a capture coefficient* ρ is defined as

$$\rho = \frac{S_p}{S_{id}} = \frac{4S_p}{A\bar{v}}, \tag{6.20}$$

where A is an area of the virtual surface at the pump entrance hole.

6.1.6 Vacuum System with a Pump

Let us consider a simple vacuum system schematically show in Figure 6.5. Pressure, P [Pa], in this vessel is defined by the total gas load, Q^* [Pa·m³/s], and total pumping speed, S [m³/s], with a so-called *Vacuum Plumber's Formula 1*:

$$P = \frac{Q^*}{S}. \tag{6.21}$$

Similarly, the gas density, n [molecules/m³], is defined by the total gas load, Q [molecules/s], and total pumping speed, S [m³/s]:

$$n = \frac{Q}{S}. \tag{6.22}$$

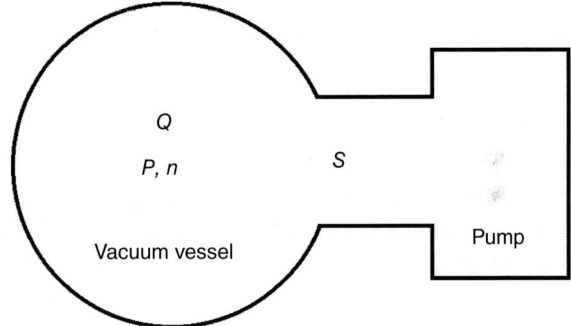

Figure 6.5 A simple vacuum system with the total gas load, Q, and total pumping speed, S.

6.1.7 Vacuum Conductance

Let us consider two large vacuum volumes with different pressures P_1 and P_2 connected with a tube with a characteristic cross-sectional dimension (for example, a tube diameter) much smaller than the characteristic dimensions of this volume (see Figure 6.6). The gas flow rate Q^* between these volumes depends on pressure difference $\Delta P = P_1 - P_2$ and vacuum conductance of connecting tube, which is defined with *Vacuum Plumber's Formula 2*:

$$U = \frac{Q^*}{\Delta P} = \frac{Q}{\Delta n}. \tag{6.23}$$

Vacuum conductance depends on tube dimensions, molecular mass, and temperature of gas. In the viscous and transitional gas flow rate regimes, the vacuum conductance also depends on pressure. In the molecular flow regime, which is

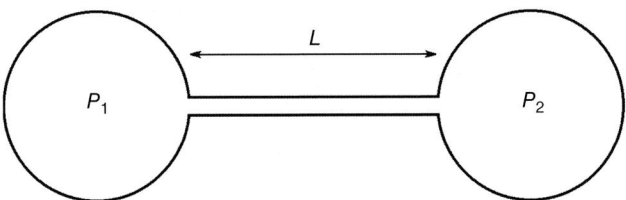

Figure 6.6 Two vacuum volumes with a connecting tube.

the most interesting for particle accelerator application, the vacuum conductance does not depend on pressure. Analytical and empirical formulas for various cross sections of vacuum tubes are well described in literature (see Refs. [1–14] in Introduction). More specific conductance can be found using TPMC (see Section 5.3) or a method of angular coefficients. Here, we will show a few examples to explain how vacuum conductances are used in the modelling of accelerator vacuum system.

6.1.7.1 Orifice

The simplest example of the connecting tube is a tube with zero length, or orifice (see Figure 6.7). In the isotropic gas in the molecular flow regime, the number of molecules hitting a vacuum vessel wall (or crossing from one side any virtual wall) of area A is defined by Eqs. (6.15) and (6.16). In the application to an orifice, it gives a simple formula for its vacuum conductance:

$$U_o = \frac{Q^*_{12} - Q^*_{21}}{P_1 - P_2} = A\frac{\bar{v}}{4} = S_{id}; \qquad (6.24)$$

6.1.7.2 Vacuum Conductance of Long Tubes

Vacuum conductance of tubes were studied by M. Knudsen, based on an earlier work by M. Smoluchowski. It was shown that vacuum conductance of a tube with a constant cross section increases with a cross-sectional area, A, and reduces with a tube length, L. However, an accurate derivation of a formula even for a simple case of a long circular tube is quite complicated.

A number of formulas for tubes with a few various cross sections and shapes are published in literature (see references in Chapter 1). However, all the variety of these formulas is usually limited to a few simple cross sections such as circular, elliptic, and rectangular. It is also important to remember that these formulas are approximate ones and, in some cases, may lead to significant errors.

Thus, the vacuum conductance of a required cross section could be obtained by a few methods:

- The vacuum conductance for any shape of vacuum chamber can accurately be calculated with TPMC method (see, for example, [7, 8]) and will be discussed in this chapter.

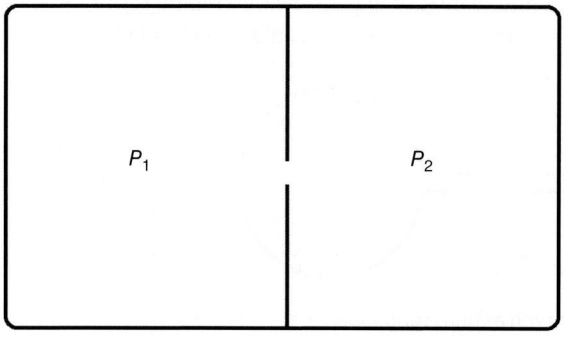

Figure 6.7 Two vacuum volumes with a connecting orifice.

- Results for tubes with a few simple cross sections (e.g. circular, elliptic, and rectangular) can obtained by the integro-moment method based on the linearised Bhatnagar–Gross–Krook (BGK) equation (see, for example, Chapter 12 in Refs. [5, 6, 9–13]).
- These results for discrete geometrical parameters can be fitted with a formula allowing to calculate the result for various parameters.

A vacuum conductance of the tube with a constant *circular cross section* with a diameter d can be calculated with a formula that provides an accuracy of 0.03% in comparison to the numerical results published in Ref. [11]:

$$U = \frac{\pi d^3 \bar{v}}{12L} = 0.2618 \frac{d^3 \bar{v}}{L}. \tag{6.25}$$

A vacuum conductance for an *elliptic cross section* with the major axis b and minor axis a (see Figure 6.8) can be calculated with a formula fitting the results in Ref. [11] as follows:

$$U = 0.52375 \frac{a^2 b^2}{(a+b)\left(1 - \dfrac{\lambda}{1.8 + 2\sqrt{1 - 1.03\lambda}}\right)} \frac{\bar{v}}{L}, \text{ where } \lambda = \left(\frac{a-b}{a+b}\right)^2. \tag{6.26}$$

In comparison to the numerical results published in Ref. [11], this formula is correct for $a = b$, which provides an accuracy $\pm 1.2\%$ for $1 < a/b \leq 100$.

For a tube with a *rectangular cross section* with sides a and b, the vacuum conductance can be described with a formula fitting the results in Ref. [12]:

$$U = 0.74327 \frac{b^4}{(a+b)} \left(\frac{a}{b}\right)^{2.13} \frac{\bar{v}}{L}. \tag{6.27}$$

In comparison to numerical results published in Ref. [12], this equation is correct for $a = b$, which provides with an accuracy $\pm 8\%$ for $1 < a/b \leq 100$.

6.1.7.3 Vacuum Conductance of Short Tubes

Vacuum conductance of short channel can be calculated with a formula:

$$\frac{1}{U} = \frac{1}{U_o} + \frac{1}{U_t} \tag{6.28}$$

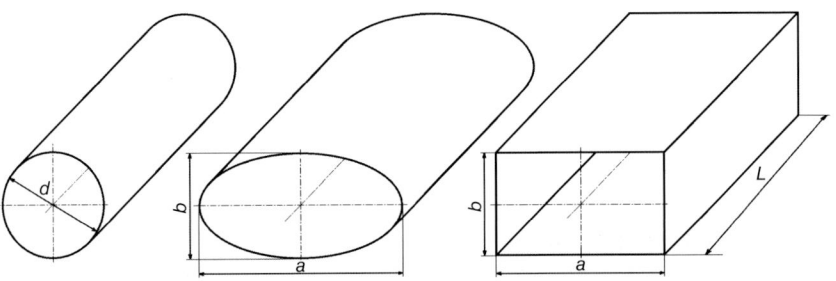

Figure 6.8 Circular, elliptic, and rectangular tubes and their geometrical characteristics.

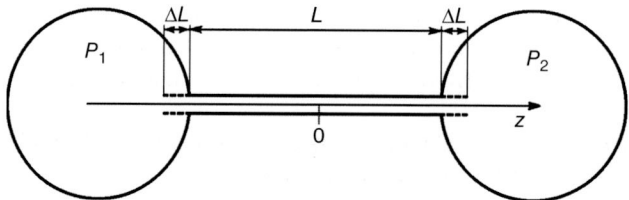

Figure 6.9 An effective length of short channel: $L_{eff} = L + 2\Delta L$.

This is an approximate formula that gives up to 10% accuracy for a circular cross section [1] and can be written as

$$U = \frac{\pi d^3 \bar{v}}{16d + 12L}. \tag{6.29}$$

Equation (6.28) for short channels can also be written in a different form:

$$U = U_t(L_{eff}), \tag{6.30}$$

where L_{eff} is an effective length of short channel: $L_{eff} = L + 2\Delta L$ [14]. For a circular cross section, $\Delta L = 0.63d/L$ [14]. A value of ΔL calculated from approximate Eq. (6.29) gives 5.8% a higher result: $\Delta L = 2d/3L \approx 0.67d/L$.

Using Eq. (6.30) for tube conductance, gas density along the channel axis z between two large vessels with pressures P_1 and P_2 (see Figure 6.9) can be calculated more accurately:

$$P(z) = \begin{cases} P_1 & \text{for } z \leq -0.5L_{eff} \\ P_1 + \dfrac{P_2 - P_1}{L_{eff}}(z + 0.5L_{eff}) & \text{for } -0.5L_{eff} < z < 0.5L_{eff}, \\ P_2 & \text{for } z \geq 0.5L_{eff} \end{cases} \tag{6.31}$$

Equation (6.31) accounts that pressure outside the tube ends is not equal to P_1 or P_2 for a distance ΔL from the tube ends.

6.1.7.4 Serial and Parallel Connections of Vacuum Tubes

A vacuum conductance of a few tubes connected in series (see Figure 6.10a) is calculated as

$$\frac{1}{U} = \sum_i \frac{1}{U_i}. \tag{6.32}$$

While in the case of parallel connection of vacuum tubes (see Figure 6.10b,c), it is

$$U = \sum_i U_i \tag{6.33}$$

6.1.8 Effective Pumping Speed

When a pump is connected to a vacuum vessel with a connecting tube, the pumping efficiency is reduced due to vacuum conductance U of the tube, the resulting

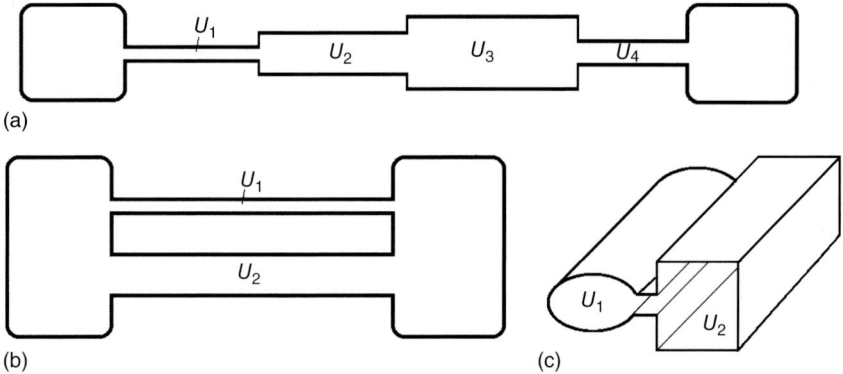

Figure 6.10 (a) Serial connection of a few vacuum tubes, (b) parallel connections of two vacuum tubes, and (c) increasing of longitudinal vacuum conductance of beam chamber, U_1, with a large antechamber conductance U_2.

pumping speed in respect to the vacuum vessel is called an effective pumping speed and defined as

$$\frac{1}{S_{eff}} = \frac{1}{S_p} + \frac{1}{U}. \tag{6.34}$$

In a case when a pump is connected with a few tubes (with vacuum conductances U_i) connected in series, the above equation can be rewritten in more general form:

$$\frac{1}{S_{eff}} = \frac{1}{S_p} + \sum_i \frac{1}{U_i}. \tag{6.35}$$

A pump effective capture coefficient ρ_{eff} could also be specified for such assembly of a pump with a connecting tube as follows:

$$\rho_{eff} = \frac{S_{eff}}{S_{id}}. \tag{6.36}$$

A normalised effective pumping speed, S_{eff}/S_p, as a function of ratio between a conductance of pumping port and a pump pumping speed, U/S_p, is shown in Figure 6.11. One can see the following:

- The effecting pumping speed is almost equal to the pump pumping speed when the vacuum conductance is much greater than the pump pumping speed: $S_{eff}(U \gg S_p) = S_p$. The condition $U \gg S_p$ provides the most efficient use of a vacuum pump. If necessary, the effecting pumping speed can be increased a few times by connecting a pump with a larger pumping speed.
- When the vacuum conductance is equal to the pump pumping speed, the effective pumping speed is reduced to 50% of the pump pumping speed: $S_{eff}(S_p = U) = 0.5S_p = 0.5U$. The condition $U \approx S_p$ provides a reasonable balance between the size of the pump and available vacuum conductance. If necessary, the effecting pumping speed can be increased by connecting a pump with a larger pumping speed, but not greater than a factor of 2: for example, replacing a pump with another one with $S_p = 2U$ will increase the effective

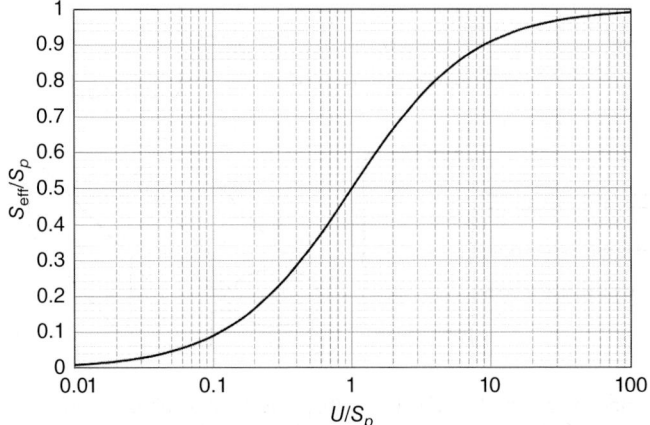

Figure 6.11 A normalised effective pumping speed, S_{eff}/S_p, as a function of ratio between a conductance of pumping port and a pump pumping speed, U/S_p.

pumping speed increased by a factor of 1.33: $S_{eff}(S_p = 2U) = 0.33S_p = 0.67U$; for $S_p = 9U$ the effective pumping speed increases by a factor of 1.95 only: $S_{eff}(S_p = 9U) = 0.1S_p = 0.9U$.

- When the vacuum conductance is much lower than the pump pumping speed, the effective pumping speed is defined by the vacuum conductance and does not depend on the pump pumping speed: $S_{eff}(U \ll S_p) = U$. This condition, when the effective pumping speed does not depend on the pump pumping speed, is called a *conductance-limited* pumping. This is the least efficient way of pumping, not economic use of pump and should be avoided if possible. Possible ways to provide a larger effective pumping speed are related to increase the vacuum conductance by shorter length, larger cross section, or increased number of connecting vacuum tubes connected in parallel.

6.2 One-Dimensional Approach in Modelling Accelerator Vacuum Systems

The diffusion model is an analytical 1D approach. It uses global and averaged parameters: pressure, pumping speed, uniform molecular velocity distribution, temperature, etc. In many cases accuracy is within 0.1–10%; however, its applicability should always be checked because in some cases (e.g. vacuum chamber with sorbing walls, beaming effect) the error may be times or even orders of magnitude. The parameters used for gas dynamics modelling are defined in respect to a longitudinal coordinate z. The non-uniform distribution of gas flow, pressure or gas density, temperature of gas or vacuum chamber walls, and gas desorption or injection is not considered at this approach; therefore the calculated results could be not accurate.

6.2.1 A Gas Diffusion Model

This approach is based on continuous flow fluid dynamics along the vacuum chamber with the uniform cross section A and the specific vacuum conductance u, which is defined as a vacuum conductance per unit of vacuum chamber length:

$$u = UL. \tag{6.37}$$

Here, L is a total length of vacuum chamber of the same cross section. Thus, for example, a specific vacuum conductance of a circular vacuum chamber can be calculated as

$$u = \frac{\pi d^3 \bar{v}}{16\frac{d}{L} + 12}. \tag{6.38}$$

It is quite common in accelerators when a length of vacuum chamber of the same cross section is much larger than its diameter: $L \gg d$. When this condition is met, Eq. (6.38) simplifies to

$$u = \frac{\pi d^3 \bar{v}}{12}. \tag{6.39}$$

Let us consider a part of a vacuum chamber between longitudinal coordinates z_i and z_{i+1} with a length $\Delta z = z_{i+1} - z_i$ and a volume $V_{\Delta z} = A \Delta z$. A gas density inside this volume changes with time as a sum of all incoming and outcoming molecules in three terms: the gas desorbed from the walls, $Q_i = q_i \, \Delta z$; the gas pumped at the walls, $Q_{p,i} = C_i \, n_i = c_i \, \Delta z \, n_i$; and gas diffusion form and to the neighbouring elements, $Q_{d,i} = U_{\Delta z}(n_i - n_{i-1})$ and $Q_{d,i+1} = U_{\Delta z}(n_{i+1} - n_i)$, correspondingly (see Figure 6.12). Note that $U_{\Delta z} = u/\Delta z$.

It is important to pay attention on units and dimension of used parameters: gas desorption flux per unit axial length q is measured in [molecules/(s·m)], a distributed pumping speed per unit axial length c is measured in in [m^2/s], and a specific vacuum chamber conductance per unit axial length u is measured in [m^4/s].

The gas balance in a vacuum chamber can be written as follows:

$$V_{\Delta z} \frac{dn_i}{dt} = Q_i - C_i n_i + U_{\Delta z}(n_{i-1} - n_i) - U_{\Delta z}(n_i - n_{i+1}) \tag{6.40}$$

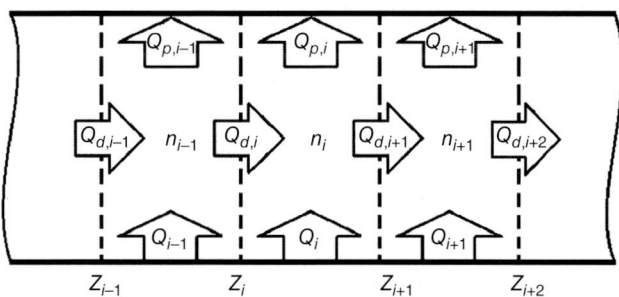

Figure 6.12 Gas balance in a vacuum chamber volume between longitudinal coordinates z_i and z_{i+1}.

or

$$A\Delta z \frac{dn_i}{dt} = q\Delta z - c_i n_i \Delta z + \frac{u}{\Delta z}(n_{i-1} - n_i) - \frac{u}{\Delta z}(n_i - n_{i+1}); \quad (6.41)$$

which leads to the second order differential equation:

$$A\frac{\partial n}{\partial t} = q - cn + u\frac{\partial^2 n}{\partial z^2}; \quad (6.42)$$

where gas desorption q in may consists of a few main sources – thermal desorption, photon stimulated desorption (PSD), and electron stimulated desorption (ESD):

$$q = \eta_t F + \eta_\gamma \Gamma + \eta_e \Theta; \quad (6.43)$$

where η_t is a specific thermal outgassing rate, F is vacuum chamber cross section circumference (or surface area per unit axial length of vacuum chamber), η_γ is PSD yield, Γ is a photon flux per unit axial length, η_e is ESD yield, and Θ is electron flux per unit axial length.

In this model, distributed pumping per unit axial length c can describe a few different types of distributed pumps:

– An non-evaporable getter (NEG)-coated vacuum chamber with sticking probability α:

$$c = \alpha F \quad (6.44)$$

– A vacuum chamber with an antechamber equipped with a distributed pump (such as getter strip, in-built sputter ion pumps, or cryopumping) with a pumping speed per unit axial length s_{pump} [m²/s] and a vacuum conductance of a slot between a beam chamber and antechamber per unit axial length u_{slot} [m²/s]:

$$\frac{1}{c} = \frac{1}{s_{\text{pump}}} + \frac{1}{u_{\text{slot}}}, \quad (6.45)$$

– A lumped pump with a pumping speed, S_p [m³/s], as a part of a vacuum chamber with a length equal to a z-axis width of a pumping port connected to it, L_p [m], and a vacuum conductance, U_p [m³/s], between the pump and the beam chamber:

$$\frac{1}{c} = \left(\frac{1}{S_p} + \frac{1}{U_p}\right) L_p = \frac{L_p}{S_{\text{eff}}}, \quad (6.46)$$

The left-hand side term of Eq. (6.42) is significant in accelerators mainly during beam injection, topping up, and dumping. Most of the times in accelerators, the beam is slowly changing, providing a so-called quasi-equilibrium condition described as

$$q - cn + u\frac{\partial^2 n}{\partial z^2} \gg A\frac{\partial n}{\partial t}.$$

When the quasi-equilibrium condition is met (which is reached within a few milliseconds after the beam injection, top-up, or dumping), one can consider $A\partial n/\partial t \approx 0$, and then Eq. (6.42) can be simplified to

$$u\frac{d^2 n}{dz^2} - cn + q = 0. \quad (6.47)$$

This second-order differential equation for the function $n(z)$ has two solutions:

$$\begin{cases} n(z) = -\dfrac{q}{2u}z^2 + C_{1a}z + C_{2a} & \text{for } c = 0, \\ n(z) = \dfrac{q}{c} + C_{1b}e^{\sqrt{\frac{c}{u}}z} + C_{2b}e^{-\sqrt{\frac{c}{u}}z} & \text{for } c > 0; \end{cases} \qquad (6.48)$$

where the constants C_1 and C_2 depend on the boundary conditions. These formulas for a gas density along a vacuum chamber are the main tools for the following analysis. The solutions for $c = 0$ corresponds to the vacuum chamber with no pumping properties and the solutions for $c > 0$ corresponds to a vacuum chamber with a distributed pumping. It is worth calculating a term $\omega = \sqrt{c/u}$, as it may simplify the following calculations. For example, in the case of a circular tube,

$$\omega = \sqrt{\dfrac{c}{u}} = \dfrac{\sqrt{3}}{d}. \qquad (6.49)$$

Note 1: In this book the model equations are written in SI units, while the units used in practical vacuum technology are often a mixture of various standards: pressure in mbar and Torr instead of Pa, pumping speed in l/s instead of m^3/s, gas flow rate in sccm and Torr l/s instead of Pa m^3/s or kg/s.

Note 2: These equations are written for fixed parameters q, u, and c: i.e. they do not depend on a coordinate z. Although it is possible to deliver the equations for variable q, u, and c, implementing this would complicate the analysis unnecessarily. The main advantage of 1D model in comparison to be more accurate as 3D models is its simplicity, which is applied for a quick optimisation of vacuum systems in the early design phase. Thus there is no reason to complicate inaccurate 1D model further. Thus, within an accuracy of 1D model, the variation of q, u, and c within 10–20% can be considered 'practically constant' and the model can be applied to a part of a vacuum chamber that meets this criteria.

6.2.2 A Section of Accelerator Vacuum Chamber in a Gas Diffusion Model

In this model, an accelerator vacuum chamber is fragmented along the beam path on N elements. Every ith element ($i = 1, \ldots, N$) of length L_i lying between longitudinal coordinates z_{i-1} and z_i (see Figure 6.13) will be described by Eqs. (6.48)

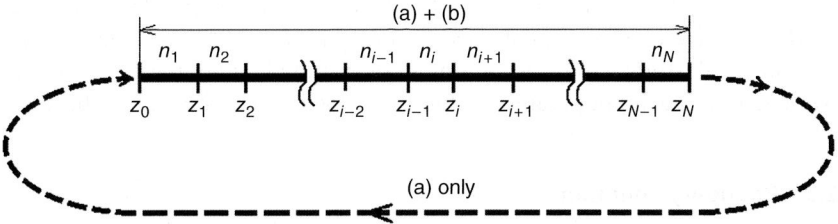

Figure 6.13 A 1D model layout for (a) a ring or repeatable sections with the closed loop conditions and (b) linear accelerators with the open end conditions.

with two unknowns $C_{1,i}$ and $C_{2,i}$ and three parameters: q, u, and c. This means that the gas flow rate q_i, the vacuum chamber shape (defining its specific vacuum conductance u_i), and distributed pumping c_i should not change within ith element; however each of these parameters could be the same or different in the neighbour components.

This suggest that the natural borders between the elements should be at the locations of

- a change of vacuum chamber cross section (for example, from a round beam pipe to a chamber with an antechamber)
- a change in gas desorption along the beam vacuum chamber (for example, a jump in synchrotron radiation (SR) intensity due to SR absorber or location of dipoles)
- a change in pumping, where the neighbour components could be of any combination of pumping properties: either one with $c = 0$ and another with $c > 0$, or both with $c = 0$, or both with $c > 0$. The latter could have two cases: $c_i = c_{i+1}$ or $c_i \neq c_{i+1}$.

In practice, one needs to make additional borders, because in a real accelerator vacuum chamber the values of q, u, and c may continuously vary with a coordinate z. To apply this method, one can choose the lengths of each model element of vacuum chamber to meet a criterion that q, u, and c vary (within the element) less than a certain limit (for example, $\pm 5\%$) and then to use their average values q_i, u_i, and c_i within the element i. Thus, the method allows varying the desorption q_i, the vacuum conductance u_i, and pumping speed c_i in a wide range of values. The only additional thing to consider is that the shorter the length of elements, the larger the number of elements N will be.

A modelled vacuum chamber is described now as a function of a coordinate z with N equations with $2N$ unknowns ($C_{1,i}$ and $C_{2,i}$):

$$n_i(z) = \begin{cases} -\dfrac{q_i}{2u_i}z^2 + C_{1,i}z + C_{2,i} & \text{if } c_i = 0, \\ \dfrac{q_i}{c_i} + C_{1,i}e^{\sqrt{\frac{c_i}{u_i}}z} + C_{2,i}e^{-\sqrt{\frac{c_i}{u_i}}z} & \text{if } c_i > 0; \end{cases} \quad z_{i-1} \leq z \leq z_i$$

$$\text{for } i = 1, 2, \ldots, N. \tag{6.50}$$

The solutions for $C_{1,i}$ and $C_{2,i}$ can be found when the boundary conditions are defined.

An example of splitting an accelerator vacuum chamber and choosing the boundaries between the elements is show in Figure 6.14. Such a layout is quite common for SR storage rings (for example, Diamond Light Source [DLS] at RAL in Didcot, UK), the input parameters for it may look as it is shown in Table 6.4.

6.2.3 Boundary Conditions

The boundary conditions between ith and $(i+1)$th elements are defined as

$$n_i(z_i) = n_{i+1}(z_i) \quad \text{and} \quad u_i \partial n_i(z_i)/\partial z = u_{i+1} \partial n_{i+1}(z_i)/\partial z. \tag{6.51}$$

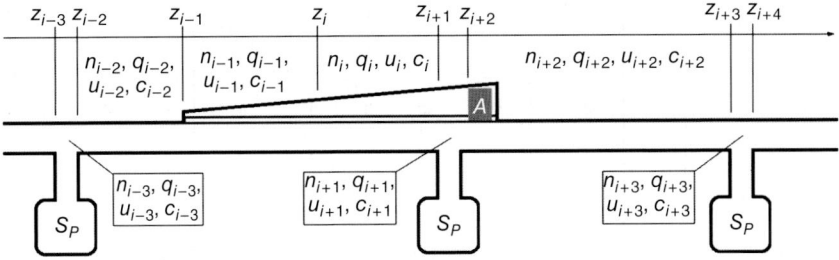

Figure 6.14 An example of splitting an accelerator vacuum chamber and choosing the boundaries between the elements. Here, A is an SR absorber placed in an antechamber and S_p is a pumping speed of a lumped pump.

The boundary conditions for two extremes z_0 and z_N can be defined by two different ways: a closed loop or an open end.

The *closed loop* boundary conditions can be used for small machines or if a model vacuum chamber is a repeatable section of the accelerator; in this case the condition at the beginning of the next sections should be the same as at the beginning of the current section. This is common for storage rings, dumping rings, and many circular machines. Thus the boundary conditions can be written as follows:

$$n_1(z_0) = n_N(z_N) \quad \text{and} \quad u_1 \partial n_1(z_0)/\partial z = u_N \partial n_N(z_N)/\partial z. \tag{6.52}$$

However, more commonly the boundary conditions at two extremes are not well defined; the *open end* boundary conditions based on reasonable assumption should be applied in this case. This assumption should cause minimum influence on average pressure along the vacuum chamber and maximum local pressure (pressure bumps). The main assumption is that there are locations in the vacuum system where the net gas flow rate along the vacuum chamber is equal to zero or at least negligible compared with that in other locations. The most likely location for such conditions is in the middle of the pumping port. Therefore, the modelled vacuum chamber should have pumping elements at both extremes: i.e. $c_1 > 0$ and $c_N > 0$. If the assumption, that gas flow rate from an undefined section to the boundary pumps (pumping elements 1 and N) is the same as than one being modelled, is used, then the following boundary conditions can be applied:

$$n_1(z_0) = n_1(z_1) \quad \text{and} \quad \partial n_1(z_0)/\partial z = -\partial n_1(z_1)/\partial z;$$
$$n_N(z_{N-1}) = n_N(z_N) \quad \text{and} \quad \partial n_N(z_{N-1})/\partial z = -\partial n_N(z_N)/\partial z. \tag{6.53}$$

These conditions mean that the gas density will have a minimum in the middle of elements 1 and N (at $z = (z_0 + z_1)/2$ and $z = (z_{N-1} + z_N)/2$ correspondingly) and the net gas flow rate at these coordinates is zero. In practice, the net gas flow rate could be not equal to zero; however, if the gas flow rate from the undefined section is less, it will reduce pressure inside the pump by a maximum factor of 2, which means the average pressure of studied elements is overestimated a little and this does not harm the performance of the accelerator. If gas flow rate from the undefined section is greater, it requires more careful consideration; the easiest way within the model is to extend the modelled vacuum section to the next pumps upstream z_0 and downstream z_N. Then the comparison of the results for

Table 6.4 Input parameters for a vacuum chamber shown in Figure 6.14.

Element	q_t	q_γ	u	c
$i-3$	$\eta_t F_{i-3}$	$\eta_b \left(\dfrac{z_{i-3}+z_{i-2}}{2}, D \right) \Gamma_t \left(\dfrac{z_{i-3}+z_{i-2}}{2}, t \right)$	$u_{i-3} = u_{bc}$	$c_{i-3} = \dfrac{S_{i-3}}{z_{i-2}-z_{i-3}}$
$i-2$	$\eta_t F_{i-2}$	$\eta_b \left(\dfrac{z_{i-2}+z_{i-1}}{2}, D \right) \Gamma_t \left(\dfrac{z_{i-2}+z_{i-1}}{2}, t \right)$	$u_{i-2} = u_{bc}$	$c_{i-2} = 0$
$i-1$	$\eta_t \dfrac{F(z_{i-1})+F(z_i)}{2}$	$q_1 + q_2$	$u_{i-1} = u_{bc} + u_a \left(\dfrac{z_{i-1}+z_i}{2} \right)$	$c_{i-1} = 0$
i	$\eta_t \dfrac{F(z_i)+F(z_{i+1})}{2}$	$q_1 + q_2 + q_3$	$u_{i-1} = u_{bc} + u_a \left(\dfrac{z_i+z_{i+1}}{2} \right)$	$c_i = 0$
$i+1$	$\eta_t \dfrac{F(z_{i+1})+F(z_{i+2})}{2}$	$q_1 + q_2 + q_3$	$u_{i+1} = u_{bc} + u_a \left(\dfrac{z_{i+1}+z_{i+2}}{2} \right)$	$c_{i+1} = \dfrac{S_{i+1}}{z_{i+1}-z_{i+2}}$
$i+2$	$\eta_t F_{i+2}$	$\eta_b \left(\dfrac{z_{i+2}+z_{i+3}}{2}, D \right) \Gamma_t \left(\dfrac{z_{i+2}+z_{i+3}}{2}, t \right)$	$u_{i+2} = u_{bc}$	$c_{i+2} = 0$
$i+3$	$\eta_t F_{i+2}$	$q_1 \left(\dfrac{z_{i-2}+z_{i-1}}{2}, t \right)$	$u_{i+3} = u_{bc}$	$c_{i+3} = \dfrac{S_{i+3}}{z_{i+3}-z_{i+4}}$

with $\kappa = 1$

two models will tell how sensitive the results are for a studied vacuum section to the boundary conditions.

Other boundary conditions could also be applied, but the two described above are sufficient in most of cases for quick optimisation of particle accelerator vacuum design.

In Eq. (6.50) the gas density within the element i is described by one of two equations depending on whether $c = 0$ or $c > 0$; therefore there could four types of boundaries between elements i and $i + 1$, which can be written for $i = 1, \ldots, N - 1$ as follows:

$$\begin{cases} \text{if } c_i = 0 \text{ and } c_{i+1} = 0 \text{ then} \\ \quad \begin{cases} \dfrac{q_i}{2u_i} z_i^2 + C_{1,i} z_i + C_{2,i} = \dfrac{q_{i+1}}{2u_{i+1}} z_i^2 + C_{1,i+1} z_i + C_{2,i+1}, \\ \dfrac{q_i}{u_i} z_i + C_{1,i} = \dfrac{q_{i+1}}{u_{i+1}} z_i + C_{1,i+1}; \end{cases} \\[4pt] \text{if } c_i = 0 \text{ and } c_{i+1} > 0 \text{ then} \\ \quad \begin{cases} \dfrac{q_i}{2u_i} z_i^2 + C_{1,i} z_i + C_{2,i} = \dfrac{q_{i+1}}{c_{i+1}} + C_{1,i+1} e^{\omega_{i+1} z_i} + C_{2,i+1} e^{-\omega_{i+1} z_i}, \\ \dfrac{q_i}{u_i} z_i + C_{1,i} = \omega_{i+1}(C_{1,i+1} e^{\omega_{i+1} z_i} - C_{2,i+1} e^{-\omega_{i+1} z_i}); \end{cases} \\[4pt] \text{if } c_i > 0 \text{ and } c_{i+1} = 0 \text{ then} \\ \quad \begin{cases} \dfrac{q_i}{c_i} + C_{1,i} e^{\omega_i z_i} + C_{2,i} e^{-\omega_i z_i} = \dfrac{q_{i+1}}{2u_{i+1}} z_i^2 + C_{1,i+1} z_i + C_{2,i+1}, \\ \omega_i(C_{1,i} e^{\omega_i z_i} - C_{2,i} e^{-\omega_i z_i}) = \dfrac{q_{i+1}}{u_{i+1}} z_i + C_{1,i+1}; \end{cases} \\[4pt] \text{if } c_i > 0 \text{ and } c_{i+1} > 0 \text{ then} \\ \quad \begin{cases} \dfrac{q_i}{c_i} + C_{1,i} e^{\omega_i z_i} + C_{2,i} e^{-\omega_i z_i} = \dfrac{q_{i+1}}{c_{i+1}} + C_{1,i+1} e^{\omega_{i+1} z_i} + C_{2,i+1} e^{-\omega_{i+1} z_i}, \\ \omega_i(C_{1,i} e^{\omega_i z_i} - C_{2,i} e^{-\omega_i z_i}) = \omega_{i+1}(C_{1,i+1} e^{\omega_{i+1} z_i} - C_{2,i+1} e^{-\omega_{i+1} z_i}); \end{cases} \end{cases}$$

(6.54)

where $\omega_i = \sqrt{c_i/u_i}$. In the case of the *closed loop* boundary conditions for $i = N$ the index '$i + 1$' should be substituted with '1'.

These equations can be rewritten in a form of linear equations:

$$\begin{cases} \alpha 1_i C_{1,i} + \alpha 2_i C_{2,i} - \alpha 3_i C_{1,i+1} - \alpha 4_i C_{2,i+1} = b_{2i-1}, \\ \beta 1_i C_{1,i} + \beta 2_i C_{2,i} - \beta 3_i C_{1,i+1} - \beta 4_i C_{2,i+1} = b_{2i}, \end{cases} \quad \text{for } i = 1, \ldots, N-1;$$

$$\begin{cases} \alpha 1_N C_{1,N} + \alpha 2_N C_{2,N} - \alpha 3_N C_{1,1} - \alpha 4_N C_{2,1} = b_{2N-1}, \\ \beta 1_N C_{1,i} + \beta 2_N C_{2,i} - \beta 3_N C_{1,1} - \beta 4_N C_{2,1} = b_{2N}, \end{cases} \quad \text{for } i = N.$$

(6.55)

for unknowns $C_{1,i}$ and $C_{2,i}$ for $i = 1, \ldots, N-1$:

$$\begin{cases} \text{if } c_i = 0 \text{ and } c_{i+1} = 0 \text{ then} \\ \begin{cases} z_i C_{1,i} + C_{2,i} - z_i C_{1,i+1} - C_{2,i+1} = \left(\dfrac{q_{i+1}}{u_{i+1}} - \dfrac{q_i}{u_i}\right)\dfrac{z_i^2}{2}, \\ C_{1,i} - C_{1,i+1} = \left(\dfrac{q_{i+1}}{u_{i+1}} - \dfrac{q_i}{u_i}\right) z_i; \end{cases} \\ \text{if } c_i = 0 \text{ and } c_{i+1} > 0 \text{ then} \\ \begin{cases} z_i C_{1,i} + C_{2,i} - e^{\omega_{i+1} z_i} C_{1,i+1} - e^{-\omega_{i+1} z_i} C_{2,i+1} = \dfrac{q_{i+1}}{c_{i+1}} - \dfrac{q_i}{2u_i} z_i^2, \\ -C_{1,i} + \omega_{i+1} e^{\omega_{i+1} z_i} C_{1,i+1} - \omega_{i+1} e^{-\omega_{i+1} z_i} C_{2,i+1} = \dfrac{q_i}{u_i} z_i; \end{cases} \\ \text{if } c_i > 0 \text{ and } c_{i+1} = 0 \text{ then} \\ \begin{cases} e^{\omega_i z_i} C_{1,i} + e^{-\omega_i z_i} C_{2,i} - z_i C_{1,i+1} - C_{2,i+1} = \dfrac{q_{i+1}}{2u_{i+1}} z_i^2 - \dfrac{q_i}{c_i}, \\ \omega_i e^{\omega_i z_i} C_{1,i} - \omega_i e^{-\omega_i z_i} C_{2,i} - C_{1,i+1} = \dfrac{q_{i+1}}{u_{i+1}} z_i; \end{cases} \\ \text{if } c_i > 0 \text{ and } c_{i+1} > 0 \text{ then} \\ \begin{cases} e^{\omega_i z_i} C_{1,i} + e^{-\omega_i z_i} C_{2,i} - e^{\omega_{i+1} z_i} C_{1,i+1} - e^{-\omega_{i+1} z_i} C_{2,i+1} = \dfrac{q_{i+1}}{c_{i+1}} - \dfrac{q_i}{c_i}, \\ \omega_i e^{\omega_i z_i} C_{1,i} - \omega_i e^{-\omega_i z_i} C_{2,i} - \omega_{i+1} e^{\omega_{i+1} z_i} C_{1,i+1} + \omega_{i+1} e^{-\omega_{i+1} z_i} C_{2,i+1} = 0. \end{cases} \end{cases}$$

(6.56)

Thus equation coefficients α and β are defined in Table 6.5. The analysis algorithm should define which of these four cases is applicable for each for each boundary with a coordinate z_i. Then two equations can be written for each boundary at coordinate z_i.

Table 6.5 Coefficients α and β in Eq. (6.55) for four cases depending in pumping on c_i and c_{i+1}.

Conditions	$c_i = 0$ and $c_{i+1} = 0$	$c_i = 0$ and $c_{i+1} > 0$	$c_i > 0$ and $c_{i+1} = 0$	$c_i > 0$ and $c_{i+1} > 0$
$\alpha 1_i$	z_i	z_i	$e^{\omega_i z_i}$	$e^{\omega_i z_i}$
$\alpha 2_i$	1	1	$e^{-\omega_i z_i}$	$e^{-\omega_i z_i}$
$\alpha 3_i$	$-z_i$	$-e^{\omega_{i+1} z_i}$	$-z_i$	$-e^{\omega_{i+1} z_i}$
$\alpha 4_i$	-1	$-e^{-\omega_{i+1} z_i}$	-1	$-e^{-\omega_{i+1} z_i}$
b_{2i-1}	$\left(\dfrac{q_{i+1}}{u_{i+1}} - \dfrac{q_i}{u_i}\right)\dfrac{z_i^2}{2}$	$\dfrac{q_{i+1}}{c_{i+1}} - \dfrac{q_i}{2u_i} z_i^2$	$\dfrac{q_{i+1}}{2u_{i+1}} z_i^2 - \dfrac{q_i}{c_i}$	$\dfrac{q_{i+1}}{c_{i+1}} - \dfrac{q_i}{c_i}$
$\beta 1_i$	1	-1	$\omega_i e^{\omega_i z_i}$	$\omega_i e^{\omega_i z_i}$
$\beta 2_i$	0	0	$-\omega_i e^{-\omega z_i}$	$-\omega_i e^{-\omega z_i}$
$\beta 3_i$	-1	$\omega_{i+1} e^{\omega_{i+1} z_i}$	-1	$-\omega_{i+1} e^{\omega_{i+1} z_i}$
$\beta 4_i$	0	$-\omega_{i+1} e^{-\omega_{i+1} z_i}$	0	$\omega_{i+1} e^{-\omega_{i+1} z_i}$
b_{2i}	$\left(\dfrac{q_{i+1}}{u_{i+1}} - \dfrac{q_i}{u_i}\right) z_i$	$\dfrac{q_i}{u_i} z$	$\dfrac{q_{i+1}}{u_{i+1}} z_i$	0

Thus there is a system of $2N$ equation defining the boundary conditions of N elements allowing to define $2N$ unknowns ($C_{1,i}$ and $C_{2,i}$). In matrix form this system of linear equations can be written as

$$\mathbf{A}\mathbf{x} = \mathbf{B} \qquad (6.57)$$

where in the case of *closed loop* boundary conditions:

$$\mathbf{A} = \begin{bmatrix} \alpha 1_1 & \alpha 2_1 & \alpha 3_1 & \alpha 4_1 & 0 & 0 & \cdots & 0 & 0 & 0 & 0 \\ \beta 1_1 & \beta 2_1 & \beta 3_1 & \beta 4_1 & 0 & 0 & \cdots & 0 & 0 & 0 & 0 \\ 0 & 0 & \alpha 1_2 & \alpha 2_2 & \alpha 3_2 & \alpha 4_2 & \cdots & 0 & 0 & 0 & 0 \\ 0 & 0 & \beta 1_2 & \beta 2_2 & \beta 3_2 & \beta 4_2 & \cdots & 0 & 0 & 0 & 0 \\ \vdots & \vdots & \vdots & \vdots & \vdots & \vdots & \ddots & \vdots & \vdots & \vdots & \vdots \\ 0 & 0 & 0 & 0 & 0 & 0 & \cdots & \alpha 1_{N-1} & \alpha 2_{N-1} & \alpha 3_{N-1} & \alpha 4_{N-1} \\ 0 & 0 & 0 & 0 & 0 & 0 & \cdots & \beta 1_{N-1} & \beta 2_{N-1} & \beta 4_{N-1} & \beta 4_{N-1} \\ \alpha 3_N & \alpha 4_N & 0 & 0 & 0 & 0 & \cdots & 0 & 0 & \alpha 1_N & \alpha 2_N \\ \beta 3_N & \beta 4_N & 0 & 0 & 0 & 0 & \cdots & 0 & 0 & \beta 1_N & \beta 2_N \end{bmatrix};$$

$$\mathbf{x} = \begin{bmatrix} C_{1,1} \\ C_{2,1} \\ \vdots \\ C_{1,i} \\ C_{2,i} \\ \vdots \\ C_{1,N} \\ C_{2,N} \end{bmatrix}; \quad \mathbf{B} = \begin{bmatrix} b_1 \\ b_2 \\ \vdots \\ b_{2i-1} \\ b_{2i} \\ \vdots \\ b_{2N-1} \\ b_{2N} \end{bmatrix}.$$

In the case of *open ends* boundary conditions with $c > 0$ and the boundary conditions for $i = 0$ and $i = N$, one can easily find that

$$\begin{cases} C_{1,1} = C_{2,1}, \\ C_{1,N} = C_{2,N}. \end{cases} \qquad (6.58)$$

Thus matrixes $\mathbf{A}$ and $\mathbf{B}$ in Eq. (6.57) will be modified to

$$\mathbf{A} = \begin{bmatrix} \alpha 1_1 & \alpha 2_1 & \alpha 3_1 & \alpha 4_1 & 0 & 0 & \cdots & 0 & 0 & 0 & 0 \\ 1 & -1 & 0 & 0 & 0 & 0 & \cdots & 0 & 0 & 0 & 0 \\ 0 & 0 & \alpha 1_2 & \alpha 2_2 & \alpha 3_2 & \alpha 4_2 & \cdots & 0 & 0 & 0 & 0 \\ 0 & 0 & \beta 1_2 & \beta 2_2 & \beta 3_2 & \beta 4_2 & \cdots & 0 & 0 & 0 & 0 \\ \vdots & \vdots & \vdots & \vdots & \vdots & \vdots & \ddots & \vdots & \vdots & \vdots & \vdots \\ 0 & 0 & 0 & 0 & 0 & 0 & \cdots & \alpha 1_{N-1} & \alpha 2_{N-1} & \alpha 3_{N-1} & \alpha 4_{N-1} \\ 0 & 0 & 0 & 0 & 0 & 0 & \cdots & \beta 1_{N-1} & \beta 2_{N-1} & \beta 4_{N-1} & \beta 4_{N-1} \\ \alpha 3_N & \alpha 4_N & 0 & 0 & 0 & 0 & \cdots & 0 & 0 & \alpha 1_N & \alpha 2_N \\ 0 & 0 & 0 & 0 & 0 & 0 & \cdots & 0 & 0 & 1 & -1 \end{bmatrix};$$

$$\mathbf{x} = \begin{bmatrix} C_{1,1} \\ C_{2,1} \\ \vdots \\ C_{1,i} \\ C_{2,i} \\ \vdots \\ C_{1,N} \\ C_{2,N} \end{bmatrix}; \quad \mathbf{B} = \begin{bmatrix} b_1 \\ 0 \\ \vdots \\ b_{2i-1} \\ b_{2i} \\ \vdots \\ b_{2N-1} \\ 0 \end{bmatrix}.$$

Solving the matrix Eq. (6.57) allows to obtain $C_{1,i}$ and $C_{2,i}$.

6.2.4 Global and Local Coordinates for Each Element

The system of equation can be solved analytically and provide an accurate answer. Then the number of equations in a system of equations is large, where one can prefer a numerical solution. There are a few problems that often met by applying the above analysis directly, and there are ways to avoid these problems.

The exponent is a very strong function, and for the large z values the product of $e^{\omega z}$ could be above a maximum number for this computer, leading to interruption or error in the program run, or too small difference of two large numbers rounded to zero. This can be demonstrated by calculating $e^{100} + 1 - e^{100}$, where analytically it is clear that the answer is $= 1$, but your computer will either over limit or give an answer $= 0$. The main reason is that the number of digits is a limited value, for example, it is a 19-digit number for a 64-bit number: from -2^{63} to $2^{63} - 1$ (or from $-9\,223\,372\,036\,854\,775\,808$ to $9\,223\,372\,036\,854\,775\,807$), while $e^{100} = 2.688 \times 10^{43}$, hence a 44-digit number.

There is a simple way to solve these problems. The large values for the product of $e^{\omega z}$ should be avoided, which could be done by avoiding large z. For this the global coordinate z is replaced with a new local coordinate $Z(z, i)$ defined for each element $i = 1, \ldots, N$ as follows:

$$Z(z, i) = z - 0.5(z_i + z_{i-1}). \tag{6.59}$$

A local coordinate $Z(z, i)$ is equal to zero in a middle of each element; thus $Z(z, i) \leq L_i/2$ and the product of $e^{\omega Z(z,i)}$ will not exceed $e^{\omega L_i/2}$. This gives us criteria for the maximum length of elements. If $e^{\omega L_i/2}$ is still too large, then this element should be split into shorter elements.

All the analysis described in Section 5.2 remains practically the same with the only difference that each boundary coordinate z_i should be replaced with two Z_i for each ith element, where

$$Z1_i = -L_i/2 \text{ and } Z2_i = L_i/2. \tag{6.60}$$

Thus, the boundary conditions between ith and $(i+1)$th elements are redefined as

$$n_i\left(\frac{L_i}{2}\right) = n_{i+1}\left(-\frac{L_{i+1}}{2}\right) \text{ and } u_i \partial n_i \left(\frac{L_i}{2}\right)/\partial z = u_{i+1} \partial n_{i+1}\left(-\frac{L_{i+1}}{2}\right)/\partial z. \tag{6.61}$$

The Eq. (6.54) can be rewritten with the boundaries at $\pm L_i/2$ for $i = 1, \ldots, N-1$ as follows:

if $c_i = 0$ and $c_{i+1} = 0$ then

$$\begin{cases} \dfrac{q_i}{2u_i}\left(\dfrac{L_i}{2}\right)^2 + C_{1,i}\dfrac{L_i}{2} + C_{2,i} = \dfrac{q_{i+1}}{2u_{i+1}}\left(\dfrac{L_{i+1}}{2}\right)^2 - C_{1,i+1}\dfrac{L_{i+1}}{2} + C_{2,i+1}, \\ \dfrac{q_i L_i}{u_i 2} + C_{1,i} = -\dfrac{q_{i+1}}{u_{i+1}}\dfrac{L_{i+1}}{2} + C_{1,i+1}; \end{cases}$$

if $c_i = 0$ and $c_{i+1} > 0$ then

$$\begin{cases} \dfrac{q_i}{2u_i}\left(\dfrac{L_i}{2}\right)^2 + C_{1,i}z_i + C_{2,i} = \dfrac{q_{i+1}}{c_{i+1}} + C_{1,i+1}e^{-\lambda_{i+1}} + C_{2,i+1}e^{\lambda_{i+1}}, \\ \dfrac{q_i}{u_i}z_i + C_{1,i} = \omega_{i+1}(C_{1,i+1}e^{-\lambda_{i+1}} - C_{2,i+1}e^{\lambda_{i+1}}); \end{cases} \qquad (6.62)$$

if $c_i > 0$ and $c_{i+1} = 0$ then

$$\begin{cases} \dfrac{q_i}{c_i} + C_{1,i}e^{\lambda_i} + C_{2,i}e^{-\lambda_i} = \dfrac{q_{i+1}}{2u_{i+1}}\left(\dfrac{L_{i+1}}{2}\right)^2 - C_{1,i+1}\dfrac{L_{i+1}}{2} + C_{2,i+1}, \\ \omega_i(C_{1,i}e^{\lambda_i} - C_{2,i}e^{-\lambda_i}) = -\dfrac{q_{i+1}}{u_{i+1}}\dfrac{L_{i+1}}{2} + C_{1,i+1}; \end{cases}$$

if $c_i > 0$ and $c_{i+1} > 0$ then

$$\begin{cases} \dfrac{q_i}{c_i} + C_{1,i}e^{\lambda_i} + C_{2,i}e^{-\lambda_i} = \dfrac{q_{i+1}}{c_{i+1}} + C_{1,i+1}e^{-\lambda_{i+1}} + C_{2,i+1}e^{\lambda_{i+1}}, \\ \omega_i(C_{1,i}e^{\lambda_i} - C_{2,i}e^{-\lambda_i}) = \omega_{i+1}(C_{1,i+1}e^{-\lambda_{i+1}} - C_{2,i+1}e^{\lambda_{i+1}}); \end{cases}$$

where $\lambda_i = \omega_i \dfrac{L_i}{2} = \sqrt{\dfrac{c_i}{u_i}}\dfrac{L_i}{2}$. These equations can be rewritten in a form of linear equations for unknowns $C_{1,i}$ and $C_{2,i}$ for $i = 1, \ldots, N-1$:

if $c_i = 0$ and $c_{i+1} = 0$ then

$$\begin{cases} \dfrac{L_i}{2}C_{1,i} + C_{2,i} + \dfrac{L_{i+1}}{2}C_{1,i+1} - C_{2,i+1} = \dfrac{q_{i+1}L_{i+1}^2}{8u_{i+1}} - \dfrac{q_i L_i^2}{8u_i}, \\ -C_{1,i} + C_{1,i+1} = \dfrac{q_{i+1}L_{i+1}}{2u_{i+1}} + \dfrac{q_i L_i}{2u_i}; \end{cases}$$

if $c_i = 0$ and $c_{i+1} > 0$ then

$$\begin{cases} \dfrac{L_i}{2}C_{1,i} + C_{2,i} - e^{-\lambda_{i+1}}C_{1,i+1} - e^{\lambda_{i+1}}C_{2,i+1} = \dfrac{q_{i+1}}{c_{i+1}} - \dfrac{q_i L_i^2}{8u_i}, \\ -C_{1,i} + \omega_{i+1}e^{-\lambda_{i+1}}C_{1,i+1} - \omega_{i+1}e^{\lambda_{i+1}}C_{2,i+1} = \dfrac{q_i L_i}{2u_i}; \end{cases} \qquad (6.63)$$

if $c_i > 0$ and $c_{i+1} = 0$ then

$$\begin{cases} e^{\lambda_i}C_{1,i} + e^{-\lambda_i}C_{2,i} + \dfrac{L_{i+1}}{2}C_{1,i+1} - C_{2,i+1} = \dfrac{q_{i+1}L_{i+1}^2}{8u_{i+1}} - \dfrac{q_i}{c_i}, \\ -\omega_i e^{\lambda_i}C_{1,i} + \omega_i e^{-\lambda_i}C_{2,i} + C_{1,i+1} = \dfrac{q_{i+1}L_{i+1}}{2u_{i+1}}; \end{cases}$$

if $c_i > 0$ and $c_{i+1} > 0$ then

$$\begin{cases} e^{\lambda_i}C_{1,i} + e^{-\lambda_i}C_{2,i} - e^{-\lambda_{i+1}}C_{1,i+1} - e^{\lambda_{i+1}}C_{2,i+1} = \dfrac{q_{i+1}}{c_{i+1}} - \dfrac{q_i}{c_i}, \\ \omega_i e^{\lambda_i}C_{1,i} - \omega_i e^{-\lambda_i}C_{2,i} - \omega_{i+1}e^{-\lambda_{i+1}}C_{1,i+1} + \omega_{i+1}e^{\lambda_{i+1}}C_{2,i+1} = 0. \end{cases}$$

Table 6.6 Coefficients α and β in Eq. (6.55) for four cases depending on pumping on c_i and c_{i+1}.

Conditions	$c_i=0$ and $c_{i+1}=0$	$c_i=0$ and $c_{i+1}>0$	$c_i>0$ and $c_{i+1}=0$	$c_i>0$ and $c_{i+1}>0$
$\alpha 1_i$	$\dfrac{L_i}{2}$	$\dfrac{L_i}{2}$	e^{λ_i}	e^{λ_i}
$\alpha 2_i$	1	1	$e^{-\lambda_i}$	$e^{-\lambda_i}$
$\alpha 3_i$	$\dfrac{L_{i+1}}{2}$	$-e^{-\lambda_{i+1}}$	$\dfrac{L_{i+1}}{2}$	$-e^{-\lambda_{i+1}}$
$\alpha 4_i$	-1	$-e^{\lambda_{i+1}}$	-1	$-e^{\lambda_{i+1}}$
b_{2i-1}	$\dfrac{q_{i+1}L_{i+1}^2}{8u_{i+1}} - \dfrac{q_i L_i^2}{8u_i}$	$\dfrac{q_{i+1}}{c_{i+1}} - \dfrac{q_i L_i^2}{8u_i}$	$\dfrac{q_{i+1}L_{i+1}^2}{8u_{i+1}} - \dfrac{q_i}{c_i}$	$\dfrac{q_{i+1}}{c_{i+1}} - \dfrac{q_i}{c_i}$
$\beta 1_i$	-1	-1	$-\omega_i e^{\lambda_i}$	$\omega_i e^{\lambda_i}$
$\beta 2_i$	0	0	$\omega_i e^{-\lambda_i}$	$-\omega_i e^{-\lambda_i}$
$\beta 3_i$	1	$\omega_{i+1} e^{-\lambda_{i+1}}$	0	$-\omega_{i+1} e^{-\lambda_{i+1}}$
$\beta 4_i$	0	$-\omega_{i+1} e^{\lambda_{i+1}}$	0	$\omega_{i+1} e^{-\lambda_{i+1}}$
b_{2i}	$\dfrac{q_{i+1}L_{i+1}}{2u_{i+1}} + \dfrac{q_i L_i}{2u_i}$	$\dfrac{q_i L_i}{2u_i}$	$\dfrac{q_{i+1}L_{i+1}}{2u_{i+1}}$	0

Thus equation coefficients α and β for the boundaries at $\pm L_i/2$ are defined in Table 6.6.

6.2.5 Using the Results

After obtaining the solution for $C_{1,i}$ and $C_{2,i}$, the gas density (or pressure) along the modelled vacuum chamber is described now with N Eq. (6.50) analytically. This allows studying the gas density (or pressure) profiles and average values along the beam path, change pumping and desorption pattern, length and cross sections of different elements, etc., with very little effort. Therefore, this method is quite a good tool for vacuum system optimisation of the location of absorbers, collimators, and antechambers as well as the location and pumping speed of the pumps. However, one should remember that this is a simplified method for quick calculations of pressure profiles in a design phase when position and size of pumps are not finalised yet. This method might lead to errors, especially when cross section varies significantly along the beam path or when there is a difference in orders of magnitude between minimum and maximum pressures. Besides, significant uncertainties should be analysed when a vessel with an antechamber is modelled:

- The vacuum conductance of an antechamber could be larger than the one of the beam chamber.
- There is some vacuum conductance between the antechamber and the beam chamber.
- The PSD molecules are desorbed from SR absorbers (preferred direction is normal to the absorber surface), the vacuum chamber walls (due to direct and reflected photons), and antechamber walls (due to photons reflected from absorbers).
- Pumps might be connected both to either the antechamber or the beam chamber or both.

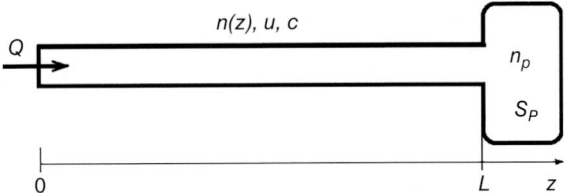

Figure 6.15 A tubular vacuum chamber with an injected gas flow rate Q.

Carefully considering these uncertainties, the 1D model can still be used. For example, it was successfully used during the storage ring vacuum system design of the DLS [15–17]. However, for the confidence of correct 1D model results, it is advisable to check complex shaped sections at an early stage and then a final design (or sections) with one of TPMC codes.

6.2.6 A Few Practical Formulas

6.2.6.1 Gas Injection into a Tubular Vacuum Chamber

Consider a tubular vacuum chamber with an injected gas flow rate Q at one end with a coordinate $z = 0$ and a pump with a pumping speed S_p at another coordinate with a coordinate $z = L$ (see Figure 6.15). The gas density as a function of coordinate z can be written with simplified Eqs. (6.48) as

$$n_i(z) = \begin{cases} C_1 z + C_2 & \text{for } c = 0, \\ C_3 \cosh\left(\sqrt{\dfrac{c}{u}} z\right) & \text{for } c > 0. \end{cases} \quad (6.64)$$

In the case of $c = 0$, the gas density in the pump is $n_p = Q/S_p$, and in the case of $c \neq 0$, the gas density in the pump is $n_p = \dfrac{dn}{dz}\dfrac{u}{S_p}$. Assuming that $n(L) = n_p$, one can find constants C_1, C_2, and C_3 and write a final formula:

$$n_i(z) = \begin{cases} Q\left(\dfrac{L-z}{u} + \dfrac{1}{S_p}\right) & \text{for } c = 0, \\ \dfrac{u}{c}\cosh\left(\sqrt{\dfrac{c}{u}} z\right)\tanh\left(\sqrt{\dfrac{c}{u}} L\right) & \text{for } c > 0. \end{cases} \quad (6.65)$$

It should be noted that Eq. (6.65) for $c > 0$ has a quite limited applicability for straight tubes because of molecular beaming effect (see Section 6.4.4).

6.2.6.2 Vacuum Chamber with Known Pumping Speed at the Ends

Consider a vacuum chamber of length L along a coordinate axis z between two pumps with a pumping speed S_p, and $z = 0$ corresponds to a middle of vacuum chamber. For example, this could be a part of experimental facility for PSD measurements from tubular samples shown in Figure 6.16. Gas density as a function of coordinate z, $n(z)$, along a vacuum chamber with distributed gas desorption (in this example, this is PSD due to SR collimated to irradiate the sample tube only) q [molecules/(s·m)], specific vacuum conductance u and without distributed pumping (i.e. $c = 0$) is described as

$$n(z) = \dfrac{q}{2u}\left(\dfrac{L^2}{4} - z^2\right) + \dfrac{qL}{2S_p}. \quad (6.66)$$

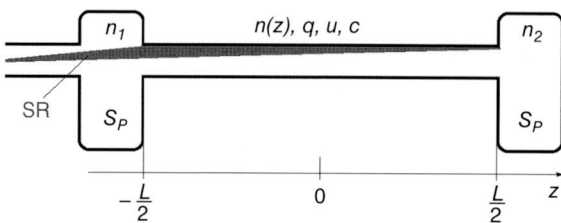

Figure 6.16 A schematic layout of PSD experiment on a tube between two pumps.

In the case of vacuum chamber with distributed pumping (i.e. $c > 0$), it is described as

$$n(z) = \frac{q}{c}\left(1 - \frac{\cosh\left(\sqrt{\frac{c}{u}}z\right)}{\cosh\left(\sqrt{\frac{c}{u}}\frac{L}{2}\right)\left(1 + \frac{\sqrt{cu}}{S_p}\tanh\left(\sqrt{\frac{c}{u}}\frac{L}{2}\right)\right)}\right). \tag{6.67}$$

The average value of the gas density defined as

$$\langle n_L \rangle = \frac{1}{L}\int_{-L/2}^{L/2} n(z)dz. \tag{6.68}$$

For the case of $c = 0$, it is

$$\langle n_L \rangle = q\left(\frac{L^2}{12u} + \frac{L}{2S_p}\right). \tag{6.69}$$

For the case of $c > 0$, it is,

$$\langle n_L \rangle = \frac{q}{c}\left(1 - \frac{\tanh\left(\sqrt{\frac{c}{u}}\frac{L}{2}\right)}{\sqrt{\frac{c}{u}}\frac{L}{2}\left(1 + \frac{\sqrt{cu}}{S_p}\tanh\left(\sqrt{\frac{c}{u}}\frac{L}{2}\right)\right)}\right). \tag{6.70}$$

For example, let us consider a tubular vacuum chamber with a diameter d and a length L irradiated by SR with an intensity $\Gamma = 10^{17}$ photons/(s·m). A distributed desorption corresponding to PSD yield $\eta_\gamma = 10^{-3}$ CO/photon is $q = \eta\Gamma = 10^{14}$ CO/(s·m). Figure 6.17 shows examples of calculated CO gas density for a tube with non-sorbing walls ($c = 0$ or $\alpha = 0$):

- A gas density n as a function of coordinate z for a tube with $L = 10$ m and $S_p = 100$ l/s significantly varies with tube diameter d.
- A gas density n as a function of coordinate z for a tube with $d = 50$ mm and $L = 10$ m significantly reduces between $S_p = 1$ and 10 l/s reduces insignificantly between $S_p = 10$ and 100 l/s and practically does not depend for $S_p > 100$ l/s, demonstrating conductance-limited pumping.
- An average gas density $\langle n_L \rangle$ for a tube with $L = 10$ m and diameter $d = 5$ mm does not depend on pumping speed S_p in the range between 1 and 100 l/s (i.e. conductance-limited pumping), while $\langle n_L \rangle$ linearly decreases pumping speed with for $d = 200$ mm. A tube with $d = 50$ mm demonstrates both types of dependences: an average gas density $\langle n_L \rangle$ decreases with pumping speed in the range between 1 and 50 l/s and conductance-limited behaviour for $S_p > 50$ l/s.

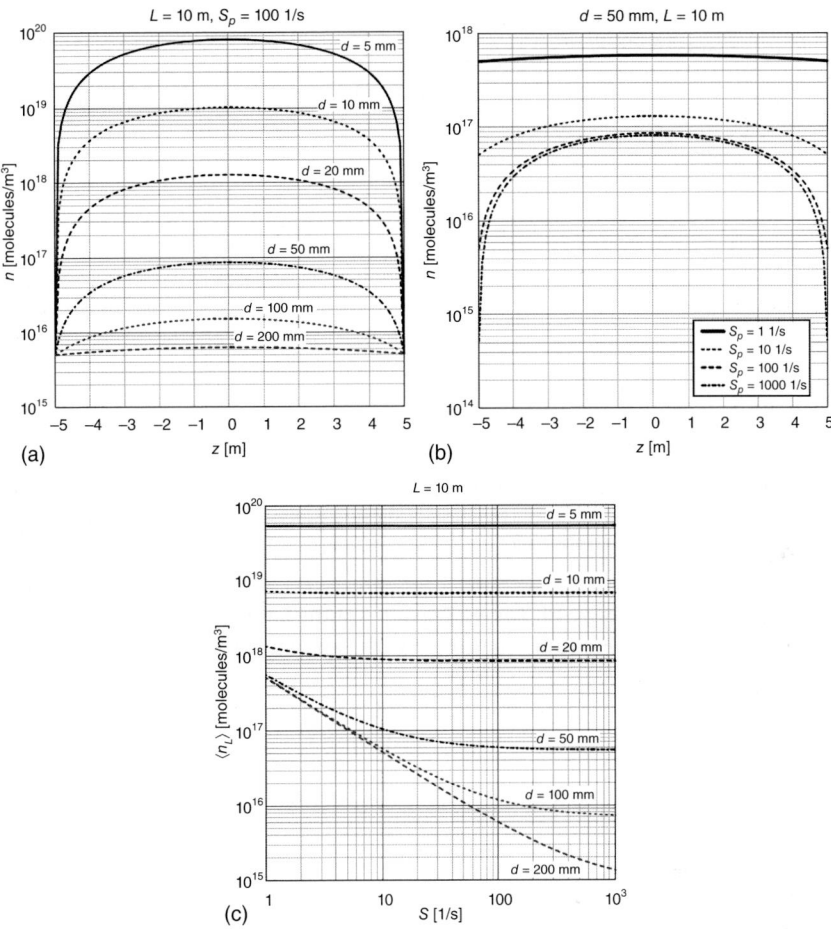

Figure 6.17 A gas density n as a function of coordinate z for (a) various tube diameters and (b) various pumping speed, (c) an average gas density $\langle n_L \rangle$ as a function of pumping speed S_p for various tube diameters.

Figure 6.18 shows examples of calculated CO gas density for a tube with sorbing walls ($c > 0$ or $\alpha > 0$):

- A gas density n as a function of coordinate z for a tube with $d = 50$ mm, $L = 10$ m, and $S_p = 100$ l/s linearly increases with α^{-1} in the range $10^{-3} \leq \alpha \leq 1$, while it is practically not dependent on $\alpha < 10^{-5}$ and almost identical to a solution for $\alpha = 0$.
- An average gas density $\langle n_L \rangle$ for a tube with $S_p = 100$ l/s: when for $d = 5$ mm, the solutions for $L = 1$ and 10 m are identical for sticking probability in the range $10^{-3} \leq \alpha \leq 1$, and when for $d = 50$ and 200 mm, the solutions for $L = 1$ and 10 m are identical for sticking probability in the range $0.05 \leq \alpha \leq 1$.

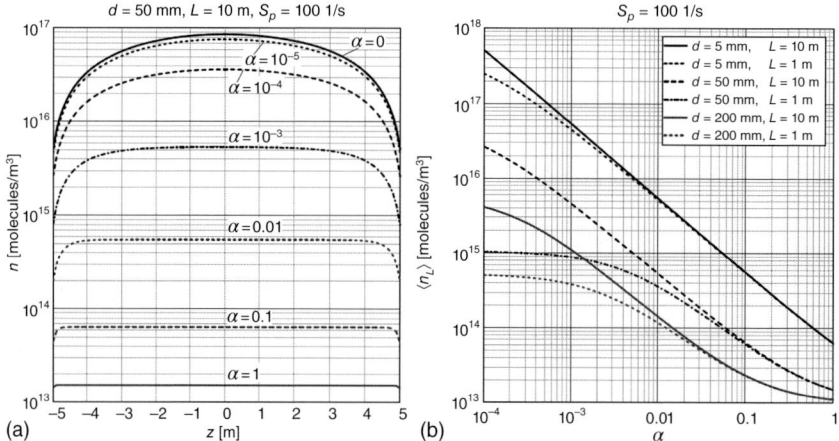

Figure 6.18 (a) A gas density n as a function of coordinate z for various sticking probabilities α, (b) an average gas density $\langle n_L \rangle$ as a function of sticking probabilities α for vacuum conductance u and pumping speed S_p.

6.2.6.3 Vacuum Chamber with Known Pressures at the Ends

Consider the same vacuum chamber with known pressures at the ends, $P(-L/2) = P_1$ and $P(L/2) = P_2$. Gas pressure as a function of coordinate z, $P(z)$, along a vacuum chamber with distributed gas desorption q^* [Pa·m²/s] and without distributed pumping (i.e. $c = 0$) is described as

$$P(z) = \frac{q^*}{2u}\left(\frac{L^2}{4} - z^2\right) + \frac{P_2 - P_1}{L}z + \frac{P_1 + P_2}{2}. \tag{6.71}$$

In the case of vacuum chamber with distributed pumping (i.e. $c > 0$), it is described as

$$P(z) = \frac{q^*}{c}\left(1 - \frac{\cosh\left(\sqrt{\frac{c}{u}}z\right)}{\cosh\left(\sqrt{\frac{c}{u}}\frac{L}{2}\right)}\right) + \frac{P_1 + P_2}{2}\frac{\cosh\left(\sqrt{\frac{c}{u}}z\right)}{\cosh\left(\sqrt{\frac{c}{u}}\frac{L}{2}\right)}$$

$$+ \frac{P_2 - P_1}{2}\frac{\sinh\left(\sqrt{\frac{c}{u}}z\right)}{\sinh\left(\sqrt{\frac{c}{u}}\frac{L}{2}\right)} \tag{6.72}$$

The average pressure for the case of $c = 0$ is

$$\langle P_L \rangle = q^*\frac{L^2}{12u} + \frac{P_1 + P_2}{2} \tag{6.73}$$

For the case of $c > 0$, it is

$$\langle P_L \rangle = \frac{q^*}{c}\left(1 - \frac{1}{L}\sqrt{\frac{u}{c}}\tanh\left(\sqrt{\frac{c}{u}}\frac{L}{2}\right)\right) + (P_1 + P_2)\frac{1}{L}\sqrt{\frac{u}{c}}\tanh\left(\sqrt{\frac{c}{u}}\frac{L}{2}\right) \tag{6.74}$$

If the gas pressure is measured at the middle of the sample tube, $P(0)$ and at the ends, P_1 and P_2, then the PSD yields can be calculated with the formulas:

$$\eta = \frac{8u}{k_B T L^2 \Gamma}\left(P(0) - \frac{P_1+P_2}{2}\right) \qquad \text{for } c = 0;$$

$$\eta = \frac{c}{k_B T \Gamma} \frac{P(0)\cosh\left(\sqrt{\frac{c}{u}}\frac{L}{2}\right) - \frac{P_1+P_2}{2}}{\cosh\left(\sqrt{\frac{c}{u}}\frac{L}{2}\right) - 1} \qquad \text{for } c > 0. \qquad (6.75)$$

These formulas are used for calculating PSD and ESD yields from the experimental measurements on tubular samples with a three-gauge method described in Chapters 4 and 5.

6.3 Three-Dimensional Modelling: Test Particle Monte Carlo

6.3.1 Introduction

Monte Carlo methods are a widely used class of computational algorithms for simulating the behaviour of various physical and mathematical systems (and many others). A 'Monte Carlo' experiment means that a random number generator was used to examine a problem. Because of the repetition of algorithms and the large number of calculations involved, Monte Carlo methods are more and more affordable with development and accessibility of powerful computers and even on modern personal computers.

TPMC method is an adaptation of the more general direct simulation Monte Carlo (DSMC) method [18] used in rarefied gas dynamics to describe the free molecular flow regime: i.e. when the molecular density is so low that intramolecular collisions can be neglected. It is based on random statistical generation of molecule initial position, velocity distribution, reflection from walls and sorption probability, and it is practically realised by adapting well-known ray-tracing techniques. The method is well described in literature (see, for example, [1, 19–21]), so we can focus on its application to the accelerator design.

Most of vacuum designers do not write their TPMC codes but use either the TPMC-based in-house programs (quite a few codes have been developed in accelerator centres) or commercially available programs. Although these programs could look very different, are optimised for different applications, and have a different level of flexibility and a different number of additional options, they are based on the same TPMC algorithm, and thus most of specific tasks that can be modelled, most of input and output parameters are common for all these codes. Therefore, in the following discussion, we will focus on TPMC modelling form user's point of view.

For a clarity, the author often refers to the parameters used in a widespread code called MOLFLOW+ [22], while other codes could employ different terminally, but functionality should be the same or at least very similar. The advantage of MOLFLOW+ is that it has been developed to meet the need of an accelerator vacuum system design. MOLFLOW+ allows to model 3D structures drawn directly in MOLFLOW or by importing the drawings of vacuum system saved in STL format [23]. This allows to significantly save time in producing 3D layout in the TPMC model. Furthermore, it is compatible with a SR ray-tracing code SYNRAD+ described in Chapter 2, allowing to accurately model a 3D distribution of PSD, the main source of gas in machines with SR.

6.3.2 A Vacuum Chamber in the TPMC Model

Although the TPMC method allows realising an algorithm for any complexity of vacuum chamber, in practice, a comparison of accuracy vs. effort and time schedules encourage a vacuum designer to accept less accurate result for a simplified geometry because vacuum chamber geometry generation is probably the most time-consuming part of the TPMC. Some TPMC codes have a number of in-built simple geometries such as cylindrical, elliptical, and right-angular tubes, trapezoidal, conical, cubes, etc., which can be used in a model to represent (with some level of simplification) a real vacuum chamber. Another approach is using flat surfaces only (this simplifies a calculation algorithm), which finally leads to a regular or irregular computing grid on engineering drawings. This grid divides the surface to flat geometrical elements called *facets*. The simplest facet is a triangle, with well-known formulas for calculating its area, boundaries, and a normal vector required for TPMC.

Only fully closed volume can be modelled in TPMC; thus all open ends leading to pumps or other part of vacuum system are represented with a virtual surfaces (facets) with a defined properties as for a real one.

6.3.3 TPMC Code Input

To build a TPMC model of vacuum chamber, one needs to not only determine the vacuum chamber shape but also know where the gas is injected and/or desorbed, what the ratio is between different gas sources, how the gas molecules interact with walls, where gas is pumped, and how to convert all this to the input data for TPMC.

From a practical point of view, the TPMC allows to study the gas flows at surfaces that could be real or virtual and have properties related to desorption, sorption, and transparency (see Figure 6.19). Thus, each facet in the TPMC model should be assigned with a number of specific properties listed below:

– Desorption:
  - Is the surface (facets) desorbing? Desorbing surface could be a real surface with thermal outgassing, ESD, PSD, and/or ISD (ion stimulated desorption); a surface reproducing gas injection; or a virtual surface reproducing a gas flow rate from other parts of vacuum chamber outside the present model.
  - If it is desorbing, what is a spatial distribution of desorbed particles? Most commonly it is a diffuse distribution (also called a cosine low distribution), but special (non-diffuse) distributions could also be used.

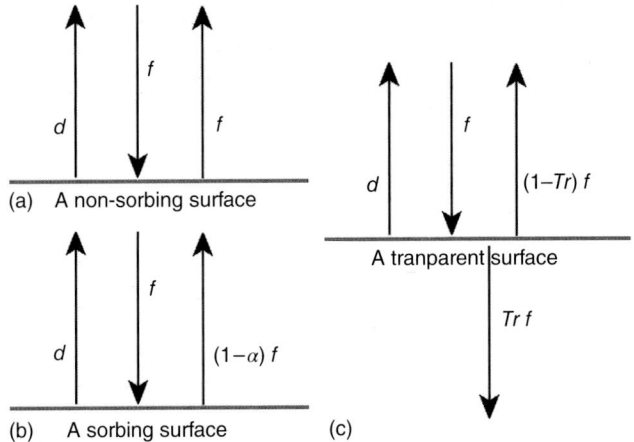

Figure 6.19 Gas flows at (a) a non-sorbing surface, (b) a sorbing surface, and (c) a transparent surface, where d is desorption and f is a flow particles hitting a surface.

- Reflection:
  - What is a directional velocity distribution of particles after interaction with a wall?

    A diffuse distribution (cosine low) is the most common for vacuum chamber surfaces.

    A specular (or mirror) reflection could take place at grazing incident on polished surfaces.

    A special reflection, for example, backscattered reflection, could be used when necessary.

    Reflection should be defined for real or semi-transparent surfaces, because it is irrelevant for fully transparent ones.
- Pumping:
- Sticking probability or capture coefficient should be defined in the range $0 \leq \alpha \leq 1$. A surface is not pumping when $\alpha = 0$ and is adsorbing all interacting particles when $\alpha = 1$.
  - A sticking probability is a defined property of sorbing surface (such as NEG or cryogenic surface) and is usually different for each gas.
  - In the case of pumps with known pumping speed for each gas, a capture coefficient should be calculated using Eq. (6.20) or (6.36) and is usually different for each gas.
- Transparency/opacity:
  - Transparency should be defined in the range $0 \leq Tr \leq 1$. A surface is fully transparent when $Tr = 1$ and fully opaque when $Tr = 0$. Opacity is defined as $k_{op} = 1 - Tr$.
  - Transparency (or opacity) is used for internal components with a complex geometry limiting free molecular travel inside the vacuum chamber, such as a mesh. This allows to significantly simplify a TPMC model and, therefore, computing time. A transparency of such complex geometry components could be provided from earlier studies, supplier's specification data, or

calculated in advance with a detailed TPMC model. In other words, transparency is a known transmission probability of this component: $Tr = w$.
- Fully transparent facets are used to study the gas flows at the locations away from the real walls.

These parameters are sufficient to model a vacuum system for the most of application in vacuum. The TPMC modelling and its results related to gas transfer do not depend neither on nature of gas species (or mass of a molecules) nor the temperature of vacuum chamber.

For completeness it is worth mentioning that when a heat transfer by moving particles should be investigated, the TPMC algorithm allows to include additional properties such as wall temperature, accommodation coefficient, and a velocity of particle. However, this is outside of the scope of this chapter.

6.3.4 TPMC Code Output

The TPMC algorithm allows to generate a particle and, based on input parameters, directions, reflections, and interaction with walls until particle absorption. After generating N particles within a TPMC program, the following results could be obtained for each facet:

- A number of hits, Mh_i.
- A number of pumped (or absorbed) particles Mp_i (for sufficiently large Mp_i one should obtain $Mp_i = \alpha_i Mh_i$).
- An area of a facet, A_i.
- Angular distribution of incident or absorbed particle at the facets.

This is sufficient to calculate a gas density and pressure near each facet, a gas flow rate to a pump or a pumping surface, and a vacuum conductance. The relations between TPMC parameters and measured values are well described in Refs. [1–23]. The most important relations are discussed as follows and summarised in Table 6.7.

6.3.4.1 Gas Flow Rate

It was already mentioned earlier that a gas flow rate in gas dynamics can be expressed in various ways in application to vacuum in particle accelerators; these are Q [molecules/s] and Q^* [Pa·m³/s] (see Table 6.1). Since a gas flow rate Q from a source and an impingement rate I_i are represented in TPMC by a number of generated molecules N and a number of hits Mh_i, respectively, the relation between them can be written as follows:

$$\frac{N[\text{particles}]A_i[\text{m}^2]}{Mh_i[\text{particles}]} = \frac{Q[\text{molecules/s}]}{I_i[\text{molecules}/[\text{m}^2 \cdot \text{s}]]}. \tag{6.76}$$

Combining with Eq. (6.14), this can be rewritten as

$$\frac{N[\text{particles}]}{Mh_i[\text{particles}]} = \frac{4Q[\text{molecules/s}]}{n[\text{molecules/m}^3]A_i[\text{m}^2]\bar{v}[\text{m/s}]}. \tag{6.77}$$

Table 6.7 Relations between TPMC parameters and measured values.

Measured parameter	Symbol and dimensions	Model parameter	Symbol	Relation
Gas flow rate	Q_n [molecules/s] Q_p [Pa·m³/s]	A number of generated particles	N	See Eq. (6.76)
Gas density and pressure	n [molecules/m³] P [Pa]	A number of hits on facet i	Mh_i	See Eq. (6.78)
Vacuum conductance	U [m³/s]$= wA\frac{\bar{v}}{4}$	Transmission probability of a tube	w	$w = \frac{Mp_{\text{outlet}}}{N}$
Ideal pumping speed	S_{id} [m³/s]$= A\frac{\bar{v}}{4}$	Maximum sticking probability	$\alpha = 1$	
Pumping speed	S [m³/s]	Sticking probability (capture coefficient)	α (or ρ)	$\alpha = \frac{S}{S_{\text{id}}}$
Effective pumping speed	S_{eff} [m³/s]$= \rho_{\text{eff}} A\frac{\bar{v}}{4}$	Effective capture coefficient	ρ_{eff}	$\frac{1}{\rho_{\text{eff}}} = \frac{1}{\rho} + \frac{1}{w}$
Pumped gas flow	Q_{pump} [molecules/s]	A number of particles pumped by facet i	Mp_i	$Q_{\text{pump}} = \frac{Q_n}{N} Mp_i$

Here A is a (effective) pumping surface.

6.3.4.2 Gas Density and Pressure

In equilibrium conditions, gas density and pressure at each facet i can be found from the results of TPMC as follows:

$$n_i \left[\frac{\text{molecules}}{\text{m}^3}\right] = 4\frac{Q[\text{molecules/s}]}{A_i[\text{m}^2]\bar{v}[\text{m/s}]}\frac{Mh_i}{N},$$

$$P_i[\text{Pa}] = nk_BT = 4\frac{Q^*[\text{Pa}\cdot\text{m}^3/\text{s}]}{A_i[\text{m}^2]\bar{v}[\text{m/s}]}\frac{Mh_i}{N}. \qquad (6.78)$$

When the conditions are not equilibrium, the gas density and pressure calculated as an average of the number of hits (or passing particles) on three surfaces orthogonal to each other (real or virtual), a, b, and c:

$$n_{abc}\left[\frac{\text{molecules}}{\text{m}^3}\right] = \frac{4}{3}\frac{Q[\text{molecules/s}]}{N\bar{v}[\text{m/s}]}\left(\frac{Mh_a}{A_a[\text{m}^2]} + \frac{Mh_b}{A_b[\text{m}^2]} + \frac{Mh_c}{A_c[\text{m}^2]}\right),$$

$$P_{abc}[\text{Pa}] = nk_BT = \frac{4}{3}\frac{Q^*[\text{Pa}\cdot\text{m}^3/\text{s}]}{N\bar{v}[\text{m/s}]}\left(\frac{Mh_a}{A_a[\text{m}^2]} + \frac{Mh_b}{A_b[\text{m}^2]} + \frac{Mh_c}{A_c[\text{m}^2]}\right). \qquad (6.79)$$

Similarly, one can use six surfaces, a pair of parallel surface with opposite normal vector parallel to each of three orthogonal surfaces.

6.3.4.3 Transmission Probability and Vacuum Conductance

When a molecule from Volume 1 enters the tube, after a number of interactions with a tube walls, it can return to Volume 1 (like molecule a in Figure 6.20) to enter to Volume 2 or (like molecule b). The ratio between a number of molecules that enters Volume 2 to a total number entered the tube is called a transmission probability w.

A transmission probability can be calculated with TPMC method for various vacuum components between the defined inlet (surface 1) and outlet (surface 2), for example, a cone tube (a) and a pumping port (b) in Figure 6.21. In this case, sticking probabilities of both inlet and outlet are set to $\alpha_{1,2} = 1$, and particle generated from inlet only is $d_1 = N$, $d_2 = 0$. After the TPMC calculations, a transmission probability between inlet and outlet is defined as

$$w = \frac{Mp_2}{N}, \qquad (6.80)$$

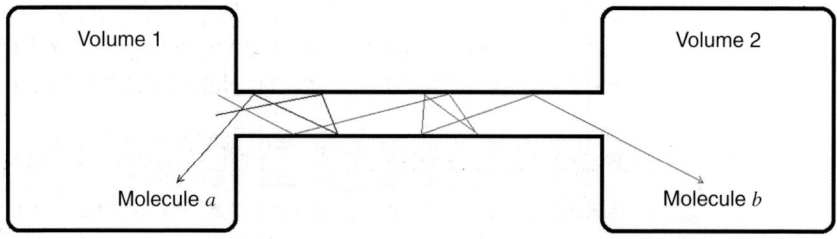

Figure 6.20 Molecules travelling between two volumes connected with a tube.

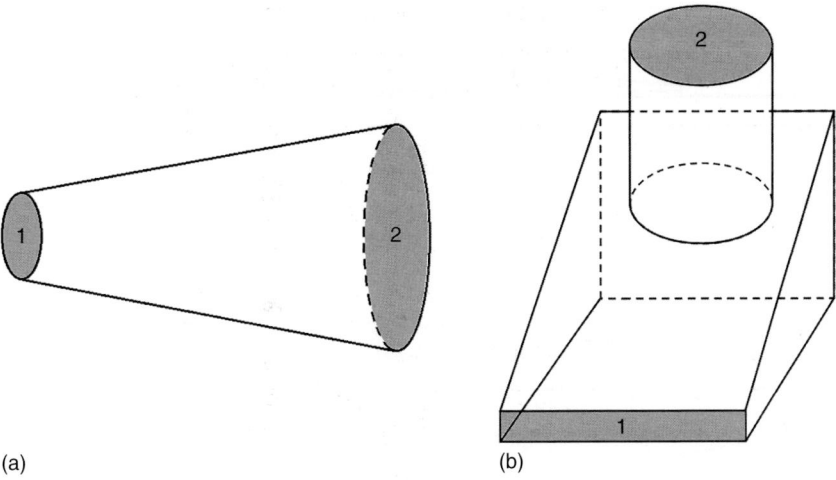

Figure 6.21 An inlet (1) and outlet (2) in a transmission probability calculations for a cone tube (a) and a pumping port (b).

and a vacuum conductance between inlet and outlet can be calculated as follows:

$$U\left[\frac{m^3}{s}\right] = wS_{id} = wA_1\frac{\bar{v}}{4}. \tag{6.81}$$

It should be noted that the transmission probability for an orifice with zero length is equal to 1, because $U_o = S_{id}$; see Eq. (6.24).

6.3.4.4 Pump-Effective Capture Coefficient
An effective pumping speed and a pump-effective capture coefficient ρ_{eff} are defined by Eqs. (6.34–6.36). A pump-effective capture coefficient could be calculated directly from a connecting tube transmission probability w and a pump capture coefficient ρ as follows:

$$\frac{1}{\rho_{eff}} = \frac{1}{\rho} + \frac{1}{w}. \tag{6.82}$$

6.3.4.5 Effect of Temperature and Mass of Molecules
Although, as it was noted earlier, the TPMC results do not depend on neither the mass of molecules nor the temperature. However the gas density n and the pressure P calculated from the TPMC results do dependent on both the mass and the temperature (see Eq. (6.78)), because the mean velocity of gas molecules, $\bar{v}(m, T)$, required for calculations of n and P, is a function of molecular mass and gas temperature and because of a temperature term in equation for pressure.

6.3.5 What Can Be Done with TPMC Results?

Let us consider a typical layout of vacuum chamber in a storage ring with SR. This layout includes a beam chamber, an antechamber with SR absorber, and lumped pumps (see Figure 6.22a).

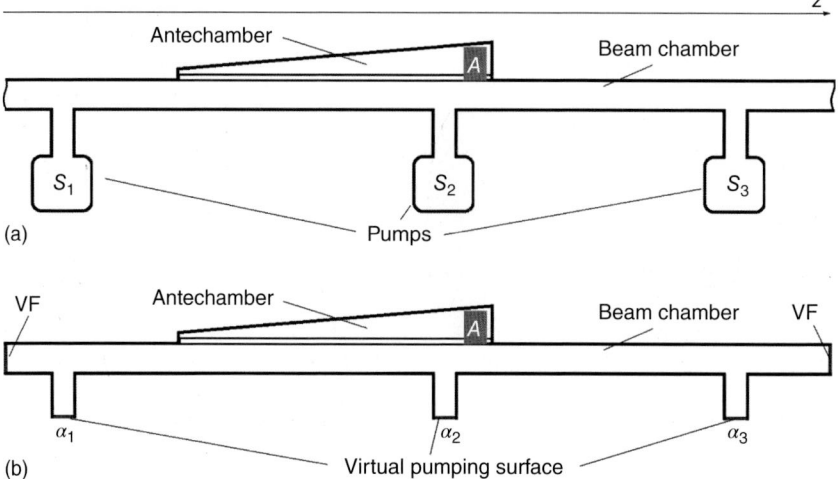

Figure 6.22 (a) A schematic vacuum layout of an accelerator section and (b) its schematic TPMC model. Here, A is a SR absorber placed in an antechamber, $S_{1,2,3}$ is a pumping speed of lumped pumps represented in TPMC model by virtual pumping surfaces (opaque with diffuse reflection and sticking probability $\alpha_{1,2,3}$, and VF are virtual surfaces (opaque with mirror reflection).

In a TPMC model (see Figure 6.22b), the pumps are represented by virtual opaque pumping surfaces with sticking probability α with a diffuse reflection.

Since the TPMC model should create a closed volume, two other virtual surfaces (VFs) are required to close the volume at the ends of the section. The set of parameters for VF depends on the conditions:

- In a case in which it is known that a gas flow rate from and to a neighbouring section is negligible, these surfaces could be set opaque with mirror reflection.
- When sections are regular, another option of boundary condition could be applied: VF absorbing a hitting particle. Then this particle is generated on another VF at the same x and y coordinates and with the same velocity vector. Mathematically it is a shift along z coordinate.
- Other conditions and set of parameters at VF are also possible.

6.3.5.1 A Direct Model with a Defined Set of Parameters

When all input parameters are defined, then the TPMC model with a single set of input parameters can provide all the necessary output results after one run of calculations.

When there is a few sources of gas, for example, gas injection, thermal desorption, and PSD from each absorber and from different parts of vacuum chamber, then a separate run of calculations with a different set of input parameters for desorption can be used for each source of gas. Since each particle travels in the model without collision with other particles (a free-molecular regime is discussed here), this allows to find the net effect as a superposition of the results for each source of gas modelled with TPMC separately:

$$n_i = \frac{4}{v} \sum_j \left(\frac{Q_j \, \mathrm{M}h_{i,j}}{N_j A_{i,j}} \right), \tag{6.83}$$

where i is a facet number and j is a gas source number. Knowing the gas flows Q_j different sources of gas one can easily calculate gas density for various operation scenarios, for example:

- Thermal outgassing is the main source of gas in accelerator vacuum chamber without a charged particle beam.
- PSD due to SR provides an additional source of gas in the presence of the beam.
- PSD from vacuum chamber and SR absorbers are conditioned differently.

Pumping speed of pumps and sorbing surfaces (and, therefore, sticking probability in TPMC model) for different gases can vary significantly; furthermore, the pumping speed may change with time, amount of pumped (or sorbed) gas, and other operation conditions. Thus it is necessary to make a separate run of calculations for each difference in a set of parameters. For example, modelling an accelerator section with SR shown in Figure 6.22, one would need to run the TPMC model for 12 sets of parameters: 3 sources of gas (thermal outgassing, PSD from SR absorbers, and PSD from a beam vacuum chamber) × 4 gases (H_2, CH_4, CO, and CO_2). Varying pumping speed for each gas would require even more runs of calculations.

6.3.5.2 Models with Variable Parameters

The following method, described and practically employed in Ref. [24, 25], allows to reduce a number of necessary TPMC runs of calculations.

Let us consider a piece of vacuum chamber with uniform thermal desorption with a pump and its TPMC model (see Figure 6.23a,b). To plot pressure distribution along the chamber for various pumping speed of the pumps, one can run a number of TPMC calculations, varying a sticking probability α between any small number and 1. Alternatively, the same result can be obtained with only two TPMC calculation runs (see Figure 6.23c). This can be done by modelling a pump as a superposition of ideal pumping and backflow from the pump. For this the pump is modelled as a surface with a sticking probability $\alpha = 1$. In Run 1, the vacuum chamber walls are desorbing and only virtual pumping surface is not desorbing (modelling an ideal pump). While in Run 2, it is opposite: the vacuum chamber walls are not desorbing and only virtual pumping surface is desorbing (modelling a backflow). The results of two runs allow to find a gas density at any facet i as follows:

$$n_i = \frac{4}{v} \left(\frac{Q_1 \mathrm{M} h_{i,1}}{N_1 A_{i,1}} + \frac{Q_2 \mathrm{M} h_{i,2}}{N_2 A_{i,2}} \right), \tag{6.84}$$

where indexes 1 and 2 correspond to Run 1 and Run 2, respectively; Q_1 is a net desorption from the vacuum chamber walls; and Q_2 is the net backflow from the pump, which equals to $Q_2 = Q_1/\alpha$. Thus, a gas density at any facet i as a function

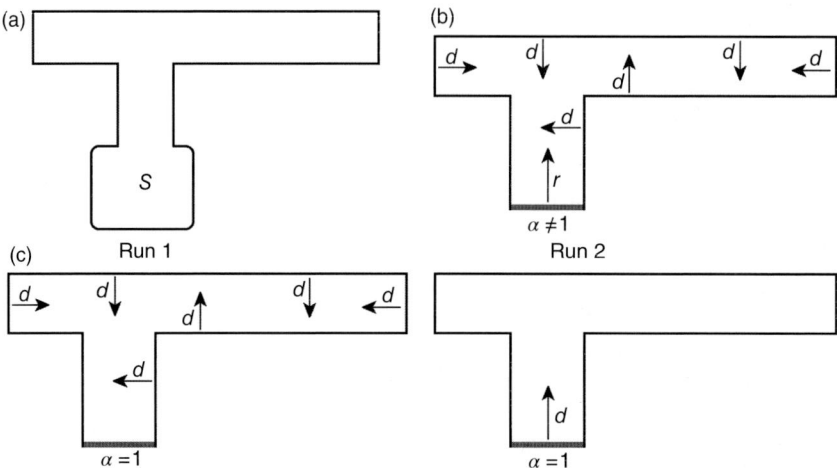

Figure 6.23 (a) A schematic layout of vacuum chamber with a pump and its schematic for (b) a direct TPMC model and (c) two runs of models with variable parameters. Here d shows desorbing surfaces and r is a pump backflow.

of sticking probability is

$$n_i(\alpha) = \frac{4Q_1}{\bar{v}} \left(\frac{Mh_{i,1}}{N_1 A_{i,1}} + \frac{Mh_{i,2}}{\alpha N_2 A_{i,2}} \right), \tag{6.85}$$

That is, having these TPMC results for two runs only, one can easily calculate a gas density distribution for any sticking probability with Eq. (6.85) without a need to perform more TPMC calculations.

Note: these formulas are correct when the gas flow is in equilibrium. When the gas flow is not in equilibrium, then the analysis should include the data from original surface at each location; see Eqs. (6.73) and (6.74).

Similarly, one can find a gas density distribution for a vacuum system with several different pumps as a function of their sticking probabilities. Let us consider a vacuum system with K pumps with sticking probabilities: $\alpha_1, \alpha_2, \alpha_3, \ldots, \alpha_K$. In TPMC model all pumps are ideal: i.e. sticking probabilities are equal to 1. In Run 0, the vacuum chamber walls are desorbing, and in the following K runs, the backflow is modelled from one of the pumps, i.e. from pump 1 in Run 1, from pump 2 in Run 2, etc. The results of $K+1$ runs allow to find a gas density at any facet i as follows:

$$n_i = \frac{4}{\bar{v}} \left(\frac{Q_0 Mh_{i,0}}{N_0 A_{i,0}} + \sum_{j=1}^{K} \frac{Q_j Mh_{i,j}}{N_j A_{i,j}} \right), \tag{6.86}$$

where Q_0 is a net desorption from the vacuum chamber walls and Q_j are the backflow from each of K pumps, which should be defined yet. To find Q_j we will consider $K+1$ elements of vacuum system: desorbing walls of vacuum chamber and K pumping surfaces. Thus, if a pump j was modelled with a few facets, a number of generated and pumped particles at pump j, N_j and Mp_j, and its surface area, A_j, are a sum of Mp_i and A_i for these facets composing pump j:

$$Q_j = \sum_{\text{pump } j} Q_i, \quad Mp_j = \sum_{\text{pump } j} Mp_i, \quad A_j = \sum_{\text{pump } j} A_i. \tag{6.87}$$

The transmission probability matrix $\mathbf{W}$ can be formed with the transmission probabilities between each pair of $K+1$ elements $w_{j,k}$ defined as a ratio between a number of particles pumped at element j to a number of particles generated at element k:

$$w_{j,k} = \frac{Mp_j}{N_k} \quad \text{for } j,k = 0,\ldots,K;$$

$$\mathbf{W} = \begin{bmatrix} 0 & 0 & 0 & \cdots & 0 & \cdots & 0 \\ w_{1,0} & w_{1,1} & w_{1,1} & \cdots & w_{1,k} & \cdots & w_{1,K} \\ w_{2,0} & w_{2,1} & w_{2,2} & \cdots & w_{2,k} & \cdots & w_{2,K} \\ \cdots & \cdots & \cdots & \cdots & \cdots & \cdots & \cdots \\ w_{j,0} & w_{j,1} & w_{j,2} & \cdots & w_{j,k} & \cdots & w_{j,K} \\ \cdots & \cdots & \cdots & \cdots & \cdots & \cdots & \cdots \\ w_{K,0} & w_{K,1} & w_{K,2} & \cdots & w_{K,k} & \cdots & w_{K,K} \end{bmatrix}. \tag{6.88}$$

The sum of transmission probabilities from element k to others should be equal to 1, and this can be used to check the correctness of obtained $w_{j,k}$ values:

$$\sum_{j=1}^{K} w_{j,k} = 1 \quad \text{for } j,k = 0,\ldots,K. \tag{6.89}$$

Another relation for checking the correctness is

$$A_j \sum_{k \neq i}^{K} w_{k,j} = \sum_{k \neq i}^{K} (A_k w_{j,k}) \quad \text{for } j = 1,\ldots,K, \tag{6.90}$$

where A_j is the area of jth pumping surface (representing pump j in TPMC model).

The net gas flow rate Q_j at the surface j consists of three components (see Figure 6.19): the incoming flow of particles hitting the sorbing surface, f_j, and two outgoing components, namely, desorption, d_j, and a backflow (or reflected flow), $(1-\alpha_j)f_j$. Considering that $\alpha_0 = 0$, one can write

$$Q_j = d_j - \alpha_j f_j \quad \text{for } j = 0,\ldots,K. \tag{6.91}$$

The flow conversation equations can be written as follows:

$$f_i = \sum_{j=0}^{K} (w_{i,j}[d_j + (1-\alpha_j)f_j]) \quad \text{for } i = 0,\ldots,K. \tag{6.92}$$

This is a system of K equations for K unknowns, f_i, which can now be written in matrix form as

$$[\mathbf{E} - \mathbf{W} \cdot \text{diag}(1-\boldsymbol{\alpha})] \cdot \mathbf{f} = \mathbf{W} \cdot \mathbf{d}, \tag{6.93}$$

where $\mathbf{E}$ is a unit matrix and $\text{diag}(1-\boldsymbol{\alpha})$ is the diagonal matrix of vector $(1-\boldsymbol{\alpha})$.

Having the solutions for f_j, the gas flows Q_j at every boundary surfaces can be calculated with Eq. (6.91) for any sticking probability at each pump and, therefore, a corresponding gas density at any facet i can now be calculated with Eq. (6.86).

This means that in a case of one defined source of gas and K pumps, the vacuum system is fully described for any pumping speed of each pump after having $K+1$ results of TPMC modelling as described earlier.

Any additional source of gas requires only one TPMC run similar to Run 0, but for this source of gas, the results are used to define a new set of the transmission probabilities $w_{j,0}$, while the other elements of the transmission probability matrix $\mathbf{W}$ remain the same. Following calculations of f_j and Q_j provide all necessary information for calculating the gas densities corresponding to this source of gas. Hence, in the case of vacuum system with K_s sources of gas and K pumps, the vacuum system is fully described for any pumping speed of each pump after having $K+K_s$ results of TPMC modelling as described earlier. For example, modelling an accelerator section with SR shown in Figure 6.22, one would need to run the TPMC model for six sets of parameters: $K_s = 3$ (sources of gas: thermal outgassing, PSD from SR absorbers, and PSD from a beam vacuum chamber) and $K = 3$ (a number of pumps). These results allow calculating analytically the gas density for any gas and any pumping speed at each pump and any ratios between three sources of gas: i.e. six sets of parameters for TPMC modelling in this approach with variable parameters allow to have much greater variety of answers in comparison with a minimum of 12 sets of parameters for direct TPMC model with a defined set of parameters. Thus this approach with variable parameters for TPMC modelling and a following analysis could save a vacuum designer significant effort and a required computing time.

6.3.6 TPMC Result Accuracy

The statistical error of the results obtained with TPMC follows a Poisson distribution and can be calculated for each facet i as follows:

$$\sigma_i = \sqrt{Mh_i \left(1 - \frac{Mh_i}{N}\right)}. \tag{6.94}$$

Therefore, the accuracy of TPMC calculations for large number of generated particles, N, decreases as the reciprocal of the square root of the number of hits, Mh_i:

$$\frac{\sigma_i}{Mh_i} = \sqrt{\frac{1}{Mh_i}}. \tag{6.95}$$

Since $Mh_i \propto N$, then the accuracy of TPMC calculations is proportional to $1/\sqrt{N}$. Thus, in calculating a transmission probability w between the inlet (1) and outlet (2), as shown in Figure 6.21 and Eq. (6.80), the accuracy is

$$\frac{\sigma_2}{Mp_2} = \sqrt{\frac{1}{Mp_2}} = \sqrt{\frac{w}{N}}. \tag{6.96}$$

When modelling a long structure (typical for accelerators) with differential pumping or with pumping walls, it might take a significant computing time to reach acceptable accuracy with the modelled results. Together with the time required to build a model, this can be a limiting factor for using TPMC in early stages of vacuum system optimisation.

6.4 Combining One-Dimensional and Three-Dimensional Approaches in Optimising the UHV Pumping System

6.4.1 Comparison of Two Methods

Both methods can be useful in different stages of vacuum system design. The main advantages and disadvantages for two methods are summarised in Table 6.8. A choice of method for modelling depends on how much time is available for modelling, what effort can realistically be applied within this time, and what accuracy is required. At the early stage of the design when many parameters are unknown, the desorption yields, pumping speed, geometry, and locations of different components are not well defined. A 1D diffusion model is applied more often for quick analysis of design modifications and optimisation; however, towards the final design when design options are much fewer and the accuracy of the model play more significant role, the intense use of TPMC models can be more beneficial.

Table 6.8 Comparison between the diffusion and TPMC models.

Model	Diffusion	TPMC
Accuracy	1D simplified model	3D accurate model
	Global averaged parameters: P, u, S, etc.	Local parameters: n, w, α, etc.
Simple shape (tubes, o-office, cone, etc.)	Simple accurate formulas	Short time accurate calculations
Complicate shape	Rough estimation	Accurate
Long structures	Short time calculations	Time-consuming calculations
Vacuum system optimisation	Easy to change geometry and vary all parameters, quick calculations	Time-consuming building and changing a model and time-consuming calculations
Molecular beaming	Does not consider at all	Accurate modelling
Time dependent processes	Easily applicable	Dramatically increases calculation time
Use	Good knowledge of gas dynamic is essential	

6.4.2 Combining of Two Methods

For modelling an accelerator vacuum chamber with a gas diffusion model, one needs to define a specific vacuum conductance along the beam chamber u and distributed pumping speed c. This can be done with formulas from vacuum handbooks, when the shape of the vacuum chamber of a pumping port is studied and reported there. However, both beam chambers and pumping ports often have a unique and complex shape (see Figures 1.3, 1.4, 6.11), so a handbook formula can provide an approximate result or could not be employed at all.

However, even at very initial design stage, the TPMC models can be built for obtaining an accurate input for 1D model, for example, to accurately calculate a transmission probability of pumping port, a vacuum conductance along a vacuum chamber, specific complex vacuum component, or other complex section of vacuum chamber, which could not be described analytically with a simple 1D model.

In an accelerator vacuum chamber, the pumping ports are often covered with various types of mesh (see examples of mesh in Figure 6.24). Considering that the dimensions of the holes and slots of the mesh could be comparable with its thickness, a TPMC model would provide the most reliable transmission probability of a mesh.

Example: A Pumping Port at Diamond Light Source

A vacuum system of the DLS was optimised by employing a 1D approach described in Section 6.2. One of the uncertainties was to correctly calculate the effective pumping speed of a pump connected to a complex shape pumping port shown in Figure 6.25. Thus, the DLS pumping ports with a few different geometries were modelled with TPMC to obtain the pumping port transmission probability w. The effective pumping speed S_{eff} can now be calculated and plotted for various gases and various pumps using only the transmission probability

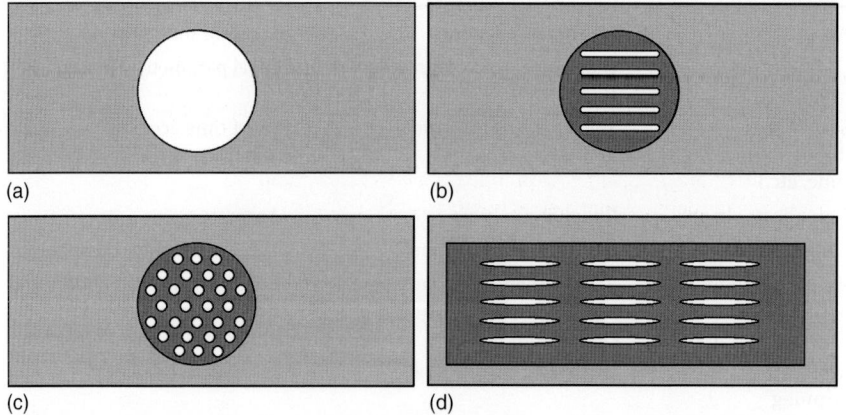

Figure 6.24 Pumping ports without a mesh and with different types of mesh. (a) A simple pumping port, (b, d) a pumping port with a mesh with slots, and (c) a pumping port with a mesh with holes.

6.4 Combining 1D and 3D Approaches

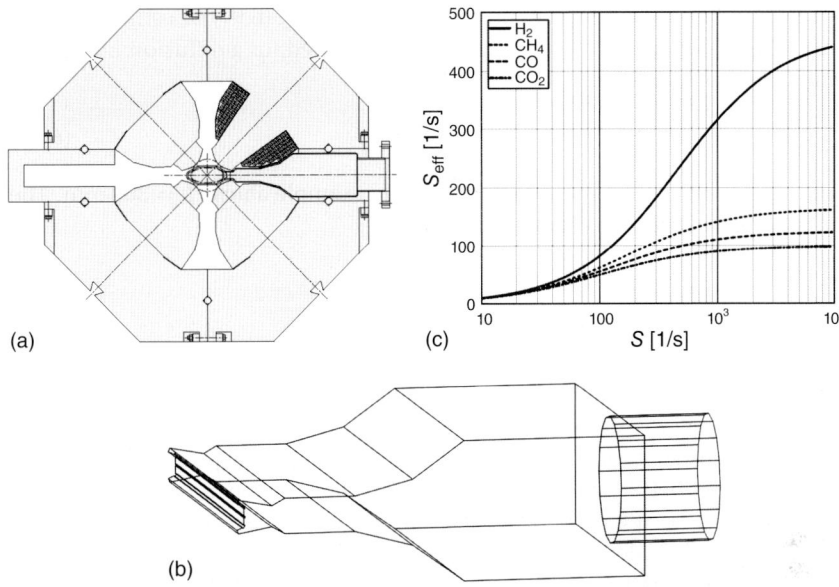

Figure 6.25 (a) A pumping ports inside the quadrupole magnets in DLS storage ring, (b) its TPMC model, and (c) the effective pumping speed as a function of a nominal pumping speed of connected pump for different gases.

value w:

$$U(M, T) = w \frac{A_e \bar{v}(M, T)}{4},$$

$$S_{\text{eff}} = \frac{U(M, T)S(M)}{U(M, T) + S(M)}, \qquad (6.97)$$

where $S(M)$ is a pumping speed of a connected pump for a gas with a molar mass M. For example, for a pumping port in an entrance dimension of 23 mm × 300 mm, the transmission probability was found to be $w = 0.15$. The calculated effective pumping speed S_{eff} is shown in in Figure 6.25c as a function of pumping speed for H_2, CH_4, CO, and CO_2. Thus if a pump has pumping speeds $S(M)$ 220 l/s for H_2, 40 l/s for CH_4, and 100 l/s for CO and CO_2, then the

Table 6.9 An example: an effective pumping speed S_{eff} of a pump with a pumping speed S connected to pumping port, and a distributed pumping speed u for the diffusion model.

Gas	S [l/s]	S_{eff} [l/s]	u [l/(s·m)]
H_2	220	150	496
CH_4	40	32	107
CO	100	55	184
CO_2	100	50	165

effective pumping speed S_{eff} can be calculated with Eq. (6.97) or estimated from Figure 6.25c. A distributed pumping speed u required for a diffusion model can be calculated as follows:

$$u = \frac{S_{\text{eff}}}{b}. \tag{6.98}$$

where b is a width of the pumping port; in the example above, $b = 300$ mm. The results for this example are shown in Table 6.9.

6.5 Molecular Beaming Effect

In particle accelerators the vacuum specification for different parts of the machine could vary in orders of magnitude. For example, the Ga/As photocathodes require a total pressure better than 10^{-10} Pa and partial pressures of oxygen containing cases better than 10^{-13} Pa, while for the other parts of the machine, it could be 10^{-6} Pa. What would be the difference between the results obtained with a diffusional model and TPMC model?

Figure 6.26 schematically shows a tube of the length L_1, a diameter d, and the wall sticking probability α. The tube is connecting two large vessels (i.e. their size is much larger than a tube diameter d) with fixed gas densities n_1 and n_2, where $n_1 \gg n_2$. A test plate is located at $z = L_2$ normal to the tube axis z. We will investigate two effects:

A gas density along the tube axis z obtained with diffusional model and TPMC.
A ratio of impingement rates at both sides of the test plate.

Figure 6.27 shows the results of TPMC calculations for a tube with a wall sticking probability $\alpha = 10^{-3}, 10^{-2}, 0.1$, and 1.0 measured for the facets parallel (X–Z and Y–Z planes) and normal (X–Y plane) to axis z. Impingement rates were normalised to one corresponding to gas density n_1 and plotted as a function of normalised length z/d. The results obtained with the parallel and normal facets are identical only for some distance from the entrance to the tube and depends on sticking probability, i.e. for $z < z_{\text{ipn}}(\alpha)$. In our example a difference of 10% in the impingement rates for the parallel and normal facets are reached at $z_{\text{ipn}}(10^{-3})/d \approx 250$, $z_{\text{ipn}}(10^{-2})/d \approx 55$, $z_{\text{ipn}}(0.1)/d \approx 4$, and $z_{\text{ipn}}(10^{-3})/d \approx 0.07$. For $z < z_{\text{ipn}}(\alpha)$ the impingement rates for the normal facets is higher than for the

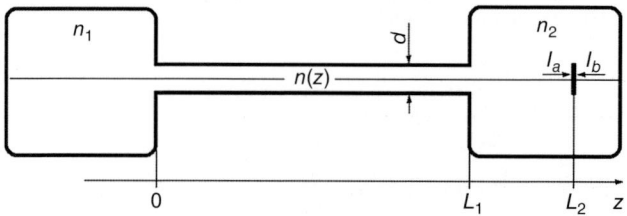

Figure 6.26 Schematics of two large vessels connected by a tube with a diameter d and a length L_1. I_a and I_b are the impingement rates at both sides of a test plate located at $z = L_2$.

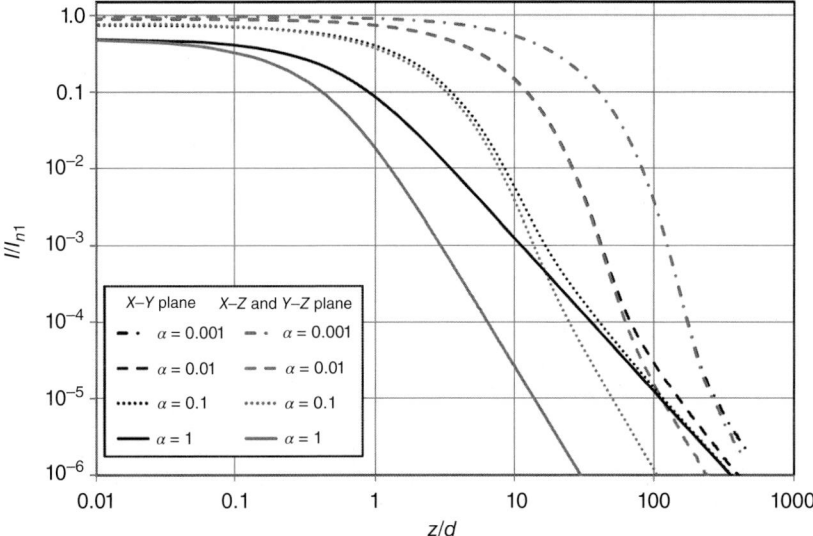

Figure 6.27 Comparison of normalised impingement rates for the facets parallel (X–Z and Y–Z planes) and normal (X–Y plane) to axis z calculated with TPMC for as a function normalised length $x = z/d$ for a tube with a wall sticking probability $\alpha = 10^{-3}, 10^{-2}, 0.1$, and 1.0.

parallel facets, which means that the angular distribution of molecular velocities is no longer uniform, and more molecules are travelling along the axis z. For example, for a tube with $\alpha = 1$, the impingement rates for the normal facets is a factor 100 higher than for the parallel facets at $z/d = 20$: i.e. practically all molecules are travelling parallel to the longitudinal axis z in its positive direction axis creating a molecular beam. This example with $\alpha = 1$ demonstrates a strong molecular beaming effect. However, the molecular beaming effect can be observed with tubes with sorbing walls, in case of differential pumping and even with non-sorbing tubes (see example as follows).

The molecular beaming effect always exists; but this effect is not considered in a diffusion model. And it is worth considering how strong it could be, referring to Figure 6.28, which shows a comparison between the results of TPMC and diffusion model for the same tube with $\alpha = 10^{-3}, 10^{-2}, 0.1$, and 1. The normalised gas density $n(z)/n_1$ was plotted as a function of normalised length z/d. The gas density for TPMC results was calculated from the average impingent rates in three orthogonal planes:

$$n_i = \frac{4Q_n}{3N\bar{v}} \left(\frac{Mh_i^{XY}}{A_i^{XY}} + \frac{Mh_i^{YZ}}{A_i^{YZ}} + \frac{Mh_i^{XZ}}{A_i^{XZ}} \right). \tag{6.99}$$

In the case of a circular tube, it can be simplified to

$$n_i = \frac{4Q_n}{3N\bar{v}} \left(\frac{Mh_i^{XY}}{A_i^{XY}} + 2\frac{Mh_i^{XZ}}{A_i^{XZ}} \right). \tag{6.100}$$

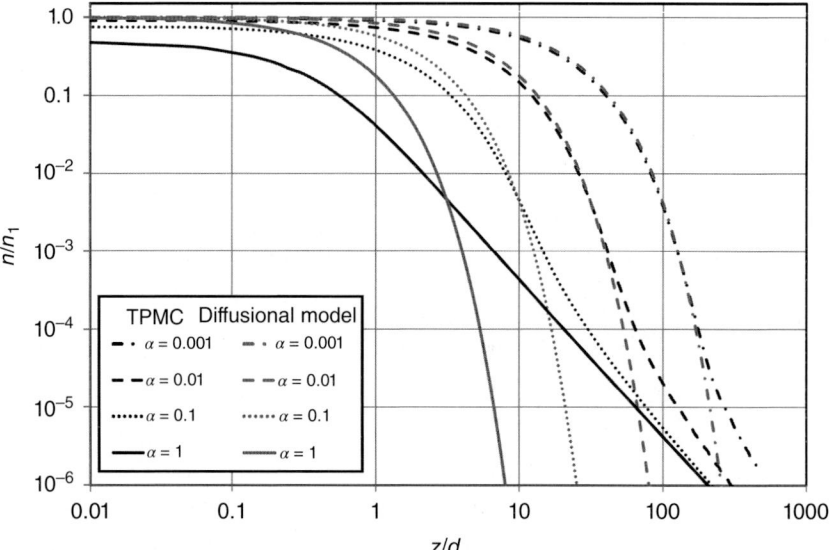

Figure 6.28 Comparison of normalised gas density calculated with TPMC and diffusional model as a function of normalised length $x = z/d$ for a tube with a wall sticking probability $\alpha = 10^{-3}$, 10^{-2}, 0.1, and 1.0.

Let us calculate the impingement rates at both sides of the test plate. A specific impingement rate calculated with a diffusion model defined as

$$I_d = \frac{n_2 \bar{v}}{4} \left[\frac{\text{molecules}}{\text{s} \cdot \text{m}^2} \right], \tag{6.101}$$

this model does not differentiate the impingement rates I_a and I_b: i.e. they are equal in this model.

In contrast, an impingement rates I_a and I_b calculated with TPMC are different:

$$I_b = I_d = \frac{n_2 \bar{v}}{4} \left[\frac{\text{molecules}}{\text{s} \cdot \text{m}^2} \right], \tag{6.102}$$

However,

$$I_a = I_b + I_{mb} = (n_1 w_{0,L2} + n_2) \frac{\bar{v}}{4} \left[\frac{\text{molecules}}{\text{s} \cdot \text{m}^2} \right], \tag{6.103}$$

where $w_{0,L2}$ is a transmission probability of molecules from the entrance to the tube at $z = 0$ to the test sample area of diameter d coaxial to the tube. One can see that I_a consists of two independent parts proportional to n_1 and n_2; in the case when $n_1 w_{0,L2} \gg n_2$, the impingement rate in vessel 2 depends mainly on the gas density in vessel 1 and practically insensitive to the gas density in vessel 2.

There are a number of examples where this effect could be very strong and must be considered.

Example 1. GaAs Photocathode Electron Gun as a Source of High Intensity Electron Beam

The advantage of GaAs photocathodes is that they can have a quite high quantum efficiency yield, a number of emitted electrons per photon, compared to other photocathodes. That is why Ga/Ar photocathode is one of the preferable sources of electron for high intensity linear accelerators. However it requires extreme high vacuum (XHV) conditions: ~0.01 ML of oxygen containing gases causes the degradation of the photocathode. At room temperature 1, ML/s may be formed at $P = 10^{-6}$ mbar. Thus, the photocathode will degrade after ~10 days at partial pressure of 10^{-14} mbar for oxygen-containing gases. So, vacuum specification for the photocathode vacuum chamber is to be below measurable with vacuum gauges, since no modern gauges can measure the pressures below 10^{-13} mbar; the lifetime of the photocathode is a best indicator of average gas density. For the other parts of linear accelerator, the required pressure could be specified as 10^{-8} mbar in N_2 equivalent.

Let us consider a vacuum chamber design near an electron gun as shown in Figure 6.29a. Pressure in the photocathode vacuum chamber $P_1 = 10^{-14}$ mbar, vacuum chambers 2 and 3 with beam position monitors (BPMs), and other instruments are equipped with pumps with pumping speeds S_2 and S_3 providing differential pumping and corresponding pressures $P_2 = 10^{-10}$ mbar and $P_3 = 10^{-8}$ mbar. Let us also consider that tubes connecting vessels 1, 2, and 3 are coated with NEG with a sticking probability $\alpha = 1$ for the oxygen-containing gases. In this case, an impingement rate on a photocathode is a sum of

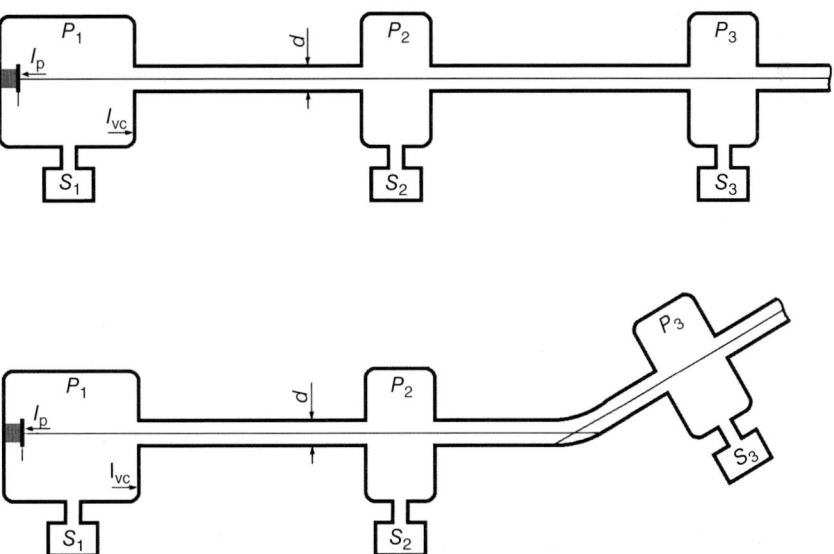

Figure 6.29 A layout of vacuum chamber design near an electron gun: (a) straight vacuum chamber and (b) vacuum chamber with a bend.

three terms:

$$I_p = I_{vc} + I_{mb21} = (n_1 + n_2 w_{21p} + n_3 w_{31p})\frac{\bar{v}}{4} = (P_1 + P_2 w_{21p} + P_3 w_{31p})\frac{\bar{v}}{4k_B T}, \quad (6.104)$$

where w_{21p} and w_{31p} are transmission probabilities of molecules from vessel 2 and vessel 3 to the photocathode in vessel 1, correspondingly. Thus, there is a possibility (which can be accurately modelled with TPMC) that $P_3 w_{31p} \gg P_1$, and/or $P_2 w_{21p} \gg P_1$. In this case increasing pumping speed S_1 (and further decreasing of P_1) would not help in increasing the photocathode lifetime because it is defined by P_3 and/or P_2.

In general, the pressures P_2 and P_3 can be reduced by increasing pumping speeds S_2 and S_3, correspondingly, the transmission probabilities w_{21p} and w_{31p} can be reduced by reducing a diameter and increasing the length of tubes. However, these methods have practical limits: a pumping speed can be increases by a few times only, a tube diameter could not be less than a beam aperture, and a tube length cannot be increased too much within the building. The best solution is to mitigate the molecular beaming effect, for this the molecular beam should hit a wall. This can be easily done by bending a tube between vessels 2 and 3 (with use of a dipole magnet for the charged particle trajectory change), as shown in in Figure 6.29b. As soon as vessel 3 is out of line of sight for the photocathode, there is no molecular beaming effect from vessel 3, i.e. $w_{31p} = 0$. Similarly, if there is still $P_2 w_{21p} \gg P_1$, then bending a tube between vessels 1 and 2 will lead to $w_{21p} = 0$.

Thus, in the case of high pressure difference between vessels 1, 2, and 3, even employing such a good vacuum technology as NEG-coated vacuum chamber would not always help to meet vacuum specifications because of molecular beaming effect. In practice, using the NEG coating vacuum chambers is not very common in the design of vacuum systems near the electron guns, so the impact of high pressure difference between vessels 1, 2, and 3 could be even stronger. However, there is no molecular beaming between the parts with no line of sight.

Example 2. ISD from the Plate and the Bottom of the Tube

Another example of molecular beaming effect was investigated for the measurements of heavy ion-stimulated gas desorption. In the experiments we wanted to compare a heavy ion-stimulated induced desorption (HISD) yields from a flat sample in normal incident as shown in Figure 6.30. Two types of samples were used: plates (20 mm × 20 mm) and tubes with a closed end (20 mm × 20 mm × 80 mm). It was calculated with TPMC model that pressure increase measured in test chamber for the same outgassing flow will be different for planar and tubular samples because the gas desorbed from the bottom of the tubular sample creates the molecular beam, which travels to the pump without interacting with the test vacuum chamber walls, so for the same desorption flux the pressure P_2 in the test chamber with a tubular sample was found to be a factor of ~1.7 lower than with a planar sample [26].

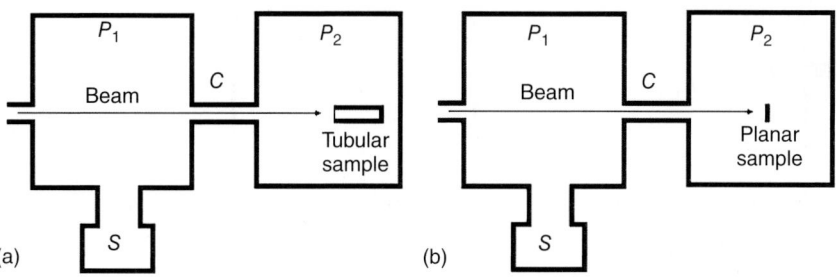

Figure 6.30 A schematic layout of experiment for measurements of HISD yields from (a) tubular and (b) planar samples.

6.6 Concluding Remarks

Modelling of gas dynamics in accelerator vacuum chamber can be done in a variety of ways:

– Simple estimation as shown in Chapter 1;
– Calculation with a 1D diffusion model;
– 3D modelling with TPMC;
– aCombined methods;
– Others, not listed above.

Each of these ways could be appropriate at different stages of the design and can be adopted to vacuum specifications, available accelerator parameters, details of vacuum chamber mechanical design and experimental input data for modelling, required accuracy, urgency, available time, software, and other resources. Therefore, knowing different modelling methods allows to make an accelerator vacuum system design and modelling in the most efficient way, i.e. in obtaining the required and reliable results with a sufficient (not excessive) effort.

Many units are used in vacuum science for various reasons (for example, Pa and mbar, m^3 and l, m^2 and cm^2). This very often results in orders of magnitude mistakes in calculating pressures, gas flows, pumping speed, etc. The best way to avoid such mistakes is to perform all calculations in SI units. If necessary, the final result can be converted to the (customer) required units at very final stage.

6.A Differential Pumping

There are a number of vacuum systems where there is a requirement to maintain a part of a system at a much higher pressure than other parts (e.g. an ion source chamber for an ion gun and a target chamber). Let us consider a vacuum chamber that consists of volume 1 pumped with a pumping speed S_1 and volume 2 pumped with a pumping speed S_2. Two volumes are connected with a vacuum conductance U (see Figure 6.A.1).

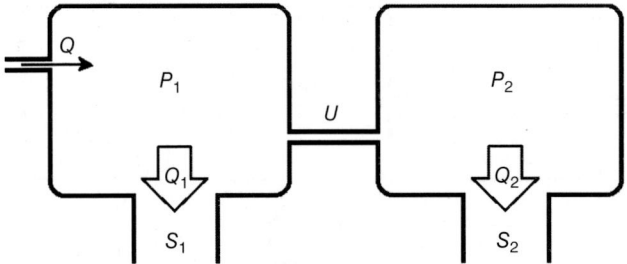

Figure 6.A.1 A layout of vacuum system with differential pumping.

We need to calculate the pumping speed S_2 required to maintain the pressure P_2. If we assume that U is small and $P_1 \gg P_2$, and the layout not allowing molecular beaming effect, then

$$S_2 = \frac{Q_2}{P_2}, \quad Q_2 = (P_1 - P_2)U \approx P_1 U, \quad P_1 = \frac{Q}{S_1 + (S_2^{-1} + U^{-1})^{-1}} \approx \frac{Q}{S_1},$$

therefore $S_2 \approx \dfrac{QU}{S_1 P}$. \hfill (A.1)

6.B Modelling a Turbo-Molecular Pump

Turbo-molecular pumps (TMPs) are often used in modern UHV systems. We would like to study the ultimate performance of TMP in the following conditions (see Figure 6.B.1): Q is total outgassing rate in a vacuum chamber; U_1 is vacuum conductance between the vacuum chamber and the TMP entrance; K is a TMP compression ratio; S_{TMP} and S_r are the pumping speeds of TMP and rough pump, respectively; U_2 is vacuum conductance between the TMP exhaust

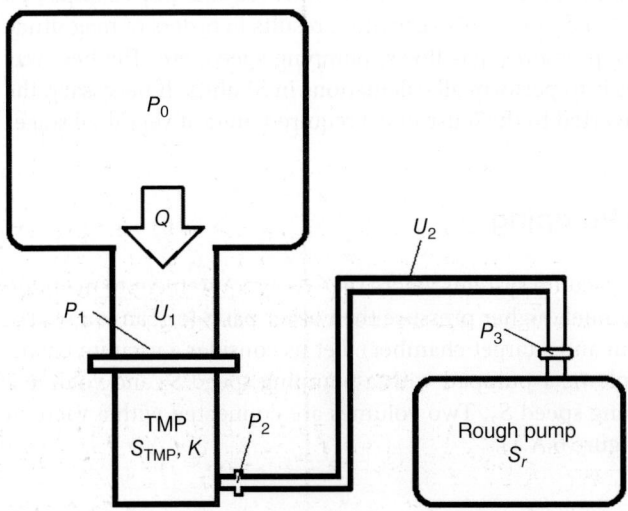

Figure 6.B.1 A vacuum chamber pumped with TMP backed with a roughing pump.

and the rough pump entrance; P_0 to P_3 are pressures in the vacuum chamber, at the TMP entrance, the TMP exhaust and the rough pump entrance, respectively. Pressures $P_0, P_1, P_2,$ and P_3 can be defined as follows:

$$P_3 = \frac{Q}{S_r} + P_{r\,ult},$$

$$P_2 = \frac{Q}{U_2} + P_3 = Q\left(\frac{1}{U_2} + \frac{1}{S_r}\right) + P_{r\,ult},$$

$$P_1 = \frac{Q}{S_{TMP}} + \frac{P_2}{K} = Q\left(\frac{1}{S_{TMP}} + \frac{1}{U_2 K} + \frac{1}{S_r K}\right) + \frac{P_{r\,ult}}{K},$$

$$P_0 = \frac{Q}{U_1} + P_1 = Q\left(\frac{1}{U_1} + \frac{1}{S_{TMP}} + \frac{1}{U_2 K} + \frac{1}{S_r K}\right) + \frac{P_{r\,ult}}{K}. \quad (B.1)$$

This formula helps optimising choice of pumps for a vacuum system, i.e. meeting vacuum specification at lowest cost. Thus, an ultimate pressure is calculated with $Q = 0$:

$$P_{0\,ult} = \frac{P_{r\,ult}}{K}. \quad (B.2)$$

When the pressure P_0 for a defined $Q \neq 0$ is defined by the lowest term between $U_1, S_{TMP}, U_2 K,$ and $S_r K$. To study how any change can affect pressure in the vacuum system, one can vary:

Vacuum conductances: U_1 and U_2.
TMP pumping speed S and compression ratio K.
Roughing pump ultimate pressure $P_{r\,ult}$ and S_r.

Acknowledgements

Author would like to acknowledge Prof. Felix Sharipov from Universidade Federal do Paraná (Curitiba, Brazil) for the many comments and suggestions.

References

1 Saksaganskii, G.L. (1988). *Molecular Flow in Complex Vacuum Systems*. Gordon and Breach Science Publisher SA.
2 Berman, A. (1992). *Vacuum Engineering Calculations, Formulas and Solved Exercises*. New York, NY: Academic Press.
3 Sharipov, F. and Barreto, Y.B. (2015). Influence of gas–surface interaction on gaseous transmission probability through conical and spherical ducts. *Vacuum* 121: 22–25.
4 Jousten, K. (ed.) (2008). *Handbook of Vacuum Technology*. Wiley. ISBN: 3527407235.
5 Sharipov, F. (2018). *Rarefied Gas Dynamics: Fundaments for Research and Practice*. Wiley. ISBN: 9783527413263.
6 Sharipov, F. and Moldover, M.R. (2016). Energy AC extracted from acoustic resonator experiments. *J. Vac. Sci. Technol., A* 34: 061604.

7 Steckelmacher, W. (1966). A review of the molecular flow conductance for systems of tubes and components and the measurement of pumping speed. *Vacuum* 16: 561–584.

8 Steckelmacher, W. (1986). Knudsen flow 75 years on: the current state of the art for flow of rarefied gases in tubes and systems. *Rep. Prog. Phys.* 49: 1083–1107.

9 Sharipov, F. and Seleznev, V. (1998). Data on internal rarefied gas flows. *J. Phys. Chem. Ref. Data* 27: 657–706.

10 Aoki, K. (1989). Numerical analysis of rarefied gas flows by finite-difference method. In: *16th International Symposium in Rarefied Gas Dynamics*, vol. 118, Washington, 1989, AIAA (eds. E.P. Muntz, D.P. Weaver and D.H. Campbell), 297–322.

11 Graur, I. and Sharipov, F. (2008). Gas flow through an elliptical tube over the whole range of the gas rarefaction. *Eur. J. Mech. B. Fluids* 27: 335–345.

12 Sharipov, F. (1999). Rarefied gas flow through a long rectangular channel. *J. Vac. Sci. Technol., A* 17: 3062–3066.

13 Sharipov, F. and Graur, I. (2014). General approach to transient flows of rarefied gases through long capillaries. *Vacuum* 100: 22–25.

14 Pantazis, S., Valougeorgis, D., and Sharipov, F. (2014). End corrections for rarefied gas flows through circular tubes of finite length. *Vacuum* 101: 306.

15 Diamond Synchrotron Light Source. (2003). Report of the design specification (Green book). CCLRC, Daresbury Laboratory, Warrington, Cheshire, UK, June 2003.

16 Herbert, J.D., Malyshev, O.B., Middleman, K.J., and Reid, R.J. (2004). Design of the vacuum system for diamond, the UK 3rd generation light source. *Vacuum* 73: 219.

17 Malyshev, O.B. and Cox, M.P. (2012). Design modelling and measured performance of the vacuum system of the Diamond Light Source storage ring. *Vacuum* 86: 1692.

18 Bird, G.A. (1994). *Molecular Gas Dynamics and the Direct Simulations of Gas Flows*. Oxford: Oxford University Press.

19 Kosumoto, Y. (2007). Reflection rules preserving molecular flow symmetry in an arbitrarily shaped pipe. *J. Vac. Sci. Technol., A* 25: 401.

20 Suetsugu, Y. (1996). Application of the Monte Carlo method to pressure calculation. *J. Vac. Sci. Technol., A* 14: 245.

21 Ady, M. (2016). Monte Carlo simulations of ultra high vacuum and synchrotron radiation for particle accelerators. Phd thesis (no. 7063). École Polytechnique Fédérale de Lausanne, https://cds.cern.ch/record/2157666.

22 Kersevan, R. and Pons, J.L. (2009). Introduction to Molflow+. *J. Vac. Sci. Technol., A* 27: 1017. https://molflow.web.cern.ch.

23 STL (file format). https://en.wikipedia.org/wiki/STL.

24 Luo, X., Day, C., Hauer, V. et al. (2006). Monte Carlo simulation of gas flow through the KATRIN DPS2-F differential pumping system. *Vacuum* 80: 864.

25 Malyshev, O.B., Day, C., Luo, X., and Sharipov, F. (2009). Tritium gas flow dynamics through the source and transport system of the Karlsruhe tritium neutrino experiment. *J. Vac. Sci. Technol., A* 27: 73.

26 Hedlund, E., Westerberg, L., Malyshev, O.B. et al. (2009). Ar ion induced desorption yields at the energies 5–17.7 MeV/u. *Nucl. Instrum. Methods Phys. Res., Sect. A* 599: 1–8.

7

Vacuum Chamber at Cryogenic Temperatures

Oleg Malyshev[1], Vincent Baglin[2], and Erik Wallén[3]

[1] ASTeC, STFC Daresbury Laboratory, Keckwick Lane, Daresbury, Warrington, WA4 4AD Cheshire, UK
[2] CERN, Organisation européenne pour la recherche nucléaire, Espl. des Particules 1, 1211 Meyrin, Switzerland
[3] Lawrence Berkeley National Laboratory, 1 Cyclotron Road, Mail Stop 15R0217, Berkeley, CA 94720, USA

7.1 Pressure and Gas Density

Modern charged particle accelerators can use superconducting dipole, quadrupole, and sextupole magnets, wigglers, and undulators. These magnets could be a single component or long section(s) occupying from a short fraction of the machine to its significant part or even the full circumference/length. The vacuum chamber inside such cryogenic magnets could have, in general, a temperature between the temperature of superconducting magnet and room temperature. Thus, the temperature of vacuum chamber along the beam path may vary in wide range.

It was already mentioned in Chapter 1 that the vacuum specifications for the machine operation are dictated by beam–gas interaction and, therefore, it is primarily defined by a residual gas density. For the machines where the whole vacuum chamber is at the same temperature T_{vc} (room temperature or any other fixed temperature), the use of pressure is equally acceptable because the pressure P and gas density n are proportional to a constant coefficient $k_B T_{vc}$:

$$P = n k_B T_{vc}. \tag{7.1}$$

However, in a cryogenic vacuum chamber with variations in the wall temperatures, the coefficient $k_B T$ is not a constant. Let us consider two vessels at temperatures T_1 and T_2 (where $T_1 > T_2$) and connected as shown in Figure 7.1.

In a viscous flow regime, there is an equal pressure in both vessels but the gas density is higher in a vessel with lower temperature:

$$P_1 = P_2 \Rightarrow n_1 = n_2 \frac{T_2}{T_1} \Rightarrow n_1 < n_2. \tag{7.2}$$

In a molecular flow regime, the condition of balance is different: a flow of molecules in both directions must be equal; thus $n_1 \bar{v}_1 = n_2 \bar{v}_2$, where $\bar{v}$ is mean

Vacuum in Particle Accelerators: Modelling, Design and Operation of Beam Vacuum Systems,
First Edition. Oleg B. Malyshev.
© 2020 Wiley-VCH Verlag GmbH & Co. KGaA. Published 2020 by Wiley-VCH Verlag GmbH & Co. KGaA.

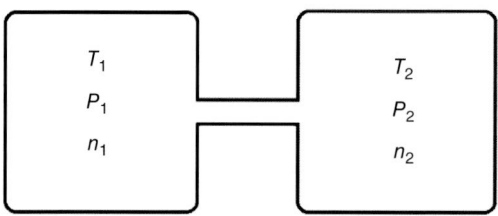

Figure 7.1 Two connected vessel at different temperatures T_1 and T_2.

molecular velocity. Therefore, pressure is higher in a warmer vessel, while gas density is higher in cooler vessel:

$$n_1 \bar{v}_1 = n_2 \bar{v}_2 \Rightarrow \begin{cases} P_1 = P_2 \sqrt{\dfrac{T_1}{T_2}} \Rightarrow P_1 > P_2, \\ n_1 = n_2 \sqrt{\dfrac{T_2}{T_1}} \Rightarrow n_1 < n_2. \end{cases} \quad (7.3)$$

Due to this thermal transpiration effect, the pressures in the two vessels located at different temperatures are different. Thus, using pressure could be misleading in the specifications and design of cryogenic vacuum systems with various temperatures; hence it is recommended to use gas density only.

It is often necessary to know the gas density and pressure inside the cryogenic vessel at temperature T_2, while the measurements are performed at room temperature T_1. In this case, one can obtain a useful formula combining Eqs. (7.1) and (7.3):

$$n_2 = n_1 \sqrt{\dfrac{T_1}{T_2}} = \dfrac{P_1}{k_B T_1} \sqrt{\dfrac{T_1}{T_2}} = \dfrac{P_1}{k_B \sqrt{T_1 T_2}} \quad (7.4)$$

This thermal transpiration effect was studied in a dedicated experimental set-up where a vacuum gauge located at room temperature was compared to a vacuum gauge immersed in liquid helium, therefore operating at 4.2 K (Figure 7.2). The two vacuum vessels operating at different temperatures are connected via a ~1 m long, 35 mm diameter tube. A Bayard–Alpert gauge is measuring the pressure at room temperature, T_1, and an extractor gauge is measuring the pressure at liquid helium temperature, T_2. A screen is placed in front of the extractor filament to avoid the direct illumination of the cryogenic surface, thereby minimising the electron-stimulated desorption (ESD).

In order to study the validity of Eq. (7.2) over several orders of magnitude, an adsorption isotherm (see next section) is measured. With the turbo-molecular pump valved off, known quantities of hydrogen are injected into the cryogenic vessel. The equilibrium pressure is then recorded, for both gauges, as a function of the surface coverage. Figure 7.3 shows the hydrogen adsorption isotherm measured by the two vacuum gauges [1]. Both pressures are given for a temperature of 4.2 K. The vacuum gauge located at room temperature has been corrected by the thermal transpiration effect according to Eq. (7.2). To take into account the fact that the power supply of the vacuum gauge located in the cryogenic vessel interprets the gas density measurement as if the vacuum gauge was operated at

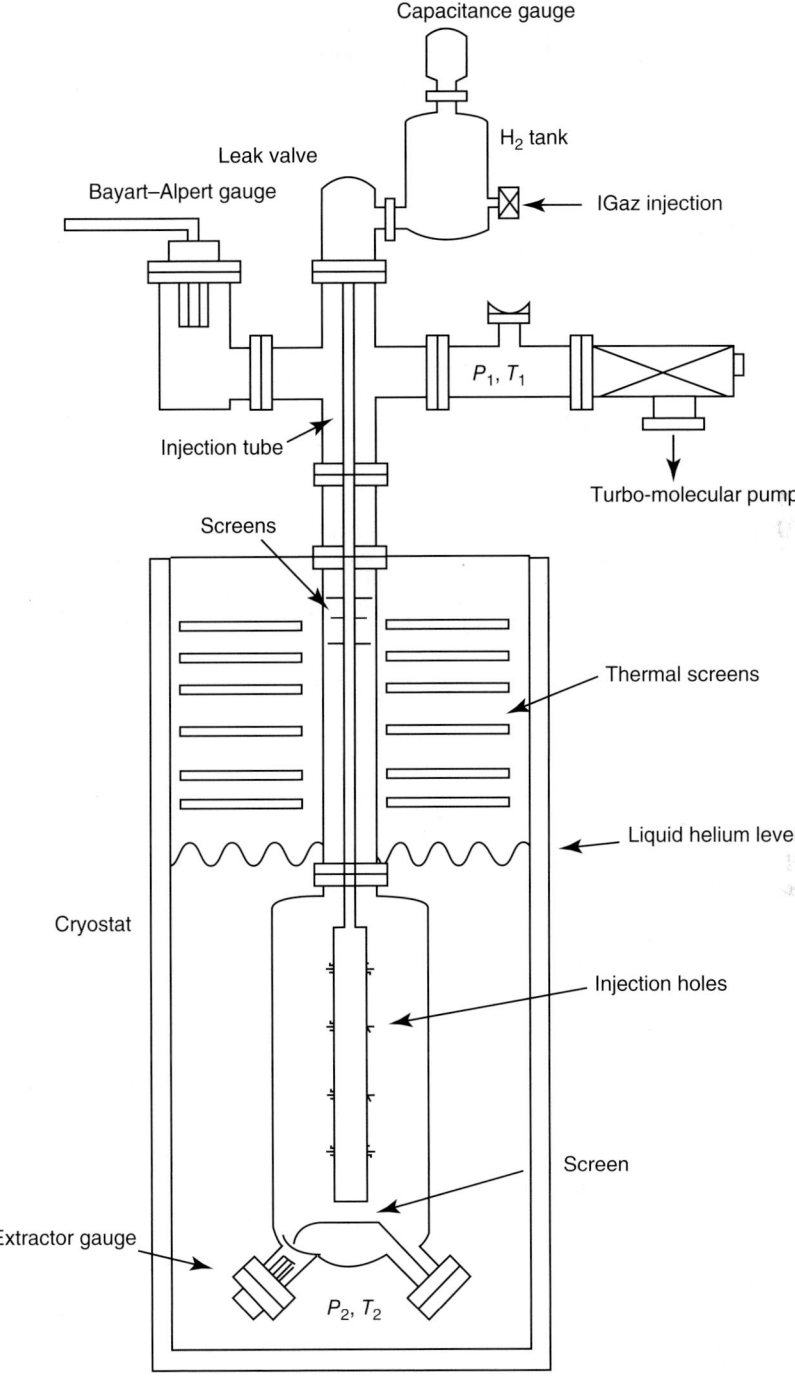

Figure 7.2 Experimental set-up to study the thermal transpiration effect.

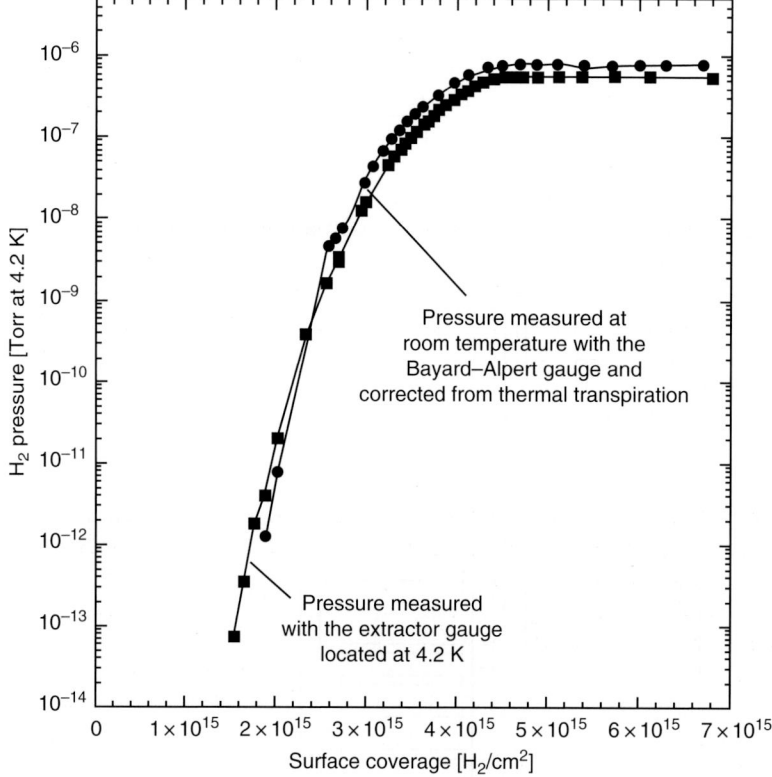

Figure 7.3 Hydrogen adsorption isotherms measured by vacuum gauges located at room temperature and cryogenic temperature. Source: Reprinted with permission from Baglin [1], Fig. 3. Copyright 2007, CERN.

room temperature, the reading, $P_{T_2,\text{read}}$, of the vacuum gauge located at 4.2 K has been corrected according to Eq. (7.4).

$$P_2[\text{Pa}] = \frac{T_2[\text{K}]}{300[\text{K}]} P_{T_2,\text{read}}[\text{Pa}]. \tag{7.5}$$

As expected, from Eq. (7.2), when corrected from thermal transpiration effects, the two pressures measured at room temperature and at cryogenic temperature are quasi-overlapped over six decades.

7.2 Equilibrium Pressure: Isotherms

The residual gas molecules in a cryogenic vacuum chamber will be attracted to the walls of the vacuum chamber due to the long-range van der Waals interaction, which stems from coupling between the dipole fluctuations in the residual gas molecules and the vacuum chamber wall. At small distances to the vacuum chamber wall, there is repulsion, which occurs when the electron cloud of the residual gas molecules starts to overlap appreciably with the electrons of the material in the vacuum chamber wall. The residual gas molecules are trapped, or physisorbed, in the potential well formed by the long-range attractive force and the short-range repulsive force between the residual gas molecules and the surface. Physisorption is sometimes called cryosorption, and it has to be distinguished from cryocondensation or just condensation. In condensation, the adsorbate is adsorbed on its own liquid or solid, while in physisorption the adsorbate is adsorbed on a substrate that consists of a material other than the adsorbate. The chemical state of the physisorbed molecules is virtually identical to the gas-phase state as shown by, e.g. [2].

Physisorption, which does not involve any change in chemical configuration, is very inefficient in adsorbing the kinetic energy of incident particles. The probability for sticking of an incident H_2 molecule with a kinetic energy of about 20 meV on a bare metal surface is of the order of a few percent. Physisorbed H_2 molecules are not tightly bound in localised sites and their freedom to translate parallel to the surface when struck by an incident particle provides an efficient means for absorption of a substantial amount of the incident kinetic energy. In fact, the sticking of H_2 on Cu is dominated by particle–particle collisions and the probability of sticking increases with the number of molecules already adsorbed on the surface. For a relative surface coverage equal to zero, the probability of sticking is about 0.1 for H_2 molecules with 15 meV of kinetic energy, and it increases linearly up to close to unity when the surface when the surface is completely covered [3]. The increase in sticking probability with increasing surface coverage in physisorption of H_2 on Cu is fundamentally different from the case of chemisorption. In chemisorption, where the pre-adsorbed species tend to block the chemisorption process, the sticking probability falls with increasing surface coverage. The physisorbed molecules are affected by the presence of other physisorbed molecules, usually in the form of an increased binding energy with increasing surface coverage. On the other hand, adsorption sites with high adsorption energy will be occupied earlier than less attractive adsorption sites, which might balance the increased binding energy with increasing surface coverage caused by lateral interactions between the physisorbed molecules. For the case of the H_2–Cu interaction, the difference between different adsorption sites is small compared to the forces introduced by the presence of other adsorbed molecules.

7.2.1 Isotherms

Although physisorption has been studied extensively for many years, it is still not completely understood and no single theory can accurately describe this phenomenon. There are, however, a number of theories for adsorption isotherms that are in use and the most frequently appearing equations are described in the following. P is the equilibrium pressure at surface coverage s.

The simplest adsorption isotherm, often called Henry's law [4]:

$$s = cP \qquad (7.6)$$

where c is a constant which depends on temperature, adsorbate, and substrate. The amount of adsorbed gas varies linearly with the pressure. Henry's law originates from observations of the solubility of gases in liquids and it builds on the assumptions that both the gas and the adsorbed phase are dilute enough to obey the perfect gas law and that there are no lateral interactions between the adsorbed gas molecules [5]. When plotted on a log-log chart, Henry's law represents a straight line with unit slope. Henry's law is the expected behaviour for all physisorption isotherms at very low surface coverage, up to some fraction of a percent of a monolayer; however, the true range of applicability of Henry's law is very short. In fact it is often below the range of experimental accessibility.

Another isotherm is the Freundlich equation [6], which, in principle, only has empirical justification:

$$s = cP^{1/n}, \tag{7.7}$$

where c and n are temperature, adsorbate and substrate dependent constants. When s is plotted against P on a log–log chart, the Freundlich isotherm is represented by a straight line with slope $1/n$. In all cases n is greater or equal to unity [7]. A large number of experimental isotherms have shown to obey the Freundlich equation, but an even larger number cannot be described by this equation.

The Langmuir isotherm is another isotherm that occupies a central position in the field of adsorption [8]. It originates from one of the first theoretical treatments of adsorption. It is based on the assumptions that the substrate has only energetically equal adsorption sites and that there is no lateral interaction between adsorbed molecules. It is further assumed that it is not possible for the adsorbed molecules to move between different sites via surface diffusion and that adsorbate molecules that impinge on a bare surface have a certain probability of being adsorbed, but those impinging on a site already occupied by an adsorbed molecule would be immediately re-evaporated; thus the Langmuir model is limited to monolayer coverage. The isotherm resulting from the assumptions above can be written as

$$s = \frac{s_m bP}{1 + bP} \tag{7.8}$$

In practice both s_m and b are determined from experiments but they have well defined physical significance. s_m is the monolayer capacity and b is an adsorption coefficient given by

$$b = \frac{\alpha_0 \exp(\Delta H_a/k_B T)}{\beta_0 (2\pi m k_B T)^{1/2}}, \tag{7.9}$$

where α_0 and β_0 are the condensation and evaporation coefficients, m is the mass of the adsorbent molecule, ΔH_a is the adsorption energy, k_B is Boltzmann's constant, and T is the absolute temperature. A linear plot of the isotherm data in P/s and P coordinates would provide a straight line with an intercept of $1/(s_m b)$ and a slope of $1/s_m$. The Langmuir isotherm is in general more applicable to chemisorption than physisorption and it has been shown to be successful in describing the isotherm of molecular hydrogen on metal surfaces at temperatures much higher than the temperatures for physisorption, while it fails at low temperatures and pressures for physisorption of H_2 on Pyrex glass [9], porous materials like silica gel [10], and gas condensates [11, 12].

With the adsorption theory from Brunauer, Emmet, and Teller [13, 14], referred to as BET, the assumptions of the Langmuir theory are expanded into the multilayer region. The basic assumption of the BET theory in its simplest form is that molecules in the first layer can serve as adsorption sites for molecules in the second layer and so on. The adsorption or desorption will always take place in the topmost layer and for the exchange of molecules between the lower layers a dynamic equilibrium is assumed. It is further assumed that the first layer will have some value for the heat of adsorption, ΔH_a, but for all succeeding layers, the adsorption energy is assumed equal to the heat of vaporisation of the pure bulk adsorbate, ΔH_L. The assumptions of the BET theory leads to the BET isotherm for multi-monolayer adsorption on a free surface, which has the form

$$s = \frac{s_m \alpha_{BET} P}{(P_0 - P)(1 + (\alpha_{BET} - 1)P/P_0)} \tag{7.10}$$

where s_m is the monolayer capacity, P_0 is the saturated vapour pressure, and α_{BET} is a dimensionless parameter: $\alpha_{BET} \approx \exp((\Delta H_a - \Delta H_L)/k_B T)$. For any particular surface the values of s_m and α_{BET} have to be determined from the measured adsorption isotherms. The BET equation may be transformed into

$$\frac{P}{s(P_0 - P)} = \frac{1}{\alpha_{BET} s_m} + \frac{(\alpha_{BET} - 1)}{\alpha_{BET} s_m} \frac{P}{P_0} \tag{7.11}$$

which would yield a straight line for the expression $P/(s(P_0 - P))$ plotted versus P/P_0. The intercept is $1/(\alpha_{BET} s_m)$ and the slope is $(\alpha_{BET} - 1)/(\alpha_{BET} s_m)$. Since α_{BET} in general is large for the case of physisorption, the slope is close to $1/s_m$.

These attempts to predict the shape of an isotherm from a few basic assumptions have had limited success in describing the real situation. An interesting theory, which does not try to predict the shape of the isotherm, is the potential theory, developed by Polanyi [15]. This theory assumes that a potential field exists near the substrate surface and the adsorbate molecules are bound to it like an atmosphere is bound to a planet. The adsorbing molecules are treated as being more compressed close to the substrate and their density decreases outwards. The adsorption potential at a point near the substrate is defined as the work done by adsorption forces in bringing in molecules from the gas phase to that point.

The adsorption potential, $\varepsilon = f(\phi)$, is a function of ϕ, which represents the adsorbate surface coverage s divided by δ_T, the liquid density of the adsorbate at the temperature T. The function $\varepsilon = f(\phi)$ decreases from its maximum value at $\phi = 0$ at $s = 0$, to its minimum value when the saturated vapour pressure is reached at ϕ_{max} at $s = s_m$. It should be noted that the interpretation of s_m is not the same in this theory as in the BET theory. The function $\varepsilon = f(\phi)$ is called the characteristic curve and it is postulated to be independent of temperature. The adsorption potential ε_i at a specific value of $\phi_i = s_i/\delta_T$ is given by the energy needed to compress the vapour from the equilibrium pressure P_i, found at the specific surface coverage s_i, to the saturated vapour pressure P_0:

$$\varepsilon_i = k_B T \ln\left(\frac{P_0}{P_i}\right). \tag{7.12}$$

From one measured isotherm it is possible to predict the isotherm for different temperatures at the same ϕ as measured, but there is no prediction for other ϕ.

An important assumption in the potential theory as described above is that the adsorption temperature is well below the critical temperature of the adsorbate gas. The characteristic curve has proved to be very successful in predicting the temperature dependence of physisorption. Experimental work done by Dubinin and Radushkevich [16–18] has shown that the low-pressure adsorption isotherms for many adsorbate–substrate combinations can often be described by the function:

$$\ln(s) = \ln(\phi_{max}\delta_T) - D\varepsilon^2 \qquad (7.13)$$

called the DR equation. D is an adsorbate–substrate-dependent constant to be fitted with experiment. Discrepancies can be seen in the high-pressure end, close to the saturated vapour pressure and in the low-pressure end where the isotherm approaches Henry's law. If the temperature dependence of the density of the liquid adsorbate is assumed to be negligible, it is possible, as shown by Kaganer [19], to use the DR equation to estimate the monolayer capacity of the substrate. In combination with the absorption potential, we arrive at the DRK equation, which is

$$\ln(s) = \ln(s_m) - D\left(k_B T \ln\left(\frac{P_0}{P}\right)\right)^2 \qquad (7.14)$$

where P is the equilibrium pressure at surface coverage s and s_m can be seen as the monolayer capacity of the surface. The DRK equation has, in many cases, been able to describe the isotherms for porous substrates and condensed gas substrates as well as very low-pressure isotherms for metal and glass substrates, situations where the BET and other theories have failed.

Comparative studies by Hobson [20] of the monolayer capacity given by the DRK equation compared with that given by the BET equation have shown that the DRK equation in general gives lower values for the monolayer capacity than the BET equation. The BET equation usually describes an isotherm well in the region $0.003 < P/P_0 < 0.3$, while the DRK equation is valid at lower pressures. Hobson suggests that the DRK monolayer coverage is the greatest coverage for which the lateral adsorbate–adsorbate interactions can be neglected.

In the case of cryogenic accelerator vacuum systems, the main concern is the hydrogen gas that is created by photon-stimulated desorption (PSD) process when the synchrotron radiation (SR) is impinging on the walls of the vacuum system and the equilibrium pressure depending on the amount of physisorbed hydrogen gas on the cold walls of the vacuum system. Figure 7.4 shows the measured H_2 isotherms at the temperature 4.2 K on Cu plated stainless steel. The measured H_2 isotherms at 4.2 K in Figure 7.4 is a multi-monolayer isotherm and it has proved difficult to find an appropriate isotherm equation that can cover the whole pressure range from the lower experimental limit of about 2.3×10^6 molecules/cm^3 up to the saturated vapour density of 1.4×10^{12} H_2/cm^3. The measured adsorption isotherm for H_2 on Cu-plated stainless at 4.2 K has been compared to the equations for Henry's law, the Freundlich isotherm, the Langmuir isotherm, the BET isotherm, and the DRK isotherm and the measured isotherm is fairly well described by the empirical Freundlich isotherm

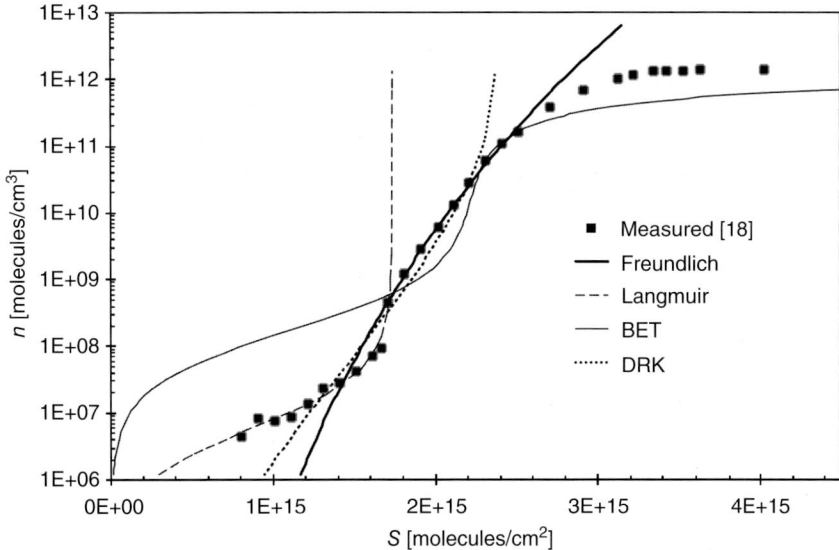

Figure 7.4 Measured H_2 isotherm on Cu plated stainless steel at 4.2 K and fits of isotherm equations to the measured data.

(see Figure 7.4). The following equation parameters have been used in the H_2 isotherms plot in Figure 7.4:

$$\text{Freundlich: } s = 4.76 \times 10^{14} n^{1/15.6}, \quad (7.15)$$

$$\text{Langmuir: } s = \frac{2.99 \times 10^8 n}{1 + 1.72 \times 10^{-7} n}, \quad (7.16)$$

$$\text{BET: } s = \frac{1.74 \times 10^{19} n}{(n_0 - n)(1 + (7.86 \times 10^3 - 1) n/n_0)}, \quad (7.17)$$

$$\text{DRK: } s = 2.39 \times 10^{15} \exp\left(-3.61 \times 10^4 \left(k_B T \ln\left(\frac{n_0}{n}\right)\right)^2\right), \quad (7.18)$$

where n represents the volume density, s the surface density, and $n_0 = 1.36 \times 10^{12}$ H_2/cm^3 is the H_2 saturated vapour density at 4.2 K.

The gas pressure was measured with an ionisation gauge at room temperature, which, with use of Eq. (7.4), gave the density above the Cu surface immersed in the liquid He bath at 4.2 K. The experimental method is described in detail in [21, 22].

The radical change in the shape of the isotherm at $s = 1.65 \times 10^{15}$ H_2/cm^2 and the later saturation at $s = 3.3 \times 10^{15}$ H_2/cm^2 lead to the conclusion that one monolayer consists of 1.65×10^{15} H_2/cm^2. There is a detectable pressure rise in the system only when the first monolayer approaches completion. As soon as the first layer of H_2 molecules is completed, there is an immediate decrease in the adsorption energy. The effect of the lower adsorption energy is seen in the very steep pressure rise for $s > 1.65 \times 10^{15}$ H_2/cm^2. From the start of the third layer at $s = 3.3 \times 10^{15}$ H_2/cm^2 and onwards, there is no change in the adsorption energy with increasing thickness of the adsorbed layer.

Even though it has shown difficult to find a theoretical isotherm that can accurately describe the full pressure range from zero surface coverage to multiple layers of adsorbed gas on the substrate surface, it is possible from a vacuum engineering point of view to establish relationships that span over orders of magnitude in equilibrium pressure. Figure 7.5 shows measured isotherms for He and H_2 on Cu-plated stainless steel at 4.2 K with fitted curves. The H_2 isotherms was fitted there with the Freundlich isotherm Eq. (7.15) and the DRK isotherm Eq. (7.18). The DRK equation and the Freundlich isotherm give good fits to the measured He data over a vast pressure region of the isotherm with the following parameters:

$$\text{Freundlich:} \quad s = 1.28 \times 10^{12} n^{1/5.08}, \tag{7.19}$$

$$\text{DRK:} \quad s = 1.84 \times 10^{15} \exp\left(-3.08 \times 10^4 \left(k_B T \ln\left(\frac{n_0}{n}\right)\right)^2\right), \tag{7.20}$$

where $n_0 = 1.75 \times 10^{21}$ He/cm^3 is the measured saturated vapour density of He at 4.2 K.

The Freundlich isotherm was successfully used for modelling of the travel speed of a He pressure wave in long vacuum tubes at 1.9 K [23] and 4.4 K [24]. In the case of the He pressure wave at 4.4 K, the DRK equation was used to estimate the isotherm that later was simplified by using the Freundlich isotherm, which is simpler to use for numerical fits.

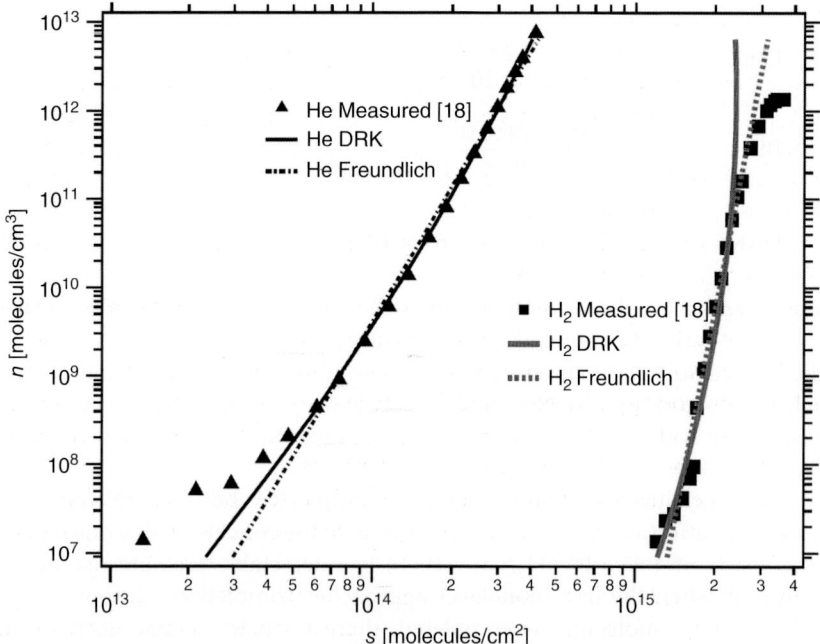

Figure 7.5 Isotherms of He and H_2 at 4.2 K on Cu-plated stainless steel.

The measurements of the He pressure wave in long vacuum tubes were carried out since there was a concern that a He leak into the cold bore vacuum system of an accelerator using superconducting magnets will not be detected by the vacuum gauges until the He gas has propagated to the pressure gauge as a pressure wave with a steep pressure front. It was found that the time to detect a He leak into a cryogenically cooled system can be considerably longer than in a room temperature system due to physisorption of He on the cold walls of the vacuum system. The experiment at 1.9 K [23] was carried out on a test bench for superconducting magnets at CERN where there was a 70-m-long cold bore tube that can be held at a stable temperature between 1.9 and 4.3 K for a long period of time (see Figure 7.6). A He leak was introduced at one end of the 75-m-long cold bore tube, as illustrated in Figure 7.7, and the time it took for the He pressure to be detected at the other end was measured. As can be seen in Figure 7.8, it took 20 hours for the He pressure wave to arrive at the pressure gauge in the other end of the 75-m-long cold bore tube. The conclusion from the measurements were that it is likely that a He leak into a long cold bore accelerator vacuum system is more likely to be discovered by Bremsstrahlung radiation monitors before it is recorded by the vacuum pressure gauges.

7.2.2 Cryotrapping

A gas like H_2 is weakly physisorbed on a cold surface even at low temperatures and it is often the only gas that can be detected by the vacuum instruments in a cryopumped vacuum system. The other gases desorbed by the SR (typically CO, CH_4, and CO_2) cannot be seen by the instruments. The other gases will however still be present on the cold substrate surface. It is not pure H_2 but a mixture of H_2 and the other gases, which will be physisorbed on the cold substrate surface. If such a gas mixture physisorbs on a cold substrate, the gas with weak physisorption may be incorporated into the condensate of the other gases, which are more strongly physisorbed, and its vapour pressure will be suppressed compared to the case of physisorption of the pure gas. The effect of pressure reduction of the gas with weak physisorption when it is coadsorbed simultaneously with other more strongly physisorbed gases is called cryotrapping.

Cryotrapping may be regarded as a special case of physisorption taking place on a continuously self-renewing substrate surface. The efficiency of the pressure reduction by cryotrapping strongly depends on the gas used as partner to the weakly physisorbed gas. Ar, NH_3, and CO_2 have been shown to be efficient in the cryotrapping of H_2, while CO and especially CH_4 have a more moderate influence on the vapour pressure of H_2. Figure 7.9 shows the adsorption isotherm for coadsorption of different mixtures of H_2 and CO plotted as a function of the surface coverage of H_2 [21]. In Figure 7.9 it can be seen that the coadsorption does not have a significant effect on the surface coverage of H_2 at which the steep pressure rise starts, but it has a strong influence on the development of the pressure with increasing surface coverage. The equilibrium pressure can be suppressed by orders of magnitude by the cryotrapping effect if the amount of CO in the coadsorption mixture is high enough.

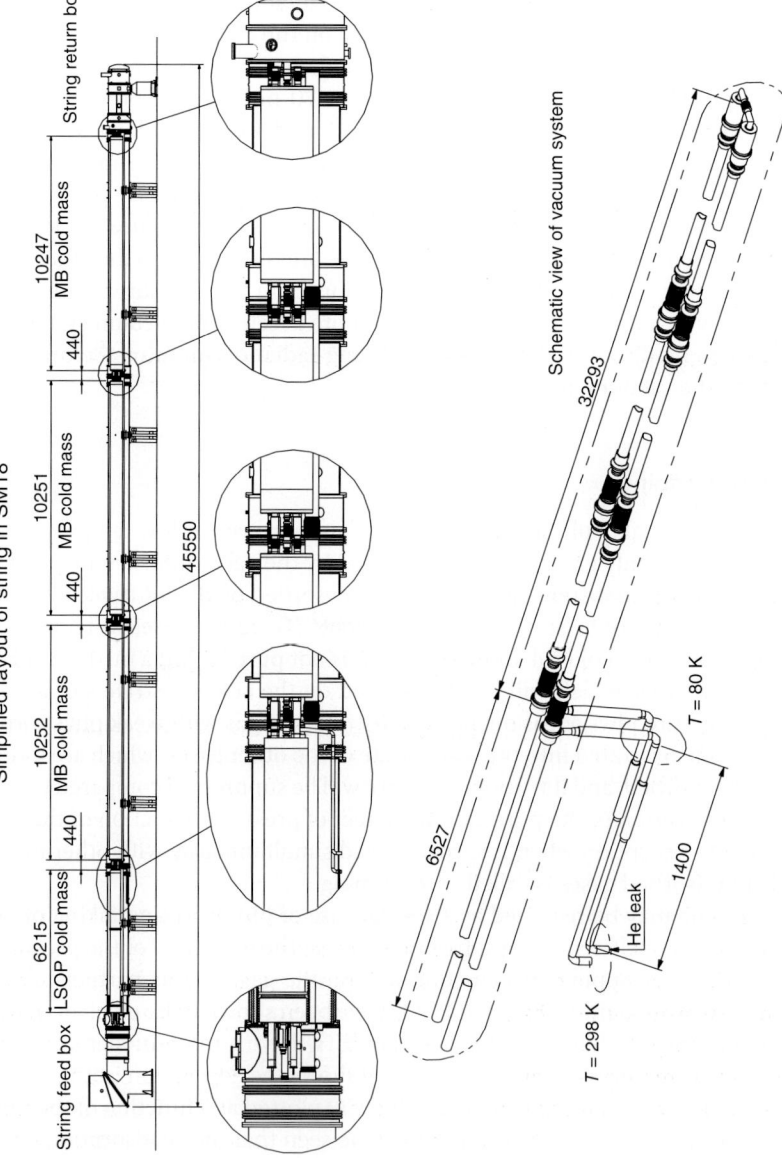

Figure 7.6 Layout of the experimental set-up for the He pressure wave measurements at CERN with one quadrupole magnet (LSQP) and three dipole magnets (MB) connected together and the cold bore vacuum system. Source: Reprinted with permission from Wallén [23], Fig. 4. Copyright 1997, American Vacuum Society.

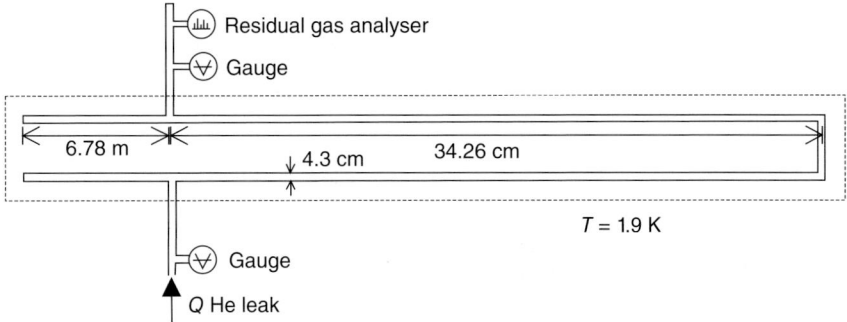

Figure 7.7 Illustration of the equivalent cold bore tube lengths used for the prediction of the time of arrival at the distant end of He pressure front propagating from the leak. Source: Reprinted with permission from Wallén [23], Fig. 5. Copyright 1997, American Vacuum Society.

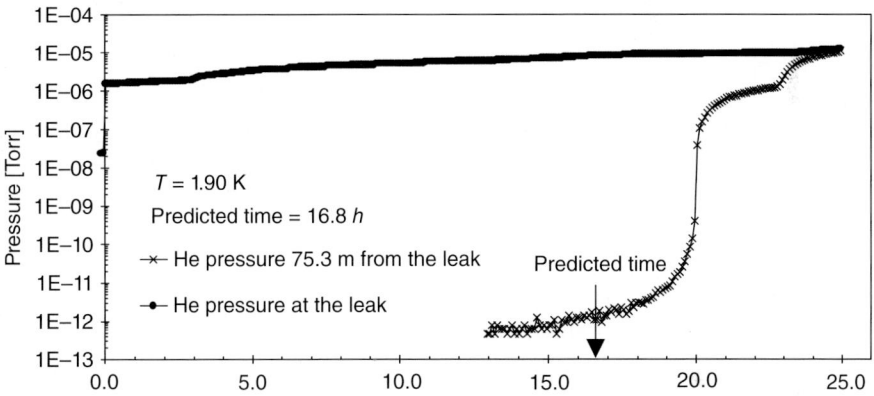

Figure 7.8 Illustration of the equivalent cold bore tube lengths used for the prediction of the time of arrival at the distant end of He pressure front propagating from the leak. Source: Reprinted with permission from Wallén [23], Fig. 8a. Copyright 1997, American Vacuum Society.

7.2.3 Physisorption on Gas Condensates

When the cold substrate surface has become covered with a certain amount of adsorbate molecules, the saturated vapour pressure is reached. In cryopumps, which usually have the design aspect of high pumping capacity between reconditioning warm-ups, the substrate is, in general, chosen to consist of materials with large adsorption capacity. Such materials are not only porous solid substrates, such as molecular sieves and activated charcoal, but also condensates of gas with a considerably higher melting point than the adsorbate. By condensation of gases, such as CO_2, polycrystalline porous adsorbents with a clean surface can be produced. The isotherm for physisorption on gas condensates, as well as on porous substrates, does, in general, show a very low, often undetectable, pressure for small amounts of physisorbed adsorbate, and the pressure slowly increases until the substrate gas condensate is saturated with adsorbate and a very steep pressure rise up to the saturated vapour pressure takes place.

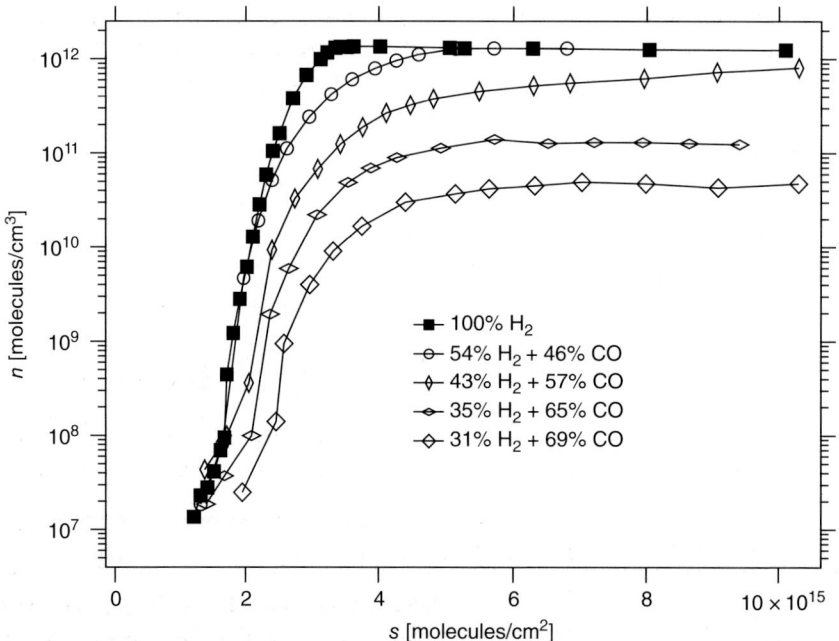

Figure 7.9 Isotherms of coadsorption of H_2 and CO on Cu-plated stainless steel at 4.2 K plotted as a function of the surface density of H_2 molecules. Source: Reprinted with permission from Wallén [21], Fig. 4a. Copyright 1996, American Vacuum Society.

The PSD yield of CO is about a factor of 10 higher than for CO_2 and CH_4 and even if CO_2 is suitable as a capacity increasing gas condensate in cryopumps, it is the behaviour of H_2 being physisorbed on top of condensed CO that is interesting for cryogenic accelerator vacuum systems. Figure 7.10 shows the adsorption isotherms of H_2 on a preconcensed layer of 1.0×10^{16} and 2.0×10^{16} CO/cm² as well as the isotherm of H_2 on the clean surface of the Cu-coated stainless steel. A surface coverage of 1.0×10^{16} CO/cm² represents roughly 6 ML of CO on the Cu surface and the isotherms show that there is no large difference in adsorbing H_2 on a gas condensate of CO compared to the metal surface. The adsorption capacity for binding H_2 before the steep pressure rise takes place has even decreased slightly with a preconcensed layer of 1.0×10^{16} CO/cm² compared to the clean Cu surface. The increase in adsorption capacity for a thicker layer of preconcensed CO may indicate that it is possible for H_2, to a small extent, to penetrate and get absorbed within the structure of the CO condensate.

7.2.4 Temperature Dependence of the H_2 Isotherms

Accelerator vacuum systems normally consist of stainless steel or copper-plated stainless steel. The vacuum chamber in an accelerator using superconducting magnets is close to liquid He temperatures, which will give rise to cryopumping of the residual gases in the vacuum chamber. Assuming that the wall temperature is 20 K or lower, it is only H_2 that has saturated vapour pressure that is high

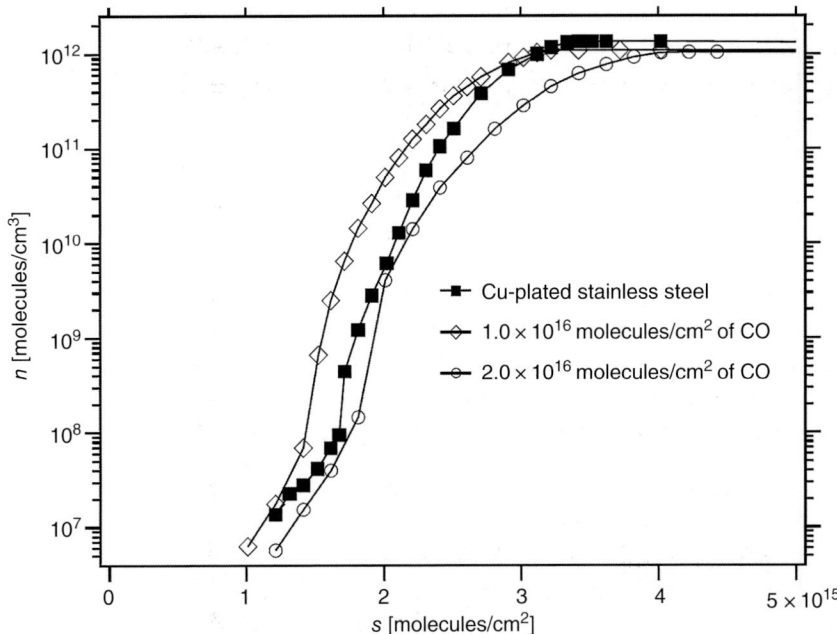

Figure 7.10 Isotherms of H_2 on a bare surface and on pre-condensed layers of 1.0×10^{16} and 2.0×10^{16} CO/cm^2 on Cu-plated stainless steel at 4.2 K. Source: Reprinted with permission from Wallén [21], Fig. 3a. Copyright 1996, American Vacuum Society.

enough to be a concern for the operation of the accelerator. When designing the vacuum system, or estimating the expected equilibrium gas density, it would be handy to have detailed knowledge through measured data of the adsorption isotherm of H_2 at the temperature of the vacuum system. Measured data for the H_2 isotherms, or the saturated vapour pressure of H_2, in the temperature region above liquid He temperatures but below 20 K is however scarce. The saturated vapour pressure H_2 varies rapidly with the temperature. At $T = 4.2$ K, it is $P_0 = 8.0 \times 10^{-5}$ Pa ($n_0 = 1.4 \times 10^{12}$ H_2/cm^3) [21] and at $T = 10$ K it is $P_0 = 2.6 \times 10^2$ Pa ($n_0 = 1.9 \times 10^{18}$ H_2/cm^3) [25], which is more than six orders of magnitude difference for the saturated vapour pressure between 4.2 and 10 K. Measurements of the adsorption isotherms of H_2 and the saturated vapour pressure of H_2 at liquid He temperatures in the range 2.30–4.17 K [26] show that the saturated vapour pressure of varies five orders of magnitude in the temperature interval 2.30–4.17 K, as shown in Figure 7.11.

The H_2 absorption capacity of the metallic surfaces in an accelerator system is hence expected to decrease rapidly with increasing temperature. If the temperature of the different parts of the vacuum system varies during operation, the physisorbed H_2 is also expected to migrate from warmer to colder parts of the vacuum system, which gives rise to pressure spikes in the accelerator vacuum during the time period of the temperature variation. The gas migration effect was, for example, observed during operation of cold bore the superconducting wiggler at the MAX II storage ring at MAX-lab in Lund, where pressure spikes

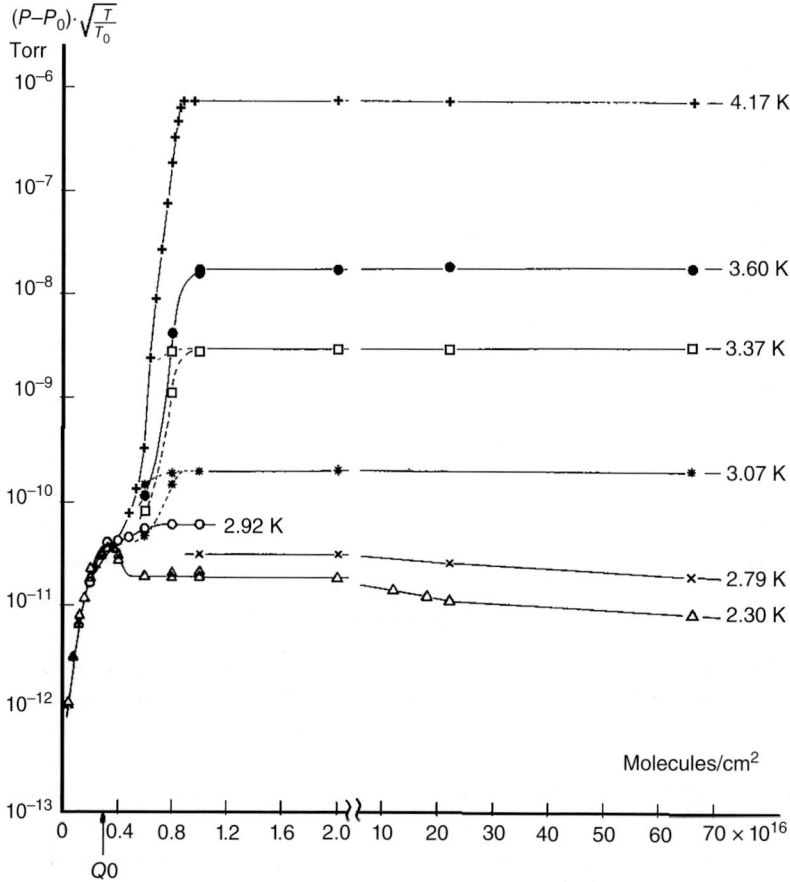

Figure 7.11 Measured adsorption isotherms of H_2 at liquid He temperatures in the range 2.30–4.17 K. Source: Reprinted with permission from Benvenuti et al. [26], Fig. 7. Copyright 1976, American Vacuum Society.

and reduced beam lifetime was observed during refills of the liquid He reservoir in the superconducting wiggler.

The DRK equation is the most suitable tool for estimating the temperature variation of the H_2 isotherm. Figure 7.12 shows measured H_2 isotherms on stainless steel and Cu-plated stainless steel at 4.2 K [22] fitted with the DRK with parameters s_m and D in Eq. (7.13) shown in Table 7.1. Table 7.2 shows the saturated vapour pressure of H_2 in the temperature range of 4.2–20 K.

When comparing the DRK fits of the He isotherms in Figure 7.5 and the H_2 isotherms in Figure 7.12, it is obvious that the DRK fit of the He curve at 4.2 works rather well, which also was confirmed at different temperatures in [22], while for the H_2 isotherms at 4.2 K the pressure range of the measurements is too close to the H_2 saturation pressure and full monolayer formation in order for the DRK to work well. It is hence doubtful if an extrapolation of the DRK parameters found at 4.2 K H_2 is an appropriate way to estimate the H_2 isotherms for metallic surfaces

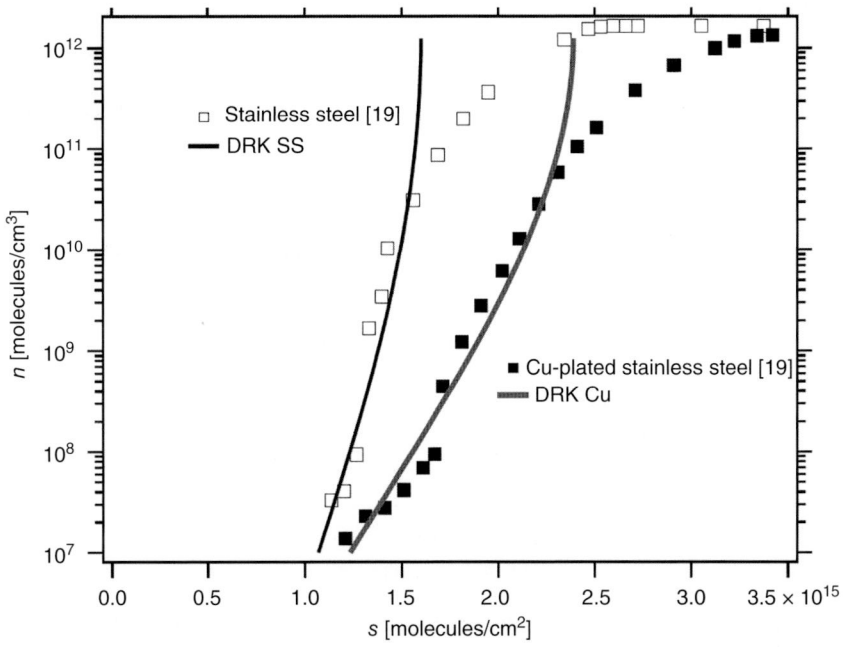

Figure 7.12 Measured H_2 isotherms on stainless steel and Cu-plated stainless steel at 4.2 K.

Table 7.1 Fitted DRK parameters s_m and D from the measured H_2 isotherms at 4.2 K shown in Figure 7.12.

	Stainless steel	Cu-plated stainless steel
s_m [H_2/cm^2]	1.60×10^{15}	2.39×10^{15}
D	22 000	36 100

Table 7.2 Saturated H_2 vapour pressure P_0 and the corresponding number density n_0.

T [K]	P_0 [Pa]	n_0 [H_2/cm^3]
20	9.32×10^4	3.38×10^{20}
15	1.34×10^4	6.46×10^{19}
10	2.57×10^2	1.86×10^{18}
4.2	8.00×10^{-5}	1.38×10^{12}

The value for 4.2 K is from Ref. [22] and the values for 10–20 K are from Ref. [25].
Source: Adapted from Wallén 1997 [22] and Hoge and Arnold 1951 [25].

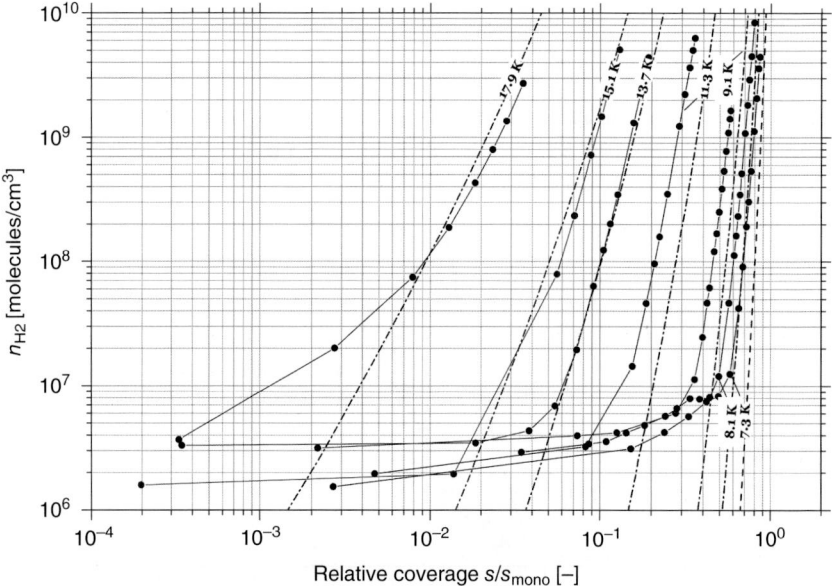

Figure 7.13 Hydrogen adsorption isotherms on an electropolished stainless steel surface in the temperature range 7.3–17.9 K. The experimental data are fitted with the DRK isotherms with the experimentally determined constants $D = 3075$ eV^{-2} and $s_m = 6.45 \times 10^{14}$ cm^{-2}. Source: Courtesy to S. Wilfert and F. Chill, GSI, Germany.

at temperatures above 4.2 K. The scarcity of measured data of H_2 isotherms on metallic surfaces in the range 4.2–20 K has recently been addressed by the GSI in Darmstadt, where the FAIR accelerator complex is under construction.

Figure 7.13 shows the H_2 isotherm measured (points) on electropolished stainless steel from 7.3 to 17.9 K fitted with the DRK model (dash lines) [27, 28]. The fitted values are $D = 3075$ eV^{-2} and $s_m = 6.45 \times 10^{14}$ H$_2$/cm^2. As compared to stainless steel, the monolayer coverage is reduced due to the electropolishing by a factor of ~2.5 and the value of the DRK parameter D is numerically closer to the value found for Cu-plated stainless steel than the bare stainless steel in Table 7.1.

7.2.5 Choice of Operating Temperature for Cryogenic Vacuum Systems

When designing a vacuum system to operate at cryogenic temperature, the choice of the operating temperature can be of primary importance for the system performance. Indeed, as shown before, the vapour pressure of the gases might vary over several decades for a variation of a few Kelvin. The knowledge of the saturated vapour pressure P_0 of some common gases is a therefore major importance since it can be used as a guideline to define possible operating temperatures.

The saturated vapour pressure of gases at any temperature can be derived from the Clausius–Clapeyron law:

$$\log P_0 = A - \frac{B}{T}, \tag{7.21}$$

where A and B are constants, which depends on the nature of the gas.

7.2 Equilibrium Pressure: Isotherms

Table 7.3 A and B constants of some gases when the saturated vapour pressure, given at 300 K, is expressed in mbar.

	He	H_2	CH_4	H_2O	Ne	N_2	CO	C_2H_6	O_2	Ar	CO_2
A	5.018	5.847	7.523	10.118	7.562	8.386	8.900	9.690	8.778	8.062	10.007
B	4.647	47.381	476.870	2576.522	107.486	372.196	433.376	1039.344	455.033	412.596	1338.981

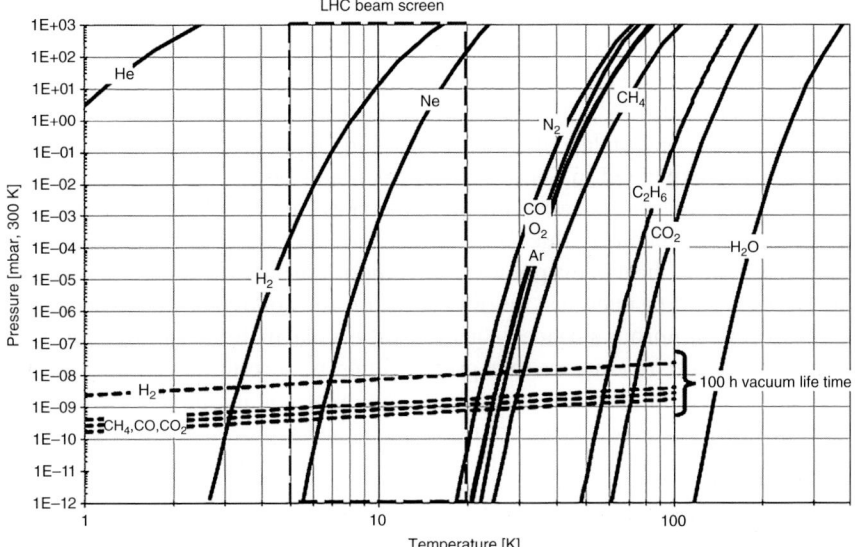

Figure 7.14 Saturated vapour pressure, given at 300 K, expressed in mbar and LHC 100 hours beam lifetime limit for some common gases.

A compilation of the saturated vapour pressure data of several gases at different temperatures can be found in [29]. Table 7.3 gives the fitted values of A and B for some common gases in vacuum technology when the saturated pressure vapour pressure, given at 300 K, is expressed in mbar.

Figure 7.14 shows the saturated vapour pressure, given at 300 K, for some common gases in vacuum technology. Also shown, is the Large Hadron Collider (LHC) 100 hour vacuum lifetime, which is a gas density limit, defined by the acceptable proton beam scattering on the residual gas (see Section 1.2.3). This design pressure is computed from the gas density limit using Eq. (7.3). The LHC vacuum system (see Section 8.5.2.3) is made of a 1.9 K cold bore into which is inserted a perforated beam screen, whose function is to intercept the beam-induced heating at a more favourable temperature than 1.9 K. For the chosen LHC beam screen temperature operating range, 5–20 K, the saturated vapour pressure of He and H_2 is very high. Therefore, these gases do not condense on the beam screen but are thermally flushed towards the cold bore, through the beam screen perforation. With the exception of Ne, which is not a gas commonly present in vacuum system, the saturated vapour pressures of all the other gases are negligible. These gases gas therefore condense into the beam screen without degrading the vacuum lifetime. However, if for some reason the beam screen temperature is increased to 24.5 K, the CO-saturated vapour pressure will reach the vacuum lifetime and, at 26 K, the vacuum lifetime will drop to 10 hours: in this last case, the proton loss during machine operation is dominated by the vacuum level! Thus, it is desirable to limit the operation temperature of the LHC beam screen to below ~25 K, for example, to 20 K, to allow design and operation margin.

It is worth underlining that the saturated vapour pressure is given when several monolayers of gas are condensed onto the surface: i.e. the molecules of the saturated vapour interact with its condensed phase (liquid or solid) and not anymore with the surface material. For example, as shown in Figure 7.11, for a surface coverage above 10^{16} H_2/cm^2, the saturated vapour pressure is reached and follows the Clausius–Clapeyron law of H_2 as depicted in Figure 7.14.

However, as explained in Section 7.2.1, gas might be also physisorbed in the sub-monolayer regime. As a result, the molecules can be adsorbed on the surface at higher temperature due to the larger binding energy of the physisorbed molecules than the heat of evaporation. For smooth surfaces, the adsorption temperatures (or binding energies) for physisorption and condensation are very similar; however, for porous surfaces, e.g. cryosorbers like material (see Section 7.5), both adsorption temperatures may greatly differ. As a result, molecules can be physisorbed up to much higher temperature than the condensation temperature. Figure 7.15 shows the pressure evolution inside the COLDEX stainless steel cold bore during the natural warm-up. The cold bore was initially cooled down at kept at 4.5 K for 3.5 days, then the LHe supply to the COLDEX cryostat was stopped, and the COLDEX stainless steel cold bore was naturally warmed from 4.5 K to room temperature at a rate of 2.2 K/h. Each physisorbed gas was desorbed at different temperatures corresponding to its (different for each gas) binding energies. The peak temperature ranges from 18 K for H_2 to 250–290 K for H_2O.

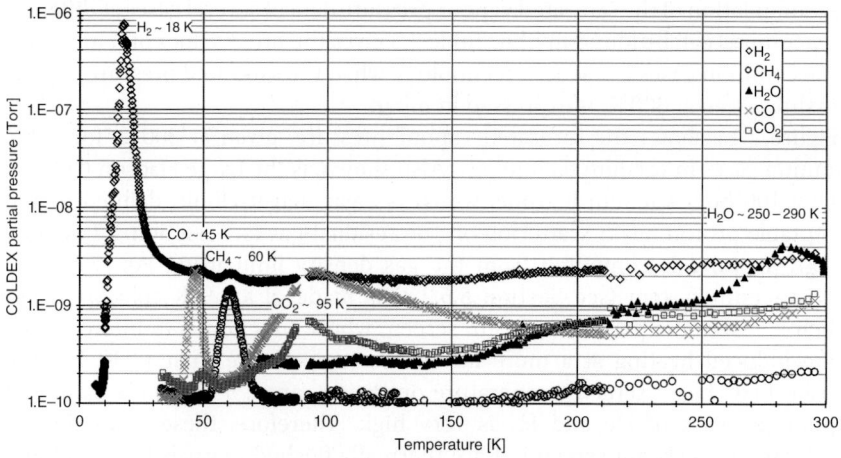

Figure 7.15 Desorption of gases when warming up to room temperature a stainless steel surface at a rate of 2.2 K/h.

Table 7.4 Surface coverage, peak temperature, activation energy, boiling point, and evaporation heat of the molecules on a stainless steel surface.

Molecules	H_2	CH_4	H_2O	CO	CO_2
Surface coverage [molecules/cm^2]	2×10^{14}	3×10^{12}	2×10^{13}	2×10^{13}	4×10^{12}
Peak temperature [K]	18	60	285	45	95
Activation energy [meV]	57	199	965	152	303
Boiling point [K]	22.3	111.7	373.2	81.7	195.1
Evaporation heat at boiling point [meV]	9.3	89.4	421.6	62.7	158.9

According to the first-order thermal desorption kinetics [30], the activation energy for desorption of the molecule, E_d, can be computed from the peak temperature, T_p, and the heating rate, β:

$$\frac{E_d}{kT_p^2} = \frac{1}{\tau_0 \beta} \exp\left(-\frac{E_d}{k_B T_p}\right), \qquad (7.22)$$

where τ_0 is the Frenkel period of vibration of the molecule ($\tau_0 = 10\text{–}13$ seconds) and k_B is the Boltzmann constant ($k = 88.2$ meV/K).

Table 7.4 gives the corresponding surface coverage, peak temperature, and activation energy to the Figure 7.15. It shows, for comparison, the boiling point and the evaporation heat of the molecules. Low surface coverages provide large binding energy as opposed to large surface coverages (i.e. liquid phase in this case), which exhibit low binding energy.

7.3 Gas Dynamics Model of Cryogenic Vacuum Chamber Irradiated by SR[1]

The cryogenic vacuum chamber under SR was intensively studied in 1990–2005 for the Superconducting Super Collider (SSC) and the LHC beam pipes [31–48]. It was discovered that gas density inside such a chamber has three sources:

(1) PSD, ESD, and ISD (ion-stimulated desorption) of tightly bounded molecules (as it happened with PSD, ESD, and ISD at room temperature) called 'primary desorption'.
(2) PSD, ESD, and ISD of cryosorbed molecules (called 'recycling' or 'secondary desorption').
(3) Equilibrium gas density n_e of cryosorbed molecules.

---

1 Before reading this section it is advisable to familiarise with Section 6.2.

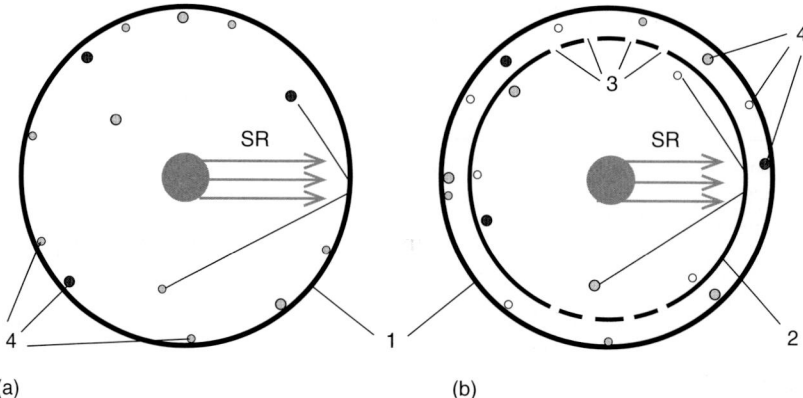

Figure 7.16 Two types of cryogenic vacuum chamber: (a) a vacuum chamber as a simple cold bore (1) and (b) a vacuum chamber consisting of a cold bore (1) and a beam screen (2) with pumping slots (3) for desorbed and cryosorbed molecules (4).

Two types of cryogenic beam vacuum chamber shown in Figure 7.16 are under consideration in this chapter chamber:

(a) A vacuum chamber as a simple cold bore with wall temperature T_{cb}.
(b) A vacuum chamber consisting of a cold bore with wall temperature T_{cb} and a beam screen (BS) (or liner) with wall temperature T_{bs} and with pumping slots for desorbed molecules.

The gas molecules desorbed with SR from a surface directly exposed to the beam, which is an inner wall of a cold bore or a beam screen (depending of the type of vacuum chamber) can be cryosorbed back onto this surface with a sticking probability α, and, if this is a beam screen with pumping slots, be pumped through these slots onto a cold bore, with a distributed pumping speed C. It is also important to point out that in all designs of cryogenic accelerator vacuum chambers known to author, $T_{cb} \leq T_{bs}$. These desorption, diffusion, and pumping processes can be described with the equations of gas dynamic balance inside a beam cryogenic vacuum chamber as follows [33, 34]. The first equation describes volumetric gas density n [molecules/cm³] in volume per unit of axial length $V/L = A$:

$$A\frac{\partial n}{\partial t} = q + q'(s) - \alpha S(n - n_e(s,T)) - Cn + u\frac{\partial^2 n}{\partial z^2}, \qquad (7.23)$$

and the second equation describes the surface density (or surface coverage) of cryosorbed gas s [molecules/cm²] per unit of axial length $A_w/L = F$:

$$F\frac{\partial s}{\partial t} = \alpha S(n - n_e(s,T)) - q'(s); \qquad (7.24)$$

where

$q(M, D)$ [molecules/(s·m)] is the primary beam-induced desorption flux.
$q'(M, s)$ [molecules/(s·m)] is secondary beam-induced desorption flux (desorption of cryosorbed molecules).
$\alpha(M, T)$ is the sticking probability.

$S(M, T)$ [m²/s] $= (1 - \rho) F \bar{v}(M, T)/4$ is the ideal wall pumping speed per unit axial length.
$C(M, T)$ [m²/s] $= \rho w_h F \bar{v}(M, T)/4$ is the distributed pumping speed of holes.
ρ is the area ratio of pumping holes to beam screen.
w_h is the transmission probability of the pumping holes.
$\bar{v}(M, T)$ [m/s] is the mean molecular velocity.
$n_e(M, T)$ [molecules/cm³] is the thermal equilibrium gas density.
$u(M, T)$ [m⁴/s] is the specific vacuum chamber conductance per unit axial length z (see Eqs. (6.35)–(6.37)).

The beam-induced desorption flux consists of PSD and ESD (the latter will take place in case of positively charged beam and beam induced electron multipacting discussed in Chapter 7):

$$q(M, D) = \eta_\gamma(M, D)\Gamma + \eta_e(M, D)\Theta,$$
$$q'(M, s) = \eta'_\gamma(M, s)\Gamma + \eta'_e(M, s)\Theta; \qquad (7.25)$$

where

η_γ and η'_γ [molecules/photon] are the primary and secondary PSD yields.
Γ [photon/(s·m)] is the photon flux per unit axial length.
η_e and η'_e [molecules/electron] are the primary and secondary ESD yields.
Θ [electron/(s·m)] is the electron flux per unit axial length.

These equations should be written for each gas with mass M present in the vacuum system. However, in this analysis each gas behaviour is considered to be independent from others; thus parameter M will be omitted. The temperature of gas is considered to the equal to the wall temperature of cold bore (in case of vacuum chamber without a beam screen) or beam screen (when it is present).

The left-hand side term of Eq. (7.23) can be significant only when the beam current in accelerators changes very quickly: for example, in the storage rings this may happen when the beam is injected or dumped. In the normal operation of the machine, there is either no beam or the beam is present and slowly changing, providing so-called quasi-equilibrium conditions described as $q + q'(s) - \alpha S(n - n_e(s, T)) - Cn + u\frac{\partial^2 n}{\partial z^2} \gg A\frac{\partial n}{\partial t}$. When these conditions are satisfied, then Eq. (7.23) can be simplified to the equation

$$q + q'(s) - \alpha S(n - n_e(s, T)) - Cn + u\frac{d^2 n}{dz^2} = 0; \qquad (7.26)$$

which mathematically is the same as Eq. (5.7):

$$u\frac{d^2 n}{dz^2} - cn + q^* = 0; \qquad (7.27)$$

where $c = \alpha S + C$ and $q^* = q + q'(s) + \alpha S n_e(s, T)$.

7.3.1 Infinitely Long Vacuum Chamber Solution

In the application to the large accelerators, the main interest is a solution for long vacuum chambers with large distance between the pumps. The 'infinitely long' vacuum chamber approximation can be applied when the pumping effect at the ends is negligible inside the chamber. In this case, there is no net axial diffusion, i.e. d^2n/dz^2; thus this approximation is only valid for $c > 0$, and the solution to Eq. (7.27) is

$$n_{\inf} = \frac{q^*}{c}. \tag{7.28}$$

7.3.1.1 Vacuum Chamber Without a Beam Screen

The gas density inside a vacuum chamber without beam screen (i.e. $C = 0$) is given by

$$n_{\inf t} = \frac{q + q'(s)}{\alpha S} + n_e(s). \tag{7.29}$$

The secondary PSD, ESD, and the thermal equilibrium gas density n_e depend implicitly on the surface density s of cryosorbed molecules, which increases with time as

$$s_{\inf t}(t) = s_0 + \frac{1}{F} \int_{t=0}^{t} q \, dt; \tag{7.30}$$

where s_0 is initial surface density. In the case of $s_0 \ll s_m$, the initial gas density is

$$n_{0 \inf t} = \frac{q}{\alpha S} \tag{7.31}$$

The H_2 gas density in a cryogenic vacuum chamber at 4.2 K shown in Figure 7.17 was calculated for primary and secondary PSD, equilibrium gas density, and a sum of these three (total gas density). For a primary PSD reducing with the accumulated photon dose, the gas density inside the vacuum chamber increases with the dose due to the secondary PSD and the H_2 isotherm (the thermal equilibrium gas density n_e), which are increasing due to accumulation of primary desorbed gas and therefore the increasing surface coverage s of cryosorbed molecules shown in Figure 7.19.

These calculations were made with the following parameters: $\Gamma \approx 10^{17}$ photons/(s·m) (similar to the LHC arcs), the diameter of tube $d = 4.5$ cm, and $T = 4.17$ K. Experimental data for $\eta(D)$ and $\eta'(s)$ are shown in Section 7.4.

7.3.1.2 Vacuum Chamber with Holes in the Beam Screen

The gas density inside a vacuum chamber with beam screen (i.e. $C > 0$) is given by

$$n_{\inf bs} = \frac{q + q'(s) + \alpha S n_e(s)}{\alpha S + C} \tag{7.32}$$

Initially, the beam screen surface has no condensed gas; therefore the secondary desorption and the equilibrium gas density equal to zero, and the initial gas density n_0 is calculated as

$$n_{0 \inf bs} = \frac{q}{\alpha S + C}. \tag{7.33}$$

Combining Eq. (7.26) with $d^2n/dz^2 = 0$ and Eq. (7.24), an expression for gas density can be written differently:

$$n_{\inf bs}(t) = \frac{q}{C} - \frac{F}{C} \cdot \frac{ds}{dt}. \tag{7.34}$$

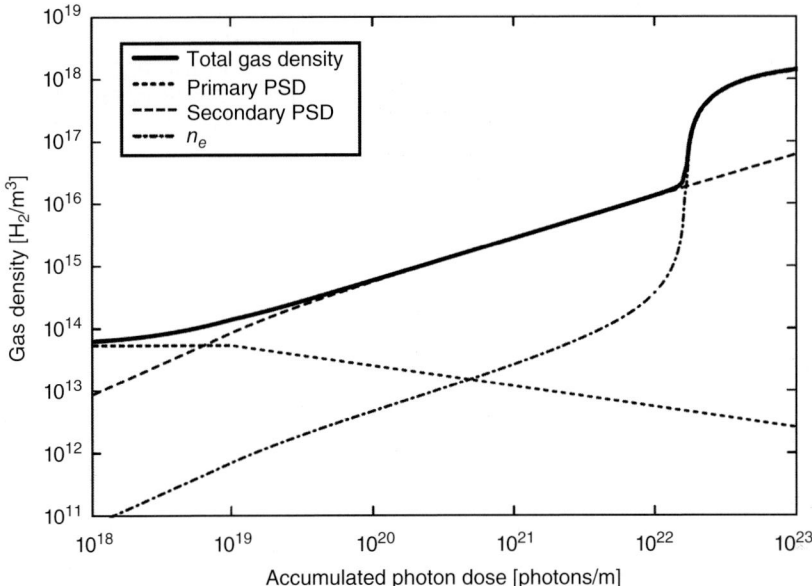

Figure 7.17 The H_2 density as a function of photon dose in a cryosorbing tube.

Since the molecules can escape the vacuum chamber through the pumping holes, the growth of the surface density $s(t)$ on the inner beam screen surface is limited by the distributed pumping C and is described by

$$s_{\text{inf bs}}(t) = s_0 + \frac{1}{F}\int_{t=0}^{t}(q - Cn(t))dt. \tag{7.35}$$

where s_0 is initial surface density.

After a certain time the quasi-static condition of the surface density $s(t)$ will be reached, in this condition for desorption flux is equal to the flux of molecules passing through the pumping holes from beam screen inner volume to the cold bore: $q = Cn$; thus no more molecules are absorbed on the inner beam screen surface, i.e. $F\,ds/dt \approx 0$; therefore the gas density depends on the primary desorption and the pumping through the pumping slots only:

$$n = \frac{q}{C}. \tag{7.36}$$

The H_2 gas density in a cryogenic vacuum chamber with a beam screen at 4.17 K shown in Figure 7.18 was calculated with the following parameters: $\Gamma \approx 10^{17}$ photons/(s·m) (similar to the LHC arcs), the diameter of beam screen $d = 4.5$ cm, and $T_{\text{bs}} = 4.17$ K, for the same experimental data as used for a simple tube shown in Figure 7.17. For a primary PSD reducing with the accumulated photon dose, the gas density inside the vacuum chamber initially increases with the accumulated photon dose due to accumulation of primary desorbed gas on a beam screen walls and, therefore, due to secondary PSD and H_2 isotherm. However, the higher gas density, the more gas pumped through the pumping holes (or slots) in the beam screen. Maximum total gas density is reached at

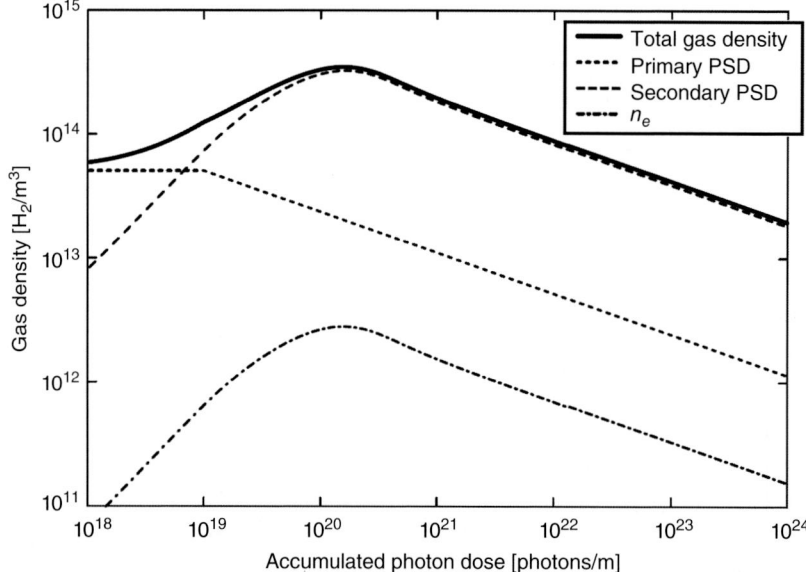

Figure 7.18 The H_2 density as a function of photon dose in a cryosorbing tube with a beam screen.

the dose of $D_{max\_n} = 1.5 \times 10^{20}$ photons/s and corresponds approaching to the condition $F\,ds/dt \approx 0$. For doses higher than D_{max_n}, the total gas density is described with Eq. (7.36). This condition was reach in relatively short time, in our example a dose D_{max_n} corresponds to 25 minutes of the LHC operation under nominal conditions.

The surface coverage is following the same trend as gas density (see Figure 7.19), initially it increases (similar to the tube without a beam screen); however, it reached its maximum and then reduces with a photon dose.

It is important to note here that in this example we used an assumption that no gas comes back from the space between cold bore and the beam screen. This is possible if equilibrium gas density at cold bore temperature, $n_e(T_{cb})$, is much lower than required gas density in the beam path. This can be achieved for all gases (except He) when $T_{cb} < 3.3\,K$. Thus, this criterion is met for typical cold bore temperature of 1.9 K. The other typical cold bore temperature is 4.5 K, and in this case this criterion can be met by using cryosorbing materials in the space between cold bore and the beam screen.

7.3.2 Short Vacuum Chamber Solution

In this book one refers to 'short' vacuum chamber when the conditions at the extremities of the chamber have an influence on the gas density along the whole length of the chamber. In this chapter the different boundary conditions, i.e. known pumping speed or gas density at the ends of the chamber, will be discussed.

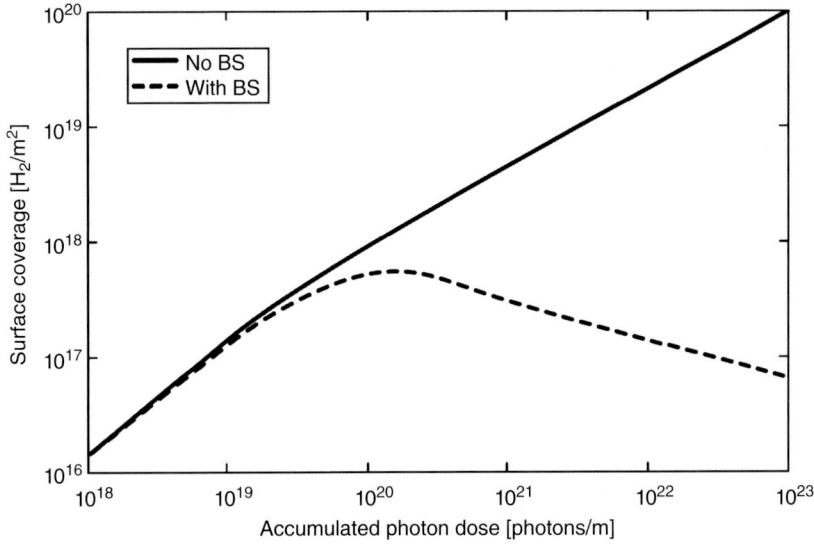

Figure 7.19 The H_2 surface coverage as a function of photon dose in a cryosorbing tube with and without a beam screen.

The second-order differential equation (6.16) for the function $n(z)$ has two solutions:

$$\text{Case (a)} \quad n(z) = \frac{q^*}{c} + C_{1b}e^{\omega z} + C_{2b}e^{-\omega z} \quad \text{for } c > 0,$$

$$\text{Case (b)} \quad n(z) = -\frac{q^*}{2u}z^2 + C_{1a}z + C_{2a} \quad \text{for } c = 0; \tag{7.37}$$

where $\omega = \sqrt{c/u}$ and the constants C_1 and C_2 depend on the boundary conditions. Similarly to the room temperature model in Section 6.2, it is worth evaluating a term $\omega = \sqrt{c/u}$: for example, for a circular tube it simplifies to $\omega = \sqrt{3/d}$ (see Eq. (6.47)).

These solutions are similar to ones obtained in Chapter 5 with expanded definitions of a few terms: distributed gas desorption $q^* = q + q'(s) + \alpha S n_e(s, T)$ and distributed pumping $c = \alpha S + C$.

Since the molecules can be pumped at the extremes of the vacuum chamber and, in the case of beam screen, escape the vacuum chamber through the pumping holes, the growth of the surface density $s(t)$ on the tube or the inner beam screen surface is limited. Similarly to Eq. (7.35), the equation for the surface density can be written as

$$s(z, t) = s_0 + \frac{1}{F}\int_{t=0}^{t}\left(q - Cn(z, t) + u\frac{d^2 n(z, t)}{dz^2}\right)dt. \tag{7.38}$$

where s_0 is initial surface density. After a certain time the quasi-static condition of the surface density $s(t)$ will be reached: $F\,\partial s/\partial t \approx 0$. No more molecules are absorbed on the inner beam screen surface; therefore the gas density depends on the primary desorption and the pumping with the pumps at the extremities

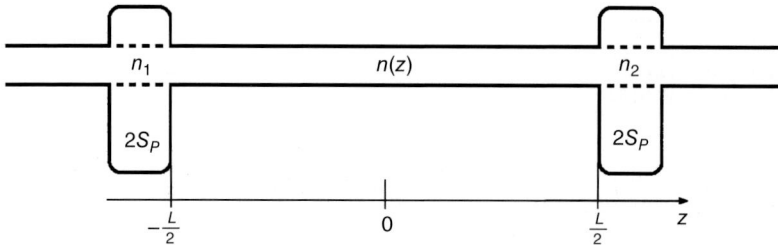

Figure 7.20 A schematic diagram of the layout for a cryosorbing vacuum chamber of length L with a given pumping.

described by Case (b) in Eq. (6.26) and, in the case of beam screen, through the pumping slots described by Case (a) in Eq. (6.26).

Let's consider a uniform cryogenic vacuum chamber without or with a beam screen with pumps with an effective pumping speed $2S_p$ connected to it. We will study the gas density of along a piece of vacuum chamber of length L centred at $z = 0$ as shown in Figures 7.20 and 7.22.

7.3.2.1 Solution for a Short Vacuum Chamber with a Given Pressure at the Ends

In this case the boundary conditions are

$$n\left(-\frac{L}{2}\right) = n_1 \quad \text{and} \quad n\left(\frac{L}{2}\right) = n_2. \tag{7.39}$$

The gas density is described as

Case (a) with $c > 0$: $n(z) = n_{\text{inf}} + (n_1 + n_2 - 2n_{\text{inf}}) \dfrac{\cosh(\omega z)}{2\cosh\left(\omega\frac{L}{2}\right)}$

$$+ (n_2 - n_1)\dfrac{\sinh(\omega z)}{2\sinh\left(\omega\frac{L}{2}\right)},$$

Case (b) with $c = 0$: $n(z) = \dfrac{q}{2u}\left(\dfrac{L^2}{4} - z^2\right) + \dfrac{n_1 - n_2}{L}z + \dfrac{n_1 + n_2}{2}.$ \qquad (7.40)

In case the pressures at both extremes are the same, i.e. $n_1 = n_2$, the expression for the gas density may be written as

Case (a) with $c > 0$: $n(z) = n_{\text{inf}} + (n_1 - n_{\text{inf}})\dfrac{\cosh(\omega z)}{\cosh\left(\omega\frac{L}{2}\right)},$

Case (b) with $c = 0$: $n(z) = \dfrac{q}{2u}\left(\dfrac{L^2}{4} - z^2\right) + n_1.$ \qquad (7.41)

It is also useful to calculate the average value of the gas density in the vacuum chamber of length L:

Case (a) with $c > 0$: $\langle n(L) \rangle = n_{\text{inf}} + (n_1 - n_{\text{inf}})\dfrac{2}{\omega L}\tanh\left(\omega\dfrac{L}{2}\right),$

Case (b) with $c = 0$: $\langle n(L) \rangle = \dfrac{q^*}{12u}L^2 + n_1.$ \qquad (7.42)

Vacuum Chamber Without a Beam Screen

Initial gas density distribution is described by Case (a) in Eqs. (7.40) and (7.41) with the distributed gas desorption $q^* = q + q'(s) + \alpha S n_e(s, T)$ and the distributed pumping $c = \alpha S$.

After reaching the quasi-static condition of the surface density $s(t)$: $F \partial s/\partial t \approx 0$, the gas density distribution is described by Case (b) in Eqs. (7.40) and (7.41). It is important to mention that this is exactly the same equation as used for particle stimulated desorption at room temperature; however the vacuum conductance is significantly lower at cryogenic temperature. Therefore the pumping at the extremities is very reduced at cryogenic temperatures.

An example of the H_2 density along a cryosorbing vacuum chamber of length $L = 4$ m was calculated with the following parameters: $\eta = 2 \times 10^{-3}$ H_2/photon, $\Gamma = 10^{17}$ photons/(s·m), $\alpha = 0.1$, $d = 4.5$ cm, $T = 4.17$ K, gas densities at the extremities $n_1 = 1 \times 10^{14}$ H_2/m^3, and $n_2 = 4 \times 10^{14}$ H_2/m^3. The result of calculations is shown in Figure 7.21 with four curves:

(1) Initial the gas density (with $\eta' \ll \eta$), where the gas density near the middle is equal to solution for an infinity long tube $n(0) = n_{\inf t} = \eta/\alpha S$ and $n_1 < n_{\inf t} < n_2$.
(2) After some time when some gas was cryosorbed leading to $\eta' = 1 \times 10^{-2}$ H_2/photon), $n(0) = n_{\inf t} = (\eta + \eta')/\alpha S$ and $n(0) > n_2 > n_1$.
(3) After longer time when more gas was cryosorbed leading to $\eta' = 5 \times 10^{-2}$ H_2/photon), where $n(0) = n_{\inf t} > n_2 > n_1$.
(4) After reaching condition of the quasi-static condition on the surface: $F \partial s/\partial t \approx 0$, the density is described by parabolic equation Case (b); in this case $n(0) \ll n_{\inf t}$.

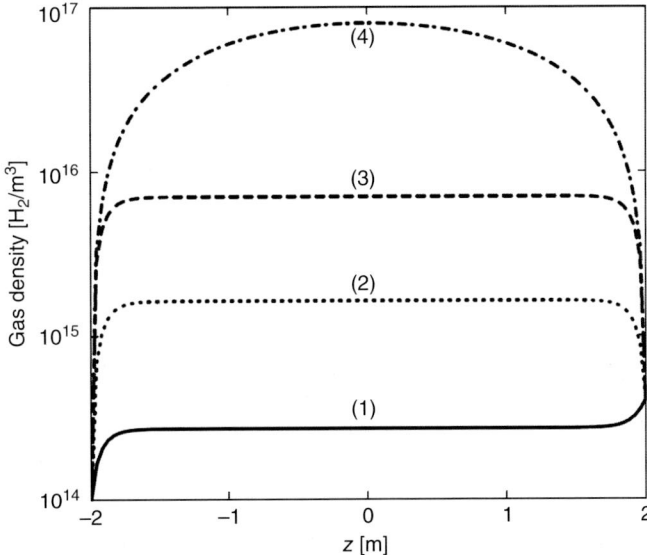

Figure 7.21 The H_2 density along a cryosorbing vacuum chamber of length L with a given gas density at the extremities.

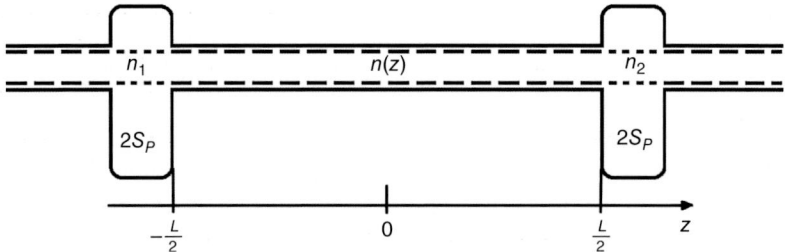

Figure 7.22 A schematic diagram of the layout for a cryosorbing vacuum chamber of length L with a beam screen and with a given pumping speed at the ends.

So, even in such a short tube the gas density in a middle of tube can reach a value which is orders of magnitudes higher than initial one, pumping at the extremes is not efficient.

Vacuum Chamber with a Beam Screen
Initial gas density distribution is described by Case (a) in Eqs. (6.28) and (6.29) with the distributed gas desorption $q^* = q + q'(s) + \alpha S n_e(s, T)$ and the distributed pumping $c = \alpha S + C$.

After reaching the quasi-static condition of the surface density $s(t)$: $F \, \partial s / \partial t \approx 0$, gas density distribution is described by Case (a) in Eqs. (7.40) and (7.41) with the distributed gas desorption $q^* = q$ and the distributed pumping $c = C$ (Figure 7.22).

An example of the H_2 density along a cryosorbing vacuum chamber of length $L = 4$ m with a beam screen was calculated with the following parameters (similar to a previous example for a simple tube): $\eta = 2 \times 10^{-3}$ H_2/photon, $\Gamma = 10^{17}$ photons/(s·m), $\alpha = 0.1$, $d_{bs} = 4.5$ cm, $T_{bs} = 4.17$ K, gas densities at the extremities $n_1 = 1 \times 10^{14}$ H_2/m^3, and $n_2 = 4 \times 10^{14}$ H_2/m^3. The result of calculations is shown in Figure 7.23 with three curves:

(1) Initial the gas density (with $\eta' \ll \eta$), where the gas density near the middle is equal to solution for an infinity long tube $n(0) = n_{\inf bs} = \eta/(\alpha S + C)$ and $n_1 < n_{\inf bs} < n_2$.
(2) After some time when some gas was cryosorbed leading to $\eta' = 3 \times 10^{-3}$ H_2/photon, the gas density near the middle is $n(0) - n_{\inf bs} = (\eta + \eta')/(\alpha S + C)$ and where $n(0) = n_{\inf bs} > n_2 > n_1$.
(3) After reaching condition of the quasi-static condition on the surface: $F \, \partial s / \partial t \approx 0$, the gas density near the middle is $n(0) = n_{\inf bs} = \eta/C$ and $n(0) = n_{\inf bs} > n_2 > n_1$.

So, even in such a short tube the gas density in a middle of tube with a beam screen is insensitive to gas density at the extremes. Due to the gas cryosorption on the inner surface of the beam screen, the gas density increase is limited by a factor $(\alpha S + C)/C$.

7.3.2.2 Solution for a Short Vacuum Chamber with a Given Pumping Speed at the Ends

In this case the boundary conditions with pumps at the two ends of pumping speed $2S_p$. The conditions at the extremities are

$$n(\pm L/2) = \mp \frac{dn(\pm L/2)}{dz} \frac{u}{S_p}. \tag{7.43}$$

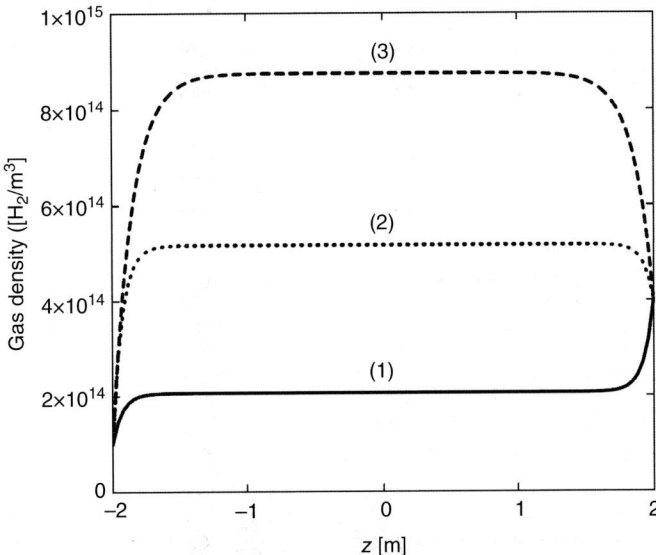

Figure 7.23 The H_2 density along a cryosorbing vacuum chamber with a beam screen of length L with a given pressure at the extremities.

The gas density is described as

Case (a) with $c > 0$: $n_t(z) = n_{\inf}\left(1 - \dfrac{\cosh(\omega z)}{\cosh\left(\omega\frac{L}{2}\right)\left(1 + \frac{\sqrt{cu}}{S_p}\tanh\left(\omega\frac{L}{2}\right)\right)}\right),$

Case (b) with $c = 0$: $n_t(z) = \dfrac{q^*}{2u}\left(\left(\dfrac{L}{2}\right)^2 - z^2\right) + \dfrac{q^*L}{2S_p}.$ (7.44)

It is also useful to calculate the average value of the gas density in the vacuum chamber of length L:

Case (a) with $c > 0$: $\langle n(L) \rangle = n_{\inf}\left(1 - \dfrac{\tanh\left(\omega\frac{L}{2}\right)}{\omega\frac{L}{2}\left(1 + \frac{\sqrt{cu}}{S_p}\tanh\left(\omega\frac{L}{2}\right)\right)}\right),$

Case (b) with $c = 0$: $\langle n(L) \rangle = q\left(\dfrac{L^2}{12u} + \dfrac{L}{2S_p}\right).$ (7.45)

Vacuum Chamber Without a Beam Screen
Initial gas density distribution is described by Case (a) in Eqs. (7.44) and (7.45) with the distributed gas desorption $q^* = q + q'(s) + \alpha S n_e(s, T)$ and the distributed pumping $c = \alpha S$.

After reaching the quasi-static condition of the surface density $s(t)$: $F\,\partial s/\partial t \approx 0$, gas density distribution is described by Case (b) in Eqs. (7.44) and (7.45).

Vacuum Chamber with a Beam Screen
Initial gas density distribution is described by Case (a) in Eqs. (7.44) and (7.45) with the distributed gas desorption $q^* = q + q'(s) + \alpha S n_e(s, T)$ and the distributed pumping $c = \alpha S + C$.

After reaching the quasi-static condition of the surface density $s(t)$: $F \partial s/\partial t \approx 0$, gas density distribution is described by Case (a) in Eqs. (7.44) and (7.45) with the distributed gas desorption $q^* = q$ and the distributed pumping $c = C$.

7.4 Experimental Data on PSD from Cryogenic Surface

An intense study of PSD processes in a cryogenic vacuum chamber was started in early 1990s. The interest to these processes was initiated by a vacuum design of two cryogenic colliders: SSC in Dallas (Texas, USA) and LHC at CERN in Geneva (Switzerland). These colliders were to be the first accelerators with SR irradiating a cryogenic vacuum chamber:

- $\Gamma \approx 10^{16}$ photons/(s·m) with $\varepsilon_c = 284$ eV in SSC arcs.
- $\Gamma \approx 10^{17}$ photons/(s·m) with $\varepsilon_c = 44$ eV in LHC arcs.

Initial considerations include the following: a beam vacuum chamber is a cold bore of superconducting magnet operating at 1.9 or 4.5 K – this is a distributed cryopump. As it was shown in Section 7.2, the equilibrium pressures are negligible for all gases except H_2 and He at 4.2–4.5 K and except He at 1.9 K. There was an experience of operating HERA in DESY (Hamburg, Germany), demonstrating that there is no vacuum related problem for a beam in a cryogenic vacuum system, no pressure rise in a presence of the beam, but HERA has no SR. Thus, the initial thought was that PSD from vacuum chamber at cryogenic temperature could be a source of hydrogen similar to room temperature, but there were no experimental data on PSD yields at cryogenic temperature as a function of photon dose, critical photon energy, and wall temperature. So, the initial aim of study was to measure the PSD yields and the amount of desorbed H_2 as a function of photon dose at chosen critical photon energy and wall temperature. Science pumping from the ends of long cryogenic vacuum chamber is negligible; the desorbed H_2 molecules will be cryosorbed on vacuum chamber walls. When H_2 surface density reaches approximately $s_m = 3 \times 10^{15}$ H_2/cm^2, the corresponding equilibrium pressure at 4.2–4.5 K rapidly increases to a non-negligible value. For the machine operation this would mean that a volumetric gas density went above the specified value, thus the operation of the machine will be interrupted for warming up the vacuum chamber and pumping away of gas accumulated on vacuum chamber walls. Time required to accumulate s_m would define a machine operation time between warming up, which should be greater than six months.

Helium equilibrium pressure could be even a greater problem, but He is not present in PSD gases. The only source of He could be LHe used for cooling superconducting magnets; thus this is a cryogenic and mechanical engineering task to build a leak-free LHe systems with a highest attention to a cold bore.

First studies were performed at the Budker Institute of Nuclear Physics in Novosibirsk (Russia) in collaborations with both SSC Lab and CERN to study cryogenic vacuum chamber under SR on two dedicated SR beamlines on the VEPP-2M. The advantage of VEPP-2M was an ability to generate the same Γ and ε_c as either in the SSC or in the LHC. The results for primary and secondary PSD measurements were reported in Refs. [31–39]. Additionally, the secondary PSD studies were reported in Refs. [44–46]. All the measurements were made with SR critical energy of either 284 eV (SSC related studies) or 44–194 eV (LHC related studies). Later the experiments were carried out on a facility called COLDEX at CERN [40–43].

7.4.1 Experimental Facility for Studying PSD at Cryogenic Temperatures

The experimental set-up for cold beam tube experiments designed and built at BINP is shown in Figure 7.24. The main idea was to reproduce and measure the processes in an accelerator cryogenic vacuum chamber. Thus the test vacuum chamber was placed between two pumps (with $S \approx 1000$ l/s and equipped with gate valves, which allow to reduce an effective pumping speed by varying the opening the valve gates), the partial pressures were measured with residual gas analysers (RGAs) not only at the ends (P_1 and P_3) but also in the middle of the test chamber (P_2); collimators and phosphor screens allow the sample tube alignment in respect to SR.

7.4.2 Discovery of Secondary PSD

The first experimental study of cryogenic vacuum chamber irradiated by SR was performed on a facility described above with an SSC beam pipe prototype: a 32 mm inner diameter and 1-m-long stainless steel tube with a cross section shown in Figure 7.16a, the inner surface was electrodeposited with 70 μm of copper. The sample tube was held at 4.2 K and exposed SR with $\Gamma \approx 10^{16}$ photons/(s·m) and

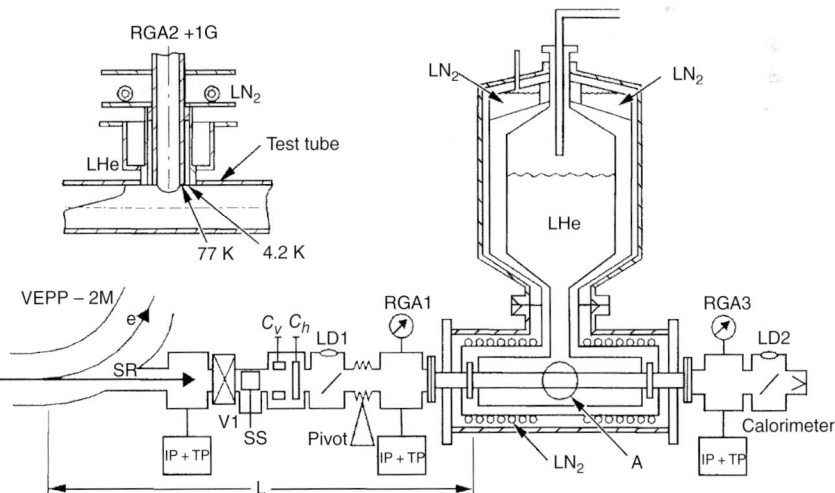

Figure 7.24 The experimental set-up for cold beam tube experiments at BINP: V1, vacuum valve; C_v, C_h, vertical and horizontal collimators; SS, safety shutter; LD1, LD2, phosphor screens; IGs, ion gauges; RGA, residual gas analysers; IP+TP, combined ion and Ti sublimation pumps. Source: Baglin 2007 [36], Fig. 1. Reprinted with permission of CERN.

$\varepsilon_c = 284$ eV at the incident angle of 10 mrad. Initially the pump gate valves were fully open providing maximum pumping speed to avoid gas condensation on the test tube, the pressures P_1 and P_3 at the extremes was in the range of 10^{-10} Torr, and the pressures in the middle was $P_{2\_e} \approx 8 \times 10^{-10}$ Torr.

Figure 7.25 shows the H_2 pressure in the middle of the test tube measured during SR irradiation (shown as Photons 'on') and without SR (Photons 'off') in in the first experiment. When SR irradiation begins, the pressure in the middle increased to $P_{2\_SR} \approx 1.3 \times 10^{-9}$ Torr. Then pressure P_2 was steadily increasing with a photon dose (unlike usually observed reduction of dynamic pressure with a photon dose at room temperature). The H_2 pressure measurements without SR (corresponding to the equilibrium pressure of cryosorbed H_2, P_e) demonstrated much smaller pressure increase with the photon dose. At the dose of 1.2×10^{21} photons/m, the temperature of the sample tube was reduced to 3.3 K. The result was the following: the equilibrium pressure of cryosorbed H_2, P_e, had reduced with temperature, while dynamic difference between measurements with and without SR $\Delta P_2 = P_{2\_SR} - P_{2\_e}$ was found to be temperature independent in the measured range of temperatures between 3.3 and 4.2 K. Then the tube was warmed up to 4.2 K and continue the SR irradiation. At the dose of 1.5×10^{21} photons/m, the pressure with SR reached a value $P_{2\_SR} \approx 10^{-8}$ Torr and did not change with a photon dose further. Since the pressure difference between the centre and the ends was approximately two orders of magnitude,

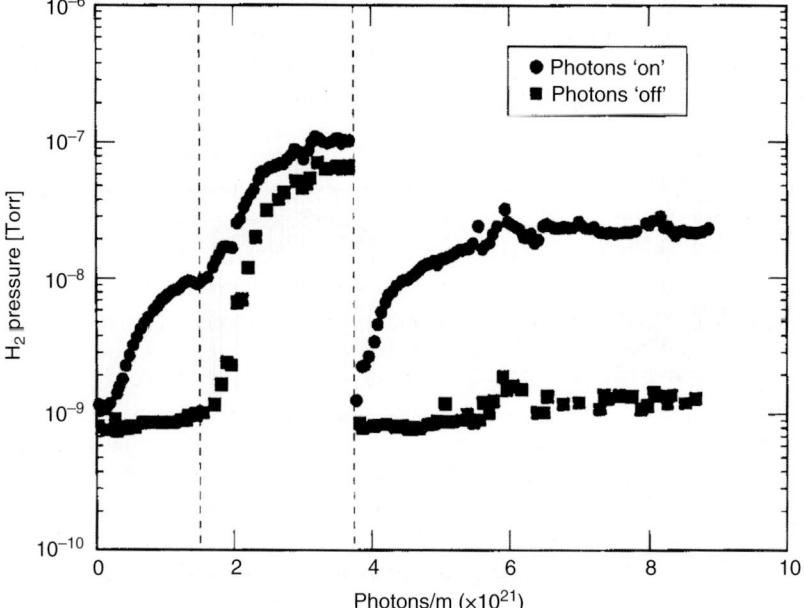

Figure 7.25 The H_2 pressure measurements in the middle of the test tube in in the first experimental study of cryogenic vacuum chamber irradiated by SR with $\Gamma \approx 10^{16}$ photons/(s·m). The vertical dashed lines correspond to features discussed in the text. Source: Reprinted with permission from Anashin et al. [34], Fig. 1. Copyright 1994, American Vacuum Society.

it was reasonable to speculate that a quasi-static condition was reached and then the photodesorbed gas flow is equal to one pumped at the end of sample tube. To check this, the gate vales at the pumps were partially closed, and the pressure then increased to a new quasi-static condition with $P_{2\_SR} \approx 10^{-7}$ Torr and $P_{2\_e} \approx 6.5 \times 10^{-8}$ Torr. The amount of gas cryosorbed on the sample tube was measured at the end of experiment by warming up to 77 K. Then the sample tube was cooled again to 4.2 K and the gas was recondensed on it, and measured pressures with and without SR were the same as before warming up. The sample tube was warmed up again and all released gas was pumped away, and then the experiment was repeated with partially closed valves.

The first experiments with cryogenic vacuum chamber irradiated by SR has demonstrated the following:

(1) The gas density inside the simple sample tube rapidly increases with a photon dose and in the case of SCC reached a maximum specified value after 2×10^{21} photons/m, which is too short as corresponds to only six hours of SSC operation.
(2) The gas density with SR can be orders of magnitude higher than equilibrium gas density.
(3) The desorption mechanism at 4.2 K differs from one at room temperature: it depends on amount of cryosorbed gas and does not depend on temperature (at least in the range of temperatures between 3.3 and 4.2 K).

Since the problem of gas density growth with photon dose is clearly related the amount of cryosorbed gas exposed to SR, the proposed solution was to make it possible for the gas to transfer to the places where is no SR. The cryostat cold bore was increased to ID = 41.9 mm and a liner (or a beam screen) with ID = 32 mm was inserted as it is shown in Figure 7.16b. The liner was perforated with 600 holes of 2 mm diameter spaced 1 cm axillary and 60° azimuthally. A liner will be irradiated with SR and create a shadow for the cold bore from SR. The desorbed gas molecules can now reach the shadow after a few interactions with the liner walls. The larger the area of holes, the fewer interactions with liner, the lower gas density should be. The results of the exposure to SR are shown in Figure 7.26 where the constant pressures without SR have been subtracted.

Initial temperature of the cold bore and a liner was the same, $T_{cb} = T_{bs} = 4.2$ K. H_2 pressure steadily increased until reaching an equilibrium value of 1.5×10^{-8} Torr at 6×10^{20} photons/m. Alike in the previous experiment without a liner, the pump valves were particle closed at 1.2×10^{21} photons/m and H_2 pressure has increased to 3×10^{-8} Torr. This pressure was the H_2 isotherm pressure of the gas cryosorbed on the bold bore and the liner.

Then the cold bore temperature was reduced reaching $T_{cb} = 3.2$ K at 1.6×10^{21} photons/m, while liner temperature remains the same $T_{bs} = 4.2$ K. The H_2 pressure in the presence of SR was reduced to 1×10^{-9} Torr, while the H_2 equilibrium pressure measured without SR was reduced to below sensitivity value; CO pressure does not change. The pressure bumps at 1.6×10^{21} and 3.8×10^{21} photons/m corresponds to LHe Dewar refilling, causing the cold bore temperature increase to 4.2 K. At 6.5×10^{21} and 9.1×10^{21} photons/m, the

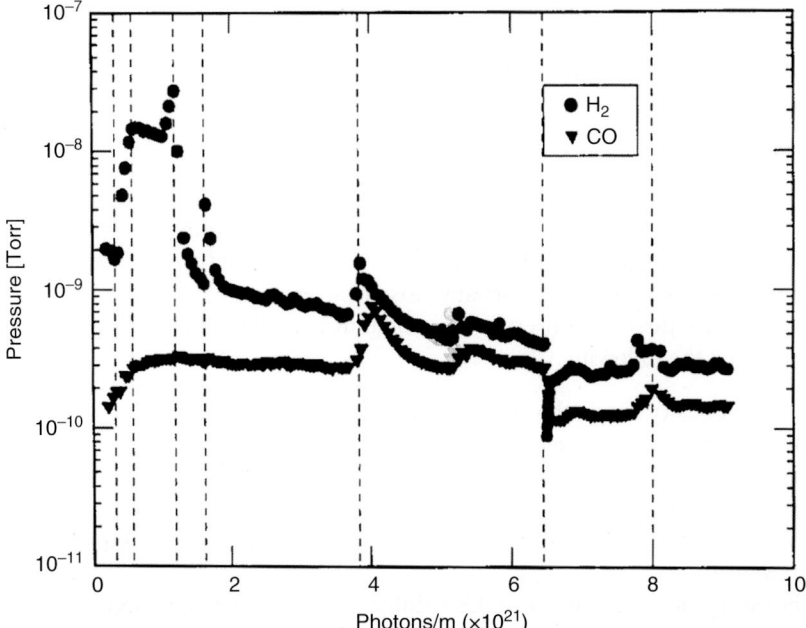

Figure 7.26 The H_2 and CO pressure measurement at the centre of a liner irradiated by SR with $\Gamma \approx 10^{16}$ photons/(s·m). The vertical dashed lines correspond to features discussed in the text. Source: Reprinted with permission from Anashin et al. [34], Fig. 2. Copyright 1994, American Vacuum Society.

cryostat was warmed up to measure the amount of cryosorbed gas. Then the experiment was repeated.

The results obtained in these experiments were plotted in terms of H_2 gas density and are shown in Figure 7.27, where A and B are vacuum chamber without a liner, C and D are experiments with a liner with pumping holes, and E is the SSC beam lifetime limit [36]. In the experiments without a liner, the initial H_2 gas density is lower than the required density (E); however the gas density increases with a photon dose and becomes higher than the required density after 2×10^{20} and 4×10^{20} photons/m in experiments (A) and (B), correspondingly; this is only 6 and 12 hours of the SSC operation. To the required gas density, the cryosorbed gas should be removed by warming the beam tube. This would be very expensive and unpractical. In the experiments with a liner (C) and (D), the initial gas density is higher than the required gas density by a factor of 2; however the gas density reduces with a photon dose and reaches the required density after 1.6×10^{21} and 4×10^{21} photons/m in experiments (C) and (D), correspondingly. This would be a quite reasonable conditioning time for the SSC.

Similar experiments were later performed in COLDEX facility at CERN [40], described in Section 7.5. The observations demonstrated a comparable behaviour under SR of a liner (also called as a *beam screen*), held at cryogenic temperature and produced with and without pumping holes (see Figure 7.28).

7.4 Experimental Data on PSD from Cryogenic Surface | 305

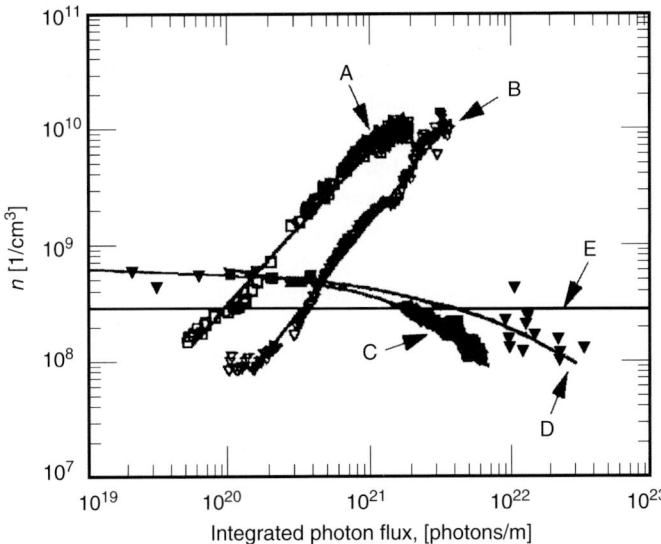

Figure 7.27 H_2 as density at the centre of the beam tube irradiated by SR with $\Gamma \approx 10^{16}$ photons/(s·m): A and B are for the simple tubes without a liner, C and D are for the cold bore with a liner with pumping holes, and E is the SSC beam lifetime limit. Source: Anashin et al. 1994 [36], Fig. 2. Reprinted with permission of CERN.

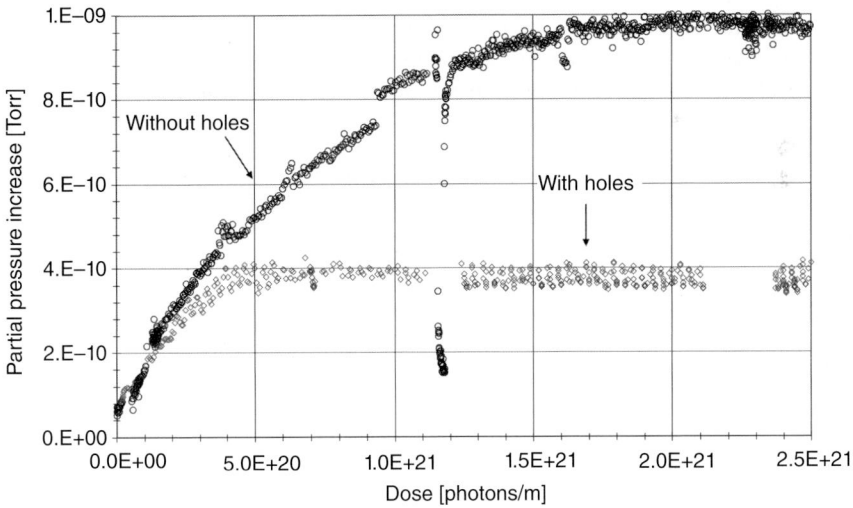

Figure 7.28 Comparison of the H_2 photodesorption between a liner with and without pumping holes. Source: Baglin et al. 2000 [40], Fig. 5. Reprinted with permission of CERN.

For a cryogenic vacuum chamber with a liner without holes, the H_2 partial pressure in the middle of the chamber increases with SR dose due primary PSD, gas accumulation in the liner surface and due to the recycling effect (secondary PSD). The pressure grows continues until the quasi-static condition is reached due to the pumping at the extremities (see Eqs. (7.41) and (7.44) with $c = 0$ at $z = 0$). For longer cryogenic tubes with smaller diameter, the pumping speed at the extremity would have been negligible and the pressure would continue increasing: in term of gas dynamics model, it will be limited by reaching maximum secondary PSD and by saturated equilibrium pressure for given wall temperature; however, in practice, the pressure will be too high to operate the accelerator well before the gas dynamics limit.

In the case of a liner perforated with pumping holes, the H_2 partial pressure increases due to the recycling desorption effect and reach an equilibrium defined by the ratio of the PSD flux to the holes conductance (see Eqs. (7.41) and (7.44) with $c \neq 0$ at $z = 0$). For very long tubes where the end pumping effect inside the chamber is negligible, the gas density is given by Eq. (7.36).

Hence, we can conclude the following:

- Low temperature does not necessary provide good vacuum in a vacuum chamber:
    o In the case of SR the gas density is rapidly grooving with an accumulated photon dose due to accumulation of cryosorbed gas.
- The solution is in creating areas with a shadow from SR where the desorbed gas can escape:
    o This can be provided by a liner with holes.
    o The equilibrium gas density should be below the required gas density (controlled by cold bore temperature or by allying cryosorbers).

These first results of experiments with cryogenic vacuum chambers allow proposing the model described in Section 6.3, which is used for both obtaining experimental results and using the results for modelling future accelerators with a cryogenic vacuum chamber.

7.4.3 Calculation of the Desorption Yields from Experimental Data

Based on this model, the PSD yields can be calculated from measured partial pressures. Thus the formulas for simple tube are

$$\eta = \frac{8u}{L^2 \Gamma} \left(n(0) - \frac{n\left(-\frac{L}{2}\right) + n\left(\frac{L}{2}\right)}{2} \right) \quad \text{for } c = 0;$$

$$\eta = \frac{c}{\Gamma} \frac{n(0) \cosh\left(\omega \frac{L}{2}\right) - \frac{n\left(-\frac{L}{2}\right) + n\left(\frac{L}{2}\right)}{2}}{\cosh\left(\omega \frac{L}{2}\right) - 1} \quad \text{for } c > 0. \tag{7.46}$$

7.4 Experimental Data on PSD from Cryogenic Surface

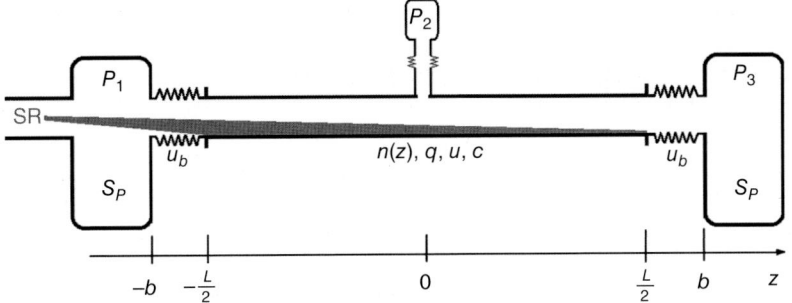

Figure 7.29 Layout of the vacuum chamber with a sample tube and bellows and pressure readings.

Now, we have to consider that pressure measurements are performed with RGAs placed at room temperature, T_{RT}, while the sample tube is at cryogenic temperature, T_t, and that there are bellows between the ends of a cryogenic sample tube and the pumping ports (see Figure 7.29).

From Eq. (7.3) one can obtain that

$$P_2 = P(0)\sqrt{\frac{T_{RT}}{T_t}} = n(0)k_B\sqrt{T_{RT}T_t} \qquad (7.47)$$

The boundary conditions at the ends of sample tube are a bit more complicated, as vacuum conductance of the bellows between the ends of a cryogenic sample tube and the pumping ports should be taken into account. The effect is proportional to the ratio between the entrance conductances of the sample tube without the bellows,

$$U_{et} = \frac{\pi d^2 \bar{v}(T_t)}{16}, \qquad (7.48)$$

and with the bellows roughly estimated using Eq. (5.9):

$$\frac{1}{U_b} = \frac{1}{\frac{\pi d_b^3}{12L_b}\frac{\bar{v}(T_{RT})+\bar{v}(T_t)}{2}} + \frac{1}{\frac{\pi d_b^2 \bar{v}(T_{RT})}{16}} + \frac{1}{\frac{\pi d^2 \bar{v}(T_t)}{16}\left(1-\frac{d_b^2}{d^2}\right)}, \qquad (7.49)$$

where d_b and L_b are the inside diameter and the length of bellows. Then in the case of the BINP experiments with the bellows of $d_b = 63$ mm and $L = 100$ mm, one can estimate:

$$\frac{U_{et}}{U_b} = 0.016. \qquad (7.50)$$

Considering that negligible end effect, one can write

$$P_{1,3} = n\left(\mp\frac{L}{2}\right)k_B\sqrt{T_{RT}T_t} \qquad (7.51)$$

In this case Eq. (7.46) can be written for measured pressures:

$$\eta = \frac{8u}{k_B L^2 \Gamma \sqrt{T_{RT} T_t}} \left(P(0) - \frac{P_1 + P_2}{2} \right) \quad \text{for } c = 0;$$

$$\eta = \frac{c}{k_B \Gamma \sqrt{T_{RT} T_t}} \frac{P(0) \cosh\left(\omega \frac{L}{2}\right) - \frac{P_1 + P_2}{2}}{\cosh\left(\omega \frac{L}{2}\right) - 1} \quad \text{for } c > 0. \tag{7.52}$$

If distributed pumping speed is not well defined and if the infinity log tube solution can be applied, then the results can be reported in a form of ratio η/α:

$$\frac{\eta}{\alpha} = \frac{\bar{v}_t}{4} \frac{P(0)}{k_B \Gamma \sqrt{T_{RT} T_t}} \quad \text{for } c > 0. \tag{7.53}$$

The following work was focuses on obtaining the experimental data required for the SCC and LHC design.

7.4.4 Primary PSD Yields

It was shown in Chapter 2 that all vacuum chamber materials contain gas atoms that can diffuse to the surface, recombine, and desorb. As diffusion is a temperature-dependent process, it should reduce the desorption process. Indeed, the thermal desorption of tightly bonded molecules is below detectable level at cryogenic temperatures.

The comparison of H_2 and CO PSD yields at room temperature and 4.2 K for the copper-laminated stainless steel was compared at $\varepsilon_c = 284$ eV (see Figure 7.30) [32]. The PSD yields at 294 K are higher than at 4.2 K. The PSD yields are reduced with an accumulated dose (except a short transition for CO at 4.2 K below 1×10^{21} photons/s).

The PSD yields for H_2, CO, CO_2, and CH_4 as a function of accumulated photon dose are shown in Figure 7.31 measured at $\varepsilon_c = 50$ eV and for the LHC beam screen prototype (copper laminated stainless steel) at $T = 77$ K [38].

The comparison of H_2 PSD yields under different conditions are shown in Figure 7.32. Initially the sample was exposed to SR with $\varepsilon_c = 50$ eV at $T = 4.2$–10 K (varied along the beam screen) in runs #1 and #2, then at $T = 77$ K in run #3, and finally the sample was exposed to SR with $\varepsilon_c = 284$ eV at $T = 77$ K in run #4. Considering a scrubbing in each run, it demonstrates insignificant difference in PSD as a function of temperature in runs #1, #2, and #3. However PSD yield increased proportionally with critical photon energy (by a factor 6) between runs #3 and #4.

It should be noted that if the slope of PSD as a function of dose at room temperature at $\varepsilon_c = 284$ eV is $a = 0.6$, it lowers at cryogenic temperatures: $a = 0.25$–0.3 at $T = 77$ K and $a = 0.1$–0.2 at $T = 4.2$ K. However, this is valid for smooth surfaces irradiated at grazing incidence. In the experiments with sawtooth surface where the incidence is quasi-perpendicular [41] (see Section 7.5), the slope measured at 7 K was $a = 0.6$.

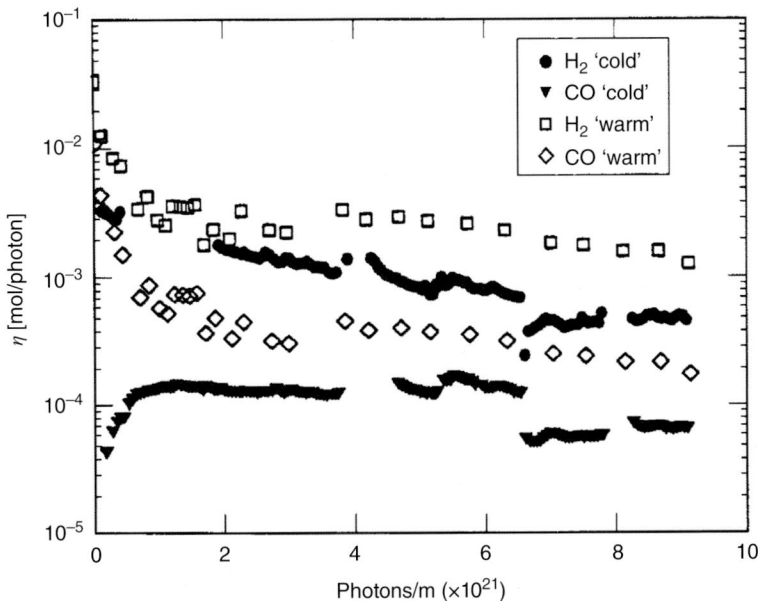

Figure 7.30 $\eta(H_2)$ and $\eta(CO)$ with $\varepsilon_c = 284$ eV versus integrated photon dose at $T = 294$ K ('warm') and $T = 4.2$ K ('cold'). Source: Reprinted with permission from Anashin et al. [32], Fig. 4. Copyright 1993, American Vacuum Society.

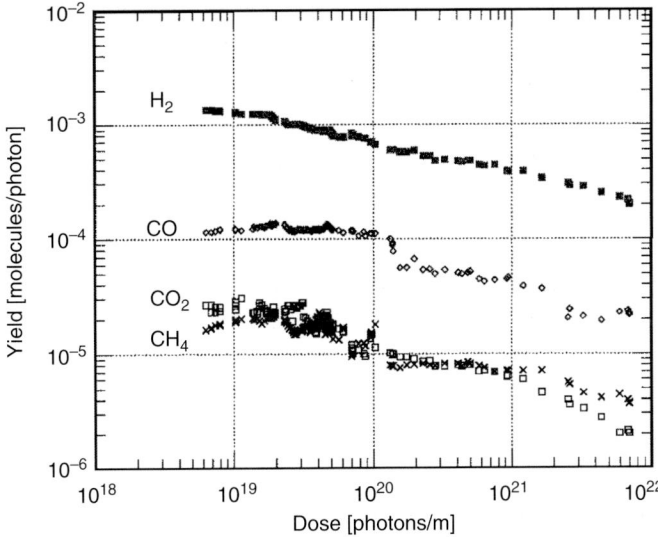

Figure 7.31 PSD yields at 77 K as a function of photon dose with $\varepsilon_c = 50$ eV. Source: Reprinted with permission from Calder et al. [38], Fig. 3. Copyright 1996, American Vacuum Society.

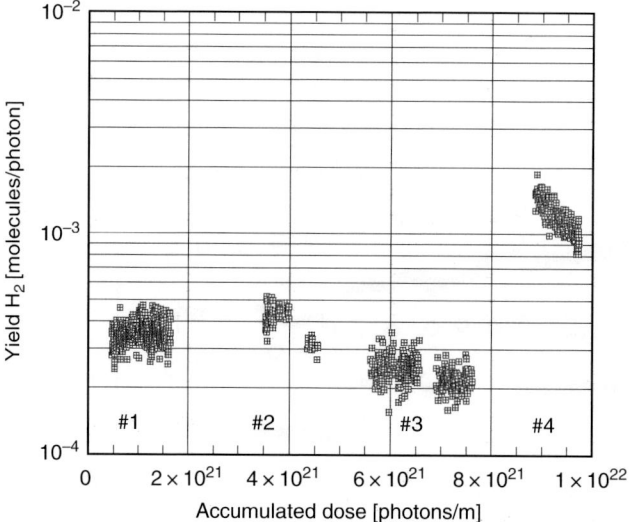

Figure 7.32 PSD yields various conditions: #1 and #2 with $\varepsilon_c = 50$ eV at $T = 4.2$–10 K, #3 with $\varepsilon_c = 50$ eV at $T = 77$ K, #4 with $\varepsilon_c = 284$ eV at $T = 77$ K. Source: Reprinted with permission from Calder et al. [38], Fig. 5. Copyright 1996, American Vacuum Society.

7.4.5 Secondary PSD Yields

First results on the secondary PSD yields (or the PSD yields for cryosorbed gas) were published in Ref. [32] and represented in Figure 7.33. The data are well represented with a linear fit:

$$\eta'(s) = \eta'(s_m) \frac{s}{s_m}. \tag{7.54}$$

where $s_m = 3 \times 10^{15}$ H$_2$/cm^2 and $\eta'(s_m) = 0.7$ in assumption of hydrogen sticking probability $\alpha_{H_2} = 0.1$.

To study the secondary PSD for higher surface density for different gases such as H$_2$, CH$_4$, CO, and CO$_2$, another research facility shown in Figure 7.34 was designed and built at BINP. This set-up allows condensing these gases at 3–68 K. In this experiment, the PSD of cryosorbed gas was calculated from accurate measurements of initial amount of gas Q_i before condensation on a substrate of area A and final amount of gas Q_f after SR irradiation with an accumulated photon dose D and warming up. To highlight the method applied, i.e. removal of gas with SR, the measured value was called the average removal coefficient $\langle \eta_r \rangle$ [molecules/photon]:

$$\langle \eta_r \rangle = \frac{Q_i - Q_f}{D}. \tag{7.55}$$

The average surface coverage of condensed gas $\langle s \rangle$ (molecules/cm^2) for each measurement was defined as

$$\langle s \rangle = \frac{Q_i + Q_f}{2A}. \tag{7.56}$$

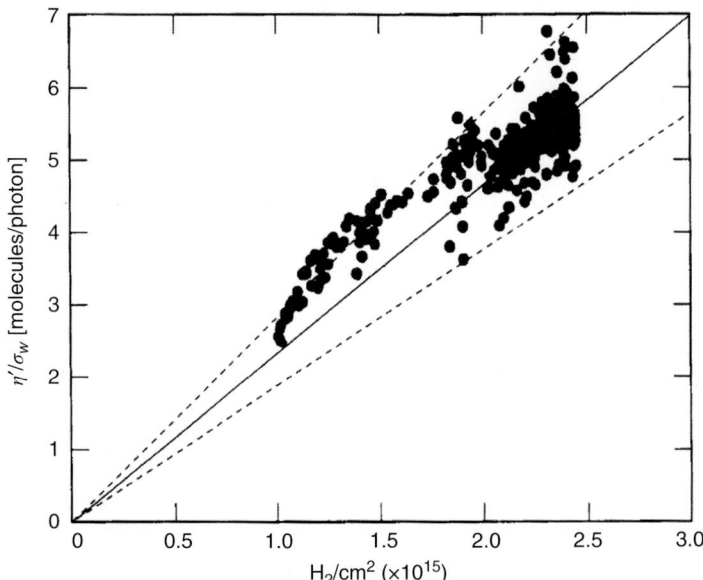

Figure 7.33 Secondary desorption η'/σ_w as a function of cryosorbed gas surface density, where σ_w is hydrogen sticking probability. Source: Reprinted with permission from Anashin et al. [32], Fig. 3. Copyright 1993, American Vacuum Society.

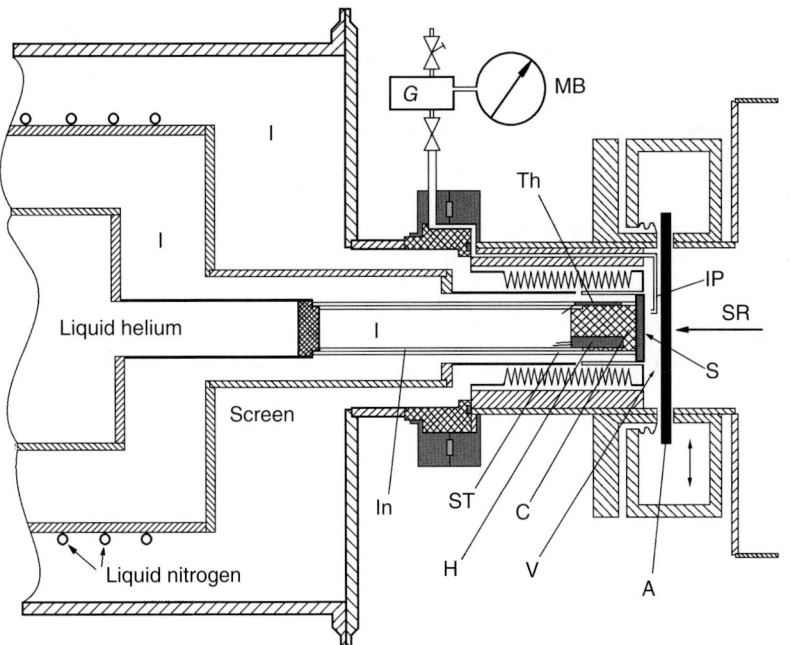

Figure 7.34 Experimental set-up for studying PSD from cryosorbed gases. A – gate valve shown in closed position. Source: Reprinted with permission from Anashin et al. [46], Fig. 1. Copyright 1999, Elsevier.

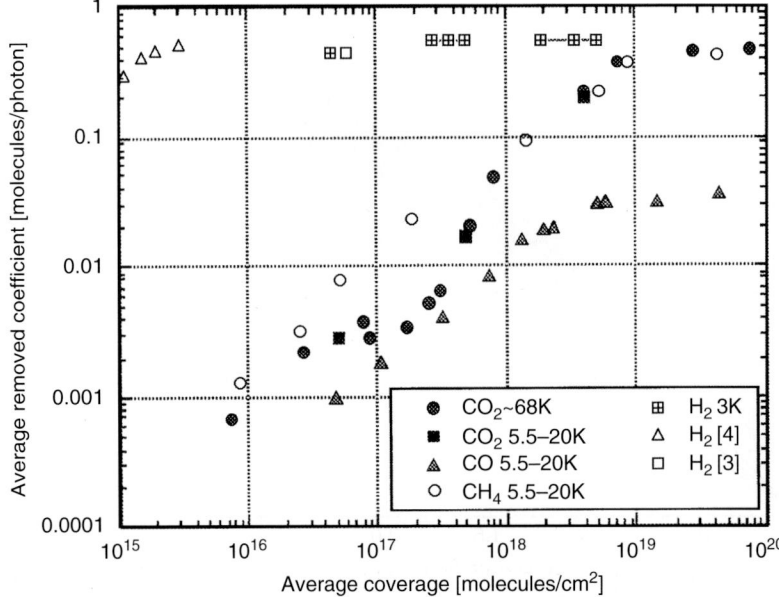

Figure 7.35 Average removal coefficient as a function of the average surface coverage. Data for H_2 shown as from Refs. [3, 4] in this figure corresponds to [32, 45] in this chapter. Source: Reprinted with permission from Anashin et al. [46], Fig. 2. Copyright 1999, Elsevier.

The results obtained with this facility as well as in earlier studies [32, 45, 46] are shown in Figure 7.35. One can see that for maximum values of η_r for H_2, CH_4, and CO_2 are practically the same – $\eta_{r\,max}(H_2, CH_4, CO_2) = 0.4$–$0.5$ molecules/photon – while maximum values of η_r for CO is in orders of magnitude lower: $\eta_{r\,max}(CO) = 0.04$ molecules/photon. However, a maximum value for H_2 was reached at $s \approx 3 \times 10^{16}$ molecules/cm², while for other gas species at orders of magnitude higher surface coverage of $s \approx 10^{18}$ molecules/cm².

Another observation was that η_r for CH_4, CO, and CO_2 is insensitive to the substrate temperature variation in the range between 5.5 and 20 K.

7.4.6 Photon-Induced Molecular Cracking of Cryosorbed Gas

7.4.6.1 Experimental Measurements

Four main photodesorbed gases in a cryogenic vacuum chamber were studied: H_2, CH_4, CO, and CO_2, two of which (CH_4 and CO_2) have shown they can be cracked by photons, $\tilde{\gamma}$, [40, 44, 46–48]. An additional amount of H_2, CO, and O_2 appears in a vacuum chamber due to photo-cracking of CH_4 and CO_2:

$$CH_4 + \tilde{\gamma} \rightarrow C + 2H_2,$$
$$2CO_2 + \tilde{\gamma} \rightarrow 2CO + O_2. \tag{7.57}$$

The efficiency of photo-cracking of CH_4 and CO_2 is approximately 10 times higher than recycling. In the case of irradiating of condensed CO_2 with coverage

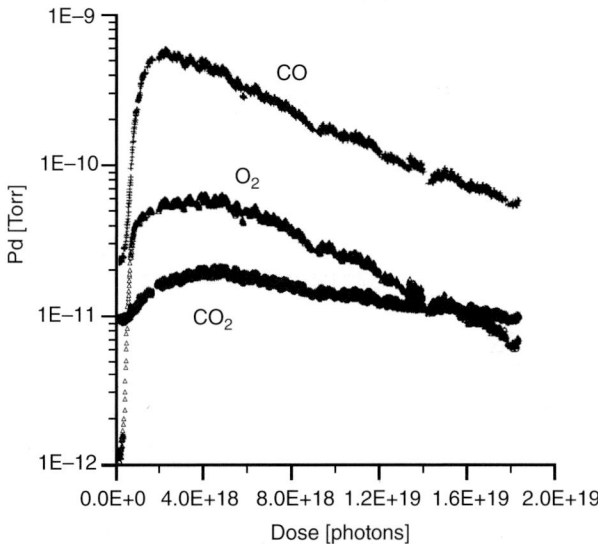

Figure 7.36 The dynamic pressure dependence on the photon dose for CO, O_2, and CO_2 in the experiment with an average CO_2 coverage of 8.2×10^{17} molecules/cm². Source: Reprinted with permission from Anashin et al. [44], Fig. 3. Copyright 1998, Elsevier.

of 8.2×10^{17} molecules/cm², the dynamic pressure is dominant by CO and O_2, as shown in Figure 7.36. In the case of irradiating of condensed CH_4, the dynamic pressure is dominant by H_2 as shown in Figure 7.38.

Similar results were observed at CERN in different experimental set-ups using not only usual gases but also isotopes. The main advantage of using isotopes was to ensure that the photo-cracking occurs only with cryosorbed gases rather than with gas molecules contained in the substrate. Studies were conducted at 4.2 and 77 K on thick coverages (10^{17} to 5×10^{17} molecules/cm²) of CH_4, CO_2, and their isotopes CD_4 and $C^{12}O_2^{18}$ irradiated by SR at perpendicular incidence and 45.3 eV critical energy [48]. Figure 7.37 shows the partial pressure of the usual gases together with the isotopes of CO, O_2, and CO_2 during the irradiation by SR of 45.3 eV critical energy at a perpendicular incidence of 2×10^{17} $C^{13}O_2^{18}$/cm² condensed onto the stainless steel surface at 4.2 K. The arrows on the vertical axis indicate the base partial pressure before irradiation. During irradiation, the partial pressure of the usual gases (H_2, CH_4, H_2O, CO, and CO_2) remains constant due to the primary photodesorption. But the partial pressure of $C^{13}O^{18}$, $C^{13}O_2^{18}$, and O_2^{18} initially increases under the recycling and photo-cracking process and then decreases towards zero during the cleaning process due to the removal of the recycled molecules by the external pumping. The main gas $C^{13}O^{18}$ is produced under $C^{13}O_2^{18}$ photo-cracking. In this experiment, the removal coefficient for $C^{13}O_2^{18}$ equals 9×10^{-3}, similar to the value of Figure 7.35.

Irradiation with 194 eV critical energy at 11 mrad of a non-perforated Cu liner held at 6 K with condensed layers of CH_4 and CO_2 exhibits also the photo-cracking effect [41]. Figure 7.38 shows the CH_4 photo-cracking into H_2

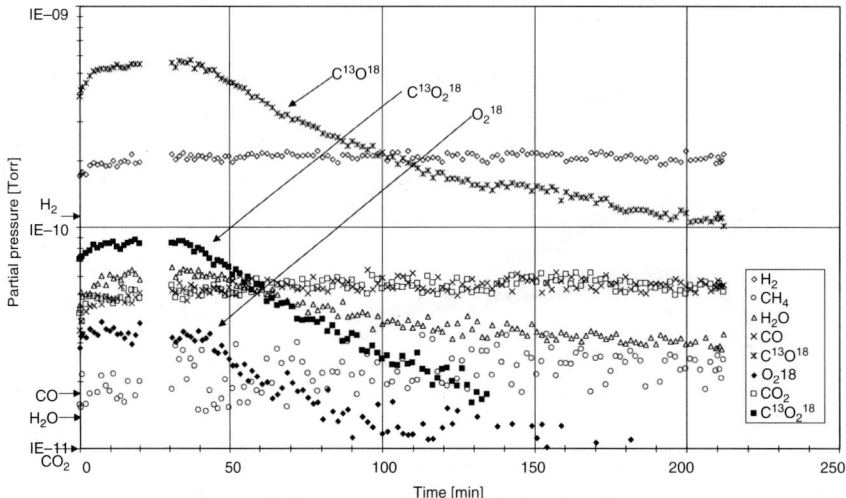

Figure 7.37 The dynamic pressure dependence on the photon dose for $C^{13}O^{18}$, O_2^{18}, and $C^{13}O_2^{18}$ in the experiment with an average $C^{13}O_2^{18}$ coverage of 2×10^{17} molecules/cm². Source: Adapted from Baglin 1997 [48].

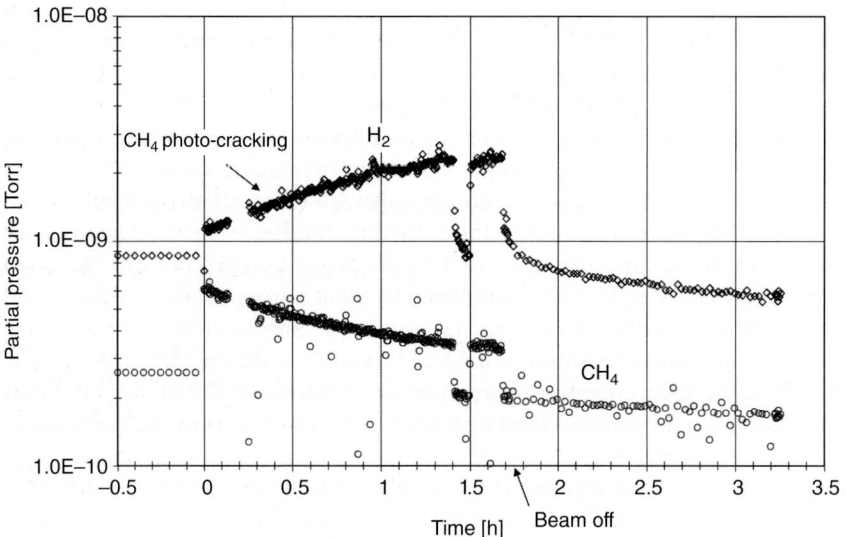

Figure 7.38 The dynamic pressure dependence on the photon dose for H_2 and CH_4 in the experiment with an average CH_4 coverage of 8×10^{15} molecules/cm². Source: Baglin et al. 2002 [41], Fig. 5. Reprinted with permission of CERN.

when 8×10^{15} CH_4/cm² is condensed on the surface prior SR irradiation. The H_2 is the main component. Under irradiation, CH_4 is photo-cracked into H_2, thereby decreasing the CH_4 partial pressure. Doing so, the H_2 surface coverage increases with time together with the secondary desorption (see Figures 7.33 and 7.34) and thus the H_2 partial pressure increases during the irradiation.

7.4.6.2 How to Include Cracking into the Model

To incorporate the cracking physical processes into the gas dynamics model, we introduce the parameter $\chi_i(s_j)$, which is the efficiency of producing type i molecules by cracking of the type j parent molecules, and the parameter $\kappa_{j \to i+n}(s_j)$, which is the cracking efficiency of type j molecules into type i and type n. The two quantities χ and κ are connected through the relation: by

$$\chi_i(s_j) = a_{i,j}\, \kappa_{j \to i+n}(s_j), \tag{7.58}$$

involving the efficiency coefficients $a_{i,j}$.

Then the balance equations for each individual gas species, i, can be written for the infinity long vacuum chamber as

$$V \frac{dn_i}{dt} = (\eta_i + \eta_i' + \chi_i)\Gamma - \alpha_i S_i(n_i - n_{e\,i}) - C_i n_i, \tag{7.59}$$

$$A \frac{ds_i}{dt} = \alpha_i S_i(n_i - n_{e\,i}) - (\eta_i' + \kappa_{i \to k+m})\Gamma. \tag{7.60}$$

Similar to analysis in other parts of this book, a rather slow evolution over many hours and even months of photon irradiation is in a consideration. Thus the following approximations for a 'quasi-static' vacuum system have been made:

$$\frac{dn_i}{dt} \approx 0 \quad \text{and} \quad \frac{ds_i}{dt} \neq 0. \tag{7.61}$$

The parameters η', χ, κ, and n_e depend on the surface coverage; thus the gas density will slowly evolve during the exposure to SR and accumulation of cryosorbed gas on the surface. Furthermore, the different cryosorbed species are transformed from one type into another due to the cracking process and their partial pressures become mutually interdependent.

A Cold Bore Without a Beam Screen

The gas density for each gas species is given by

$$n_i \approx \frac{(\eta_i + \eta_i'(s_i) + \chi_i(s_j))\Gamma}{\alpha_i S_i} + n_{e\,i}. \tag{7.62}$$

The slowly changing surface density of the cryosorbed molecules, $s(t)$, can be computed starting from the initial value $s_i(0)$ as

$$s_i(t) = s_i(0) + \frac{1}{A}\int_{t=0}^{t} (\eta_i + \chi_i(s_j) - \kappa_i(s_i))\Gamma\, dt \tag{7.63}$$

To estimate how including of cracking can affect the maximum gas density in the beam pipe, we can assume that η', χ, κ increase with surface coverage until they reach their respective maximum values $\eta_{\max}$, $\kappa_{\max}$, and $\chi_{\max}$ [34, 44, 46, 47]. Then in a general case the gas density would have an upper bound:

$$n_i \leq \frac{\left(\eta_i + \eta'_{i\max}(s_i) + \sum_{j \neq i} \chi_{i\max}(s_j)\right)\Gamma}{\alpha_i S_i} + n_{e\,i}(s_i). \tag{7.64}$$

A more constraint estimate can be made when there is no pre-condensed gas at the beginning of the SR irradiation. In this case the rate of cracking of molecules can never exceed the primary production rate: $\kappa_{j \to i+k} = a_{i,j}\,\chi_i(s_j) \leq \eta_j$. From this condition follows

$$n_i \leq \frac{\left(\eta_i + \eta'_{i\max} + \sum_{j \neq i} \frac{\eta_j}{a_{i,j}}\right)\Gamma}{\alpha_i S_i} + n_{e\ i}. \tag{7.65}$$

A Cold Bore with a Beam Screen

The gas density for each gas species is given by

$$n_i = \frac{(\eta_i + \eta'_i + \chi_i)\Gamma + \alpha_i S_i n_{e\ i}}{\alpha_i S_i + C_i} \tag{7.66}$$

In this case, the gas density and the surface density on the beam screen will always be limited by the distributed pumping C of the holes/slots in the beam screen. The surface density of the cryosorbed molecules on the beam screen can be computed with the expression:

$$s_i(t) = s_i(0) + \frac{1}{A}\int_{t=0}^{t}[(\eta_i + \chi_i(s_i) - \kappa_i(s_j))\Gamma - C_i n_i]dt. \tag{7.67}$$

Under conditions where the thermal equilibrium density n_e can be neglected, the slowly varying gas density can be expressed as

$$n_i(t) = \frac{(\eta_i + \chi_i - \kappa_i)\Gamma}{C_i} - \frac{A}{C_i}\frac{ds_i}{dt}. \tag{7.68}$$

When the surface coverage has reached a condition $ds_i/dt = 0$ and a constant value, which in turn implies a constant gas density independent of the wall pumping speed, the gas density is

$$n_i = \frac{(\eta_i + \chi_i - \kappa_i)\Gamma}{C_i}. \tag{7.69}$$

Since, the gas density has an upper limit

$$n_i \leq \frac{\left(\eta_i + \sum_{j \neq i} \frac{\eta_j}{a_{i,j}}\right)\Gamma}{C_i}. \tag{7.70}$$

7.4.6.3 Example

The evolution of the gas density in the LHC vacuum system with a cold bore at 1.9 K and a beam screen at 5 K has is shown in Figure 7.39 [47]. The H_2 density increases to its maximum value after 10^{20} photons/m (about 15–20 minutes of LHC operation) and decreases for further irradiation. The initial increase is due to the finite pumping capacity of the beam screen. The maximum values of gas density n, surface coverage s, and secondary desorption yield η' were reached when a condition $ds_i/dt = 0$ was satisfied.

The CH_4 and CO_2 gas densities are at their maximum at the start of irradiation and decreases as a function of accumulated dose due to photon scrubbing. The surface coverage is limited by cracking and the recycling of these gases is negligible.

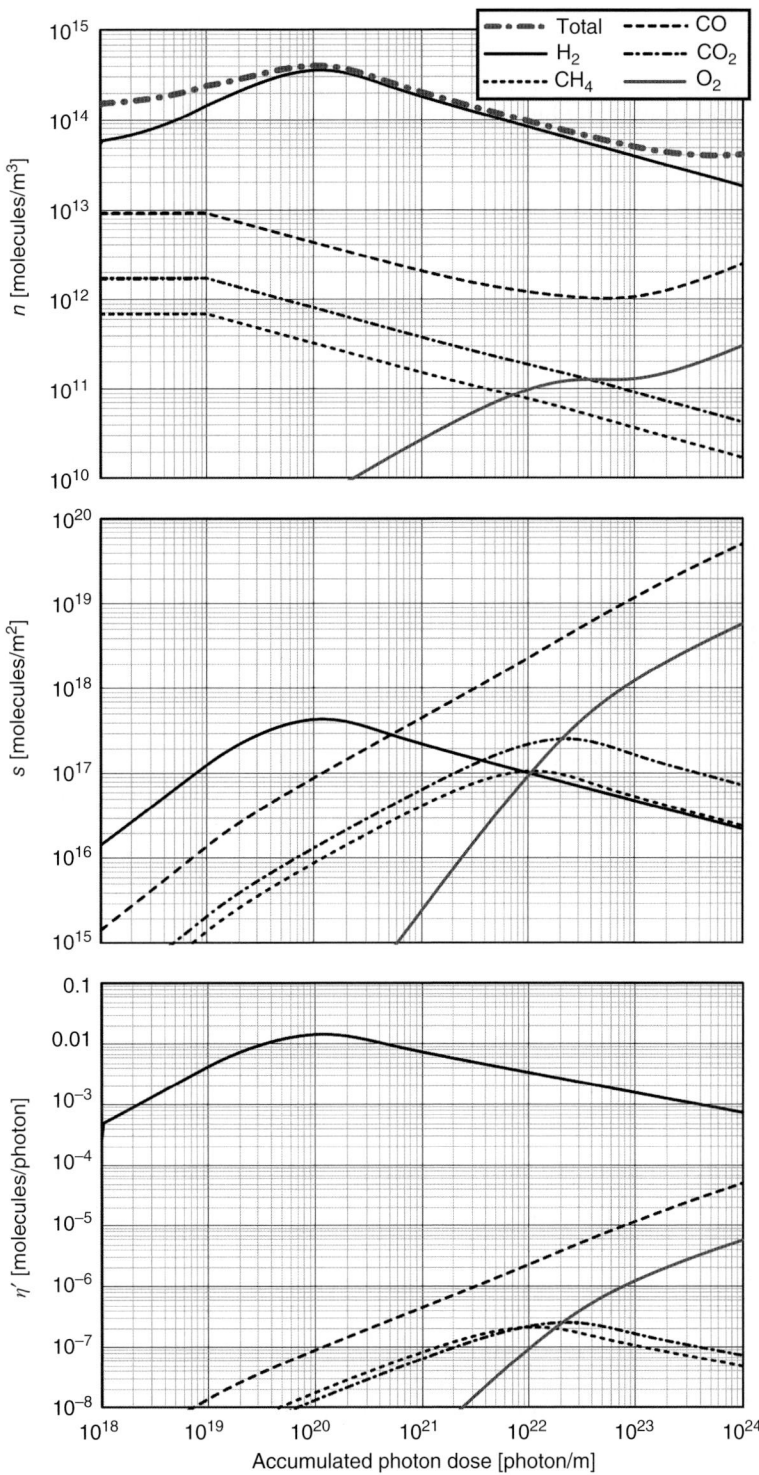

Figure 7.39 Gas density n, surface coverage s, and secondary desorption yield η' for H_2, CH_4, CO, O_2, and CO_2 as a function of the accumulated photon dose in the LHC with a beam screen at 5 K. Hydrogen-equivalent gas density is labelled as Total. Source: Anashin et al. 2001 [47]. Replotted with permission of Elsevier.

The CO density initially decreases due to the photon scrubbing of the surface. However, from a photon dose of about 10^{22} photons/m, it increases again because of the additional contribution from CO_2 cracking, which increases the surface coverage s and results in a higher secondary desorption yield η'. Nevertheless, since the CO secondary desorption yield remains relatively low up to a rather high surface coverage of many 1000 ML, the gas density does not reach a saturation level even at an accumulated photon dose of 10^{24} photons/m (which corresponds to approximately four months of LHC operation).

Total hydrogen-equivalent gas density is shown in Figure 7.39 and labelled as Total. In the beginning of irradiation, the CO gas density is significant in the total hydrogen-equivalent gas density. At higher photon doses (10^{20}–10^{23} photons/m), the total gas density is practically dominated to hydrogen gas density; however CO contribution increases ~50% to at 10^{23} photons/m.

Another interesting result is the presence of oxygen in gas density spectrum, which is very unusual for room temperature machines.

7.4.7 Temperature of Desorbed Gas

In the gas dynamics calculation, the temperature of gas is one of the key parameters. The temperature of gas is often considered the same as wall temperature. This is correct for a vacuum chamber with all the walls at the same temperature T_w and with a gas in equilibrium state. In some cases it could be considered correct: for example, in a long vacuum chamber (i.e. with $d/L \ll 1$) with temperature transition at the ends. However, it is not obvious what is a gas temperature in a cryogenic vacuum chamber irradiated by SR. Let us consider that molecules are desorbed with temperature T_d, which is higher than the wall temperature T_w: i.e. $T_d > T_w$.

Heat transfer in a gas–wall interaction is described with the thermal accommodation coefficient defined as follows:

$$\psi(T_a, T_w) = \frac{T_a - T_b}{T_a - T_w} \tag{7.71}$$

where T_a and T_b is a temperature of a molecule before and after an interaction with a wall. After each interactions with walls the gas temperature is

$$T_b = T_a + (T_w - T_a)\psi(T_a, T_w). \tag{7.72}$$

Thus, the temperature of gas molecules can be calculated after each infraction with a wall when the thermal accommodation coefficient $\psi(T_a, T_w)$ is known in the temperature range $T_w \leq T_a \leq T_d$.

If we consider the thermal accommodation coefficient is a constant, then using Eq. (7.72), one can write

$$T_b - T_w = (T_a - T_w)(1 - \psi). \tag{7.73}$$

Thus, after N gas–wall interactions, the gas temperature can be calculated from

$$\frac{T_N - T_w}{T_g - T_w} = (1 - \psi)^N \tag{7.74}$$

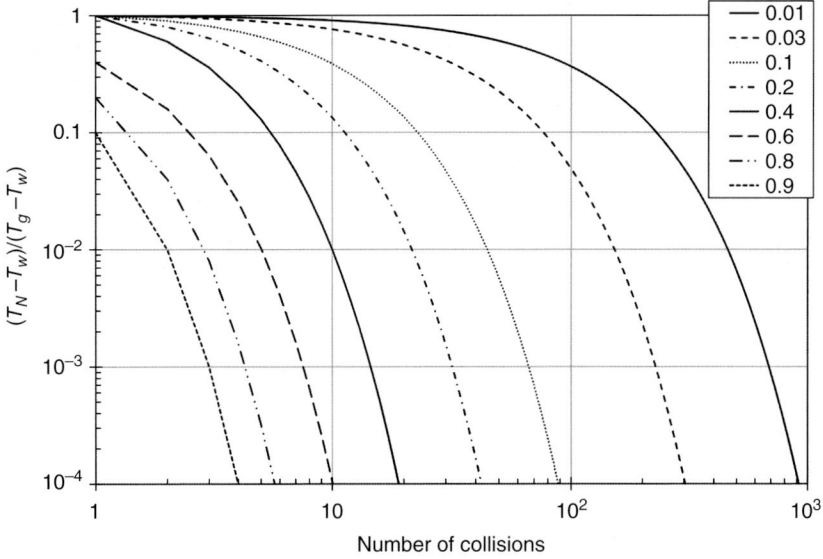

Figure 7.40 The normalised temperature of gas molecules as a function of a number of molecule collisions with the wall for various thermal accommodation coefficients: 0.01, 0.03, 0.1, 0.2 0.4, 0.6, 0.8, and 0.9.

The normalised temperature of gas molecules, $(T_N - T_w)/(T_g - T_w)$, is shown in Figure 7.40 as a function of a number of interactions with the walls for the thermal accommodation coefficients in the range: $0.01 \leq \psi \leq 0.9$. One can see that the smaller the accommodation coefficient, the more gas–wall interactions required for thermalisation of molecules.

In general, the thermal accommodation coefficient, $\psi(T_a, T_w)$, depends not only on temperatures of an incident molecule and a wall but also on gas species, incident angle, wall surface material, and structure (smooth, rough, porous) [49–52]. Unfortunately, there are not many data available, but for this section it is sufficient to know the following:

- For industrial metal surfaces the thermal accommodation coefficient could vary in the range from 0.01 to 1.
- The thermal accommodation coefficient for light gas species (H_2 and He) are lower whang or heavier species. Therefore, they may require a more interactions with vacuum chamber walls for a full temperature accommodation.
- The accommodation coefficient for heavier gas species can be considered equal to 1: i.e. gas molecules are fully thermalised after one collision with vacuum chamber walls.

On another side, the sticking probability defines a number of gas–wall interaction before the molecule is cryosorbed:

- The H_2 sticking probability at $T = 4.2\,K$ lies in the range $0.1 \leq \alpha \leq 0.5$, if an initial temperature of photodesorbed molecules is high: $T_d > T_w$. Then at each collision, some molecules are cryosorbed, while others are partially

thermalised. Since a fraction of gas molecules, which are just desorbed, could be significant (10–50%), other molecules could be not fully thermalised, the average temperature of these molecules could be higher than the wall temperature.
- The sticking probabilities for heavier gas species are higher: in the range $0.5 \leq \alpha \leq 1$. In case of $\alpha = 1$, the temperature of gas is equal to T_d. Thus, the average temperature of heavy gas species could also be higher.

To study how much the average temperature of gas could be higher than the vacuum chamber wall temperature, the facility at BINP (Novosibirsk, Russia) shown Figure 7.24 has been updated to measure the gas density inside the sample tube using an effect of charge exchange of 20 keV H$^+$ beam with gas molecules (see Figure 7.41) [35, 37]. The gas density measurements are based on two processes:

$$H^+ + A^0 = H^0 + A^+,$$
$$H^+ + A^0 = H^- + A^{++}. \tag{7.75}$$

The cross section values of these processes is known and the amounts of H^0 and H$^-$ are proportional to the gas density. The H$^+$ beam is measured with a Faraday Cup, while H^0 and H$^-$ beams can be measured with secondary electron multipliers: SEM1 and SEM2. The four-pole superconducting magnet (SM) installed at the middle of the sample tube allows to separate and measure the intensity of H^0 and H$^-$ created on the 20-cm-long section in the middle of the sample tube (see Figure 7.42).

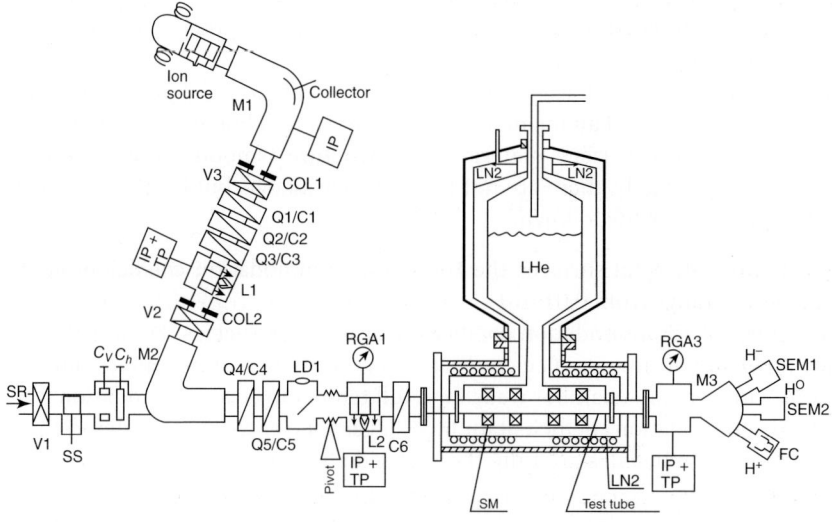

Figure 7.41 The experimental set-up for direct measurements of gas density inside a 4.2 K beam tube by the H$^+$ beam method at BINP: M1–M3, bending magnets; Q1–Q5, quadrupoles; C1–C4, correctors; SM, superconducting magnet; FC, Faraday cup; SEM1 and SEM2, secondary electron multipliers; V1, vacuum valve; C_v, C_h, vertical and horizontal collimators; SS, safety shutter; LD1, LD2, phosphor screens; IG, ion gauges; RGA, residual gas analysers; IP + TP, combined ion and Ti sublimation pumps. Source: Adapted from Alinovsky et al. 1994 [37].

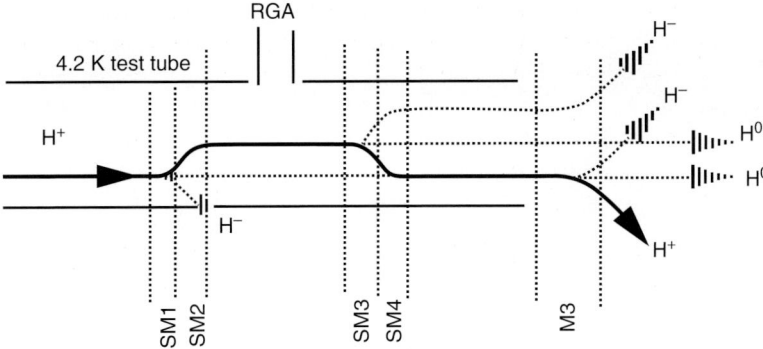

Figure 7.42 Illustration of secondary particle beam separation inside the sample tube with the help of superconducting magnet poles SM1–SM4. Source: Adapted from Alinovsky et al. 1994 [37].

The results obtained in the experiments allow to conclude that average speed of H_2 molecules during SR irradiation of sample tube held at 4.2 K with photon critical energy of $\varepsilon_c = 284$ eV and intensity of $\Gamma = 10^{16}$ photons/(s·m) is equal to 800 ± 150 m/s and corresponds to an effective temperature of 60 ± 20 K.

This result demonstrates that from one side the calculations of PSD yields from the cryogenic experiments could be underestimated due to the difference between gas and vacuum chamber wall temperature. From another side, the same effect results in overestimating the gas density in calculation for future machines. However, if the vacuum chamber prototypes were tested under the conditions similar to the future machine, and these results are used for gas dynamics modelling, the possible gas temperature errors in the measurements and modelling will cancel each other; thus this is the safest way to avoid such a mistake.

7.5 In-Depth Studies with COLDEX

In the framework of the LHC construction, in-depth studies of PSD on beam tube held at cryogenic temperature were conducted at CERN with the COLD bore Experiment (COLDEX) [40–43] that is an advanced version of the facility shown in Figure 7.24.

7.5.1 COLDEX Experimental Facility

The COLDEX experimental facility was installed and operated either in one SR beamline (Synchrotron Light Facility 92) of the Electron Positron Accumulator (EPA) or in the EPA storage ring itself (Figure 7.43) [53]. EPA is a part of the Large Electron Positron pre-injector complex. When EPA is operated with electron beam with energies from 200 to 600 MeV, it could generate SR in the UV range with critical energy from 12 to 335 eV. Although, when operated at 308 MeV, EPA generated an SR with critical energy of 45.3 eV very similar

Figure 7.43 COLDEX installed in a SR beam line (a) and in the EPA ring (b).

to that of LHC, the machine was, most of the time, operated at its nominal energy of 500 MeV with corresponding critical energy of 194 eV and a photon flux of $\sim 3.4 \times 10^{16}$ photons/(m·s), i.e. 1/3 of the LHC photon flux. The SR light impinged the tested tube at a grazing angle of 11 ± 2.7 mrad over the length of 2112 mm (see Figure 7.44). The light source was located at 4.34 m from the middle of the 2.8-m-long COLDEX cryostat. The SR fan dimension was defined by a fixed collimator of 7.5 mm × 11 mm horizontal and vertical aperture, with 5.3 mrad × 7.8 mrad opening angle. The radiation low energy cut-off, set by the collimator vertical opening, equalled 4.2 eV, below which the photon energy spectrum is attenuated. Thus 66% of the photon energy spectrum was not attenuated by the vertical collimation. The beam size dimension in the middle of the COLDEX cryostat was 23.0 mm × 33.8 mm in the horizontal and vertical planes. The total and partial pressure were monitored in the centre and at the extremities of the apparatus. A combination of ion and Ti sublimation pumps was installed at each extremity of the COLDEX cryostat. To measure the gas flux from/to the test chamber, the conductances of 72 l/s (in nitrogen equivalent) were installed on both upstream and downstream sides to the COLDEX cryostat.

The large cold bore (113 mm) of the COLDEX allows an easy exchange of the tested beam tubes with 47 mm inner diameter and 2232 mm length. The cold bore is a part of LHe cryostat which can be filled with liquid helium, its temperature can be controlled from 2.5 to 4.2 K (below 3 K, the saturated vapour pressure of hydrogen is negligible, $<10^{-10}$ mbar). The beam screen is cooled with gaseous helium circulating inside cooling channels brazed on each side of it. The beam screen temperature can be controlled from 5 to ~ 150 K. It is monitored with three calibrated temperature sensors. Two port were produced in the middle of the beam screen (Figure 7.45). The warm chimney allows the collected molecules desorbed from the cold surface to be measured at room temperature with BA4 and RGA3 total pressure and partial pressure gauges. An extractor gauge is also placed in the centre of the beam screen to measure the desorbed molecules. This gauge operates at ~ 90 K. The distance between the warm chimney, the extractor gauge shield, and the two beam screen ports is set to 1 mm by design to optimise the collection of the desorbed molecules. Known quantities of gas can be admitted into the system via a leak valve. The cryostat can be isolated from the

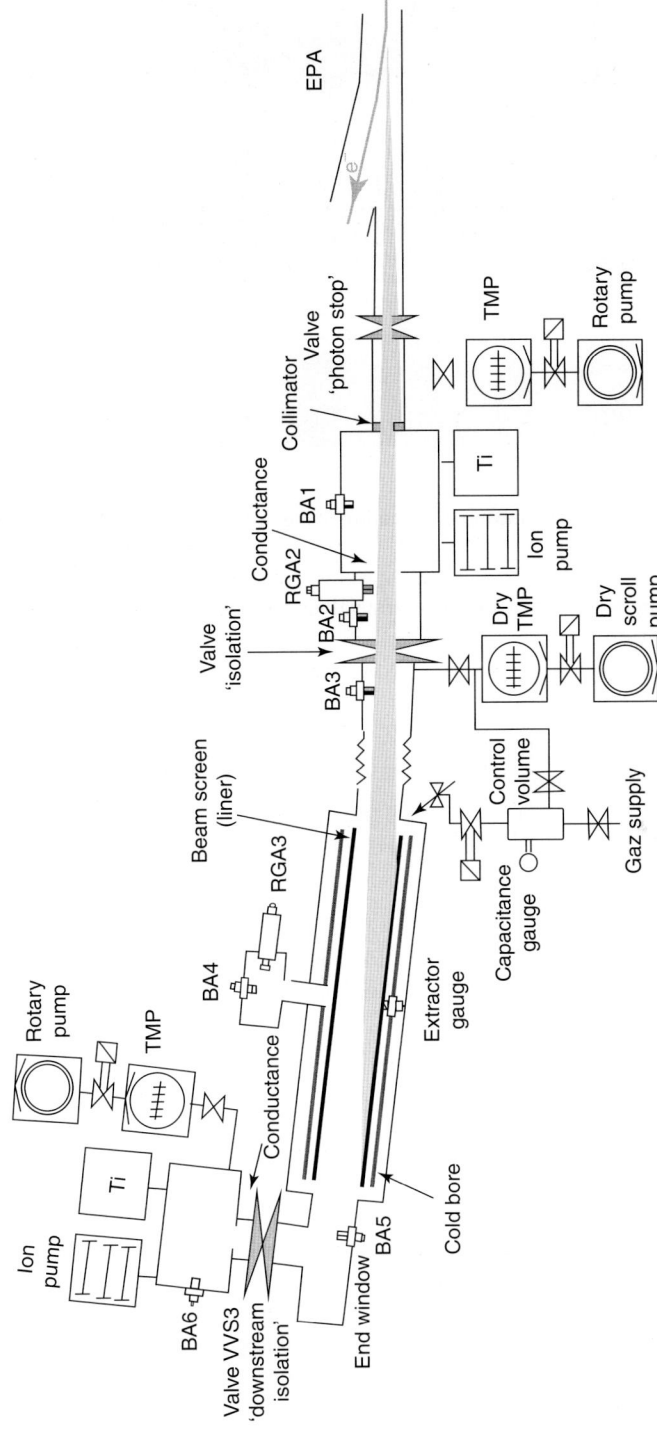

Figure 7.44 COLDEX experimental facility layout (Fig. 1 in [42]). BA, Bayard–Alpert gauge; RGA, residual gas analyser; TMP, turbo-molecular pump; Ti, Ti sublimation pump. Source: Baglin et al. 2000 [40], Fig. 1. Reprinted with permission of CERN.

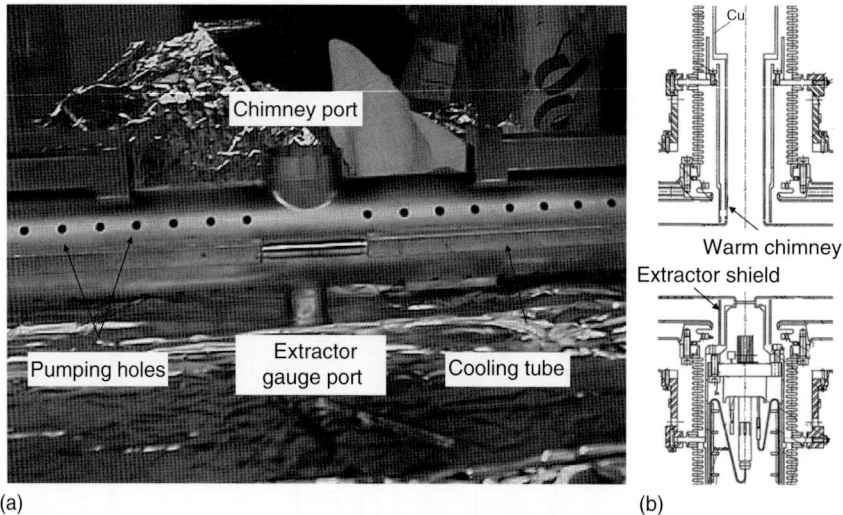

Figure 7.45 (a) Picture and (b) drawing of the centre of the beam screen and cryostat assembly.

external pumping system with two sector valves to study adsorption isotherms and gas propagation.

During the LHC design phase, the following beam screens where tested at the SR facility until December 2001:

- A stainless steel liner without holes.
- A Cu liner without holes.
- A perforated Cu beam screen with 1% transparency.
- A perforated Cu beam screen equipped with cryosorber (activated charcoal).
- A perforated Cu-co-laminated beam screen with a sawtooth structure, i.e. an LHC-like beam screen.

Then the COLDEX facility has then re-installed in a bypass of the CERN Super Proton Synchrotron (SPS) ring to study the interaction of the electron cloud with cryogenic walls (see Chapter 8).

7.5.2 PSD of Cu as a Function of Temperature

As shown previously, when the end pumping effect inside the chamber is negligible, the primary PSD yield can be measured with a perforated beam screen using Eq. (7.36). For the temperature dependence study, the COLDEX cryostat was equipped with a perforated Oxygen Free Electronic Grade (OFE) Cu beam screen [42]. The beam screen was perforated with 264 holes of 4 mm diameter each equally distributed over the full length and representing 1% of its surface area. The equivalent pumping speed of the holes, taking into account the Clausing factor, equals 122 l/(s·m) for air at room temperature.

To minimise the cleaning effect under photon bombardment during the measurement at each temperature, the beam tube was pre-exposed to a dose of

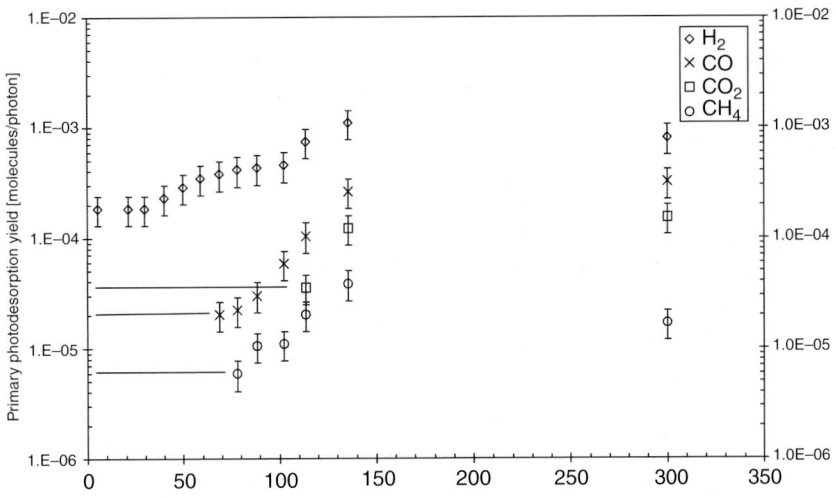

Figure 7.46 Primary PSD for an OFE copper BS irradiated with 194 eV critical energy SR after an accumulated dose of 3×10^{22} photons/m. Source: Reprinted with permission from Baglin et al. [42], Fig. 2. Copyright 2002, Elsevier.

3×10^{22} photons/m. At cryogenic temperature (<80 K), the primary PSD yield, η, is dominated by H_2 ($\eta_{H_2} = 2 \times 10^{-4}$ H_2/photon) being about 1 order of magnitude larger than CO, CO_2, and CH_4 (Figure 7.46). The observed reduction from room temperature to cryogenic temperature is about 1 order of magnitude far all the gases except methane. The straight lines are an upper limit of the primary PSD yields of CO, CO_2, and CH_4.

The sticking probability, α, can be measured under SR by combining Eqs. (7.32) and (7.36). Below 20 K, the sticking probability of an OFE Cu 'bare surface' (<10^{14} molecules/cm^2), i.e. in physisorption regime, lies in the range 0.01–0.1 (Figure 7.47). Carbon dioxide has the largest sticking probability ($\alpha = 0.1$–0.2) and can be physisorbed up to ~100 K with $\alpha = 0.01$. In other words, at 100 K, the beam screen surface pumping speed is equivalent to the perforations pumping speed. Methane, carbon monoxide, and hydrogen are physisorbed with $\alpha > 0.01$ up to 60, 40, and 20 K, respectively.

7.5.3 Secondary PSD Yields

The secondary PSD yield, η', can be measured with a non-perforated liner ($C = 0$) under SR by using Eq. (7.29). For this purpose, gas was pre-condensed onto the non-perforated liner before SR irradiation. Known amount of gas were injected with the non-perforated liner held at 5 K and the isolation valves closed. The temperature of the non-perforated liner was increased till the pressure at BA3, BA4, and BA5 reached a few 10^{-5} mbar, allowing a complete redistribution of the gas along the tube. Then, the temperature was slowly decreased to 5 K while keeping the same pressure across the gauges. Tables 7.5 and 7.6 give the secondary PSD yields of some gases condensed on the OFE Cu held at 5 K when irradiated with 194 eV critical energy SR.

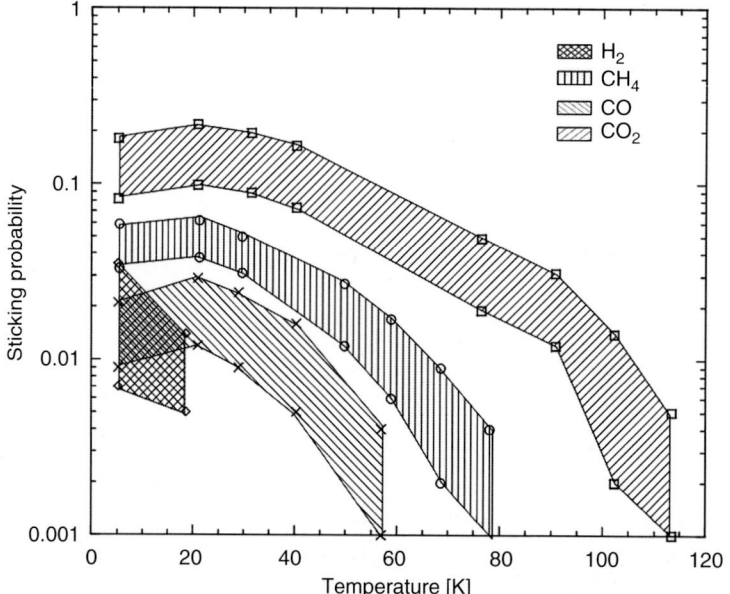

Figure 7.47 Sticking probabilities for an OFE copper BS irradiated with 194 eV critical energy SR after an accumulated dose of 3×10^{22} photons/m. Source: Reprinted with permission from Baglin et al. [42], Fig. 3. Copyright 2002, Elsevier.

Table 7.5 Hydrogen secondary PSD yields over sticking probability at two surface coverages, s_1 and s_2, when irradiated with 194 eV critical energy SR.

Surface coverage	s_1	s_2
s [H_2/cm^2]	1.6×10^{15}	3.2×10^{15}
η'_{H_2}/α [H_2/photon]	0.7	3.3

Table 7.6 Secondary PSD yields over sticking probability of some common gases at a surface coverage of 8.5×10^{15} molecules/cm^2 when irradiated with 194 eV critical energy SR.

Gas	CH_4	CO	N_2	CO_2
η'/α [molecules/photon]	8×10^{-3}	3×10^{-2}	3×10^{-2}	1×10^{-3}

7.5.4 PSD of a BS with Sawtooth for Lowering Photon Reflectivity and PEY

Most of the modern storage rings are prone to electron cloud build-up; thus the accelerator's vacuum systems require a specific design to cope with the potential detrimental effects associated with this electron cloud build-up (see Chapter 8). The LHC vacuum system was also designed in this perspective [54]. In the LHC,

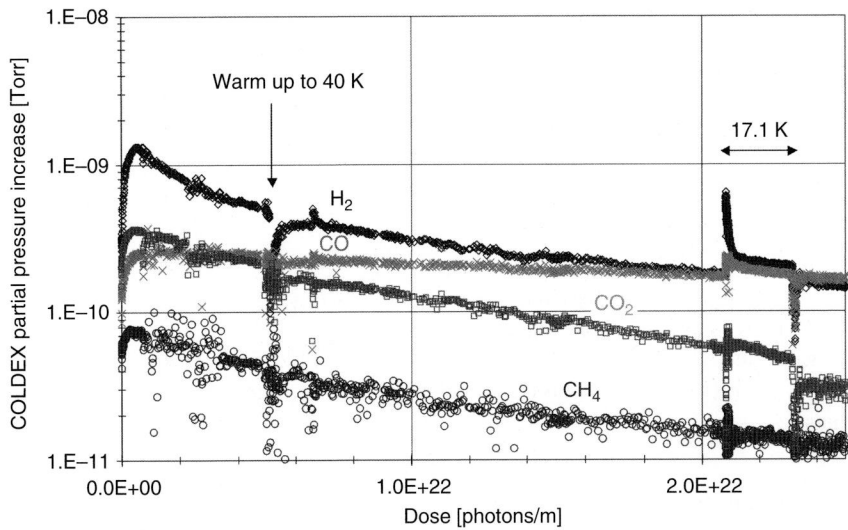

Figure 7.48 Cu-co-laminated stainless sawtooth beam screen irradiated with 194 eV critical energy SR. Source: Baglin et al. 2002 [41], Fig. 1. Reprinted with permission of CERN.

the SR is intercepted on the vacuum chamber at a perpendicular incidence by a sawtooth structure (see Figure 3.4) for lowering the photon reflectivity and the photoelectron yield (PEY).

Figure 7.48 shows the partial pressure behaviour of an LHC-like beam screen (of 1% transparency) as a function of photon dose when irradiated with 194 eV critical energy SR [41]. During the complete study, the cold bore was held at 2.7 K for which the saturated vapour pressure of all the gases, except He, is well below 10^{-12} mbar. At the start of the irradiation, the H_2 partial pressure increases due to the secondary PSD from 2×10^{-10} Torr up to a quasi-equilibrium at 1.3×10^{-9} Torr. Then, it decreases due to the cleaning effect induced by the SR. This behaviour corresponds to the one computed for Figure 7.18. At a dose of 5×10^{21} photons/m, the BS was warmed up to 40 K to flush all the gas, except CO_2, toward the cold bore. With the BS at 7 K, SR was resumed showing again the effect of the secondary PSD. At a dose of 2.1×10^{22} photons/m, irradiation was stopped and the BS temperature was raised to 17.1 K. When resuming irradiation, a vacuum transient due to the excess of gas on the BS surface was seen (see Section 7.5.5). Along the SR irradiation, all the partial pressures, except CO, decrease with time. The origin is attributed to the photo-cracking of CO_2 into CO (see Figure 7.39).

The primary and secondary PSD yields and the conditioning rate were measured from the above data and are shown in Table 7.7. The primary and secondary PSD yields are given in Table 7.7 at a dose of 10^{22} photons/m. The primary PSD yield reduces with photon dose D with a conditioning rate, a, which was obtained with commonly used formula in relation to the initial primary PSD yield η_0 at the photon dose D_0:

$$\eta(D) = \eta_0 \left(\frac{D}{D_0}\right)^{-a} \tag{7.76}$$

Table 7.7 Primary and secondary PSD yields and conditioning rate of a Cu-co-laminated stainless steel sawtooth beam screen when irradiated with 194 eV critical energy SR.

	H_2	CH_4	CO	CO_2
η	2×10^{-4}	$<6 \times 10^{-6}$	$<3 \times 10^{-5}$	$<2 \times 10^{-5}$
$(\eta + \eta')/\alpha$	2×10^{-2}	6×10^{-4}	3×10^{-3}	2×10^{-3}
a	0.6	0.6	0.2	0.8

Source: Baglin et al. 2002 [41]. Reproduced with permission of CERN.

Due to the low conditioning rate and low primary PSD yield, more than 10 years of LHC operation with design parameter is required to desorb a total of 100 ML of gas.

7.5.5 Vacuum Transient

When operating with a perforated beam screen, the initial gas density inside a beam screen without an absorbed gas is described with Eq. (7.33). With accumulation of cryosorbed gas, it can be described with Eq. (7.34) and then (when a quasi-equilibrium state is reached, i.e. $F\, ds/dt \approx 0$) the gas density given by Eq. (7.36). The surface coverage is described with Eq. (7.35), it is increasing when $\eta\Gamma > Cn$ (i.e. more gas primary desorbed than removed through the beam screen pumping holes), and a quasi-equilibrium state corresponds to $\eta\Gamma = Cn$ (surface coverage does not change).

This is illustrated in Figure 7.18 and Figure 7.19 where gas is adsorbed on the beam screen surface until an accumulated photon dose of $\sim 10^{20}$ photons/m. Above this dose, the effective pumping speed of the surface equals zero, and the gas decrease corresponds to reduction of primary PSD yield due to the SR conditioning (scrubbing) effect. From another side, the reduction of primary PSD means that more gas removed through the beam screen pumping holes than primary desorbed, i.e. $\eta\Gamma - Cn < 0$, and the access gas is flushed away towards the cold bore surface until the quasi-static condition of surface coverage is reached again; therefore, the surface coverage decreases together with primary PSD yield.

The H_2 vacuum transient effect was demonstrated in two consequent experiments shown in Figure 7.49. In these experiments the OFE beam screen held at 5 K was exposed to SR. The first experiment started with a bare surface (i.e. no cryosorbed gas) of OFE beam screen, while prior to irradiation in the second experiment, the beam screen was covered with approximately a monolayer of H_2 ($3.2 \times 10^{15}\, H_2/cm^2$) [55]. When no H_2 is condensed on the beam screen, the pressure increases due to the secondary PSD. After five hours of irradiation, a quasi-equilibrium was reached at $\sim 3 \times 10^{-10}$ Torr: i.e. the primary PSD is balanced by the pumping speed through the holes. In the second experiment with a monolayer of H_2 condensed on the beam screen, at the beginning of SR irradiation the pressure increases steeply to 3×10^{-8} Torr due to the secondary PSD.

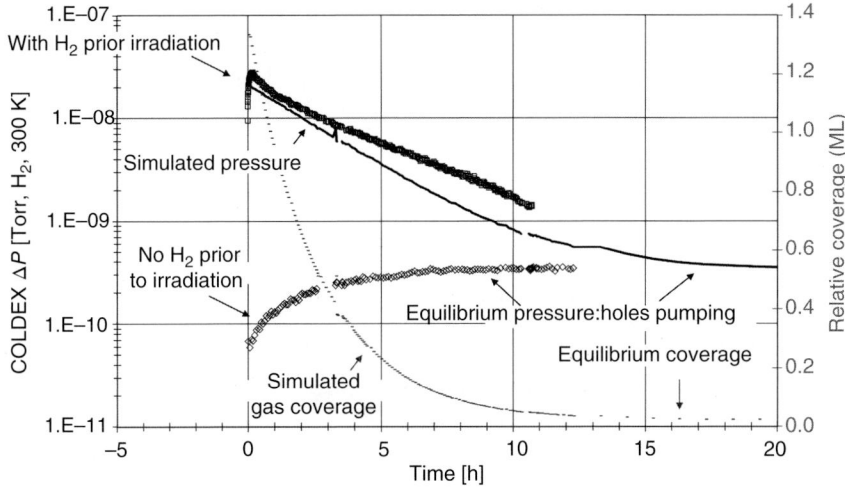

Figure 7.49 Observed and simulated H_2 vacuum transients of a bare and a monolayer covered OFE Cu beam screen surface during irradiation with SR of 194 eV critical energy. Source: Reprinted with permission from Sharipov and Moldover [51], Fig. 3. Copyright 2016, CERN.

During the following few hours of irradiation, the gas was slowly flushed though the pumping holes towards the cold bore until the quasi-equilibrium pressure is reached. This vacuum transient can be simulated using Eqs. (7.34) and (7.35) with $\eta = 3 \times 10^{-4}$ H_2/photon, $\eta' = 0.5$ H_2/(photon ML) and $\alpha = 0.5$. The simulated pressure and gas coverage shown in Figure 7.49 reach a quasi-equilibrium after more than 15 hours of irradiation.

In order to avoid vacuum transient in the LHC, the beam screen surface must remain free of physisorbed/condensed molecules. For this reason, an appropriate scenario decoupling the beam screen and cold bore cool down was set. Moreover, in the event an excess of gas is physisorbed or condensed on the inner surface of the LHC beam screen, e.g. following a magnet quench, beam screen heaters allow warming up the beam screen up to ~80 K to flush the gas from the beam screen towards the cold bore before operation with beams [51].

That vacuum transients have been observed due to an excess of surface coverage and SR irradiation. However, when designing vacuum system operating at cryogenic temperature, one should bear in mind that other sources of stimulated desorption (electron cloud, ion bombardment, particle losses) can also lead to vacuum transients.

7.5.6 Temperature Oscillations

Another type of vacuum transient can occur even in the absence of stimulated desorption. Indeed, temperature oscillations of a cryogenic surface might lead to pressure excursions. Depending on the nature of the substrate, the nature of the gas, the surface coverage, and the temperature range, the vapour pressure might become non-negligible and alter the performance of an accelerator.

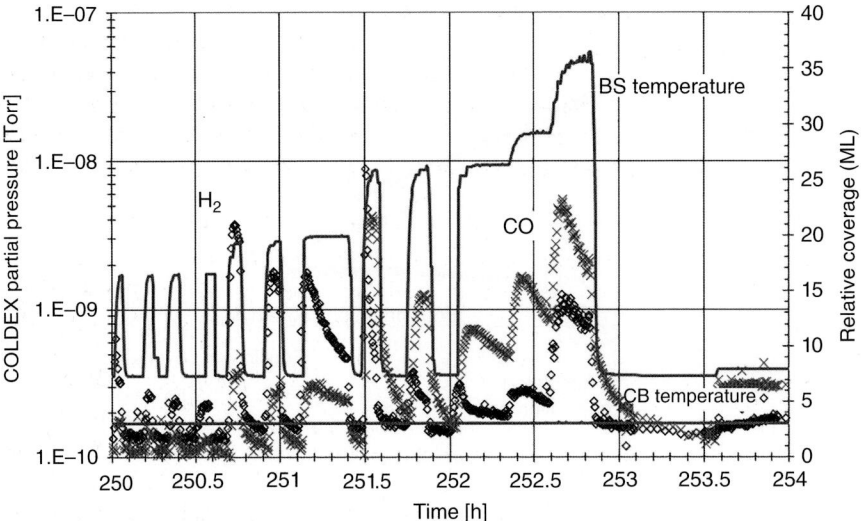

Figure 7.50 Temperature oscillations after accumulation of 2.3×10^{22} photons/m. Source: Baglin et al. 2002 [41], Fig. 2. Reprinted with permission of CERN.

Figure 7.50 shows the H_2 and CO pressures during temperature oscillation in the range between 15 and 35 K after an accumulated photon dose of 2.3×10^{22} photons/m [41]. Around 20 and 25 K, H_2, and CO, accumulated during the SR irradiation, are desorbed from the cryogenic surface. The pressure reaches $\sim 10^{-8}$ Torr. In the CO case, the reached level of pressure was above the LHC design pressure ($\sim 10^{-9}$ Torr) (see Section 7.2.5 and Figure 7.14). For this reason, during operation, the temperature of the LHC beam screen shall remain below 25 K.

The impact of temperature oscillations on the machine operation might be acceptable for some gas since their presence in the beam screen volume is reduced by the large secondary desorption yield, which flushes away the molecules towards the cold bore. However, gas with low secondary desorption yield remains for a longer period in the beam screen volume.

Figure 7.51 shows the H_2 and CO vacuum levels during SR irradiation when 8.5×10^{15} CO/cm^2 are condensed onto the beam screen [41]. During daily operation, such surface coverage on the beam screen might be the result of a magnet quench. Although the secondary desorption yield of CO is low, the level of pressure remains above the LHC design pressure (10^{-9} mbar) for several hours. When temperature oscillations up to 25 K occurs, the vapour pressure adds up to the molecular stimulated desorption, increasing further the vacuum level approaching the magnet quench limit (2.5×10^{-7} mbar).

Such vacuum transient is avoided by a careful control of the cooling scheme. In the LHC, a feedforward control loop taking into account the impact of the proton beams (impedance increase during the machine filling, heat load due to electron cloud) is applied.

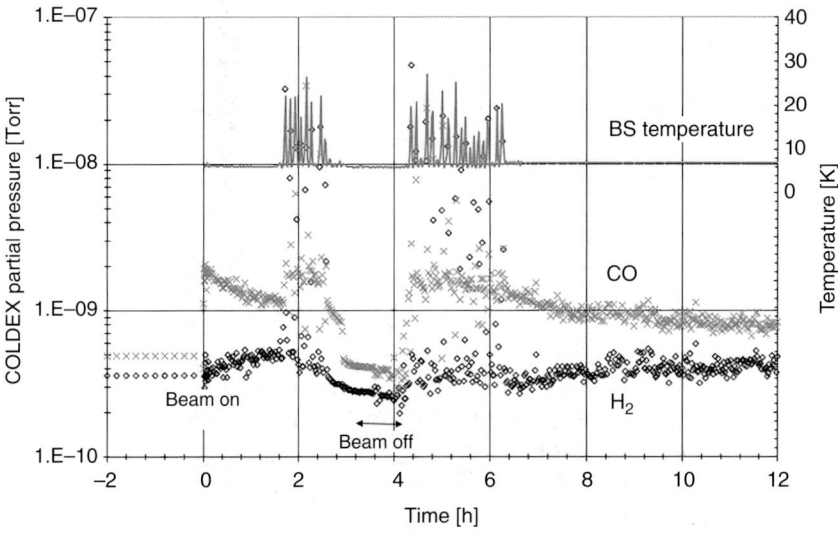

Figure 7.51 Temperature oscillations when 8.5×10^{15} CO/cm² are condensed on the BS. Source: Baglin et al. 2002 [41], Fig. 4. Reprinted with permission of CERN.

7.6 Cryosorbers for the Beam Screen at 4.5 K

The accelerator vacuum chamber is cold usually when it is inserted into a superconducting magnet, which operates at either 1.9 or 4.2 K. At 1.9 K equilibrium gas density of all gases (except He) is negligible; thus sorption capacity is sufficient to operate for years. In contrast, at 4.2 K the H_2 equilibrium gas density is too high for coverages greater than $\sim 3 \times 10^{15}$ H_2/cm²; see hydrogen adsorption isotherm at 4.2 K for stainless steel and copper shown in Figure 7.52 [56]. Thus hydrogen surface density should be kept below a monolayer. Ti–Zr–V non-evaporable getter (NEG) coating has a columnar structure, which allows to increase the surface area, and after activation the binding energy is higher. However the net effect on cryosorption capacity is still within a factor of 2.

BET adsorption isotherms (see Section 7.2) are often used to measure the roughness factor of a material. For this purpose, Xe adsorption isotherms are measured at 77 K (see Figure 7.53) [57]. As an inert gas, Xe can only be adsorbed on a surface by physisorption. Since the saturated vapour pressure of a gas molecule is due to the adsorption/desorption mechanism on the liquid/solid phase of the same molecule, the appearance of the saturated vapour pressure in an isotherm is proportional to the surface capacity, hence the roughness of the material. Metallic surfaces, which are smooth, have a surface capacity of $\sim 10^{15}$ Xe/cm² before reaching the saturated vapour pressure of Xe (3×10^{-3} Torr at 300 K). Other material such as, e.g. an NEG strip or sealed and unsealed anodised Al have a surface capacity up to $\sim 3 \times 10^{17}$ Xe/cm².

The roughness factor, R, of a material can be derived from the BET multi-monolayer theory. As explained in Section 7.2, a plot of the above

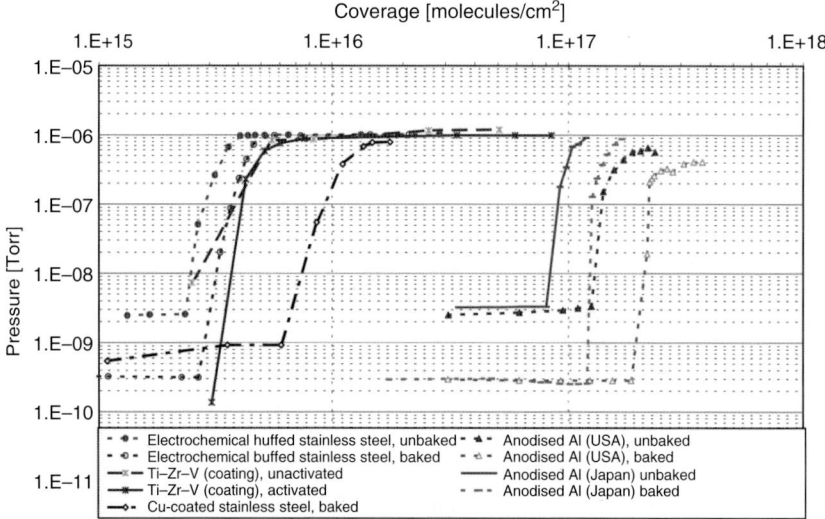

Figure 7.52 Hydrogen adsorption isotherm at 4.2 K for various samples. Source: Reprinted with permission from Sharipov and Bertoldo [52], Fig. 3. Copyright 2006, Elsevier.

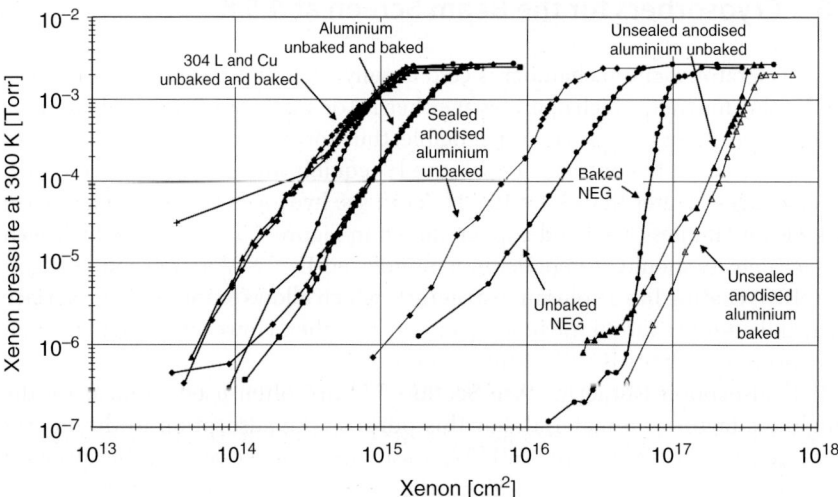

Figure 7.53 Xenon adsorption isotherm at 77 K for various samples. Source: Reprinted with permission from Potier and Rinolfi [53], Fig. 2. Copyright 1998, CERN.

isotherm in the so-called BET coordinates $[P/P_0, P/(s(P_0 - P))]$ in the range P/P_0 [0.01, 0.3] yield a straight line, whose slope is inversely proportional to the BET monolayer s_m (see Eq. (7.11)):

$$\frac{P}{s(P_0 - P)} = \frac{1}{\alpha_{BET} s_m} + \frac{\alpha_{BET} - 1}{\alpha_{BET} s_m} \frac{P}{P_0} \cong \frac{1}{\alpha_{BET} s_m} + \frac{1}{s_m} \frac{P}{P_0} \quad (7.77)$$

where α is a dimensionless parameter.

Table 7.8 Roughness factor of some common material in unbaked and baked states.

Technical surface	Unbaked	Baked at 150 °C
Copper Cu-DHP[a] acid etched	1.4	1.9
Stainless steel 304L vacuum fired	1.5	1.5 (at 300 °C)
Aluminium degreased	3.5	3.5
Sealed anodised aluminium at 12 V	24.9	—
Unsealed anodised aluminium at 12 V	537.5	556.0
NEG St 707	70.3	156.3

a) Cu-DHP (also called C12200) is a 99.9% pure Cu deoxidised with phosphorus, leaving relatively high residual phosphorus content.

Knowing the surface area of the molecule, $A = 25 \text{ Å}^2$ for Xe, the roughness factor, R, which is the ratio of the real surface, A_r, to the geometrical surface A_g, is given by

$$R = \frac{A_r}{A_g} = \frac{As_m}{A_g} \tag{7.78}$$

Table 7.8 give the roughness factor of some common material in unbaked and baked states. Metallic surfaces have a roughness factor of a few units; it can be increased up to a few hundred for specially treated surfaces.

The natural choice for the cryogenic system is using the cryosorbers. These cryosorbers should meet a number of criteria:

– Its sorption capacity should be sufficient for accelerator operation for at least a few months (preferably years).
– It should not produce particulates (it should be dust free).
– It should be easy to install in a confined space between a beam screen and a cold bore and attach to either.
– It should be maintenance free.
– It should have a lifetime of the accelerator or greater.

The following are a few materials that were considered in the application for the LHC.

7.6.1 Carbon-Based Adsorbers

7.6.1.1 Activated Charcoal

Activated charcoal is a well-known cryosorber and has quite a high adsorption capacity. This is due to a high bound energy of carbon and technologies allowing to produce high porosity materials in various forms such as globules, powder, particles, and granules. Activated charcoal is widely used for cryopumps [58, 59] and has huge pumping capacity even at 20 K, but can be used as a cryosorber up to 33 K [43, 60]. One of the advantages of activated charcoal that it can also pumps He [61], which may appear in a cryogenic vacuum chamber through micro cracks, especially for the systems operation with superfluid He at 1.9 K. However, when large cryopumping capacity required in space-constrained area such as interconnect or beam screen of colliders [50, 62, 63], the activated charcoal has some disadvantages: the grain size might be too large for the available space, it is difficult to attach to surfaces and to provide sufficient thermal contact, and finally it is a source of dust that can lead to unidentified falling object (UFO) problem in particle accelerator [64]. Furthermore, using a glue to fix charcoal inside the vacuum chamber could lead to presence of high mass organic molecules, the most unwanted species in the ultrahigh vacuum (UHV) accelerator vacuum chamber. Thus, in spite of excellent pumping properties, the activated charcoal should be used with a great caution.

Pros:
- Huge pumping capacity.
- Wide range operation temperatures (up to 33 K).

Cons:
- Difficult to attach without using a glue.
- Produces dust particles.

7.6.1.2 Carbon Fibre

A new carbon fibre material was used for a first time on the LHC beam screens placed inside the magnets operating at 4.5 K [65, 66]. This woven material manufactured fibre was developed at the Institute of Solid State Chemistry and Mechanochemistry SB RAS (Novosibirsk, Russia). As shown in Figure 7.54, the ~1 mm diameter carbon fibre wires are weaved together. The fibres are ~10 μm diameter with pores from 50 to 500 nm diameter. These pores represent about 15% of the total surfaces. Molecules are trapped within these pores providing pumping speed and adsorption capacity.

Figure 7.55 shows the H_2 adsorption isotherm and the sticking probability of the carbon fibre in the 6–30 K temperature range [62]. The saturated vapour pressure is reached for surface coverage larger than 10^{20} H_2/cm^2. The capacity, defined here as the maximum acceptable pressure, i.e. 10^{-8} Torr corresponding to

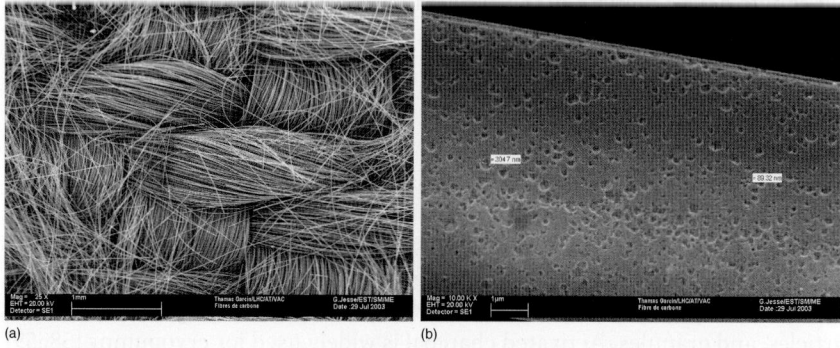

Figure 7.54 Electronic microscope photographs of a carbon fibre cryosorber with a magnification factor (a) ×25 and (b) ×10 000. Source: Hseuh et al. 1999 [62], Fig. 2. Reprinted with permission of Elsevier.

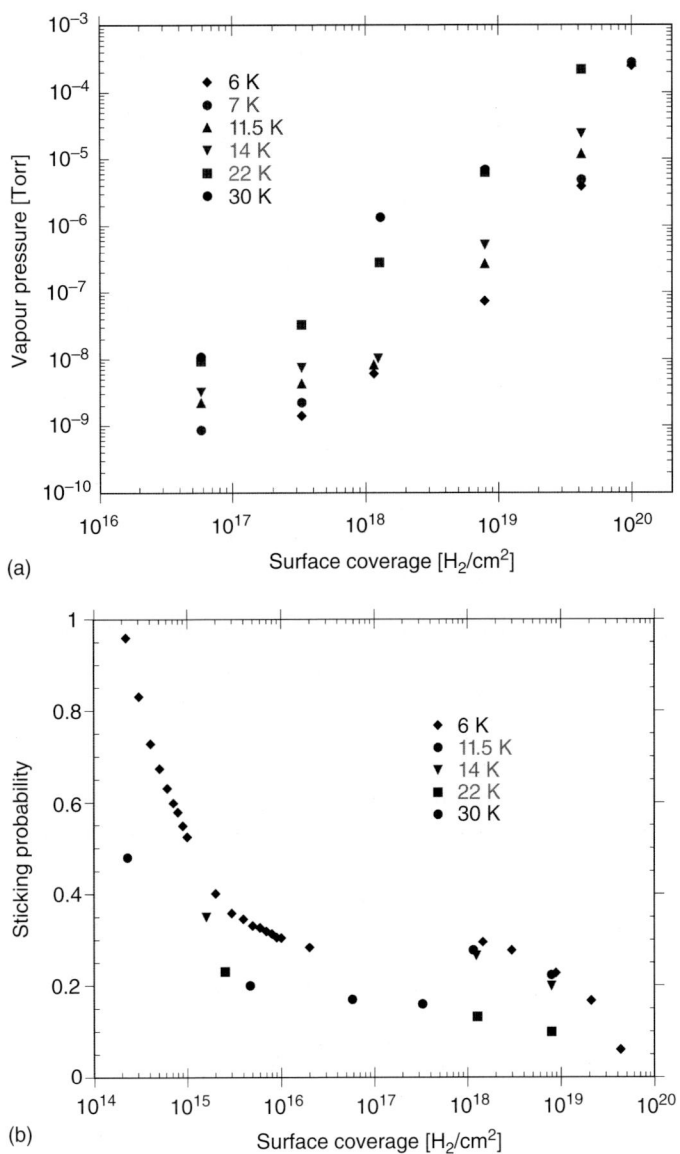

Figure 7.55 (a) H_2 adsorption isotherms and (b) H_2 sticking probabilities of the carbon fibre cryosorber in the temperature range between 6 and 30 K. Source: Hseuh et al. 1999 [62], Figs. 3 and 4. Reprinted with permission of Elsevier.

100 hours vacuum lifetime for LHC, decreases when increasing the temperature. For the LHC operating temperature (5–20 K), the capacity ranges from 6×10^{17} to 2×10^{18} H_2/cm^2. In the same parameter domain, the H_2 sticking probability ranges from 0.1 to 0.3.

The carbon fibre material has similar bond energy to activated charcoal and large porosity: the specific surface area of such material is 7.7×10^5 m^2/kg.

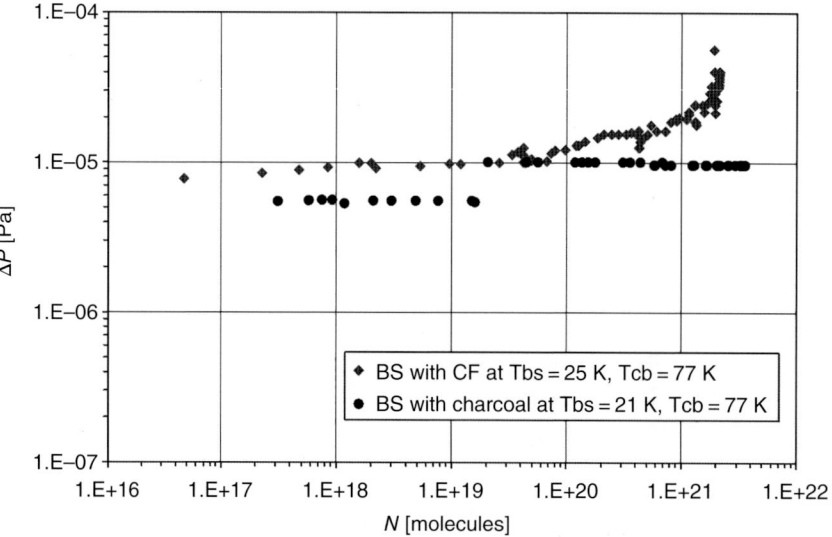

Figure 7.56 A comparison of H_2 dynamic pressures in vacuum chamber for beam screen with granulated charcoal [57] and for beam screen with carbon fibre (CF) at close beam screen temperatures. Source: Reprinted with permission from Anashin et al. [61], Fig. 4. Copyright 2004, Elsevier.

However, being less dense than activated charcoal, it provides less sorption capacity per unit of volume. In the application to the LHC, the pumping speed and sorption capacity were studied in LHC beam screen and cold bore prototypes only. Figure 7.56 shows the results of measurements in a 1-m-long LHC beam screen and cold bore prototype. Dynamic pressure in a vacuum chamber corresponds uniformly injected gas flow of 10^{15} H_2/s. One can see that that the performance of both beam screens with granulated charcoal and carbon fibre is comparable (with a factor 2) until the amount of absorbed molecules reached $\sim 10^{21}$ molecules.

Carbon fibre can be produced as thick as 0.5 mm: i.e. it can be used in narrow gaps. This material has all advantages of fabric: i.e. it can be cut to any shape, sewed, pierce, stick, bend, wrap up, etc. Thus due to these properties it can be attached to a surface not only by glue (alike activated charcoal) but also with clips, clamps, stiches, staples or simply by wrapping around.

Pros:
- Large pumping capacity and sticking probability.
- High operation temperature (up to 25 K).
- Can fit a narrow gaps, even shapes.
- Can be attached with variety of means: clips, clamps, stiches, staples, glue, etc.

Cons:
- Sorption capacity is not as high as for activated charcoal.
- Hard to provide a good thermal contact.
- Produces dust particles.

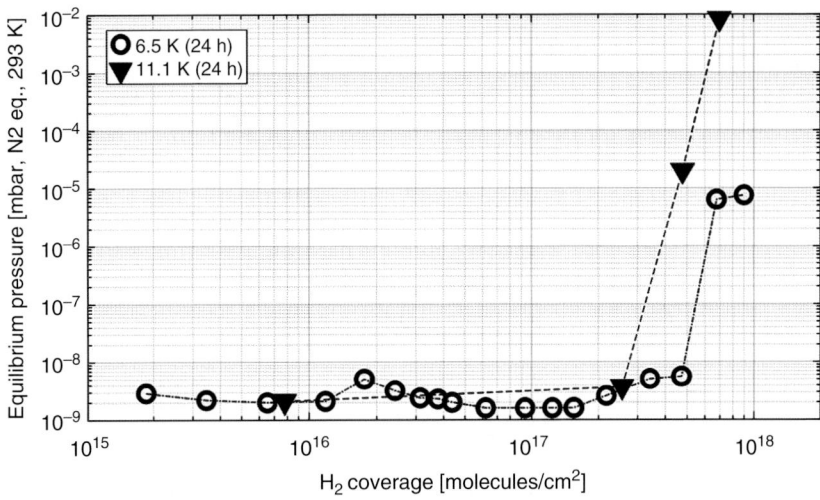

Figure 7.57 H_2 adsorption isotherm on a-C coating at 6.5 and 11.1 K. Source: Turner 1995 [63], Fig. 2. Reprinted with permission of CERN.

7.6.2 Amorphous Carbon Coating Absorption Properties

Amorphous carbon (a-C) coating has been developed to provide low secondary electron yield (SEY) surface to mitigate electron cloud build-up in the CERN SPS (see Chapter 8). Its application to cryogenic beam pipe requires the knowledge of its properties at cryogenic temperature. Figure 7.57 shows the H_2 adsorption isotherm measured on 400 nm thick a-C coated in a 2-m-long beam screen installed in the COLDEX experimental set-up [67]. The H_2 capacity of $\sim 2 \times 10^{17}$ H_2/cm^2 is larger than metallic surface. Thus, the a-C coating acts also as a cryosorber.

For application in storage rings, an appropriate choice of the operating temperature compatible with the cryosorber properties is needed to avoid pressure excursion induced by temperature oscillations. Figure 7.58 shows the thermal desorption spectroscopies of H_2, CO, and N_2 for low and large surface coverages. H_2 is physisorbed on a-C coating up to ~ 35 K and fully desorbed above ~ 65 K. Other gases, such as CO and N_2, are physisorbed up to ~ 85 K and desorb up to 140 K. Operation of a machine in e.g. the 40–60 K region is not UHV compatible in the presence of H_2.

For machine application, it is interesting to study the physisorption properties of the material as a function of the gas species and the surface coverage. Figure 7.59 shows the hydrogen thermal desorption spectroscopies of a-C coating for several surface coverage [68]. At low surface coverage ($< 3.4 \times 10^{15}$ H_2/cm^2), the hydrogen is physisorbed on the a-C coated surface up to 40 K. When increasing the surface coverage, the hydrogen starts to desorb at lower temperature. It reach ~ 8 K for 1.3×10^{18} H_2/cm^2 condensed on the surface. The activation energy is estimated from first order thermal desorption kinetics (see Eq. (7.22)).

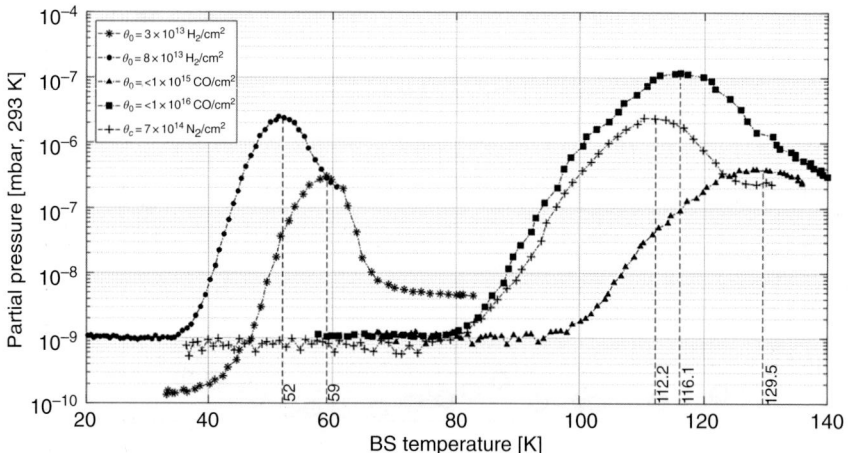

Figure 7.58 Thermal desorption spectroscopies of a-C coating for several gas, surface coverages, and heating rates. Source: Turner 1995 [63], Fig. 1. Reprinted with permission of CERN.

During the adsorption process, the pores are first filled at an activation energy of $\sim$180 meV. Above $\sim 10^{18}$ H_2/cm^2, when all the pores are filled, hydrogen starts to condense with an activation energy of $\sim$60 meV, similarly to metallic surface (see Table 7.4 and Ref. [52]). The desorption peaks around $\sim$10 K in Figure 7.59 are a typical signature of hydrogen condensation; therefore, the monolayer capacity of 400-nm-thick a-C coating can be estimated to be in the range $(5-10) \times 10^{17}$ H_2/cm^2.

Pros:
- Efficient for electron cloud suppression due to low SEY.
- High operation temperature (up to $\sim$30 K).
- As a thin film (50–500 nm), it does not reduce beam aperture.
- Can be deposited on Al, stainless steel, and Cu surfaces of vacuum chambers.

Con:
- Sorption capacity is not as high as for activated charcoal.

7.6.3 Metal-Based Absorbers

Metal-based absorbers [52, 69, 70] have good thermal conductivity, can be welded or brazed to the cooling surface, or can even be produced directly on vacuum chamber wall. Therefore, there is an interest to the cryosorption properties of porous metal structures produced with different technologies on different metals.

7.6.3.1 Aluminium-Based Absorbers

A process of aluminium anodising creates a 100-μm-deep rough and porous surface that demonstrates up to two orders of magnitude higher sorption capacity at

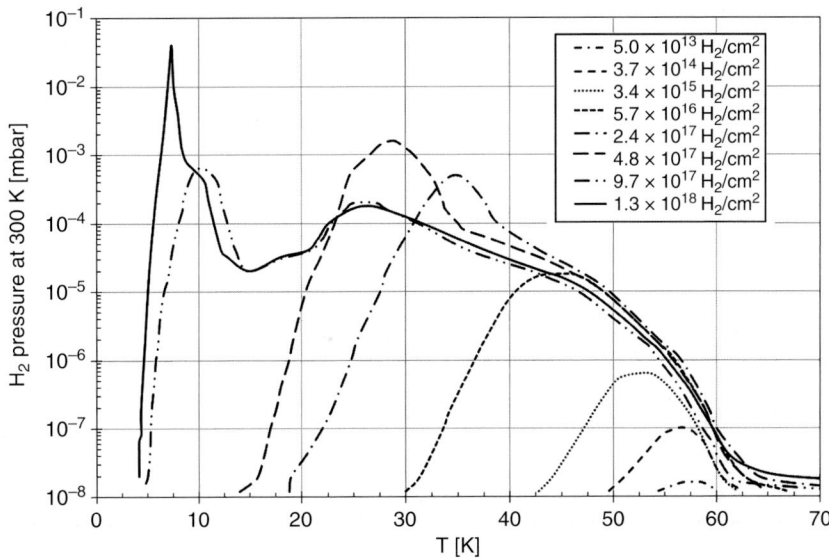

Figure 7.59 Hydrogen thermal desorption spectroscopies of a-C coating for several surface coverages.

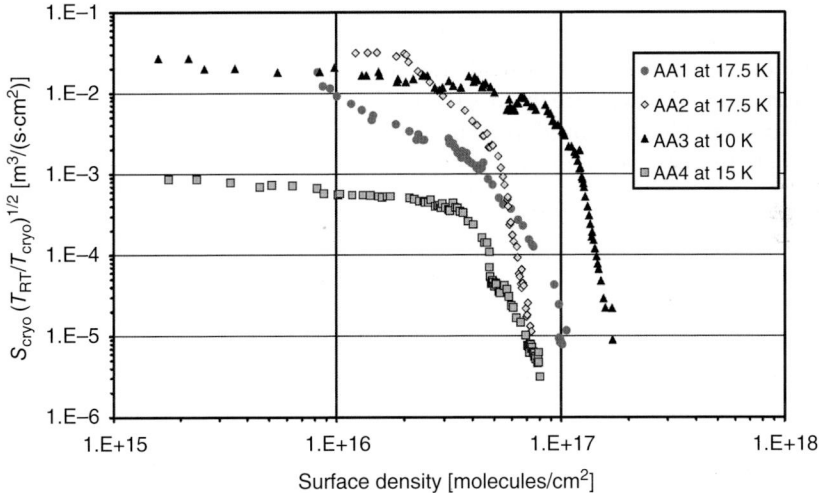

Figure 7.60 Room temperature specific hydrogen pumping speed as a function of the number of the adsorbed molecules for the anodised aluminium samples at different temperatures. Source: Reprinted with permission from Anashin et al. [65], Fig. 2. Copyright 2004, Elsevier.

4.2 K than untreated metal surface (see Figure 7.52). At higher temperatures the sorption pumping speed and capacity rapidly reduces (see Figure 7.60).

Pros:
- Pumping capacity up to 100 times higher than untreated surface.
- Operation temperature up to ~10–15 K.
- Can be done directly on aluminium vacuum chamber walls and does not need space.
- High thermal conductivity.

Cons:
- Sorption capacity is not as high as for activated charcoal.
- Lower range of temperatures than carbon based cryosorbers.

7.6.3.2 Copper-Based Absorbers

There are a number of industrially produced porous copper materials such as sponges, filters, mould pressed copper powder, etc. An example of pumping properties study is shown in Figure 7.61. At the temperature of 12–13 K, the sorption capacity of these samples is up to an order of magnitude higher than the sorption capacity of untreated metal surface at 4.2 K. The advantage of these samples is that the degradation of pumping speed as a function of amount of absorbed gas is not as sharp as for aluminium, but inverse proportional to amount of absorbed gas.

Pros:
- Pumping capacity up to 10 times higher than untreated surface.
- Some pumping is still provided after sorbing 10^{18} H_2/cm^2.
- Operation temperature up to ~12–13 K.
- Can be welded or brazed to vacuum chamber walls.
- High thermal conductivity.

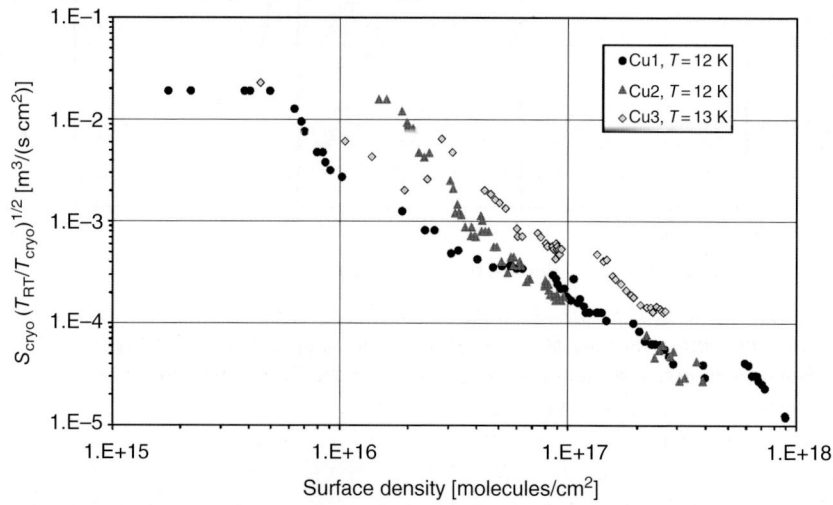

Figure 7.61 Room temperature hydrogen pumping speed as a function of the number of the adsorbed molecules for the porous copper samples at the different temperatures. Source: Reprinted with permission from Anashin et al. [65], Fig. 3. Copyright 2004, Elsevier.

Cons:
- Sorption capacity is not as high as for activated charcoal.
- Lower range of temperatures than carbon-based cryosorbers.

7.6.3.3 LASE for Providing Cryosorbing Surface

Laser Ablation Surface Engineering (LASE) is discussed in Chapter 8 as a low SEY surface for e-cloud mitigation. At cryogenic temperatures, a LASE surface with a developed surface area could provide cryosorption properties similar to the copper- and aluminium-based cryosorbers described in the previous section. An ongoing study has already demonstrated that LASE can significantly increase the surface area and sorption capacity at cryogenic temperatures [71]. Cryosorption properties of only a few sample has been measured up to now, and a variety of different surfaces should be investigated in the future to find optimum laser parameters for high sorption capacity at various operation temperatures. However it is already clear that this is a very promising technology of increasing sorption capacity of metal surfaces.

Pros:
- Pumping capacity should be higher than for untreated surface.
- Expected operation temperature up to ~12–15 K.
- LASE can be made directly on vacuum chamber walls.
- LASE can be applied to any metal surface.
- High thermal conductivity.

Cons:
- Expected sorption capacity is not as high as for activated charcoal.
- Lower range of temperatures than carbon-based cryosorbers.

7.6.4 Using Cryosorbers in a Beam Chamber

There are a number of different things that should be considered for choosing the cryosorber that would suit the best way to meet specifications for each application such as:

- Operation temperature or temperature range.
- Sorption capacity and equilibrium pressure at the operation temperature (or at the highest operation temperature, if the temperature is not stabilised).
- Required space for the cryosorber.
- How to fit and thermalise at the required location.
- Thermal conductivity.

Direct exposure of a cryosorber to a surfaces at much higher temperature (such as room temperature) may affect their sorption characteristics and increase the equilibrium pressure. So the cryosorber should ideally be surrounded by walls at the temperatures close or below its temperature.

Special attention should be paid to protect (or screen) the cryosorbers from SR, multipacting electrons, and other energetic particle bombardment. Although there is too little data on behaviour of cryosorbers under photon, electron, or ion bombardment, it is reasonable to expect that PSD, ESD, and ISD from cryosorbers will increase with the amount of sorbed gas, similarly to secondary PSD from smooth surfaces; therefore it could potentially be a serious problem.

7.7 Beam Screen with Distributed Cryosorber

This is an example of implementation of beam screen design inside a cold bore with operating temperature above 3 K. In this case, the saturated vapour pressure of hydrogen is not negligible. Based on crysorbing studies described earlier, the equilibrium pressure of hydrogen can be reduced by using cryosorbers in the range of temperature up to 10–30 K (depending on the crysorber).

Such beam screen design with distributed crysorber was successfully adopted for the stand-alone magnets of the LHC matching sections. The cold bore of these magnets operates at 4.5 K for which the hydrogen saturated vapour pressure equals 2×10^{-5} mbar. Figure 7.62 shows, on the left side, a ribbon of a carbon fibre cryosorber produced by BINP during the LHC construction and shows, on the right side, the cryosorber attached on the electron shield clamped on the back of the LHC beam screen cooling capillary. This assembly allows protecting the inner beam tube from carbon dust and controlling the temperature of the cryosorber with the beam screen cooling circuit. About 200 cm^2/m of such crysorber was installed to guarantee the LHC vacuum performances. During beam shutdown, if needed, these cryosorbers are regenerated by increasing the beam screen temperature above 80 K (activation energy ~236 meV). In the meantime, the cold mass must be emptied in a way the cold bore temperature is held above 20 K to allow the removal of the hydrogen molecules towards the external pumping system.

For illustration, Figure 7.63 shows the pressure measured with the COLDEX experiment equipped with a beam screen with distributed activated charcoal crysorber subjected to SR [43]. For this experiment, the cold bore was held at 70 K and 2×10^{19} H$_2$/cm^2 was adsorbed on the cryosorber prior irradiation. This quantity is equivalent to ~100 ML desorbed from the BS, i.e. several years of LHC operation. At the start of irradiation, with the BS held at 15 K, the H$_2$ pressure increased to 4×10^{-9} Torr. Although the BS temperature was further increased to 20 K while applying temperature oscillation, the H$_2$ pressure was still kept below the LHC design pressure (10^{-8} Torr) demonstrating the efficiency of the activated charcoal to provide H$_2$ pumping speed and capacity.

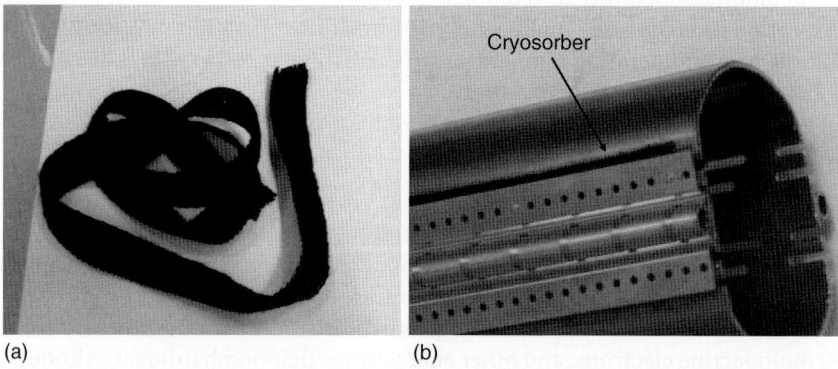

Figure 7.62 (a) Ribbon of a carbon fibre cryosorber. (b) The cryosorber attached on the electron shield clamped on the back of the LHC beam screen cooling capillary. Source: Baglin 2007 [1], Fig. 13. Reprinted with permission of CERN.

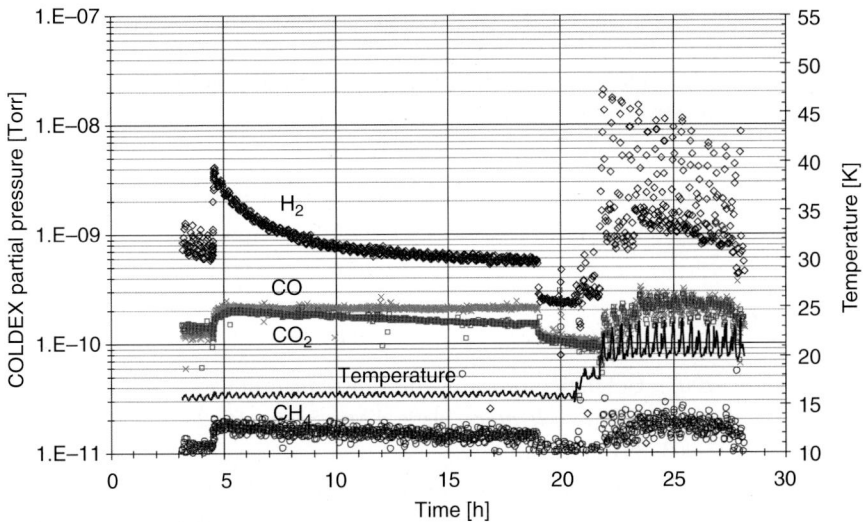

Figure 7.63 SR-induced pressure increase in a BS with distributed cryosorber when 2×10^{19} H_2/cm^2 is adsorbed on the activated charcoal.

7.8 Final Remarks

Gas dynamic models for accelerator cryogenic vacuum chamber are sufficiently developed to predict behaviour of gas density at different conditions such as:

- Temperature.
- Cryosorbed gases.
- With and without SR (or other particle-induced gas desorption).
- Different beam vacuum chamber geometries (with and without liner).
- Different condition at the extremes (known pressure, pumping speed, another chamber with different geometry, etc.).
- With and without cryosorbers.

The PSD experimental data are available for a limited range of photon critical energies of SR (40–300 eV), a few materials of vacuum chamber. and a few temperatures (3.5–25 K, 78 K). Cryosorption data allows to predict (model) equilibrium pressure of main gases of the interest and some of their mixtures on copper, stainless steel. and aluminium. Secondary PSD phenomenon is well understood, and its dependence on surface coverage has been investigated. Sticking probability of gas molecules on different materials were measured.

However, there is still a lack of experimental data, which will be required for new machines with different parameters, in particular:

- PSD for SR with photon critical energies in the keV range, at the temperatures 3.5–100 K.
- ESD and ISD data are scarce but are essential to estimate gas load in the presence of electron multipacting (see Chapter 8) and to ensure vacuum stability (see Chapter 9).

- New materials, surface treatments, and coatings could affect both gas desorption and absorption (sticking probability, equilibrium pressure).
- New cryosorbers and cryosorbing surfaces could be of practical interest.
- Some new designs will require experimental testing.

Thus, future machine parameters will highlight what exactly is missing and what experiments should be performed for obtaining new data.

References

1 Baglin, V. (2007). Cold/sticky systems. In: *Proceedings of CERN Accelerator School, Vacuum in Accelerators*, Platja d'Aro, Spain, 16–24 May 2006, CERN-2007-003, 351. CERN.
2 Andersson, S. and Harris, J. (1982). Observation of rotational transitions for H_2, D_2, and HD adsorbed on Cu(100). *Phys. Rev. Lett.* 48: 545.
3 Andersson, S., Harris, J., Persson, M., and Wilzén, L. (1989). Sticking in the quantum regime: H_2 and D_2 on Cu(100). *Phys. Rev. B* 40: 8146.
4 Carroll, J.J. (1993). Henry's law: a historical view. *J. Chem. Educ.* 70 (91) https://doi.org/10.1021/ed070p91.
5 Young, D.M. and Crowell, A.D. (1962). *Physical Adsorption of Gases*, 104–106. London: Butterworth.
6 Freundlich, H. (1930). Kapillarchemie. *Leipzig* I: 153–172.
7 Dushman, S. and Lafferty, J.M. (1962). *Scientific Foundations of Vacuum Technique*, 385–389. New York, NY: Wiley.
8 Dushman, S. and Lafferty, J.M. (1962). *Scientific Foundations of Vacuum Technique*, 390–394. New York, NY: Wiley. Original paper: Langmuir, I. (1918). The adsorption of gases on plane surfaces of glass, mica and platinum. *J. Am. Chem. Soc.* 40: 1361.
9 Hobson, J.P. (1961). Physical adsorption of nitrogen on Pyrex at very low pressures. *J. Chem. Phys.* 34: 1850.
10 Kaganer, M.G. (1961). Surface area determination from the adsorption of nitrogen, argon and krypton. *Dokl. Akad. Nauk. SSSR* 138: 405.
11 Templemeyer, K.E. (1971). Correlation of the equilibrium adsorption isotherms of low temperature cryodeposits. *Cryogenics* 11: 120.
12 Tölle, V. (1971). *Adsorptionsgleichgewichte von Wasserstoff and kondensiertem Methan, Äthan und Propan*. Berlin, Technical University, Dissertation 1971, 990003533550302884.
13 Dushman, S. and Lafferty, J.M. (1962). *Scientific Foundations of Vacuum Technique*, 395–400. New York, NY: Wiley. Original paper: Brunauer, S., Emmett, P.H., and Teller, E. (1938). Adsorption of gases in multimolecular layers. *J. Am. Chem. Soc.* 60: 309.
14 Young, D.M. and Crowell, A.D. (1962). Multilayer adsorption on uniform surfaces: the BET equation, refined treatments. In: *Physical Adsorption of Gases*, 147–170. London: Butterworth.

15 Young, D.M. and Crowell, A.D. (1962). Multilayer adsorption on uniform surfaces: earlier theories. In: *Physical Adsorption of Gases*, 137–146. London: Butterworth. Original paper: Polanyi, M. (1920). *Z. Elektrochem.* 26: 371.

16 Dubinin, M.M. and Radushkevich, L.V. (1947). Evaluation of microporous materials with a new isotherm. *Proc. Acad. Sci. USSR* 55: 327.

17 Dubinin, M.M. (1965). *Russ. J. Phys. Chem.* 39: 487.

18 Dubinin, M.M. (1967). Adsorption in micropores. *J. Colloid Interface Sci.* 23: 487.

19 Kaganer, M.G. (1957). A method for the determination of specific surfaces from the adsorption of gases. *Proc. Acad. Sci. USSR* 116: 603.

20 Hobson, J.P. (1969). Physical adsorption isotherms extending from ultrahigh vacuum to vapor pressure. *J. Phys. Chem.* 73: 2720.

21 Wallén, E. (1996). Adsorption isotherms of H_2 and mixtures of H_2, CH_4, CO, and CO_2 on copper plated stainless steel at 4.2 K. *J. Vac. Sci. Technol., A* 14: 2916–2929. https://doi.org/10.1116/1.580245.

22 Wallén, E. (1997). Adsorption isotherms of He and H_2 at liquid He temperatures. *J. Vac. Sci. Technol., A* 15: 265–274.

23 Wallén, E. (1997). Experimental test of the propagation of a He pressure front in a long, cryogenically cooled tube. *J. Vac. Sci. Technol., A* 15: 2949–2958. https://doi.org/10.1116/1.580890.

24 Hseuh, H.C. and Wallén, E. (1998). Measurements of the helium propagation at 4.4 K in a 480 m long stainless steel pipe. *J. Vac. Sci. Technol., A* 16: 1145–1150.

25 Hoge, H.J. and Arnold, R.D. (1951). Vapor pressure of hydrogen, deuterium, and hydrogen deuteride and dew-point pressures of their mixtures. *J. Res. Nat. Bur. Stand.* 47: 63, Research Paper 2228.

26 Benvenuti, C., Calder, R.S., and Passardi, G. (1976). Influence of thermal radiation on the vapor pressure of condensed hydrogen (and isotopes) between 2 and 4.5 K. *J. Vac. Sci. Technol., A* 13: 1172. https://doi.org/10.1116/1.569063.

27 Bozyk, L., Chill, F., Kester, O., and Spiller, P. (2015). Pumping properties of cryogenic surfaces in SIS100. In: *Proceedings of IPAC 2015*, Richmond, VA, USA, 3696. DRK fit by S. Wilfert, GSI, Darmstadt, Germany.

28 Wilfert, S. and Pongrac, I. (2017). The vacuum system of SIS100 at FAIR. A Talk at EUCARD-2 Workshop Beam Dynamics Meets Vacuum, Collimations, and Surfaces, 8–10 March 2017. Karlsruhe Institute of Technology (KIT), Karlsruhe, Germany. https://indico.gsi.de/event/5393/session/7/contribution/14/material/slides/0.pdf

29 Redhead, P.A., Hobson, J.P., and Kornelsen, E.V. (1993). *The Physical Basis of Ultrahigh Vacuum*. American Institute of Physics.

30 Redhead, P.A. (1962). Thermal desorption of gases. *Vacuum* 12: 203–211.

31 Anashin, V., Evsigneev, A., Malyshev, O., et al. (1993). Summary of resent photodesorption experiments at VEPP-2M. SSCL-N-825, June 1993.

32 Anashin, V., Malyshev, O., Osipov, V., et al. (1993). Cold beam tube photodesorption experiments for SSCL 20 TeV proton collider. 40th National Symposium AVS, Orlando, November 1993.

33 Turner, W. (1994). Beam tube vacuum in future superconducting proton colliders. SSCL-Preprint-564, October 1994.

34 Anashin, V.V., Malyshev, O.B., Osipov, V.N. et al. (1994). Investigation of synchrotron radiation-induced photodesorption in cryosorbing quasiclosed geometry. *J. Vac. Sci. Technol., A* 12: 2917.

35 Anashin, V.V., Derevyankin, G.E., Dudnikov, V.G. et al. (1994). Cold beam tube photodesorption and related experiments for SSCL 20 TeV proton collider. *J. Vac. Sci. Technol., A* 12: 1663.

36 Anashin, V., Malyshev, O., Osipov, V. et al. (1994). Experimental investigation of dynamic pressure in a cryosorbing beam tube exposed to synchrotron radiation. In: *Proceedings of EPAC-94*, London, 27 June to 1 July 1994, 2506.

37 Alinovsky, N., Anashin, V., Beschastny, P. et al. (1994). A hydrogen ion beam method of molecular density measurement inside a 4.2 K beam tube. In: *Proceedings of EPAC-94*, London, 27 June to 1 July 1994, 2509.

38 Calder, R., Gröbner, O., Mathewson, A.G. et al. (1996). Synchrotron radiation induced gas desorption from a prototype Large Hadron Collider beam screen at cryogenic temperatures. *J. Vac. Sci. Technol., A* 14: 2618.

39 Malyshev, O.B. and Collins, I.R. (2001). Dynamic gas density in the LHC interaction regions 1&5 and 2&8 for optics version 6.3. LHC-PROJECT-NOTE-274, December 2001, CERN, 34 pages.

40 Baglin, V., Collins, I.R., Gröbner, O. et al. (2000). First results from COLDEX applicable to the LHC cryogenic vacuum system. In: *Proceedings of EPAC 2000*, Vienna, Austria, 2283.

41 Baglin, V., Collins, I.R., Gröbner, O. et al. (2002). Synchrotron radiation studies of the LHC dipole beam screen with COLDEX. In: *Proceedings of EPAC 2002*, Paris, France, 2535.

42 Baglin, V., Collins, I.R., Gröbner, O. et al. (2002). Molecular desorption by synchrotron radiation and sticking coefficient at cryogenic temperatures for H_2, CH_4, CO and CO_2. *Vacuum* 67: 421.

43 Baglin, V., Collins, I.R., Gröbner, O. et al. (2002). Cryosorber studies for the LHC long straight section beam screen with COLDEX. In: *Proceedings of EPAC 2002*, Paris, France, 2538.

44 Anashin, V.V., Malyshev, O.B., Calder, R. et al. (1998). The study of photodesorption processes for cryosorbed CO_2. *Nucl. Instrum. Methods Phys. Res., Sect. A* 405: 258.

45 Anashin, V.V., Malyshev, O.B., Calder, R. et al. (1997). Photon induced molecular desorption from condensed gases. *Vacuum* 48: 785.

46 Anashin, V.V., Malyshev, O.B., Calder, R., and Gröbner, O. (1999). A study of the photodesorption process for cryosorbed layers of H_2, CH4, CO or CO_2 at various temperatures between 3 and 68 K. *Vacuum* 53: 269.

47 Anashin, V.V., Collins, I.R., Gröbner, O., and Malyshev, O.B. (2001). Photon stimulated desorption and effect of cracking of condensed molecules in a cryogenic vacuum system. *Vacuum* 60: 15–24.

48 Baglin, V. (1997). Etude de la photo-désorption de surfaces techniques aux températures cryogéniques. PhD thesis. University Denis Diderot, Paris.

49 Zweerink, G.L. and Roach, D.V. (1970). The thermal accommodation coefficients of helium, neon, and argon on an ice surface. *Surf. Sci.* 19: 249–254.

50 Goodman, F.O. (1980). Thermal accommodation coefficients. *J. Phys. Chem.* 84: 1431–1445.

51 Sharipov, F. and Moldover, M.R. (2016). Energy accommodation coefficient extracted from acoustic resonator Experiments. *J. Vac. Sci. Technol., A* 34: 061604.
52 Sharipov, F. and Bertoldo, G. (2006). Heat transfer through a rarefied gas confined between two coaxial cylinders with high radius ratio. *J. Vac. Sci. Technol., A* 24: 2087–2093.
53 Potier, J.P. and Rinolfi, L. (1998). The LEP pre-injector as a multipurpose facility. *Proceedings of EPAC'98*, Stockholm, Sweden, 22–26 June 1998.
54 LHC Design Report (2004). The LHC main ring. In: *CERN-2004-003*, vol. I (eds. O. Brüning, P. Collier, P. Lebrun, et al.). Geneva: CERN.
55 Baglin, V. (2004). Vacuum transients during LHC operation. *Proceedings of LHC Project Workshop*, Chamonix, France, February 2004.
56 Moulard, G., Jenninger, B., and Saito, Y. (2001). Industrial surfaces behaviour related to the adsorption and desorption of hydrogen at cryogenic temperature. *Vacuum* 60: 43–50.
57 Baglin, V. (1997). Mesure de la rugosité de surfaces techniques à l'aide de la méthode BET. Vacuum Technical Note 97-03, January 1997, CERN, Geneva.
58 Haeffer, R.A. (1989). *Cryopumping – Theory and Practice*. Oxford Science Publications, Clarendon Press Oxford.
59 Welch, K.M. (2001). *Capture Pumping Technology*. North-Holland, Elsevier Science BV.
60 Day, C. (2001). The use of active carbons as cryosorbent. *Colloids Surf., A* 187–188 (187–206).
61 Anashin, V.V., Dostovalov, R.V., Krasnov, A.A. et al. (2004). Vacuum performance of a beam screen with charcoal for the LHC Long Straight Sections. *Vacuum* 72: 379–383.
62 Hseuh, H.C., Davis, R., Pate, D. et al. (1999). RHIC vacuum systems. *Vacuum* 53: 347–352.
63 Turner, W.C. (1995). Model of an 80 K liner vacuum system for the 4.2 K cold bore of the superconducting super collider 20 TeV proton collider. *J. Vac. Sci. Technol., A* 13: 2241.
64 Baer, T., Barnes, M., Goddard, B. et al. (2011). UFOs in the LHC. In: *Proceedings of IPAC 2011*, San Sebastián, Spain, 4–9 September 2011, 1347.
65 Anashin, V.V., Collins, I.R., Dostovalov, R.V. et al. (2004). Vacuum performance of a carbon fibre cryosorber for the LHC LSS beam screen. *Vacuum* 75: 293–299.
66 Baglin, V., Dupont, H., and Garcin, T. (2004). Vacuum characterisation of a woven carbon fiber cryosorber in presence of H_2. In: *Proceedings of EPAC 2004*, Lucerne, Switzerland, 5–9 July 2004, 1603.
67 Salemme, R., Baglin, V., Bregliozzi, G., and Chiggiato, P. Vacuum performances of amorphous carbon coating at cryogenic temperature with presence of proton beams. In: *Proceedings of IPAC 2016*, Busan, Korea, 3663.
68 Lamure, A.-L., Baglin, V., Chiggiato, P., Henrist, B. (2017). Adsorption/desorption from amorphous carbon coatings at cryogenic temperatures. *AVS 64th International Symposium and Exhibition*, 29 October to 3 November 2017, Tampa, FL, USA.

69 Anashin, V.V., Dostovalov, R.V., Krasnov, A.A. et al. (2004). Molecular cryosorption properties of porous copper, anodised aluminium and charcoal at temperatures between 10 and 20 K. *Vacuum* 76: 23–29.

70 Rao, M.G., Kneisel, P., Susta, J. (1994). Cryosorption pumping of H_2 and He with metals and metals oxides at 4.3 K. *Proceedings of the 15th International Cryogenic Engineering Conference*, Genova, Italy, 6–10 June 1994.

71 Spallino, L., Angelucci, M., Larciprete, R., and Cimino, R. (2019). On the compatibility of porous surfaces with cryogenic vacuum in future high-energy particle accelerators. *Appl. Phys. Lett.* 114: 153103. https://doi.org/10.1063/1.5085754.

8

Beam-Induced Electron Multipacting, Electron Cloud, and Vacuum Design

Vincent Baglin[1] and Oleg B. Malyshev[2]

[1] CERN, Organisation européenne pour la recherche nucléaire, Espl. des Particules 1, Meyrin, 1211, Switzerland
[2] ASTeC, STFC Daresbury Laboratory, Keckwick Lane, Daresbury, Warrington WA4 4AD, Cheshire, UK

8.1 BIEM and E-Cloud

8.1.1 Introduction

Beam-induced electron multipacting (BIEM) and electron cloud (e-cloud) are two coupled effects that can compromise the performance of modern high intensity machines with positively charged beams. E-cloud was first observed on the proton storage ring (PSR) at the Institute of Nuclear Physics (Novosibirsk, Russia) in 1965 [1–3]. A few months later the e-cloud-related instability was observed at the ZGS in Argonne National Laboratory [4] and at AGS in Brookhaven National Laboratory [5] in the United States. In the following years, the e-cloud has been detected and investigated in a number of other machines [6, 7]:

– At Bevatron in Lawrence Berkeley National Laboratory in 1971 [8], at PSR in Los Alamos in 1986 [9], at the AGS booster [10] in 1998 and at RHIC [11] in 2001 in the Brookhaven National Laboratory, at PEP-II in 2000 in SLAC [12], at Main Injector in Fermi National Laboratory in 2005 [13], at SNS in 2006 in Oak Ridge National Laboratory [14], and at CesrTA in 2007 in Cornell University [15, 16] in the United States.
– At ISR in 1972 [17], at PS in 2000 [18], at SPS in 1999 [19], and at the Large Hadron Collider (LHC) in 2010 [20, 21] in CERN (Switzerland).
– At KEK PF in 1988 [22] and at KEKB in 2000 and 2017 [23, 24] in Japan.
– At DAΦNE in LNF/INFN in 2003 [25] in Italy.
– At PETRA-III in DESY in 2009 [26] in Germany.
– At ISIS in RAL in 2008 [27] in the United Kingdom.

The e-cloud problem has also been intensively studied for future machines such as CLIC [28], HL-LHC [29], ILC [30, 31], and FCC [32, 33].

Such a strong and long-term interest to the e-cloud problem highlights its importance for the beam dynamics:

– E-cloud can drive very fast, is often destructive, and has beam instabilities (both single and multi-bunch).

Vacuum in Particle Accelerators: Modelling, Design and Operation of Beam Vacuum Systems,
First Edition. Oleg B. Malyshev.
© 2020 Wiley-VCH Verlag GmbH & Co. KGaA. Published 2020 by Wiley-VCH Verlag GmbH & Co. KGaA.

– A negative space charge of the e-cloud focuses on the positively charged beam, thus leading to the betatron tune shift and energy spread.

The BIEM process can qualitatively be described as follows. The initial electrons, which appear in the beam chamber due to photoelectron emission (PEE) induced by synchrotron radiation (SR) from vacuum chamber walls or in-vacuum components or due to beam-induced gas ionisation, can be accelerated in the electric field of the passing bunches, acquire kinetic energies of up to several hundreds of electron volt, and then collide with the beam pipe walls and produce secondary electrons (secondary electron emission [SEE]). When the electromagnetic field generated by the bunch train creates the resonant conditions, the electron multipacting can be triggered.

All these electrons create a negative space charge along the beam path and form an e-cloud; thus the beam bunches are circulating under modified conditions:

– Interacting with a space charge may cause the beam emittance growth above a tolerable level.
– This in conjunction with the collisions between beam particles and e-cloud electrons may increase the beam particle loss rate.

Furthermore, electron multipacting causes the following negative impacts:

– The BIEM transfers energy from the beam to the vacuum chamber walls, increasing
   o the beam energy loss, requiring more power for the RF cavities to compensate the loss;
   o the heat of vacuum chamber – the power of such additional heat loads on a cryogenic vacuum chamber could become critical for its cryogenic system.
– Electron-stimulated desorption (ESD) due to multipacting electrons will increase the gas density, which can in its turn increase:
   o The beam particle loss rate.
   o The rate of electron production due to residual gas ionisation.

Figure 8.1 illustrates the electron cloud build-up in the LHC beam pipe. SR emitted in the dipole magnets by the proton beams generates photoelectrons from the photon interaction with the vacuum chamber wall. Since these

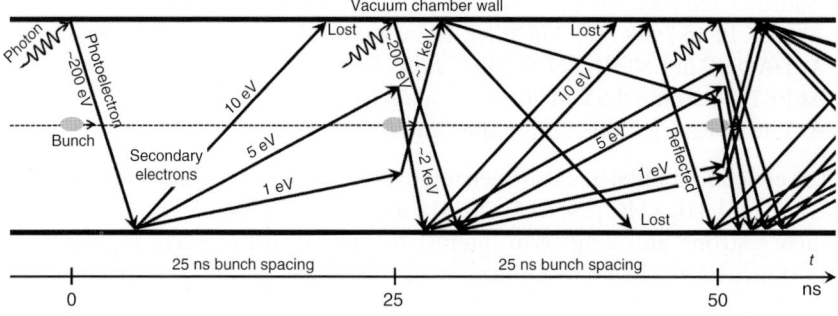

Figure 8.1 Illustration of BIEM and electron cloud build-up in the LHC beam pipe.

photoelectrons are in quasi-synchronism with the proton bunch, they can be accelerated by it towards the opposite vacuum chamber wall. In the LHC, taking into account the pipe diameter and the bunch population, the photoelectrons bombard the vacuum chamber wall at ~200 eV. As a consequence, secondary electrons are created and emitted into the beam pipe. If these secondary electrons have not been absorbed by the vacuum chamber wall when the second bunch passes, they can be accelerated towards the wall. In this illustration, the low energy secondary electrons (1 and 5 eV) are in the vicinity of the bunch when it arrives 25 ns later. Therefore, these secondary electrons receive a larger kick from the bunch, bombarding the vacuum chamber wall in the kiloelectron volt range. Secondary electrons are then produced and emitted into the beam pipe. When cumulated with the following bunches, this process, called BIEM, results in a multiplication of the electrons into the beam pipe.

In this example, the seed of electrons originates from the photoelectron production, which is the dominant process in LHC. However, the seed of electrons can also originate from beam gas ionisation or particle loss at the vacuum chamber wall.

Note 1: In general, e-cloud could appear without BIEM, when the main sources of electrons are gas ionisation and/or photoemission. From another side, BIEM would not necessarily create the e-cloud with an electron density above a tolerable level. It is also worth mentioning that the RF cavities and waveguides may also suffer from electron multipacting, while e-cloud is not an issue there.

Note 2: Theoretically, BIEM could be triggered in the very high intensity machines with negatively charged bunches [34]; the e-cloud effects for an electron beam were studied both theoretically and experimentally in some machines [35–37].

Since the discovery of e-cloud and its impact on beam dynamics, there was a significant effort in looking for various techniques of e-cloud mitigation. This activity was intensified in last 20–25 years with the design and operation of high intensity and low emittance colliders such as LHC, ILC, KEK-B, SuperKEKB, DAΦNE, PEP-II, CESR, etc. These studies are covering a number of activities:

- Photoelectron yield (PEY) and secondary electron yield (SEY) measurements of various materials after different treatments, coatings, conditioning, etc.
- Photon reflectivity of these materials.
- BIEM and e-cloud modelling.
- BIEM and e-cloud studies in the machines.
- Developing various BIEM and e-cloud mitigation techniques.

8.1.2 E-Cloud Models

Approximated equations, based on a stationary electron model, can be derived to evaluate the possible existence of BIEM and e-cloud in a machine [38].

A *first condition* to allow wall-to-wall electron multipacting is that the electrons cross the beam pipe of radius r_p, between the successive bunch passages spaced in time by t_{bb}. Assuming a bunch of uniform transverse density, the minimum electric field required to provide enough speed to the electrons can be computed. This translates in a minimum bunch population, N_b, given by

$$N_b \geq \frac{r_p^2}{r_e c\, t_{bb}} \qquad (8.1)$$

where $r_e = 2.8 \times 10^{-15}$ m is the classical electron radius and c the speed of light.

A *second condition* for electron multipacting is that the energy of the electron bombarding the wall shall be large enough to allow the production of secondary electrons. For this reason, the energy gain by the electron from the passing beam bunch shall be computed. In the simple kick approximation, the energy gain to the electron, $\Delta W(r)$, received by the kick from the bunch field as a function of the radial position r, is given by

$$\Delta W(r) = 2mc^2 r_e^2 \left(\frac{N_b}{r}\right)^2, \qquad (8.2)$$

where $m = 511$ keV/c^2 is the electron mass.

In the previous calculation, the transverse bunch density was assumed uniform. Therefore, the electric field outside the bunch falls off proportionally to $1/r$. In a real machine, the transverse r.m.s. beam size of a beam σ is described by the Gaussian distribution modifying the electric field within the beam pipe. However, at a distance larger than 2σ, the difference between the simple kick approximation and the real electric field is negligible. Thus, the simple kick approximation is valid for electrons, which are located at the distance of a few beam radii away from the bunch. In particular, the previous equation is valid near the vacuum chamber wall.

Figure 8.2a shows the minimum bunch population required to allow a wall to wall multipacting as a function of bunch spacing for several vacuum chamber radii (see Eq. (8.2)). It is seen that a bunch population in the range of 10^{11} particles per bunch will generally trigger multipacting for almost any bunch spacing and medium size (ID[1] ≈ 80 mm) vacuum chambers. Here, the bunch population is computed for a multipacting to occur between each bunch.

Figure 8.2b shows the energy gain received during the passage of the bunch by a stationary electron at several locations in the beam vacuum chambers (Eq. (8.2)). For a bunch population in the range of 10^{11} particles per bunch, the energy gain received by the electron is 50 eV for electrons located at 40 mm from the bunch and increases to 800 eV for electrons located at 10 mm from the bunch. For positions closer to the beam (i.e. <10 mm), the movement of the electron during the passage of the beam shall be taken into account. A more accurate evaluation of the energy gain can be then obtained by integrating the equation of motion.

The average gain in kinetic energy of an e-cloud stationary in time and uniformly distributed is given by

$$\langle \Delta W \rangle = 6mc^2 r_e^2 \left(\frac{N_b}{r_p}\right)^2, \qquad (8.3)$$

---

1 ID is an inner diameter of a beam pipe.

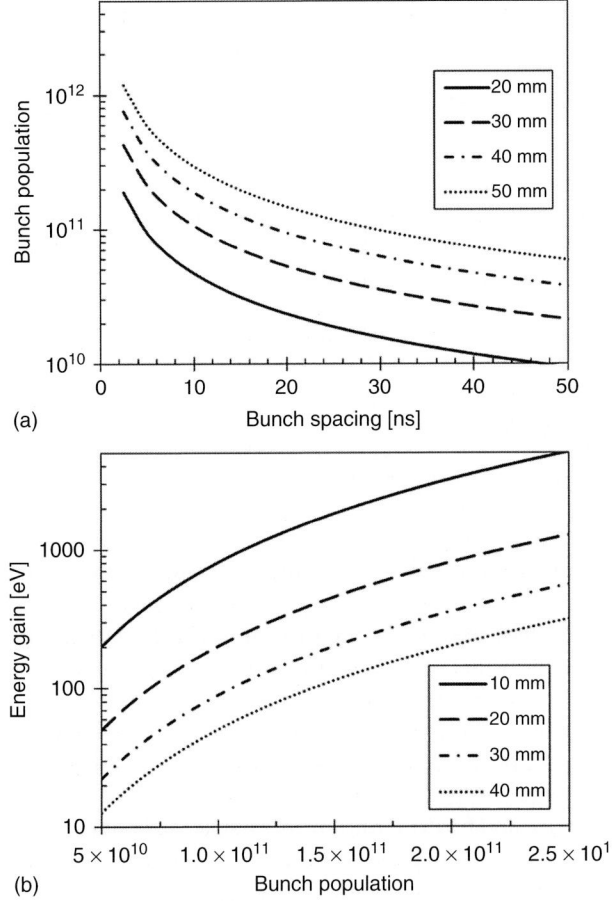

Figure 8.2 (a) Minimum bunch population for wall-to-wall multipacting as a function of bunch spacing for several beam vacuum chambers radii. (b) Energy gain received by the electron from the bunch kick as a function of the bunch population for several electron positions in the beam vacuum chamber.

This handy formula can be used to evaluate the ESD yields (which are a function of the energy) of a beam vacuum chamber subjected to BIEM or to evaluate the deposited heat load if the electron flux to the wall is known.

Figure 8.3 shows the average kinetic energy gain of a stationary and uniformly distributed e-cloud for several beam chamber radii. For a bunch population of 10^{11} particles per bunch, the average kinetic energy varies from 30 to 200 eV when decreasing the vacuum chamber inner diameter from 100 to 40 mm.

When dealing with magnetic fields, complex vacuum chamber shapes, or specific beam structures, a better evaluation of the BIEM and e-cloud parameters can only be achieved by simulation codes, e.g. PyECLOUD, CLOUDLAND, CSEC, POSINST, Factor2, etc. [39, 40]. Figure 8.4 shows the simulated heat load and electron current at the beam vacuum chamber wall as a function of the SEY in

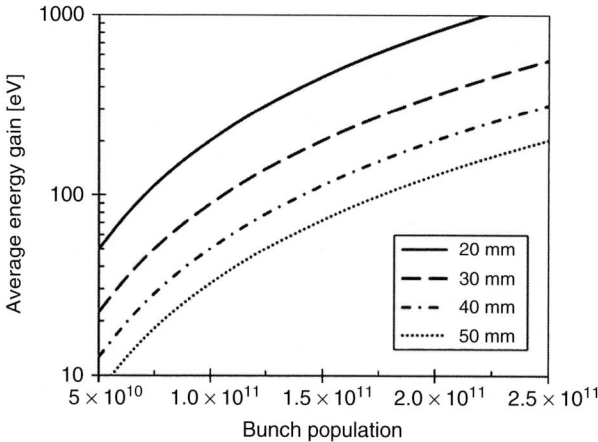

Figure 8.3 Average kinetic energy gain of a stationary and uniformly distributed electron cloud for several beam vacuum chamber radii.

a beam screen dipole of the LHC. These data were computed with PyECLOUD simulation code [41].

In the absence of a magnetic field, the e-cloud interacts only with the circulating bunches and the beam chamber wall. In this case, the motion of the electrons is not guided, and the e-cloud occupies the full cross section of the beam chamber. An evolution of e-cloud distribution in the beam chamber cross section is shown in Figure 8.5. In the presence of a positively charged bunch, the electrons from the walls of beam chamber are attracted towards the beam in the centre (left side), occupying more and more space and finally filling an entire space inside the beam chamber (middle picture). After the beam passage, the electrons continue drifting towards the opposite walls and kick it with energy gained from the beam. When the kicked energy is large enough, secondary electrons are produced and drift inside the beam chamber, ready to be accelerated by the following bunch (right side) [42].

In the presence of a magnetic field, the electron motion is guided by the magnetic field lines. Figure 8.6 shows an example of the transverse distribution of the e-cloud density for dipole, quadrupole, sextupole, and solenoid fields in the LHC. In the case of a dipole field, the electron motion can form two stripes. The position of the stripes is a function of the bunch density and dipole field. The e-cloud interacts only with the upper part and lower part of the vacuum chamber. For the cases of the quadrupolar and sextupolar fields, the electron motion follows the field lines and interacts with the vacuum chamber wall at the pole positions. Applying a solenoid field maintains the e-cloud close to the vacuum chamber walls along its perimeter. In this case, the kick received from the bunch is minimised and, in the occurrence, the electrons receive enough energy to produce secondary electrons, the energy of the secondary electrons being very low, and they remain in the close vicinity of the wall chamber [43, 44].

Input parameters for the BIEM and e-cloud phenomena depend on the shape of the vacuum chamber and its surface properties. BIEM in the vacuum chamber

Figure 8.4 Simulated heat load (a) and electron current (b) as a function of the SEY in a beam screen dipole of the LHC when operated at 7 TeV with various bunch populations. Source: Courtesy of Dr. G. Iadarola (CERN, Geneva, Switzerland).

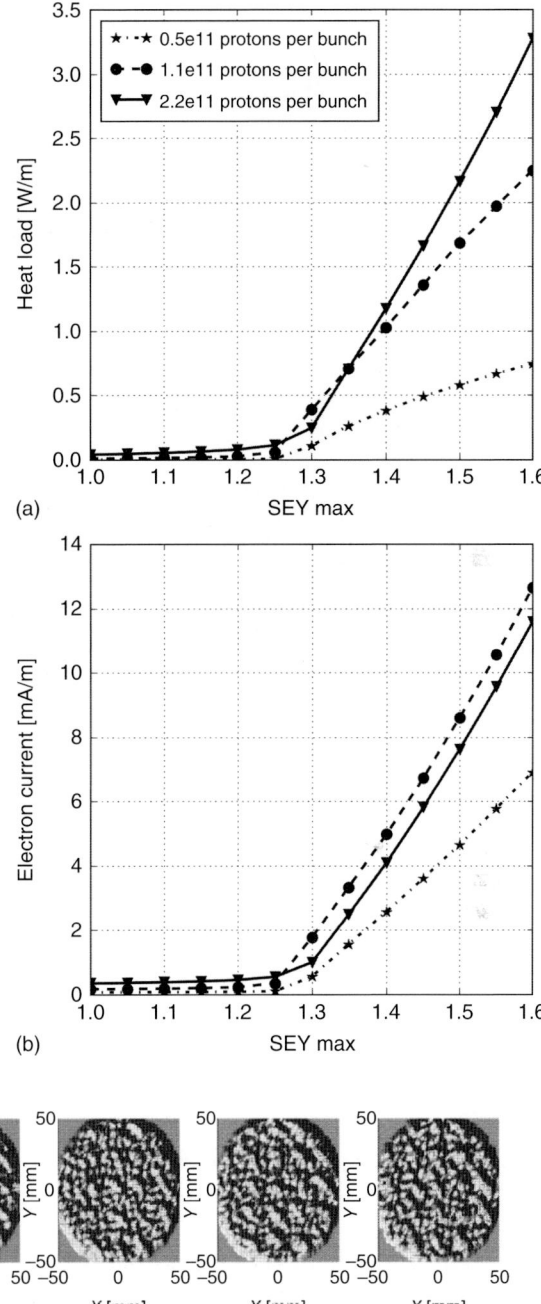

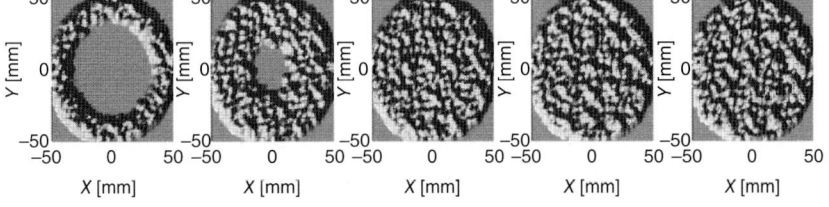

Figure 8.5 An evolution of electron cloud cross-sectional distribution in the absence of magnetic field during and after the bunch passage (from left to right). Source: Courtesy of Dr. Lanfa Wang (SLAC, CA, USA).

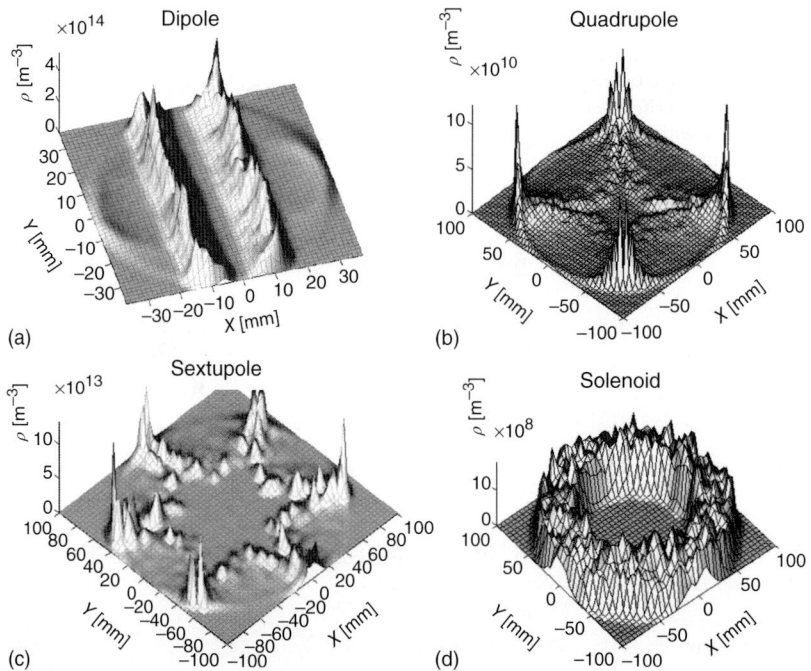

Figure 8.6 Effect of the magnetic field configuration on the electron cloud transverse distribution. (a) Dipole, (b) quadrupole, (c) sextupole, and (d) solenoid. Source: Wang et al. 2004 [43], Fig. 10 and Wang et al. 2004 [44], Figs. 25 and 26. Reprinted with permission of CERN.

causes ESD and may require modification of the vacuum system to deal with it. These input characteristics for BIEM and e-cloud modelling will be discussed the following sections of this chapter.

8.2 Mitigation Techniques and Their Impact on Vacuum Design

There are three main sources of electrons for BIEM and e-cloud in vacuum chamber [45]:

1. Electrons that appear due to *residual gas ionisation* by the beam particles.
2. *Photoelectrons* that are emitted from vacuum chamber walls due to SR (see Chapter 3).
3. *Secondary electrons* emitted due to electrons multipacting (see Section 8.3).

Apart from these three main sources, there could be a few additional ones that may contribute to the process:

4. Electrons that appear due to *residual gas ionisation*
   o By multipacting electrons
   o By SR

5. Bremsstrahlung photons, which appear due to beam–gas collisions, can contribute to
   - residual gas ionisation and
   - PEE.
6. Beam particles hitting vacuum chamber, collimators, and tapers may cause
   - SEE,
   - Bremsstrahlung radiation and, therefore, photoelectron production.

To lower the BIEM intensity and the e-cloud density, these sources of electrons must be minimised or even suppressed. Thus, the production of electrons due to residual gas ionisation can be mitigated by specifying the maximum gas density, which requires using vacuum chambers with low photon stimulated desorption (PSD) and ESD yields and providing sufficient (preferably distributed) pumping speed. Production of PEE due to SR is proportional to the photon flux, so it can be reduced by reducing the photon flux irradiating *the beam vacuum chamber walls* using photon traps, photon absorbers, antechambers, etc. Production of both photoelectrons and secondary electrons could be reduced by reducing their production rates: PEY and SEY, defined as the number of photoelectrons or secondary electrons, respectively, per impact of photon or electron.

A large number of BIEM and e-cloud mitigation methods have been developed over recent years. These mitigation methods can be divided into two groups: passive (i.e. they require no controllers or power sources after installation or implementation) and active (requiring controllers, power sources, feedback electronics, etc.).

Various solutions can be applied, and the choice is dictated by a few criteria. First of all, the chosen solution must solve the e-cloud and/or BIEM problem. Among working solutions, the most preferable ones should be simple, cost effective, and very reliable in the long term. Ideally, the preferred solution should work for the duration of the machine lifetime, take no space, and not require controllers and/or feedthroughs.

8.2.1 Passive Methods

Passive methods are implemented as an integrated part of the vacuum chamber, and they do not require controllers, power sources, feedthroughs, feedback electronics, etc. Once implemented they should work for the full lifetime of the machine without maintenance. If a passive method works (i.e. reduces BIEM and e-cloud below a tolerable level) for the whole machine or its sections, then it is a preferable solution. Thus, a significant effort was made to find and develop a number of passive mitigation solutions [46–49]:

(1) The PEY and SEY are related to surface material and chemical state; thus significant interest was attracted to materials and surface coating, which have a low intrinsic SEY:
   - *Materials* have different PEY/SEY, for example, Cu has a lower SEY than Al (see Figures 8.16 and 8.17):
     - The PEY/SEY of as-received materials with a natural oxide layer [50–53].

- o The PEY/SEY of specially treated surfaces, for example, etched and controlled oxidation, nitration, or graphitisation [54–59].
- o The PEY/SEY of pure materials after scrubbing with SR, electrons, or ions in vacuum [14–17, 22].
- o Effect of temperature and cryosorbed gases [60, 61].
  - *Pro*: a simple choice from a short list of materials and low cost treatments.
  - *Con*: a limited choice of materials compatible with vacuum chamber specifications and beam impedance issues.
- – *Coating* any material with low PEY/SEY materials (see Figure 8.18):
  - o TiN coating is a well-known technique with low PEY/SEY coating used in various applications [62–66].
  - o Non-evaporable getter (NEG) coating provides similar low PEY/SEY [62, 67–69] combined with distributed pumping, i.e. a very good combined solution to suppress all three sources of electrons; thus, when a bakeout is acceptable, NEG coating is a preferred solution as an ideal for vacuum design due to low gas desorption and distributed pumping (see Chapter 5).
  - o Amorphous carbon (a-C) coating [56, 70, 71]) leads to significant reduction of SEY ($\delta_{max} < 1$) and suppression of electron multipacting.
  - o Other low SEY coatings such as Cr, CrN, and TiCr [72].
    - *Pro*: It does not affect a choice of vacuum chamber materials, and low SEY coating can be done on any vacuum chamber material.
    - *Con*: It requires using a vacuum-based deposition technology, therefore increasing the cost of the vacuum chamber.

(2) Surface geometry may help to reduce the net SEY:
- – Modifying a surface geometry by machining different shape grooves [46, 73–75] (see Figures 8.7–8.9, and 8.19). Although the intrinsic SEY remains the same as for a flat surface, the reduction of net SEY happens due to multiple interactions of the electrons with the grooved walls, thus increasing the probability of electron energy loss and electron absorption. The net SEY reduces with an increase of the ratio between height h and distance between grooves (a and b in Figure 8.7) and with the decrease of the rib width.
  - *Pros*:
    - o It works.
    - o To reduce the cost, vacuum chamber with groves can be produced by extrusion.
  - *Cons*:
    - o It requires precise machining, increasing the cost considerably.
    - o A less accurate sharpness of the groove top edges of extruded vacuum chamber results in reducing the efficiency of the SEY mitigation in comparison to precise machining.
- – Coating with low SEY microstructure (e.g. copper black, gold black; columnar NEG) [69, 76–78]:
  - o Materials that form as columns, pyramids, nanotubes, and flakes could be very efficient in reducing PEY and SEY.

8.2 Mitigation Techniques and Their Impact on Vacuum Design | 359

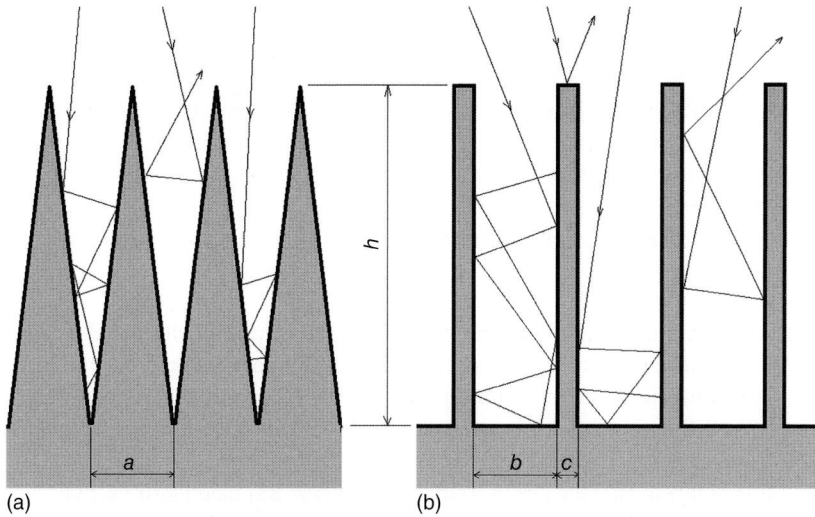

Figure 8.7 Two types of grooves for reducing the net SEY: (a) sawtooth-shaped grooves (h is a height and a is a distance between grooves) and (b) rib-shaped grooves (h is a height, b is a distance between the ribs, and c is a rib width).

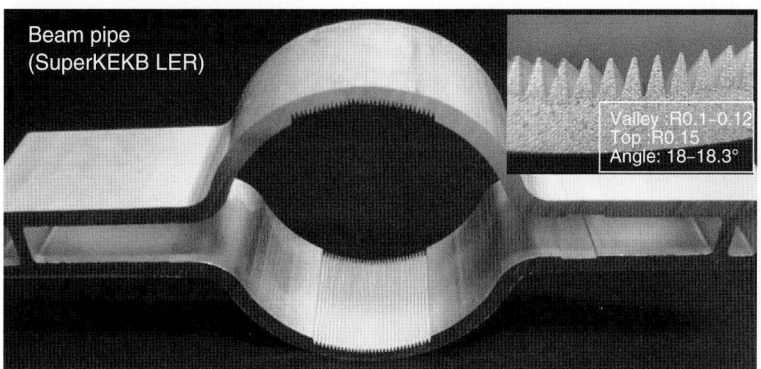

Figure 8.8 KEK vacuum chamber with grooves and an antechamber. Source: Shibata 2012 [78], Fig. 6. Reprinted with permission of CERN.

Figure 8.9 A layout of wiggler vacuum chamber with grooves and an antechamber equipped with NEG strips for distributed pumping in the ILC positron damping ring. Source: Malyshev et al. 2010 [74], Fig. 3. Reprinted with permission of CERN.

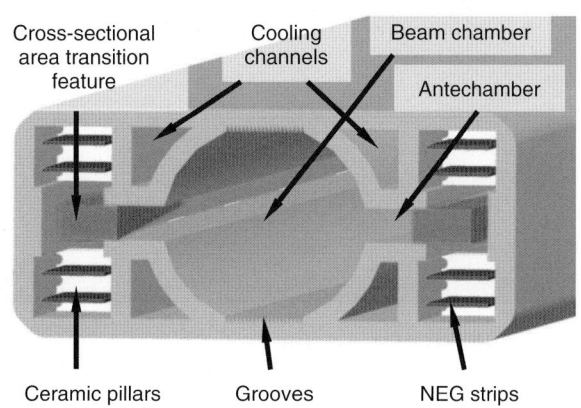

- *Pro*: It works.
- *Con*: The created features may be fragile; cleaning with solvents could be difficult; a possibility of increasing thermal outgassing, PSD, ESD, and ISD (ion stimulated desorption) has to be investigated; and there is a risk of particulate generation and increase in surface resistance.
– Surface micro-engineering by ion etching, chemical etching, etc. [76] (see Figure 8.10):
  ○ A technology that allows to roughen the surface and reduce PEY and SEY.
    - *Pro*: It works.
    - *Cons*: The created features may be fragile; cleaning with solvents may be difficult; after this wet technology, the surface is saturated with hydrogen thus thermal outgassing; PSD, ESD, and ISD may increase; and there is a risk of particulate generation and increase in surface resistance.
– Laser ablation surface engineering (LASE) [79–83] (see Figure 8.11):
  ○ The most recently invented technology for reducing PEY and SEY allows to obtain low SEY with ($\delta_{max} < 1$) for as-received metal surfaces modified by a nano- or picosecond pulsed laser. Bakeout or bombardment with electrons led to even lower values of $\delta_{max} < 1$.
    - *Pros*: In comparison with other technologies used for surface engineering, the laser treatment has a few advantages such as simple equipment for production, the treatment does not require vacuum, it can be done in air under atmospheric pressure, and it reduces PSD, ESD, and ISD.

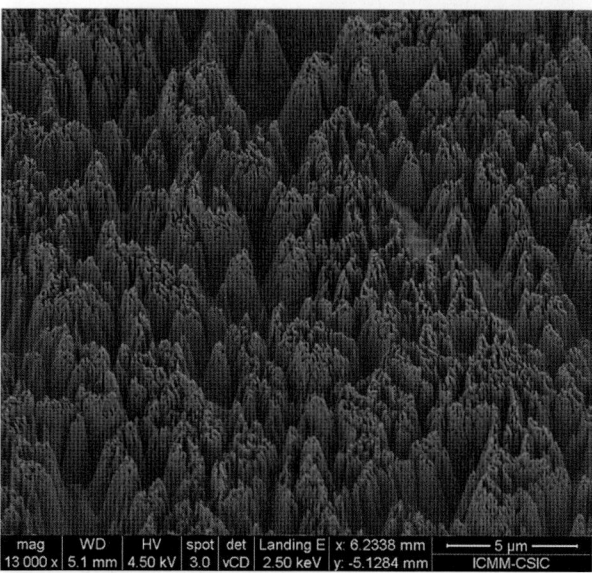

Figure 8.10 Ag plating, ion etched with Mo mask. Source: Montero et al. 2012 [75], Fig. 3. Reprinted with permission of CERN.

Figure 8.11 (a)–(c) Low to high resolution scanning electron microscopy (SEM) images and (d) a graph of SEY as a function of primary electron energy of LASE samples treated by ASTeC. Source: Courtesy of Dr. Bhagat-Taaj Sian (CI, Warrington, UK).

- *Cons*: It requires particulate control, and there is a risk of increase in surface resistance.

Note: All surface engineering technologies may increase the surface resistance, which, in turn, could potentially affect on the beam wakefield impedance.

(3) Vacuum chamber shape may help in e-cloud suppression:
  - Using antechamber and shadowing critical components from SR is a very efficient means to reduce photoelectron production in the beam chamber. The photoelectrons emitted in the antechamber do not play a role in BIEM and e-cloud [24, 74, 84, 85]:
    o In comparison to photoelectrons emitted from the beam chamber walls, the photoelectrons produced in the antechamber are exposed to a much weaker beam electric field. They have a strongly reduced probability to enter the beam chamber through a narrow gap between the chamber and antechamber to participate in BIEM and e-cloud.
    o SR absorbers in the antechamber can have a special design to minimise PEY, for example, absorbing photons at normal incidence, photoelectron traps similar to Faraday cups, using low PEY materials, etc.
      - *Pro*: It takes away one of the main sources of electrons in the beam chamber and it allows to efficiently use the benefits of an antechamber when it is already considered in a vacuum chamber design (e.g. for SR beamline, for higher vacuum conductance, or for distributed pumping);
      - *Con*: A more complicated design and a vacuum chamber more expensive than a simple (e.g. round or elliptic) beam chamber.

(4) Use of permanent magnet field for trapping electrons [49, 86]:
   ○ Permanent magnets are applied to suppress e-cloud at KEK-B (see Figure 8.12).
      - *Pros*: It works, it is quite simple and low cost, it does not require power controllers and cables, it can be installed at any stage of the machine lifetime, and it does not affect the vacuum system as it is outside of the vacuum chamber.
      - *Cons*: It can only mitigate e-clouds in magnetic field-free areas, and it may affect operations of beam position monitors (BPMs), vacuum gauges, and other sensitive equipment. Maximum bakeout temperature of these permanent magnets should be checked. It could be below the vacuum temperature bakeout temperature of 250–300 °C. In the latter case these magnets should be removed before and reinstalled after each vacuum chamber bakeout.

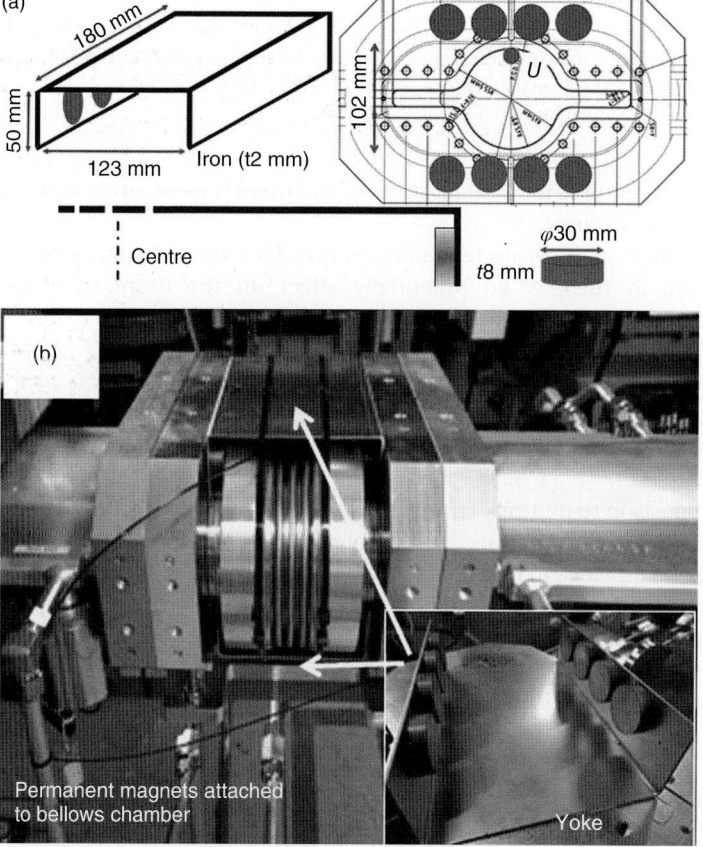

Figure 8.12 (a) Schematics and (b) a photo of yokes with permanent magnets attached to the bellow chambers at SuperKEKB low energy ring. Source: Suetsugu et al. 2016 [86], Figs. 12 and 11(b). https://journals.aps.org/prab/abstract/10.1103/PhysRevAccelBeams.19.121001. Licensed Under CC BY 3.0.

(5) Various combinations of all of the above can be applied as well [49, 74] (see Figures 8.8 and 8.9):
   a. Antechamber combined with grooves in the beam chamber.
   b. Antechamber combined with TiN coating.
   c. NEG coating on rough surfaces.
   d. Any other possible combinations.

8.2.2 Active Methods

Active methods are used in many machines, and they could be used as a single solution or in combination with others. Some examples of such methods, with their pros and cons, are shown below:

(1) As shown in Figure 8.13a, a weak solenoid field (~50 G) can be applied along the drift chambers [87, 88]. Magnetic field traps photoelectrons and secondary electrons near the surface and away from the beam path, thereby reducing BIEM and e-cloud as shown by the reduction of the pressure increase in Figure 8.13b. Usually it is used in combination with TiN or NEG coating of such vacuum chambers.
   - *Pro*: A coil can be added after the beginning of the machine operation to solve unexpected issues. It is can be easily implemented only on simple straight sections.

Figure 8.13 (a) KEKB low energy ring solenoids. (b) Observation of pressure rise with and without solenoid field as a function of the positron beam current. The non-linear behaviour, signature of electron multipacting, is suppressed when a solenoid field is applied. Source: (a) Courtesy of Y. Suetsugu, KEK, Tsukuba, Japan. (b) Suetsugu 2001 [87], Fig. 7. Reprinted with permission of IEEE.

(a)

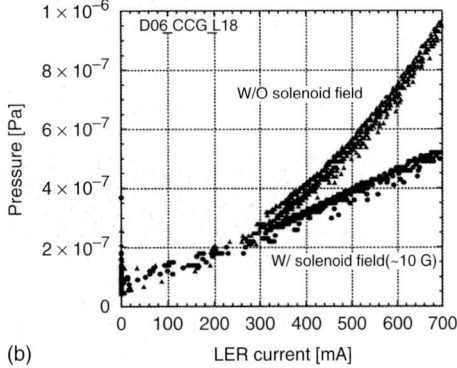

(b)

- *Con*: It requires winding the coils, cables, controllers, and power supplies. Such solution might be not compatible with a bakeout system. It can be used in magnetic field-free regions only.
(2) Clearing electrodes (biased up to ±1000 V) in wigglers and dipoles do not allow electrons to multipact [48, 89–93]. As shown in Figure 8.14, the clearing electrodes in the DAΦNE dipole chambers strongly reduces the horizontal instability even at large beam current.
  - *Pro*: It works. It can be installed inside dipole, quadrupole and wigglers
  - *Con*: It requires designing and manufacturing of these electrodes and redesigning and adopting a vacuum chamber to accommodate these electrodes, their holder, and electric feedthroughs. Electrodes and insulating materials may dramatically increase the gas density in a vacuum chamber due to thermal, photon-, electron-, and ion-induced gas desorption. Choice of material for electrodes and insulating layer as well as in-vacua design must be ultra high vacuum (UHV) compatible: i.e. it requires additional vacuum studies and testing. Feedthroughs increase the chance of vacuum leaks to air. It requires cables and controllers/power supplies, which significantly increase the cost for large machines.
(3) Optimising the beam train parameters to avoid high intensity resonant conditions by varying the beam charge and the beam spacing, introducing satellite beams, etc. [94, 95].
  - *Pro*: It also works. It requires no change in vacuum design, no additional cables, controllers and power supplies in the machine tunnel.

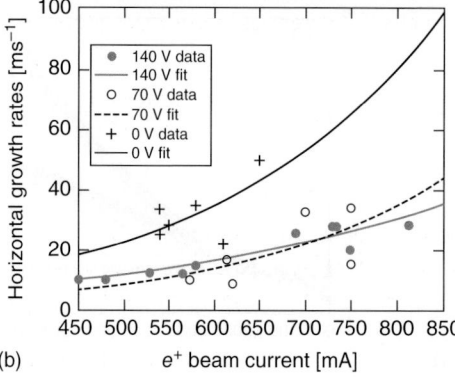

Figure 8.14 (a) Clearing electrode in a DAΦNE dipole vacuum chamber. (b) Electron intensity grow rate. Source: Wang et al. 2006 [89], Figs. 5 and 14. Reprinted with permission of Creative Commons Attribution 3.0 license (CC-BY 3.0).

- *Con*: It reduces the flexibility during machine operation. In high energy particle colliders, it may result to more collisions per crossing, thereby limiting the particle detector's efficiency.

8.2.3 What Techniques Suit the Best

It is important to mention that all the active and passive methods are working well and can be employed on their own or in various combinations with each other.

In general, different techniques could be most efficient for different components and different sections of the machine. Thus, a machine design team should balance effort, efficiency, impact on various machines systems, costs, etc., and choose the techniques, which will suit the best for the whole machine, for a sector, or for a particular component.

- First of all, the chosen BIEM and e-cloud mitigation technique must be explicitly *fitted for the purpose*.
- A *simple solution* is preferable as it is usually more robust.
- The *cost* of mitigation could vary for different techniques; thus an increase in the cost of vacuum chamber should be taken into consideration during the design phase.
- The mitigation technique should be *vacuum compatible*, and vice versa, and a solution for beam vacuum chamber should be compatible with the BIEM and e-cloud mitigation technique.
- The mitigation techniques employed inside beam vacuum chambers (surface roughening, coatings, asymmetric vacuum chamber shape (e.g. due to an antechamber), clearing electrodes, etc.) may affect the longitudinal and/or transversal beam impedance. The implementation of such mitigation techniques would be recommended when the impact on beam impedance is below a tolerable level.
- Therefore, a mitigation solution should meet all the requirements for BIEM and e-cloud mitigation, beam impedance, particulate generation, vacuum specification, and machine risk analysis.

It must be noted here that an implementation of mitigation techniques in a design phase is much easier, while solving the BIEM and e-cloud problem on an already existing and operating machine has a shorter list of possible solutions.

8.3 Secondary Electron Emission (Laboratory Studies)

8.3.1 SEY Measurement Method

When a primary electron interacts with matter, it may cause the emission of secondary electrons in the typical energy interval for the e-cloud build-up. Electrons can penetrate into the solid along 1–10 nm and produce secondary electrons. Those electrons can be scattered or diffused into the solid and might be consequently emitted into the vacuum. The total SEY (or δ) is defined as the average

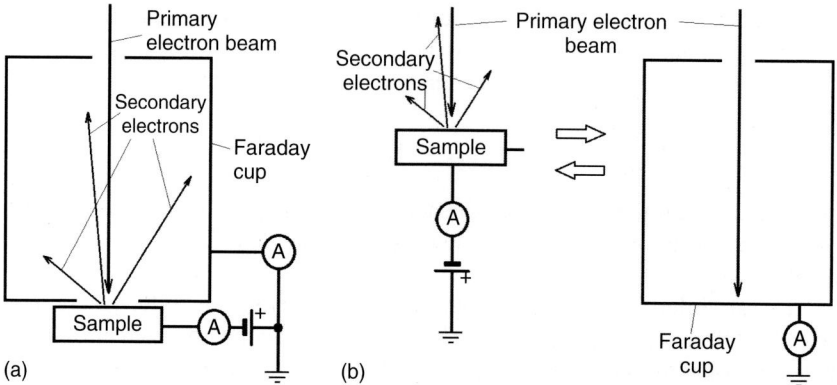

Figure 8.15 Typical layouts for secondary yield measurements: (a) simultaneous measurements of currents from a sample and a Faraday cup with a passing through primary electron beam and (b) two-step measurements of currents from a sample only and a close Faraday cup only.

number of electrons emitted from the surface per incident electron:

$$\delta = \frac{I_S}{I_P}; \quad (8.4)$$

where I_S is the secondary electron current (including both elastic and inelastic processes) and I_P is the primary electron current.

A typical layout of secondary yield measurements is shown in Figure 8.15. During SEY measurements, the sample is bombarded by electrons with energy in the range from 10 eV to 3 keV (depending on the operation range of connected electron gun). The sample can be biased between 0 and −100 V to repel secondary electrons towards the Faraday cap. The SEY measured with a bias are higher than the ones with no bias. The SEY increases with a bias and reaches its saturation at certain bias value, which depends on the gap between the sample and a Faraday cap. In the experiments reported in the following, the bias was always applied at saturation values. Two currents can be measured simultaneously in the experiment: a sample drain current, I_D, and a current at the Faraday cap, I_F. The total SEY can be calculated as

$$\delta = \frac{I_S}{I_P} = \frac{I_F}{I_F + I_D}. \quad (8.5)$$

An alternative method used to measure the SEY consists of doing the measurement in two steps. In the first step, the sample is biased negatively to repel the secondary electrons and the sample drain current, I_D, is measured. In the second step, the sample is replaced by a Faraday cup from which the primary electron beam current, I_P, can be measured. The amount of secondary electrons is then derived from the subtraction of the primary electrons current minus the sample drain current. In this case, the SEY is given by the following equation:

$$\delta = \frac{I_S}{I_P} = \frac{I_F - I_D}{I_F} = 1 - \frac{I_D}{I_F}. \quad (8.6)$$

8.3 Secondary Electron Emission (Laboratory Studies)

In this section we summarise the main properties of and tendencies in SEY under various conditions.

8.3.2 SEY as a Function of the Incident Electron Energy

Figure 8.16 shows SEY curves of typical materials used for the construction of accelerator vacuum systems as a function of the incident electron energy. The values are given for 'as-received', i.e. cleaned according to UHV standards and unbaked material, therefore without any specific treatment to reduce the yield [96]. The curves have a 'bell' shape with a maximum in the energy range between 200 and 300 eV. The maximum of the SEY, δ_{max}, ranges from $\delta_{max} = 1.5$ for TiN-coated sample to $\delta_{max} = 3.5$ for aluminium. Typical values for 'as-received' titanium, copper, and stainless steel are $\delta_{max} = \sim 2$. Any electrons bombarding a surface with a primary energy up to 2 keV will subsequently produce secondary electrons, increasing the total amount of electron in the accelerator.

8.3.3 Effect of Surface Treatments by Bakeout and Photon, Electron, and Ion Bombardment

As mentioned previously, several techniques are used to lower the SEY of technical surfaces. This is illustrated in Figure 8.17a, where, for instance, the maximum SEY of copper surface reduces by baked at 300 °C to $\delta_{max} \approx 1.7$. Cleaned copper surface, i.e. bombarded with Ar ions in a way to sputter away the native oxide, hydrocarbon, and contaminants layer, has a maximum SEY of $\delta_{max} \approx 1.4$ at a primary kinetic energy of ~650 eV [96]. Electron bombardment can also reduce the SEY of technological surfaces. This effect is often used in various equipment, e.g. radio frequency cavities or accelerators, to mitigate the electron multipacting. Figure 8.17b shows the evolution of δ_{max} of an as-received Cu surface as a function

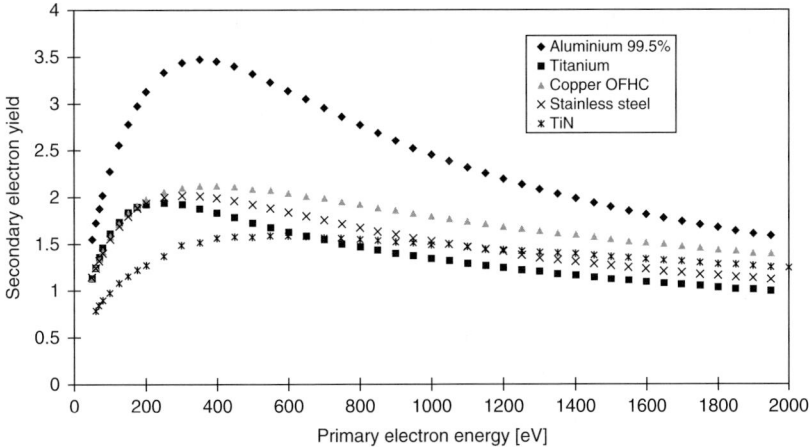

Figure 8.16 Secondary electron yield (SEY) curves of some technological materials used for the construction of accelerators. Source: Hilleret et al. 2000 [96], Fig. 3. Reprinted with permission of CERN.

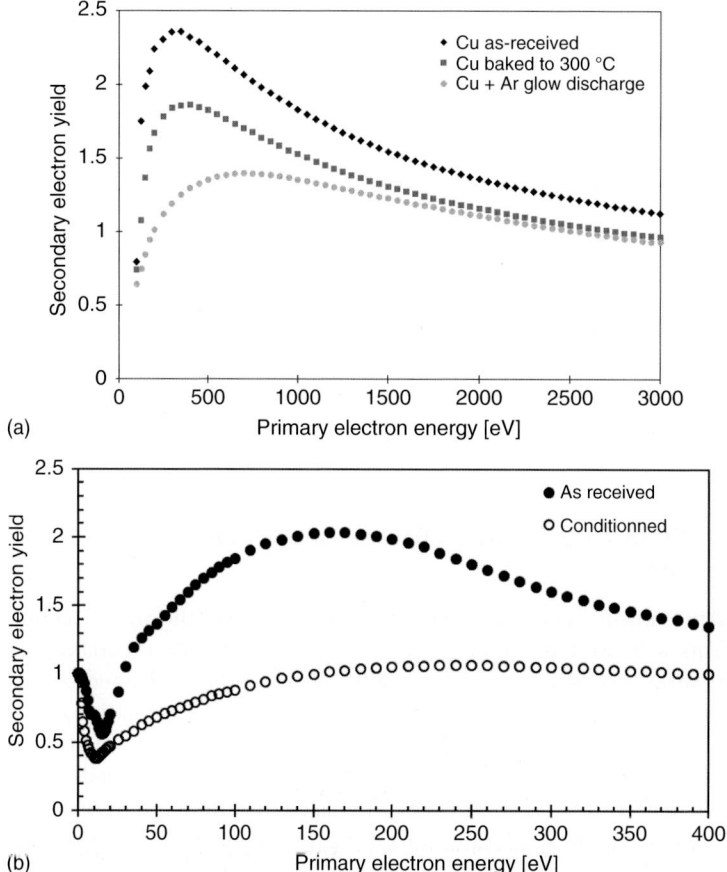

Figure 8.17 Secondary electron yield curves of Cu as a function of surface treatment. (a) Bakeout, Ar glow discharge. (b) Electron bombardment. Source: (a) Hilleret et al. 2000 [96], Fig. 2. Reprinted with permission of CERN. (b) Reprinted with permission from Cimino et al. [55], Fig. 2. Copyright 2012, American Physical Society.

of the electron dose. After bombardment with electron of kinetic energy above 50 eV, δ_{max} can be reduced to ~1.1. However, for lower kinetic energy and the same electron dose, the conditioning is less efficient and the value of δ_{max} reaches ~1.35. The origin of this reduction phenomenon (called 'scrubbing' or 'conditioning' in accelerator scientist's jargon) is ascribed to removing oxides (by ESD) and the graphitisation of the surface [55, 69].

8.3.4 Effect of Surface Material

Even cleaned following UHV standards, technological surfaces are covered by a surface layer consisting of oxides and physisorbed/chemisorbed gases, which modify the SEE of the pure material. An ion bombardment can sputter the contaminants, removing them from the surface. As already shown in Figure 8.17, Cu has $\delta_{max} \approx 1.4$ when sputter cleaned by ion bombardment.

Table 8.1 Maximum SEY of some chemical components.

Material	Ag	Al	Au	Polished C	Cu	Fe	Nb	Ti
δ_{max}	1.5	1	1.5	1	1.4	1.3	1.2	0.9
E_{max} [eV]	800	300	700	300	600	400	375	280

Source: Adapted from Arianer 2006 [97].

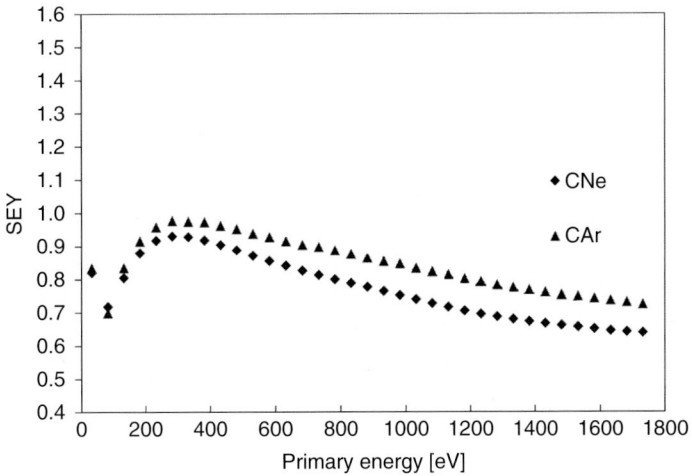

Figure 8.18 Secondary electron yield curves of amorphous carbon coating. Source: Reprinted with permission from Costa-Pinto et al. [99], Fig. 1. Copyright 2013, Elsevier.

Table 8.1 shows the maximum SEY, δ_{max}, and the corresponding impinging energy, E_{max}, of some common pure materials [97]. These values are much different than 'as-received' metal surfaces with an oxide layer.

A reduction of the secondary electron production can also be achieved by coating the surface with a low SEY material.

The TiN coating on technological surfaces such as copper, aluminium, or stainless steel substrate allows reducing δ_{max} to 1.5 (see Figure 8.16).

The NEG films coated on the same substrates can also reduce the maximum SEY, as it was described in Chapter 5 (see Figures 5.25 and 5.26). Thus, SEY was reduced to $\delta_{max} \approx 1.1$ with a 1-μm-thick Ti–Zr–V getter film coated on copper and activated at 200 °C [98].

Carbon coating also reduces the production of secondary electrons. Figure 8.18 shows the SEY curve of stainless steel with amorphous carbon coated with Ne and Ar discharge. A remarkable value of $\delta_{max} \leq 1$ is achieved: multipacting does not occur with such surface [99].

8.3.5 Effect of Surface Roughness

The morphological aspect of a surface reduces also the secondary electron production. When the secondary electrons, produced from the surface, perform several collisions with the material, their chance of being absorbed is increased. This results in a reduction of the net SEY.

Thus, the low SEY of as-received TiN coating resulted not only due to low SEY of TiN but also due to pyramidal structure created by TiN grains (i.e. due to the roughness of its surface). Indeed, thin TiN coating has a goldish colour with $\delta_{max} \approx 2.4$, whereas the surface-coated thick layer of TiN is black with $\delta_{max} \approx 1.5$, the black aspect being attributed to the roughness of the material [100]. Thick amorphous carbon films reduce as well the SEY due to its surface roughness [99].

Various techniques can be employed for producing macroscopic or microscopic structures: mechanical, chemical, or ion etching, or laser ablation.

As illustrated in Figures 8.7, 8.10, and 8.19, a groove, rib, peak, or honeycomb structure can trap the secondary electrons to reduce the effective SEY of the material [101, 102].

A TiN-coated grooved surface with 40° opening angle and depth of 1 mm provides $\delta_{max} \approx 1.2$, as was demonstrated at SLAC (CA, USA) [103].

A surface of copper, 75% of which was occupied by drilled holes with a depth of 0.5 mm and a diameter of 1 mm (see Figure 8.19), provides $\delta_{max} \approx 1.2$, measured in experiments at CERN [102].

Reducing SEY with LASE is a promising technology currently under development [79–83]. The laser treatment of metal surfaces (such as stainless steel, copper, aluminium and their alloys, and other metals) with a fluence above the ablation threshold results in the formation of hierarchy of microstructure, submicron structure, and nanostructures. The LASE process can produce the grooves with tens of microns in depth (microstructures) and the groove walls are covered with submicron and nanospheres (see Figure 8.20) or/and nanowire structures

Figure 8.19 Surface roughness produced by drilled means.

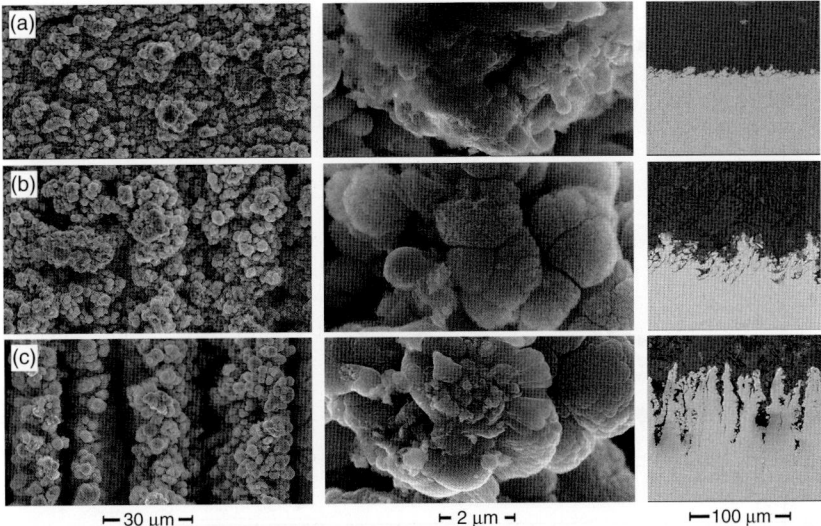

Figure 8.20 Low (on the left) and high (in the middle) resolution planar and X section (on the right) SEM micrographs of 1-mm-thick copper samples treated with laser using different scan speeds: (a) 180 mm/s, (b) 90 mm/s, and (c) 30 mm/s. Source: Reprinted with permission from Valizadeh et al. [83], Fig. 3. Copyright 2017, Elsevier.

(see details in Ref. [83]). These structures on a surface could provide SEY below 1: multipacting does not occur with such a surface.

Initially, the main emphasis was on creating microstructures (in the form of pyramids and grooves) [79]; however, these structures have a large RF surface resistance. Later, it was shown that submicron structures and nanostructures also play a role in reducing SEY; thus, an effort was focused on LASE, which should provide not only low SEY but also a low surface resistance [81, 83, 104, 105]. It has been demonstrated that LASE surfaces can be produced with a marginally visible microstructure (and much lower resurface resistance), while the submicron structures and nanostructures can still provide $\delta_{max} < 1$ (see Figures 8.20 and 8.21). As depicted in Figure 8.21, the SEY of all laser-treated Cu samples is below 1 for impinging electron energies up to 480 eV, which is sufficient for e-cloud mitigation in machines like LHC or FCC where multipacting electrons have energies up to 300 eV.

The results shown in Figure 8.11 illustrated another example where all measured samples have $\delta < 1$ in the whole range of used electron energies (i.e. up to 1 kV). It must be noted that these are initial data obtained after 4–10 hours of pumping; electron or photon beam conditioning, and bakeout reducing SEY further down to as low as $\delta_{max} < 0.6$.

8.3.6 'True' Secondary Electrons, Re-Diffused Electrons, and Reflected Electrons

It must be stressed that the secondary electrons taken into account for the evaluation of the SEY in the figures above are the sum of 'true' secondary

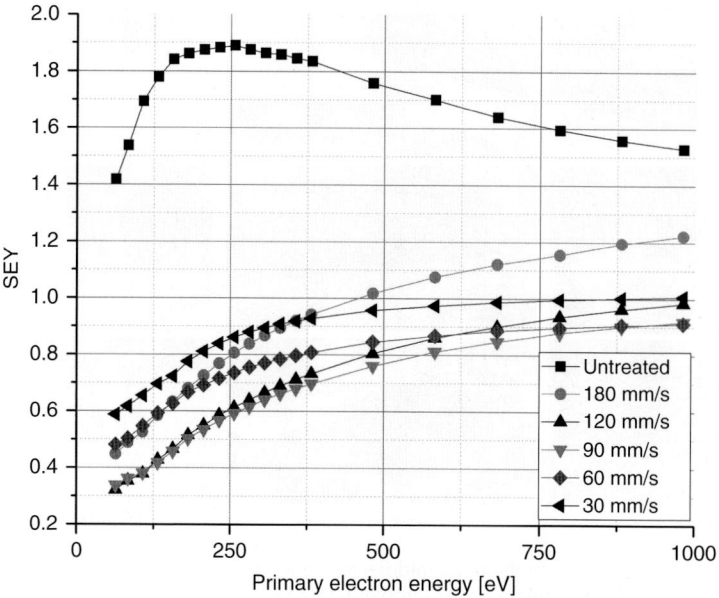

Figure 8.21 Comparison of SEY of copper samples as a function of primary electron energy for an untreated sample and the samples treated with LASE with varying scan speed after 10-hour pumping. Source: Reprinted with permission from Valizadeh et al. [83], Fig. 4. Copyright 2017, Elsevier.

electrons, re-diffused electrons, and reflected electrons [106, 107]. Thus, for each point measured at a given incident primary electron kinetic energy, the emitted electrons are not mono-energetic but have rather a distribution in energy. Such electron distribution curves are highly peaked towards low energy below ~5 eV. However, there is always a contribution from reflected electrons. This contribution increases when decreasing the incident electron kinetic energy. For a technical surface such as copper, reflectivity above ~60% is measured for impinging electrons with kinetic energy below ~20 eV. This is illustrated in Figure 8.22 where the electron energy distribution curves are measured for primary electron energy ranging from 4 to 200 eV [108]. These curves are typically measured with semi-hemispherical electron analysers or low energy electron diffraction devices. The reflectivity at low electron energy of clean polycrystalline copper (i.e. sputtered copper) is ~20% [109].

For the purpose of modelisation of the BIEM and e-cloud, the SEY curve can be described by the sum of true secondary electrons and reflected electrons as a function of energy of the primary impinging electrons, E_p [110–112]:

$$\delta(E_p) = \delta_{true}(E_p) + \delta_{elastic}(E_p) \tag{8.7}$$

where the contribution of 'true' secondary electrons is given by Eq. (8.8) with $s \approx 1.35$, δ_{max} as the maximum of the SEY curve, and E_{max} as the primary energy

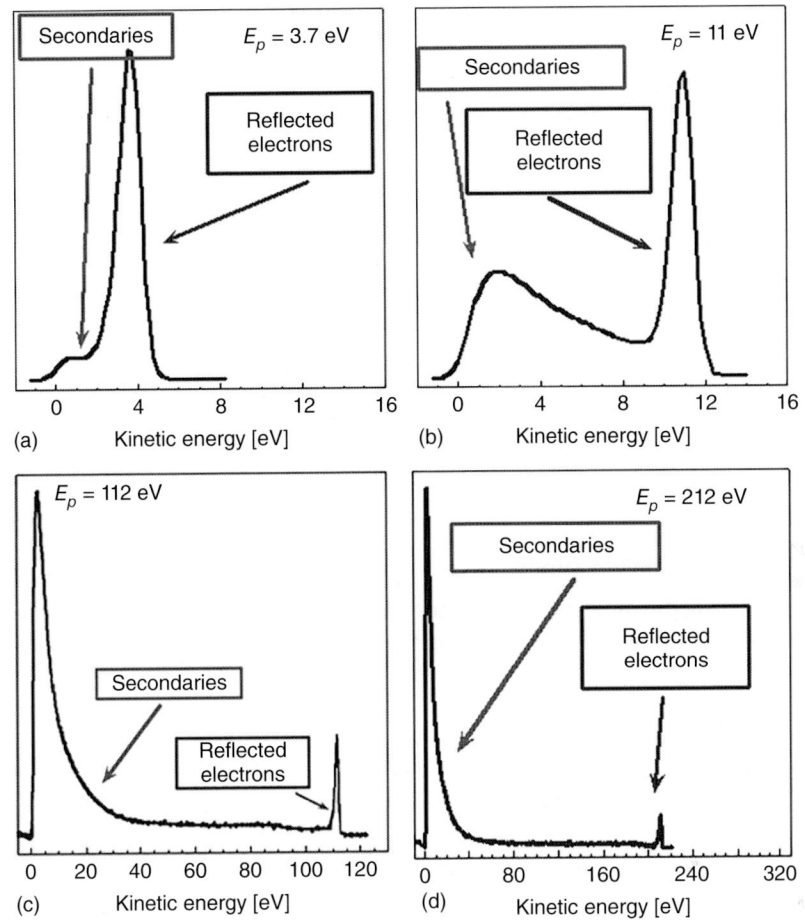

Figure 8.22 Electron energy distribution curves for different primary electron energies, E_p: (a) 3.7 eV, (b) 11 eV, (c) 112 eV and (d) 212 eV. Source: Reprinted with permission from Cimino and Collins [108], Fig. 1. Copyright 2004, Elsevier.

corresponding to δ_{max}:

$$\delta_{true}(E_p) = \delta_{max} \frac{s \frac{E_p}{E_{max}}}{s - \left(1 + \frac{E_p}{E_{max}}\right)^s}. \tag{8.8}$$

Figure 8.23 shows the curves for Cu obtained with Eq. (8.8) for $E_{max} = 250$ eV and $\delta_{max} = 2.2, 1.7,$ and 1.2. As shown, when the maximum SEY is decreased, not only the amount of multiplied electrons is decreased, but also the energy range of the primary electrons that could be multiplied. When $\delta_{max} = 2.2$, the SEY is larger

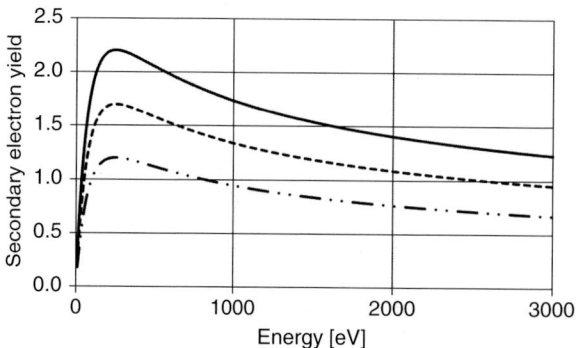

Figure 8.23 SEY curves of Cu, computed with Eq. (8.8), for several values of δ_{max}.

than 1 for primary energy range up to 3 keV; however, when $\delta_{max} = 1.2$, the energy range of multiplied primary electrons is reduced to the range of 100–700 eV.

8.3.7 Effect of Incidence Angle

When decreasing the incidence angle, the SEY increases due to the electron production closer to the surface. Introducing the angle θ with respect to the surface normal, the parameters δ_{max}, E_{max}, and s may vary [113]:

$$\delta_{max}(\theta) = \delta_{max} e^{0.4(1-\cos\theta)} \quad (8.9)$$
$$E_{max}(\theta) = E_{max} \times (1 + 0.7(1 - \cos\theta)) \quad (8.10)$$
$$s(\theta) = s \times (1 - 0.18(1 - \cos\theta)) \quad (8.11)$$

In magnetic field-free regions, the azimuthal distribution of the true secondary electrons follows the Beer–Lambert law but the reflection of the elastic electrons is specular.

The contribution of the 'elastically' reflected electrons as a function of impinging electron energy is given by [112]:

$$\delta_{elastic}(E_P) = R_0 \left(\frac{\sqrt{F_P} - \sqrt{F_P + \varepsilon_0}}{\sqrt{E_P} + \sqrt{E_P + \varepsilon_0}} \right)^2 \quad (8.12)$$

with R_0, the reflectivity for electron impinging the surface close to zero energy ($R_0 \sim 0.6$), and ε_0 the negative step potential onto which the plane wave–electron wave function is incident ($\varepsilon_0 \sim 150$ eV). Figure 8.24 shows the relative contribution to the total SEY (Eq. (8.7)) of 'true' and 'elastic' secondary electrons. The contribution of reflected electrons is important for primary energies below 10 eV and negligible above 50 eV.

8.3.8 Insulating Materials

Insulating materials are widely used in vacuum chambers. Their SEY values are much larger than for metals (see Table 8.2). The bakeout may result in the SEY increase [114].

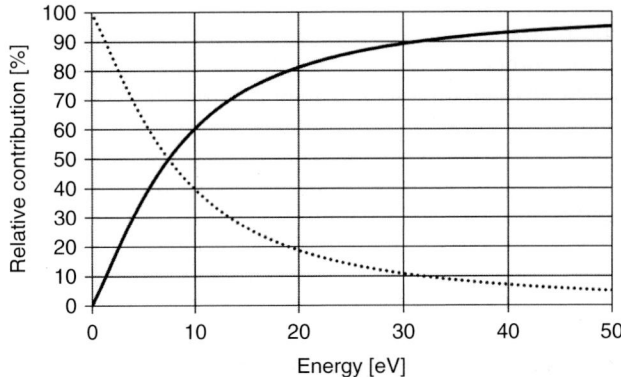

Figure 8.24 Relative contribution to the SEY of the 'true' secondary electrons (plain) and the 'elastic' secondary electrons (dot) computed from Eqs. (8.7), (8.8), and (8.12).

Table 8.2 Maximum SEY of some insulators.

Material	As received		Baked to 350 °C		References
	δ_{max}	E_{max} [eV]	δ_{max}	E_{max} [eV]	
Quartz	3.0	370	3.2	405	[114]
Alumina 97.6%	5.7	935	8.2	1150	[114]
Alumina 97.6%	3.6	695	5.75	1000	[114]
Alumina 100%	4.6	1090	—	—	[114]
Sapphire	4.2	755	—	—	[114]
Alumina	4.7	600	—	—	[115]
	7.2	850	—	—	[116]
	7.8	1400	—	—	[117]
	9	2000	—	—	[118]
Polytetrafluoroethylene (PTFE)	1.5	400	—	—	[117]
	2.25	375	—	—	[119]
	3.0	300	—	—	[120]
Polyethylene (PE)	1.6	500	—	—	[117]
	2.4	270	—	—	[119]
	2.9	250	—	—	[121]
Machinable ceramics (SiO_2–B_2O_3–Al_2O_3–ZnO–MgO–F system)	2.3	800	—	—	[117]

It should be noted that isolators are very sensitive to both production process and experimental conditions and, therefore, the results are varying significantly from one research group to another.

Insulating materials can be charged under electron bombardment. This makes studying their SEY quite difficult. While in the machines the charged insulator can be acting as a potential source of electrons and stimulating beam instabilities.

Therefore, ceramic components should be avoided in locations of possible electron multipacting or at least to be screened from SR and primary electron bombardment.

8.4 How the BIEM and E-Cloud Affect Vacuum

It was already mentioned earlier that electrons can be accelerated by the bunch charge, gain energy of up to a few hundred electron volt, and hit the vacuum chamber wall, not only producing other secondary electrons but also inducing ESD.

Therefore, the gas dynamics balance equations in Chapters 5 and 6 include the gas desorption term q, which consists of three main sources, thermal desorption, PSD, and ESD:

$$q\left[\frac{\text{molecules}}{\text{s·m}}\right] = \eta_t F + \eta_\gamma \Gamma + \eta_e \Theta \tag{8.13}$$

To estimate the BIEM and e-cloud effects on vacuum, one needs to obtain

- the electron flux Θ [e$^-$/(s·m)],
- the (average) energy of hitting electrons $\langle E_e \rangle$ [eV], which is required to estimate the ESD yields (η_e),
- how well the surface was conditioned with SR and photoelectrons before the electron multipacting has been triggered on.

8.4.1 Estimation of Electron Energy and Incident Electron Flux

The e-cloud modelling code input includes a selected set of parameters such as PEY, SEY, gas density, and charge particle beam characteristics. The results of the modelling provides an intensity of BIEM and an e-cloud density distribution.

To calculate ESD for gas dynamics, we can use either the electron energy distribution, $\theta(E)$:

$$q_{\text{ESD}} = \sum_{i=1}^{N} (\eta_e(E_i)\Theta_i(E_i)), \tag{8.14}$$

or the average power dissipation of multipacting electrons to the walls of vacuum chamber, P [W/m], and the average energy of hitting electrons $\langle E_e \rangle$. The use of the average electron energy, $\langle E_e \rangle$, is sufficient to calculate the effective electron flux Θ from the electron power P. Considering a linear dependence of ESD yields with the electron energy in the range 50 eV $< E <$ 1 keV (see Chapter 4).

8.4 How the BIEM and E-Cloud Affect Vacuum

Since the power is

$$P\left[\frac{W}{m}\right] = I_e\left[\frac{A}{m}\right] U[V] = \Theta\left[\frac{e^-}{s\cdot m}\right] \langle E_e \rangle [eV], \tag{8.15}$$

then the effective electron flux per meter of vacuum chamber length can be calculated as

$$\Theta\left[\frac{e^-}{s\cdot m}\right] = \frac{P\left[\frac{W}{m}\right]}{\langle E_e \rangle [eV] q_e [C]}. \tag{8.16}$$

Then the ESD per meter of vacuum chamber can be calculated:

$$q_{ESD} = \eta_e(\langle E_e \rangle)\Theta. \tag{8.17}$$

If the average electron energy is not provided, it can be estimated as follows. The time-averaged electric field $\mathcal{E}$ of the beam with a Gaussian profile can be given in SI units by

$$\mathcal{E}(r) = \frac{I}{2\pi\varepsilon_0 cr}\left(1 - e^{-\left(\frac{r}{\sigma_r}\right)^2}\right) \tag{8.18}$$

where I is the proton beam current; $\varepsilon_0 = 8.85 \times 10^{-12}$ [F/m] is the permittivity of free space; c is the speed of light in vacuum; σ_r is the r.m.s. beam size, $\sigma_r = \sqrt{\beta\varepsilon_n/\gamma}$; and r is the distance from the centre of beam. However, a peak electric field (in the presence of bunch of particles) is a factor T/τ higher than an average electric field value (here τ is the bunch length and T is the bunch spacing):

$$\mathcal{E}_{peak}(r) = \frac{I}{2\pi\varepsilon_0 cr}\left(1 - e^{-\left(\frac{r}{\sigma_r}\right)^2}\right)\frac{T}{\tau} \tag{8.19}$$

The electrons are accelerated towards the beam when bunch is present and are drifting between bunches. Thus, the electron can gain the energy up to

$$E_{max} = \int_0^a \mathcal{E}_{peak}(r)dr = \frac{I}{2\pi\varepsilon_0 c}\frac{T}{\tau}\int_0^a\left(\frac{1 - e^{-\left(\frac{r}{\sigma_r}\right)^2}}{r}\right)dr. \tag{8.20}$$

The electron incident energy varies in the range $0 < E_e < E_{max}$.

Integration of Eq. (8.20) can be performed numerically (see, for example, the procedure employed in Chapter 9) or analytically with Euler gamma functions:

$$E_{max} = \int_0^a \mathcal{E}_{peak}(r)dr = \frac{I}{4\pi\varepsilon_0 c\sigma_r}\frac{T}{\tau}$$

$$\times\left(\log\left(\left(\frac{a}{\sigma_r}\right)^2\right) + \Gamma\left(0, \left(\frac{a}{\sigma_r}\right)^2\right) + \gamma\right); \tag{8.21}$$

here, $\Gamma(0, z)$ is the incomplete Euler gamma function, $\Gamma(0, z) = \int_z^\infty x^{-1}e^{-x}dx$ and γ is Euler's constant, $\gamma \approx 0.577\,216$.

Then the average electron energy, $\langle E \rangle$, can be calculated from numerical result or artificially set to $\langle E_e \rangle = E_{max}/2$.

8.4.2 Estimation of Initial ESD

When there is no SR in a designed particle accelerator, then the ESD results from Chapter 4, from literature, or from new dedicated experiments are applicable directly.

In the presence of SR, most commonly, the start-up scenario of a machine allows sufficient time for tuning and commissioning. This means that the machine operates with low beam current, with SR emitted in the bends, irradiating and conditioning (or scrubbing) the beam pipe walls, but operates with beam parameters below the BIEM threshold. Thus, the beam pipe is conditioned with direct and reflected SR (see Figure 8.25). The SR conditioning will reduce both PSD and ESD yields (see Chapters 4 and 5). Furthermore, the SR conditioning reduces PEY and SEY. The conditioning effect can also be enhanced by photoelectrons that bombard the area of the beam pipe, which is not irradiated by direct SR; however the magnetic field can redistribute or trap these photoelectrons along the field lines [51, 122]. Therefore, the beam chamber conditioning is non-uniform around the beam pipe cross section but more efficient on a directly irradiated area than where the reflected photons and photoelectrons can reach. Also, the conditioning effect is non-uniform along the beam path (see Figures 4.31–4.33), and the intensity of SR directly irradiating the vacuum chamber reduces with distance from the source (a dipole, a wiggler, an undulator, or a quadrupole); therefore, the vacuum chamber, at and near, a dipole is conditioned more intensively than the parts of straight vacuum chamber at a larger distance from a dipole. For a detailed study, the use of ray and electron-tracing codes (see Section 2.5) are mandatory to compute and estimate the pre-conditioning effect during the start-up of a machine.

Thus, during the machine start-up, when reaching the BIEM threshold in the machines with SR, the initial ESD must be lower than in the machines without SR. The main problem is to estimate how strong this effect is. The ESD yield data were discussed in Chapter 4. The ESD is proportional to ESD yield η_e, the number of electrons hitting the walls in unit of time (the electron flux Θ [e$^-$/(s·m)]). The ESD yield increases with incident electron energy E_e. Like the PSD, the ESD reduces with both an integral photon and electron dose.

As a very simplistic estimation, one can use the assumption that the ESD yield (for any fixed electron energy) will be reduced with SR photon dose proportionally to the PSD yield: i.e. when PSD yield was reduced with SR conditioning by a factor 100, then it is reasonable to expect that ESD yield may also be reduced by a factor 100. That means that the lower the SR photon flux at the surface, the

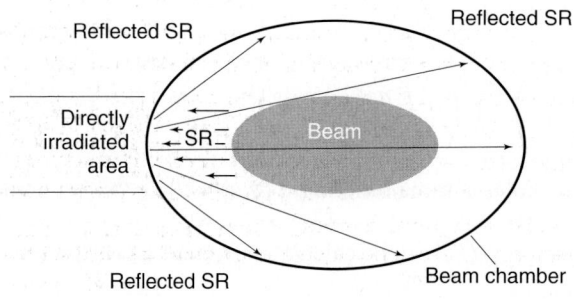

Figure 8.25 A beam chamber irradiated with direct and reflected SR.

higher the initial ESD yields will be when the BIEM is triggered: i.e. the highest ESD yields will be expected at the locations where there is the lowest SR intensity or no SR.

Then the terms in Eq. (8.13) can be compared to each other. If ESD is comparable or larger than the thermal and PSD, then in the gas dynamic model the significant attention should be paid to the ESD term, $\eta_e \Theta$, to check how much the gas density will increase due to BIEM. The pressure profile may dramatically change, and optimum pumping arrangement for PSD may become not ideal to cope with ESD.

One should pay attention that the electron energy depends on beam and beam train parameters. Beam current may increase by increasing the number of bunches or/and by increasing the number of particles in each bunch.

- Increasing the number of bunches increases the electron flux Θ but does not affect the average energy of hitting electrons E_e; thus the beam current will cause a linear increase in ESD.
- Increasing the number of particles in each bunch increases the average energy of hitting electrons E_e, which leads to higher ESD yield and may also result in higher SEY (for $E_e < E_{max}$, E_{max} = 150–650 eV for flat surfaces and E_{max} = 900–1000 eV for rough surfaces), which, in turn, lead to a higher electron flux Θ. In this case, the ESD increase with beam current may be to a power larger than 1.

The ESD can significantly increase the gas density in the beam vacuum chamber; a higher gas density, in turn, will increase the gas ionisation rate by the beam particles, contributing into electron production for the e-cloud build-up and change e-cloud density to unacceptably high value.

That should be considered during the vacuum system design (to provide better pumping or apply different BIEM and e-cloud mitigation techniques) and machine conditioning scenario. Such work was reported, for example, for LHC [123] and ILC [45].

8.5 BIEM and E-Cloud Observation in Machines

8.5.1 Measurements in Machines

The study and characterisation of BIEM and e-cloud in machines is obviously of great importance during the design phase of a machine and during its commissioning and operating periods. Many observables, which can be used by the engineers and scientists for the analysis of BIEM and electron cloud, are available in the control room. They can be related to the measurement of beam properties such as the beam and bunch current lifetimes, emittance growth, beam instabilities (usually in the vertical plane since the majority of a synchrotron is built of dipole magnet), energy and particle losses, etc. This information is mainly available via specific beam instrumentation systems. The impact of BIEM and e-cloud can also be observed with other technical systems such as pressure rise

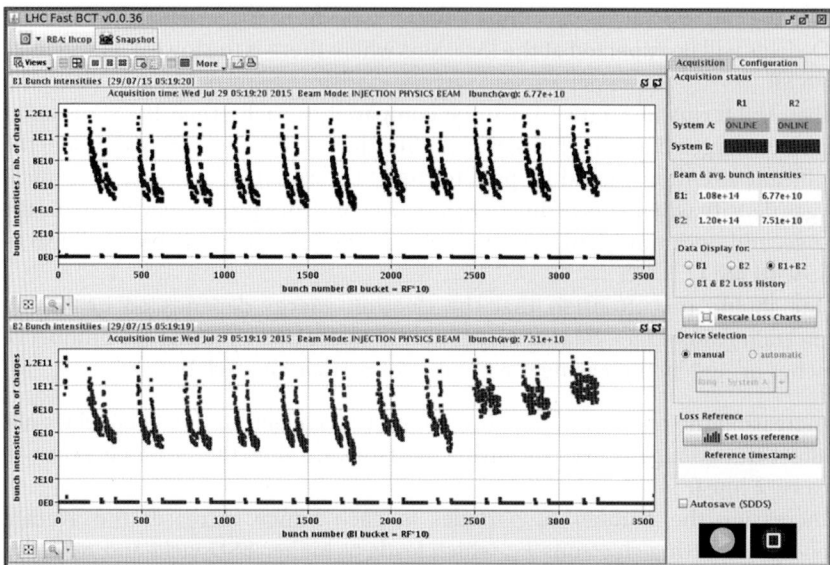

Figure 8.26 A typical signature of electron cloud observed on a bunch intensity fixed display of the CERN Large Hadron Collider (fill 4088, 29 July 2015).

in a vacuum system, energy deposition on a cryogenic system, beam energy loss compensated by the radio frequency system, etc.

Figure 8.26 shows a typical signature of e-cloud observed with a bunch current monitor at the CERN LHC with injection energy of 450 GeV during a beam conditioning campaign. During this run, the machine was filled with 1596 bunches in both beams, Beam 1 and Beam 2. After extraction from the Proton Synchrotron (PS) injector, the proton bunches are separated by 25 ns and grouped into a batch of 72 bunches each. Two batches are then injected from the PS into the Super Proton Synchrotron (SPS) to form a train. Then, 12 bunches are injected into the LHC followed by 11 trains of 144 bunches each. As shown by the fast beam current transformer, the bunch intensity is drastically reduced along each batch from 1.2×10^{11} protons per bunch (the nominal value) to 5×10^{10} protons per bunch. This reduction is due to the presence of an e-cloud into the LHC ring as supported by the measurement of a heat load onto the 5–20 K cryogenic system of ~0.5 W/m per aperture (~22 kW around the ring!). To be noticed, in the lower plot for Beam 2, is the smaller reduction of bunch intensity in the last three freshly injected trains from bunch number 2500 onwards.

The impact of BIEM and e-cloud can be seen also on the beam emittance as shown in Figure 8.27. Due to the dipole field, the emittance blow-up is limited to the vertical plane (lower plots). Along the batch, the first bunches are not affected by the e-cloud, but rather the ones at the end producing a typical triangular shape along each train. To be noticed also is the absence of vertical emittance blow-up for last three freshly injected trains into Beam 2.

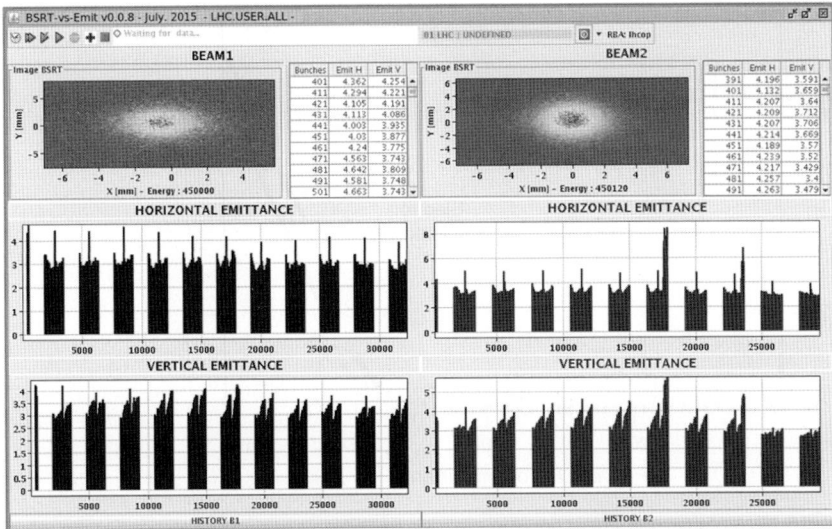

Figure 8.27 A typical signature of electron cloud observed on a horizontal and vertical emittance fixed display of the CERN Large Hadron Collider (fill 4088, 29 July 2015).

8.5.1.1 Vacuum Pressure

The observation of BIEM and electron cloud is usually associated with a pressure increase. Depending on the electron flux to the wall, the kinetic energy of the bombarding electrons, and the nature of the surface, the observed pressure increase can significantly vary. Moreover, since the BIEM is a non-linear mechanism, the pressure increase usually changes with the beam structure from which a threshold is a typical signature. Figure 8.28 shows the pressure threshold observed in PEP-II due to BIEM in the vacuum chamber [124].

The BIEM and e-cloud mechanism is a strong function of the beam structure; therefore strong impact on the associated pressure rise can be observed [21]. Below a given bunch intensity, BIEM cannot be triggered due to the low energy of the electrons. So, in this case, the gas dynamics is not dominated by other desorption processes (induced by, e.g. photons or ions), and there is no or little pressure increase. Increasing further the bunch intensity, above a given threshold, a sudden pressure increase appears due to ESD induced by the BIEM process. If the bunch intensity is kept constant but the amount of bunches is increased, the pressure increases linearly with the number of bunches. On the other hand, if the distance between bunches is increased, the pressure rise associated with the BIEM is reduced. However, a large time lag between bunches (a few 10 μs) is needed to clear completely the BIEM, thereby the associated pressure rise. A much better mean than increasing the bunch distance to clear the associated pressure rise in magnetic field-free region is the wrapping of a solenoid around the vacuum chamber. Figure 8.29a shows a solenoid wrapped around the LHC beam pipe. During the LHC commissioning, as shown on Figure 8.29b, a pressure

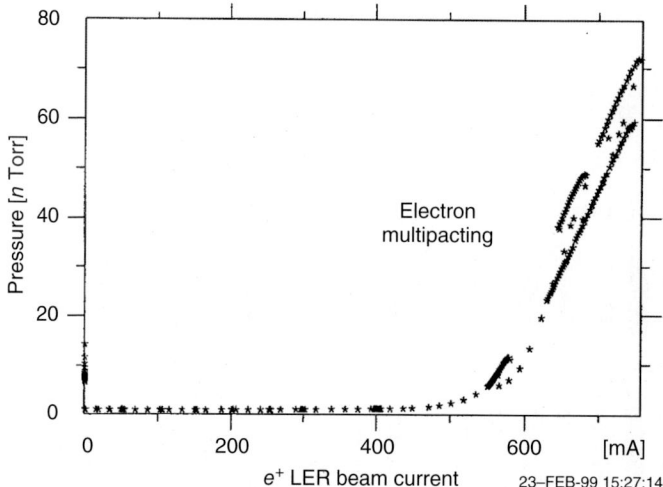

Figure 8.28 A typical non-linear pressure rise due to BIEM in a vacuum chamber. Source: Seeman et al. 2000 [124], Fig. 4. Reprinted with permission of CERN.

increase up to 10^{-8} mbar was observed around the experimental areas despite the distance between the bunches was 150 ns. Since the two beams circulate in opposite direction in the common beam pipe, the effective distance between the bunches is reduced at specific locations. When applying a solenoidal field of 20 G, this pressure increase was reduced in the 10^{-10} mbar range. The observed pressure reduction was due to the suppression of BIEM. However, it must be stressed that, despite this method is highly effective to suppress the BIEM, it does not affect the surface state of the vacuum chamber itself. Therefore, if the solenoidal field is switched off, the BIEM and the associated pressure rise occur again.

8.5.1.2 Vacuum Chamber Wall Properties

In the last decades, several specific diagnostics tools have been designed to study in detail not only the impact of BIEM and e-cloud on the beam itself but also the interaction of the phenomena with the vacuum surface itself in a real machine environment.

The exposure of different types of samples to the irradiation of direct SR or electrons was realised in several laboratories around the world [62, 125–129]. These studies allowed monitoring in detail the variation of important surface parameters for the understanding of the BIEM and e-cloud such as the SEY, the photon reflectivity, and the PEY.

Figure 8.30 shows a typical arrangement of such a system for the *in situ* measurement of SEY [125]. During the machine operation, the Cu sample was placed at the bottom of the beam pipe, while SR was irradiating the side of the wall. When the irradiation stopped, the sample can be moved upwards the SEY monitor for the SEY measurement. During each SEY measurement, electrons from the gun probe the sample with scanning kinetic energy from 10 eV to 3 keV. The accumulated dose is of the order of 1 nC/cm² in such a way that any reduction of the SEY due to electron gun bombardment during the

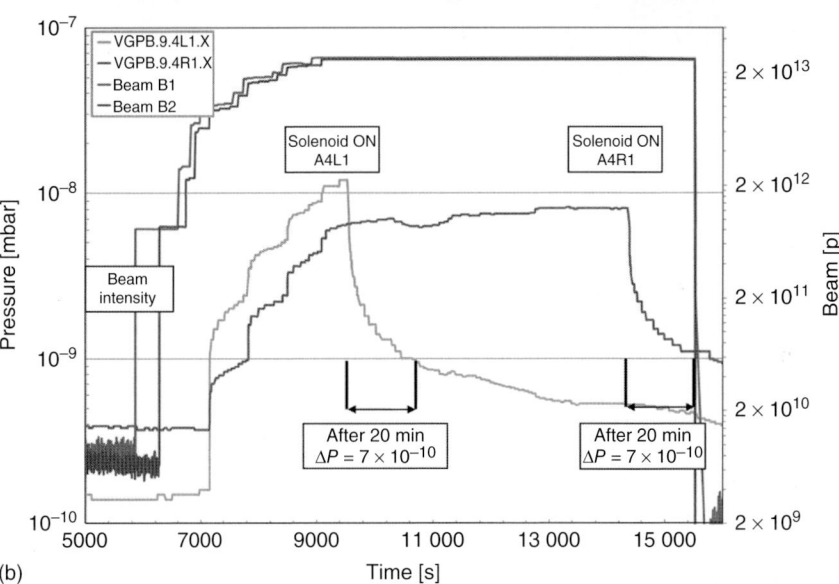

Figure 8.29 (a) A solenoid wrapped around the LHC beam pipe. (b) Observed pressure reduction when a solenoidal field of 20 G is applied. Source: Bregliozzi et al. 2011 [21], Fig. 7. Reprinted with permission of CERN.

measurement process can be neglected. With this device, the SEY is obtained by the simultaneous measurement of the sample drain current I_D and by the collected secondary electron current of a Faraday cup (or Cage) placed in front of the sample, I_F. The sample is negatively biased (−50 to −90 V) to expel the secondary electrons and the caged positively biased to attract these electrons. The SEY is then given by Eq. (8.5).

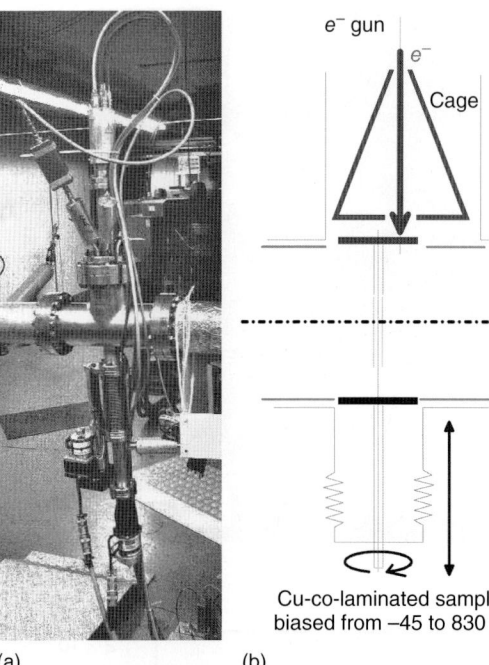

Figure 8.30 Picture (a) and schematic (b) of a set up for SEY *in situ* measurement. A translator exposes the sample to the electron cloud or to the SEY measurement device located on the upper part. Source: Reprinted with permission from Baglin et al. [125], Fig. 3. Copyright 2001, CERN.

Figure 8.31 shows the result of beam exposure to technical surfaces of accelerator machines. Figure 8.31a presents a network of SEY curves as a function of the dose of the incident particle. In this case, a CERN LHC Cu co-laminated-type sample was exposed to photoelectrons produced by SR and accelerated towards the sample to 100 eV [125]. Figure 8.31b shows the SEY curves of a TiN-coated sample exposed to the photoelectrons and electron cloud generated during two months by the positron beam of the PEP-II low energy ring (LER). The accumulated electron dose was estimated to be $\sim$40 mC/mm^2 [127]. In both cases, a reduction of the SEY curve under conditioning is noticed. In the laboratory, similar modifications of the SEY curve were also observed when directly bombarding the sample with electrons produced with a gun [55, 96].

A typical plot of δ_{max} during beam conditioning is shown in Figure 8.32 for an unbaked Cu co-laminated sample obtained from LHC beam screen material. In this particular case, the photoelectrons produced on the horizontal plane by the incident SR were attracted towards the sample by biasing it. Doing so, any artefact, due to the presence of an electron gun irradiating directly a sample (which might be the case when performing a similar measurement in the laboratory), is avoided. It is shown that the maximum of the SEY curve decreases as a function of electron dose. A value of $\delta_{max} \approx 1.2$ being reached after an electron dose of $\sim$10 mC/mm^2. A closer look at this figure indicates that the conditioning efficiency varies with the beam energy. The reduction of the maximum SEY is attributed to the graphitisation of the native oxide and carbide layers, as demonstrated in a dedicated X-ray photoelectron spectroscopy study. In this study the conversion of the C^{1s} bond from sp^3 to sp^2 was pointed out together

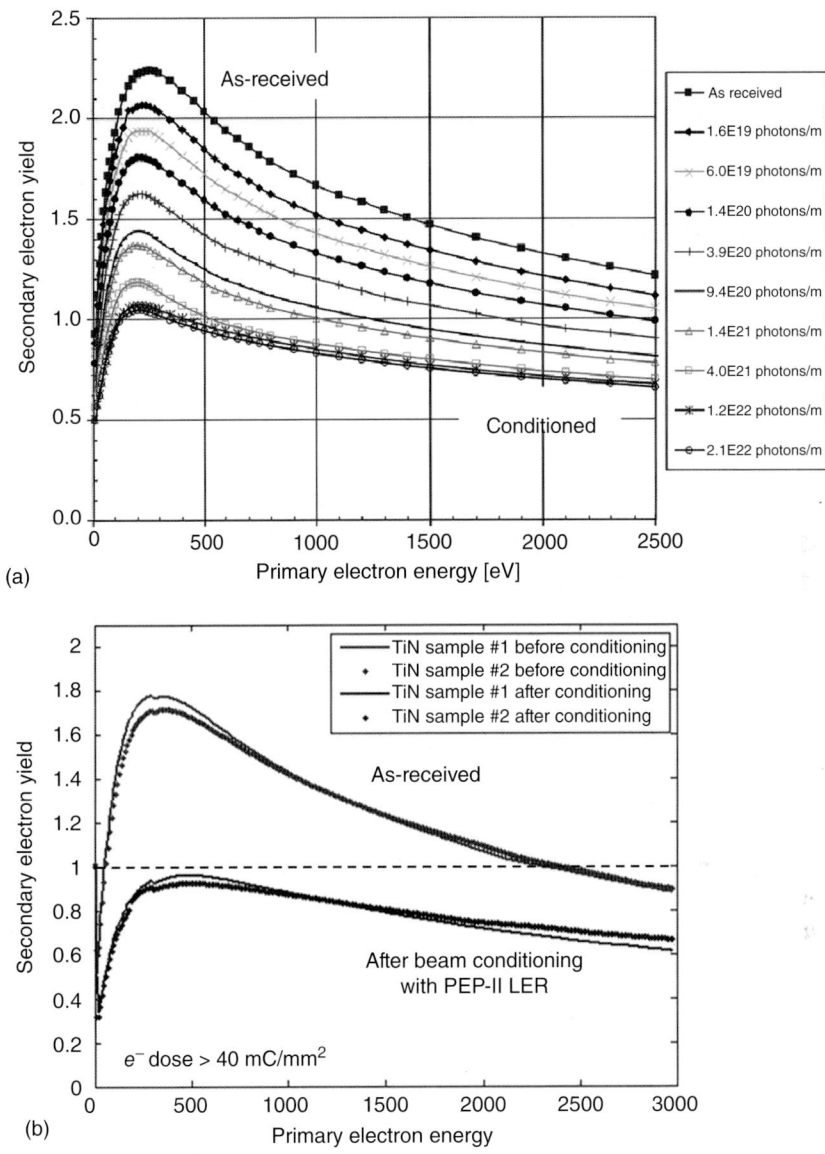

Figure 8.31 (a) Reduction of the SEY curve as a function of photon dose when photoelectrons are accelerated to 100 V. (b) Reduction of the SEY curve after two months operation of the PEP-II LER. Source: Reprinted with permission from Pivi et al. [127], Fig. 8. Copyright 2010, Elsevier.

with a higher conditioning efficiency for electrons with kinetic energy larger than 50 eV [55, 130]. Indeed, pure sp^2 carbon (or highly oriented pyrolytic graphite) has $\delta_{max} \approx 1$.

From these measurements, the SEY curves were fitted by Eq. (8.14) according to Eqs. (8.5) and (8.2). The result is depicted in Figure 8.33 for Cu-co-laminated

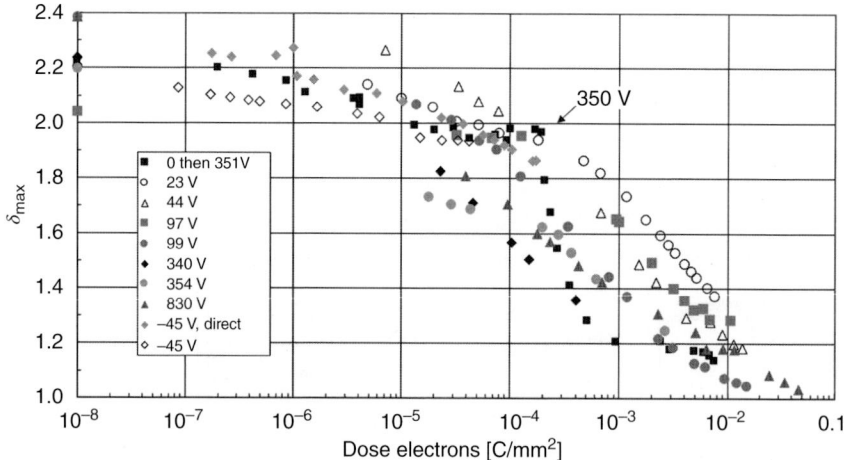

Figure 8.32 Reduction of the maximum of the SEY curve as a function of electron dose. Source: Reprinted with permission from Baglin et al. [125], Fig. 4. Copyright 2001, CERN.

Table 8.3 Parameters for the description of the SEY curve of as-received and conditioned Cu-co-laminated material.

State	δ_{max}	E_{max} [eV]	s	a	σ [eV]
As-received	2.24	220	1.401	0.560	36.8
Conditioned	1.18	200	1.329	0.360	27.0

where the as-received and conditioned SEY curves are shown on (a) and (b), respectively. The error bars are defined by 3σ across the different measured samples, where σ is a standard deviation:

$$\delta(E) = \delta_{max} \frac{E}{E_{max}} \frac{s}{s - 1 + \left(\frac{E}{E_{max}}\right)^2} + a \exp\left(-\frac{E^2}{2\sigma^2}\right). \tag{8.22}$$

Table 8.3 shows the result of the fitting process across five samples measured in the as-received and conditioned states. It can be seen that the conditioning process could induce a slight modification of the parameters possibly due to the graphitisation of the surface.

8.5.1.3 Specific Tools for BIEM and Electron Cloud Observation

Besides the characterisation of SEY and PEY surface parameters with dedicated devices, instruments designed to observe the BIEM and e-cloud have been developed and installed in numerous machines around the world [129, 131–141]. Most of the instruments are based on electron flux measurement at the vacuum chamber wall in a way to derive properties of the e-cloud. In this section, we will discuss only a few fundamental instruments, which can be easily integrated in a vacuum system.

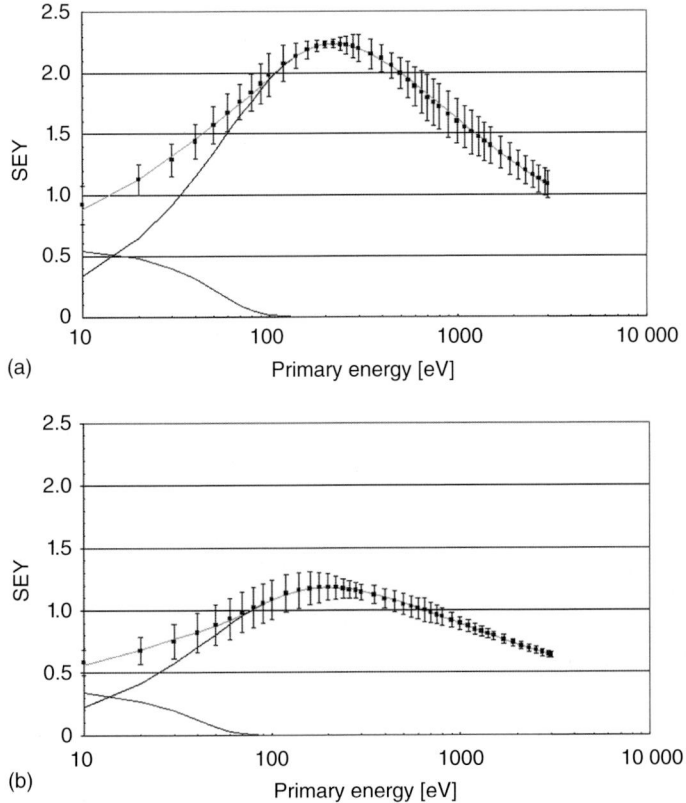

Figure 8.33 (a) Fit to as-received Cu-colaminated sample. (b) Fit to conditioned Cu-co-laminated sample. Source: Reprinted with permission from Baglin et al. [125], Fig. 5. Copyright 2001, CERN.

The simplest detector is a button pick-up (of a few 10 mm in diameter), which allows the measurement of an electron current. When directly facing the beam, the pick-up is non-shielded. A wideband beam signal can be recorded with a scope. This signal indicates the presence of a beam and can be used as a trigger if needed. When shielded from the beam, the pick-up measures a current proportional to the electron cloud density. Such pick-ups are installed behind a grounded RF shield made of slots or round holes with size in the mm range. Typical transparency of the RF shield is in the range 10–40%, a compromise between the electron collection efficiency and induced perturbation on the BIEM and e-cloud. The button pick-up is positively biased in order to collect the secondary electrons produced at the button by the incoming electrons originating from the e-cloud. About +50 V was applied in practice to cancel the effect on the measured current due to secondary electrons produced at the button pick-up. For illustration, Figure 8.34a shows a shielded button pick-up and, Figure 8.34b, a signal induced by the passage of the beam in front of a non-shielded pick-up (strip line) and a shielded pick-up (BPU2) signal induced by an e-cloud [137, 142]. In this example, the distance between successive bunches can be measured (25 ns) and

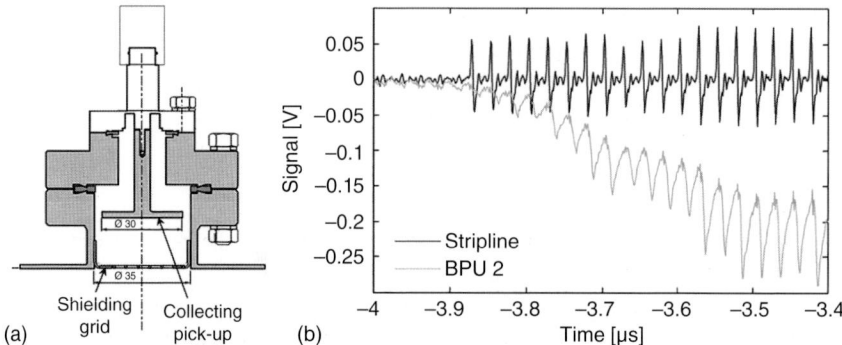

Figure 8.34 (a) Shielded button pick-up. (b) Typical signal measured with a scope at an unshielded (strip line) and shielded (BPU2) button pick-up. Source: (a) Reprinted with permission from Jimenez et al. [142], Fig. 4a. Copyright 2002, CERN. (b) Mahner et al. 2008 [137], Fig. 5a. https://journals.aps.org/prab/abstract/10.1103/PhysRevSTAB.11.094401. Licensed Under CC BY 3.0.

the number of bunches needed to reach a quasi-equilibrium for the e-cloud is 15. In steady state, collected electron current is typically in the range 0.1–10 µA.

Another important instrument is the Retarding Field Analyser (RFA) (see Figure 8.35a) [143]. This instrument allows measuring the energy spectra of the electrons extracted from the electron cloud. In this case, a retarding voltage is inserted between the grounded RF shield connected to the vacuum chamber wall and the collector held at ∼+50 V. When the voltage of the analysing grid is set at a potential V_g, only the electrons whose energy is larger than V_g can cross the grid and reach the collector. During the measurement, the voltage of the analysing grid is sweep from 0 to ∼−500 V and the collected current is differentiated with respect to the grid voltage to obtain the electron energy spectra. In order to reduce the noise induced by the capacitive coupling between the variable voltage grid and the collector, a grid held at ground can be placed

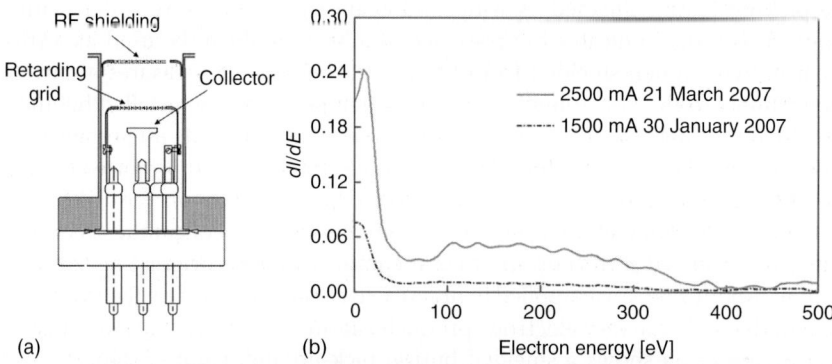

Figure 8.35 (a) Retarding Field Analyser. (b) Electron energy distribution of an electron cloud. Source: (a) Reprinted with permission from Laurent and Iriso Ariz [143], Fig. 2a. Copyright 2003, CERN. (b) Reprinted with permission from Pivi et al. [127], Fig. 16. Copyright 2010, Elsevier.

around the collector. For further noise reduction, another grounded grid can be placed between the vacuum chamber's RF shield and the retarding grid. Typical transparencies of mesh grids range from 50 to 90%. In upgraded versions, the collector is replaced by a multichannel plate to increase the detection sensitivity. The energy resolution of the electron spectra can also be upgraded by adding a fifth grid and using an etherodine technique similar to the one used in LEED technology to perform Auger studies [144]. In all cases, when all the grids are grounded, the instrument can be used in electron current detection mode in a similar way to a shielded button pick-up. Figure 8.35b shows a typical electron energy distribution measured by an RFA exposed to e-cloud. The distribution is peaked at low energy ($\sim$5 eV) and most of the electrons have an energy below 50 eV. Finally, when operating the RFA at large negative voltage (~ -1 kV), i.e. detecting electrons kicked near the beam, the device can also be used to estimate the e-cloud density around the beam [139]. One must stress that this last mode of detection is of paramount importance for the machine designer since above a given e-cloud density threshold, in the range of 10^{10}–10^{12} e/m^3, an accelerator machine can be strongly unstable.

As explained previously, the BIEM and e-cloud are sensitive to magnetic fields. For this reason, specific diagnostic tools have been designed to collect electron current with a spatial resolution. A typical detector is made of $\sim$1-mm-wide Cu strips. These strips were deposited on an insulating substrate, e.g. Macor™, or a lithographed Kapton foil with $\sim$1 mm pitch. Typical collected current ranges from 20 nA to 20 µA [132]. Figure 8.36 shows typical electron current signals obtained with a multi-strip detector [129, 142]. In this case, the detectors are placed in the lower part of the vacuum chamber. As shown, a spatial resolution in the mm range can be achieved. The signal measured in a vertical dipole field is shown on Figure 8.36a for 9×10^{10} particles per bunch. Above $\sim$20 Gs, the electrons are guided along the magnetic field lines. At low bunch current, the electron signal is centred in the vacuum chamber. Above $\sim 5 \times 10^{10}$ particles per bunch, two stripes appears. The distance between the two signals is a function of the bunch current. The larger the bunch current, the larger the kick received by the electron. Increasing further the bunch density above $\sim 1.3 \times 10^{10}$ particles per bunch, a third stripe appears in the centre of the vacuum chamber. Therefore, in a dipole field, the electron bombards the vacuum chambers in the horizontal and vertical planes at four distinct positions, which are a function of the bunch density. In the CERN LHC, the two stripes are separated by 20 mm with nominal bunch density (1.1×10^{10} protons per bunch). The signal measured in a quadruple field is shown on Figure 8.36b where only the lower part of the vacuum chamber is shown. In this case, the electrons are also guided along the magnetic field lines. Thus, the electrons bombard the vacuum chamber walls at four pole locations.

Other tools are available for specific studies to understand the interplay of the e-cloud with the vacuum chamber wall. An RFA coupled with a 'sweeping' electrode can monitor the electron density during the bunch passage and after its passage [136, 145]. Calorimetry is performed by pick-ups or liner chambers [146, 147]. The study of the multipacting mechanism can be performed with resonant standing wave coaxial set-up [148, 149].

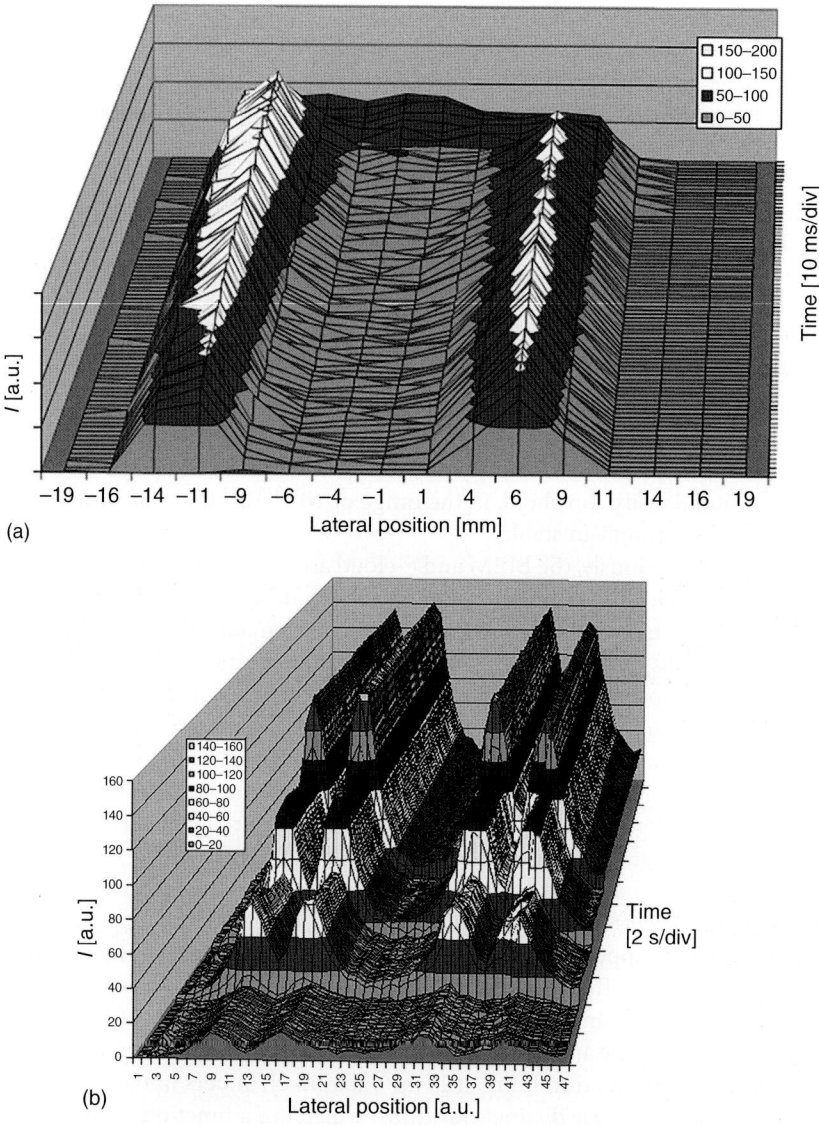

Figure 8.36 Typical electron current signals obtained with a multi-strip detector. (a) The signals measured in a dipole field. (b) The signal measured in a quadrupole field. Source: (a) Reprinted with permission from Jimenez et al. [142], Fig. 22b. Copyright 2002, CERN. (b) Reprinted with permission from Jimenez et al. [129], Fig. 7. Copyright 2005, American Vacuum Society.

8.5.2 Machines Operating at Cryogenic Temperature

With the increasing use of superconducting magnets or RF cavities in machines around the world, there is more and more interest to understand the interplay between the vacuum system operating at cryogenic temperature and the BIEM and e-cloud. This is the case of, e.g. RHIC, LHC, and future machines such as SIS100, FCC-hh, or CPPS.

8.5.2.1 Surface Properties at Cryogenic Temperature

At cryogenic temperature, the vacuum chamber wall is usually held at temperature close to the liquid helium boiling temperature (i.e. 4.2 K). In this regime, as discussed in Chapter 7, the vacuum chamber wall acts as a pumping surface. Thus, gas molecules are physisorbed or condensed on the wall. Thereby the behaviour of the vacuum system is significantly modified; see, e.g. [150, 151]. The nature of the adsorbed molecules onto the surface is a function of the vacuum chamber material and its temperature. For smooth and metallic surfaces, hydrogen is adsorbed below 20 K; methane, nitrogen, and carbon monoxide, below 40 K; carbon dioxide, below 80 K; and water, below 190 K. The sticking probability of the molecules is a function of the vacuum chamber material, its temperature, the surface coverage, and the kinetic energy of the molecule. As it will be shown below, the surface temperature, with the presence of adsorbate, has significant impact on the BIEM and e-cloud through the modification of the photoelectrons and secondary electron properties.

For surface coverage below 10^{16}–10^{17} molecules/cm^2, the PEY of a stainless steel surface held at cryogenic temperature irradiated at perpendicular incidence with SR of 200 eV critical energy does not differ as compared to room temperature (i.e. $\sim 10^{-2}$ e$^-$/photon). For much larger surface coverage, the PEY decreases by 1 order of magnitude. However, the photo-interaction with thick layers of CH$_4$ and CO induces a charging of the condensate, which results in a slight enhancement of the PEY [152].

At cryogenic temperature, the SEY of an as-received Cu surface is not modified as compared to room temperature. Figure 8.37 shows the SEY behaviour of an LHC Cu-co-laminated beam screen sample held at room temperature and at 9 K. As shown on Figure 8.37a, the SEY at 200 eV as a function of electron dose is very similar for cryogenic and room temperatures. Similarly, to Figure 8.32, the SEY value reduces to 1–1.2 for an accumulated dose of about 10 mC/mm^2 [153]. The SEY curve at 9 K is very similar to the one at room temperature with a maximum around 250 eV [58].

However, when thick layers of gas are condensed onto the surface, i.e. above a monolayer, the SEY is strongly modified. Figure 8.38 shows the evolution of the SEY when water is adsorbed at 77 K onto a sputter-cleaned Cu-baked sample [154]. The left curve shows that the water adsorption on a baked sample modifies drastically the SEY. At 200 ML, i.e. a thickness of ~ 80 nm, the benefit of the bakeout (Figure 8.17) is fully lost and the SEY properties of an unbaked sample is recovered (Figure 8.16). The right curve shows the maximum SEY saturates at 2.3 above 160 ML of condensed water. About 10 ML of adsorbed water are sufficiently large to modify drastically the BIEM and e-cloud phenomena, thereby deeply perturbing the operation of a machine.

In fact, as illustrated in Figure 8.39, only a few monolayers of any condensable gases modify the SEY curve of a surface [155]. The left curve shows the evolution of the maximum SEY of a Cu surface held at 4.7 K as a function of the amount of CO$_2$ and N$_2$ monolayers. When a monolayer is adsorbed on the surface, the

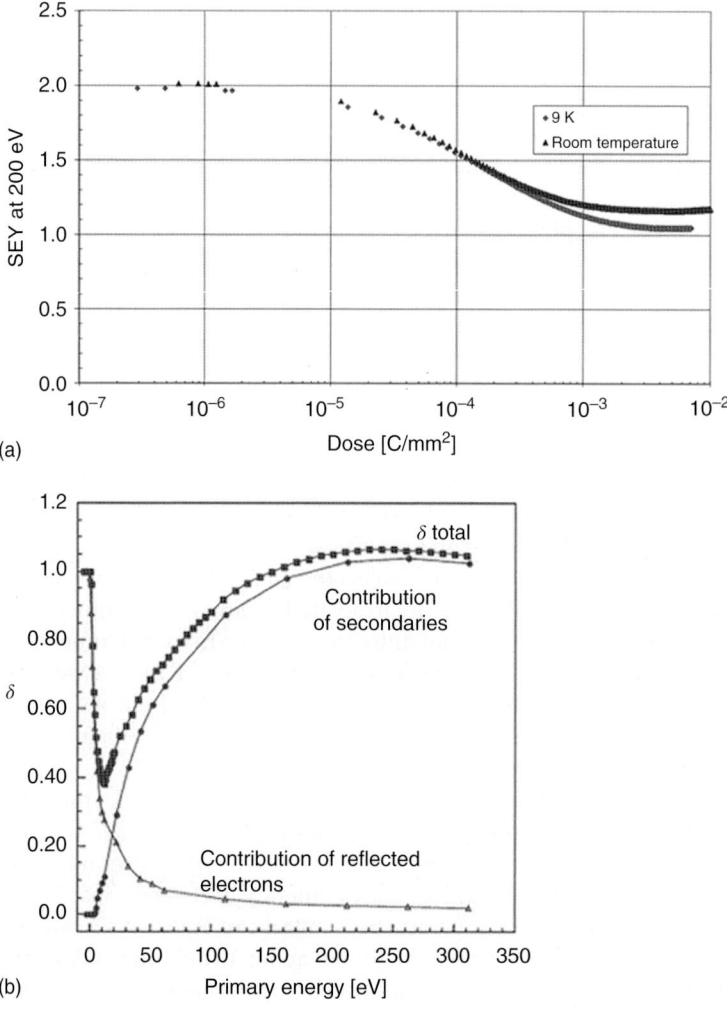

Figure 8.37 (a) SEY at 200 eV as a function of electron dose of a Cu sample held at room temperature and 9 K. (b) SEY curve of a Cu sample held at 9 K. Source: (a) Reprinted with permission from Baglin [153], Fig. 2. Copyright 2017, Elsevier. (b) Reprinted with permission from Cimino and Collins [108], Fig. 3. Copyright 2004, Elsevier.

maximum SEY, δ_{max}, of the as-received surface is reduced by 0.4–0.6 unit. Above 10 ML, the value starts to level off around 1.7–1.9 for N_2 and CO_2. The right curve shows the evolution of the maximum SEY of condensed CO for different substrates (aluminium, copper, and electropolished copper). Again, the as-received SEY, $\delta_{max} = 1.7$–3, for the different technical surfaces, is modified as soon as a monolayer is adsorbed on the surface. Above 10 ML, the maximum SEY of condensed CO saturates at $\delta_{max} = 1.3$. For CH_4 (not shown), the SEY saturates at $\delta_{max} = 1.5$.

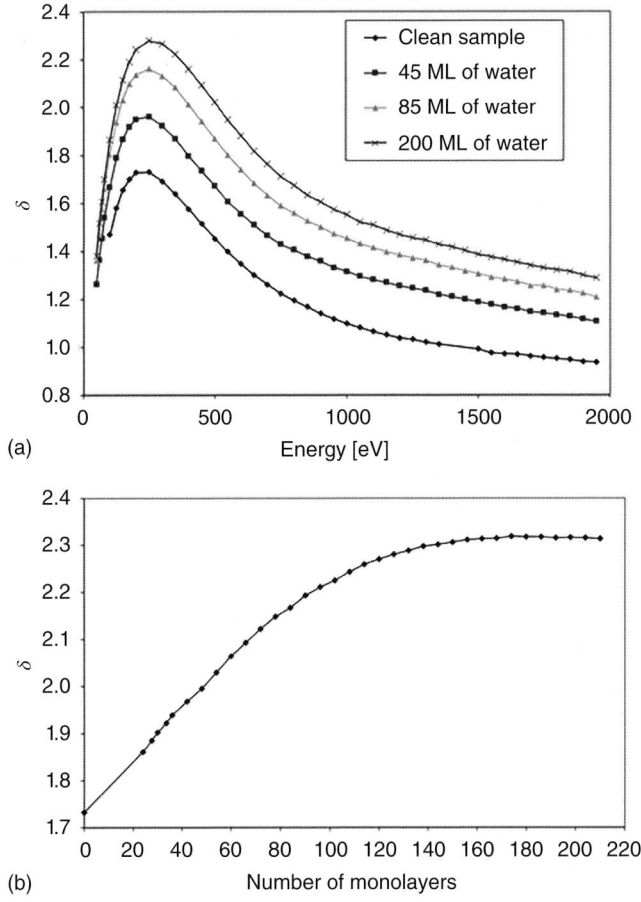

Figure 8.38 (a) SEY curve of water condensed on a sputter cleaned Cu baked sample held at 77 K. (b) Maximum SEY as a function of the number of water monolayer. Source: Reprinted with permission from Baglin et al. [154], Figs. 6 and 7. Copyright 2000, CERN.

The presence of adsorbed gas onto a surface strongly modifies also the electron-stimulated molecular desorption yield [156]. Figure 8.40 gives the ESD yield of H_2 and CO condensed on a baked Cu sample held at 4.2 K and bombarded by 300 eV electrons [157]. For a few monolayers of gas condensed on the surface, the molecular desorption yield equals 400 H_2/e and 5 CO/e. Values for CH_4 and N_2 are similar to CO, whereas CO_2 is constant along the studied range and equals 0.3 CO_2/e. When bombarded with electrons of 40 eV, the yields are roughly 1 order of magnitude lower.

Similarly to the case of photon irradiation, the values of electron desorption yields of condensed gases are much larger than the intrinsic desorption yield of the surface.

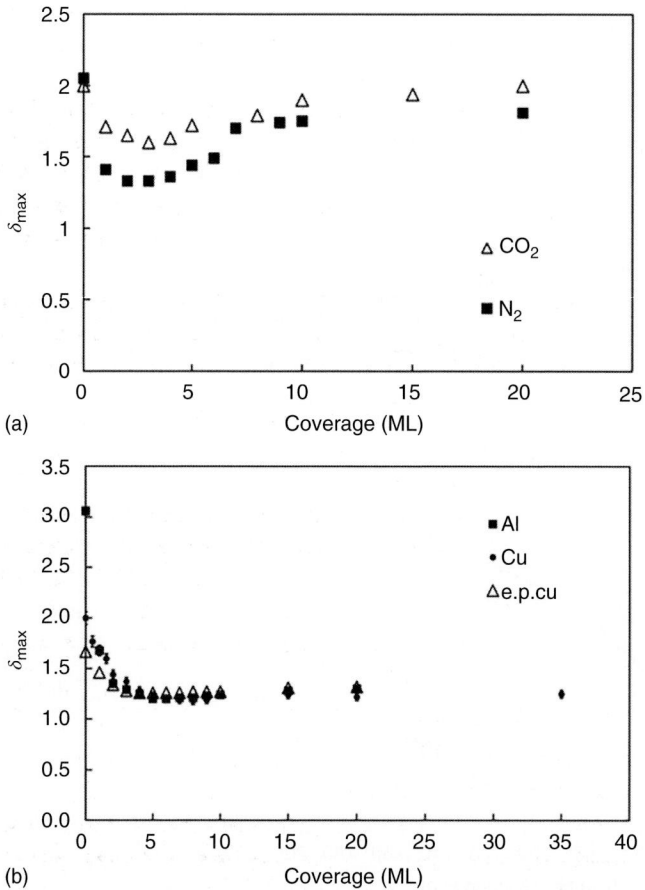

Figure 8.39 (a) Maximum SEY of CO_2 and N_2 condensed on a Cu sample held at 4.7 K. (b) Maximum SEY of technical surfaces as a function of the number of CO monolayers. Source: Reprinted with permission from Kuzucan et al. [155], Figs. 9 and 12. Copyright 2012, American Vacuum Society.

Thus, the adsorption of gas on a cryogenic surface can significantly modify the surface parameters of a vacuum system in such a way that the pressure level, the BIEM, and e-cloud phenomena are affected as illustrated in the next paragraph. When designing a vacuum system to operate at cryogenic temperature, it is therefore of primary importance that all these elements are taken into account and that all the required precautions are taken to guarantee a nominal operation of the vacuum system.

8.5.2.2 Observations with Beams

To study the impact of the e-cloud on an LHC-type vacuum system, the COLD bore EXperiment (COLDEX) was installed in a bypass of the CERN Super Proton Synchrotron. The cryostat is made of a 2.2-m-long unbaked beam screen

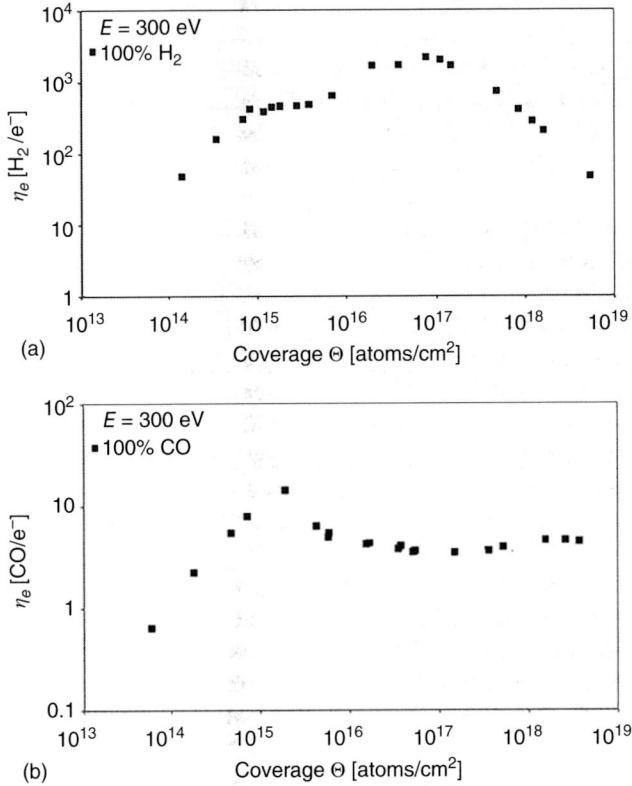

Figure 8.40 ESD yield of H_2 (a) and CO (b) condensed on a baked Cu sample held at 4.2 K. Source: Reprinted with permission from Tratnik et al. [157], Figs. 5 and 7. Copyright 2007, Elsevier.

inserted into a cold bore (Figure 8.41). To mimic an LHC-type beam screen, 1% of the 67 mm inner diameter Cu beam screen is perforated with slots. The slots are shielded to protect the cold bore from the heat load due to the e-cloud. During dedicated studies, the experimental set-up can be moved IN to let the LHC type beams circulating through the device. In the OUT position, the system can be prepared prior the study: the temperature of the beam screen and the cold bore can be selected and any condensable gas can be adsorbed onto the beam screen sample [158].

The interaction of the 25 ns spaced proton bunches with the test system is monitored with total and partial pressure gauges, electron collector, and calorimeters. The vacuum gauges are located in the middle of the cryostat at the top extremity of a room temperature chimney. The bottom extremity of the chimney is placed at less than one mm from the middle port of the beam screen. It allows the molecules desorbed from the cryogenic part to be measured by the vacuum gauges. Electron collectors are placed inside the chimney and behind the beam screen perforation. These collectors are shielded from the beam by a grid and by

(a) (b)

Figure 8.41 (a) The COLD bore EXperiment (COLDEX) installed in a bypass of the CERN SPS. (b) The Cu beam screen.

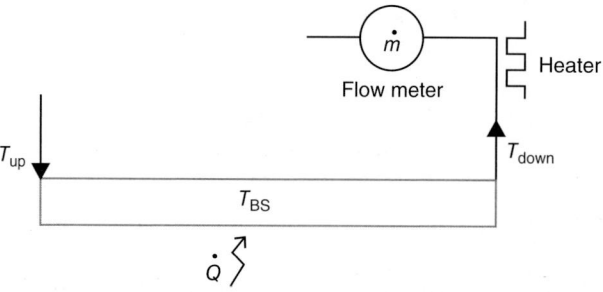

Figure 8.42 Schematic of the COLDEX beam screen calorimetric system.

the slots of the beam screen. As shown in Figure 8.42, the heat load onto the beam screen (operating at T_{BS}) is measured by temperature sensors, T_{up} and T_{down} increase, and a flow meter. Calibrated instruments are used. A heater is placed in front of the flow meter to warm up to room temperature the gaseous helium that circulates trough the beam screen. A heater wire (not shown) extended along the beam screen is used to check the calibration. With this method, beam-induced heat load above 100 mW/m can be measured.

The heat load induced by the proton beam onto the beam screen, $\dot{Q}$, is given by the following equation:

$$\dot{Q} = \dot{m}(h_{He}(T_{down}) - h_{He}(T_{up})) \tag{8.23}$$

with $\dot{m}$ as the helium mass flow and $h_{He}(T_{down})$ and $h_{He}(T_{up})$ as the helium enthalpies at the temperatures of the downstream and the upstream temperature sensors, respectively.

To evidence the BIEM and e-cloud phenomena, a typical test is used in studying the impact of the bunch current. As shown in Figure 8.43, when scanning the number of protons per bunch, a heat load is measured above a given threshold. This heat load is associated with a pressure increase and currents measured at the

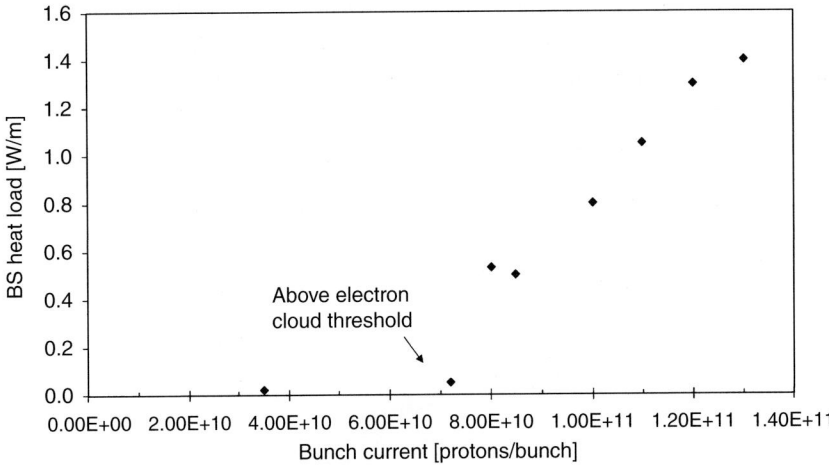

Figure 8.43 Evidence of a BIEM and electron cloud mechanism by observation of a threshold effect.

electron collectors. At 1.1×10^{11} protons per bunch, a typical pressure increase is the 10^{-8} mbar range with a collected current of 20 µA, i.e. an electron activity inside the beam screen of 25 mA/m. Using simulation codes, it is shown that the SEY corresponding to the observed heat load is in the range $\delta_{max} = 1.1$–1.2.

Other tests such as applying a solenoid field to mitigate multipacting or modifying the beam structure can also be done to demonstrate the presence of BIEM and e-cloud in a machine [21].

As explained in Chapter 7, two desorption process are observed at cryogenic temperature. The primary desorption is due to the desorption of molecules from the surface of the material, while the secondary (or recycling) desorption is due to the desorption of molecules physisorbed (or condensed) on the surface of the material.

During the irradiation process, when an equilibrium pressure P_e is reached, the primary desorption yield, η, is measured from the gas flux passing through the beam screen holes:

$$\eta = \frac{GC\Delta P_e}{\Theta} \tag{8.24}$$

where $G = 2.4 \times 10^{19}$ is a constant converting mbar l to a number of molecules, C [l/s] is the conductance of the beam screen holes, ΔP_e [mbar] is the pressure increase at equilibrium, and Θ [electron/s] is the electron flux.

During the whole electron bombardment process, the combination of the primary and recycling effects are observed. The sum of the primary and recycling desorption yields, η', divided by the sticking probability, α, is given by Eq. (8.17):

$$\frac{\eta + \eta'}{\alpha} = \frac{GS\Delta P}{\Theta} \tag{8.25}$$

where S [l/s] is the pumping speed of the beam screen surface and ΔP [mbar] the pressure increase.

If the sticking probability is known, the recycling yield can be derived from Eqs. (8.16) and (8.17). In the absence of data, it is usually assumed that $\alpha \approx 1$.

Figure 8.44 shows the measured yields when the beam screen operates at 12 K and the cold bore at 3 K [158]. The left curve shows the primary electron desorption yield of hydrogen as a function of the electron dose. The yield is comparable with the results obtained at room temperature. At a dose of 10^{19} e$^-$/cm^2 (i.e. 16 mC/mm^2), the primary desorption yield of hydrogen equals 10^{-3} H$_2$/e$^-$.

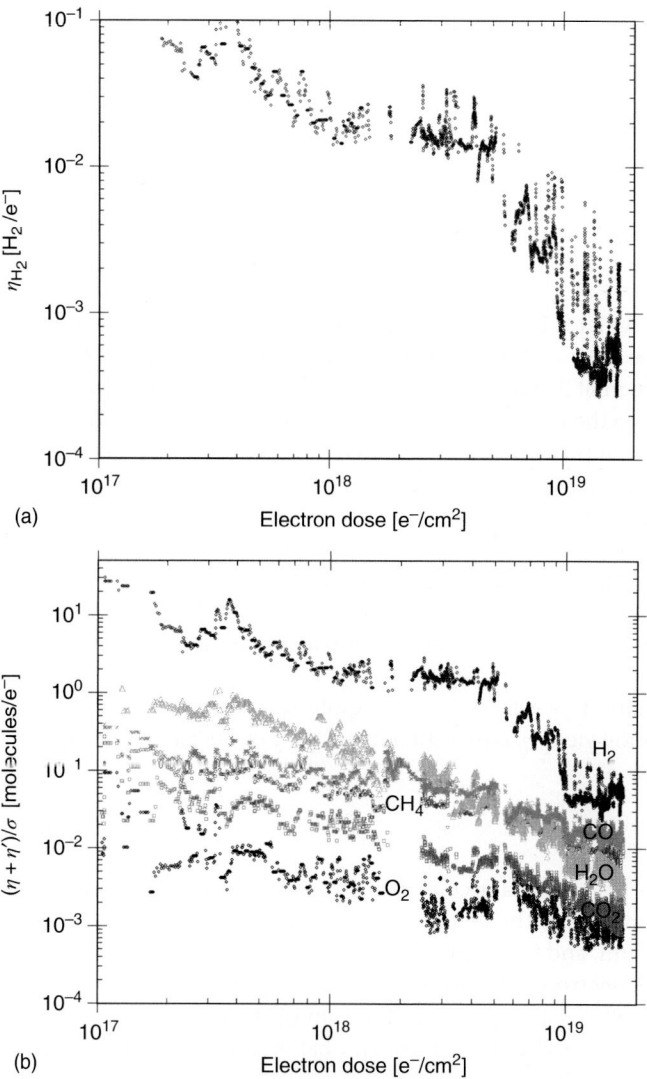

Figure 8.44 Primary ESD of hydrogen (a) and sum of the primary and recycling desorption yields divided by the sticking probability of the desorbed gases (b) as a function of the electron dose when the beam screen operates at 12 K. Source: Reprinted with permission from Baglin and Jenninger [158], Figs. 7 and 8. Copyright 2004, CERN.

8.5 BIEM and E-Cloud Observation in Machines

The right curve shows the sum of the primary and recycling desorption yields divided by the sticking probability. At a dose of 10^{19} e$^-$/cm^2, the sum of the primary and recycling desorption yields divided by the sticking probability is in the range 10^{-3} to 10^{-1} molecules/e$^-$, which corresponds to a dynamic pressure in the range 10^{-9} mbar.

Figure 8.45 shows the pressure increase due to the recycling effect when 10^{15} H$_2$/cm^2 (a) and 5×10^{15} CO/cm^2 (b) are condensed on the beam screen's

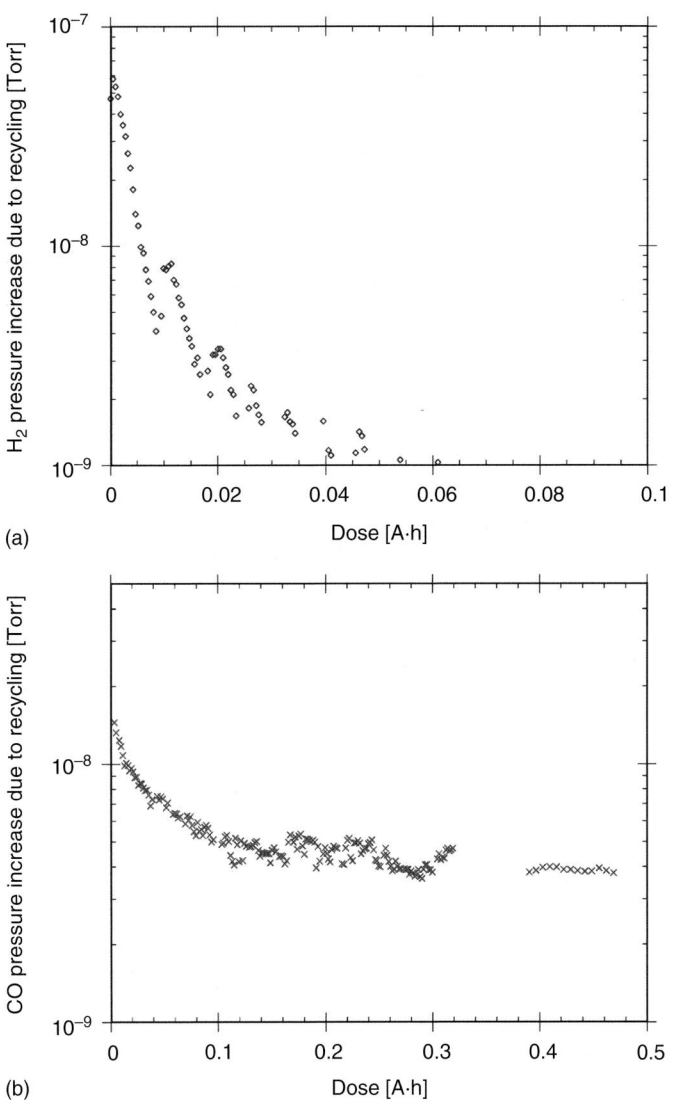

Figure 8.45 Pressure increase due to the recycling effect when 10^{15} H$_2$/cm^2 (a) and 5×10^{15} CO/cm^2 (b) are condensed on the beam screen held at 5 K. Source: Baglin and Jenninger 2004 [159], Figs. 2 and 3. Reprinted with permission of CERN.

inner surface held at 5 K [159]. Large pressure increases, in the range 10^{-8} to 10^{-7} mbar, were observed. Following Eq. (8.17), one can calculate that $\eta'_{H_2}/\alpha = 3\,H_2/e^-$ and $\eta'_{CO}/\alpha = 0.4\,CO/e^-$. A fast flushing of H_2 from the beam screen towards the cold bore within less than 0.01 A h is indeed seen, whereas a much slower flushing (more than 0.5 A h) for CO with a pressure level of $\sim 5 \times 10^{-9}$ mbar takes place. As compared to a bare surface, the heat load increase due to such surface coverage is less than 0.1 W/m.

However, increasing further the surface coverage of CO to $6.0 \times 10^{16}\,CO/cm^2$, the heat load dissipated on the beam screen rose to 6 W/m. Due to the slow flushing under electron bombardment of the CO towards the cold bore, this heat load level was maintained for a beam dose of at least 1.5 A h. The observations reveal that thick coverage of CO results in large heat load associated with a slow flushing of gas towards the cold bore.

Figure 8.46 shows the case of $1.5 \times 10^{16}\,CO_2/cm^2$ condensed onto the beam screen exposed to the electron bombardment due to BIEM and e-cloud [160]. Under electron bombardment, the condensate is cracked into CO and O_2 molecules. As shown, the gas composition is dominated by CO, whose pressure level is about 7 times larger than O_2 and CO_2. It was found that $\eta'_{CO_2}/\alpha = 0.01\,CO_2/e^-$. As compared to a bare surface, the heat load increase due to such surface coverage is less than 0.1 W/m [159].

Figure 8.47 shows the consequence of water condensed on the beam screen when exposed to BIEM and e-cloud [161]. Due to the continuous electron bombardment, the total pressure inside the beam screen decreases during the study from 10^{-4} to 10^{-6} Pa. However, during the first phase (before 100 hours), the heat load increases up to 8 W/m, while the beam screen was maintained in the range 8–20 K. This increase is attributed to the electron desorption of water from the

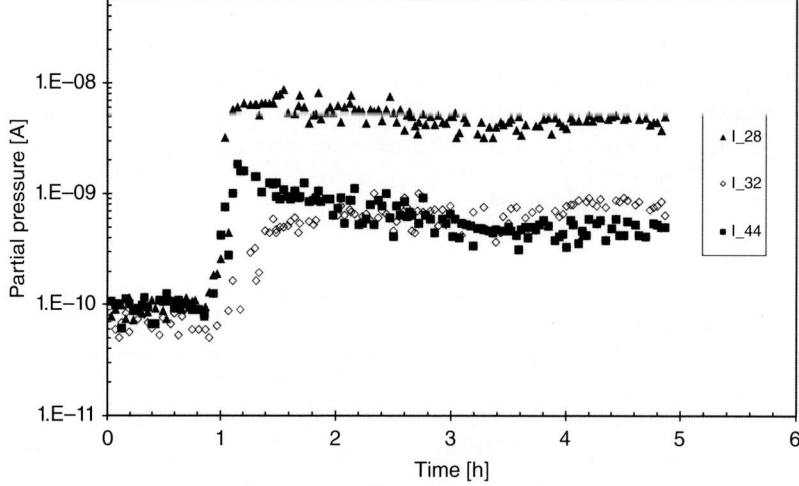

Figure 8.46 CO_2 cracking into CO and O_2 induced by BIEM and electron cloud. I_28, I_32, and I_44 in the legend correspond to the residual gas analyser (RGA) reading at 28, 32, and 44 amu.

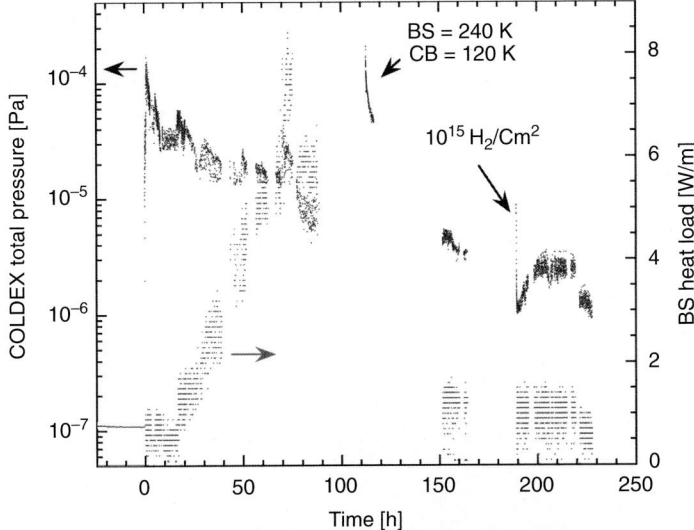

Figure 8.47 Total pressure and heat load due to BIEM and electron cloud. Source: Reprinted with permission from Baglin et al. [161], Fig. 2. Copyright 2004, Elsevier.

unbaked Cu beam screen, which is subsequently physisorbed on the surface modifying the apparent SEY as shown in Figure 8.38. The accumulation of water on the surface results from a low recycling yield, i.e. a slow flushing of the desorbed molecules by the BIEM and e-cloud towards the cold bore. Indeed, a warm-up to 240 K to remove the condensed water from the beam screen followed by a cool down (at time = 150 hours in Figure 8.47) to nominal value results in the much lower heat load of 1 W/m.

The condensation of gas can therefore strongly modify the performance of a cryogenic vacuum system subjected to BIEM and e-cloud phenomena. It can be associated with pressure rise and heat load increase. During the design and operation phase of a cryogenic vacuum system, the vacuum engineer shall aim to minimise the amount of physisorbed (or condensed) gas on the surfaces potentially exposed to electron bombardment. The next section describes the design and operational choices made the CERN LHC cryogenic vacuum system to mitigate the BIEM and e-cloud.

8.5.2.3 The CERN Large Hadron Collider Cryogenic Vacuum System

The CERN LHC is a storage ring bringing into collisions the protons at 14 TeV in their centre of mass. The superconducting machine is located underground, in a 27 km circumference tunnel, where 1232 bending dipoles of 8.33 T nominal field and 392 focusing quadrupoles of 223 T/m nominal gradient are installed. The two counter-rotating beams are located inside two-in-one magnets made of NbTi/Cu Rutherford-type cables, which operates with superfluid helium at 1.9 K. The ring comprises 8 arcs of 2.7 km each, made of basic repetitive half-cell of 53.4 m in length [162].

8 Beam-Induced Electron Multipacting, Electron Cloud, and Vacuum Design

The vacuum system consists of a seamless stainless steel cold bore at 1.9 K into which is inserted a beam screen. This system is the result of several years of studies [163–165]. The following describes the main items (Figure 8.48) functions. The beam screen intercepts the beam induced heat loads from impedance, SR, and electron cloud. As a result, the beam screens temperature along a half-cell increase from 5 to 20 K, the temperature being controlled with supercritical helium gas circulating in the cooling tubes. The racetrack shape optimises the beam aperture while leaving space for the cooling system. Sliding rings, placed every 0.75 m, ease the insertion of the 16-m-long tube into the cold bore. Pumping slots are located on top and bottom of the beam screen. The 4.4%

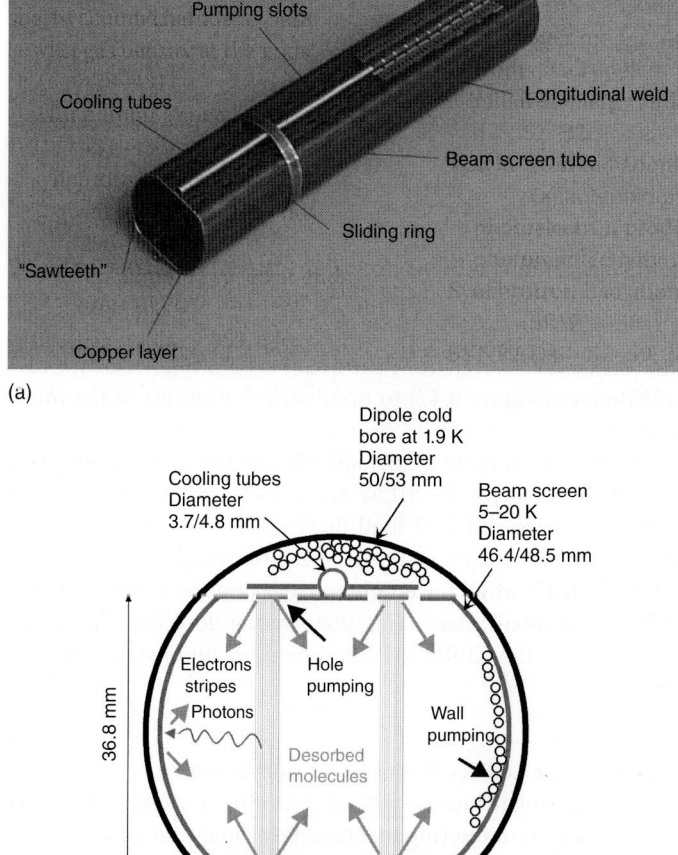

Figure 8.48 (a) Picture of an LHC beam screen tube and (b) schematic cross section of the LHC cold bore/beam screen assembly. Source: Reprinted with permission from Baglin [153], Fig. 1. Copyright 2017, Elsevier.

beam screen transparency allow to control the gas density level at the desired value. When desorbed, the molecules can be either cryosorbed on the beam screen wall or pumped through the holes towards the cold bore where the saturated vapour pressure of all gas (except helium) is negligible ($<10^{-19}$ mbar for H_2). The transverse and longitudinal random distribution size of the rounded pumping slots minimises the electromagnetic leakage towards the cold bore. The 75-μm-thick Cu layer of the inner surface of the beam screen minimises the impedance seen by the beam. In case of a transition from the superconducting to the resistive state, Eddy current circulates into the Cu material. The resulting 'quench' force of a few tons are sustained by the non-magnetic stainless steel onto which Cu is co-laminated. A 'sawtooth' pattern on the outer side of the beam screen intercepts the SR at quasi-perpendicular incidence, thereby reducing the photoelectron production and the forward scattering of the photons minimising the amount of electron available for BIEM. In dipole magnets, electron shields clamped on the cooling tube intercept the electrons, which circulate along the vertical field lines, protecting the cold bore from BIEM and e-cloud heat load.

The LHC operation started on 10 September 2008 [166]. Unfortunately, on 19 September 2008, an electrical fault provoked the damage of several magnets, whose repair took 14 months. After a thorough repair and consolidation of the vacuum system, the LHC reached its design luminosity on 26 June 2016 [21, 153, 167–172].

The first signs of BIEM and e-cloud in the LHC vacuum system were observed in autumn 2010 during physics operation. As already explained above (Figure 8.29), a solenoid wrapped around the vacuum chamber demonstrated the presence of BIEM and mitigated the e-cloud build-up [21]. Until the end of 2012, the machine mainly operated with 50 ns bunch spacing, which limited the appearance of BIEM and e-cloud in the arc to dedicated studies during which the bunch spacing was reduced to 25 ns [170, 173].

After the 2013–2014 shutdown, from 2015 onwards, the LHC operated with 25 ns bunch spacing. Therefore, BIEM and e-cloud was routinely observed in the LHC arcs. Figure 8.49 shows the reduction of the dynamic pressure (a) and of the maximum SEY (b) observed during the first years of LHC operation when increasing the total beam intensity. Thanks to this beam conditioning, the machine could reach its design luminosity in June 2016 ready for full operation [153].

However, the e-cloud phenomenon in a cryogenic machine remains highly sensitive to the surface state. In particular, the growth of gas on the cryogenic surface shall be controlled at any time. As illustrated in Figure 8.50 for the LHC case, pressure excursions along the cold beam pipe can occur under some circumstances. Figure 8.50a shows the impact of an air leak ($\sim 10^{-8}$ mbar·l/s) into the system during a month. During this period, some air can accumulate onto the beam screen resulting in an increase of the N_2 surface coverage up to a monolayer. In this example, operating the LHC with 1.5 W/m dissipated on the beam screen by the e-cloud results in a pressure rise above the magnet quench limit. Figure 8.50b illustrates the consequence of an excess of CO coverage on the beam screen. For instance, after a magnet quench, the LHC cold bore temperature is increased up to 30–40 K leading to a redistribution of the condensed gas on the surface

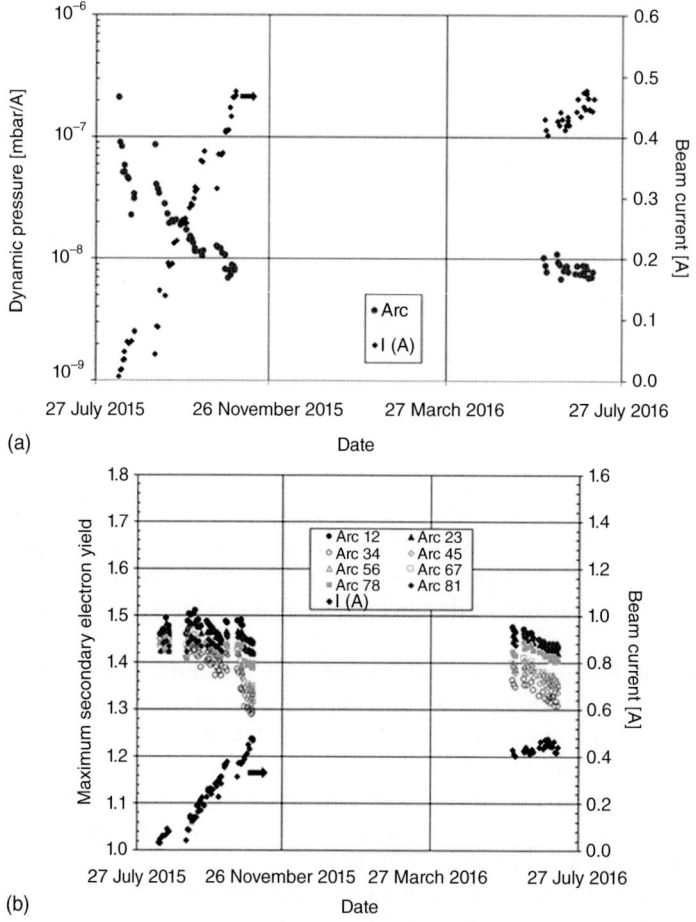

Figure 8.49 Dynamic pressure (a) and maximum SEY evolution with time (b). Source: Reprinted with permission from Baglin [153], Figs. 6 and 7. Copyright 2017, Elsevier.

exposed to the BIEM. Thus, without taking any particular precaution, successive quenches can occur when the machine is operated at its maximum performances. Indeed, an e-cloud dissipating 1.5 W/m on the beam screen cryogenic can trigger a magnet quench. As shown, a mitigation occurs to reduce the electron flux on the vacuum chamber wall. This allows flushing slowly the excess of gas towards the cold bore without the risk of triggering a quench. In this example, 20 hours is needed to flush ∼25 ML of gas towards the cold bore. A better solution consists in warming up the beam screen to thermally flush the excess of gas towards the cold bore or an external pumping system [174, 175].

This type of event was indeed observed in the sector 16L2 of LHC where an excess of condensed gas, following an uncontrolled maintenance of the vacuum

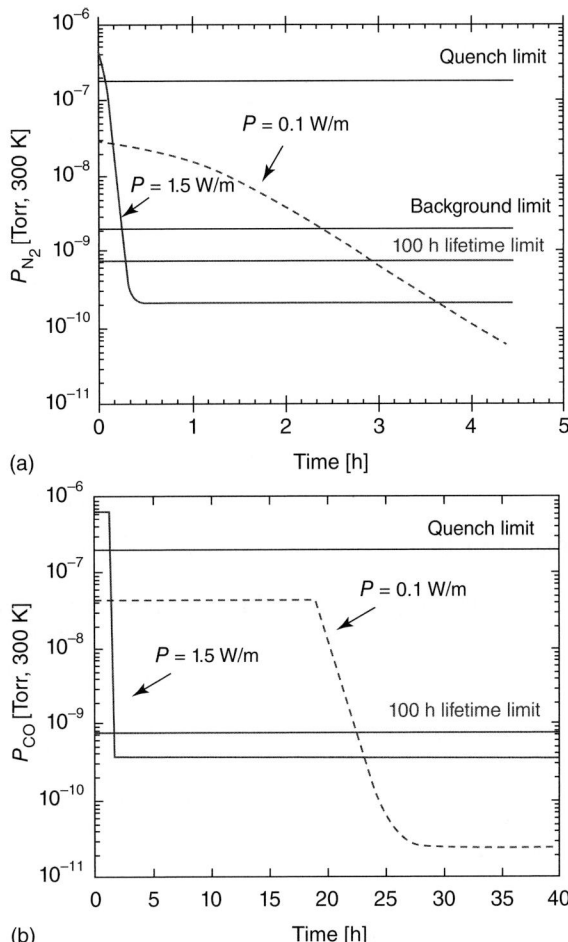

Figure 8.50 Vacuum transients due to an air leak (a) and due to large coverage of carbon monoxide (b). Source: (a) Reprinted with permission from Baglin [174], Fig. 1. Copyright 2005, CERN. (b) Reprinted with permission from Baglin [175], Fig. 7. Copyright 2004, CERN.

system, triggered magnet quenches and beam dumps. The solution consisted in using specific beams with longer spaces between bunches to reduce the BIEM and warming up the sector to allow a proper maintenance of the vacuum system [176].

8.6 Contribution of BIEM to Vacuum Stability

The effect of ion-induced pressure instability is discussed in Chapter 9. Here we would like to demonstrate how BIEM may enhance this effect.

The interaction of the e-cloud with the residual gas causes the production of ions. Well known from the vacuum expert, the hot cathode gauge, which operates in extreme high vacuum (XHV) and UHV regimes, uses this mechanism

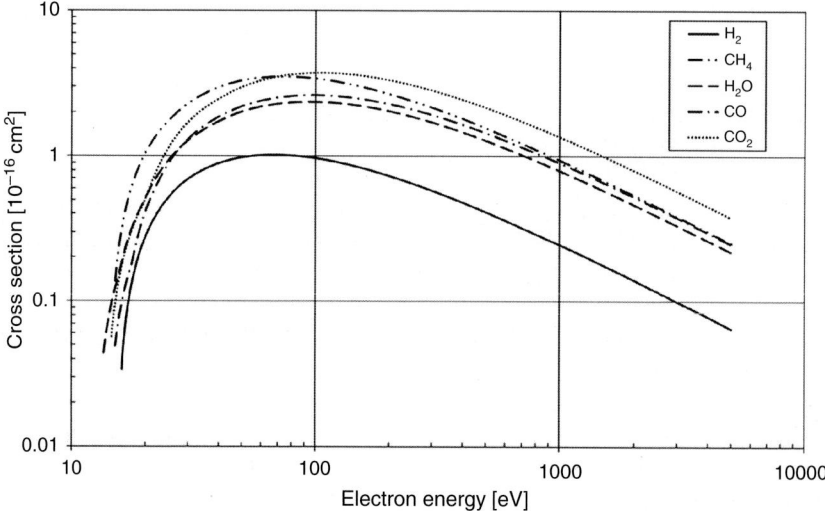

Figure 8.51 Electron ionisation cross section of some common gases.

to measure the pressure. In this gauge, the residual gas is ionised by the emitted electrons from a filament. This gauge, for which the electron kinetic energy is set to ~100 eV, is designed to optimise its sensitivity by maximising the ion production cross section according to Figure 8.51 based on data from the National Institute of Standards and Technology from atomic and molecular databases [177].

At this stage, it must be underlined that a similar range of electron kinetic energy is achieved within accelerator vacuum system where BIEM occurs!

The ion flux, I_e^+, from the residual gas ionisation process due to the electron cloud is given by

$$I_e^+ = \sigma_e \frac{P}{k_B T} L_e I_e \qquad (8.26)$$

where σ_e is the electron ionisation cross section of a given gas species, P is the pressure, T is the temperature, L_e is the ionisation path length, and I_e is the linear e-cloud flux.

In a hot cathode gauge, the typical ionisation path length is of the order of a few centimeters. However, in a synchrotron, this length might be significantly increased by several orders of magnitude and shall be computed for each case.

This ion production rate shall be compared to the ion flux produced by the circulating beam for, e.g. a proton beam. In this case, the ion flux, I_p^+, is given by Eq. (8.19):

$$I_p^+ = \sigma_p \frac{P}{k_B T} I \qquad (8.27)$$

where σ_p is the proton ionisation cross section of a given gas species and I is the proton beam current.

Figure 1.1 shows the electron and proton ionisation cross sections for typical gas species as a function of the beam energy. This cross section depends only on the velocity of the ionising particle but neither on its charge nor on its mass. Thus,

$$\frac{I_e^+}{I_p^+} = \frac{\sigma_e L_e I_e}{\sigma_p I} \tag{8.28}$$

Proton ionisation cross sections are in the 10^{-22} m² range, whereas electron ionisation cross sections are ~100 times larger. Therefore, in the case the electron ionisation path length is 1 m long and assuming a typical e-cloud flux of ~0.01 A/m, the contribution of the ion production rate by the e-cloud is very similar to the ion production rate of a 1 A circulating proton beam.

In the next chapter, we will develop in detail the interaction of the ions with the vacuum chamber wall and its interplay with the circulating beam. However, we will simply give here the result of the analysis including the ionisation rate by the e-cloud.

Under the combination of the (e-cloud and beam) ionisation processes and the ion bombardment of the vacuum chamber wall, the pressure, P, in a vacuum system obtained from the gas balance equation, is given by Eq. (8.21):

$$P = \frac{Q_0 + \eta_e k_B T \frac{I_e}{q_e}}{S_{\text{eff}} \left(1 - \frac{\eta_{\text{ion}}(\sigma_p I + \sigma_e L_e I_e)}{q_e}\right)} \tag{8.29}$$

where Q_0 is the thermal desorption rate, η_e is the ESD yield, η_{ion} is the ISD yield, and S_{eff} is the effective pumping speed.

This equation illustrates the feedback mechanism on the residual gas following the molecular desorption stimulated by the ion bombardment due to the beam–gas and e-cloud–gas ionisation processes on the residual gas. When the denominator of Eq. (8.21) approach zero, i.e. when the effective pumping speed cannot cope anymore with the ion production rate, a pressure runaway appears. Assuming a linear increase of the e-cloud flux on the vacuum chamber wall with the beam current, $I_e = aI$, where a is a coefficient, then the critical current, I_c, for which the pressure runaway becomes unstable is given by Eq. (8.22):

$$I_c = \frac{S_{\text{eff}} q_e}{\eta_{\text{ion}}(\sigma_p + \sigma_e L_e)} \tag{8.30}$$

Thus, the presence of the e-cloud reduces the critical current of a given vacuum system. In the simple case of beam stacking in a circular machine, $\alpha = 1$; therefore, the critical current is divided by ~10 for an electron ionisation path length of 10 cm.

This phenomenon was observed in RHIC [67].

8.7 Past, Present, and Future Machines

Since BIEM and e-cloud phenomena can severely limit the operation of a storage rings, the detailed understanding of the interplay of the BIEM and e-cloud with the vacuum chamber wall is of paramount importance for the vacuum scientist. Significant progresses have been achieved over the last decades, thanks

Table 8.4 Overview of BIEM and electron cloud key parameters for some past, present, and future synchrotrons around the world.

	PEP-II low energy ring	KEKB low energy ring	DAΦNE	LHC	HL-LHC	SuperKEKB	ILC damping ring	FCC-hh
Particle	e+	e+	e+	p	p	e+	e+	p
Energy [GeV]	3.1	3.5	0.51	7000	7000	4	5	50 000
Intensity [A]	2.2	1.7	1.4	0.58	1.12	3.6	0.4	0.5
Luminosity [Hz/cm^2]	3×10^{33}	2×10^{34}	5×10^{32}	1×10^{34}	5×10^{34}	80×10^{34}	n/a	5×10^{34}
Circumference [km]	2.2	3.02	0.10	26.66	26.66	3.02	3.2	97.75
Nb of bunches	1658	1284	120	2808	2748	2500	1312	10 426
Bunch population	6×10^{10}	9×10^{10}	2×10^{10}	1.15×10^{11}	2.2×10^{11}	9×10^{10}	2×10^{10}	1×10^{11}
Bunch spacing [ns]	4.2	7	2.7	25	25	4	554	25
Bunch length [ns]	0.05	0.02	0.1	0.25	0.25	0.02	0.02	0.25
Horizontal/vertical normalised emittance [μm·rad]	0.049/0.002	0.018/1.8×10^{-4}	1/0.01	3.75	2.5	0.003/10^{-5}	5.5/0.02	2.2
Electron cloud density instability threshold [m^{-3}]	1×10^{12}	4×10^{11}	1×10^{13}	5×10^{11} at 450 GeV	1×10^{12} at 450 GeV	3×10^{11}	4×10^{10}	4×10^{10} at 3 TeV
Vacuum chamber material	Al	Cu	Al	Cu co-laminated on stainless steel	Cu co-laminated on stainless steel	Cu/Al	Cu	Cu co-laminated on stainless steel
BIEM and electron cloud mitigation technique	Antechamber, TiN coating, solenoid windings	Solenoid windings	Antechamber, clearing electrodes, solenoids, beam scrubbing, machine feedback	Beam scrubbing, sawtooth, TiZrV coating	Beam scrubbing, sawtooth, TiZrV coating, a-C coating	TiN coating, grooves with TiN coating, clearing electrodes, solenoid windings, permanent magnets, antechamber	TiN coating, grooves with TiN coating, NEG coating, clearing electrodes, solenoid windings, antechamber	Laser treated surface, antechamber, sawtooth

to detailed studies, simulations, and the development of new technologies and mitigations techniques. Table 8.4 gives an overview of the BIEM and e-cloud key parameters of several machines around the world, which have dealt, are dealing, or will deal with BIEM and electron effects. And although further studies on e-cloud mitigation are continued, the modern state of knowledge and technologies provides sufficient confidence that the new machine can be designed and built to meet the specifications on the following:

- PEY and SEY to ensure operation of accelerator below the e-cloud thresholds on the e-cloud density (or the space charge).
- The BIEM power dissipation limit to ensure that cooling capacity of beam chamber of the beam screen is sufficient for keeping temperature within a design range and for the stable operation of the cryogenics.
- The gas density due to ESD is below a specified value.

Acknowledgements

The authors would like to acknowledge Dr. Mikhail Zobov and Dr. Roberto Cimino from LNF-INFN (Frascati, Italy) and Dr. Jim Crittenden from the Cornell University (Ithaca, NY, USA) for the many comments and suggestions. Authors would also like to thank Dr. Bhagat-Taaj Sian from CI (Warrington, UK) for providing the SEM images and the SEY results for a LASE sample in Figure 8.11.

References

1 Budker, G., Dimov, G., and Dudnikov, V. (1966). Experiments on production of intense proton beam by charge exchange injection method. *Proceedings of the International Symposium on Electron and Positron Storage Rings*, Saclay, France, Article no. VIII-6-1.
2 Budker, G., Dimov, G., and Dudnikov, V. (1967). Experimental investigation of the intense proton beam accumulation in storage ring by charge-exchange injection method. *Sov. At. Energy* 22: 384.
3 Budker, G., Dimov, G., Dudnikov, V., and Shamovsky, V. (1967). *Proceedings of the International Conference on High Energy Accelerators*. Cambridge, MA: CEA.
4 Martin, J.H., Winje, R.A., Hilden, R.H., and Mills, F.E. (1966). *Proceedings of the 5th International Conference on High Energy Accelerators*, Frascati. Rome: CNEN, p. 347.
5 Raka, E.C. (1967). *Proceedings of the International Conference on High Energy Accelerators*, Cambridge, MA. Cambridge, MA: CEA, p. 428.
6 Zimmermann, F. (2004). Review of single bunch instabilities driven by an electron cloud. *Phys. Rev. Spec. Top. Accel. Beams* 7: 124801.
7 Zimmermann, F. (2013). Electron-cloud effects in past & future machines – walk through 50 years of electron-cloud studies. *Proc.*

ECLOUD'12, La Biodola, Isola d'Elba, Italy, p. 9-17 https://doi.org/10.5170/CERN-2013-002.9

8 Grunder, H.A. and Lambertson, G.R. (1971). *Proceedings of the 8th International Conference on High Energy Accelerators*, Geneva. Geneva: CERN.

9 Neuffer, D., Colton, E., Fitzgerald, D. et al. (1992). Observations of a fast transverse instability in the PSR. *Nucl. Instrum. Methods Phys. Res., Sect. A* 321: 1–12.

10 Blaskiewicz, M. (1999). *Workshop on Instabilities of High Intensity Hadron Beams in Rings*, AIP Conference Proceedings No. 496 (eds. T. Roser and S.Y. Zhang). New York: AIP.

11 Fischer, W., Brennan, J.M., Blaskiewicz, M., and Satogata, T. (2002). Electron cloud measurements and simulations for the Brookhaven Relativistic Heavy Ion Collider. *Phys. Rev. Spec. Top. Accel. Beams* 5: 124401.

12 Kulikov, A., Fisher, A.S., Heifets, S. et al. (2001). The electron-cloud instability at PEP-II. *Proceedings of 2001 Particle Accelerator Conference*, p. 1903.

13 Zhang, X., Chen, A.Z., Chou, W. et al. (2007). Electron cloud studies at Tevatron and main injector. *Proceedings of 2007 Particle Accelerator Conference*, p. 3501.

14 Danilov, V. and Cousineau, S. (2006). Accumulation of high intensity beam and first observations of instabilities in the SNS accumulator ring. *Proceedings of the ICFA HB 2006 Workshop*, Tsukuba, Japan.

15 Palmer, M.A., Codner, G., Rice, D. et al. (2007). Electron cloud studies at CESR-c and Cesr-TA. *Proceedings of ECLOUD 2007, 2007-10*. Tsukuba, Japan: KEK, p. 108–113.

16 Palmer, M.A., Alexander, J., Billing, M. et al. (2009). The conversion and operation of the Cornell electron storage ring as a test accelerator (CesrTA) for damping ring research and development. *Proceedings of PAC'09 (4–8 May 2009)*, Vancouver, British Columbia, Canada, p. 4200.

17 Calder, R., Fischer, E., Gröbner, O., and Jones, E. (1974). *Proceedings of the 9th International Conference on High Energy Accelerators*, Stanford, CA, USA. Washington, DC: A.E.C., 1975, p. 70–74.

18 Cappi, R., Giovannozzi, M., Métral, E. et al. (2002). Electron cloud buildup and related instability in the CERN protron synchrotron. *Phys. Rev. Spec. Top. Accel. Beams* 5: 094401.

19 Arduini, G., Cornelis, K., Gröbner, O. et al. (2000). Electron cloud: observations with LHC-type beams in the SPS. *Proceedings of EPAC'00*, Vienna, Austria (26–30 June 2000), p. 939.

20 Dominguez, O., Li, K., Cuna, H.M. et al. (2013). First electron-cloud studies at the large Hadron Collider. *Phys. Rev. Spec. Top. Accel. Beams* 16: 011003.

21 Bregliozzi, G., Baglin, V., Chiggiato, P. et al. (2011). Observations of electron cloud effects with the LHC vacuum system. *Proceedings of IPAC'11*, San Sebastián, Spain, p. 1560.

22 Izawa, M., Sato, Y., and Toyomasu, T. (1995). The vertical instability in a positron bunched beam. *Phys. Rev. Lett.* 74: 5044.

23 Fukuma, H. (2002). 3D simulation of photoelectron cloud in KEKB LER. *Proceedings of ECLOUD'02*, Geneva (Report No. CERN-2002-001).

24 Suetsugu, Y., Shibata, K., Ishibashi, T. et al. (2017). Achievements and problems in the first commissioning of superKEKB vacuum system. *J. Vac. Sci. Technol., A* 35: 03E103.

25 Cimino, R., Drago, A., Vaccarezza, C. et al. (2005). Electron cloud build-up study for DAΦNE. *Proceedings of PAC'05*, Knoxville, TN, USA (16–20 May 2005), p. 779.

26 Wanzenberg, R. (2012). Observations of electron cloud phenomena at PETRA III. *Proceedings of ECLOUD12*, La Biodola, Isola d'Elba (5–9 June 2012), p. 89.

27 Pertica, A. and Payne, S.J. (2012). Electron cloud observations at the ISIS proton synchrotron. *Proceedings of ECLOUD12*, La Biodola, Isola d'Elba (5–9 June 2012), p. 237.

28 Zimmermann, F., Korostelev, M., Schulte, D. et al. (2005). Collective effects in the CLIC damping rings. *Proceedings of PAC'05*, Knoxville, TN, USA (16–20 May 2005), p. 1312.

29 Cuna, H.M., Contreras, J.G., and Zimmermann, F. (2012). Simulations of electron-cloud heat load for the cold arcs of the CERN Large Hadron Collider and its high-luminosity upgrade scenarios. *Phys. Rev. Spec. Top. Accel. Beams* 15: 051001.

30 Crittenden, J.A., Conway, J.V., Dugan, G.F. et al. (2014). Investigation into electron cloud effects in the International Linear Collider positron damping ring. *Phys. Rev. Spec. Top. Accel. Beams* 17: 03002.

31 Pivi, M., Raubenheimer, T., Wang, L. et al. (2006). Simulation of the electron cloud for various configurations of a damping ring for the ILC. *EPAC'06, Edinburgh*, Scotland, p. 2958.

32 Zimmermann, F., Bartmann, W, Benedikt, M et al. (2016). Beam dynamics issues in the FCC. *Proceedings of HB2016*, Malmö, Sweden (3–8 July 2016), p. 373.

33 Belli, E., Costa Pinto, P., Rumolo, G. et al. (2018). Electron cloud studies in FCC-ee. *Proceedings of IPAC2018*, Vancouver, Canada (April-29 to May 4, 2018), p. 374.

34 Zimmermann, F. (2003). Horizontal instability in the KEKB HER. *KEK* (19 October to 1 November 2002 and 16–29 March 2003).

35 Feng, B., Huang, C., Decyk, V. et al. (2009). Simulation of electron cloud effects on electron beam at ERL with pipelined QuickPIC. *AIP Conference Proceedings*, Volume 1086, p. 340. https://doi.org/10.1063/1.3080929.

36 Crittenden, J.A., Calvey, J.R., Dugan, G. et al. (2009). Studies of the effects of electron cloud formation on beam dynamics at CesrTA. *Proceedings of PAC'09*, Vancouver, British Columbia, Canada (4–8 May 2009), p. 106.

37 Iriso, U., Casalbuoni, S., Rumolo, G., and Zimmermann, F. (2009). Electron cloud simulations for ANKA. *Proceedings of PAC'09*, Vancouver, British Columbia, Canada (4–8 May 2009), p. 3321.

38 Gröbner, O. (1997). Beam induced multipacting, *Proceedings of PAC 1997*, Vancouver, Canada, p. 3589–3591.

39 Zimmermann, F. (2006). Electron-cloud benchmarking and CARE-HHH codes. *Proceedings of HB2006*, Tsukuba, Japan, THBW02.

40 Bruns, W., Schulte, D., and Zimmermann, F. (2006). Factor2: a code to simulated collective effects of electrons and ions. *EPAC'06*, Edinburgh, Scotland, p. 2242.

41 Dijkstal, P., Iadarola, G., Mether, L., and Rumolo, G. (2017). Simulation studies on the electron cloud build-up in the elements of the LHC Arcs at 6.5 TeV. *CERN-ACC-NOTE-2017-0057*. Geneva: CERN.

42 Wang, L., Chao, A., and Fukuma, H. Short bunch. *Presented at ECLOUD'04*, Napa, CA, USA, slide 11.

43 Wang, L., Chao, A., and Fukuma, H. (2004). Energy spectrum of an electron cloud with short bunch. *Proceedings of ECLOUD'04*, Napa, CA, USA, CERN-2005-001.

44 Wang, L., Blaskiewicz, M., Hseuh, H. et al. (2004). Multipacting and remedies of electron cloud in long bunch proton machine. *Proceedings of ECLOUD'04*, Napa, CA, USA, CERN-2005-001.

45 Malyshev, O.B. and Bruns, W. (2008). ILC DR vacuum design and e-cloud. *Proceedings of EPAC 2008*, Genova, Italy, p. 673–675.

46 Cimino, R. and Demma, T. (2014). Electron cloud in accelerators. *Int. J. Mod. Phys. A* 29: 1430023.

47 Cimino, R., Rumolo, G., and Zimmermann, F. (eds). (2013). *"Ecloud'12: Joint INFN-CERN-EuCARD-AccNet Workshop on Electron-Cloud Effects*, La Biodola, Isola d'Elba, Italy (5–9 June 2012), CERN–2013–002, Geneva.

48 Pivi, M.T.F., Wang, L., Demma, T. et al. (2010). Recommendation for the feasibility of more compact LC damping rings. *Proceedings of IPAC'10*, Kyoto, Japan (23–28 May 2010), p. 3578.

49 Suetsugu, Y., Shibata, K., Hisamatsu, H. et al. (2006). R&D on copper beam ducts with ante-chambers and related vacuum components. *Proceedings of EPAC'06*, Edinburgh, Scotland, p. 1438.

50 Anashin, V.V., Collins, I.R., Gröbner, O. et al. (2000). Reflection of photons and azimuthal distribution of photoelectrons in a cylindrical beam pipe. *Nucl. Instrum. Methods Phys. Res., Sect. A* 448: 76–80.

51 Anashin, V.V., Collins, I.R., Dostovalov, R.V. et al. (2001). Magnetic and electric field effect on the photoelectron emission from prototype LHC bean screen material. *Vacuum* 60: 255–260.

52 Boon, L. and Harkay, K. (2012). Chamber surface roughness and electron cloud for the Advanced photon source superconducting undulator. *Proceedings of ECLOUD12*, La Biodola, Isola d'Elba (5–9 June 2012), p. 95.

53 Anashin, V.V., Baglin, V., Cimino, R. et al. (1999). Experimental Investigations of the Electron Cloud Key Parameters. *LHC Project Report 313*. CERN, p. 8.

54 Nishiwaki, M. and Kato, S. (2010). Graphitization of inner surface of copper beam duct of KEKB positron ring. *Vacuum* 84: 743.

55 Cimino, R., Commisso, M., Grosso, D.R. et al. (2012). Nature of the decrease of the secondary electron yield by electron bombardment and its energy dependence. *Phys. Rev. Lett.* 109: 064801.

56 Yin Vallgren, C., Arduini, G., Bauche, J. et al. (2010). Amorphous carbon coatings for mitigation of electron cloud in the CERN SPS. *Proceedings of IPAC'10*, paper TUPD048.

57 Nishiwaki, M. and Kato, S. (2007). Influence of electron irradiation and heating on secondary electron yields from non-evaporable getter films observed with in situ X-ray photoelectron spectroscopy. *J. Vac. Sci. Technol., A* 25: 675.

58 Calliari, L., Filippi, M., and Laidani, N. (2004). Electron beam irradiation of hydrogenated amorphous carbon films. *Surf. Interface Anal.* 36: 1126.

59 Larciprete, R., Grosso, D.R., Commisso, M. et al. (2012). The chemical origin of SEY at technical surfaces. *Proceedings of ECLOUD12*, La Biodola, Isola d'Elba (5–9 June 2012), p. 99.

60 Anashin, V.V., Malyshev, O.B. and Pyata, E.E. (1999). Photoelectron current from the substrate with cryosorbed gases. *Vacuum Technical Note 99-04*. CERN, p. 13.

61 Anashin, V.V., Krasnov, A.A., Malyshev, O.B., and Pyata, E.E. (1999). The effect of the temperature and of a thick layer of condensed CO_2 on the photoelectron emission and on the photon reflection. *Vacuum Technical Note 99-05*. CERN, p. 10.

62 Suetsugu, Y., Kanazawa, K., Shibata, K., and Hisamatsu, H. (2006). Continuing study on the photoelectron and secondary electron yield of TiN coating and NEG (Ti-Zr-V) coating under intense photon irradiation at the KEKB positron ring. *Nucl. Instrum. Methods Phys. Res., Sect. A* 556: 399.

63 He, P., Hseuh, H.C., Mapes, M. et al. (2001). Development of titanium nitride coating for SNS ring vacuum chambers. *Proceedings of PAC'01*, p. 2159.

64 Michizono, S., Saito, Y., Suharyanto et al. (2004). Secondary electron emission of sapphire and anti-multipactor coatings at high temperature. *Appl. Surf. Sci.* 235: 227–230.

65 Suharyanto, Michizono, S., Saito, Y. et al. (2007). Secondary electron emission of TiN-coated alumina ceramics. *Vacuum* 81: 799–802.

66 Ruiz, A., Román, E., Lozano, P. et al. (2007). UHV reactive evaporation growth of titanium nitride thin films, looking for multipactor effect suppression in space applications. *Vacuum* 81: 1493–1497.

67 Fischer, W., Blaskiewicz, M., Brennan, J.M. et al. (2008). Electron cloud observations and cures in the Relativistic Heavy Ion Collider. *Phys. Rev. Spec. Top. Accel. Beams* 11: 041002.

68 Henrist, B., Hilleret, N., Scheuerlein, C., and Taborelli, M. (2001). The secondary electron yield of TiZr and TiZrV non-evaporable getter thin film coatings. *Appl. Surf. Sci.* 172: 95–102.

69 Wang, S. (2016). Secondary electron yield measurements of anti-multipacting surfaces for accelerators. PhD thesis. Loughborough University.

70 Costa Pinto, P., Calatroni, S., Chiggiato, P. et al. (2012). Carbon coating of the SPS dipole chambers. *Proceedings of ECLOUD12*, La Biodola, Isola d'Elba (5–9 June 2012), p. 141.

71 Bundaleski, N., Candeias, S., Santos, A. et al. (2012). Study of SEY degradation of amorphous carbon coatings. *Proceedings of ECLOUD12*, La Biodola, Isola d'Elba (5–9 June 2012), p. 149.

72 Fuentes, G.G., Rodríguez, R.J., García, M. et al. (2007). Spectroscopic investigations of Cr, CrN and TiCr anti-multipactor coatings grown by cathodic-arc reactive evaporation. *Appl. Surf. Sci.* 253: 7627–7631.

73 Stupakov, G. and Pivi, M. (2004). Suppression of the effective secondary emission yield for a grooved metal surface. *Proceedings of ECLOUD'04*, p. 139.

74 Shibata, K. (2012). SuperKEKB vacuum system. *Proceedings of ECLOUD12*, La Biodola, Isola d'Elba (5–9 June 2012), p. 67.

75 Malyshev, O.B., Lucas, J.M., Collomb, N. et al. (2010). Mechanical and vacuum design of the wiggler section of the ILC damping rings. *Proceedings of IPAC'10*, Kyoto, Japan (23–28 May 2010), p. 3563–3565.

76 Montero, I., Aguilera, L., Dávila, M.E. et al. (2012). Novel types of anti-e-cloud surfaces. *Proceedings of ECLOUD12*, La Biodola, Isola d'Elba (5–9 June 2012), p. 153.

77 Nistor, V., González, L.A., Aguilera, L. et al. (2014). Multipactor suppression by micro-structured gold/silver coatings for space applications. *Appl. Surf. Sci.* 315: 445–453.

78 Ye, M., He, Y.N., Hu, S.G., and Wang, R. (2013). Suppression of secondary electron yield by micro-porous array structure. *J. Appl. Phys.* 113: 074904.

79 Valizadeh, R., Malyshev, O.B., Wang, S. et al. (2014). Low secondary electron yield engineered surface for electron cloud mitigation. *Appl. Phys. Lett.* 105: 231605.

80 Valizadeh, R. and Malyshev, O. (2015). Apparatus and methods relating to reduced photoelectron yield and/or secondary electron yield. Patent publication number WO2015189645 A1.

81 Valizadeh, R., Malyshev, O.B., Wang, S. et al. (2017). Reduction of secondary electron yield for e-cloud mitigation by laser ablation surface engineering. *Appl. Surf. Sci.* 404: 370–379. https://doi.org/10.1016/j.apsusc.2017.02.013.

82 Valizadeh, R., Malyshev, O.B., Wang, S. et al. (2016). Low secondary electron yield of laser treated surfaces of copper, aluminium and stainless steel. *Proceedings of IPAC 2016*, Busan, Korea, p.1089.

83 Calatroni, S., Garcia-Tabares Valdivieso, L., Neupert, H. et al. (2017). First accelerator test of vacuum components with laser-engineered surfaces for electron-cloud mitigation. *Phys. Rev. Accel. Beams* 20: 113201.

84 Suetsugu, Y., Kanazawa, K., Shibata, K. et al. (2005). R&D of copper beam duct with antechamber scheme for high-current accelerators. *Nucl. Instrum. Methods Phys. Res., Sect. A* 538: 206–217.

85 Zolotarev, K., Malyshev, O.B., Korostelev, M. et al. (2010). SR power distribution along wiggler section of ILC DR. *Proceedings of IPAC'10*, Kyoto, Japan (23–28 May 2010), p. 3569–3571.

86 Suetsugu, Y., Shibata, K., Ishibashi, T. et al. (2016). First commissioning of the SuperKEKB vacuum system. *Phys. Rev. Accel. Beams* 19: 121001.

87 Suetsugu, Y. (2001). Observation and simulation of the nonlinear dependence of vacuum pressures on the positron beam current at KEKB. *Proceedings of PAC 2001*, Chicago, USA, p. 2180–2182.

88 Suetsugu, Y., Tanimoto, Y., Hori, Y. et al. (2001). Effects of external magnetic fields on the photoelectron emission from a copper beam chamber. *Proceedings of PAC 2001*, Chicago, USA, p. 2180–2182.

89 Wang, L., Fukuma, H., Kurokawa, S., et al. (2006). A perfect electrode to suppress secondary electrons inside the magnets. *EPAC'06*, Edinburgh, Scotland, p. 1489.

90 Zobov, M., Alesini, D., Drago, A. et al. (2012). Operating experience with electron cloud clearing electrodes at DAΦNE. *Proceedings of ECLOUD12*, La Biodola, Isola d'Elba (5–9 June 2012), p. 259.

91 Alesini, D., Drago, A., Gallo, A. et al. (2013). Dafne operation with electron-cloud-clearing electrodes. *Phys. Rev. Lett.* 110: 124801.

92 Suetsugu, Y., Fukuma, H., Wang, L. et al. (2009). Demonstration of electron clearing effect by means of a clearing electrode in a high-intensity positron ring. *Nucl. Instrum. Methods Phys. Res., Sect. A* 598: 372–378.

93 Suetsugu, Y., Kanazawa, K., Shibata, K. et al. (2012). Design and construction of the SuperKEKB vacuum system. *J. Vac. Sci. Technol., A* 30: 031602.

94 Zimmermann, F. (2000). Electron-cloud simulations for SPS and LHC. *Proceedings of Chamonix X*, Chamonix, France, CERN-SL-2000-007 DI, Geneva 2000.

95 Bartosik, H. and Rumolo, G. (2016). Beams from the injectors. *Proceedings of 7th Evian Workshop on LHC Beam Operation*, Evian Les Bains, France.

96 Hilleret, N., Baglin, V., Bojko, J. et al. (2000). The secondary electron yield of technical material and its variation with surface treatments. *Proceedings of the EPAC'00*, CERN LHC Project Report 433, Geneva, Vienna, Austria.

97 Arianer, J. (2006). To cite this version: Joël Arianer. Les sources de particules chargées. DEA. HAL Id: cel-00092960.

98 Henrist, B., Hilleret, N., Scheuerlein, C., and Taborelli, M. (2001). The secondary electron yield of TiZr and TiZrV non-evaporable getter thin film coating. *Appl. Surf. Sci.* 172: 95–102.

99 Costa-Pinto, P., Calatroni, S., Neupert, H. et al. (2013). Carbon coatings with low secondary electron yield. *Vacuum* 98: 29–36. https://doi.org/10.1016/j.vacuum.2013.03.001.

100 He, P., Hseuh, H.C., Todd, R. et al. (2004). Secondary electron emission measurements for TiN coating on the stainless steel of SNS accumulator ring vacuum chamber. *Proceedings of EPAC'04*, Lucerne, Switzerland (5–9 July 2004), p. 1804.

101 Krasnov, A.A. (2004). Molecular pumping properties of the LHC arc beam pipe and effective secondary electron emission from Cu surface with artificial roughness. *Vacuum* 73: 195–199.

102 Baglin, V. and Kos, H. (1997). Unpublished, CERN.

103 Pivi, M., King, F.K., Kirby, R.E. et al. (2008). Sharp reduction of the secondary electron emission yield from grooved surfaces. *J. Appl. Phys.* 104: 104904.

104 Sian, T., Valizadeh, R., and Malyshev, O. (2018). LASE surfaces for mitigation of electron cloud in accelerators. *Proceedings of IPAC'18*, Vancouver,

British Columbia, Canada (29 April – 4 May 2018), paper THPAL133, p. 3958.
105 Malyshev, O.B. and Valizadeh, R. (2018). Complex technological solutions for particle accelerators. *Proceedings of ECLOUD'18* (3–7 June 2018), Elba, Italy.
106 Bruining, H. (1954). *Physics and Applications of Secondary Electron Emission*. Pergamon Press.
107 Hachenberg, O. and Brauer, W. (1959). Secondary electron emission from solids. *Adv. Electron Phys.* 11: 413–499.
108 Cimino, R. and Collins, I.R. (2004). Vacuum chamber surface electronic properties influencing electron cloud phenomena. *Appl. Surf. Sci.* 235: 231–235. https://doi.org/10.1016/j.apsusc.2004.05.270.
109 Cimino, R., Gonzalez, L.A., Larciprete, R. et al. (2015). Detailed investigation of the low energy secondary electron yield of technical Cu and its relevance for the LHC. *Phys. Rev. Spec. Top. Accel. Beams* 18: 051002.
110 Furman, M.A. (1998). The Electron-Cloud Effect in the Arcs of the LHC. CERN LHC Project Report 180, 1998 or LBNL-41482/CPB note 247.
111 Rumolo, G., Ruggiero, F., and Zimmermann, F. (2001). Simulation of the electron-cloud build up and its consequences on heat load, beam stability, and diagnostics. *Phys. Rev. Spec. Top. Accel. Beams* 4: 012801.
112 Cimino, R., Collins, I.R., Furman, M.A. et al. (2004). Can low-energy electrons affect high energy physics accelerators? *Phys. Rev. Lett.* 93: 014801.
113 Kirby, R.E. and King, F.K. (2001). Secondary electron emission yields from PEP-II accelerator materials. *Nucl. Instrum. Methods Phys. Res., Sect. A* 469: 1–12.
114 Barnard, J., Bojko, I., and Hilleret, N. (1997). Measurements of secondary electron emission of some insulators. Vacuum Technical Note 97-xx, August1997, CERN, Geneva, Switzerland. arXiv:1302.2333 [physics.acc-ph].
115 Dawson, P.H. (1966). Secondary electron emission yields of some ceramics. *J. Appl. Phys.* 37: 3644–3645.
116 Hopman, H., Alberda, H., Attema, I. et al. (2003). Measuring the secondary electron emission characteristic of insulators. *J. Electron. Spectrosc. Relat. Phenom.* 131: 51–60.
117 Song, B.-P., Shen, W.-W., Mu, H.-B. et al. (2013). Measurements of secondary electron emission from dielectric window materials. *IEEE Trans. Plasma Sci.* 41: 2117.
118 Michizono, S. (2007). Secondary electron emission from alumina RF windows. *IEEE Trans. Dielectr. Electr. Insul.* 14: 583–592.
119 Gross, B. and Hessel, R. (1991). Electron emission from electron-irradiated dielectrics. *IEEE Trans. Electron. Insul.* 26: 18–25.
120 Willis, R.F. and Skinner, D.K. (1973). Secondary electron emission yield behaviour of polymers. *Solid State Commun.* 13: 685–688.
121 Burke, E.A. (1980). Secondary emission from polymers. *IEEE Trans. Nucl. Sci.* 27: 1759–1764.
122 Maslennikov, I., Turner, W., Anashin, V. et al. (1993). Photodesorption experiments on SSC collider beam tube configurations. *Proceedings of 1993*

IEEE Particle Accelerators Conference, Volume 5, Washington, DC (17–20 May 1993), p. 3876–3878.

123 Malyshev, O.B. and Collins, I.R. (2001). Dynamic Gas Density in the LHC Interaction Regions 1&5 and 2&8 For Optics Version 6.3. LHC-PROJECT-NOTE-274, December 2001, CERN, p. 34.

124 Seeman, J., Cai, Y., Clendenin, J. et al. (2000). Status report on PEP-II performance. *Proceedings of EPAC 2000*, Vienna, Austria, p. 38–42.

125 Baglin, V., Collins, I.R., Gröbner, O. et al. (2001). Measurement at EPA of vacuum and electron cloud related effect. *Proceedings of the LHC Performance Workshop 2001*, CERN-SL-2001-003, Geneva 2001, Chamonix, France.

126 Suetsugu, Y., Kanazawa, K., Shibata, K., and Hisamatsu, H. (2007). Recent study on photoelectron and secondary electron yields of TiN and NEG coating using the KEKB positron ring. *Nucl. Instrum. Methods Phys. Res., Sect. A* 578: 470–479.

127 Pivi, M.T.F., Collet, G., King, F. et al. (2010). Experimental observations of in situ secondary electron yield reduction in the PEP-II particle accelerator beam line. *Nucl. Instrum. Methods Phys. Res., Sect. A* 621: 47–56.

128 Hartung, W.H., Asner, D.M., Conway, J.V. et al. (2015). In-situ measurements of the secondary electron yield in an accelerator environment: instrumentation and methods. *Nucl. Instrum. Methods Phys. Res., Sect. A* 783: 95–109.

129 Jimenez, J.M., Henrist, B., Hilleret, N. et al. (2005). LHC and SPS electron cloud studies. *AIP Conference Proceedings*, Volume 773, p. 211.

130 Larciprete, R., Grosso, D.R., Commisso, M. et al. (2013). Secondary electron yield of Cu technical surfaces: dependence on electron irradiation. *Phys. Rev. Spec. Top. Accel. Beams* 16: 011002.

131 Rosenberg, R.A. and Harkay, K.C. (2000). A rudimentary electron energy analyser for accelerator diagnostics. *Nucl. Instrum. Methods Phys. Res., Sect. A* 453: 507–513.

132 Arduini, G., Collier, P., Dehning, B. et al. (2002). Measurement of the electron cloud properties by means of a multi-strip detector in the CERN SPS. *Proceedings of the EPAC'02*, Paris, France.

133 Harkay, K.C. and Rosenberg, R.A. (2003). Properties of the electron cloud in a high-energy positron and electron storage ring. *Phys. Rev. Spec. Top. Accel. Beams* 6: 034402.

134 Wang, J.Q., Guo, Z.Y., Liu, Y.D. et al. (2004). Electron cloud instability studies in the Beijing Electron Positron Collider. *Phys. Rev. Spec. Top. Accel. Beams* 7: 094401.

135 Iriso, U. and Fischer, W. (2005). Electron induced molecular desorption from electron clouds at the Relativistic Heavy Ion Collider. *Phys. Rev. Spec. Top. Accel. Beams* 8: 113201.

136 Macek, R.J., Browman, A.A., Ledford, J.E. et al. (2008). Electron cloud generation and trapping in a quadrupole magnet at the Los Alamos proton storage ring. *Phys. Rev. Spec. Top. Accel. Beams* 11: 010101.

137 Mahner, E., Kroyer, T., and Caspers, F. (2008). Electron cloud detection and characterization in the CERN proton synchrotron. *Phys. Rev. Spec. Top. Accel. Beams* 11: 094401. https://doi.org/10.1103/PhysRevSTAB.11.094401.

138 Pivi, M.T.F., Ng, J.S.T., Cooper, F. et al. (2010). Observation of magnetic resonances in electron clouds in a positron ring. *Nucl. Instrum. Methods Phys. Res., Sect. A* 621: 33–88.

139 Kanazawa, K., Fukuma, H., and Puneet, J. (2010). Analysis of the electron cloud density measurement with RFA in a positron ring. *Proceedings of ECLOUD10 Workshop*, Ithaca, NY, USA, p. 184.

140 Crittenden, J.A., Billing, M.G., Li, Y. et al. (2014). Shielded button electrodes for time resolved measurements of electron cloud buildup. *Nucl. Instrum. Methods Phys. Res., Sect. A* 749: 42–46.

141 Calvey, J.R., Hartung, W.H., Li, Y. et al. (2015). Measurements of electron cloud growth and mitigation in dipole, quadrupole, and wiggler magnets. *Nucl. Instrum. Methods Phys. Res., Sect. A* 770: 141–154.

142 Jimenez, J.M., Arduini, G., Collier, P. et al. (2002). Electron cloud with LHC-type beams in the SPS: a review of three years of measurements. *Proceedings of ECLOUD'02*, Geneva, Switzerland, CERN-2002-01.

143 Laurent, J.-M. and Iriso Ariz, U. (2003). Particle Collectors for Electron Cloud Studies. CERN Vacuum Technical Note 03-05, EDMS: 374712. Geneva, Switzerland: CERN.

144 Commisso, M., Demma, T., Giuducci, S. et al. (2008). A retarding field detector to measure the actual energy of electrons participating in e-cloud formation in accelerators. *Proceedings of EPAC'08*, Genoa, Italy, TUPC019.

145 Macek, R.J., Browman, A.A., Borden, M. et al. (2003). Electron cloud diagnostics in use at the Los Alamos PSR. *Proceedings of PAC'03*, Portland, OR, USA.

146 Henrist, B., Hilleret, N., and Jimenez, J.M. (2002). Electron cloud observations in the SPS. Presented at the "Mini-workshop on SPS scrubbing run results and implications for the LHC", CERN, Geneva.

147 Baglin, V. and Jenninger, B. (2003). CERN SPS electron cloud heat load measurements and simulations. *Phys. Rev. Spec. Top. Accel. Beams* 6: 063201.

148 Brüning, O., Caspers, F., Laurent, J.-M. et al. (1998). Multpacting tests with magnetic field for the LHC beam screen. *Proceedings of the EPAC'98*, Stockholm, Sweden.

149 Laurent, J.-M. and Iriso Ariz, U. (2002). Characterisation of Multipacting with a 100 MHz Resonant Cavity. CERN Vacuum Technical Note 02-12, EDMS: 3488701, Geneva, Switzerland: CERN.

150 Welch, K. (2001). *Capture Pumping Technology*. North Holland.

151 Baglin, V. (2006). Cold/sticky systems. *Proceedings of CERN Accelerator School, Vacuum in Accelerators*, Platja d'Aro, Spain, CERN-2007-003.

152 Anashin, V.V., Malyshev, O.B., and Pyata, E.E. (1999). Photoelectron Current from the Substrate with Cryosorbed Gases. CERN Vacuum Technical Note 99-04, EDMS: 678255. Geneva, Switzerland: CERN.

153 Baglin, V. (2017). The LHC vacuum system: commissioning up to nominal luminosity. *Vacuum* 138: 112–119.

154 Baglin, V., Henrist, B., Hilleret, N. et al. (2000). Ingredients for the understanding and the simulation of multipacting. *Proceedings of Chamonix X*, Chamonix, France, CERN-SL-2000-007 DI, Geneva 2000.

155 Kuzucan, A., Neupert, H., Taborelli, M., and Störi, H. (2012). Secondary electron yield on cryogenic surfaces as a function of physisorbed gases. *J. Vac. Sci. Technol., A* 30: 051401.

156 Tratnik, H. (2005). Electron stimulated desorption of condensed gas on cryogenic surfaces. PhD thesis. University Vienna, CERN-THESIS-2006-038.

157 Tratnik, H., Hilleret, H., and Störi, H. (2007). The desorption of condensed noble gas and gas mixtures from cryogenic surfaces. *Vacuum* 81: 731–737.

158 Baglin, V. and Jenninger, B. (2004). Pressure and heat load in a LHC type cryogenic vacuum system subjected to electron cloud. *Proceedings of ECLOUD'04*, Napa, CA, USA, CERN-2005-001.

159 Baglin, V. and Jenninger, B. (2004). Gas condensates onto a LHC type cryogenic vacuum system subjected to electron cloud. *Proceedings of EPAC'04*, Lucerne, Switzerland, p. 126.

160 Baglin, V. (2003). Data collected during COLDEX run #93. CERN.

161 Baglin, V., Collins, I.R., and Jenninger, B. (2004). Performance of a cryogenic vacuum system (COLDEX) with a LHC type beam. *Vacuum* 73: 201–206.

162 CERN (2004). LHC Design Rport. CERN-2004-003-V-1. Geneva.

163 Gröbner, O. (1995). Vacuum system for LHC. *Vacuum* 46: 797–801.

164 Gröbner, O. (1995). Technological problems related to the cold vacuum system of the LHC. *Vacuum* 47: 591–595.

165 Gröbner, O. (2001). Overview of the LHC vacuum system. *Vacuum* 60: 25–34.

166 Jimenez, J.M. (2010). LHC: the world's largest vacuum systems being operated at CERN. *Vacuum* 84: 2–7.

167 Baglin, V., Henrist, B., Jimenez, J.M. et al. (2010). Recovering about 5 km of LHC beam vacuum system after sector 3-4 incident. *Proceedings of IPAC'10*, Kyoto, Japan, p. 3870.

168 Bregliozzi, G., Baglin, V., and Jimenez, J.M. (2010). Summary of beam vacuum activities held during the LHC 2008-2009 shutdown. *Proceedings of IPAC'10*, Kyoto, Japan, p. 3864.

169 Baglin, V., Bregliozzi, G., Jimenez, J.M., and Lanza, G. (2011). Synchrotron radiation in the LHC vacuum system. *Proceedings of IPAC'11*, San Sebastián, Spain, p. 1563.

170 Baglin, V., Bregliozzi, G., Jimenez, J.M., and Lanza, G. (2012). Vacuum performances and lessons for 2012. *Proceedings of Chamonix X*, Chamonix, France, CERN-ATS-2012-069, Geneva 2012, p. 74.

171 Lanza, G., Baglin, V., Bregliozzi, G., and Jimenez, J.M. (2012). LHC vacuum system: 2012 review and 2014 outlook. *Proceedings of 4th Evian Workshop on LHC Operation*, Evian-les-bains, France, CERN-ATS-2013-045, Geneva 2013, p. 139.

172 Baglin, V., Bregliozzi, G., Chiggiato, P., et al. (2014). CERN vacuum system activities during the long shutdown 1: the LHC beam vacuum. *Proceedings of IPAC'14*, Dresden, Germany, p. 2375.

173 Iadarola, G., Arduini, G., Baglin, V. et al. (2013). Electron cloud and scrubbing studies for the LHC. *Proceedings of IPAC'13*, Shanghai, China, p. 1331.

174 Baglin, V. (2005). How to deal with leaks in the LHC beam vacuum. *Proceedings of Chamonix 2012 LHC Performance Workshop*, Chamonix, France, CERN-AB-2005-014, Geneva 2005, p. 105.

175 Baglin, V. (2004). Vacuum transients during LHC operation. *Proceedings of Chamonix XIII*, Chamonix, France, CERN-AB-2004-014 ADM, Geneva 2004, p. 275.

176 Jimenez, J.M., Antipov, S., Arduini, G. et al. (2018). Observations, analysis and mitigation of recurrent LHC beam dumps caused by fast losses in arc half-cell 16L2. *Proceedings of IPAC2018*, Vancouver, Canada (29 April to May 4 2018), p. 228.

177 NIST. Electron-impact cross sections for ionisation and excitation database. Data of the National Institute of Standards and Technology from atomic and molecular databases. https://www.nist.gov/pml/electron-impact-cross-sections-ionisation-and-excitation-database (accessed 15 December 2016).

9

Ion-Induced Pressure Instability

Oleg B. Malyshev[1] and Adriana Rossi[2]

[1] ASTeC, STFC Daresbury Laboratory, Keckwick Lane, Daresbury, Warrington WA4 4AD, Cheshire, UK
[2] CERN, Organisation européenne pour la recherche nucléaire, Espl. des Particules 1, 1211, Meyrin, Switzerland

9.1 Introduction

The ion-induced pressure instability and ion-stimulated desorption (ISD) have been studied since the start-up of the Ion Storage Ring (ISR) at CERN in 1971. The circulating high energy particle (protons, positrons, hadrons) can ionise the residual gas molecules inside the beam vacuum chamber. These ions, accelerated by the beam space charge, bombard the vacuum chamber walls and stimulate gas desorption leading to pressure increase (see Figure 9.1). The higher the pressure, the higher the gas ionisation rate, and consequently, the higher the intensity of ion bombardment and the further pressure increase. This self-sustaining process can lead to uncontrollable pressure rises (pressure runaway) that then cause proton (or positrons or hadrons) beam losses and possible magnet overheating.

The ion-induced pressure instability can be demonstrated by comparing pressure values in the identical electron and positron rings of the ILC damping rings as a function of the beam current (see Figure 9.2). The residual gas pressure in the electron ring is proportional to the beam current due to increasing intensity of synchrotron radiation (SR) and, therefore, due to photon stimulated desorption (PSD). The same happens in the positron ring at very low beam current. However, as the current increases, while the pressure remains proportional to the current in the electron ring, in the positron ring it rapidly increases due to ion-induced desorption until it runs away at a specific beam current called *critical current*, I_c.

The phenomenon of pressure instability in proton machines was reported in Refs. [1–5]. Very interesting cold bore experiments (COLDEXs) were performed at ISR [6, 7]. A complete study was made by W. Turner concerning the ion desorption stability in the superconducting high energy physics proton colliders [8]. The numerical estimations of ion desorption stability are given for the number of cases relevant to the Superconducting Super Collider (SSC) and Large Hadron Collider (LHC); later the studies were continued for the LHC [9–14] and the International Linear Collider Positron Damping Ring [15].

In this chapter we will demonstrate the gas dynamic model describing ion-induced pressure instability, experimental results that can be used as

Vacuum in Particle Accelerators: Modelling, Design and Operation of Beam Vacuum Systems,
First Edition. Oleg B. Malyshev.
© 2020 Wiley-VCH Verlag GmbH & Co. KGaA. Published 2020 by Wiley-VCH Verlag GmbH & Co. KGaA.

9 Ion-Induced Pressure Instability

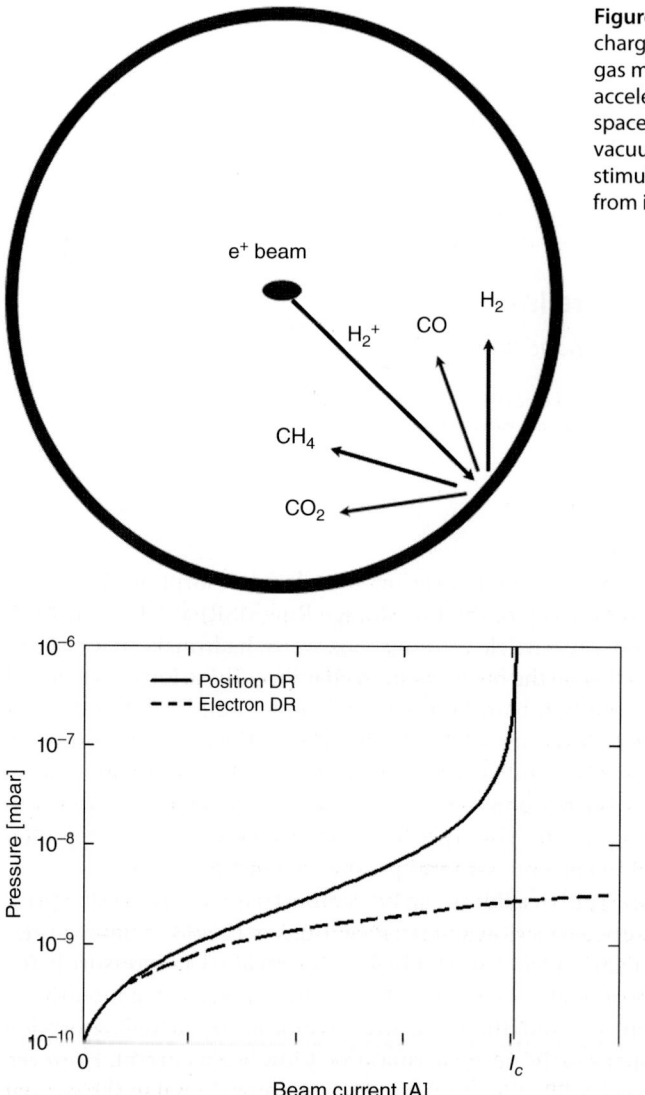

Figure 9.1 Positively charged beam can ionise a gas molecule, which is accelerated by the beam space charge, bombard the vacuum chamber walls, and stimulate the gas desorption from it.

Figure 9.2 Comparison of electron and positron dumping rings in the ILC.

the input data for modelling, and some examples for room temperature and cryogenic machines.

9.2 Theoretical

9.2.1 Basic Equations

The basic equations describing the evolution of the volumetric and surface gas density in an accelerator vacuum chamber exposed to SR were discussed in Chapter 6 for room temperature vacuum chambers and in Chapter 7 for cryogenic vacuum chambers. These equations are rewritten here to include the gas ionisation (with a cross section σ) and the ISD (χ and χ' for the primary and secondary ISD yields, correspondingly), electron-stimulated desorption (ESD) from, for example, BIEM[1] (ξ and ξ' for the primary and secondary ESD yields, correspondingly,[2] and Θ for the electron flux to the wall), and thermal desorption for room temperature systems q_t.

Thus, the equation for the volumetric gas density inside a room temperature vacuum chamber reads

$$A\frac{\partial n}{\partial t} = \chi\frac{I\sigma}{q_e}n - (\alpha S + C)n + \eta\Gamma + \xi\Theta + q_t + u\frac{\partial^2 n}{\partial z^2}, \tag{9.1}$$

and the equations for the volumetric and surface gas density inside a cryogenic temperature vacuum chamber are

$$A\frac{\partial n}{\partial t} = (\chi + \chi'(s))\frac{I\sigma}{q_e}n - (\alpha S + C)n + (\eta + \eta'(s))\Gamma + (\xi + \xi'(s))\Theta$$
$$+ \alpha S n_e(s, T) + u\frac{\partial^2 n}{\partial z^2}, \tag{9.2}$$

$$F\frac{\partial s}{\partial t} = \alpha S(n - n_e(s, T)) - \eta'(s)\Gamma - \xi'(s)\Theta - \chi'(s)\frac{I\sigma}{q_e}n, \tag{9.3}$$

where the first term on the right-hand side of Eqs. (9.1) and (9.2) accounts for the ion-induced desorption from the wall substrate and from the gas condensed on the surface, the second term accounts for the distributed pumping along the wall (where α is the sticking probability either at room temperature on surfaces like non-evaporable getter (NEG) coatings, or at cryogenic temperature, due to cryosorbing), the third term accounts for PSD from the wall and the condensed gas (we recall that symbols with apex correspond to the desorption stimulated on the condensed gas layers), the fourth term accounts for ESD from the wall and the condensed gas, the fifth term (in Eq. (9.1)) accounts for thermal surface desorption at room temperature, the fifth term (in Eq. (9.2)) indicates that the volumetric density is limited by the thermal equilibrium density n_e, and the last term accounts for the axial diffusion of gas molecules. The terms on the right-hand side of Eq. (9.3) similarly account for cryosorbing, equilibrium gas density and photon-, electron-, and ion-stimulated gas desorption of cryosorbed gas. Equations (9.1)–(9.3) are valid in an axisymmetric geometry. It was assumed that the parameters are constant in time and do not depend on the axial coordinate z.

In the quasi-static conditions with $V(\partial n/\partial t) \approx 0$ and $A(\partial s/\partial t) \neq 0$, the gas density is described (similarly to analysis in Chapters 6 and 7) by a second-order

---

1 Beam induces electron multipacting in the beam chamber; see a previous Chapter 8 for more details.
2 For the sake of clarity, in this chapter we use the ξ symbol for ESD yield, and not n_e as in Chapters 5 and 6, since we will require several apexes and indexes when going to the multigas system and multi-element vacuum chamber.

differential equation for $n(z)$:

$$u\frac{d^2n}{dz^2} - cn + q = 0; \qquad (9.4)$$

where

c is the net effect between the wall-distributed pumping speed and the ion-induced desorption:

- $c = \alpha S + C - \chi\frac{I\sigma}{q_e}$ at room temperature.
  - Here, α is an NEG coating sticking probability, S is an ideal pumping speed of cryosorbing vacuum chamber per meter of length, and C is a distributed pumping speed of pumping holes or slots per metre of length.
- $c = \alpha S + C - (\chi + \chi'(s))\frac{I\sigma}{q_e}$ at cryogenic temperature.
  - Here, α is a sticking probability of a cryogenic surface, S is an ideal pumping speed of cryosorbing vacuum chamber per metre of length, and C is a distributed pumping speed of pumping holes or slots in a beam screen per metre of length.

q is the PSD, ESD, and thermal desorption term:
- $q = \eta\Gamma + \xi\Theta + q_t$ at room temperature.
- $q = (\eta + \eta'(s))\Gamma + (\xi + \xi'(s))\Theta + \alpha S n_e(s, T)$ at cryogenic temperature.

For the following analysis, we have to consider that u is always positive; c can be positive, negative or equal to zero; and q can be positive or equal to zero. The second-order differential equation (9.4) for the function $n(z)$ has three solutions:

$$\begin{aligned} \text{Case (a) with } c > 0 \quad & n(z) = \frac{q}{c} + C_1 e^{\sqrt{\frac{c}{u}}z} + C_2 e^{-\sqrt{\frac{c}{u}}z}, \\ \text{Case (b) with } c = 0 \quad & n(z) = -\frac{q}{2u}z^2 + C_3 z + C_4 \\ \text{Case (c) with } c < 0 \quad & n(z) = \frac{q}{c} + C_5 e^{i\sqrt{\frac{|c|}{u}}z} + C_6 e^{-i\sqrt{\frac{|c|}{u}}z} \end{aligned} \qquad (9.5)$$

where the constants depend on the boundary conditions.

In order to study the evolution of the gas density within an accelerator vacuum chamber, the simplest case of the infinitely long tube is studied in Section 9.2.2. Subsequently, the boundary conditions for a finite length chamber are taken into account and their influence on the gas density analysed.

The solutions for different geometry configurations are presented in the successive sections with comments on the vacuum stability. The solutions for room and cryogenic temperatures are the same, but with different definitions of c and q. Thus we are interested in studying the vacuum chambers in the following conditions:

- In a room temperature vacuum chamber
  - With a distributed pumping speed
    - NEG-coated vacuum chamber without pumping slots (i.e. $\alpha \neq 0$, $C = 0$)
    - Vacuum chamber with pumping slots (i.e. $\alpha = 0$, $C \neq 0$)
  - Without a distributed pumping speed (i.e. $\alpha = 0$, $C = 0$)

- In a cryogenic vacuum chamber
  o Without a beam screen (i.e. $\alpha \neq 0$, $C = 0$):
    - $A\frac{ds}{dt} \neq 0$
    - $A\frac{ds}{dt} \approx 0$ (a quasi-static case)
  o With a beam screen (i.e. $\alpha \neq 0$, $C \neq 0$):
    - $A\frac{ds}{dt} \neq 0$
    - $A\frac{ds}{dt} \approx 0$ (a quasi-static case)

9.2.2 Solutions for an Infinitely Long Vacuum Chamber

The 'infinitely long' vacuum chamber approximation can be applied when the pumping effect at the ends is negligible inside the chamber. In this case, there is no net axial diffusion; thus $d^2n/dz^2 = 0$ and the solution to Eq. (9.4) is

$$n_{inf} = \frac{q}{c}; \tag{9.6}$$

which is independent of the coordinate z. Since the gas density can only be positive (or equal to zero) and finite, and the term q is positive or equal to zero, then the solution exists only for $c > 0$. Therefore we consider solution with $\alpha \neq 0$ and/or $C \neq 0$.

9.2.2.1 Room Temperature Vacuum Chamber

The gas density inside an *NEG-coated vacuum chamber* without pumping slots (i.e. $\alpha \neq 0$, $C = 0$) is given by

$$n_{inf} = \frac{\eta\Gamma + \xi\Theta + q_t}{\alpha S - \chi\frac{I\sigma}{q_e}}. \tag{9.7}$$

The effective pumping speed is decreased by the ion-induced desorption $c = \alpha S - \chi\frac{I\sigma}{q_e}$ and the gas density approaches infinity as $c \to 0$, i.e. the density is finite and positive only as long as the following condition is verified:

$$\alpha S > \chi\frac{I\sigma}{q_e}. \tag{9.8}$$

Therefore, the vacuum will be stable as long as the beam current is lower than the critical current, I_c:

$$I_c = \frac{\alpha S q_e}{\chi\sigma}. \tag{9.9}$$

The gas density inside *a vacuum chamber with pumping slots* (i.e. $\alpha = 0$, $C \neq 0$) is given by

$$n_{inf} = \frac{\eta\Gamma + \xi\Theta + q_t}{C - \chi\frac{I\sigma}{q_e}}. \tag{9.10}$$

The gas density stability in this case is

$$C > \chi\frac{I\sigma}{q_e}. \tag{9.11}$$

The critical current is given by

$$I_c = \frac{Cq_e}{\chi\sigma}. \tag{9.12}$$

9.2.2.2 Cryogenic Vacuum Chamber

The gas density inside *a vacuum chamber without beam screen* (i.e. $\alpha \neq 0$, $C = 0$) is given by

$$n_{\inf} = \frac{(\eta + \eta'(s))\Gamma + (\xi + \xi'(s))\Theta + \alpha' Sn_e(s,T)}{\alpha S - (\chi + \chi'(s))\dfrac{I\sigma}{q_e}}. \tag{9.13}$$

The secondary photon and ion-induced desorption and the thermal equilibrium gas density depend implicitly on the surface density s of cryosorbed molecules:

$$s(t) = s_0 + \frac{1}{A}\int_{t=0}^{t}\left(\eta\Gamma + \chi\frac{I\sigma}{q_e}n(t)\right)dt. \tag{9.14}$$

Hence, the gas density inside the vacuum chamber will increase with the accumulated photon dose due to both photon and ion-induced desorptions.

The gas density stability in this case is

$$\alpha S > (\chi + \chi')\frac{I\sigma}{q_e}. \tag{9.15}$$

The critical current, I_c, is

$$I_c = \frac{\alpha S}{(\chi + \chi')\dfrac{\sigma}{q_e}}. \tag{9.16}$$

The gas density inside *a vacuum chamber with a beam screen* (i.e. $\alpha \neq 0$, $C \neq 0$) is given by

$$n_{\inf} = \frac{(\eta + \eta'(s))\Gamma + (\xi + \xi'(s))\Theta + \alpha Sn_e(s,T)}{\alpha S + C - (\chi + \chi'(s))\dfrac{I\sigma}{q_e}}. \tag{9.17}$$

Since the molecules can escape from the beam chamber through the pumping holes, the growth of the surface density, s, on the beam screen surface is limited by the distributed pumping, C, and is described by

$$s(t) = s_0 + \frac{1}{A}\int_{t=0}^{t}\left[\eta\Gamma + \xi\Theta + \left(\chi\frac{I\sigma}{q_e} - C\right)n(t)\right]dt. \tag{9.18}$$

As in the previous case, the effective pumping speed of the vacuum chamber with a beam screen is decreased by the ion-induced desorption and the stability condition is

$$\alpha S + C > (\chi + \chi')\frac{I\sigma}{q_e}. \tag{9.19}$$

The critical current is given by

$$I_c = \frac{(\alpha S + C)q_e}{(\chi + \chi')\sigma}. \tag{9.20}$$

9.2 Theoretical

The expression for the gas density inside an infinitely long vacuum chamber with a beam screen can also be written in another form:

$$n_{\text{inf}} = \frac{\eta\Gamma + \xi\Theta - A\frac{ds}{dt}}{C - \chi\frac{I\sigma}{q_e}}. \tag{9.21}$$

In the *quasi-static case*, when the condition $A(\partial s/\partial t) \approx 0$ is satisfied, the gas density depends only on η and χ:

$$n_{\text{inf}} = \frac{\eta\Gamma + \xi\Theta}{C - \chi\frac{I\sigma}{q_e}}. \tag{9.22}$$

This equation is the same as Eq. (9.7); therefore, the stability condition and the critical current in this case are also described with Eqs. (9.8) and (9.9), respectively.

9.2.2.3 Summary for an Infinitely Long Vacuum Chamber

The results for the infinity long tube are summarised in Table 9.1. The higher the critical current, the safer the operation of the accelerators. In all studied conditions the critical current I_c increases with a distributed pumping speed ($\alpha S + C$) and decreases with gas ionisation cross section and ISD yield. This naturally suggests that mitigation of ion-induced instability go by choosing materials and surface treatments of vacuum chamber with lowest ISD yield and increasing pumping speed.

Table 9.1 Gas density, stability criteria, and critical current for an infinitely long vacuum chamber.

Conditions	n_{inf} [molecules/m³]	Stability criteria	Critical current, I_c
At room temperature			
$\alpha \neq 0$, $C = 0$	$\dfrac{\eta\Gamma + \xi\Theta + q_t}{\alpha S - \chi\frac{I\sigma}{q_e}}$	$\alpha S > \chi\dfrac{I\sigma}{q_e}$	$I_c = \dfrac{\alpha S q_e}{\chi\sigma}$
$\alpha = 0$, $C \neq 0$	$\dfrac{\eta\Gamma + \xi\Theta + q_t}{C - \chi\frac{I\sigma}{q_e}}$	$C > \chi\dfrac{I\sigma}{q_e}$	$I_c = \dfrac{C q_e}{\chi\sigma}$
At cryogenic temperature			
$\alpha \neq 0$, $C = 0$, $A\dfrac{ds}{dt} \neq 0$	$\dfrac{(\eta + \eta'(s))\Gamma + (\xi + \xi'(s))\Theta + \alpha Sn_e(s, T)}{\alpha S + C - (\chi + \chi'(s))\frac{I\sigma}{q_e}}$	$\alpha S > (\chi + \chi')\dfrac{I\sigma}{q_e}$	$I_c = \dfrac{\alpha S q_e}{(\chi + \chi')\sigma}$
$\alpha = 0$, $C \neq 0$, $A\dfrac{ds}{dt} \neq 0$	$\dfrac{(\eta + \eta'(s))\Gamma + (\xi + \xi'(s))\Theta + \alpha Sn_e(s, T)}{\alpha S + C - (\chi + \chi'(s))\frac{I\sigma}{q_e}}$	$\alpha S + C > (\chi + \chi')\dfrac{I\sigma}{q_e}$	$I_c = \dfrac{(\alpha S + C)q_e}{(\chi + \chi')\sigma}$
$\alpha = 0$, $C \neq 0$, $A\dfrac{ds}{dt} \approx 0$	$\dfrac{\eta\Gamma + \xi\Theta}{C - \chi\frac{I\sigma}{q_e}}$	$C > \chi\dfrac{I\sigma}{q_e}$	$I_c = \dfrac{C q_e}{\chi\sigma}$

From the stability criteria ($c > 0$), a particle accelerator could not operate with a beam current $I \geq I_c$. However, in a design of the vacuum system for an accelerator, it is important to consider that in an infinitely long vacuum chamber even at $I = 0.5\, I_c$, the ion-induced desorption increases the gas density by a factor of 2. Moreover, the ISD yield measured in experiments could be different from one on real vacuum chamber wall. Thus, the ideal operational condition is $I \ll I_c$. If this is impossible to reach, then the authors would advise to aim the vacuum system design with a criterion at $I_{max} \leq 0.5\, I_c$.

9.2.3 Short Vacuum Chamber

In this book, one refers to 'short' vacuum chamber when the conditions at the extremities (boundary conditions) of the chamber have an influence on the gas density along the whole length of the chamber. In this chapter, solutions for gas density, stability criteria, and critical current are studied for different boundary conditions, i.e. known gas density or pumping speed at the ends. The schematic diagram of the layout of vacuum chamber is shown in Figure 9.3. The solutions will be shown for the most common case of q and c, while the specific cases are summarised in the tables.

9.2.3.1 Solution for a Short Vacuum Chamber with a Given Gas Density at the Ends

This case can be used for both the operations of the LHC or in the case of a laboratory experiment like the COLDEX experiment [16]. Consider a vacuum chamber centered at $z = 0$ and length L and density $n(-L/2) = n_1$ and $n(L/2) = n_2$.

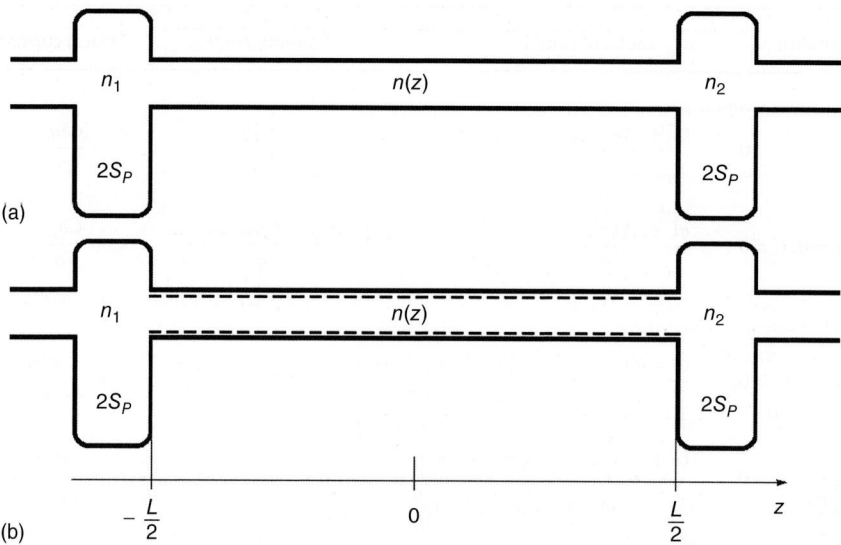

Figure 9.3 A layout of a vacuum chamber without a beam screen (a) and with a beam screen (b) between two pumps with known pumping speeds and known gas density at the extremes.

As discussed previously, there are three different solutions for the gas density, depending on the value of the parameter c, given by Eq. (9.5):

Case (a): $c > 0$

$$C_1 = \frac{n_1 + n_2 - 2n_{\text{inf}}}{4\cosh(\omega_a L/2)} + \frac{n_2 - n_1}{4\sinh(\omega_a L/2)};$$

$$C_2 = \frac{n_1 + n_2 - 2n_{\text{inf}}}{4\cosh(\omega_a L/2)} + \frac{n_1 - n_2}{4\sinh(\omega_a L/2)}$$

For $n_1 = n_2$ the expression for the gas density may be written as

$$n(z) = n_{\text{inf}} - (n_{\text{inf}} - n_1)\frac{\cosh(\omega_a z)}{\cosh(\omega_a L/2)}. \tag{9.23}$$

where
- n_{inf} is given by Eqs. (9.7) and (9.13) at room temperature and by Eq. (9.17) at cryogenic temperature

$$\omega_a = \sqrt{\left(\alpha S + C - (\chi + \chi')\frac{I\sigma}{q_e}\right)\frac{1}{u}}$$

- in the quasi-static case with $A\frac{ds}{dt} \approx 0$ at cryogenic temperature n_{inf} is given by Eq. (9.22) and $\omega_a = \sqrt{\left(C - \chi\frac{I\sigma}{q_e}\right)\frac{1}{u}}$

This solution is always stable in the interval of existence: i.e. $I < I_c$ for $c > 0$.

Case (b): $c = 0$

The expression for the gas density is

$$n(z) = \frac{(\eta + \eta')\Gamma + (\xi + \xi')\Theta + \alpha S n_e}{2u}\left(\left(\frac{L}{2}\right)^2 - z^2\right) + \frac{n_1 - n_2}{2L}z + \frac{n_1 + n_2}{2}. \tag{9.24}$$

At cryogenic temperature in the quasi-static case and at room temperature, it can be simplified to

$$n(z) = \frac{\eta\Gamma + \xi\Theta + q_t}{2u}\left(\left(\frac{L}{2}\right)^2 - z^2\right) + \frac{n_1 - n_2}{2L}z + \frac{n_1 + n_2}{2}. \tag{9.25}$$

This solution is always stable and corresponds to a parabolic pressure profile.

Case (c): $c < 0$

$$C_5 = \frac{n_1 + n_2 - 2n_{\text{inf}}}{4\cos(\omega_b L/2)} + \frac{n_2 - n_1}{4\sin(\omega_b L/2)};$$

$$C_6 = \frac{n_1 + n_2 - 2n_{\text{inf}}}{4\cos(\omega_b L/2)} + \frac{n_1 - n_2}{4\sin(\omega_b L/2)}$$

For $n_1 = n_2$ the expression for the gas density may be written as

$$n(z) = (n_{\text{inf}} + n_1)\frac{\cos(\omega_b z)}{\cos(\omega_b L/2)} - n_{\text{inf}}, \tag{9.26}$$

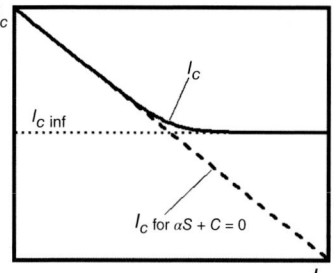

Figure 9.4 Critical current as a function of the length of vacuum chamber (shown with log–log axis).

where $\omega_b = \sqrt{\left((\chi + \chi')\frac{I\sigma}{q_e} - (\alpha S + C)\right)\frac{1}{u}}$. The gas density stability in this case is

$$\cos\left(\frac{\omega_b L}{2}\right) > 0 \Rightarrow \left|\frac{\omega_b L}{2}\right| < \frac{\pi}{2} \Rightarrow |\omega_b| < \frac{\pi}{L}. \tag{9.27}$$

This leads to the critical current in the common form given by

$$I_c = \left(\alpha S + C + \frac{\pi^2 u}{L^2}\right) \frac{q_e}{(\chi + \chi')\sigma}. \tag{9.28}$$

Schematically, the dependence of the critical current from the length of vacuum chamber, $I_c(L)$, is shown in log–log scale in Figure 9.4. One can see that the critical current decreases with the square of the length of vacuum chamber. However, for vacuum chambers with distributed pumping speed ($\alpha S + C > 0$), the critical current could not be lower than the critical current of infinity long tubes.

For a specified maximum operation beam current in a machine, I_s, we can define a maximum length of vacuum chamber, L_{max}:

$$L_{max} = \pi \sqrt{\frac{u}{\frac{(\chi + \chi')\sigma I_s}{q_e} - (\alpha S + C)}}. \tag{9.29}$$

The gas density (or pressure) remains stable when the vacuum chamber length is below L_{max}. For vacuum chambers with distributed pumping speed ($\alpha S + C > 0$), if a criterion $(\chi + \chi')\sigma I/q_e < (\alpha S + C)$ is met then $L_{max} = \infty$.

The specific solutions for the critical current in various conditions are shown in Table 9.2. The critical current I_c increases with a distributed pumping speed $\alpha S + C$ and vacuum conductance of the beam chamber u, and it decreases with gas ionisation cross section σ, the ISD yields $(\chi + \chi')$, and the length of vacuum chamber L. Thus, in the solution for a short vacuum chamber with a given pressure at the ends, the critical current I_c is higher than for an infinitely long vacuum chamber due to the gas diffusion to the end of vacuum chamber. The stability of vacuum chamber without distributed pumping (when $\alpha = 0$ and $C = 0$) is provided by vacuum conductance of the beam chamber.

Table 9.2 Critical current I_c for a short vacuum chamber with a given gas density at the ends or $\hat{I}_c$ for the infinitely large pumping speed at the ends.

Conditions	Critical current, I_c or $\hat{I}_c$	
	Room temperature	Cryogenic temperature
$\alpha \neq 0,\ C \neq 0,\ A\dfrac{\partial s}{\partial t} \neq 0$	$\left(\alpha S + C + \dfrac{\pi^2 u}{L^2}\right)\dfrac{q_e}{\chi\sigma}$	$\left(\alpha S + C + \dfrac{\pi^2 u}{L^2}\right)\dfrac{q_e}{(\chi + \chi')\sigma}$
$\alpha = 0,\ C \neq 0,\ A\dfrac{\partial s}{\partial t} \neq 0$	$\left(C + \dfrac{\pi^2 u}{L^2}\right)\dfrac{q_e}{\chi\sigma}$	$\left(C + \dfrac{\pi^2 u}{L^2}\right)\dfrac{q_e}{(\chi + \chi')\sigma}$
$\alpha \neq 0,\ C = 0,\ A\dfrac{\partial s}{\partial t} \neq 0$	$\left(\alpha S + \dfrac{\pi^2 u}{L^2}\right)\dfrac{q_e}{\chi\sigma}$	$\left(\alpha S + \dfrac{\pi^2 u}{L^2}\right)\dfrac{q_e}{(\chi + \chi')\sigma}$
$\alpha = 0,\ C = 0,\ A\dfrac{\partial s}{\partial t} \neq 0$	$\dfrac{\pi^2 q_e u}{\chi\sigma L^2}$	$\dfrac{\pi^2 q_e u}{(\chi + \chi')\sigma L^2}$
$\alpha = 0,\ C \neq 0,\ A\dfrac{\partial s}{\partial t} = 0$	—	$\left(C + \dfrac{\pi^2 u}{L^2}\right)\dfrac{q_e}{\chi\sigma}$
$\alpha = 0,\ C = 0,\ A\dfrac{\partial s}{\partial t} = 0$	—	$\dfrac{\pi^2 q_e u}{\chi\sigma L^2}$

9.2.3.2 Solution for a Short Vacuum Chamber with a Given Pumping Speed at the Ends

Consider a vacuum chamber of length L centered at $z = 0$, with pumps at the two ends of pumping speed S_p. The conditions at the ends are

$$n(\pm L/2) = \mp \frac{dn(\pm L/2)}{dz}\frac{u}{S_p}$$

In the **case (a)** when $c > 0$, the gas density $n(z)$ is given by

$$n(z) = n_{\inf}\left[1 - \frac{\cosh(\omega_a z)}{\cosh\left(\dfrac{\omega_a L}{2}\right)\left(1 + \dfrac{u}{S_p}\omega_a \tanh\left(\dfrac{\omega_a L}{2}\right)\right)}\right]. \quad (9.30)$$

It is useful to calculate the average value of the gas density in the vacuum chamber of length L:

$$\langle n(L)\rangle = n_{\inf}\left[1 - \frac{2\tanh\left(\dfrac{\omega_a L}{2}\right)}{\dfrac{\omega_a L}{2}\left(1 + \dfrac{u}{S_p}\omega_a \tanh\left(\dfrac{\omega_a L}{2}\right)\right)}\right]. \quad (9.31)$$

The gas density is always finite in range of values of c (i.e. $c > 0$).

In the **case (b)** when $c = 0$, the gas density $n(z)$ is given by

$$n(z) = ((\eta + \eta')\Gamma + (\xi + \xi')\Theta + \alpha S n_e)\left[\frac{1}{2u}\left(\left(\frac{L}{2}\right)^2 - z^2\right) + \frac{L}{2S_p}\right]. \tag{9.32}$$

At cryogenic temperature in the quasi-static case and at room temperature, it can be simplified to

$$n(z) = (\eta\Gamma + \xi\Theta + q_t)\left[\frac{1}{2u}\left(\left(\frac{L}{2}\right)^2 - z^2\right) + \frac{L}{2S_p}\right]. \tag{9.33}$$

This solution is always stable and corresponds to a parabolic pressure profile.

In the **case (c)** when $c < 0$, the gas density $n(z)$ is given by

$$n(z) = n_{\inf}\left(\frac{\cos(\omega_b z)}{\cos\left(\frac{\omega_b L}{2}\right)\left(1 - \frac{u}{S_p}\omega_b \tan\left(\frac{\omega_b L}{2}\right)\right)} - 1\right). \tag{9.34}$$

The average value of the gas density is

$$\langle n(L) \rangle = n_{\inf}\left(\frac{\tan\left(\frac{\omega_b L}{2}\right)}{\frac{\omega_b L}{2}\left(1 - \frac{u}{S_p}\omega_b \tan\left(\frac{\omega_b L}{2}\right)\right)} - 1\right). \tag{9.35}$$

The stability conditions in the case (c) are

$$\begin{cases} \left|\frac{\omega_b L}{2}\right| < \frac{\pi}{2} \\ \frac{u}{S_p}\omega_b \tan\left(\frac{\omega_b L}{2}\right) < 1. \end{cases} \tag{9.36}$$

and the critical current I_c is the smallest root of two equations:

$$|\omega_b| = \frac{\pi}{L} \tag{9.37}$$

and

$$\omega_b \tan\left(\frac{\omega_b L}{2}\right) = \frac{S_p}{u} \tag{9.38}$$

Schematically, the dependence of the critical current from pumping speed at the extremes, $I_c(S_p)$, is shown in log–log scale in Figure 9.5. One can see that for small pumping speeds, the higher pumping speed of the pumps, the higher critical current, but the critical current is limited by a value $\hat{I}_c$ corresponding to the infinity large pumping speed:

$$I_c \xrightarrow{S_p \to \infty} \hat{I}_c. \tag{9.39}$$

The critical current for the infinity large pumping speed $\hat{I}_c$ corresponds to a first condition in Eq. (9.36), which is exactly the same as Eq. (9.27). Thus its solution

Figure 9.5 Critical current as a function of pumping speed at the end of pumping speed (shown with log–log axis).

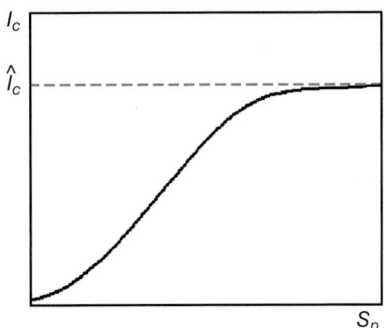

similar to Eq. (9.28):

$$\hat{I}_c = \frac{\frac{\pi^2 u}{L^2} + \alpha S + C}{(\chi + \chi')\sigma} q_e. \tag{9.40}$$

The specific solutions for the critical current for the infinity large pumping speed $\hat{I}_c$ in various conditions are shown in Table 9.2.

It is more complicated to find a solution for Eq. (9.38), because it is a transcendental equation for ω_b that can be solved either numerically or by applying the power series such as the Taylor or Maclaurin series. Thus, the Maclaurin series for $\tan(x)$ is

$$\tan(x) = x + \frac{1}{3}x^3 + \frac{2}{15}x^5 + \frac{17}{315}x^7 + \frac{62}{2835}x^9 + \cdots \quad \text{for } |x| < \frac{\pi}{2}. \tag{9.41}$$

Only the first term of the Maclaurin series can be used for a narrower domain: $\tan(x) = x$ with an accuracy of 0.8% for $|x| < \pi/20$, 3.3% for $|x| < \pi/10$, and 9.3% for $|x| < \pi/6$.

Considering that experimental accuracy of ISD yield is greater than 20%, an approximation $\tan(\omega_b L/2) \approx \omega_b L/2$ provides tolerable accuracy of 10% for a domain $|\omega_b L/2| \le 0.5$, which is written for the common form of c as

$$(\chi + \chi')\frac{I\sigma}{q_e} < \alpha S + C + \frac{u}{8L^2}. \tag{9.42}$$

One can see from this inequality that the first term of the Maclaurin series can be used for large distributed pumping ($\alpha S + C$) or for large vacuum conductance u and short vacuum chambers with length L. It can be easily shown that the condition (8.41) is always met when $u/L > 16S_p$, i.e. for large vacuum conductance and small pumping speed and the ends.

Having in mind these conditions, Eq. (9.38) can be simplified to

$$\omega_b^2 = \frac{2S_p}{uL} \tag{9.43}$$

In the common form of c, this equation is written as

$$(\chi + \chi')\frac{I\sigma}{q_e} - (\alpha S + C) = \frac{2S_p}{L} \tag{9.44}$$

Table 9.3 Critical current for a short vacuum chamber with a given pumping speed at the ends at the condition $u/L > 16S_p$.

Conditions	Critical current, I_c	
	Room temperature	Cryogenic temperature
$\alpha \neq 0, C \neq 0, A\dfrac{\partial s}{\partial t} \neq 0$	$\left(\alpha S + C + u\dfrac{2S_p}{L}\right)\dfrac{q_e}{\chi\sigma}$	$\left(\alpha S + C + \dfrac{2S_p}{L}\right)\dfrac{q_e}{(\chi + \chi')\sigma}$
$\alpha = 0, C \neq 0, A\dfrac{\partial s}{\partial t} \neq 0$	$\left(C + \dfrac{2S_p}{L}\right)\dfrac{q_e}{\chi\sigma}$	$\left(C + \dfrac{2S_p}{L}\right)\dfrac{q_e}{(\chi + \chi')\sigma}$
$\alpha \neq 0, C = 0, A\dfrac{\partial s}{\partial t} \neq 0$	$\left(\alpha S + \dfrac{2S_p}{L}\right)\dfrac{q_e}{\chi\sigma}$	$\left(\alpha S + \dfrac{2S_p}{L}\right)\dfrac{q_e}{(\chi + \chi')\sigma}$
$\alpha = 0, C = 0, A\dfrac{\partial s}{\partial t} \neq 0$	$\dfrac{2S_p q_e}{\chi \sigma L}$	$\dfrac{2S_p q_e}{(\chi + \chi')\sigma L}$
$\alpha = 0, C \neq 0, A\dfrac{\partial s}{\partial t} = 0$	—	$\left(C + \dfrac{2S_p}{L}\right)\dfrac{q_e}{\chi\sigma}$
$\alpha = 0, C = 0, A\dfrac{\partial s}{\partial t} = 0$	—	$\dfrac{2S_p q_e}{\chi\sigma L}$

Therefore the critical current is given by

$$I_c = \frac{\alpha S + C + \dfrac{2S_p}{L}}{(\chi + \chi')\sigma} q_e. \tag{9.45}$$

The specific solutions for the critical current when $u/L > 16S_p$ in various conditions are shown in Table 9.3.

9.2.3.3 Solution for a Short Vacuum Chamber Without a Beam Screen Between Two Chambers With a Beam Screen

Consider a vacuum chamber centered at $z = 0$, of length L. The conditions at the ends are (Figure 9.6):

$$n(\pm L/2) = n_{bs}(\pm L/2), \quad dn(\pm L/2)/dz = dn_{bs}(\pm L/2)/dz$$

In the **case (a)** when $c > 0$ the gas density $n(z)$ is given by

$$n(z) = n_{t\,inf} + \frac{(n_{bs\,inf} - n_{t\,inf})\cosh(\omega_{ta}z)}{\cosh\left(\dfrac{\omega_{ta}L}{2}\right)\left(1 + \dfrac{\omega_{ta}}{\omega_{bs}}\tanh\left(\dfrac{\omega_{ta}L}{2}\right)\right)}. \tag{9.46}$$

where $n_{t\,inf}$ is the solution for an infinitely long vacuum chamber without the beam screen (i.e. $C = 0$) in Eq. (9.13) and $n_{bs\,inf}$ is the solution for an infinitely long vacuum chamber with a beam screen in Eq. (9.17),

$$\omega_{ta} = \sqrt{\left(\alpha_t S_t - (\chi + \chi'(s_t))\frac{I\sigma}{q_e}\right)\frac{1}{u_t}} \text{ and}$$

$$\omega_{bs} = \sqrt{\left(\alpha_{bs} S_{bs} + C_{bs} - (\chi + \chi'(s_{bs}))\frac{I\sigma}{q_e}\right)\frac{1}{u_{bs}}},$$

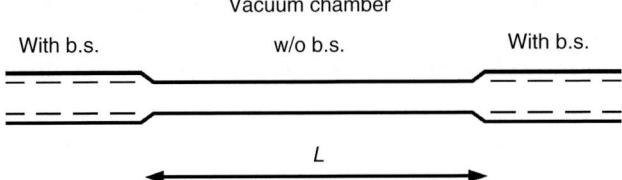

Figure 9.6 A schematic diagram of the layout for a vacuum chamber without a beam screen between two vacuum chambers with a beam screen.

where the indexes t and bs indicate the parameters for the vacuum chamber without and with the beam screen, respectively.

The average value of the gas density in the vacuum chamber with length L is given by:

$$\langle n(L) \rangle = n_{t\,\text{inf}} + \frac{(n_{bs\,\text{inf}} - n_{t\,\text{inf}})\tanh\left(\frac{\omega_{ta}L}{2}\right)}{\frac{\omega_{ta}L}{2}\left(1 + \frac{\omega_{ta}}{\omega_{bs}}\tanh\left(\frac{\omega_{ta}L}{2}\right)\right)}. \qquad (9.47)$$

The gas density is always finite in the range of values of the parameters (i.e. $c > 0$) for case (a).

In the **case (c)** when $c < 0$ the gas density $n(z)$ is given by:

$$n(z) = \frac{(n_{t\,\text{inf}} - n_{bs\,\text{inf}})\cos(\omega_{tb}z)}{\cos\left(\frac{\omega_{tb}L}{2}\right)\left(1 - \frac{\omega_{tb}}{\omega_{bs}}\tan\left(\frac{\omega_{tb}L}{2}\right)\right)} - n_{t\,\text{inf}}. \qquad (9.48)$$

where $\omega_{td} = \sqrt{\left((\chi + \chi'(s_t))\frac{I_\sigma}{q_e} - \alpha_t S_t\right)\frac{1}{u}}$.

The average value of the gas density in the vacuum chamber with length L is

$$\langle n(L) \rangle = \frac{(n_{t\,\text{inf}} - n_{bs\,\text{inf}})\tan\left(\frac{\omega_{td}L}{2}\right)}{\frac{\omega_{td}L}{2}\left(1 - \frac{\omega_{td}}{\omega_{bs}}\tan\left(\frac{\omega_{td}L}{2}\right)\right)} - n_{t\,\text{inf}}. \qquad (9.49)$$

The stability condition in the case (c) is

$$\begin{cases} \left|\frac{\omega_{td}L}{2}\right| < \frac{\pi}{2} \\ \frac{\omega_{td}}{\omega_{bs}}\tan\left(\frac{\omega_{td}L}{2}\right) < 1. \end{cases} \qquad (9.50)$$

and the critical current I_c is the smallest root of two equations:

$$|\omega_{td}| = \frac{\pi}{L} \qquad (9.51)$$

and

$$\omega_{td}\tan\left(\frac{\omega_{td}L}{2}\right) = \omega_{bs}. \qquad (9.52)$$

The solutions for Eq. (9.50) were already discussed and are shown in Table 9.2.

Solutions for Eq. (9.51) can be obtained similar to what was done for Eq. (9.38), applying the first term of the Maclaurin series with a narrower domain. Then Eq. (9.51) can be simplified to

$$\omega_{td}^2 = \frac{2\omega_{bs}}{L} \tag{9.53}$$

In the common form of c, this equation is written as

$$\frac{1}{u_t}\left((\chi + \chi'(s_t))\frac{I\sigma}{q_e} - \alpha_t S_t\right)$$
$$= \frac{2}{L}\sqrt{\left((\chi + \chi'(s_{bs}))\frac{I\sigma}{q_e} - (\alpha_{bs}S_{bs} + C_{bs})\right)\frac{1}{u_{bs}}} \tag{9.54}$$

That can be rewritten in a form of square equation:

$$\left((\chi + \chi'(s_t))\frac{\sigma}{q_e}\right)^2 I^2 - \left[2\alpha_t S_t(\chi + \chi'(s_t)) + \frac{4u_t^2}{u_{bs}L^2}(\chi + \chi'(s_{bs}))\right]\frac{\sigma}{q_e}I$$
$$+ \left[(\alpha_t S_t)^2 + (\alpha_{bs}S_{bs} + C_{bs})\frac{4u_t^2}{u_{bs}L^2}\right] = 0 \tag{9.55}$$

Therefore the critical current is the smallest positive root given by

$$I_c = \frac{\alpha_t S_t + \frac{2u_t^2}{u_{bs}L^2}\frac{(\chi + \chi'(s_{bs}))}{(\chi + \chi'(s_t))} \pm \sqrt{\left(\alpha_t S_t + \frac{2u_t^2}{u_{bs}L^2}\frac{(\chi + \chi'(s_{bs}))}{(\chi + \chi'(s_t))}\right)^2 - (\alpha_t S_t)^2 - (\alpha_{bs}S_{bs} + C_{bs})\frac{4u_t^2}{u_{bs}L^2}}}{(\chi + \chi'(s_t))\sigma} q_e. \tag{9.56}$$

The specific solutions of our interest are for the following.

The critical current for *a cryogenic vacuum chamber* in quasi-equilibrium state (i.e. $A\, \partial s_t/\partial t - 0$ and $A\, \partial s_{bs}/\partial t - 0$) is given by

$$I_c = \left(\frac{2u_t^2}{u_{bs}L^2} \pm \sqrt{\left(\frac{2u_t^2}{u_{bs}L^2}\right)^2 - C_{bs}\frac{4u_t^2}{u_{bs}L^2}}\right)\frac{q_e}{\chi\sigma} \tag{9.57}$$

The critical current for a *room temperature uncoated vacuum chamber* between two NEG coated vacuum chambers is given by

$$I_c = \left(\frac{2u_t^2}{u_N L^2} \pm \sqrt{\left(\frac{2u_t^2}{u_N L^2}\right)^2 - \alpha_N S_N \frac{4u_t^2}{u_N L^2}}\right)\frac{q_e}{\chi\sigma}; \tag{9.58}$$

where index 'N' corresponds to NEG-coated vacuum chamber parameters.

The critical current for *a room temperature NEG-coated vacuum chamber* between NEG coated vacuum chambers with a pumping mesh or slots (leading to a lumped or distributed pump):

$$I_c = \left(\alpha_N S_N + \frac{2u_t^2}{u_m L^2} \pm \sqrt{\left(\alpha_N S_N + \frac{2u_t^2}{u_m L^2} \right)^2 - (\alpha_N S_{NEG})^2 - (\alpha_N S_m + C_m)\frac{4u_t^2}{u_m L^2}} \right) \frac{q_e}{\chi \sigma} \qquad (9.59)$$

where index 'N' corresponds to NEG coated vacuum chamber parameters and index 'm' corresponds to NEG-coated vacuum chamber with a pumping mesh. It is assumed here that sticking probability of NEG coating is the same for all parts.

9.2.3.4 Some Remarks to Solutions for Short Tubes

One important conclusion follows from the earlier discussion: increasing the lumped pumping speed at the ends of vacuum chamber will effectively improve the vacuum stability only as long as $I_c \ll \frac{uq_e}{\chi\sigma}\left(\frac{\pi}{L}\right)^2$. However, when $I_c \leq \frac{uq_e}{\chi\sigma}\left(\frac{\pi}{L}\right)^2$, the vacuum stability could be effectively improved by reducing the length L and/or increasing the vacuum conductance u (i.e. increasing the cross section A) of the vacuum chamber.

And finally it is important to mention that the analysis above is done for given ISD yields χ and χ'. However, reducing ISD yields χ and χ' is an effective way to achieve stable vacuum. The ISD yield χ depends on material, surface preparation and cleaning, coating, and treatments as described in Section 2.8. The ISD yield χ' depends on surface density of cryosorbed gas. Thus using the relevant and reliable data is very critical in the design of vacuum system without beam induced vacuum instability.

9.2.4 Multi-Gas System

In the previous chapter, the vacuum stability was studied assuming that only one single gas species is present in the system. In reality, several species coexist, as observed experimentally from the dominating peaks in the mass spectra measurements. The generic gas species can desorb other gas species. Therefore, the equilibrium equations for the gas density of each species will be cross-correlated to those of other species. Thus, one can write a system of equations for N gas species, A_i ($i = 1, 2, \ldots, N$); for volumetric gas density inside a room temperature vacuum chamber (instead of Eq. (9.1)):

$$A_c \frac{\partial n_i}{\partial t} = \sum_{j=1}^{N} \frac{\chi_{A_i,A_j^+} I \sigma_j}{q_e} n_j - (\alpha_i S_i + C_i) n_i + \eta_i \Gamma + \xi_i \Theta + q_{ti} + u_i \frac{\partial^2 n_i}{\partial z^2}; \qquad (9.60)$$

and for volumetric and surface gas density inside a cryogenic temperature vacuum chamber (instead of Eqs. (9.2) and (9.3)):

$$A_c \frac{\partial n_i}{\partial t} = \sum_{j=1}^{N} \frac{(\chi_{A_i,A_j^+} + \chi'_{A_i,A_j^+}(s_i)) I \sigma_j}{q_e} n_j - (\alpha_i S_i + C_i) n_i$$

$$+ (\eta_i + \eta'_i(s_i))\Gamma + (\xi_i + \xi'_i(s_i))\Theta + + \alpha_i S_i n_{ei}(s_i, T) + u_i \frac{\partial^2 n_i}{\partial z^2}; \qquad (9.61)$$

$$F\frac{\partial s_i}{\partial t} = \alpha_i S_i(n_i - n_{ei}(s_i, T)) - \eta_i'(s_i)\Gamma - \xi_i'(s_i)\Theta - \sum_{j=1}^{N} \frac{\chi'_{A_i,A_j^+}(s_i)I\sigma_j}{q_e} n_j; \quad (9.62)$$

where A_c is a vacuum chamber cross-section area, σ_j is the ionisation cross section of the residual gas molecules A_j by beam particles, and q_i and $q_i'(s_i)$ are primary and secondary PSD and ESD.

Solving the system of equation analytically for three or more gases is quite complicated, but it can done numerically. An example of realisation of numerical solution is described in Section 9.3.

9.2.5 Two-Gas System

Solving the system of equations for two-gas model with gases A and B can be delivered. In quasi-static conditions, where $V(dn/dt) \approx 0$ and $A(ds/dt) \neq 0$, the gas densities of two gases system, $n_A(z)$ and $n_B(z)$, are described by a system of two differential equation of the second order:

$$\begin{cases} u_1 \dfrac{d^2 n_A}{dz^2} - c_1 n_A + q_1 + d_1 n_B = 0; \\ u_2 \dfrac{d^2 n_B}{dz^2} - c_2 n_B + q_2 + d_2 n_A = 0. \end{cases} \quad (9.63)$$

The solution is

$$\begin{cases} n_A(z) = \dfrac{q_2 d_1 + c_2 q_1}{c_1 c_2 - d_1 d_2} + C_1 e^{\sqrt{\omega_1} \cdot z} + C_2 e^{-\sqrt{\omega_1} \cdot z} + C_3 e^{\sqrt{\omega_2} \cdot z} + C_2 e^{-\sqrt{\omega_2} \cdot z}; \\ n_B(z) = \dfrac{q_1 d_2 + c_1 q_2}{b_1 b_2 - d_1 d_2} + K_1 e^{\sqrt{\omega_1} \cdot z} + K_2 e^{-\sqrt{\omega_1} \cdot z} + K_3 e^{\sqrt{\omega_2} \cdot z} + K_2 e^{-\sqrt{\omega_2} \cdot z}; \end{cases}$$

$$(9.64)$$

where

$$\omega_{1,2} = \frac{1}{2}\left(\frac{c_1}{u_1} + \frac{c_2}{u_2} \pm \sqrt{\left(\frac{c_1}{u_1} - \frac{c_2}{u_2}\right)^2 + 4\frac{d_1 d_2}{u_1 u_2}}\right)$$

with

$$c_1 = \alpha_A S_A + C_A - (\chi_{A,A^+} + \chi'_{A,A^+})\frac{I\sigma_A}{q_e};$$

$$c_2 = \alpha_B S_B + C_B - (\chi_{B,B^+} + \chi'_{B,B^+})\frac{I\sigma_B}{q_e};$$

$$q_1 = (\eta_A + \eta_A')\Gamma + (\xi_A + \xi_A')\Theta + \alpha_A S_A n_{eA};$$

$$q_2 = (\eta_B + \eta_B')\Gamma + ((\xi_B + \xi_B'))\Theta + \alpha_B S_B n_{eB};$$

$$d_1 = (\chi_{A,B^+} + \chi'_{A,B^+})\frac{I\sigma_B}{q_e};$$

$$d_2 = (\chi_{B,A^+} + \chi'_{B,A^+})\frac{I\sigma_A}{q_e}.$$

When $\chi > 0$, the parameters c_1 and c_2 can be positive, negative, or equal to zero; u_1, u_2, d_1, and d_2 are always positive; and q_1 and q_2 can be positive or equal to zero. Therefore, the parameters ω_1 and ω_2 can be positive, negative, or equal to zero.

The constants C_i and K_i ($i = 1, \ldots, 4$) are dependent on the conditions at the ends of vacuum chamber.

9.2.5.1 Solutions for an Infinitely Long Vacuum Chamber

In the case of an infinitely long vacuum chamber, there is no net axial diffusion, i.e. $d^2 n_A/dz^2 = 0$ and $d^2 n_B/dz^2 = 0$. The gas density of the two components of the gas mixture is given by the system of two linear equations:

$$\begin{cases} -c_1 n_A + q_1 + d_1 n_B = 0; \\ -c_2 n_B + q_2 + d_2 n_A = 0; \end{cases} \tag{9.65}$$

for which the solution is

$$\begin{cases} n_{\inf A} = \dfrac{q_2 d_1 + c_2 q_1}{c_1 c_2 - d_1 d_2}; \\ n_{\inf B} = \dfrac{q_1 d_2 + c_1 q_2}{c_1 c_2 - d_1 d_2}. \end{cases} \tag{9.66}$$

The stability conditions in this case are

$$c_1 c_2 - d_1 d_2 > 0, \quad \text{with } c_1 > 0 \text{ and } c_2 > 0. \tag{9.67}$$

The critical current can be calculated as

$$I_c = \frac{(\alpha_1 + \alpha_2) \pm \sqrt{(\alpha_1 - \alpha_2)^2 + 4\beta_1 \beta_2}}{2(\alpha_1 \alpha_2 - \beta_1 \beta_2)}; \tag{9.68}$$

where

$$\alpha_1 = \frac{1}{I_c(A, A^+)} = \frac{(\chi_{A,A^+} + \chi'_{A,A^+}) \sigma_A}{\alpha_A S_A + C_A} \cdot \frac{1}{q_e}; \quad \alpha_2 = \frac{1}{I_c(B, B^+)} = \frac{(\chi_{B,B^+} + \chi'_{B,B^+}) \sigma_B}{\alpha_B S_B + C_B} \cdot \frac{1}{q_e};$$

$$\beta_1 = \frac{1}{I_c(A, B^+)} = \frac{(\chi_{A,B^+} + \chi'_{A,B^+}) \sigma_B}{\alpha_A S_A + C_A} \cdot \frac{1}{q_e}; \quad \beta_2 = \frac{1}{I_c(B, A^+)} = \frac{(\chi_{B,A^+} + \chi'_{B,A^+}) \sigma_A}{\alpha_B S_B + C_B} \cdot \frac{1}{q_e}.$$

9.2.5.2 Solution for a Short Vacuum Chamber in the Equilibrium State

Consider a vacuum chamber centered at $z = 0$ of length L and then taking in mind the following: $\chi > 0$, $\omega_1 < 0$, ω_2 is positive when $c_1 c_2 - d_1 d_2 < 0$ and negative when $c_1 c_2 - d_1 d_2 > 0$. Assuming the same conditions at the ends, the solution described by Eq. (9.64) can be simplified to

$$\begin{cases} n_A(z) = \dfrac{q_2 d_1 + c_2 q_1}{c_1 c_2 - d_1 d_2} + C_5 \cos(\sqrt{|\omega_1|} \cdot z) + C_6 \cos(\sqrt{|\omega_2|} \cdot z), \\ n_B(z) = \dfrac{q_1 d_2 + c_1 q_2}{c_1 c_2 - d_1 d_2} + K_5 \cos(\sqrt{|\omega_1|} \cdot z) + K_6 \cos(\sqrt{|\omega_2|} \cdot z); \end{cases} \text{for } \omega_2 < 0;$$

$$\tag{9.69}$$

and

$$\begin{cases} n_A(z) = \dfrac{q_2 d_1 + c_2 q_1}{c_1 c_2 - d_1 d_2} + C_5 \cos(\sqrt{|\omega_1|} \cdot z) + C_6 \cosh(\sqrt{|\omega_2|} \cdot z), \\ n_B(z) = \dfrac{q_1 d_2 + c_1 q_2}{c_1 c_2 - d_1 d_2} + K_5 \cos(\sqrt{|\omega_1|} \cdot z) + K_6 \cosh(\sqrt{|\omega_2|} \cdot z); \end{cases} \quad \text{for } \omega_2 < 0.$$

(9.70)

When the pumping speed at the ends of the vacuum chamber is sufficiently high, that is, when increasing S_p does not improve appreciably the stability margin, as discussed in Section 9.2.3.2, we can use for the stability condition the inequalities:

$$\begin{cases} \sqrt{|\omega_1|} \dfrac{L}{2} < \dfrac{\pi}{2}; \\ \sqrt{|\omega_2|} \dfrac{L}{2} < \dfrac{\pi}{2}; \end{cases} \quad (9.71)$$

and since it is always $|\omega_1| > |\omega_2|$, the first condition is the most stringent one. Therefore, the critical current is the root of the equation:

$$-\frac{1}{2}\left(\frac{c_1}{u_1} + \frac{c_2}{u_2} + \sqrt{\left(\frac{c_1}{u_1} - \frac{c_2}{u_2}\right)^2 + 4\frac{d_1 d_2}{u_1 u_2}}\right) = \left(\frac{\pi}{L}\right)^2$$

and is given by

$$I_c = \frac{2}{(\alpha_1 + \alpha_2 + \sqrt{(\alpha_1 - \alpha_2)^2 + 4\beta_1 \beta_2})} \quad (9.72)$$

where

$$\alpha_1 = \frac{1}{I_c(A, A^+)} = \frac{\chi_{A,A^+} \sigma_A}{u_A q_e} \left(\frac{L}{\pi}\right)^2; \alpha_2 = \frac{1}{I_c(B, B^+)} = \frac{\chi_{B,B^+} \sigma_B}{u_B q_e} \left(\frac{L}{\pi}\right)^2;$$

$$\beta_1 = \frac{1}{I_c(A, B^+)} = \frac{\chi_{A,B^+} \sigma_B}{u_A q_e} \left(\frac{L}{\pi}\right)^2; \beta_2 = \frac{1}{I_c(B, A^+)} = \frac{\chi_{B,A^+} \sigma_A}{u_B q_e} \left(\frac{L}{\pi}\right)^2.$$

9.2.6 Some Comments to the Analytical Solutions

The estimation of the gas density and the stability criteria in the equilibrium state gives the upper limit for the gas density distribution and the lowest values for the critical current for a machine with quasi-static or static parameters. It could be demonstrated that for a system in equilibrium conditions, if the beam current and, hence, the photon flux, are suddenly changed, the system will undergo to a fast transition, i.e. $A_c(dn/dt) \neq 0$. A new equilibrium gas density and surface coverage will be reached, for which the value for the critical current must be estimated. If, during the transition phase, the gas density is greater than the equilibrium density, then the critical current estimated at the equilibrium might be too high. Therefore, it is necessary to have some safety margin at the highest beam current in the machine, i.e. the critical current for all elements of the machine should be **at least a factor of 2–3** higher than the highest beam current.

9.2.7 Effect of the Ion-Stimulated Desorption on the Gas Density

It is useful to estimate the effect of the ion-induced desorption on the gas density, in order to know when the ISD can be neglected. To this purpose, we will calculate the ratio of gas density with $\chi \neq 0$ and $\chi' \neq 0$ to the gas density with $\chi = 0$ and $\chi' = 0$: $n_{\chi \neq 0}/n_{\chi = 0}$.

9.2.7.1 Infinitely Long Vacuum Chamber (One Gas)

For an infinitely long vacuum chamber, the ratio can be estimated using Eq. (9.13) as

$$\frac{n_{\inf \chi \neq 0}}{n_{\inf \chi = 0}} = \frac{\alpha S + C}{\alpha S + C - (\chi + \chi'(s))\dfrac{I\sigma}{q_e}} = \frac{1}{1 - \dfrac{(\chi + \chi'(s))I\sigma}{(\alpha S + C)q_e}} = \frac{1}{1 - \dfrac{I}{I_c}}$$

Or using Eq. (9.22),

$$\frac{n_{\inf \chi \neq 0}}{n_{\inf \chi = 0}} = \frac{C}{C - \chi \dfrac{I\sigma}{q_e}} = \frac{1}{1 - \dfrac{\chi I\sigma}{q_e C}} = \frac{1}{1 - \dfrac{I}{I_c}}$$

Hence, for the infinitely long vacuum chamber, with or without a beam screen, the ratio can be expressed as a function of ratio I/I_c:

$$\frac{n_{\inf \chi \neq 0}}{n_{\inf \chi = 0}} = \left(1 - \frac{I}{I_c}\right)^{-1} \tag{9.73}$$

One can conclude that the effect of the ISD on the gas density is negligible when $I \ll I_c$.

9.2.7.2 Vacuum Chamber with a Given Pumping Speed at the Ends (One Gas)

The same estimation can be done for the vacuum chamber with a given pumping speed at the ends using the Eqs. (9.23) and (9.26):

$$\frac{\langle n_{\chi \neq 0}\rangle}{\langle n_{\chi = 0}\rangle} = \frac{\alpha S + C}{\alpha S + C - (\chi + \chi'(s))\dfrac{I\sigma}{e}} \cdot \frac{1 - \dfrac{\omega L}{2}\left(\dfrac{1}{\tan\left(\dfrac{\omega L}{2}\right)} - \dfrac{u}{S_p}\omega\right)}{1 - \dfrac{\beta L}{2}\left(\dfrac{1}{\tanh\left(\dfrac{\beta L}{2}\right)} + \dfrac{u}{S_p}\beta\right)}^{-1};$$

$$\tag{9.74}$$

here

$$\beta = \sqrt{\frac{(\alpha S + C)}{u}}, \text{ and } \omega = \sqrt{\left((\chi + \chi')\frac{I\sigma}{q_e} - (\alpha S + C)\right)\frac{1}{u}} = \beta\sqrt{\frac{I}{I_{c\,\inf}} - 1}.$$

This formula can be rewritten as

$$\frac{\langle n_{\chi\neq 0}\rangle}{\langle n_{\chi=0}\rangle} = \frac{1}{1-\dfrac{I}{I_{c\,\mathrm{inf}}}} \cdot \frac{1 - \dfrac{\beta L}{2}\sqrt{\dfrac{I}{I_{c\,\mathrm{inf}}}-1}\left(\tan\left(\dfrac{\beta L}{2}\sqrt{\dfrac{I}{I_{c\,\mathrm{inf}}}-1}\right)\right)^{-1}\left(-\dfrac{u\beta}{S_p}\sqrt{\dfrac{I}{I_{c\,\mathrm{inf}}}-1}\right)}{1 - \dfrac{\beta L}{2}\left(\dfrac{1}{\tanh\left(\dfrac{\beta L}{2}\right)} + \dfrac{u\beta}{S_p}\right)^{-1}}. \quad (9.75)$$

When $S_p > \tilde{S}_p$, the terms with S_p can be neglected:

$$\frac{\langle n_{\chi\neq 0}\rangle}{\langle n_{\chi=0}\rangle} = \frac{1}{1-\dfrac{I}{I_{c\,\mathrm{inf}}}} \cdot \frac{1 - \dfrac{1}{\dfrac{\beta L}{2}\sqrt{\dfrac{I}{I_{c\,\mathrm{inf}}}-1}}\tan\left(\dfrac{\beta L}{2}\sqrt{\dfrac{I}{I_{c\,\mathrm{inf}}}-1}\right)}{\left(1 - \dfrac{2}{\beta L}\tanh\left(\dfrac{\beta L}{2}\right)\right)}.$$

In the equilibrium case for a vacuum chamber without the beam screen, this becomes

$$\frac{\langle n_{\chi\neq 0}\rangle}{\langle n_{\chi=0}\rangle} = \frac{\sqrt{\dfrac{\chi I\sigma}{uq_e}\dfrac{L}{2}}\left(1 - \dfrac{u}{S_p}\sqrt{\dfrac{\chi I\sigma}{uq_e}}\tan\left(\sqrt{\dfrac{\chi I\sigma}{uq_e}\dfrac{L}{2}}\right)\right)^{-1} \tan\left(\sqrt{\dfrac{\chi I\sigma}{uq_e}\dfrac{L}{2}}\right) - 1}{\dfrac{\chi I\sigma}{q_e}\left(\dfrac{L^2}{12u} + \dfrac{L}{2S_p}\right)} \quad (9.76)$$

For a very short tube, the conductance of vacuum chamber is much higher than the pumping at the ends, i.e. $u/L \gg S_p$, and the critical current can be written as $I_c = 2S_p q_e/(L\chi\sigma)$. In this case the ratio can be rewritten as

$$\frac{n_{\mathrm{inf}\,\chi\neq 0}}{n_{\mathrm{inf}\,\chi=0}} = \left(1 - \frac{I}{I_c}\right)^{-1} \quad (9.77)$$

which is the same as in Eq. (9.73).

When the conductance of the vacuum chamber is much lower than the pumping at the ends, i.e. $u/L \ll S_p$, the critical current can be written as $I_c \approx \dfrac{ue}{\chi\sigma}\left(\dfrac{\pi}{L}\right)^2$, and the ratio is

$$\frac{\langle n_{\chi\neq 0}\rangle}{\langle n_{\chi=0}\rangle} = \frac{I_c}{I}\frac{12}{\pi^2}\left[\frac{\tan\left(\dfrac{\pi}{2}\sqrt{\dfrac{I}{I_c}}\right)}{\dfrac{\pi}{2}\sqrt{\dfrac{I}{I_c}}} - 1\right]. \quad (9.78)$$

9.2.7.3 Two-Gas System

The gas density for an infinitely long tube for two gases system is written in Eq. (9.66). The ratio between the gas density values estimated taking or not into account the ISD is

$$\begin{cases} \dfrac{n_{\inf A}(\chi \neq 0)}{n_{\inf A}(\chi = 0)} = \dfrac{\dfrac{n_{\inf B}(\chi = 0)}{n_{\inf A}(\chi = 0)} \dfrac{I}{I_c(A, B^+)} + \left(1 - \dfrac{I}{I_c(B, B^+)}\right)}{\left(1 - \dfrac{I}{I_c(A, A^+)}\right)\left(1 - \dfrac{I}{I_c(B, B^+)}\right) - \dfrac{I}{I_c(A, B^+)}\dfrac{I}{I_c(B, A^+)}}; \\[2ex] \dfrac{n_{\inf B}(\chi \neq 0)}{n_{\inf B}(\chi = 0)} = \dfrac{\dfrac{n_{\inf A}(\chi = 0)}{n_{\inf B}(\chi = 0)} \dfrac{I}{I_c(B, A^+)} + \left(1 - \dfrac{I}{I_c(A, A^+)}\right)}{\left(1 - \dfrac{I}{I_c(A, A^+)}\right)\left(1 - \dfrac{I}{I_c(B, B^+)}\right) - \dfrac{I}{I_c(A, B^+)}\dfrac{I}{I_c(B, A^+)}}. \end{cases}$$

(9.79)

The analytical solutions for the vacuum chamber of limited length is more complicated, but even this solution shows that the estimated gas density is higher when the ISD is taken into account.

9.2.8 Some Numeric Examples from the LHC Design

The ion-induced gas density instability was intensively studied during the LHC design. The parameters shown in Table 9.4 have been used.

The ISD yields from condensed gas depends on the amount of cryosorbed gas. In the worst-case scenario, they can reach its maximum value shown in Table 9.4 (see Chapter 4). Due to the highest PSD, the main cryosorbed gas on the beam chamber wall will be H_2, and the second gas will be CO. The amount of CH_4

Table 9.4 The beam–gas ionisation cross section at 7.0 TeV and the ion-stimulated desorption yields for different gases and ion energy of 300 eV.

			ISD yield χ from bare walls [molecules/ion]					Maximum value of secondary ISD yield χ' from condensed gases [molecules/ion]	
Ion impact energy		500 eV	300 eV					500 eV	
Gas	σ [cm²]	H_2^+	H_2^+	He⁺	CH_4^+	CO⁺	CO_2^+	H_2^+	CO⁺
H_2	3.14×10^{-19}	0.8	0.47	0.52	2.6	4.3	5.2	8×10^3	1.5×10^4
He	3.41×10^{-19}	—	—	—	—	—	—	—	—
CH_4	6.76×10^{-18}	0.045	0.03	0.03	0.17	0.3	0.39	—	—
CO	5.36×10^{-18}	0.28	0.17	0.25	1.4	2.8	3.9	20	25
CO_2	1.214×10^{-17}	0.09	0.05	0.08	0.45	0.9	1.26	—	—

and CO_2 is limited due to their cracking under SR (see Chapter 7); thus their secondary ISD yields χ' will be much lower than for H_2 and CO and can be neglected.

9.2.8.1 The Critical Current for an Infinitely Long Vacuum Chamber

The solution for an infinitely long vacuum chamber gives the lowest limit for the critical current (i.e. worst case) in comparison with the short vacuum chambers. The results of the critical current calculation are shown in Table 9.5 for the single gas model – $I_{c(A,A+)}$ (where A is H_2, CH_4, CO, and CO_2) – and two-gas model: I_c.

In the case of vacuum chamber without a beam screen, the calculations were made for a vacuum chamber with inner diameters of 40, 50, and 60 mm at two temperatures, 1.9 and 4.5 K. As it was shown in Chapter 7, in vacuum chamber irradiated by SR, an amount of cryosorbed H_2 and CO molecules increase with a photon dose, while an amount of cryosorbed CH_4 and CO_2 molecules is limited due to SR-induced cracking. Therefore, critical currents were calculated with the maximum value of secondary ISD yields χ' for H_2 and CO (see Table 9.4) and for primary ISD yield χ for CH_4 and CO_2. Sticking probabilities $\alpha(H_2, CO) = 0.1$ and $\alpha(CH_4, CO_2) = 1$ are assumed here. The lowest critical current for single gas model $I_{c(A,A+)}$ were obtained for H_2 and CO (in bold in Table 9.5), and these two gases were further explored in two-gas model and the results are shown in column under I_c. All the I_c results are much lower than the LHC maximum design beam current of 0.85 A: i.e. a vacuum chamber without a beam screen will suffer from the ion-induced gas density instability. Increasing the vacuum chamber diameter and temperature help to increase the wall pumping speed and therefore, to increase the critical current, but it would still be insufficient for safe operation of the machine.

Table 9.5 The beam-gas ionisation cross-section at 7.0 TeV and the ion-stimulated desorption yields for different gases and ion energy of 300 eV.

T [K]	ID [mm]	I_c [A]	$I_{c(H_2,H_2^+)}$ [A]	$I_{c(CH_4,CH_4^+)}$ [A]	$I_{c(CO,CO^+)}$ [A]	$I_{c(CO_2,CO_2^+)}$ [A]
				$\alpha(H_2, CO) = 0.1$		
				Without a beam screen		
1.9	Ø40	0.22	0.28	2200	**1.5**	75
	Ø50	0.28	0.25	2740	**1.72**	94
	Ø60	0.34	0.43	3290	**2.2**	112
4.5	Ø40	0.34	0.44	3370	**2.2**	116
	Ø50	0.43	0.55	4220	**2.8**	145
	Ø60	0.52	0.66	5060	**3.4**	173
				With a beam screen		
5.5	Ø45	2.7	1600	70	7.3	4.2
20	Ø45	5.1	3060	134	14.0	8.1

In the case of vacuum chamber with a beam screen, the amount of cryosorbed gas on the inner walls of beam screen irradiated by SR is negligible (see Chapter 7); thus the primary ISD yield χ from bare walls are used in the calculation. The beam screen transparency was set to 4.4%. In this case, the lowest critical current for single gas model $I_{c(A,A+)}$ was obtained for CO and CO_2 (in bold in Table 9.5); thus these two gases were further explored in the two-gas model. The results shown in Table 9.5 demonstrate that using a beam screen with a temperature between 5.5 and 20 K leads to solving the ion-induced instability problem with a safety margin factor from 3.2 at 5.5 K to 6.0 at 20 K.

It is important to note that the critical current calculated with the two-gas model is always lower than the ones for the single gas model.

9.2.8.2 Short Vacuum Chambers

The critical currents of all components along the beam path of the LHC storage ring was calculated. It has been shown that some components require special attention to increase the critical currents.

For example, the interconnects between the dipole magnets and the bellows between two magnet cold bores have a higher temperature due to pure thermal conductivity. Thus, although the perforated beam screen is exactly the same as in the dipole, it does not provide distributed pumping through the pumping holes (see Section 9.2.3.3). To solve this, either a cryosorber should be attached to the outside of the beam screen (see Chapter 7), or the length of the interconnect should be reduced below the critical length.

Standing alone dipoles in Long Straight Section would be unstable without a beam screen with sufficient bean screen transparency (area of pumping holes) (see Section 9.2.5). Magnets at 4.5 K also require using a cryosorber attached to the outside of the beam screen.

Room temperature part require the length of vacuum chambers between the pumps to be much less than the critical length and the pumping speed of the pump to be sufficiently large (see Sections 9.2.3.1–9.2.3.2). Since a vacuum chamber bakeout reduces ISD yields, it helps to increase the critical length or reduce the critical current. Providing a distributed pumping speed either by NEG coating or with NEG strips in the antechamber is also a good solution.

9.2.8.3 Effect of the Ion-Stimulated Desorption on the Gas Density

Table 9.6 shows the ratio $n_{\chi \neq 0}/n_{\chi = 0}$ of the average gas density estimated taking into account the ISD (i.e. with $\chi \neq 0$ and $\chi' \neq 0$) to the average gas density estimated without taking into account the ISD (i.e. with $\chi = 0$ and $\chi' = 0$).

Table 9.6 The ratios $n_{\chi \neq 0}/n_{\chi = 0}$ of gas densities with and without taking into account the ion-stimulated desorption for an infinitely long vacuum chamber.

I/I_c	0.1	1/3	0.5	2/3	0.8	0.9
$n_{\chi \neq 0}/n_{\chi = 0}$	1.1	1.5	2.0	3.0	5.0	10.0

From these numerical results for the *single gas estimation* obtained with Eq. (9.77) for an infinitely long vacuum chamber and with Eq. (9.78) for a finite vacuum chamber with sufficiently high pumping ($S_p > \tilde{S}_p$), one can deduce that the effect of the ISD will be practically negligible (10% gas density increase) when the machine current is 10 times less than the critical current. However, when $I/I_c = 0.5$ the gas density increases by a factor 2, i.e. the ISD contribution is practically the same as all other sources of gas. For higher I/I_c, the ISD is the main source of gas.

The *two gases system* studied with Eq. (9.79) considers that gas A is the same as in the single gas model. It is interesting to study how gas B affects the picture. The ratio between the average gas densities of gases A and B is defined as: $r_{BA} = n_{\inf B}/n_{\inf A}$. This ratio estimate without taking into account the ISD (i.e. with $\chi = 0$ and $\chi' = 0$) is defined as $r_{BA}(\chi = 0) = n_{\inf B}(\chi = 0)/n_{\inf A}(\chi = 0)$. Although the gas composition in the cryogenic accelerator vacuum chamber irradiated by SR is dominated by H_2 and CO, the ion-induced gas density instability in the beam screen case depends on CO and CO_2 (see above).

The pressure ratio $n_{\inf \chi \neq 0}/n_{\inf \chi = 0}$ calculated with the parameters from Table 9.4 for two pairs of gas H_2 and CO with $r_{CO,H_2}(\chi = 0) = 0.1$, and CO and CO_2 with $r_{CO_2,CO}(\chi = 0) = 0.1$ shown on the left and right side respectively. In case of modelling H_2 and CO, the ISD of H_2 is negligible in comparison to PSD, while ISD for CO is comparable to PSD at $I = 2$ A, and CO becomes a dominant gas when $I > 10$ A. In case of modelling CO and CO_2, the ISD of CO is higher than in a previous example due to a contribution from high ISD yield $\chi_{CO,CO_2}{}^+$, while ISD for CO_2 growing with a beam current even faster, CO_2 becomes a dominant gas when $I > 2.5$ A.

This examples qualitatively demonstrate the effect which was experimentally observed in machines: the gas density composition can significantly change (strong increase in CO and CO_2) with beam current when approaching the ion-induced pressure instability (Figure 9.7).

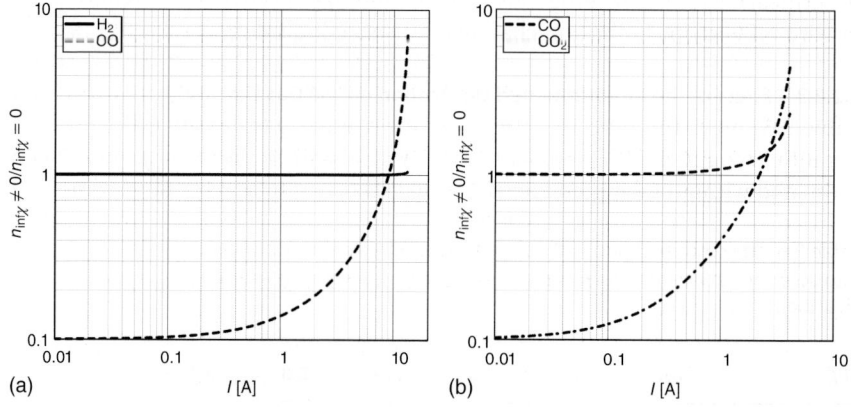

Figure 9.7 The ratios $n_{\chi \neq 0}/n_{\chi = 0}$ of gas densities as a function of the beam current in two-gas models: (a) $H_2 + CO$ and (b) $CO + CO_2$.

9.3 VASCO as Multi-Gas Code for Studying the Ion-Induced Pressure Instability

The VASCO code [17] was written to estimate the residual gas pressure and critical current in a hadron machine, taking into account the multi-gas model and complex vacuum system. The code gives a numerical answer, subdividing the vacuum system into 'elements' with constant properties (diameter, material and desorption yields, temperature, etc.) as explained in the following.

9.3.1 Basic Equations and Assumptions

The basic equation used in the VASCO code merges Eqs. (9.2) and (9.3) as follows:

$$A_c \frac{\partial n_i}{\partial t} = \sum_{j=1}^{N} \frac{(\chi_{A_i,A_j^+} + \chi'_{A_i,A_j^+}(s_i))I\sigma_j}{q_e} n_j - (\alpha_i S_i + C_i)n_i$$

$$+ (\eta_i + \eta'_i(s_i))\Gamma + (\xi_i + \xi'_i(s_i))\Theta + q_{ti} + \alpha_i S_i n_{ei}(s_i, T) + u_i \frac{\partial^2 n_i}{\partial z^2} \approx 0$$

(9.80)

where all terms related to cryogenic desorption or equilibrium density (n_e) are identically zero for chambers at room temperature and thermal desorption (q_t) is negligible at cryogenic temperature.

It should be noted that in the model described by Eqs. (9.2), (9.3), and (9.80) ions are created only by interaction of the residual gas with the running beam. To be complete, one should include ionisation by photons (mainly from SR) and electrons (especially in the presence of electron cloud).

In the following we describe how to resolve Eq. (9.80) with the assumption of the following:

- *Time invariant parameters*: This implies that the residual gas density estimation is relevant to a specific moment in time, since some of the parameters are indirectly time dependent such as the following: (i) Induced gas desorption yields, which depend on the surface history (wall pre-conditioning, particle bombardment) and on the cryogenic surfaces, change with the amount of gas condensed on the surface, and (ii) NEG-distributed pumping, which varies with the quantity of gas already pumped and the activation history.
- *Cylindrical geometry*: This allows for one-dimensional approximation (along the beam axis), assuming radial symmetry. Chambers with non-cylindrical cross sections (i.e. conical or elliptical) are approximated with cylinders having the same conductance. When a parameter is a function of the wall surface area (e.g. thermal outgassing and wall pumping), it is scaled to take the real area into account.
- *Finite elements*: A series of 'short' vacuum chamber, for each element (identified by the index $k = 1, \ldots, N$), we define the following equations:

$$A_{c,k} \frac{\partial n_{i,k}}{\partial t} = \sum_{j=1}^{N} \frac{(\chi_{(A_i,A_j^+),k} + \chi'_{(A_i,A_j^+),k})I\sigma_j}{q_e} n_{j,k} - (\alpha_{i,k} S_{i,k} + C_{i,k})n_{i,k}$$

$$+ (\eta_{i,k} + \eta'_{i,k})\Gamma_k + (\xi_{i,k} + \xi'_{i,k})\Theta_k$$

$$+ \alpha_{i,k} S_{i,k} n_{ei,k}(s_{i,k}, T_k) + q_{t,k} + u_{i,k}\frac{\partial^2 n_{i,k}}{\partial z^2} \approx 0 \qquad (9.81)$$

each characterised by a specific set of parameters (for example, a certain cross section diameter, material properties, lumped pumping at the extremities, etc.). The boundary conditions are chosen to ensure the continuity of the gas density and flux functions between elements.

9.3.2 Multi-Gas Model in Matrix Form and Fragmentation in Several Vacuum Chamber Elements

Equation (9.81) represents a set of systems of linearly dependent equations, one system per chamber element k and one equation per gas species considered i. In order to numerically solve such equations, we are going to express them into a more compact matrix form. From now on we will only consider the gas species generally dominating the residual gas in a ultra high vacuum (UHV) system: H_2, CH_4, CO, and CO_2, but any number and type of gas species will bring the same matrix form. We define a set of vectors and matrices as follows (note that ionisation cross section in independent of the chamber element k):

$$\mathbf{n_k} = \begin{bmatrix} n_{H_2,k} \\ n_{CH_4,k} \\ n_{CO,k} \\ n_{CO_2,k} \end{bmatrix}, \quad \boldsymbol{\eta_k} = \begin{bmatrix} \eta_{H_2,k} \\ \eta_{CH_4,k} \\ \eta_{CO,k} \\ \eta_{CO_2,k} \end{bmatrix}, \quad \boldsymbol{\eta'_k} = \begin{bmatrix} \eta'_{H_2,k} \\ \eta'_{CH_4,k} \\ \eta'_{CO,k} \\ \eta'_{CO_2,k} \end{bmatrix},$$

$$\mathbf{q_{t,k}} = \begin{bmatrix} q_{tH_2,k} \\ q_{tCH_4,k} \\ q_{tCO,k} \\ q_{tCO_2,k} \end{bmatrix}, \quad \boldsymbol{\xi_k} = \begin{bmatrix} \xi_{H_2,k} \\ \xi_{CH_4,k} \\ \xi_{CO,k} \\ \xi_{CO_2,k} \end{bmatrix}, \quad \boldsymbol{\xi'_k} = \begin{bmatrix} \xi'_{H_2,k} \\ \xi'_{CH_4,k} \\ \xi'_{CO,k} \\ \xi'_{CO_2,k} \end{bmatrix},$$

$$\boldsymbol{\chi_k} = \begin{bmatrix} \chi_{(H_2,H_2^+),k} & \chi_{(H_2,CH_4^+),k} & \chi_{(H_2,CO^+),k} & \chi_{(H_2,CO_2^+),k} \\ \chi_{(CH_4,H_2^+),k} & \chi_{(CH_4,CH_4^+),k} & \chi_{(CH_4,CO^+),k} & \chi_{(CH_4,CO_2^+),k} \\ \chi_{(CO,H_2^+),k} & \chi_{(CO,CH_4^+),k} & \chi_{(CO,CO^+),k} & \chi_{(CO,CO_2^+),k} \\ \chi_{(CO_2,H_2^+),k} & \chi_{(CO_2,CH_4^+),k} & \chi_{(CO_2,CO^+),k} & \chi_{(CO_2,CO_2^+),k} \end{bmatrix},$$

$$\boldsymbol{\chi'_k} = \begin{bmatrix} \chi'_{(H_2,H_2^+),k} & \chi'_{(H_2,CH_4^+),k} & \chi'_{(H_2,CO^+),k} & \chi'_{(H_2,CO_2^+),k} \\ \chi'_{(CH_4,H_2^+),k} & \chi'_{(CH_4,CH_4^+),k} & \chi'_{(CH_4,CO^+),k} & \chi'_{(CH_4,CO_2^+),k} \\ \chi'_{(CO,H_2^+),k} & \chi'_{(CO,CH_4^+),k} & \chi'_{(CO,CO^+),k} & \chi'_{(CO,CO_2^+),k} \\ \chi'_{(CO_2,H_2^+),k} & \chi'_{(CO_2,CH_4^+),k} & \chi'_{(CO_2,CO^+),k} & \chi'_{(CO_2,CO_2^+),k} \end{bmatrix};$$

$$\mathbf{u_k} = \begin{bmatrix} u_{H_2,k} & 0 & 0 & 0 \\ 0 & u_{CH_4,k} & 0 & 0 \\ 0 & 0 & u_{CO,k} & 0 \\ 0 & 0 & 0 & u_{CO_2,k} \end{bmatrix}, \quad \boldsymbol{\sigma} = \begin{bmatrix} \sigma_{H_2} & 0 & 0 & 0 \\ 0 & \sigma_{CH_4} & 0 & 0 \\ 0 & 0 & \sigma_{CO} & 0 \\ 0 & 0 & 0 & \sigma_{CO_2} \end{bmatrix},$$

$$\alpha_k = \begin{bmatrix} \alpha_{H_2,k} & 0 & 0 & 0 \\ 0 & \alpha_{CH_4,k} & 0 & 0 \\ 0 & 0 & \alpha_{CO,k} & 0 \\ 0 & 0 & 0 & \alpha_{CO_2,k} \end{bmatrix}, \quad S_k = \begin{bmatrix} S_{H_2,k} & 0 & 0 & 0 \\ 0 & S_{CH_4,k} & 0 & 0 \\ 0 & 0 & S_{CO,k} & 0 \\ 0 & 0 & 0 & S_{CO_2,k} \end{bmatrix},$$

$$C_k = \begin{bmatrix} C_{H_2,k} & 0 & 0 & 0 \\ 0 & C_{CH_4,k} & 0 & 0 \\ 0 & 0 & C_{CO,k} & 0 \\ 0 & 0 & 0 & C_{CO_2,k} \end{bmatrix}.$$

Thus, Eq. (9.4) can be rewritten in matrix form as follows:

$$u_k \frac{d^2 n_k}{dz^2} - C_k n_k + q_k = 0 \tag{9.82}$$

where

c_k is the net effect between the wall distributed pumping speed and the ion-induced desorption:

- $c_k = \alpha_k S_k + C_k - \chi_\kappa \frac{I\sigma}{q_e}$ at room temperature
- $c_k = \alpha_k S_k + C_k - (\chi_k + \chi'_k)\frac{I\sigma}{q_e}$ at cryogenic temperature

q_k is the photon-stimulated, electron-stimulated, and thermal desorption term:

- $q_k = \eta_k \Gamma_k + \xi_k \Theta_k + q_{t,k}$ at room temperature
- $q_k = (\eta_k + \eta'_k)\Gamma_k + (\xi_k + \xi'_k)\Theta_k + \alpha_k S_k n_{ek}$ at cryogenic temperature

9.3.2.1 Boundary Conditions

At the boundary between two segments, the continuity of the density function must be guaranteed. Moreover, the sum of flow of molecules coming from the two side of one boundary must be equal to the amount of molecules pumped (S_p) or generated by a local source (g) as pictured in Figure 9.8.

In the specific case of the VASCO code, for the first and the last elements, where there are no lumped pumps, 'mirror' boundary conditions are adopted. In other words, it is supposed that the sector considered is open and continues with the same cross section, and molecules travelling in the outwards direction are the

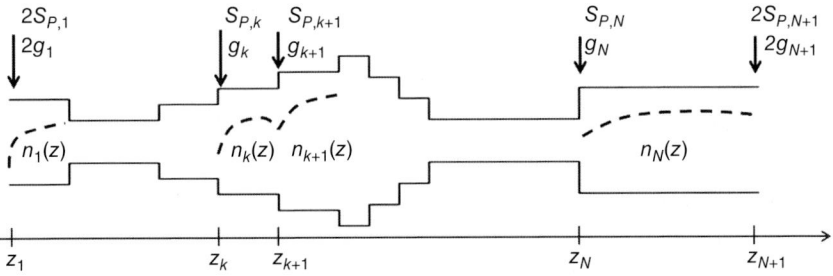

Figure 9.8 Schematic of a beam pipe subdivided in several elements. z_k represents the distance of the segment from an arbitrary reference point (distance included in the input file for convenience of the code user). The sketched light grey lines show how the gas density of a specific gas species might evolve along the beam pipe, due to outgassing and pumping.

same as those travelling back into the system (as if reflected). If there are lumped pumps or gas sources, the pumping speed and gas fluxes are halved.

$$\mathbf{n}_{k-1}(z_k) = \mathbf{n}_k(z_k)$$

$$-\mathbf{u}_{k-1}\frac{d\mathbf{n}_{k-1}}{dz}\bigg|_{z_k} + \mathbf{u}_k\frac{d\mathbf{n}_k}{dz}\bigg|_{z_k} = \mathbf{S}_{p,k}\mathbf{n}_k(z_k) - \mathbf{g}_k \quad \text{for } k = 2, \ldots, N. \quad (9.83)$$

Assuming an open pipe, where the pumping speed and gas sources are equally shared between the two parts of the pipe, only half of the pumping speed and gas flux at the extremities of the considered beam pipe section are available[3]:

$$\mathbf{u}_1\frac{d\mathbf{n}_1}{dz}\bigg|_{z_1} = \mathbf{S}_{p,1}\mathbf{n}_1(z_1) - \mathbf{g}_1$$

$$-\mathbf{u}_N\frac{d\mathbf{n}_N}{dz}\bigg|_{z_{N+1}} = \mathbf{S}_{p,N+1}\mathbf{n}_N(z_{N+1}) - \mathbf{g}_{N+1} \quad \text{for } k = 1 \text{ and } k = N+1 \quad (9.84)$$

where the matrices $\mathbf{S}_{p,k}$ is the lumped pumping speed and $\mathbf{g}_k$ is the localised gas source at the intersection between the segments $(k-1)$ and (k).

$$\mathbf{S}_{p,k} = \begin{bmatrix} S_{pH_2,k} & 0 & 0 & 0 \\ 0 & S_{pCH_4,k} & 0 & 0 \\ 0 & 0 & S_{pCO,k} & 0 \\ 0 & 0 & 0 & S_{pCO_2,k} \end{bmatrix} \quad \text{and} \quad \mathbf{g}_k = \begin{bmatrix} g_{H_2,k} \\ g_{CH_4,k} \\ g_{CO,k} \\ g_{CO_2,k} \end{bmatrix} \quad (9.85)$$

9.3.3 Transformation of the Second-Order Differential Linear Equation into a System of First-Order Equations

A linear differential equation of the second order can be transformed into two equations of the first order, with a change of variable. Let us call

$$\begin{cases} \mathbf{y}_{1,k} = \mathbf{n}_k \\ \mathbf{y}_{2,k} = \dfrac{d\mathbf{n}_k}{dz} \end{cases} \quad \text{for } k = 1, \ldots, N \quad (9.86)$$

Equation (9.81) becomes the system of equations:

$$\begin{cases} \dfrac{d\mathbf{y}_{1,k}}{dz} = \mathbf{y}_{2,k} \\ \dfrac{d\mathbf{y}_{2,k}}{dz} = -|\mathbf{u}_k|^{-1}\mathbf{c}_k\mathbf{y}_{1,k} - |\mathbf{u}_k|^{-1}\mathbf{q}_k \end{cases} \quad (9.87)$$

If **I** is the identity matrix, and **0** the identically *zero* matrix, we can now define the new matrices:

$$\mathbf{Y}_k = \begin{bmatrix} \mathbf{y}_{1,k} \\ \mathbf{y}_{2,k} \end{bmatrix} \quad (9.88)$$

$$\mathbf{M}_k = \begin{bmatrix} \mathbf{0} & \mathbf{I} \\ -|\mathbf{u}_k|^{-1}\mathbf{c}_k & \mathbf{0} \end{bmatrix} \quad \text{and} \quad \mathbf{b}_k = \begin{bmatrix} \mathbf{0} \\ -|\mathbf{u}_k|^{-1}\mathbf{c}_k \end{bmatrix} \quad (9.89)$$

---

[3] With the above boundary conditions, only the solution of the density for known pumping speed or gas source at the pipe extremities can be found.

9.3 VASCO as Multi-Gas Code for Studying the Ion-Induced Pressure Instability

Let z be the variable running along each segment $z \in [0\ L_k]$ where L_k represents the length of the segment kth. $\mathbf{Y}_{0k} = \mathbf{Y}(z = 0)$ represents the set of initial conditions.

We define:

$$\mathbf{P}_k(z) = \exp(\mathbf{M}_k z);$$

$$\mathbf{Q}_k(z) = \int_0^z \exp\{\mathbf{M}_k(z - \tau)\} \mathbf{b}_k\, d\tau. \tag{9.90}$$

We can write the solution as a function of the initial conditions:

$$\mathbf{Y}_k(z) = \mathbf{P}_k(z) \mathbf{Y}_{0,k} + \mathbf{Q}_k(z). \tag{9.91}$$

9.3.3.1 Boundary Conditions

The boundary conditions in Eq. (9.84) must now be expressed in terms of the new matrices:

$$\mathbf{H}_{k-1} \mathbf{Y}_{k-1}(z_k) - (\mathbf{H}_k + \Xi_k) \mathbf{Y}_k(z_k) = \mathbf{G}_k \quad \text{for } k = 2, \ldots, N \tag{9.92}$$

$$\mathbf{F}_1 \mathbf{Y}_1(z_1) = -\mathbf{g}_1 \quad \text{for } k = 1$$
$$\mathbf{F}_N \mathbf{Y}_N(z_{N+1}) = \mathbf{g}_{N+1} \quad \text{for } k = N + 1 \tag{9.93}$$

where

$$\mathbf{H}_k = \begin{bmatrix} I & 0 \\ 0 & -\mathbf{u}_k \end{bmatrix}, \quad \Xi_k = \begin{bmatrix} 0 & 0 \\ S_k & 0 \end{bmatrix}, \quad \text{and} \quad \mathbf{G}_k = \begin{bmatrix} 0 \\ -\mathbf{g}_k \end{bmatrix}. \tag{9.94}$$

$$\mathbf{F}_1 = \begin{bmatrix} -S_1 & \mathbf{u}_1 \end{bmatrix} \quad \text{and} \quad \mathbf{F}_N = \begin{bmatrix} S_{N+1} & \mathbf{u}_N \end{bmatrix}. \tag{9.95}$$

9.3.4 Set of Equations to be Solved

The set initial conditions, $\mathbf{Y}_{0,k}$, in the solution Eq. (9.91) can be calculated from the set of equations imposed by the boundary conditions. Therefore, we want to express such equations in terms of $\mathbf{Y}_{0,k}$ for each segment.

Taking into account that at $z = L_{k-1}$, $\mathbf{Y}_{k-1}(L_{k-1}) = \mathbf{P}_{k-1}(L_{k-1}) \mathbf{Y}_{0,k-1} + \mathbf{Q}_{k-1}(L_{k-1})$, we finally get to the set of equations to be solved:

$$\mathbf{H}_{k-1}[\mathbf{P}_{k-1}(L_{k-1})\mathbf{Y}_{0,k-1} + \mathbf{Q}_{k-1}(L_{k-1})]$$
$$- (\mathbf{H}_k + \Xi_k)\mathbf{Y}_{0,k} = \mathbf{G}_k \quad \text{for } k = 2, \ldots, N \tag{9.96}$$

$$\mathbf{F}_1 \mathbf{Y}_{0,1} = -\mathbf{g}_1 \quad \text{for } k = 1$$
$$\mathbf{F}_N[\mathbf{P}_N(L_N)\mathbf{Y}_{0,N} + \mathbf{Q}_N(L_N)] = \mathbf{g}_{N+1} \quad \text{for } k = N + 1 \tag{9.97}$$

We can compact Eqs. (9.96) and (9.97) together and find a system of $(2 \times N_{\text{gas}} \times N)$ equations for the $(2 \times N_{\text{gas}} \times N)$ variables:

$$\begin{bmatrix} \mathbf{F}_1 & 0 & 0 & \cdots & \cdots & 0 & 0 \\ \mathbf{H}_1 \mathbf{P}_1(L_1) & -(\mathbf{H}_2 + \Xi_2) & 0 & & & 0 & 0 \\ 0 & \mathbf{H}_2 \mathbf{P}_2(L_2) & -(\mathbf{H}_3 + \Xi_3) & \cdots & \cdots & 0 & 0 \\ 0 & \vdots & \vdots & & & 0 & 0 \\ \cdot\cdot & \vdots & \vdots & \ddots & \ddots & \cdots & \cdots \\ 0 & 0 & 0 & \cdots & \cdots & \mathbf{H}_{N-1}\mathbf{P}_{N-1}(L_{N-1}) & -(\mathbf{H}_N + \Xi_N) \\ 0 & 0 & 0 & \cdots & \cdots & 0 & \mathbf{F}_N \mathbf{P}_N(L_N) \end{bmatrix}$$

$$\begin{bmatrix} Y_{0,1} \\ Y_{0,2} \\ Y_{0,3} \\ \vdots \\ Y_{0,N-1} \\ Y_{0,N} \end{bmatrix} = \begin{bmatrix} -g_1 \\ G_2 - H_1 Q_1(L_1) \\ G_3 - H_2 Q_2(L_2) \\ \vdots \\ G_N - H_{N-1} Q_{N-1}(L_{N-1}) \\ g_{N+1} - F_N Q_N(L_N) \end{bmatrix} \tag{9.98}$$

$$\text{SolvMatrix} = \begin{bmatrix} F_1 & 0 & 0 & \cdots & \cdots & 0 & 0 \\ H_1 P_1(L_1) & -(H_2 + \Xi_2) & 0 & \cdots & \cdots & 0 & 0 \\ 0 & H_2 P_2(L_2) & -(H_3 + \Xi_3) & \cdots & \cdots & 0 & 0 \\ 0 & \vdots & \vdots & \cdots & \cdots & 0 & 0 \\ \vdots & \vdots & \vdots & \ddots & \ddots & \vdots & \vdots \\ 0 & 0 & 0 & \cdots & \cdots & H_{N-1} P_{N-1}(L_{N-1}) & -(H_N + \Xi_N) \\ 0 & 0 & 0 & \cdots & \cdots & 0 & F_N P_N(L_N) \end{bmatrix} \tag{9.99}$$

$$\mathbf{w} = \begin{bmatrix} -g_1 \\ G_2 - H_1 Q_1(L_1) \\ G_3 - H_2 Q_2(L_2) \\ \vdots \\ G_N - H_{N-1} Q_{N-1}(L_{N-1}) \\ g_{N+1} - F_N Q_N(L_N) \end{bmatrix} \tag{9.100}$$

$$\text{SolvMatrix} \cdot \mathbf{Y}_0 = \mathbf{w} \quad \Rightarrow \quad \mathbf{Y}_0 = \text{SolvMatrix}^{-1} \cdot \mathbf{w} \tag{9.101}$$

9.3.5 'Single Gas Model' Against 'Multi-Gas Model'

The VASCO code has been validated against the analytical solution, where possible. Moreover, in Ref. [17] the code results have been successfully compared to a different code called PRESSURE, used previously for vacuum calculations [18], which uses the single gas model.

The sensitivity of the models used in the VASCO program (single and multi-gas models) to parameter variation was also studied in [17]. In this section we report only the analysis of the effect of the matrix c_k parameters in Eq. (9.82), since the major novelty of the VASCO code is the introduction of the multi-gas model, i.e. the cross interaction between gas species expressed by this matrix.

The vacuum system geometry for this analysis consists of two (arbitrarily chosen) segments with the same diameter (91.63 mm) and length (3 m), as for the Ion Store Ring machine, with three lumped pumps with pumping speed independent of gas species, two at the extremities ($S_{z=0} = S_{z=6\,m} = 1.2\,\text{m}^3/\text{s}$) and one between segments ($S_{z=3\,m} = 0.6\,\text{m}^3/\text{s}$) at room temperature (300 K). The parameters were estimated from lab measurements, as discussed in detail in [19]. In particular, the ionisation cross section of residual gas interacting with protons are extrapolated to 7 TeV protons in Ref. [20], electron-induced desorption measured for baked copper in Ref. [21], photon-induced desorption yields as in [22], and thermal outgassing from [23]. The values of electron and photon flux are typical values for the LHC [24]:

$$\sigma = \begin{bmatrix} 4.45 \times 10^{-23} & 0 & 0 & 0 \\ 0 & 3.18 \times 10^{-22} & 0 & 0 \\ 0 & 0 & 2.75 \times 10^{-22} & 0 \\ 0 & 0 & 0 & 4.29 \times 10^{-22} \end{bmatrix}$$

$$\xi = \begin{bmatrix} 1.73 \times 10^{-3} & 6.46 \times 10^{-5} & 4.52 \times 10^{-4} & 3.87 \times 10^{-4} \end{bmatrix} T$$

$$\eta = \begin{bmatrix} 1.50 \times 10^{-4} & 4.00 \times 10^{-6} & 1.50 \times 10^{-5} & 2.50 \times 10^{-5} \end{bmatrix} T$$

$$q_t = \frac{1.33 \times 10^{-5}}{k_B T} \begin{bmatrix} 1 \times 10^{-12} & 5 \times 10^{-15} & 1 \times 10^{-14} & 5 \times 10^{-15} \end{bmatrix} T$$

$$\Theta = 1.2 \times 10^{14}$$
$$\Gamma = 3 \times 10^{15}$$

For the multi-gas model, the ion-induced desorption yields come from measurements on baked Cu samples performed in [25] as a function of the incident ion at 5 keV ion incident energy and scaled at 300 eV (ion energy as estimated in field free regions of the LHC Long Straight Section [12]) using the scaling factor from [20]:

$$\chi = \begin{bmatrix} 0.54 & 0.54 & 0.54 & 0.54 \\ 0.04 & 0.05 & 0.07 & 0.11 \\ 0.25 & 0.29 & 0.29 & 0.33 \\ 0.14 & 0.14 & 0.14 & 0.14 \end{bmatrix}$$

For the single gas model, the ion-induced desorption yields for each gas species (corresponding to elements on the diagonal) have been chosen as the average between all the desorption yields for that species from the different ions weighted by the respective ionisation cross section:

$$\chi_i = \frac{\sum_j \{\chi_{(A_i, A_j^+), k} \sigma_j\}}{\sum_j \sigma_j}.$$

The results of density calculated with the multi-gas model are shown in Figure 9.9. In this case, the critical current, i.e. the value of beam current above which pressure runaway is expected, has been estimated to be 65.4 A. The hydrogen thermal outgassing and the hydrogen contributions from photon and electron desorptions are much higher than for the other gas species. Thus H_2 density at current much lower than the critical current is the highest (Figure 9.9a,b). As the current is increased, however, the net production of CO increases faster, given the high ion-induced desorption yield and the lower conductance. The gas composition changes and is dominated by CO and H_2 (Figure 9.9c). When the critical current is approached, CO species is the major component of the residual gas and determines the critical current (Figure 9.9d). The evolution of the density and composition with beam current reproduces what was observed in the ISR when pressure runaway occurred [2].

If the equivalent *single gas model* is used, the value of critical current results 148.6 A, determined by CO. As shown in Figure 9.10, the gas density at low

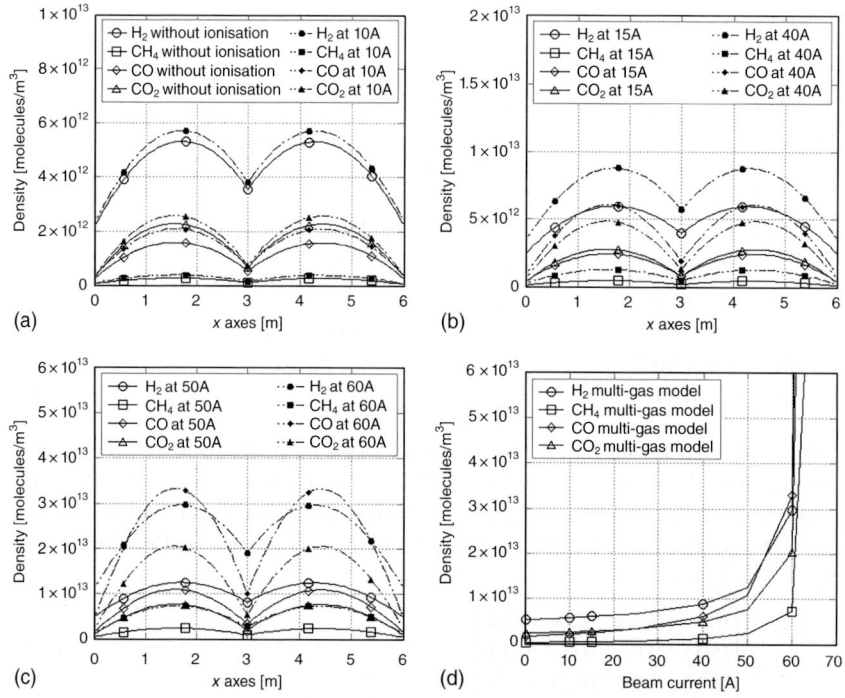

Figure 9.9 Multi-gas model: (a) density profile along the pipe axes calculated at 0 and 10 A beam current; (b) at 15 and 40 A; (c) at 50 and 60 A. (d) Gas density at 1.7 m location as a function of the beam current. Note that the vertical scale changes from figure to figure.

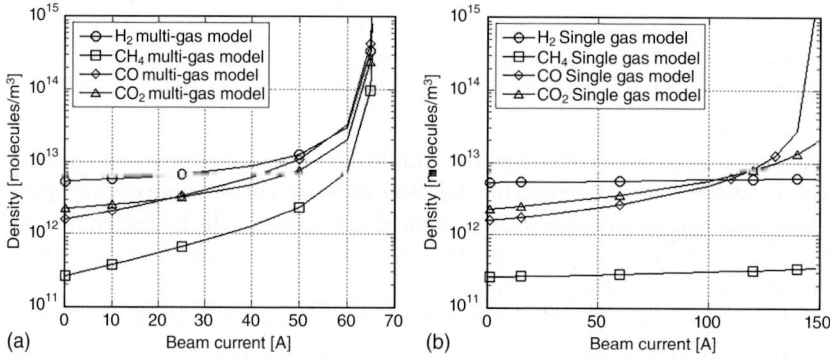

Figure 9.10 Gas density at $z = 1.7$ m with location as a function of the beam current for (a) multi-gas model and (b) single gas model.

beam current is the same for the two cases, since the ion-induced desorption contribution is negligible with respect to other gas sources. When increasing the beam current, the gas composition changes and the single gas model are dominated by CO and CO_2, contrary to the experimental observations [2].

In conclusion, the multi-gas model seems to better reproduce experimental results. The model can be used when designing an accelerator for positrons or hadrons, and to calculate the expected critical current and determine the parameters (material, vacuum conditioning, vacuum chamber diameter, pumping, distributed pumping, etc.) of the vacuum system. A factor of 2 minimum should nevertheless be taken as safety margin, given that ion-induced desorption yield are measured with large error bars (see Section 9.5) and that, with the VASCO code, the critical current is calculated by increasing the beam current progressively until the results diverge. This of course does not take into account any time evolution and progressive increase of pressure, and therefore ions, which will probably cause a pressure runaway at lower current values.

The model and the VASCO code have been used to design the vacuum chambers in the experimental regions and the warm long straight sections of the LHC [26] as well as to estimate the residual gas pressure to expect in the machine [27], in order to then assess the background noise to the LHC experiments due to beam–gas interactions.

9.4 Energy of Ions Hitting Vacuum Chamber

The ISD yields depend on the energy and mass of ions bombarding the vacuum chamber walls (see Chapter 4); thus calculation of the ion energy for each type of ions is essential for analysis of ion-induced pressure instability. The electron loss ionisation has the highest cross section; thus the positively charged ions are under consideration here. Immediately after ionisation, the newly born ions are exposed to the electric field of a positive space charge of the positron, proton, or ion beam, and, therefore, the ions are repelled from the beam and accelerated in this field towards the vacuum chamber wall. Thus the ions impact with vacuum chamber walls with an energy that depends on the beam current, on the beam r.m.s., bunch spacing, on the ion mass and the position where the residual gas molecule was ionised, and the cross-sectional dimension of the vacuum chamber [8, 9, 13, 28, 29].

9.4.1 Ion Energy in the Vacuum Chamber Without a Magnetic Field

9.4.1.1 Circular Beams

Let us consider a circular beam with a Gaussian profile. The time-averaged electric field of the beam can be given in SI units by

$$E = \frac{I}{2\pi\varepsilon_0 cr}\left(1 - e^{-\left(\frac{r}{\sigma_r}\right)^2}\right);\qquad(9.102)$$

where

I is the proton beam current
$\varepsilon_0 = 8.85 \times 10^{-12}$ [F/m] is the permittivity of free space
c is the speed of light in vacuum
σ_r is the r.m.s. beam size, $\sigma_r = \sqrt{\beta\varepsilon_n/\gamma}$
r is the distance from the centre of beam to the ion

In the estimation with a continuous (unbunched) beam, the ions arrive at the vacuum chamber wall with a kinetic energy equal to the difference in potential between the point of the ionisation and the wall:

$$W(a) = \int_a^R E(r)dr; \tag{9.103}$$

where a is the radial position where the molecule was ionised and R is the internal radius of the vacuum chamber. The probability of ionisation $\rho(a)$ for the residual gas molecules is proportional to the Gaussian distribution of beam particles and the initial radial position, a, of the ion (i.e. where a gas molecule was ionised):

$$\rho(a) \propto 2\pi r e^{-\left(\frac{a}{\sigma_r}\right)^2} \tag{9.104}$$

Then numerical integrating of equation (9.103) with K different initial radial positions, a_k (for example, $a_k = 3\sigma_r/k$, $k = 1, 2, \ldots, K$), gives the average value of ion energies for unbunched beam:

$$\langle W_u \rangle = \sum_{k=1}^{K} w_k W(r_k) \tag{9.105}$$

where w_k is weight of $W(a_k)$: $w_k = \frac{\rho(a_k)}{\sum_{j=1}^{N} \rho(a_j)}$.

The estimation described above does not take into account the effect of a bunched beam. For example, the bunch length in the LHC is $\tau = 0.257$ ns and the bunch spacing is $T = 24.95$ ns, i.e. $T/\tau \approx 97$. Thus, a peak electric field (in the presence of bunch of particles) is a factor T/τ higher than an average electric field value:

$$E_{peak} = \frac{I}{2\pi\varepsilon_0 cr}\left(1 - e^{-\left(\frac{r}{\sigma_r}\right)^2}\right)\frac{T}{\tau} \tag{9.106}$$

A newly born ion is accelerated by the peak electric field during the bunch passage and then it drifts with a constant velocity until the next bunch arrives. An estimation of its final velocity can be obtained by numerical integration. The iteration formulae for ion velocity and the radial position in the presence of a bunch can be written as

$$\begin{cases} v_n = v_{n-1} + E_{peak}\dfrac{q}{m} \cdot \Delta t; \\ r_n = r_{n-1} + v_n \cdot \Delta t; \end{cases} \tag{9.107}$$

where $\Delta t = \tau/N$ is the time interval, $n = 1, 2, \ldots, N$. The time interval should be small enough so as not to influence the final result. This requirement was found to be satisfactory for $N = 1000$. Between two bunches the electric field is equal to zero; thus the ions are drifting with energies gained before and their velocities can be described as

$$\begin{cases} v_d = v_N; \\ r_d = r_N + v_N T. \end{cases} \tag{9.108}$$

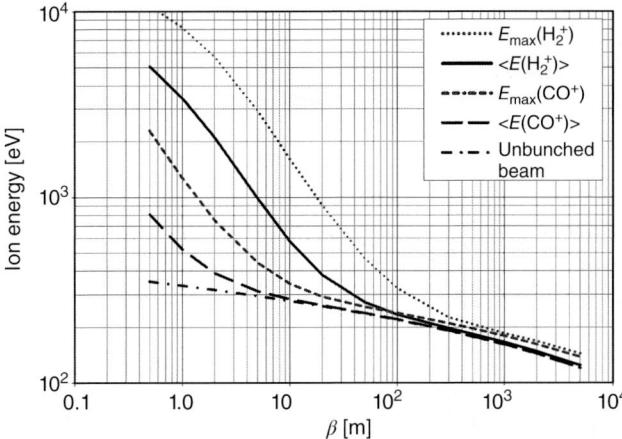

Figure 9.11 The ion energy in the LHC beam chamber as a function of β-function for a beam current $I = 0.85$ A in a vacuum chamber with $R = 25$ mm.

The ion can be born at a different radial positions and anywhere along the length of the bunch: in the head, in the middle part, or at the tail. The position along the bunch can be described in terms of time, i.e. the duration of acceleration of the ion by the first bunch: $\tau_1 = \frac{m}{M}\tau$, $m = 1, 2, \ldots, M$. The formula for the average ion energy of bunched beam is

$$\langle W_b \rangle = \frac{1}{M} \sum_{m=1}^{M} \sum_{k=1}^{K} w_k W_k \left(a_k, \frac{m}{M}\tau\right) \tag{9.109}$$

It is obvious that the ion energy calculated for a bunched beam with Eq. (9.109) has upper and lower limits:

- The ion energy calculated for a bunched beam could not be lower than for unbunched beam with the same beam current and β-function calculated with Eq. (9.105).
- The maximum ion energy calculated for a bunched beam could not be higher than for unbunched beam calculated for peak electric field E_{peak}, i.e.:

$$\langle W_u \rangle \leq \langle W_b \rangle \leq \frac{T}{\tau} \langle W_u \rangle \tag{9.110}$$

For example, the results of ion energy calculation for H_2^+ and CO^+ ions as a function of β-function in the LHC magnetic-field-free sections are presented in Figure 9.11 for the following parameters: $\varepsilon_n = 3.75 \times 10^{-6}$ m·rad, $\gamma = 7460.6$, $\tau = 0.257$ ns, $T = 24.95$ ns, and $I = 0.85$ A. For the large values of β-function (i.e. $\beta > 300$ m for H_2^+ and $\beta > 10$ m for CO^+), the average ion energy of bunched beam calculated with Eq. (9.109) is practically the same as for an unbunched beam calculated with Eq. (8.4). For smaller values of β-function, Eq. (8.4) significantly underestimates the ion energy. It is also worth mentioning that H_2^+ ions gain higher energy than CO^+ ions due to a smaller mass.

The maximum values of ion energies $W_{\text{max}}(H_2^+)$ and $W_{\text{max}}(CO^+)$ calculated with Eq. (9.109) and shown in Figure 9.11 could be up to a factor 2.8 larger than

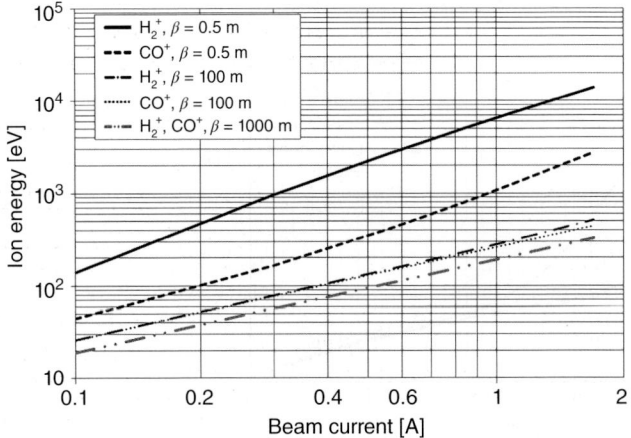

Figure 9.12 The average ion energy in the LHC beam chamber as a function of beam current in a vacuum chamber with $R = 25$ mm.

the average values. However, $W_{max}(H_2^+)$ and $W_{max}(CO^+)$ are still much below the upper limit shown in Eq. (9.110). That means that no ions reached the wall of vacuum chamber within a bunch passing time τ. Thus, no simple formula can be used for accurate calculation of average ion energies, and numerical integration procedure described above should be applied.

The H_2^+ and CO^+ average ion energies as a function of beam current are shown in Figure 9.12 for three β-function values: 0.5, 100, and 1000 m. The ion energy linearly increases with a beam current when $\beta = 100$ and 1000 m, while the ion energy is approximately proportional to $I^{1.6}$ when $\beta = 0.5$ m.

The H_2^+ and CO^+ average ion energy increases with a beam chamber radius as shown in Figure 9.13 for two β-function values, 100 and 1000 m, and for $I = 0.85$ A. However, when $\beta = 0.5$ m, the ion energy increases insignificantly with a beam chamber radius: $5.07 \text{ keV} \leq W_{max}(H_2^+) \leq 5.11 \text{ keV}$ and $788 \text{ eV} \leq W_{max}(CO^+) \leq 845 \text{ eV}$.

In particle colliders, two beams may coexist in the same beam pipe at the locations close to the interaction region. In this case, the ion energy is not uniform along the beam chamber. It depends on whether or not the beams arrive at a fixed cross section simultaneously or with a time delay. The highest average ion energy is reached in places where the beams arrive simultaneously and then in the lowest one where the beams arrive in anti-phase.

9.4.1.2 Flat Beams

Flat beams are the beams with one transversal beam size much greater than another one. Let us consider $\sigma_x \gg \sigma_y$, where σ_x and σ_y are the horizontal and vertical transversal r.m.s. beam sizes. In this case, the probability of ionisation $\rho(x, y)$ of the residual gas molecules is proportional to a Gaussian distribution of

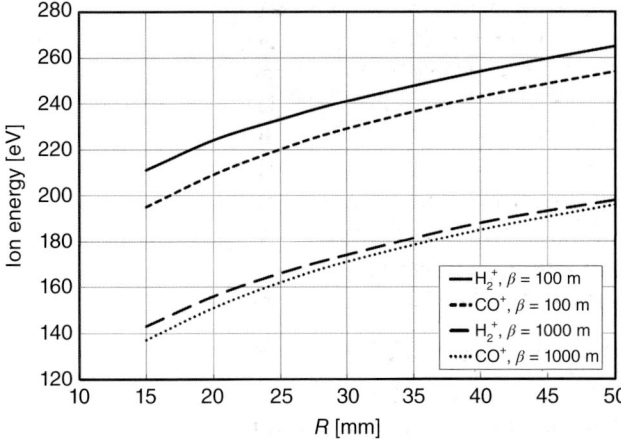

Figure 9.13 The average ion energy in the LHC beam chamber as a function of a beam chamber radius.

charges particles in the bunch and the molecule position (x, y) at the ionisation time:

$$\rho(x, y) \propto 2\pi r e^{-xy/(\sigma_x \sigma_y)}; \tag{9.111}$$

The electric field, E_b, of the flat bunch can be described with the Bassetti–Erskine formula (see, e.g. [30]) adopted for flat beams as follows [31]:

$$E(x, y) = \frac{q_b}{4\pi\varepsilon_0 l_b \sigma_x}$$

$$\left| \sqrt{\frac{8}{\pi}} \cdot \frac{y\sigma_x}{x^2 + y^2} \left[1 - \frac{(y^2 - 3x^2)\sigma_x^2}{(x^2 + y^2)^2} \right] \quad \text{for } x, y > 3\sigma_x; \right.$$

otherwise :

$$\times \left| \exp\left[-\left(\frac{x-iy}{\sqrt{2\sigma_x}}\right)^2\right] \operatorname{erf}\left(\frac{y\frac{\sigma_x}{\sigma_y} + ix\frac{\sigma_y}{\sigma_x}}{\sqrt{2\sigma_x}}\right) \right.$$

$$- \exp\left[-\left(\frac{x+iy}{\sqrt{2\sigma_x}}\right)^2\right] \operatorname{erf}\left(\frac{y - ix}{\sqrt{2\sigma_x}}\right) \tag{9.112}$$

$$+ \exp\left[-\left(\frac{x+iy}{\sqrt{2\sigma_x}}\right)^2\right] \operatorname{erf}\left(\frac{y\frac{\sigma_x}{\sigma_y} - ix\frac{\sigma_y}{\sigma_x}}{\sqrt{2\sigma_x}}\right)$$

$$- \exp\left[-\left(\frac{x-iy}{\sqrt{2\sigma_x}}\right)^2\right] \operatorname{erf}\left(\frac{y + ix}{\sqrt{2\sigma_x}}\right) \left|\right.$$

The following analysis is similar to the one described for the circular beams only with the difference that instead of varying an initial radial position of the ion, two

Table 9.7 Ion energy calculated in the ILC positron damping ring with $\tau = 0.02$ ns, $T = 3.08$ ns, and $I = 0.4$ A for different beam sizes and different components.

		Straights (magnetic field free)	Arcs (dipole magnetic field)	Wigglers (alternating magnetic field)
σ_x [m]	max	1.3×10^{-3}	1.3×10^{-3}	2.7×10^{-3}
	min	2.7×10^{-4}	6.5×10^{-4}	1.9×10^{-4}
σ_y [m]	max	1.0×10^{-5}	8.9×10^{-6}	5.5×10^{-6}
	min	5.6×10^{-6}	5.6×10^{-6}	3.8×10^{-6}
E [eV]	max	320	265	340
	min	220	220	320

coordinates x_0 and y_0 are varied. For example, the ion energies were calculated for different parts of the ILC positron damping ring: arcs, long straights, and wigglers [15]. Beam parameters were $\tau = 0.02$ ns, $T = 3.08$ ns, and $I = 0.4$ A. For the beam sizes σ_x and σ_y varying along the ring, one can find their maximum and minimum values.

The results of calculations in Table 9.7 demonstrate that the ion energy varies with the beam sizes between 220 eV for largest σ_x and σ_y and 340 eV for smallest in ILC DR; however no difference in energy was found for H_2^+, CO^+, and CO_2^+. It was also found that variation of the bunch length between 6 and 9 mm does not affect the result.

9.4.2 Ion Energy in a Vacuum Chamber with a Magnetic Field

Along most of the length of the LHC, the vacuum chambers can be located inside magnetic elements: dipoles, quadrupoles, or solenoids. The magnetic field will bend the ion, accelerated by the electric field, and the energy of ion bombarding in the vacuum chamber wall can then differ from the estimation made without magnetic field. Furthermore, the angle of incidence may go from perpendicular to grazing, which corresponds to larger ISD yields.

To include the effect of magnetic field, the iteration formulas (9.107) and (9.108) for the ion velocity and the radial position in magnetic field $\vec{B} = (B_x, B_y, B_z)$ should be rewritten as

$$\begin{cases} \vec{v}_n = \vec{v}_{n-1} + \dfrac{q}{m}(\vec{E} + \vec{v}_{n-1} \times \vec{B}) \cdot \Delta t, \\ \vec{r}_n = \vec{r}_{n-1} + \vec{v}_n \cdot \Delta t; \end{cases} \tag{9.113}$$

where $\vec{E} = \vec{E}_{peak}$ during the bunch passage and $\vec{E} = 0$, in another case.

Three cases were studied for vacuum chamber in the following:

(1) The dipole magnetic field (all formulation is also applicable to wigglers).
(2) The quadrupole magnetic field.
(3) Solenoid magnetic field (could be applied in particle detectors).

9.4.2.1 Vacuum Chamber in a Dipole Magnetic Field

The dipole magnetic field strength in the accelerators may vary significantly from $B = 1.4$ T in room temperature machines to $B = 8.4$ T in the LHC arcs and 16–20 T in the FCC. The Eq. (8.13) for a dipole magnetic field $\vec{B} = (0, B, 0)$ can be written in a more detailed form such as g

$$\begin{cases} vx_n = vx_{n-1} + \dfrac{q}{m}(E\cos\alpha - vz_{n-1}B) \cdot \Delta t; \\ vy_n = vy_{n-1} + \dfrac{q}{m}E\sin\alpha \cdot \Delta t; \\ vz_n = vz_{n-1} + \dfrac{q}{m}vx_{n-1}B \cdot \Delta t; \\ x_n = x_{n-1} + vx_n \cdot \Delta t; \\ y_n = y_{n-1} + vy_n \cdot \Delta t; \\ z_n = z_{n-1} + vz_n \cdot \Delta t. \end{cases} \qquad (9.114)$$

The results of estimations of ion energies in a dipole magnetic field with the LHC beam parameters used above shows that H_2^+ ion energy increase by 10% in $B = 1.4$ T and by 15% in $B = 8.4$ T, while the energy of CO^+ does not change. The results of calculations for the ILC DR in Table 9.7 demonstrate that the ion energy was mainly affected by the beam size at different locations and the effect of magnetic field was found to be insignificant.

It should be noted that since the ions drift along the magnetic field lines, they will bombard only the top and bottom of a vacuum chamber in a dipole (two strips along the beam path). The incident angle of ions varies between normal and very greasing angles.

9.4.2.2 Vacuum Chamber in a Quadrupole Magnetic Field

From the point of view of vacuum stability, the vacuum chambers inside the quadrupoles are the most critical elements. The quadrupole magnetic field can be described as

$$\begin{cases} B_x = Gr\sin\alpha, \\ B_y = Gr\cos\alpha; \end{cases} \qquad (9.115)$$

where G is the gradient of the quadrupole magnetic field, α is the angle of radius vector $\vec{r}: x = r\cos\alpha, y = r\sin\alpha$.

Equation (9.114) can be rewritten in a more detailed form in this case such as

$$\begin{cases} vx_n = vx_{n-1} + \dfrac{q}{m}(E - vz_{n-1}B)\cos\alpha \cdot \Delta t; \\ vy_n = vy_{n-1} + \dfrac{q}{m}(E + vz_{n-1}B)\sin\alpha \cdot \Delta t; \\ vz_n = vz_{n-1} + \dfrac{q}{m}(vx_{n-1}B\cos\alpha - vy_{n-1}B\sin\alpha) \cdot \Delta t; \\ x_n = x_{n-1} + vx_n \cdot \Delta t; \\ y_n = y_{n-1} + vy_n \cdot \Delta t; \\ z_n = z_{n-1} + vz_n \cdot \Delta t. \end{cases} \qquad (9.116)$$

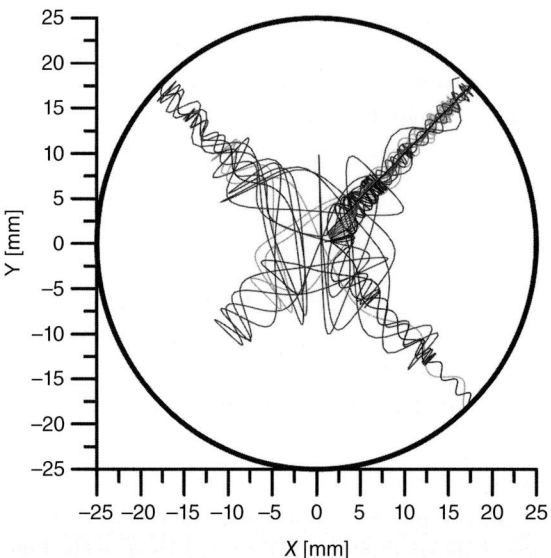

Figure 9.14 Examples of ion trajectories in a quadrupole field.

The estimations of the ion energy in the LHC quadrupoles were made for the maximum gradient of the quadrupole magnetic field $G = 240$ T/m. It was found that the H_2^+ energy is higher by 1.3–1.7 times in the presence of the quadrupole magnetic field, while CO^+ ions have practically the same energy in both cases.

Figure 9.14 shows the examples of H_2^+ and CO^+ trajectories with the initial angle of the radius vector between 0° and 45°. The axes X and Y correspond to the vacuum chamber cross section. The ions bombard four ~4 mm strips along a 50 mm diameter vacuum chamber in a quadrupole, i.e. about 10% of vacuum chamber surface. The ion migration along the vacuum chamber axis Z does not exceed the diameter of the vacuum chamber.

9.4.2.3 Vacuum Chamber in a Solenoid Magnetic Field

A strong solenoid field can be used in particle detectors at the interaction regions of colliders, for example, the strength of magnetic fields the LHC detectors is 2 T in ATLAS and 4 T in CMS. The gas density and the vacuum stability of these elements are very important parameters, which also depend on energy of ions bombarding the walls of vacuum chamber.

The iteration formula (9.114) for the ion velocity and the radial position in a solenoid magnetic field can be rewritten in that case as

$$\begin{cases} vx_n = vx_{n-1} + \dfrac{q}{m}(E \cos \alpha + vy_{n-1} B) \cdot \Delta t; \\ vy_n = vy_{n-1} + \dfrac{q}{m}(E \sin \alpha - vx_{n-1} B) \cdot \Delta t; \\ vz_n = vz_0; \\ x_n = x_{n-1} + vx_n \cdot \Delta t; \\ y_n = y_{n-1} + vy_n \cdot \Delta t; \\ z_n = z_0 + vz_0 \cdot \Delta t \cdot n. \end{cases} \quad (9.117)$$

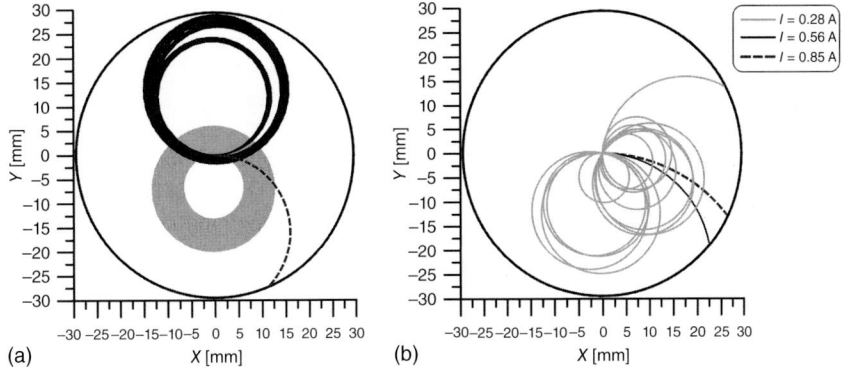

Figure 9.15 Examples of (a) H_2^+ and (b) CO^+ ion trajectories in a 2-T solenoid field.

The estimations of the ion energy in solenoid field were performed for the magnetic field with $B = 2$ T and $B = 4$ T in a vacuum chamber with diameter of 58 mm. The computation will stop when either the ion reaches the wall or after 1000 bunches passed. This means that the ion having an initial energy of about 1 eV can move longitudinally no more than 0.25 m: i.e. the ion remains practically with the same coordinate z. Few examples of the ion trajectories near the interaction point with $\beta = 0.5$ m are shown in Figure 9.15 for ions born at $r_0 = \sigma_r$.

Studying the ion energies with iteration calculations requires long computing time; however a simple estimation of energy range is made analytically. The ions bend in the solenoid magnetic field and may reach a vacuum chamber wall if the bending radius is larger than half radius of the vacuum chamber; otherwise the ion will return to the centre of the vacuum chamber. During their numerous circulation, the ions strongly accelerate or decelerate when they travel close to the beam (inside the beam with size of about σ_r). They will circulate between the centre and the wall of the vacuum chamber until the ion gains sufficient energy that its bending radius will be larger than half radius of vacuum chamber. In a strong magnetic field or for large β, the ions circulates around a beam. The ion energy and bending radius increase slowly and the ions may not reach the vacuum chamber walls even after 1000 bunches.

The bending radius r of the ion depends on its velocity v as

$$r = \frac{mv}{qB} \tag{9.118}$$

The period T_{ion} is

$$T_{ion} = \frac{2\pi r}{v} = \frac{2\pi m}{qB} \tag{9.119}$$

which can be compared with the bunch spacing T.

The minimum energy of ions reaching a wall in a solenoid magnetic field can be estimated as

$$E_1 = \frac{(qBr)^2}{2m} \approx \frac{(qBR)^2}{8m} \tag{9.120}$$

Table 9.8 The minimum energy of ions reaching a wall in a solenoid magnetic field in the LHC beam chamber.

	B = 2 T		B = 4 T	
[keV]	H_2^+	CO^+	H_2^+	CO^+
E_0	13.6	2.7	13.6	2.7
E_1	20	1.45	80	5.8
E_2	36	12	100	12
E_3	80	5.8	325	23
E_{min}	20	2.7	80	5.8
E_{max}	80	12	325	23
E'_{max}	~35	~8	~270	~8

The minimum impact energy can be estimated as a maximum of two values E_1 and is the energy, estimated in case of without the magnetic field, E_0:

$$E_{min} = \max\{E_0, E_1\} \quad (9.121)$$

The maximum impact energy can be estimated as a maximum of two values:

$$E_{max} = \max\{E_2, E_3\} \quad (9.122)$$

First value is the energy E_2 that the ion will get if it arrives about the centre of a vacuum chamber simultaneously with a bunch. This energy can be estimated with numerical integration. The second value, E_3 comes from maximal possible bending radius in a vacuum chamber, which is about the same as a radius of vacuum chamber:

$$E_3 \leq \frac{(qBR)^2}{2m} \quad (9.123)$$

In general, the case when $E_0 > E_{max}$ means that magnetic field has no effect on the ion energy.

The results of the analytical estimations of H_2^+ and CO^+ ion energies E_0, E_1, E_2, E_3, and minimum and maximum impact energies, E_{min} and E_{max}, is shown in Table 9.8 in comparison with a result obtained with iteration numerical estimation, E'_{max}.

It is important to point out two important results for ions in a solenoid field:

- The ions will bombard a wall at grazing incident angles, which leads to higher ISD yields than at normal incident.
- The ions with energy in a few kiloelectron volts circulating in a magnetic field with a lifetime of ~1000 bunch passes can also ionise the residual gas molecules. This effect was not considered in the models discussed earlier.

9.5 Errors in Estimating the Critical Currents I_c

Each of the experimental or calculated parameters used for estimations of the critical current is affected by errors, and these errors propagate in the

calculations. The parameters to which the critical current values are most sensitive are:
- Beam–gas ionisation cross sections
- Ion impact energy
- ISD yields
  o From a bare surface
  o From condensed or cryosorbed gases.
- Pumping
  o Vacuum conductivity of vacuum chamber
  o Sticking probability of cryogenic or getter-coated vacuum chamber
  o Lumped pumping speed

9.5.1 Beam–Gas Ionisation

Beam–gas ionisation cross sections can be calculated with Eq. (1.7) (see Chapter 1) under the assumption that the model is correct. However, the results reported in Ref. [32] show that the ionisation cross sections could be smaller by a factor of 1.2–1.7. As not experts on beam–gas interaction, the authors prefer to use a higher value obtained with Eq. (8.23), as it is either a correct or overestimated value leading to a safety margin in the critical current calculations.

In the machines with a significant flux of SR, the photons have a finite probability of ionising the residual gas in the chamber. To take this effect into account, the ionisation cross section in the equation describing the gas density evolution has to be included by an amount equal to the photo-ionisation probability. This intensifies the 'gain' in the ion desorption stability problem. The gas density corresponding to a certain value of the beam current will be therefore higher. It is estimated that the photo-ionisation probability can be up to a factor of 10 times larger than the probability of ionisation from protons. This effect could therefore be of importance and should be studied in a future work.

The high intensity machines may also be suffering the BIEM (see Chapter 8). The multipacting electrons can have energies in the range between 0 and 1000 eV, these electrons have a high ionisation cross section, and their addition impact on gas ionisation might be non-negligible. To include this effect, one needs to know the electron energy distribution and electron density spatial distribution. Although this effect was not included in the model yet, it could also be of importance and should also be studied in a future work.

9.5.2 Ion Impact Energy

Ion impact energy can be calculated accurately for the defined beam parameters. Since the transverse r.m.s. beam sizes vary along the beam path, the ion impact energy varies as well. Thus, the ion impact energy distribution along the beam chamber should be carefully considered.

9.5.3 Ion-Stimulated Desorption Yields

ISD yields have two sources of error: the ion impact energy discussed in the preceding text and the ISD yield experimental data. For the latter, the error can be estimated to be a factor of 0.5–2.

An additional uncertainty in the ISD yields are associated with lack of experimental data for surfaces irradiated with SR. Although it is clear that exposure to SR should reduce ISD yields, there is no data on ISD yields as a function of photon dose to put into the model. Using the available ISD data (i.e. without SR conditioning) leads to some unknown safety margin in the critical current calculations.

9.5.4 Pumping

Vacuum conductivity of a vacuum chamber can be calculated with test particle Monte Carlo (TPMC) method with an accuracy of ~0.1%. The source of errors is the difference between modelled and real vacuum chamber geometries. The difference of 1% in vacuum chamber cross section can lead to ~3% in vacuum conductivity, which is negligible in comparison to other uncertainties.

Sticking probability of cryogenic surfaces may vary with surface roughness, surface condition, vacuum chamber temperature, etc. It would be safe to consider a factor of 0.5–2.

Sticking probability of getter surfaces may vary with surface roughness, grain size, surface condition, activation temperature, and amount of adsorbed gas. It should also be considered that NEG coating sticking probability may reduce with a number of exposures to air, so the model should take into consideration a possible machine operation scenario, NEG activation temperature, and possible number of air vents.

Lumped pumping speed accuracy depends on a pumping principle, model, age, and condition of the pump. Thus, turbo-molecular pump (TMP) are quite stable and their pumping speed can be considered quite accurate ($\pm 10\%$). The sputter ion pump (SIP) pumping speed may vary between +50% to nominal pumping speed immediately after bakeout and −50% after long operation; it is also worth mentioning that the SIP pumping speed reduces at UHV by ~25–30% at 10^{-9} mbar and ~40–50% at 10^{-10} mbar. The titanium sublimation pumps (TSPs) are usually used in combination with SIP; thus 100 l/s SIP can provide with freshly activated TSP a combined pumping speed of 1000 l/s; this means that a pumping speed of such TSP + SIP may vary by a factor of 10. NEG cartridges can be used on their own or in combination with SIP. Degradation of NEG cartridges depends on the conditions of operation such as the total amount of gas absorbed, frequency of air vents, and activation temperature.

9.5.5 Total Error in Critical Current

The sources of error listed in the preceding text indicate how careful these errors should be considered. One can see that since critical current cannot be calculated accurately, a minimum error of a factor of 0.5–2 should be considered. However, in some cases the error may reach a factor of 0.1–8. Based on these considerations and the author's experience, it is advisable to design a machine with a safety margin of at least a factor of 3; i.e. the calculated critical current should be at least a factor of 3 greater than machine operation current.

9.6 Summary

Ion-induced pressure instability, when the avalanche-like pressure (or gas density) increase could rapidly reach a level above a specified value, could be a serious limiting factor for operation of high intensity, positively charged machines and has to be taken into consideration for their vacuum systems design. The effect of ion-induced pressure instability should be hardly visible when the system is properly designed, but may cause severe limitations in beam intensity, emittance, and energy spread, as well as a background in the detectors, if neglected.

Present gas dynamics models and available codes provide sufficient instruments to estimating critical currents and stability conditions, provided a margin is taken, for safe operations of future particle accelerators.

The model and the VASCO code presented here were used for LHC, ILC, and other machine designs. Successful LHC operations, with no pressure runaway occurrence (when vacuum was properly conditioned) proves their validity.

References

1 Gröbner, O.. (1976). The dynamic behavior of pressure bumps in the ISR. CERN/ISR-VA/76-25, 8 June 1976.
2 Calder, R.S. (1974). Ion induced gas desorption problems in the ISR. *Vacuum* 24: 437.
3 Fisher, E. and Zankel, K.. (1973). The stability of the residual gas density in the ISR in presence of high intensity proton beam. CERN-ISR-VA/73-52, 9 November 1973.
4 Gröbner, O. and Calder, R.S.. (1973). Beam induced desorption in the CERN Intersecting Storage Ring. CERN/ISR-VA/73-15, 27 February 1973.
5 Gröbner, O.. (1972). Dynamic pressure behavior in presence of beam induced gas desorption. ISR Performance Report, CERN, 19 December 1972.
6 Benvenuti, C., Calder, R., and Hilleret, N. (June 1977). A vacuum cold bore test section at the CERN ISR. *IEEE Trans. Nucl. Sci.* 24 (3): 1373–1375.
7 Benvenuti, C. and Hilleret, N. (June 1979). Cold bore experiments at CERN ISR. *IEEE Trans. Nucl. Sci.* 26 (3): 4086–4088.
8 Turner, W. (1996). Ion desorption stability in superconducting high energy physics proton colliders. *J. Vac. Sci. Technol., A* 14: 2026–2038.
9 Gröbner, O. (1995). Vacuum system for LHC. *Vacuum* 46: 797–801.
10 Collins, I.R., Gröbner, O., Lepeule, P., and Veness, R. (1998). Mechanical and vacuum stability design criteria for the LHC experimental vacuum chambers. In: *Proceedings of EPAC-98*, Stockholm, 22–26 June 1998, 2202–2204.
11 Malyshev, O.B. and Rossi, A.. (1999). Ion desorption stability in the LHC. Vacuum Technical Note 99-20, December 1999, CERN, 76pp.
12 Malyshev, O.B. and Rossi, A. (2000). Ion desorption vacuum stability in the LHC (The multigas model). In: *Proceedings of EPAC-2000*, Vienna, 22–26 June 2000, 948–950.

13 Malyshev, O.B. (2000). The ion impact energy on the LHC vacuum chamber walls. In: *Proceedings of EPAC-2000*, Vienna, 26–30 June 2000, 951–953.

14 Collins, I.R., Gröbner, O., Malyshev, O.B. et al. (1999). Vacuum stability for ion induced gas desorption. LHC Project Report 312, October 1999, CERN, 7pp.

15 Malyshev, O.B. (2010). Ion induced pressure instability in the ILC positron DR. In: *Proceedings of IPAC'10*, Kyoto, Japan, 23–28 May 2010, 3566–3568.

16 Baglin, V.. The experimental program for vacuum stability study with COLDEX at the SPS operating with the LHC beam parameters in May 2002–Nov 2004. (2004). Private communication.

17 Rossi, A. (2004). VASCO (VAcuum Stability COde): multi-gas code to calculate gas density profile in a UHV system. LHC-Project-Note-341, March 2004, CERN.

18 Code PRESSURE developed by Poncet, A. and Pace, A. at CERN. (1989). Monte-Carlo simulations of molecular gas flow: some applications in accelerator vacuum technology using a versatile personal computer program. CERN PS/89-49(ML), CERN 1989, presented at the *11th International Vacuum Congress (IVC-11)* and *7th International Conference on Solid Surface (ICSS-7)*.

19 Rossi, A. and Hilleret, N. (2003). Residual gas density estimations in the LHC experimental interaction regions. LHC Project Report 674, CERN, Geneva, 18 September 2003.

20 Mathewson, A.G.. (1976). Ion induced desorption coefficient for titanium alloy, pure aluminium and stainless steel. CERN-ISR-VA 76-05, CERN, Geneva, 1976.

21 Gómes-Goñi, J. and Mathewson, G. (1997). Temperature dependence of the electron induced gas desorption yields on stainless steel, copper, and aluminum. *J. Vac. Sci. Technol., A* 15: 3093.

22 Gómes-Goñi, J., Gröbner, O., and Mathewson, A.G. (1994). Comparison of photodesorption yields using synchrotron radiation of low critical energies for stainless steel, copper, and electrodeposited copper surfaces. *J. Vac. Sci. Technol., A* 12: 1714.

23 Mathewson, A.G., C. Reymermier, S. Zhang et al. (1995). The ALICE vacuum system. LHC Project Note 19, CERN, Geneva, 27 November 1995.

24 Zimmermann, F. and Rossi, A. (2003). Synchrotrotron Radiation in the LHC Experimental Insertions, CERN-LHC-Project-Report-675. *CERN* http://cds.cern.ch/record/645173/files/lhc-project-report-675.pdf.

25 Lozano, M.P. (2002). Ion induced desorption yield measurements from copper and aluminium. *Vacuum* 67: 339–345.

26 Collins, I., Knaster, J.R., Lepeule, P. et al. (2001). Vacuum calculations for the LHC experimental beam chambers. LHC Project Report 492, CERN, Geneva, 6 Aug 2001. Also in *Proceedings of 19th IEEE Particle Accelerator Conference*, Chicago, IL, USA, 18–22 June 2001, p. 3153.

27 Rossi, A. (2004). Residual gas density estimation in the LHC insertion regions IR1 and IR5 and the experimental regions of ATLAS and CMS for different beam operations. CERN LHC Project Report 783.

28 Collins, I.R., Gröbner, O., Lepeule, P., and Veness, R.. (1998). Mechanical and vacuum stability design criteria for the LHC experimental vacuum chamber. LHC Project Report 205, CERN, Geneva, 27 July 1998.
29 Malyshev, O.. (1999). The energy of the ions bombarding the vacuum chamber walls. Vacuum Technical Note 99-17, CERN, Geneva, November 1999.
30 Chao, A.W. and Tigner, M. (eds.) (1999). Beam dynamics. In: *Handbook of Accelerator Physics and Engineering*, 128. World Scientific Publishing Co. Pte. Ltd.
31 Tzenov, S.. (2008). The electric field of flat bunches. Private communication.
32 Mathewson, A.G. and Zhang, S.. (1996). The beam-gas ionization cross section at 7.0 TeV. Vacuum Technical Note 96-01, CERN, January 1996.

10

Pressure Instabilities in Heavy Ion Accelerators
Markus Bender

GSI Helmholtzzentrum für Schwerionenforschung, Planckstr. 1, Darmstadt 64291, Germany

10.1 Introduction

In 1973, for the first time, a pressure increase was observed during operation of the Intersecting Storage Rings (ISR) at CERN [1]. With increasing number of protons circulating in the ring, the pressure increased and the vacuum conditions became unstable. The underlying phenomenon was the ionisation of residual gas by the proton beam and subsequent acceleration of the gas ions within the beam potential towards the beam pipe [2]. Here, the release of adsorbed gas, the so-called desorption, was stimulated. It was found that the cleanliness of the vacuum chamber walls strongly affected the amount of desorbed gas. Finally the ISR vacuum system could be improved and stabilised by dedicated surface treatments, cleanings, and coatings. The dynamic pressure increase was reduced significantly, from some 10^{-5} Pa to less than 10^{-9} Pa [3].

Another beam-induced pressure instability, this time for heavy ions, was observed in 1997, when Pb^{54+} ions were stored in the Low Energy Antiproton Ring (LEAR) at CERN [4] and 1998 at the AGS booster at Brookhaven National Laboratory (BNL) with Au^{31+} ions [5, 6]. In 2001, also at the heavy ion synchrotron SIS18 at GSI, a pressure instability with increasing beam intensity was observed, which led to lifetime and intensity limitation of the beam [7]. It was found that beam ions that were lost due to charge exchange in collisions with residual gas hit the vacuum chamber wall at their actual velocity. There, they stimulated the release of gas [8], and the local pressure increase, in turn, increased the beam loss. Hence, beam loss-induced desorption is self-amplifying and appears to be a general intensity limitation for heavy ion ring accelerators. The situation as found at the turn of the millennium was mended significantly by improving the static pressure and increasing the pumping speed by non-evaporable getter (NEG) coatings and by the installation of dedicated beam catchers, so-called collimators, aiming to stop charge-exchanged ions after the mass-to-charge-separating dipole magnets [9]. These collimators were optimised in line with their desorption characteristic and the pumping speed in that area was increased.

Vacuum in Particle Accelerators: Modelling, Design and Operation of Beam Vacuum Systems,
First Edition. Oleg B. Malyshev.
© 2020 Wiley-VCH Verlag GmbH & Co. KGaA. Published 2020 by Wiley-VCH Verlag GmbH & Co. KGaA.

In the Relativistic Heavy Ion Collider (RHIC) at BNL, a limitation of the Au^{79+} beam current was observed in 2001 with an obvious link to the dynamic pressure rise upon operation. In this case it was finally concluded that the electron cloud effect is responsible for the pressure increase [10]. Finally, the mitigation in that case was the installation of so-called antigrazing rings to raise the threshold for electron cloud formation [11] and reduce the dynamic pressure rise.

Nowadays, heavy ion accelerators are under construction, which deliver orders of magnitude higher beam current than present machines [12–14]. For example, the Facility for Antiproton and Ion Research (FAIR) in Germany is expected to deliver 10^{12} uranium ions/s [12]. Spiral2 at GANIL in France even delivers up to 5×10^{14} medium heavy ions/s [13]. The heavy ion accelerator facility (HIAF) in China will deliver 5×10^{11} uranium ions/s and 4×10^{13} protons/s [14]. Even if the overall beam losses of these machines were in the order of a few percent, the number of lost ions is larger than the primary beam intensities of existing machines. Usually, the primary beam is intended to hit a target, e.g. for the production of rare isotopes [15, 16]. In that collision, called direct beam loss, the amount of released gas is of utmost importance for the operability of an accelerator. Due to the high beam current, a high gas load from desorption is to be expected, requiring also high pumping speed in the respective areas. Unfortunately, the desorbed gas will further be activated and therefore must not be stored in getter pumps or on cryogenic surfaces.

With the development of new accelerators, superconducting magnets have become more and more important. Larger areas of the beam pipe surface have cryogenic temperatures of few Kelvin like the Large Hadron Collider (LHC) beam screen [17]. This leads to a high sticking probability for any kind of gas on the surface. Hence, a huge amount of pumping speed is available, resulting in low operating pressures in the 1×10^{-10} Pa regime or even lower. On the other hand, the high sticking factor of gas on the surface leads to a notable surface coverage over time that acts as gas source for ion-induced desorption.

For future accelerators, mitigation techniques to overcome the dynamic pressure increase and thereby increase in the beam lifetime or intensity have been developed or are still under development.

10.2 Pressure Instabilities

In heavy ion accelerators, the gas balance is enhanced by sources that have partially been discussed in earlier chapters and sources that are exclusively present in heavy ion machines. Figure 10.1 summarises the mechanisms that may trigger pressure instabilities and related beam lifetime reductions in bunched-beam heavy ion accelerators below the space charge limit. The dominating source of gas is heavy ion-induced desorption stimulated by the loss of beam ions that collide with accelerator entities. This can be either an aperture-limiting device or production target where the ion beam is partially or completely dumped, usually under perpendicular incidence or it can occur by a modified trajectory due to charge exchange, whereas the loss of the beam ions is under grazing

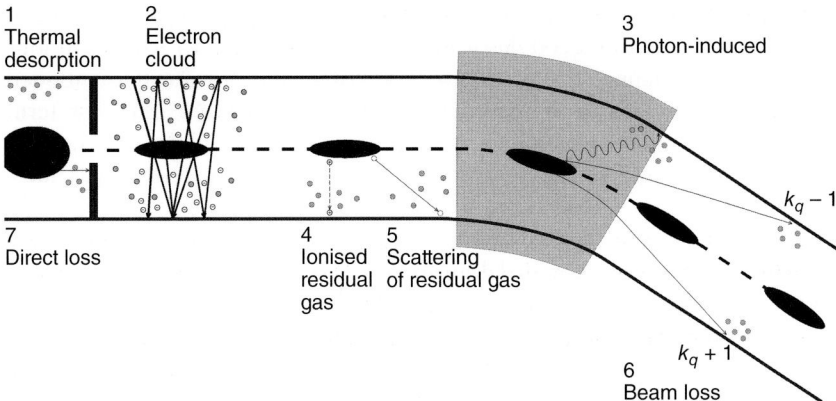

Figure 10.1 Processes of the dynamic vacuum in bunched-beam heavy ion accelerators (beam direction from left to right): (1) thermal desorption, (2) beam induced electron multipacting (BIEM) and electron cloud formation with subsequent electron-induced desorption, (3) photon-induced desorption by synchrotron radiation, (4) residual gas ionisation and acceleration of the gas ions by the electric field normally to vacuum chamber walls, (5) kinematic scattering of residual gas, (6) charge exchange of beam ions with subsequent loss, and (7) direct beam loss on aperture-limiting device. k_q is the charge state of the reference ion.

incidence. The charge exchange can occur in collision with residual gas and can lead to electron loss or electron capture. Moreover, in a collision with residual gas, the gas molecule can be ionised and subsequently accelerated within the beam potential towards the beam pipe wall where it encounters a perpendicular collision. In contrast to beam loss, where desorption is triggered by heavy ions at primary energy, the acceleration of ionised residual gas leads to a low energy desorption at maximum few kilo-electron volts collision energy. The process has been discussed in detail in Chapter 9. Also, elastically forward scattered residual gas molecules or atoms can collide with the vacuum vessel. Further, electron cloud build-up, as described in Chapter 8, can occur in heavy ion accelerators, depending on the bunch structure as well as geometrical and surface properties. With increasing energy of the ions, also synchrotron radiation can become a stimulation for gas desorption in ion accelerators though presently it is not an issue, if at all at the LHC during proton acceleration [18]. It is evident that in each case the lowest achievable amount of released gas is desired for best performance of the machine.

With ions propagating in the ring, the above discussed mechanisms can lead to desorption. Hence, the gas dynamics balance equation has to be augmented by the terms arising from these mechanisms and reads now as

$$q_{\Delta z} = (\eta_t F + \xi \Theta + \eta_\gamma \Gamma + \chi H + \chi_k H_k + \chi_g H_g) \Delta z + \chi_n H_n \tag{10.1}$$

The first term on the right side describes thermal desorption; the second, electron-induced desorption from electron cloud (Chapter 8); the third, photon-induced desorption from synchrotron radiation; and the fourth, ion-induced desorption (the effect described in Chapter 9). The fifth term describes desorption from elastically scattered residual gas molecules, which have incurred an impulse transfer in a close collision with a beam ion. The sixth term is the beam loss-induced desorption that is stimulated by lost beam ions due to charge exchange. The production rate of all the previously mentioned occasions depends on the observed fraction of the ring, Δz. The last term describes the direct beam loss on aperture-limiting devices and does not depend on Δz. $H_{(index)}$ quantifies the specific flux of stimulating particles (gas ions, atoms, and molecules as well as beam ions).

In the following, the contributions related to ion-induced desorption are described in detail, except thermal desorption and electron- and photon-induced desorption that were described in earlier chapters.

$$\chi H = \chi \left[\frac{\text{molecules}}{\text{ion}}\right] H \left[\frac{\text{ions}}{\text{s} \cdot \text{m}}\right] = \chi \frac{I\sigma}{q_e k_q \Delta z} \tag{10.2}$$

expresses ion-induced desorption, where gas molecules are ionised and subsequently accelerated by the beam potential towards the beam pipe wall. This perpendicular, low energy collision stimulates the release of gas with a rather low efficiency χ. The flux of produced residual gas ions per length unit Δz in the accelerator depends on the number of beam ions passing, calculated by $I/q_e k_q$ and the cross section σ for a beam ion ionising a residual gas molecule.

$$\chi_k H_k = \chi_k \left[\frac{\text{molecules}}{\text{atom/molecule}}\right] H_k \left[\frac{\text{atoms/molecules}}{\text{s} \cdot \text{m}}\right] = \chi_k \frac{I\sigma_k}{q_e k_q \Delta z} \tag{10.3}$$

is the amount of released gas by a residual gas molecule that was kinematically scattered in a collision. This amount is also strongly related to the number of beam ions. However, the probability σ_k for this collision channel is different. Depending on the collision parameter, the impact can be nearly perpendicular to grazing incidence and has a broad energy range. As a consequence, the efficiency χ_k of gas release from that impact is unpredictable.

$$\chi_g H_g = \chi_g \left[\frac{\text{molecules}}{\text{ion}}\right] H_g \left[\frac{\text{ions}}{\text{s} \cdot \text{m}}\right] = \chi_g \frac{I\sigma_g}{q_e k_q \Delta z} \tag{10.4}$$

describes the quantity of gas from a high energy grazing impact of a primary beam ion that gets lost in the course of a bending magnet after charge exchange in a collision with residual gas. The amount of released gas per impacting ion (χ_g) is of the highest experimentally found yields. Again, the number of impacting beam ions depends on the beam intensity and the cross section for charge exchange σ_g in a collision with residual gas.

The last term,

$$\chi_n H_n = \chi_n \left[\frac{\text{molecules}}{\text{ion}}\right] H_n \left[\frac{\text{ions}}{\text{s}}\right] = \chi_n \frac{I}{q_e k_q} f(\sigma_x, \sigma_y, a_x, a_y) \tag{10.5}$$

denotes the amount of released gas when the ion beam is dumped on a target or an aperture-limiting device such as a collimator. χ_n is the yield of gas production within this collision. The total amount of released gas depends on the fraction f of the beam that is dumped, which, in turn, depends on the size of the beam σ

and the size of the aperture a, both in x and y planes. The total amount of gas is produced at the collision point ($\Delta z = 0$) and can be zero if this situation does not encounter.

The desorption yield χ will be explained in Section 10.3. Equation (10.1) is simplified in a way that intrinsic beam processes, like emittance growth, are neglected. These processes become significant for beam intensities close to the space charge limit and are not related to the gas density.

The equation for the volumetric gas density inside a room temperature accelerator vacuum chamber reads now as

$$A\frac{\partial n(z)}{\partial t} = \eta_t F + \xi\Theta + \eta_\gamma \Gamma - (\alpha S + C)n(z) + u\frac{\partial^2 n(z)}{\partial z^2}$$

$$+ \left(\frac{\chi\sigma + \chi_k\sigma_k + \chi_g\sigma_g}{b(z) - a(z)} \int_{a(z)}^{b(z)} n(x)dx + \chi_n f(\sigma_x, \sigma_y, a_x, a_y)\right)\frac{I}{q_e k_q} \quad (10.6)$$

and the equations for the volumetric and surface gas density inside a cryogenic vacuum chamber are

$$A\frac{\partial n(z)}{\partial t} = \alpha S n_e(s, T) + (\xi + \xi'(s))\Theta + (\eta_\gamma + \eta'_\gamma(s))\Gamma - (\alpha S + C)n(z) + u\frac{\partial^2 n(z)}{\partial z^2}$$

$$+ \left(\frac{(\chi + \chi'(s))\sigma + (\chi_k + \chi'_k(s))\sigma_k + (\chi_g + \chi'_g(s))\sigma_g}{b(z) - a(z)} \int_{a(z)}^{b(z)} n(x)dx\right.$$

$$\left. + (\chi_n + \chi'_n(s))f(\sigma_x, \sigma_y, a_x, a_y)\right)\frac{I}{q_e k_q} \quad (10.7)$$

and

$$F\frac{\partial s(z)}{\partial t} = \alpha S(n(z) - n_e(s, T)) + \xi'(s)\Theta + \eta'_\gamma(s)\Gamma$$

$$+ \left(\frac{\chi'(s)\sigma + \chi'_k(s)\sigma_k + \chi'_g(s)\sigma_g}{b(z) - a(z)} \int_{a(z)}^{b(z)} n(x)dx + \chi'_n(s)f(\sigma_x, \sigma_y, a_x, a_y)\right)\frac{I}{q_e k_q} \quad (10.8)$$

where the first term on the right-hand side of Eqs. (10.6)–(10.8) represents thermal desorption (Eq. (10.6)) or thermal equilibrium density n_e (Eq. (10.7)). The second and third terms describe electron- and photon-induced desorption, respectively. For the volumetric gas density, the terms for conductance-limited distributed pumping speed and axial diffusion of gas molecules follow. Written in brackets and depending on the beam intensity are the terms for ion-induced, kinematic scattered residual gas-induced and beam loss-induced desorption as well as beam-induced desorption on aperture-limiting devices.[1] α is the sticking probability on either cryogenic surfaces or getter surfaces like NEG coatings. For a cryogenic vacuum chamber, the symbols with apex (Eqs. (10.7) and (10.8)) denote desorption stimulated on condensed gas layers.

---

1 Note that the terms for electron- and photon-induced desorption also depend on the beam intensity. For the sake of simplicity, this was omitted here. For details, see Chapters 8 and 9.

10.2.1 Model Calculations of the Dynamic Pressure and Beam Lifetime

For the simulation of the pressure-dependent beam lifetime in a ring accelerator, the average static pressure value of the system is of importance:

$$\langle P \rangle = \frac{1}{L} \int_0^L \frac{q(z)}{S_{\text{eff}}(z)} dz \tag{10.9}$$

with z denoting the position in the ring [8] and L the total circumference. Precisely, $q(z)$ represents the gas flux at position z and $S_{\text{eff}}(z)$ the effective pumping speed at this position, resulting from attached pumps and conductances. Then, the mean particle density is given by $\langle n \rangle = \langle P \rangle / k_B T$, where k_B denotes the Boltzmann constant and T the gas temperature.[2] For the dynamic case, with an ion beam propagating in the ring, the equation was supplemented by a dynamic gas flux term q_{dyn}, which is composed out of stimulated desorption from lost ions due to charge exchange and accelerated residual gas. Finally, the complete equation including the dynamic term but under non-consideration of electron clouds is [19]

$$\frac{d\langle P \rangle}{dt} = -\frac{1}{\tau_P}(\langle P \rangle - \langle P_e \rangle) + \frac{k_B T}{S_{\text{eff}}}(\dot{H}\chi_{\text{loss}} + H\chi) \tag{10.10}$$

By implementing Eq. (10.10), several calculations and measurement benchmarks were performed. After all, with the assumption of boundary conditions and start parameters such as initial pressure P_0, outgassing flux q, and pumping speed S as well as initial beam intensity I_0 and appropriate cross sections for ionisation and loss, time-dependent pressure and beam intensity could be modelled. The latter exhibits the beam lifetime as $t(I_0/e)$ in seconds. Replacing either parameter by a measured value, especially the beam intensity or pressure, the modelling results benchmark other values, e.g. the predicted cross sections. Some example results will be discussed.

10.2.1.1 Closed System (Vessel)

Using Eq. (10.10), a complete ring accelerator is modelled as a single vessel with an absolute volume (the total accelerator volume) and a mean pressure or gas density averaged over the ring [8]. Hence, also ionisation and loss rates are summed up over the ring orbit. Figure 10.2 shows the calculation of the mean pressure and its evolution versus time in SIS18 at GSI with stored beam of 1.4 mA (a) and 5.4 mA (b) U^{28+} beam current at the injection energy $E = 8.9$ MeV/u. Some findings can be excavated from the plots. First, the $\frac{1}{e}$-lifetime of the beam is much longer for the lower injection intensity (1 s for 1.4 mA and 0.25 s for 5.4 mA). Second, the pressure bump is higher for higher injection currents ($P_{\text{max}} = 1.5 \times 10^{-8}$ Pa for 1.4 mA and 3.3×10^{-8} Pa for 5.4 mA). Finally the time-dependent evolution of the values can be explained as follows: after injection the beam loss rate is high due to the high beam intensity. This leads to the pressure increase in the beginning, which would run exponentially towards an equilibrium value defined

---

2 T can depend on the position in the ring, if cold–warm transitions are present, like in cryogenic accelerators.

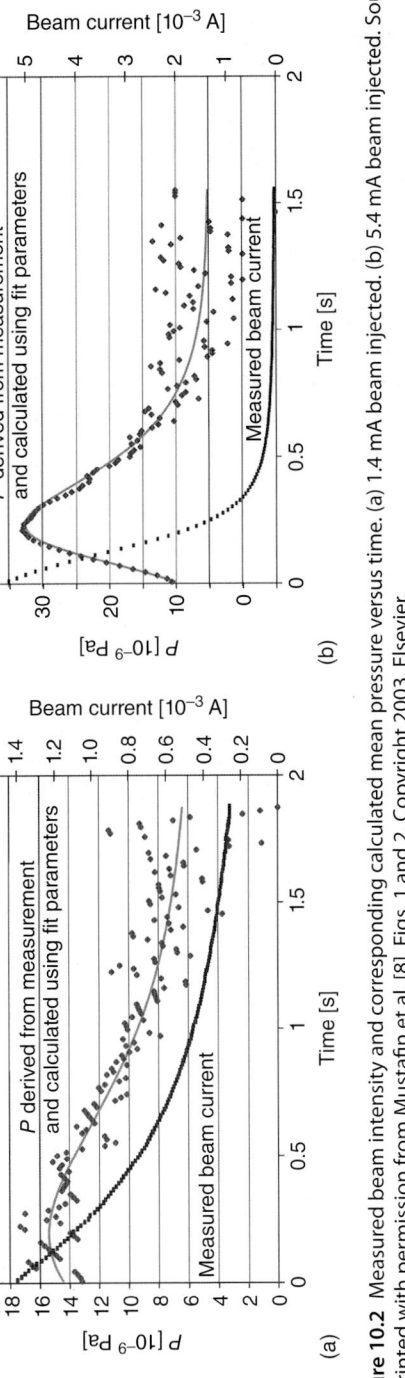

Figure 10.2 Measured beam intensity and corresponding calculated mean pressure versus time. (a) 1.4 mA beam injected. (b) 5.4 mA beam injected. Source: Reprinted with permission from Mustafin et al. [8], Figs. 1 and 2. Copyright 2003, Elsevier.

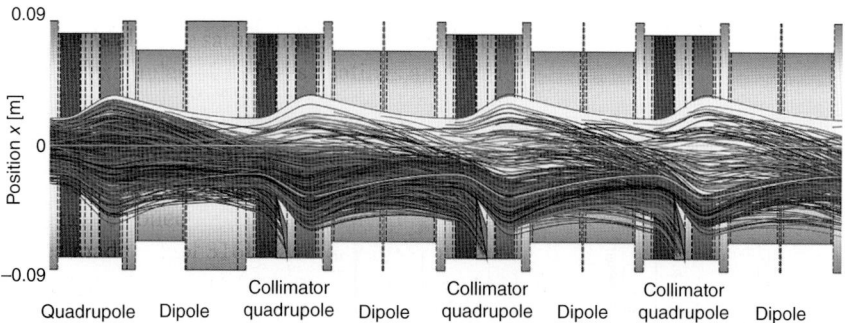

Figure 10.3 Ion-optical lattice of a section of SIS100: the original ion beam in white, and charge-exchanged particles in black. Source: Courtesy of Dr. C. Omet 2008 [21], Fig. 31.

by the pumping speed of the system. Meanwhile the beam intensity and, hence, the loss rate have decreased, whereby the pressure also decreases again.

10.2.1.2 Vessel Including Collimation

A combination of vacuum and ion optic simulation is included in the StrahlSim code [20]. Here, the vacuum system is represented as lattice, observing pumping speed and conductance by position as the accelerator ring is a conductance-limited device. The calculation of the pressure profile is done also here as mean value. The code provides the exact positions of beam losses from which one can determine the position of a collimation system. Such a system is able to capture lost beam ions in a controlled manner, i.e. with lowest possible desorption yield as compared to uncontrolled grazing incidence. However, only a fraction of lost beam ions is collected in that way and the collimation efficiency describes the number of collected ions over the number of lost ions and hence is ≤1.

Figure 10.3 shows an ion-optical simulation of an U^{28+} beam inside SIS100 including charge-exchanged and subsequently lost beam ions and collimators. From this lattice, also vacuum conductances can be obtained using the geometry of the magnets and tubes (x is the diameter in metres and z the length of the respective vacuum entity).[3]

The simulation includes cross sections for charge exchange and target ionisation that either have been measured or have been scaled from measurements. The used desorption yields have been scaled from experimental data using $(dE/dx)^2$-scaling; compare Section 10.3.4.

10.2.1.3 Longitudinal Profile

More sophisticated modelling has been performed since 2010 using an enhancement of the StrahlSim code. The principle is basically the same as before but now, the accelerator is discretised into small segments with a longitudinal raster of

---

3 In Figure 10.3 the collimation efficiency is close to 1 for the charge exchange $U^{28+} \rightarrow U^{29+}$. In upcoming systems like SIS100, even a broader range of charge exchange channels can be covered. However, in existing systems where collimators are a subsequent installation, the collimation efficiency will be < 1.

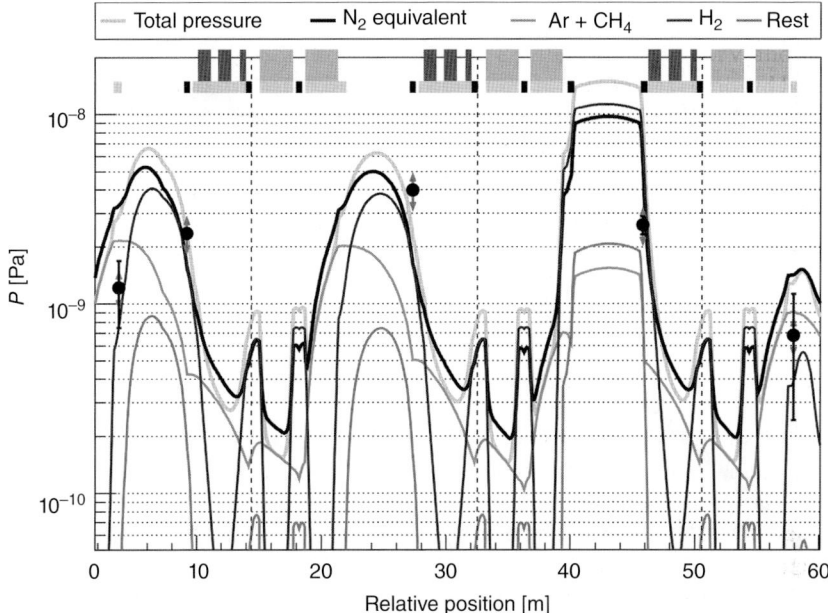

Figure 10.4 Pressure profile of three sections in SIS18 during U^{28+} lifetime measurements as calculated with StrahlSim. The pressure is calculated with individual gas components and summed up to the N_2-equivalent pressure. The calculation was fitted in good agreement to measured pressure values (points). Source: Reprinted with permission from Bozyk et al. [24], Fig. 8. Copyright 2016, Elsevier.

a few centimetre. The calculation is done by means of transfer matrices similar to the vactrac or vacdyn code [22, 23]. This now finally allows to include getter-coated and cryo surfaces in more detail and allows to respect local effects like systematic beam loss or pressure build-up. The result displays a longitudinal vacuum profile as a function of orbit position and time (see Figure 10.4) [24].

After all, it is possible to calculate the number of extracted ions out of the ring as a function of the number of injected ions, respecting loss processes and related dynamic vacuum behaviour, including several upgrade measures like collimation and getter coating.

Figure 10.5 shows measurements of the number of extracted ions out of SIS18 at GSI as a function of injected ions. Measurements carried out in 2003 and 2007 exhibit a maximum value, indicating that the dynamic vacuum limits the beam intensity. More injected beam even leads to less extracted beam. The values of 2010 and 2011 are much larger and denote an intensity world record for extracted intermediate charge state heavy ions [25]. The measures that led to the intensity increase were coating of large areas of the ring by NEG and the installation of collimation systems.

10.2.2 Consequences

From Eqs. (10.1) and (10.10), a stability criterion can be defined. A stable operation of a ring accelerator is granted in the following situations:

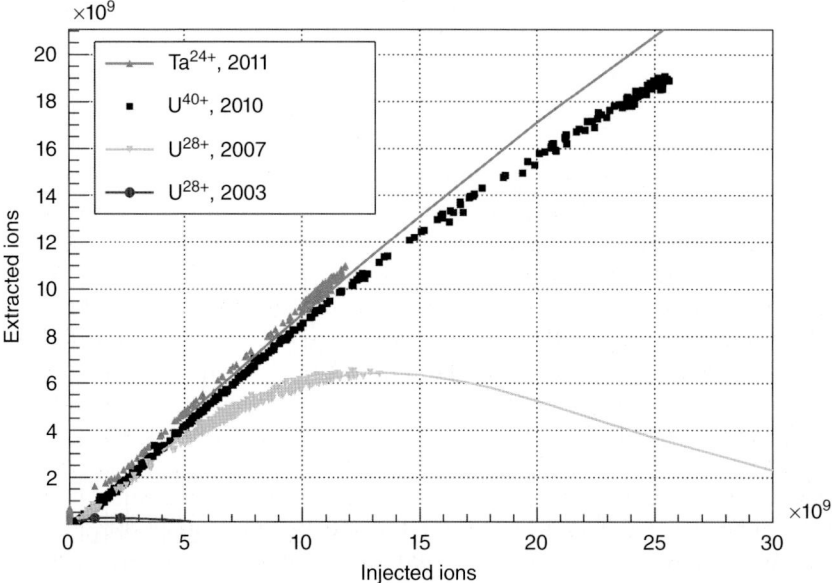

Figure 10.5 Number of extracted ions as function of injected ions in SIS18. Shown are simulated (solid line) and measured values (points). Source: Spiller et al. 2011 [25], Fig. 7. Reprinted with permission of CERN.

- The pump-down time is short (τ_p small, e.g. by large S_{eff}).
- The loss and/or ionisation rates are low.
- Low desorption yields χ.

Mathematically, these demands can be summarised as the inequation

$$(\chi\sigma + \chi_k\sigma_k + \chi_g\sigma_g)I < \frac{VeZ_p}{L\tau_p} \tag{10.11}$$

Typical measures to meet this requirements are, e.g. getter coating of large surfaces inside the accelerator, such as magnet chambers. The gained pumping speed reduces the mean pressure significantly, as the pressure inside the magnet chamber is no longer limited by the conductance. Further, a possible measure is to have the ions for short time on low energy, e.g. to minimise the injection plateau and accelerate the ions in short time, as the charge exchange cross sections decrease with increasing energy. Hence, the steeper the acceleration ramp, the lower the number of overall lost ions. Finally, the materials in loss regions should be chosen for low desorption. An intense research was undertaken to understand and minimise the desorption yield of dedicated materials as will be described in the following section.

10.3 Investigations on Heavy Ion-Induced Desorption

First of all, it is indispensable to understand the process of heavy ion-induced desorption in detail to mitigate pressure instabilities in particle accelerators.

10.3 Investigations on Heavy Ion-Induced Desorption

To quantify the amount of desorbed gas, the number of desorbed particles (molecules or atoms) per incident ion is defined as the so-called desorption yield χ[4] [26]:

$$\chi = \frac{\text{released particles}}{\text{incident ion}} \quad (10.12)$$

This definition is similar to the sputter yield. However, the sputtering process has to be distinguished from desorption, even though both contribute to the particle density inside the vacuum vessel. In some literature, all release of molecules into the vacuum including sputtering is quite simply called desorption. When regarding pressure instabilities in particle accelerators, it is useful to separate the definitions. In SPUTTERING we consider the release of target atoms such as metal atoms that have a very high sticking probability [27] and, thus, contribute to a small amount to the pressure instability, as they stick quickly to the chamber wall. Sputter yields are in the range of few atoms per incident ion [28, 29] and, frequently, an angular distribution in terms of $\cos^n$ is observed [29]. In contrast, DESORPTION is the release of volatile, predominately molecular species, such as H_2, CO, and CO_2 but also noble (atomic) gases like Ar. These species have a low sticking probability at room temperature and contribute the major fraction to the pressure instability. Further, there is no specific angular distribution visible. The released gas is emitted in 2π solid angle from the surface.

In order to get a complete picture, experimental investigations must include both measurements of the desorption yield and its conjunction to the material and surface properties. The performed investigations cover a broad variety of materials and surface treatments as well as different beam parameters. Finally, it is desirable to describe expected desorption yields by means of a theoretic model.

10.3.1 Desorption Yield Measurements

Yields of heavy ion-induced desorption have been intensively studied since the turn of the millennium [30–34]. For that purpose, the PRESSURE RISE METHOD was exclusively used [35], that is, measuring the pressure increase in a vessel of defined volume or with defined pumping speed upon irradiation of a target at a given projectile ion flux. Schematically, pressure rises are displayed in Figure 10.6 [32]. Figure 10.6a, after 0.4 seconds, a single beam pulse hits the target. An instantaneous pressure increase is visible, followed by an exponential decay whose slope is defined by the pumping speed of the system. Figure 10.6b, a pressure rise due to continuous bombardment is shown. The slope of the exponential decay in this case is much more flat as it is defined by the target cleaning with the ion beam, the so-called beam scrubbing [36]. The plot shows the total pressure evolution, which is composed of several gas species in the partial pressure spectrum, which all have different heights and decay slopes [34].

---

4 In most literature, the desorption yield is denoted as η. Within this book, however, the *ion-induced* desorption yield is denoted as χ to clearly distinguish it from electron-induced (here denoted as ξ), photon-induced, and thermal desorption; see Eq. (10.1).

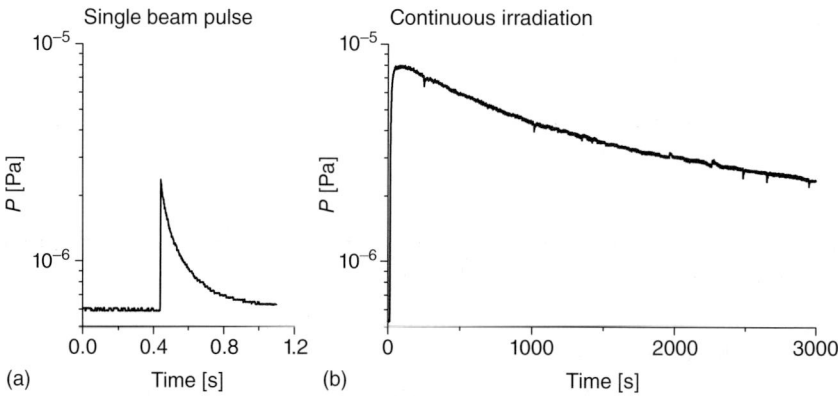

Figure 10.6 Pressure during the irradiation of a sample. (a) Single shot – one single pulse of ions hits the target. (b) Continuous irradiation – the target is hit by a beam pulse every second.

At the regarded pressure ranges, the molecules can be understood as ideal gas, moving forceless until a collision occurs. Therefore the number of particles per volume can be written as

$$PV = nk_B T \tag{10.13}$$

with the pressure P and the number density of the gas n. Desorption changes the number of particles per volume and, hence, the pressure rise in a vessel without pumping is related to an increased number of particles. The relation ΔP, respectively Δ (number of particles) to beam ions in an irradiating pulse delivers the desorption yield as [37]

$$\chi = \frac{\Delta P V}{H k_B T} \tag{10.14}$$

In a set-up with finite pumping speed, gas molecules are continuously pumped out of the volume. Therefore, a steady-state condition can only be reached by re-delivering molecules, e.g. from desorption. Continuous irradiation instead of a single pulse grants the required gas desorption. With that, Eq. (10.14) becomes time dependent in volume and in the number of ions:

$$\chi = \frac{\Delta P \dot{V}}{\dot{H} k_B T} = \frac{\Delta P S}{\dot{H} k_B T} \tag{10.15}$$

where $\dot{H}$ denotes the ion flux [37].

With the first observations of beam lifetime limitations upon increasing beam intensities, also the measured pressure rises and corresponding desorption yields in the accelerator rings were reported. However, measured desorption yields have to be considered with skepticism, as the effective pumping speed on the beam loss point is hard to determine and also, due to the grazing impact, the collision is undefined.

At LEAR pressure rises in the order of 1×10^{-7} Pa after the injection of 1×10^8 lead ions are reported [4], corresponding to a desorption yield of 2×10^4 released molecules per incident ion [38].

In SIS18 the measured pressure increase was some 10^{-8} Pa already during the injection of 5×10^7 uranium ions. Calculated desorption yields were in the order of few 10^4 molecules per uranium ion [7].

Pressure increases of almost 4 orders of magnitude have been observed at RHIC at BNL during the acceleration of gold ions at a base pressure of few 10^8 Pa [11]. However, as the phenomenon was uncovered as an electron cloud effect, desorption yields in the sense of released molecules per incident ion are not reported. Rather, electron-induced desorption yields in the order of less than 0.1 released molecules per electron were calculated [11].

10.3.2 Materials Analysis

To address the question how the materials properties influence the desorption yield, materials analysis were performed. At CERN, X-ray photoelectron spectroscopy (XPS) measurements were carried out during the bakeout of stainless steel samples with different coatings. Some information about the content of C and O at the surface of the samples and the evolution of their amount were gathered and a correlation to the subsequently measured desorption yields was visible [34, 39]. However, this is the only reported investigation on XPS.

To make use of the irradiating ion beam for materials analysis, a dedicated set-up for Rutherford Back Scattering (RBS) and Elastic Recoil Detection Analysis (ERDA) *in situ* with desorption yield measurements was built up at GSI [40]. Both techniques facilitate Rutherford kinematics for elastic scattering [41]. However, in RBS, a light projectile ion is scattered elastically on heavier target atoms. Thus, RBS is suited to detect the concentration of heavier elements in the target such as metallic components [42, 43]. In contrast, for ERDA, the target is irradiated with heavy ions and target atoms are scattered elastically in forward direction [43–45]. In both cases the energy transfer to the ejected particle is only dependent on the mass ratio of the scattering partners (projectile ion/target atom) and the scattering angle. Analysed particles are caught by the detector at a defined scattering angle ϕ. The energy of a scattered particle is

$$E_2 = kE_1 = \frac{4M_1 M_2}{(M_1 + M_2)^2} \cos^2 \phi E_1 \qquad (10.16)$$

It can be seen that for, e.g. in an ERDA configuration with a projectile ion much heavier than the target atoms, all ejectiles will have almost the same velocity after scattering. Therefore, they can be detected Z-separated in terms of their energy loss in a gas-filled ionisation chamber, called the ΔE-part (see also Eq. (10.20)) [43, 46]. The ejectiles are finally stopped in the second part of the detector, called E_{rest}; compare Figure 10.7, right side. This leads to the total energy of the particle and is a qualitative measure of the scattering depth, as the projectile as well as the ejectile suffer from energy loss on their way through the sample [47–51]. Plotting the number of events as intensity on the z-axis versus ΔE and E_{rest}, a raw spectrum as shown in Figure 10.8 can be obtained.

In this raw plot, a qualitative statement can be already given: A cumulation of events on the oxygen line at channel numbers $\Delta E = 75$ and $E_{rest} = 450$ eV is visible. Qualitatively, this result can be interpreted by the presence of an oxide

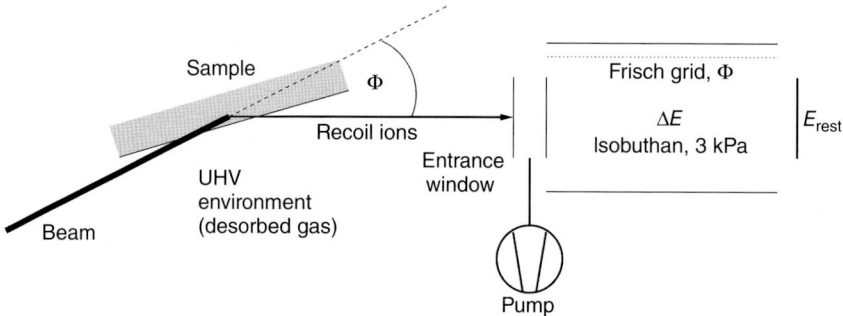

Figure 10.7 Schematic sketch of the combined desorption yield measurement and surface analysis set-up at GSI. On the right the ERDA detector, represented by a $\Delta E - E_{rest}$ telescope is shown.

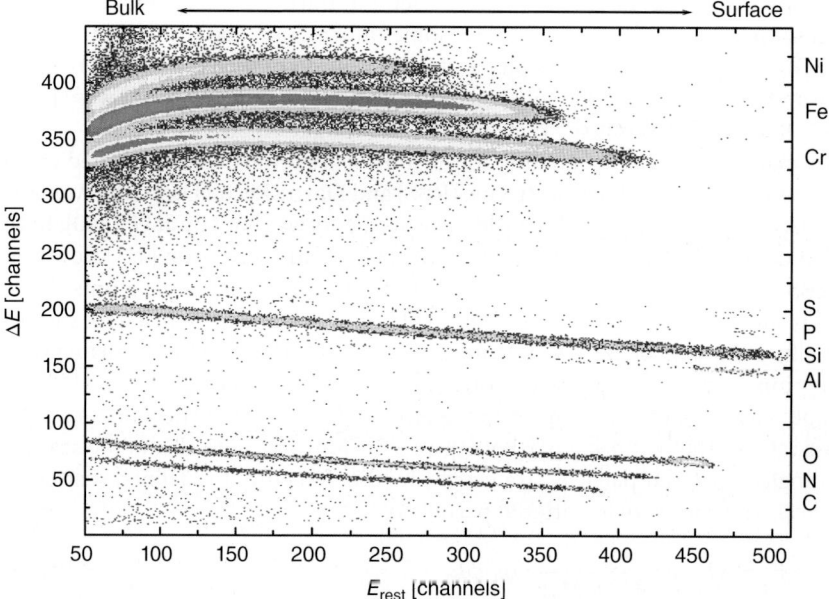

Figure 10.8 $\Delta E - E_{rest}$ spectrum of a stainless steel sample. The x-axis represents the total energy (function of the depth of origin) of a particle. Surface and bulk can be distinguished qualitatively. The y-axis represents the energy loss in the detector gas (function of the atomic number). The related elements are written to the right.

layer on the stainless steel sample. Focusing further on, e.g. oxygen within this raw plot, the energy spectrum for the element is obtained as displayed in Figure 10.9. In this example again a cumulation of intensity is visible at high channel numbers (equivalent to the surface), which states the oxide layer.

The intensity represented on the y-axis in this plot is the number of recorded recoil events for the respective element. It is determined by the concentration of the element in the sample and the cross section for a scattering event of the projectile with the observed element under a defined angle (the detector position)

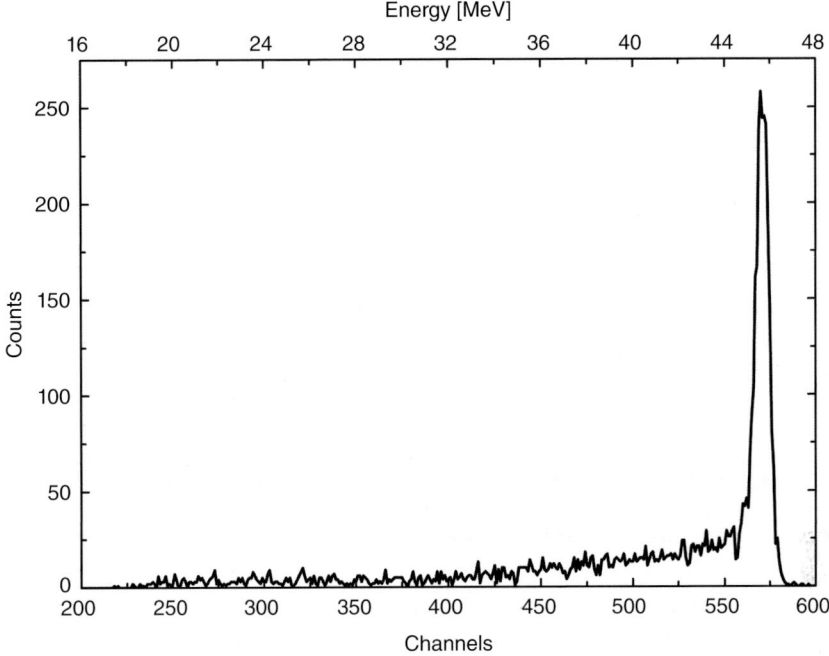

Figure 10.9 Energy spectrum of the oxygen line, carved out of Figure 10.8.

as [52]

$$\frac{d\sigma}{d\Omega} = \left(\frac{Z_1 Z_2 q_e^2 (M_1 + M_2)}{8\pi\epsilon_0 M_2 E_1}\right)^2 \frac{1}{\cos^3\phi} \tag{10.17}$$

Applying this differential (Rutherford) cross section to the scattering of the particle, the elements can be compared concerning their concentration in a certain depth, which finally leads to an element-specific depth profiling of the samples, derived, e.g. by RUMP [53, 54], KONZERD [55], or data furnace [56, 57]. The final plot of element concentration versus target depth is shown in Figure 10.10 [35, 55].

The ultimate observable depth of the used configuration is roughly 1 μm and the depth resolution for the used set-up is about 10 nm [52].

10.3.3 Dedicated Set-ups to Measure Ion-Induced Desorption Yields

For systematic desorption yield measurements, different dedicated set-ups have been used, each based on the employment of the above discussed Eqs. (10.14) and (10.15). Each set-up consists of some differential pumping stages to adapt to the vacuum conditions of the respective accelerator environment. The last pumping post usually operates in deep ultra high vacuum (UHV). The measurement chamber is connected to that last pumping post by means of a defined conductance tube, which, in combination with the installed pumps, represents the applied effective pumping speed S in Eq. (10.15). Inside the experimental chamber, the

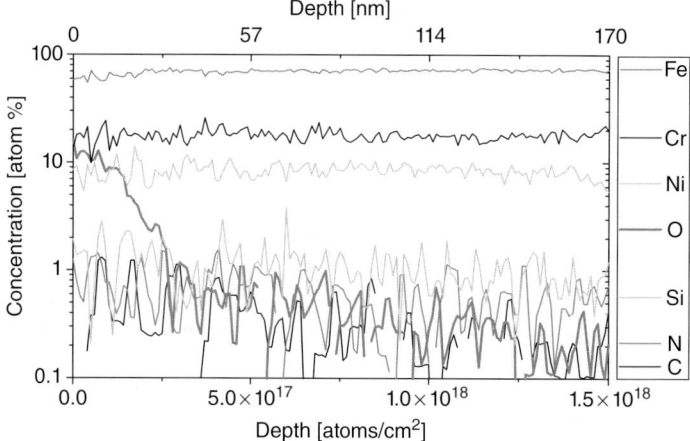

Figure 10.10 De-convoluted element concentration of Figure 10.8 versus depth. In this representation the surface is on the left. Due to the ambiguity of the lines in grey style, the elements are written as close to the corresponding line as possible. Note the thick line to be oxygen.

samples are mounted such that they can be irradiated by the beam. The pressure is measured by UHV gauges and a residual gas analyser (RGA) measures the distribution of the partial gases. The accuracy of the pressure measurement and pumping speed defines the error for the desorption yield, as they reveal the highest uncertainty. The temperature and ion flux can be measured with higher accuracy. The total error within the desorption yield measurements is between 25% and 50%, depending on the literature.

The used set-ups are listed below in chronological order:

CERN SET-UP: The first dedicated set-up for quantitative measurements of desorption yields was installed at LINAC 3 at CERN [31, 32, 58]. Figure 10.11 shows a schematic plot of the set-up [32].

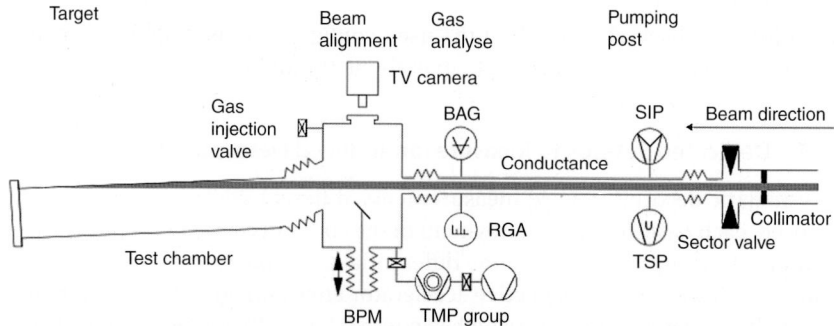

Figure 10.11 CERN set-up, used at LINAC3 with 4.2 MeV/u lead ions. The target is represented by the tube called test chamber on the left side. Source: Mahner et al. 2002 [31], Fig. 1. Reprinted with permission of CERN.

10.3 Investigations on Heavy Ion-Induced Desorption

The set-up consists of a test chamber, which in itself acts as target. Basically this test chamber is a tube of a dedicated material or with a dedicated coating or cleaning procedure [34]. A bellow allows the tilting of that test chamber, enabling the irradiation of the end flange under perpendicular incidence as well as the tube under grazing angle. A beam positioning monitor (BPM) is installed in front of the test chamber. For pressure measurement, a calibrated Bayard–Alpert gauge (BAG) and a calibrated RGA (Balzers QMA 420) are installed within the test chamber volume. The test chamber including the analytic devices is pumped through a defined conductance that is attached to a pumping post out of a sputter ion pump (SIP) and a Ti sublimation pump (TSP), resulting in an effective pumping speed in the test chamber of 7 l/s for N_2. At the pumping post, the whole set-up is connected to the accelerator. Experiments at CERN LINAC 3 have been conducted with predominately stainless steel, either pure, cleaned by argon glow discharge, or coated with NEGs. The irradiation was done with 4.2 MeV/u lead ions of charge states 27+ with an intensity of 1×10^{10} ions per pulse and 53+ with an intensity of 1.5×10^{9} ions per pulse, either in single shot mode or continuously with 0.8 Hz.

GSI SET-UP: A similar set-up of that described earlier has been used at GSI since 2003 [37]. The main difference is that here, samples of different sizes could be mounted on a linear and rotation feedthrough. This mounting allowed the installation of up to 20 targets at a time with the drawback that only perpendicular irradiation was possible.

The set-up is shown in Figure 10.12. It also consisted of the irradiation chamber with the installed targets and measurement devices. The latter have been an extractor gauge IE 514 and a Pfeiffer Prisma RGA. The test chamber was attached to the pumping post via a defined conductance of 82.5 cm in length, resulting in 5.5 l/s pumping speed for N_2. The pumping post in turn was attached to the accelerator. With the GSI set-up, experiments were carried out at different beam energies and thus in different locations over the GSI facility where the set-up could be installed with minor modifications. At the high charge state injector at

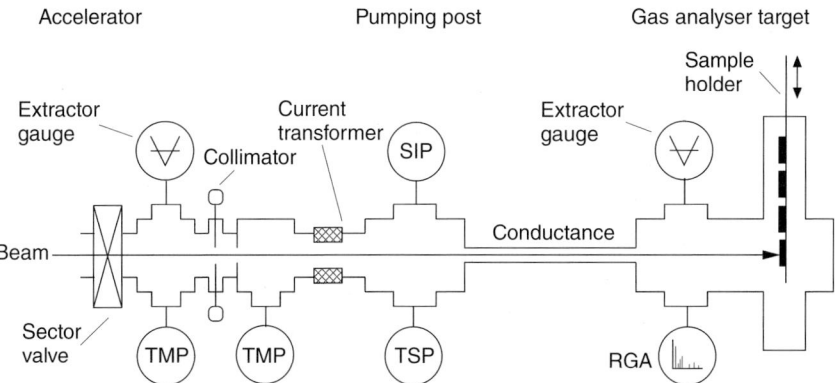

Figure 10.12 GSI set-up used with various ion species at different energies. Up to 20 targets are mounted on a target holder. Source: Reprinted with permission from Kollmus et al. [37], Fig. 1. Copyright 2005, American Institute of Physics.

GSI, experiments have been performed with C, Ca, Cr, Zn, Xe, and Pb beams, all at 1.4 MeV/u. Further experiments with higher energies have been conducted at the heavy ion synchrotron SIS18. Here, Ar and U beams with energies between 15 and 100 MeV/u were available.

CERN SPS SET-UP: Collimator materials such as graphite and stainless steel were tested in the SPS at CERN with indium ions at 158 GeV/u under grazing incidence. The measured pressure increases were rather small, in the order of some 10^{-11} Pa at a base pressure of about 8×10^{-10} Pa. Due to the high effective pumping speed of more than 1000 l/s, the calculated desorption yields were, however, in the range of some 10^4 to more than 1×10^5 released molecules per incident ion [33].

GSI SET-UP WITH IBA: In 2005, a new set-up for desorption yield measurements was installed at the high charge state injector at GSI [59]. The principle of the desorption yield measurements is identical with the other set-ups described previously, but with an effective pumping speed of 300 l/s for N_2. The main difference compared with the former used set-up was the capability of *in situ* surface treatment and analysis of the samples. For surface cleaning, a Xe sputter gun was installed, which was operated at 5 kV. For analysis purpose, ion beam analytic methods 'ERDA' and 'RBS' were performed. Both methods were briefly described in Section 10.3.2. The additional methods were applied to investigate how the surface properties of the material, such as a present oxide layer, influences the desorption yield.

TSL SET-UP: In the gamma cave of TSL in Sweden, a group of Uppsala University installed a set-up for measuring ion-induced desorption yields [60]. It made use of the same principal as the previously described set-ups. The conductance was 4 l/s for N_2. The available beam for the Uppsala experiment was Ar at 5, 9.7, and 17.7 MeV/u.

GSI MATERIALS RESEARCH SET-UP: Since 2010, a new beamline called M-branch has been in operation at GSI [61, 62]. This branch is devoted to materials research and offers different spectroscopy methods for *in situ* and online surface and bulk investigations. The line M1 is a dedicated surface physics line, operating at a pressure of 1×10^{-8} Pa or better. Samples can be introduced using a load lock system. The conductance to the pumping post is about 215 l/s for N_2. In this set-up desorption yield measurements have been performed since 2014 with Ca and Au beams, both at 3.6 and 4.8 MeV/u. One advantage at the M1 beamline is that all beams can be delivered in different charge states without changing any other parameter. A scheme of this set-up is shown in Figure 10.13.

CRYOGENIC TARGETS AT GSI: In the synchrotron SIS100 of the upcoming FAIR facility, catcher systems are planned to eliminate charge-exchanged ions in the ring. To measure the desorption yield of this cryocatcher, a prototype set-up has been installed at the existing synchrotron SIS18 [63]. The set-up consisted of the catcher block inside of a liquid helium-cooled vacuum chamber. The block was mounted electrically insulated to measure the current of the ion beam, and a dedicated heating system allowed to measure desorption yields in a temperature range between 30 and 90 K [64]. To determine the pressure rise, an extractor gauge was used, which was shielded to repress thermal load to the block from heat

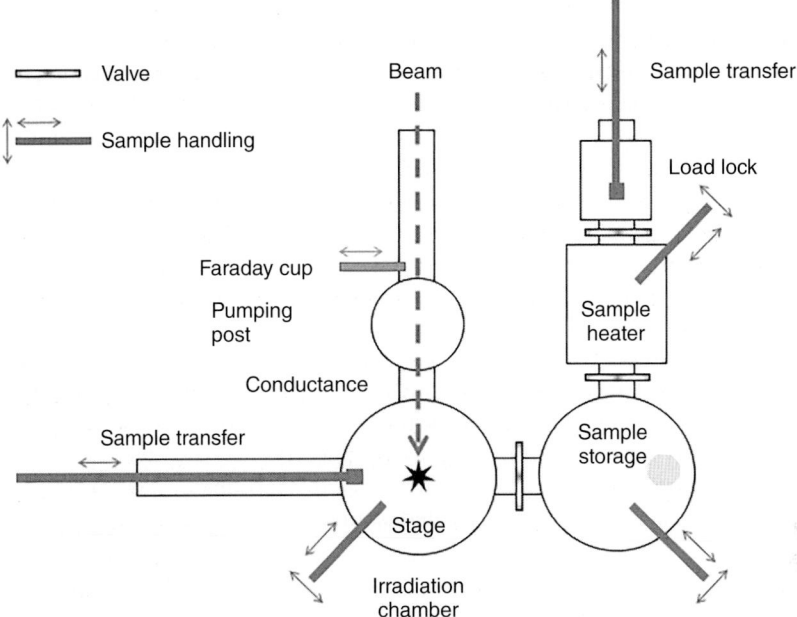

Figure 10.13 Schematic sketch of the GSI set-up at beamline M1. Vacuum analysis such as total and residual gas pressure measurements are available in the irradiation chamber.

radiation. Experiments have been conducted with Au, Ta, and Bi beams between 100 and 800 MeV/u.

Subsequently, a dedicated set-up to investigate the desorption yield of different materials at cryogenic temperatures was built [65]. This set-up follows the same principle as the above described set-ups using a conductance tube between experimental chamber and pumping post. The samples were cooled by a closed-cycle cold head to an ultimate temperature of 38 K. Special care was taken on the pumping speed of the cryostat itself that adds on the pumping speed provided by the conductance [66]. Further the samples were also investigated with variable temperature up to room temperature. Measurements were carried out with U and Bi beams between 50 and 350 MeV/u.

IMP SET-UP: The most recent set-up was installed at the high voltage platform at the Institute of Modern Physics (IMP) in Lanzhou, China [67]. Here, oxygen-free copper targets were irradiated with xenon 10+ and oxygen 1+ ions at total energies of few 100 keV for oxygen and few mega-electron volts for xenon. As all other installations, this set-up makes use of the pressure rise method. The experimental chamber is direct-pumped by a turbomolecular pump, resulting in about 350 l/s pumping speed at an ultimate pressure in the lower 10^{-6} Pa regime.

10.3.4 Results

All experimental results were obtained from the irradiation of accelerator-relevant materials, such as different stainless steel grades, as received and vacuum fired as well as copper [36, 40]. In particular cases, also aluminium, tantalum, tungsten, rhenium, and palladium were investigated [68]. Further, different coatings were applied, e.g. gold and rhodium coating for oxidation prevention and NEGs as thin film vacuum pumps [69, 70]. Although the materials altered upon the experimental series, relevant questions were addressed, like the origin of the desorbed gas, whether it is a surface or bulk effect, and how impurities, especially the oxide layer, influence the desorption behaviour.

Hundreds of samples have been tested under different conditions. A comparison of many of them is given in Figure 10.14 [33, 36]. Qualitatively, it can be concluded that desorption yields for grazing incidence are higher as for perpendicular impact and the numbers seem further to be energy dependent. There are some results spread over 3 orders of magnitude, even all recorded at perpendicular impact with the same ion species and energy. This is a hint that the target itself or its preparation plays a dominant role for the quantity of desorbed gas.

All irradiations have in common that the desorption yield decreases with increasing irradiation fluence. This proves that the ion beam itself cleans the target and this conditioning effect is called beam scrubbing [32, 35].

10.3.4.1 Materials

Stainless Steel Two types of stainless steel are commonly used in accelerators, 304L and 316LN (DIN 1.4301 and 1.4429, respectively). In general, steel contains besides metallic components (predominately iron, chrome, and nickel) intermediate heavy components such as aluminium and silicon as well as light components like carbon, nitrogen, and oxygen approximately up to the percent range [71, 72]. Further, hydrogen is incorporated in stainless steel like in most other materials and is volatile. Hence, to lower the outgassing of hydrogen, steel components are usually vacuum fired. In this procedure the components are heated in a vacuum furnace up to 950 °C to reduce the amount of incorporated hydrogen and the subsequent outgassing [73].

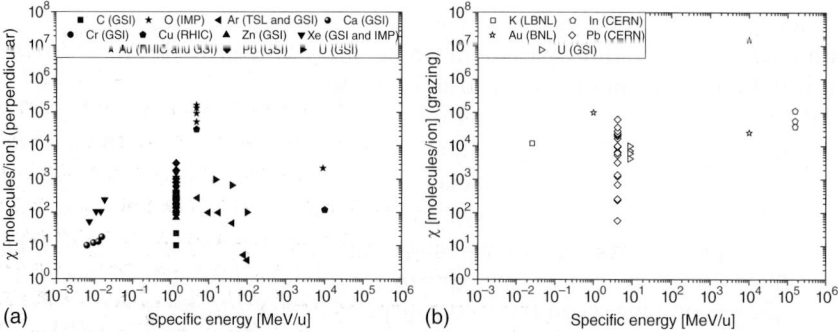

Figure 10.14 Overview plot of all published desorption yield data (in 2017, only room temperature data). Plot according to [36] but completed in published data to date. (a) Desorption yields from perpendicular impact. (b) Desorption yields from grazing incidence ($< 1°$). For clarity no error bars are displayed but they range from 25% to 50% according to the respective publication.

The fairly high amount of 'impurities', especially the oxide layer, was originally suspected to be the main contribution of the desorbed gas. To pursue this suspicion, stainless steel samples, as received and vacuum fired, have been investigated with ERDA complementary to the desorption yield measurements. Another approach was the investigation of different cleaning or coating procedures to remove or cover the oxide layer.

Upon irradiation with 1.4 MeV/u xenon ions under perpendicular incidence, the initial desorption yield after starting the irradiation was 400 and 520 molecules per incident ion for 304L and 316LN stainless steel. In case of vacuum-fired steel, the values were 330 and 265, respectively. Due to irradiation, the samples were cleaned and the desorption yield reduced to an ultimate value of 25 and 41 for 304L and 316LN as received and 46 to 52 for the vacuum-fired samples. The residual gas analysis showed that in the case of vacuum-fired steel, the desorption of hydrogen is lowered by 70%, but the desorption of CO is increased by almost a factor of 2 [40]. In summary, vacuum firing can reduce the initial desorption with regard to hydrogen. On the other hand, the ultimate value after an irradiation of 5×10^{13} ions/cm² was slightly higher for the fired samples due to CO [40].

A comparison between the thickness of the oxide layer measured with ERDA and the desorption yield versus irradiation time showed a different slope; see Figure 10.15 [74]. This proves that the sputtered content of the oxide layer is not the dominating source of the desorbed gas. However, the aforesaid is a hint that the oxide layer influences the desorption behaviour, as vacuum-fired stainless steel had a thicker oxide layer with two times more oxygen in the ERDA spectrum.

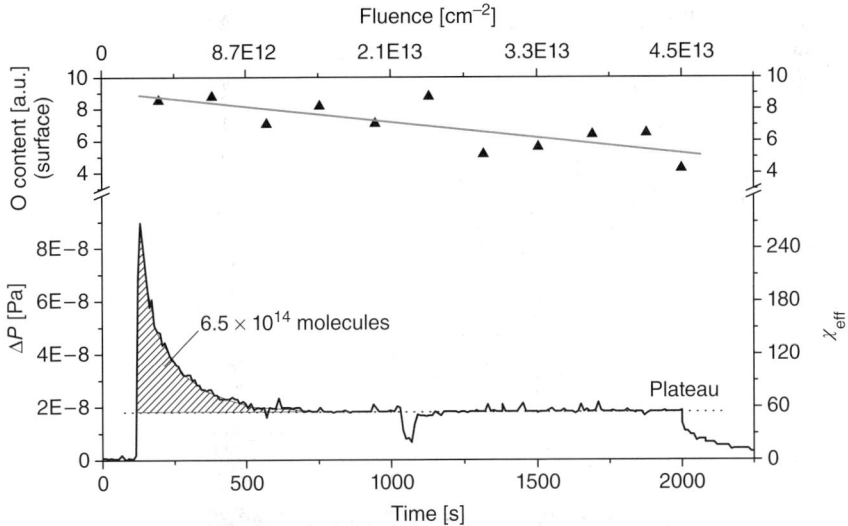

Figure 10.15 Evolution of the oxide layer thickness in arbitrary units (top) and the desorption yield (bottom), both versus irradiation fluence and irradiation time, respectively. Source: Bender et al. 2006 [74], Fig. 2. Reprinted with permission of CERN.

Mahner et al. measured cleaned and coated stainless steel vacuum chambers with 4.2 MeV/u lead ions [34]. The irradiation was performed under grazing incidence of 89.2° to the surface normal. This suspends the results to be compared to the above described results. However, this geometry represents the beam loss in the accelerator better. The results range from $\chi \geq$ some 10 molecules/ion for *activated* TiZrV-coated sample tubes to 10 000 molecules/ion for glow discharge-cleaned tubes [32, 34]. Electropolishing of the uncoated sample tubes reduced the desorption yield to 5000 molecules/ion, still significantly higher than for the TiZrV-coated tubes [34]. Also here, vacuum firing and glow discharge increase the thickness of the surface oxide layer as measured with ERDA [36, 40, 75].

Copper In contrast to stainless steel, copper is easily available and machinable at a purity of 99.99%. With this oxygen-free high-conductivity copper, the influence of the oxide layer can be studied in detail without a contribution of the bulk. Two samples with different surface properties have been studied, one with a pure copper surface, obtained by etching in HNO_3 and subsequent storing in an argon atmosphere before installation with minimal exposure to air. Another sample was polished and cleaned in propanol with a subsequent heating in atmosphere at 200 °C. This sample had a shiny dark surface. From the ERDA measurements, it was perceptible that this sample had a 100 nm layer of Cu_2O, in contrast to the pure copper sample, which had less than 3% oxygen in the first 20 nm from the surface [35, 59]. The pure sample, however, contained much hydrogen, which arose most probably from the acid treatment. A quantification of hydrogen was not possible with the available ERDA set-up.

The desorption yield of these two samples measured with 1.4 MeV/u Xe^{21+} ions resulted in $\chi = 360$ molecules/ion for pure copper. The desorbed gas consisted mainly of hydrogen. The sample with the oxide layer showed a desorption yield of $\chi = 1530$ molecules/ion with predominately CO and CO_2, which were both 1 order of magnitude higher than for the pure copper sample. In contrast, for the oxide sample, hydrogen was similar in height. After an irradiation fluence of slightly above 1×10^{13} ions/cm², the desorption yield decreased to 25 molecules/ion for the pure and 80 molecules/ion for the oxidised copper

This result is a clear indication that the electrical and, hence, the thermal conductivity of the material influences its desorption behaviour.

High-Melting Materials Since the beginning of dedicated desorption yield studies around year 2000, it has been discussed that beam ions might transfer sufficient energy to melt the target along their flight path. Therefore, high-melting-point materials were investigated concerning their desorption behaviour. The obtained data is still unpublished; however, a brief overview can be given.

Targets have been made out of niobium, molybdenum, tantalum, tungsten, and rhenium, which were irradiated perpendicular with 1.4 MeV/u zinc ions. Some of the targets have been coated by thin films of silver, gold, palladium, and NEG (consisting of titanium, zirconium, and vanadium). The results showed that there is no evidence of an effect of the high melting point on the ion-induced desorption yield. All measured yields were in the range of ±25% identical with the yields of stainless steel.

Among the coatings, however, all NEG-coated samples (*saturated*) showed a factor of 2 higher yield and the palladium-coated samples even had a factor of six to seven higher yield [68]. In both cases the higher yield was only determined by hydrogen desorption, which points to the high hydrogen gettering capability of the respective coating material.

10.3.4.2 Surface Coatings

Noble Metal Coatings To prevent oxidation and preserve high conductivity, noble metal coatings were applied to the copper samples. Gold and rhodium were tested and applied on etched oxygen-free, high conductivity copper with thicknesses between 250 nm and 1 µm, prepared by vapour deposition, sputtering, and galvanic deposition. As gold showed a high diffusion into the copper substrate upon the vacuum bakeout, a diffusion blocking layer of roughly 200 nm nickel was applied underneath the gold coating. In the case of rhodium, diffusion was not visible and, hence, a diffusion blocking layer was not applied. The rhodium coating of roughly 150 nm was sputter coated.

In the ERDA spectra, a difference in the purity of the different applied gold coatings was visible [74]. Both of the physical vapour deposition (PVD) methods resulted in very pure gold layers. On the other hand, the galvanic gold layer contained a total amount of 10% organic components such as C, N, and O. The first samples without diffusion barrier showed an intermixing of gold and copper [76], stemming from the UHV bakeout. The desorption yields for gold-coated copper, measured with 1.4 MeV/u Xe^{21+} ions, were 220 molecules/ion for evaporated, 155 molecules/ion for sputtered, and 240 molecules/ion for galvanic gold. Later the samples with a nickel diffusion barrier had a desorption yield of 90 molecules/ion initially and 25 molecules/ion after 1×10^{13} ions/cm^2 [74].

The intermixing of rhodium and copper was much less pronounced as compared to the gold samples. Only after heating the rhodium-coated samples up to 400 °C an inter-diffusion and a subsequent oxidation of the copper was visible with even up to 15% oxygen within 50 nm from the surface. The initial desorption yield was about 1050 for the rhodium-coated samples. After roughly 6×10^{12} ions/cm^2, the desorption yield dropped to 350 molecules/ion.

Getter Coatings Getter coatings such as NEG consisting of titanium, zirconium, and vanadium provide very high pumping speeds for chemical active gases. However, the gettering capability also increases the amount of available gas molecules that are bound at surface-close regions of the sample and are suspected to be released upon irradiation.

Only one qualitative measurement was performed where activated and saturated getter films were compared concerning their desorption behaviour [34]. The set-up consisted out of an NEG-coated tube and end flange that were irradiated. To determine the desorption yield, the pumping speed of the getter has to be considered and is extensively estimated in the cited work. As an example, the pumping speed for activated TiZrV getters is few 10 m^3/(s·m^2) [69].

Generally, the obtained results suggest that the desorption yield of saturated NEG is two to four times higher than for freshly activated NEG. The increase of the desorption yield within this comparison is represented exclusively by CO, which was also the saturation gas. This states the presumption mentioned earlier. Some examples of surfaces coatings are included in Table 10.1.

10.3.4.3 Cleaning Methods

To obtain clean surfaces, different techniques have been tested. The most representative collection of tested cleaning methods is published by Mahner [36]. Tubes made out of 316LN were tested untreated, polished, and cleaned by glow discharge. The chemical polished and electropolished tubes all showed a similar pressure rise of $4.1\text{--}6.7 \times 10^{-6}$ Pa when irradiated with 4.2 MeV/u lead ions under grazing incidence, corresponding to $\chi = 6200\text{--}10\,000$ molecules/ion. This is about a factor of 2 better than the untreated tube. However, the argon glow discharge-cleaned tubes showed a pressure increase between 1.5 and 3×10^{-5} Pa (χ up to 62 000 molecules/ion), while the He–O_2 glow discharge-cleaned tube behaved similar to the untreated tube.

Table 10.1 Pressure rise and related desorption yields of differently cleaned and coated target tubes, irradiated with 4.2 MeV/u lead ions under grazing incidence

Sample number	Vacuum chamber	Δp [Pa]	χ_{total} [molecules/ion]
1	TiZrV (1.5 µm sputtered) on 316 LN	3.87×10^{-8}	58
2	St707 (getter strips)	1.60×10^{-7}	240
3	Pd (0.6 µm sputtered) on 316 LN	1.73×10^{-7}	260
4	TiZrV (1.5 µm sputtered) on 316 LN	1.73×10^{-7}	260
5	Ag (2 µm galvanic) on 316 LN	8.27×10^{-7}	1 240
6	Au (30 µm galvanic) on 316 LN	9.07×10^{-7}	1 360
7	304 L	2.13×10^{-6}	3 200
8	316 LN (vented after scrubbing)	3.60×10^{-6}	5 400
9	316 LN (50 µm electropolished)	4.40×10^{-6}	6 600
10	316 LN (50 µm chem. polished)	4.13×10^{-6}	6 200
11	316 LN (as No. 10, getter purified)	4.67×10^{-6}	7 000
12	316 LN (150 µm electropolished)	6.27×10^{-6}	9 400
13	316 LN (50 µm electropolished)	6.67×10^{-6}	10 000
14	316 LN (LEIR type, not polished)	1.33×10^{-5}	20 000
15	316 LN (He–O_2 glow discharged)	1.12×10^{-5}	16 800
16	Al	1.33×10^{-5}	20 000
17	Cu	1.47×10^{-5}	22 000
18	Mo (127 µm foil)	1.60×10^{-5}	24 000
19	316 LN (Ar–O_2 glow discharged)	1.87×10^{-5}	28 000
20	Si (0.4 µm evaporated) on 316 LN	2.40×10^{-5}	36 000
21	316 LN (Ar–O_2 glow discharged)	4.13×10^{-5}	62 000

Source: Mahner 2008 [36]. Reproduced with permission of American Physical Society.

With the same experimental campaign, a technique called getter purifying was tested [36, 77]. That is, the sample is coated by a getter film, e.g. TiZrV, and subsequently the getter is activated. Afterwards the getter thin film is removed by etching. The idea behind this procedure is that the getter wrests gas form surface-close layers of the sample and thus cleans the sample, which leads to lower amounts of available gas when the getter film is removed afterwards. The results are summarised in Table 10.1. For more details and partial pressure distribution, refer to [32, 34, 36].

10.3.4.4 Energy Loss Scaling

At different energies, stainless steel and oxygen-free copper samples have been irradiated with different ion beams in a wide energy range. With exception of the irradiation with K ions, all experiments were performed under perpendicular incidence. Within the respective experimental series, the energy loss dE/dx inside of the material was varied via the ion energy (better velocity, compare also Eq. (10.20)). It was found that the desorption yields could be plotted in relation to the energy loss to the power of 2 [37, 78, 79]. Within the reported error bars of up to 30%, stemming mainly from the pressure measurement technique, all values obey the $(dE/dx)^2$ scaling. By implementing a scaling value k and writing the desorption yield as

$$\chi = k \left(\frac{dE}{dx} \right)^n \tag{10.18}$$

with $n = 2$, all values can be plotted together as shown in Figure 10.16 [67, 79, 80].

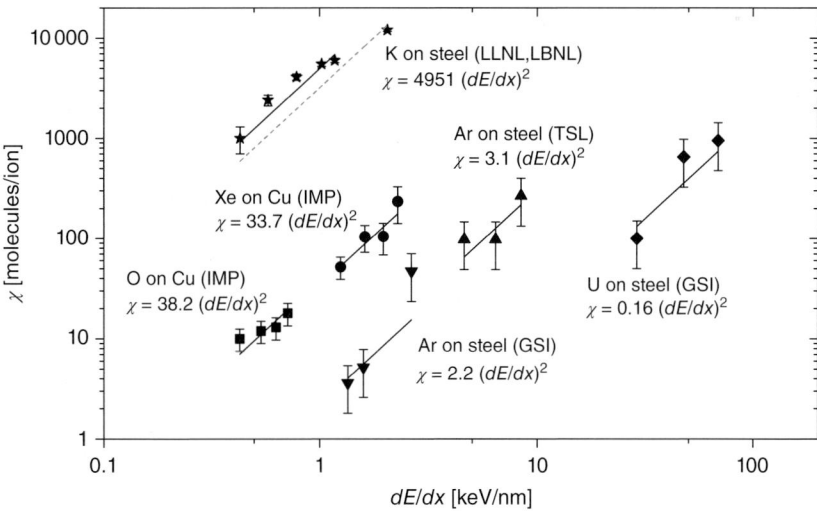

Figure 10.16 Quadratic scaling of the measured desorption yields with the electronic energy loss of the ions inside of the material. The dashed line on the K data is a fit including all data points. The best fit (solid line) is obtained excluding the value at $dE/dx = 2$ keV/nm as it was recorded in a separate experiment.

Also at low ion energies, a quadratic scaling with the electronic energy loss was found, even though, at least for xenon and potassium projectiles, the stopping power is dominated by elastic nuclear scattering; refer to Section 10.3.5.1. This finding states the significance of the electronic stopping mechanism for ion-induced desorption.

Maurer et al. also measured a quadratic energy loss scaling with uranium ions of few 100 MeV/u on gold-coated copper [65]. Here the scaling holds true for the target at ambient temperature, but for cryogenic targets no evidence for an energy loss scaling could be found so far [63].

10.3.4.5 Angle Dependence

In all irradiation experiments with ions in the mega-electron volts to giga-electron volts range, it was reported that grazing incidence leads to higher desorption yields as perpendicular impact [32, 37, 79]. Grazing incidence depicts the situation of beam loss ions impinging on a beam pipe under less than a few degrees. On the other hand, steeper angles like, 19° as used for the ERDA measurements, do not lead to higher desorption yields as compared with perpendicular incidence. Even though a dedicated investigation of the angular dependence of the desorption yields for ions impinging at grazing incidence was not performed so far with the exception of the reported experiments, three aspects have to be taken into consideration when discussing the angular dependence.

First, a technical sample has a remarkable rough surface with mean roughness values R_z of up to some micrometres. In the case of an ion hitting under grazing incidence, the ion can enter and leave the target several times on its flight path through the roughness hills and valleys as depicted in Figure 10.17a [81]. This leads to the desorption of gas from many surfaces with one and the same ion at one and the same time and also the pathway for diffusion is short within that structure. Second, regarding an ion beam of round shape with finite size, the irradiated area is increasing like the tangent of the irradiation angle α ($\Delta A \propto \tan \alpha$). The increased area offers an increase in surface adsorbates that can be released

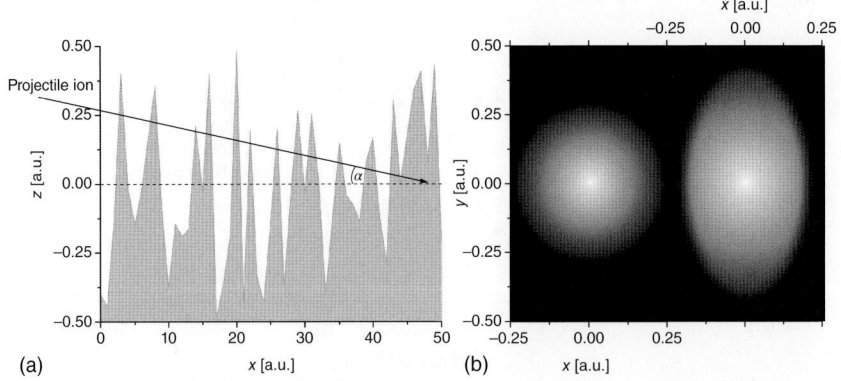

Figure 10.17 Effect of grazing incidence of the ion beam. (a) Due to rough surface, several centers of desorption contribute to the pressure increase. (b) Temperature (in arbitrary units) at surface as seen from above. The circular temperature distribution corresponds to perpendicular ion impact, while the elliptical distribution stems from a 45° impact (test simulation for 10 keV/nm energy loss on copper sample).

by the beam. However, this is also the case for steeper angles and should therefore lead to increased desorption yields whenever tilting the target away from perpendicular incidence. Moreover, with increasing area, the area density of ion impacts decreases and this, in turn, leads to lower energy density deposited into the target that should decrease the desorption yield.

Third, on grazing incidence, all of the ion energy is deposited into a surface-near region of the target. This means that a major area is heated up by the ion and a larger surface for desorption and a larger surface-near volume for diffusion are addressed; compare Figure 10.17b. Considering a transient heated region radially around the ion track, a tilted entrance angle of the projectile beam leads to an elliptically shaped temperature distribution on the sample surface.

It also has to be taken into account that, depending on the collision system, at grazing irradiation incidence a large percentage of projectile ions is scattered elastically in forward direction and collides with the environment in an uncontrolled manner [79].

10.3.4.6 Conditioning

The interaction of energetic ions with production targets or cryogenic surfaces requires the conditioning of the components prior to their application. Different conditioning methods have been tested and compared such as cleaning by the ion beam also called beam scrubbing, sputter cleaning of the surface by a 5 keV argon sputter gun, and heating of the target (Figure 10.18). It was found that cleaning by the ion beam itself is most ineffective. However, the data only consists of few

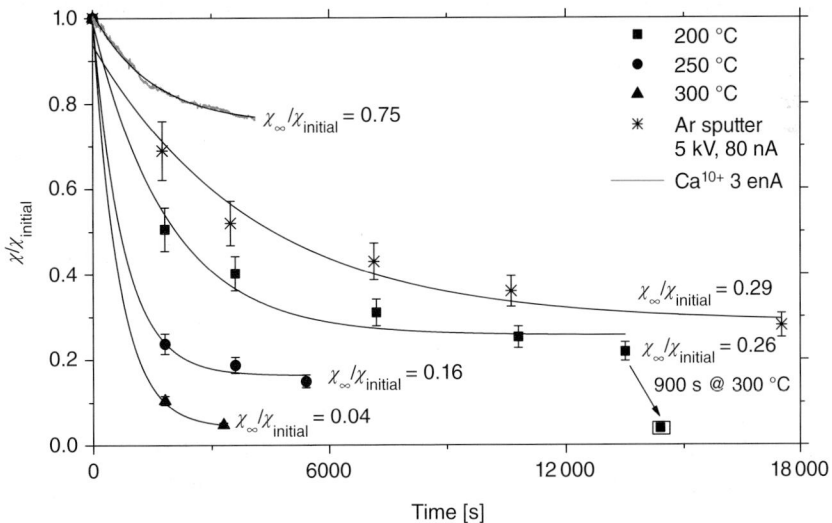

Figure 10.18 Comparison of different target conditioning methods to reduce ion-induced desorption, including beam scrubbing, sputter cleaning, and thermal annealing. enA is a notation used for multiple charged ions. Here, Ca is charged 10+, hence 3 enA equals 0.3 pnA (particle nA). For the Ar sputter, in the line above it is not necessary, as Ar is charged 1+ and enA equals pnA and is just denoted as nA. Source: Warth et al. 2016 [82], Fig. 1. Reprinted with permission of CERN.

irradiations, 1.4 MeV/u xenon on stainless steel [35, 74], 4.8 MeV/u calcium on gold-coated copper [82], and 4.2 MeV/u lead on coated steal tubes [36].

Sample cleaning by kilo-electron volts Ar ions can reduce the desorption yield. However, this procedure is complex and not very effective. Moreover, the surface of the treated component is changed significantly; in the case of the tested gold-coated copper, the gold coating was partially removed and after five hours of sputtering the surface appeared cuprous.

The most effective target conditioning can be achieved by thermal annealing. This procedure was already tested in 2007 with the rhodium-coated copper samples [40]. These samples were mounted on UHV-compatible sample heaters. One of the samples was heated up to 400 °C *in situ* before irradiation. The temperature was kept on 400 °C for roughly five minutes; however the complete heating cycle from room temperature and back took two hours. After heating, the sample was irradiated and a desorption yield of 260 was found, four times less as compared to the unbaked rhodium-coated sample.

A more detailed study was performed in 2015 with different annealing temperatures and treatment times. In this study the samples were annealed *in situ* in the UHV environment of the irradiation set-up [82]. It stated the obvious, that higher temperatures reduce the desorption yield faster and more effectively and longer annealing times lead to lower desorption yields. However, it was also found that the ultimate reduction of the desorption yield is dependent on the annealing temperature (see Figure 10.18). This means that in the example, the reduction that is achieved by 300 °C cannot be reached by 200 °C annealing in reasonable annealing time. For example, the desorption yield of a 200 °C annealed sample was reduced to 26% of the initial value after four hours of annealing. Another 15 minutes of subsequent annealing of the same sample at 300 °C reduced the yield further down to 4% of the initial value.

10.3.4.7 Cryogenic Targets

To investigate how important the surface coverage of cryogenic surfaces is, desorption experiments have been performed with cold targets at different temperatures, including dedicated application of different amounts of gas to be frozen on the surface.

The targets consisted of oxygen-free copper, pure, gold-coated, and coated with amorphous carbon. In all cases, the target was mounted onto a cold head closed-cycle cryostat where temperatures down to less than 10 K could be achieved, for what a thermal screen was needed. Assisted by DC-heaters any desired temperature between T_{min} and ambient temperature could be regulated. In the case of copper, one tubular target existed, called cold bore, that could be tilted to obtain grazing impact angles [83]. In all other cases, the irradiation was only possible under perpendicular incidence.

The pressure rise and therefore the desorption yield at T_{min} was in all cases roughly a factor of 15 lower than at ambient temperature, related to all residual gas species, except hydrogen, which only decreased by a factor of two [84]. As numbers, the desorption yield from the copper target ranged from $\chi = 500$ to 8000 molecules/ion between 6.3 K and ambient temperature and for gold-coated copper from $\chi = 400$ to 6000 molecules/ion when irradiated with 4.2 MeV/u

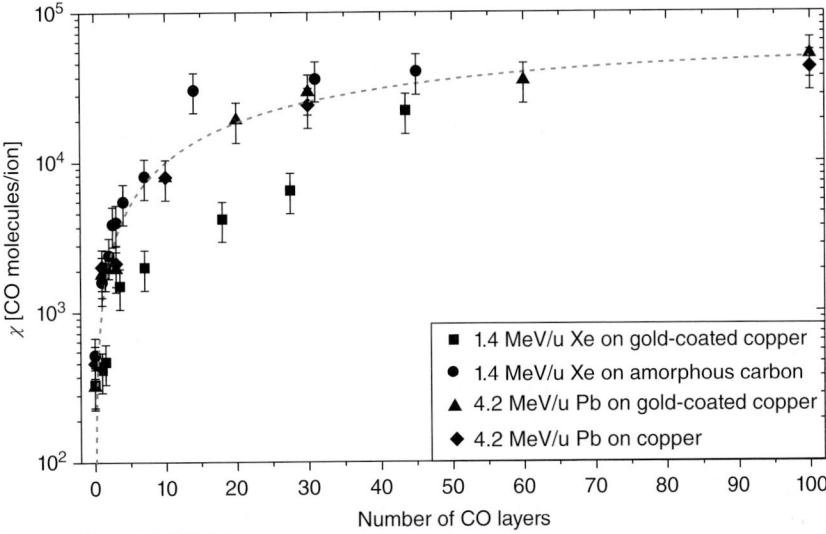

Figure 10.19 Dependence of the desorption yield on the CO surface coverage of a cryogenic sample. In all investigations, the yield saturates for high number of monolayers. Line to guide the eye. Source: Holzer et al. 2013 [85]. Reproduced with permission of American Physical Society.

lead ions [84]. An increase of the desorption yield with temperature was also observed in [65], during the irradiation of gold-coated copper with uranium ions at few 100 MeV/u. Here, the desorption yield increased from $\eta \approx 250$ to ≈ 3000 molecules/ion between 50 K and ambient temperature.

At the lowest achievable temperature, 6.3 K, an increasing number of CO layers was grown on the cryo target to test the impact of the amount of adsorbed gas on the desorption yield [85]. The results are displayed in Figure 10.19. Starting from 1 ML, for each irradiation a thicker layer was grown by the introduction of gaseous CO and the new layer was irradiated again. Finally, up to 300 ML were grown. The result was that the increasing number of CO layers increases the desorption yield from $\chi = 500$ CO/ion for 1 ML to $\chi = 50.000$ CO/ion for 40 ML regardless if irradiated with 1.4 MeV/u xenon or 4.2 MeV/u lead ions. The desorption yield does not further increase when the surface coverage exceeds about 40 ML. This result is of importance for the start-up of new cryogenic accelerators, e.g. to determine the maximum pressure at which the cool-down can be started.

10.3.5 Theoretic

10.3.5.1 Interaction of Ions with Matter

The interaction of ions with matter is characterised by their slowing down due to energy loss while penetrating the target. Basically, two slowing down mechanisms must be distinguished, nuclear and electronic stopping, whereas the type of mechanism is dependent on the ion velocity. The kinetic energy is transferred

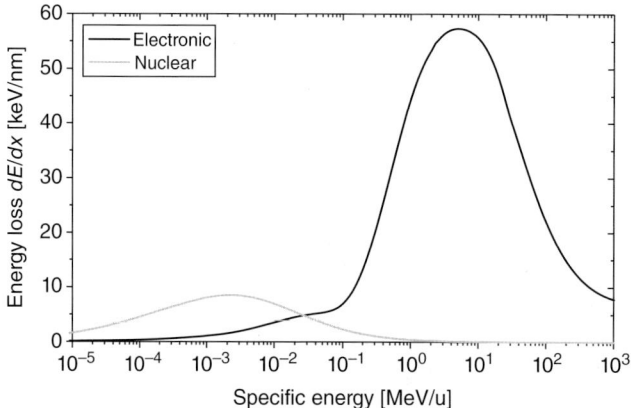

Figure 10.20 Energy loss of gold ions in stainless steel as a function of their specific energy.

to the target where several reactions can occur, depending on the stopping mechanism and the target material. Figure 10.20 shows exemplarily the amount of transferred energy per path length [keV/nm] versus the ion energy for gold ions in stainless steel [48]. It can be seen that lower energetic ions are slowed down in the nuclear stopping regime, whereas higher energetic ions are slowed down in the electronic energy loss regime. A rough comparison shows that the energy transfer in the electronic energy loss regime is 1 order of magnitude higher than in the nuclear energy loss.

Nuclear stopping is actually an improper chosen term. This mechanism occurs at collision energies below 1 MeV/u, where nuclear reactions are virtually impossible [86]. The term rather originates from the fact that only elastic two-body collisions between the ion and the target nuclei occur. The energy transfer in a two-body collision can be described by Rutherford mechanics at the basis of Coulomb forces. It is only dependent on the ratio between the ions mass M_1 and the mass of the target nucleus M_2 as well as on the scattering angle ϕ as can be seen in Eq. (10.19), which describes the ratio of the energy of a projectile before (E_0) and after (E_1) the collision. The scattering angle depends on the smallest distance of the projectile ion from the target nucleus, the so-called impact parameter, and is of statistic nature [41, 44]:

$$\frac{E_1}{E_0} = \left(\frac{\sqrt{(M_2^2 - M_1^2 \sin^2 \phi)} + M_1 \cos \phi}{M_1 + M_2} \right)^2 \tag{10.19}$$

In reality, many two-body collisions with the ion occur and subsequent collisions of accelerated target atoms with other target atoms result in collision cascades that distribute the kinetic energy along and around the ion path. The result of nuclear stopping inside the target is the creation of point defects, e.g. voids, where an atom has been removed from its lattice site, or interstitials, where the ion is stopped outboard a lattice site. Point defects change the properties of matter in many ways. For example, they lead to mechanical stress within the lattice planes, which is applied material hardening [87–89]. Other examples are the modification of electrical and optical properties, such as the conductance or coloring of the material. If a collision cascade is directed towards the surface and the remaining kinetic energy of the cascade atoms is sufficiently high, target atoms can leave the surface towards the vacuum. This mechanism is typically called sputtering and leads to erosion of the target over high irradiation fluences [90]. Sputtering yields are of the amount of ≤10 [28, 29]. They contribute to the particle stream into the vacuum system similar to gas flow, e.g. out of a vacuum leak. However, this contribution is hard to quantify as most sputtered target atoms are predominately of a non-volatile kind and their sticking probability to the vessel wall is very high [91]. Hence, their residence time in the vacuum is rather short. Nevertheless, these particles can lead to scattering and charge exchange of the primary beam in a particle accelerator.

Electronic stopping is the dominating stopping mechanism for ions at energies above few 100 keV/u [48]. The term stands for the fact that the ion beam predominately interacts with the electrons of the target due to Coulomb forces. The deposited energy is transferred by excitation of the target electrons and ionisation of the target atoms. The amount of deposited energy per path length dE/dx can be described by the Bethe–Bloch formula, which is given here in a very basic form [46]:

$$-\frac{dE}{dx} = \frac{4\pi n Z_{\text{eff}}^2}{m_e v^2} \left(\frac{e^2}{4\pi \epsilon_0} \right)^2 \ln \left(\frac{2 m_e v^2}{I} \right) \tag{10.20}$$

where n denotes the electron density and I the mean ionisation potential of the material wherein the ion is slowed down. From the equation, it can be found that the energy deposition depends mainly on the velocity v and the effective charge Z_{eff} of the projectile [92].

10.3.5.2 Inelastic Thermal Spike Model

After an ion collision in the electronic stopping regime, the electrons carry the energy radially away from the track, leaving a cylindrical space charge zone around the ion trajectory. At that time of about 1×10^{-14} s, all energy is stored in the electronic subsystem of the target, whereas the lattice of the material is still cold. Two competing models describe the subsequent behaviour of the target material. In the Coulomb explosion model, damage can occur if the space charge is not moderated fast enough [93]. This might be the leading damage creation process in insulators where electrons are immobile, but for metals it is unlikely due to the high electron mobility. However, there exist only few quantitative approaches for Coulomb explosion calculation. The other model describes the moderation of energy in the electronic subsystem via electron–phonon coupling. Coupling of energy from the electrons to the lattice leads to fast heating of the track region, eventually to temperatures higher than the melting or even evaporation temperature, depending on the material. The molten region freezes subsequently, which leads to a latent track, equivalent to radiation damage. Even below melting point, the enhanced temperature can lead to desorption of gas and diffusion out of the track several nanometres in radius around the ion

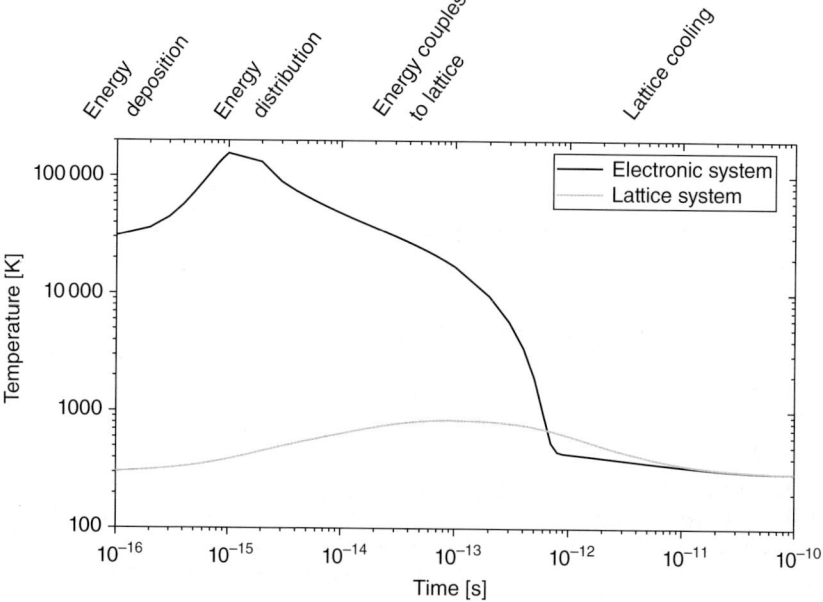

Figure 10.21 Temperature of the electronic and lattice system versus time. This plot shows the temperature in the very centre of the track, at radius = 0 nm. The calculation was done for 1.4 MeV/u xenon in copper.

impact. This contributes to the particle density inside the vacuum vessel and increases the pressure.

The earlier described mechanism of a single ion heating a track of several nanometres is called 'inelastic thermal spike model' [94]. In summary, the processes as shown in Figure 10.21 can be written as [95]:

- 1×10^{-17} to 1×10^{-16} s: Energy deposition, generation of delta electrons. Lattice cold.
- 1×10^{-15} to 1×10^{-14} s: Energy distribution by electron cascade. Lattice still cold.
- 1×10^{-13} to 1×10^{-12} s: Electron–phonon coupling. Lattice heating.
- 1×10^{-12} to 1×10^{-10} s: Lattice cooling. Cooling speed depends on the material.

Mathematically, the above described processes are a two-temperature model, which treats the electronic and the lattice subsystem independently, whereas the energy is transferred by a coupling term from one to the other system [96]. Both of the equations, for the electrons as well as for the lattice, are basically the classical heat flow equation in radial geometry at a constant volume [97], extended by a source term ($A(r, t)$ and $B(r, t)$, respectively) which is determined by the energy loss and the coupling term $g(\Delta T)$. They read as

$$C_e(T_e)\frac{\partial T_e(r,t)}{\partial t} = \frac{1}{r}\frac{\partial}{\partial r}\left(rK_e(T_e)\frac{\partial T_e(r,t)}{\partial r}\right) - g(T_e - T_a) + A(r,t)$$

(10.21a)

and

$$C_a(T_a)\frac{\partial T_a(r,t)}{\partial t} = -\frac{1}{r}\frac{\partial}{\partial r}\left(rK_a(T_a)\frac{\partial T_a(r,t)}{\partial r}\right) + g(T_e - T_a) + B(r,t)$$

(10.21b)

where T_e and T_a represent the temperature of the electronic and the lattice system, respectively. $K_{e/a}$ denotes the thermal conductivity and $C_{e/a}$ the specific heat. If the electronic system has a higher temperature than the lattice, energy is coupled to the lattice system, heating it up. If, in contrast, the lattice system has a high temperature, energy is coupled onto the electrons which thereby moderate the energy and cool the lattice. Solving the equations results in a time- and position-dependent temperature profile radially around the ion impact. The temperatures for the electronic subsystem and for the lattice system are shown versus radius and time in Figure 10.22. Note that the temperature of the electronic subsystem is orders of magnitude higher than the temperature of the lattice system.

The transient increase of the temperature leads to enhanced thermal desorption of gas of the amount s bound to the surface with the desorption energy E_{des} in terms of [98]:

$$\chi = 2\pi \int_0^{t_{max}} \int_0^{r_{max}} v_0(T(r,t))s(r,t)e^{\left(-\frac{E_{des}}{k_B T(r,t)}\right)} r\, dr\, dt$$

(10.22)

with the radius r, whereas the integration limits for the time t_{max} and for the space r_{max} are chosen according to the simulation results. Figure 10.23 shows a differential desorption yield calculation using Eq. (10.22) for a 1.4 MeV/u xenon impact on copper. Integrating the calculated result over space (in circular geometry) and time leads to the total amount of desorbed gas per incident ion.

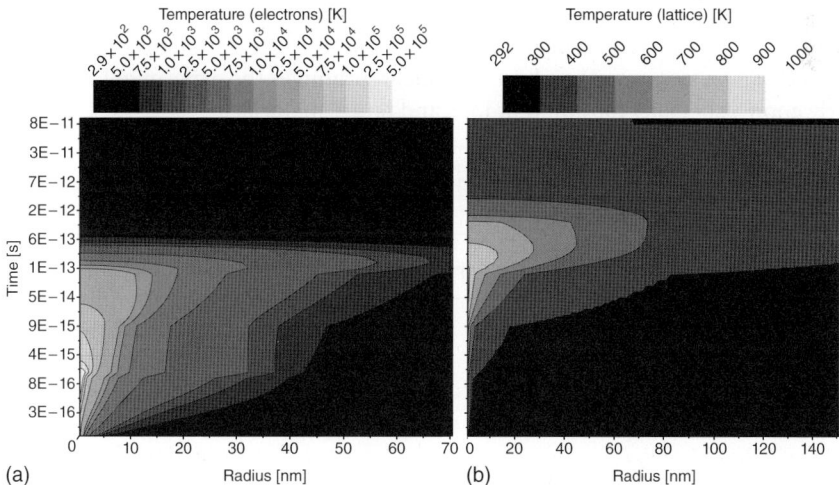

Figure 10.22 (a) Temperature of the electronic subsystem versus radius and time. (b) Temperature of the lattice system versus radius and time. The calculation was done for 1.4 MeV/u xenon in copper. Source: Reprinted with permission from Bender et al. [98], Figs. 1 and 2. Copyright 2009, Elsevier.

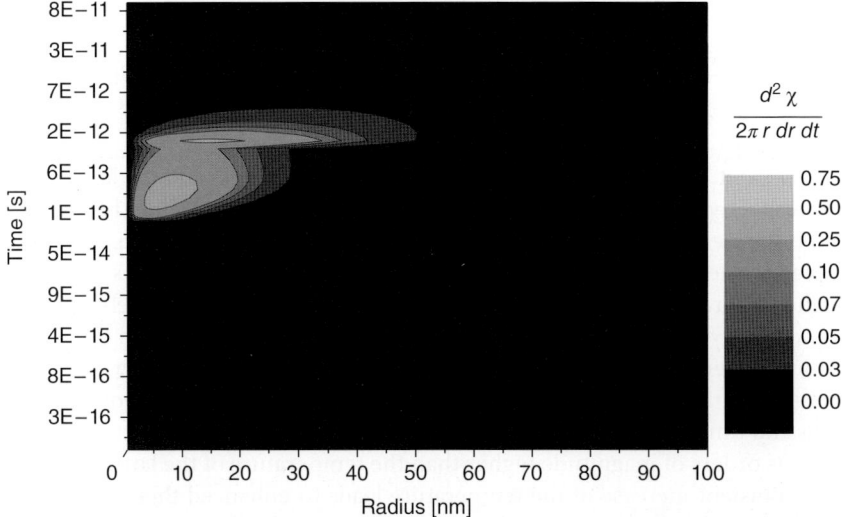

Figure 10.23 Differential desorption yield of the calculation in Figure 10.22 versus radius and time. The total amount of desorbed gas per incident ion can be gathered by integrating over space and time. Source: Reprinted with permission from Bender et al. [98], Fig. 3. Copyright 2009, Elsevier.

The model as described above has been used to calculate different experimental collision systems. With constant boundary conditions such as 1 ML gas coverage ($s(r, 0) = 10^{19}\,\mathrm{m}^{-2}$) and $E_{\mathrm{des}} = 0.4$ eV binding energy, different ions, impacting on different targets with different energies have been calculated. The results are displayed in Table 10.2. In fact, the model is not *ab initio* and some uncertainties have to be admitted. These are the electron-phonon-coupling, the binding energy which influences the results very sensitively and the absolute surface coverage. However, these values have been kept constant for the calculation to compare different collision systems. Thereby, the model reproduces the experimental results fairly good, especially for different ions and target materials.

Table 10.2 Comparison of experimental and theoretic desorption yield values of different collision systems

Projectile	E [MeV/u]	Target	$\chi_{\mathrm{experiment}}$ [molecules/ion]	$\chi_{\mathrm{calculated}}$ [molecules/ion]
Xenon	1.4	Copper	290–360	185
Xenon	1.4	Gold	90	165
Xenon	1.4	Rhodium	915–1 286	3 400
Xenon	1.4	Cu_2O	1 530	10 000
Carbon	1.4	Copper	10	5
Chrome	1.4	Copper	150	40
Lead	1.4	Copper	800	525
Lead	4.2	Gold	800	675

10.4 Conclusion: Mitigation of Dynamic Vacuum Instabilities

From the gathered modelling results and experimental findings, some guidelines for the design of ion-accelerator vacuum systems can be given to minimise the effect of vacuum instabilities during operation. These suggestions should be considered as state of technology in 2019 and can possibly be completed or slightly modified in future. However, from the view of a vacuum or surface scientist, they seem to be reasonable.

As a common solution in heavy ion ring accelerators, dedicated catchers for lost ions have been used since 2004. These collimators catch charge-exchanged ions after the dipole magnet instead of allowing them to hit the vacuum chamber wall in an uncontrolled way. The beam loss occurs on a block that is installed such that charge-exchanged ions are captured under perpendicular incidence with a collimation efficiency close to 1, while the aperture for the main ion beam is unaffected. As constituted in all experiments with swift heavy ions, the desorption yield for irradiation under perpendicular incidence is lower as compared to grazing incidence. However, the effect of increasing the desorption yield is evident for irradiation angles of few degrees or less to the surface. As an advice to the vacuum designer, the impact angle should be as steep as possible, ideally perpendicular. This can possibly be achieved by structuring the surface or by the installation of catcher devices, such as collimators, located in the loss regions. For heavy ion accelerators, the loss regions are in a distance after a bending magnet where ions with modified m/q hit the vessel wall. Depending on the charge state of the ion, electron loss as well as electron capture can occur in different intensities. However, the predominating charge state modification encounters one electron ($k_q \pm 1$) and higher modifications are less intense.

In contrast to any other imaginable collision point in the accelerator, the collimators are tailored exclusively for lowest desorption yields. In experiments between 2003 and 2008, a clear indication was found that high conducting metallic targets show the lowest desorption yields and this finding could be confirmed by thermal spike calculations. Between metals, e.g. copper has a much higher conductivity as stainless steel and tungsten, and the conductivity of gold is similar to that of copper. As gold is a noble metal, an oxidation in vacuum over a long time or even on atmosphere can be excluded, in contrast to pure copper. Copper is chemically active and therefore it changes its surface properties over time, at least in poor vacuum environments. Hence, the best material for collimators and other devices interacting with the beam would be gold. For cost reduction, a gold layer of ≈ 1 µm on top of high-conductivity copper is adequate: The conductivity remains high and gold prevents the copper from growing an oxide layer. The disadvantage of a gold layer on top of copper is the diffusion of the metals into each other. Dedicated diffusion blocking layers of chrome or nickel (≈ 0.5 µm) can reduce the inter-diffusion. Another possibility is to have a gold layer that is thick enough to stop the ion inside this layer. Depending on the ion beam energy this can charge a gold layer of few 10 µm and has to be well reflected if it should be preferred. However, in that case the block itself can be made out of stainless steel instead of copper.

All collimator blocks are heatable independent of the vacuum system bakeout. Thereby, the blocks can be kept always on higher temperature than the rest of the vacuum system, especially during bakeout and cool-down, which shall keep the surface coverage low. Recent measurements have shown that dedicated pretreatments, most effectively vacuum bakeouts, can reduce the desorption yields significantly. If high temperature is not employable, a similar effect can be achieved over time, but below the result of higher temperature. After all, a proper pretreatment seems to be more important than the choice of the material or coating; hence also stainless steel components might be optimised for low desorption.

The distributed pumping speed of the vacuum system and especially the pumping speed in the vicinity of the collimator blocks must have the maximum achievable value. If restrictions in space impede the installation of booster pumps, getter films should be taken into account. Especially NEGs are recommendable due to their outstanding performance concerning pumping speed for active gases, gettering capacity and secondary electron yield. An activation of the residual or desorbed gas by the ion beam has to be considered, and the risk of storing activated gas in getter thin films has to be calculated. The same applies for cryogenic surfaces.

Another approach tested at GSI was a wedge-shaped collimator where the ions were captured under a flat angle of 11° [9]. The idea in that conception was to keep the desorbed gas away from the beam axis as the collision and subsequent desorption occurs on the far side of the collimator with a huge pumping speed facing that side. However, this test ended without success, because the fabrication and support of the collimator block was much more complicated and, furthermore, the measured desorption yield was about five times higher as compared to the block geometry, eventually due to the flat incident angle. Figure 10.24 shows a

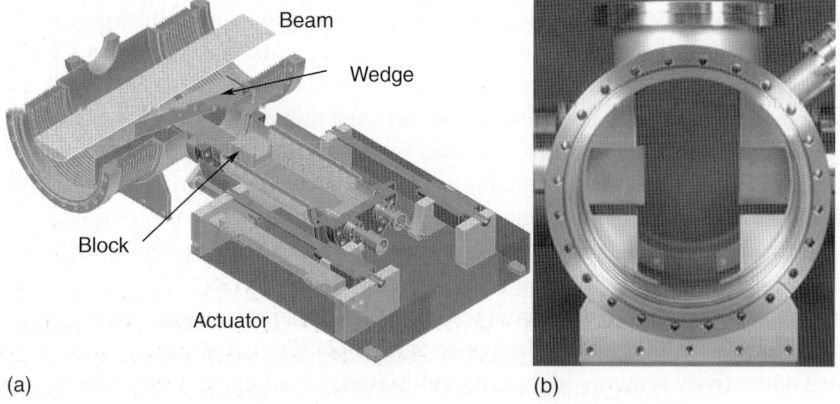

Figure 10.24 (a) Sketch drawing of a testing prototype of the collimators used at SIS18 at GSI, including a wedge and a block-shaped collector. Source: Omet et al. 2008 [9], Fig. 2. Reprinted with permission of CERN. (b) Photography of the finally installed collimators. Blocks are installed on both sides of the beam pipe to collect ions of both increased and decreased charge states.

schematic sketch of a test set-up and a photograph of the finally built beam loss collimators (block geometry) that were installed in SIS18 at GSI and in a similar form in LEIR at CERN. After the installation the achievable ion beam intensity was increased by over 1 order of magnitude [99].

Acknowledgement

The author gratefully acknowledges co-working with many people on a productive and pleasant basis. Edgar Mahner from CERN and the staff offered and organised numerous joint experiments at CERN and GSI including many fruitful discussions. Holger Kollmus, Andreas Krämer, Cristina Bellachioma, and Hartmut Reich-Sprenger from GSI supported and co-worked on a multitude of experiments and discussions. The valuable contribution of Stefan Wilfert to this manuscript is gratefully acknowledged. Being a GSI researcher himself, the author acknowledges the work of GSI, to mention the accelerator operation staff, Bettina Lommel and Birgit Kindler from the target laboratory, and the materials research group under Christina Trautmann as well as the technical staff, Arne Siegmund, and José Cavaco da Silva. The presented results are partially based on irradiations at the GSI Helmholtzzentrum fuer Schwerionenforschung, Darmstadt (Germany), in the frame of FAIR Phase-0.

The experimentation and expertise of Lars Westerberg and Emma Hedlund, TSL Schweden and Oleg Malyshev, Daresbury Laboratory, United Kingdom, is appreciated very much. Walter Assmann of Ludwig Maximilians University, Munich, provided the expertise in ion beam analysis. Marcel Toulemonde, Ganil, France, assisted with the inelastic thermal spike calculations. Discussions and data for modelling of the dynamic vacuum have been provided by Lars Bozyk and Edil Mustafin and are gratefully acknowledged. Finally the collegiate works of Alexander Warth, Verena Velthaus, and Leon Kirsch have expedited the research to complete the understanding of ion-induced desorption to the actual level.

References

1 Gröbner, O. and Calder, R.S. (1973). Beam induced gas desorption in the CERN intersecting storage rings. *Proceedings of the 1973 Particle Accelerator Conference*, San Francisco, CA, USA [(1973). IEEE Trans. Nucl. Sci. NS-20: 760]. https://doi.org/10.1109/TNS.1973.4327235.

2 DeLuca, W.H. (1969). Beam detection using residual gas ionization. *Proceedings of the 1969 Particle Accelerator Conference*, Washington, DC, USA [(1969). IEEE Trans. Nucl. Sci. NS-16: 813]. https://doi.org/10.1109/TNS.1969.4325373.

3 Calder, R., Fischer, E., Gröbner, O., and Jones, E. (1974). Vacuum Conditions for proton storage rings. *Proceedings of the 9th International Conference on High Energy Accelerators*, Stanford, CA, p. 70; (1974). Report No. CERN/ISR-VA/74-26.

4 Bosser, J., Carli, C., Chanel, M. et al. (1999). Experimental investigation of electron cooling and stacking of lead ions in a low energy accumulation ring. *Part. Accel.* 63: 171.

5 Zhang, S.Y. and Ahrens, L.A. (1998). Booster gold beam injection efficiency and beam loss. *Proceedings of the 6th European Particle Accelerator Conference*, Stockholm. London: IOP, p. 2149.

6 Zhang, S.Y. and Ahrens, L.A. (1999). Gold beam losses at the AGS booster injection. *Proceedings of the 18th Particle Accelerator Conference*, New York [New York, IEEE, 1999], p. 3294.

7 Krämer, A., Boine-Frankenheim, O., Hoffmann, I. et al. (2002). Measurement and calculation of U^{28+} beam lifetime in SIS. *Proceedings of the 8th European Particle Accelerator Conference*, Paris [EPS-IGA and CERN, Geneva, 2002], p. 2547.

8 Mustafin, E., Boine-Frankenheim, O., Hoffmann, I. et al. (2003). A theory of the beam loss-induced vacuum instability applied to the heavy-ion synchrotron SIS18. *Nucl. Instrum. Methods Phys. Res., Sect. A* 510: 199. https://doi.org/10.1016/S0168-9002(03)01811-4.

9 Omet, C., Kollmus, H., Reich-Sprenger, H., and Spiller, P. (2008). Ion catcher system for the stabilisation of the dynamic pressure in SIS18. *Proceedings of the 11th European Particle Accelerator Conference*, Genoa, 2008 [EPS-AG, Genoa, Italy, 2008], p. 295.

10 Fischer, W., Bai, M., Brennan, J.M. et al. (2002). Vacuum pressure rise with intense ion beams in RHIC. *Proceedings of the 8th European Particle Accelerator Conference*, Paris, p. 1485.

11 Fischer, W., Blaskiewicz, M., Brennan, J.M. et al. (2008). Electron cloud observations and cures in the relativistic heavy ion collider. *Phys. Rev. Spec. Top. Accel. Beams* 11: 041002. https://doi.org/10.1103/PhysRevSTAB.11.041002.

12 *FAIR Baseline Technical Report*. https://repository.gsi.de/record/54062/files/GSI-2013-04785.pdf (accessed 27 March 2018).

13 Lewitowicz, M. (2011). The SPIRAL2 project and experiments with high-intensity rare isotope beams. *J. Phys.: Conf. Ser.* 312: 052014. https://doi.org/10.1088/1742-6596/312/5/052014.

14 Yang, J.C., Xia, J.W., Xiao, G.Q. et al. (2013). High intensity heavy ion accelerator facility (HIAF) in China. *Nucl. Instrum. Methods Phys. Res., Sect. B.* 317: 263. https://doi.org/10.1016/j.nimb.2013.08.046.

15 Folger, H., Geißel, H., Hartmann, W. et al. (1991). Targets and degraders for relativistic heavy ions at GSI. *Nucl. Instrum. Methods Phys. Res., Sect. A* 303: 24. https://doi.org/10.1016/0168-9002(91)90759-J.

16 Kindler, B., Antalic, S., Burkhard, H.-G. et al. (2001). Status of the target development for the heavy element program. *AIP Conference Proceedings*, Volume 576, p. 1148. https://doi.org/10.1063/1.1395508.

17 Martinez-Darve, J., Artoos, K., Cruikshank, P. et al. (2001). Measurement of the mechanical behaviour of the LHC beam screen during a quench. *Proceedings of the 2001 Particle Accelerator Conference*, Chicago, p. 2162. https://doi.org/10.1109/PAC.2001.986467.

18 Gröbner, O. (2000). *The impact of synchrotron radiation in the LHC*, CERN-LHC/VAC, VLHC Workshop.

19 Omet, C. (2005). *Auslegung eines Kollimationssystems zur Lokalisierung von Umladungsverlusten und Beseitigung von Desorptionsgasen hochenergetischer, intensiver Schwerionenstrahlen in Ringbeschleunigeranlagen*. Diploma thesis. TU Darmstadt.

20 Puppel, P., Ratzinger, U., Bozyk, L., and Spiller, P. (2010). StrahlSim, A computer code for the simulation of charge exchange beam loss and dynamic vacuum in heavy ion synchrotrons. *Proceedings of the 1st International Particle Accelerator Conference*, Kyoto, Japan, [MOPEC058], p. 594.

21 Omet, C. (2008). *Kollimatorsystem zur Stabilisierung des dynamischen Restgasdruckes im Schwerionensynchrotron SIS18*. PhD thesis. TU Darmstadt.

22 Ziemann, V. (1992). *Vacuum Tracking*. Tech. Rep. SLAC-PUB-5962. Stanford, CA: Stanford University.

23 Ziemann, V. (2007). Vakdyn, a program to calculate time dependent pressure profiles. *Vacuum* 81: 866. https://doi.org/10.1016/j.vacuum.2006.10.003.

24 Bozyk, L., Chill, F., Litsarev, M.S. et al. (2016). *Nucl. Instrum. Methods Phys. Res., Sect. B* 372: 102. https://doi.org/10.1016/j.nimb.2016.01.047.

25 Spiller, P., Bozyk, L., and Puppel, P. (2011). SIS18 intensity record with intermediate charge state heavy ions. *Proceedings of the 2nd International Particle Accelerator Conference*, San Sebastian, Spain [WEPS003].

26 Hilleret, N. (1980). Variation of the ion induced desorption yields with temperature and the nature of incident ions. *Proceedings of the 8th International Vacuum Congress*, Cannes, CERN/ISR-VA/80-17.

27 Nafarizal, N. and Sasaki, K. (2007). Sticking probability of Ti atoms in magnetron sputtering deposition evaluated from the spatial distribution of Ti atom density. *J. Vac. Sci. Technol., A* 25: 308. https://doi.org/10.1116/1.2539256.

28 Assmann, W., Toulemonde, M., and Trautmann, C. (2007). Electronic sputtering with swift heavy ions. In: *Sputtering by Particle Bombardment*, Topics in Applied Physics, vol. 110 (ed. R. Behrisch and W. Eckstein), 401. Berlin: Springer-Verlag. https://doi.org/10.1007/978-3-540-44502-9_7.

29 Mieskes, H.D., Assmann, W., Grüner, F. et al. (2003). Electronic and nuclear thermal spike effects in sputtering of metals with energetic heavy ions. *Phys. Rev. B* 67: 155414. https://doi.org/10.1103/PhysRevB.67.155414.

30 Lozano, M.P. (2001). *Ion Induced Desorption Yield Measurements for Copper and Aluminium*. Vacuum Technical Note 01-05. LHC Vacuum Group.

31 Mahner, E., Hansen, J., Küchler, D. et al. (2002). Ion-stimulated gas desorption yields and their dependence on the surface preparation of stainless steel. *Proceedings of 2002 EPAC*, p. 2568.

32 Mahner, E., Hansen, J., Laurent, J.-M., and Madsen, N. (2003). Molecular desorption of stainless steel vacuum chambers irradiated with 4.2 MeV/u lead ions. *Phys. Rev. Spec. Top. Accel. Beams* 6: 013201. https://doi.org/10.1103/PhysRevSTAB.6.013201.

33 Mahner, E., Efthymiopoulos, I., Hansen, J. et al. (2004). Beam-loss induced pressure rise of Large Hadron Collider collimator materials irradiated with 158 GeV/u In^{49+} ions at the CERN super proton synchrotron. *Phys. Rev. Spec. Top. Accel. Beams* 7: 103202. https://doi.org/10.1103/PhysRevSTAB.7.103202.

34 Mahner, E., Hansen, J., Küchler, D. et al. (2005). Ion-stimulated gas desorption yields of electropolished, chemically etched, and coated (Au, Ag, Pd, TiZrV) stainless steel vacuum chambers and St707 getter strips irradiated with 4.2 MeV/u lead ions. *Phys. Rev. Spec. Top. Accel. Beams* 8: 053201. https://doi.org/10.1103/PhysRevSTAB.8.053201.

35 Kollmus, H., Bender, M., Assmann, W. et al. (2008). Heavy ion-induced desorption investigations using UHV-ERDA. *Vacuum* 82: 402. https://doi.org/10.1016/j.vacuum.2007.07.062.

36 Mahner, E. (2008). Review of heavy-ion induced desorption studies for particle accelerators. *Phys. Rev. Spec. Top. Accel. Beams* 11: 104801. https://doi.org/10.1103/PhysRevSTAB.11.104801.

37 Kollmus, H., Bender, M., Bellachioma, M.C. et al. (2005). Measurements on ion-beam loss induced desorption at GSI. *AIP Conference Proceedings*, Volume 773, p. 207. https://doi.org/10.1063/1.1949529.

38 Mahner, E. (2005). *The Vacuum System of the Low Energy Ion Ring at CERN: Requirements, Design, and Challenges.* Report Number CERN-AT-2005-013-VAC. Geneva, Switzerland: CERN.

39 Mahner, E., Hansen, J., Küchler, D. et al. (2003). *Ion-Stimulated Gas Desorption Yields of Coated (Au, Ag, Pd) Stainless Steel Vacuum Chambers Irradiated with 4.2 MeV/u Lead Ions.* Divisional Report No. CERN AT/2003-6. Geneva, Switzerland: CERN.

40 Bender, M. (2008). *Untersuchung der Mechanismen Schwerioneninduzierter Desorption an Beschleunigerrelevanten Materialien.* PhD thesis. Germany: Frankfurt University.

41 Rubin, S., Passell, T.O., and Bailey, E. (1957). Chemical analysis of surfaces by nuclear methods. *Anal. Chem.* 29: 736. https://doi.org/10.1021/ac60125a001.

42 Chu, Chu, and Nicolet, Nicolet (eds.) (1978). *Backscattering Spectrometry.* New York, San Francisco, CA, London: Academic Press. https://doi.org/10.13140/RG.2.1.1948.0807.

43 Assmann, W., Hartung, P., Huber, H. et al. (1994). Setup for materials analysis with heavy ion beams at the Munich MP tandem. *Nucl. Instrum. Methods Phys. Res., Sect. B* 85: 726. https://doi.org/10.1016/0168-583X(94)95911-0.

44 L'Ecuyer, J., Brassard, C., Cardinal, C., and Terreault, B. (1978). The use of ^{6}Li and ^{35}Cl ion beams in surface analysis. *Nucl. Instrum. Methods* 149: 271.

45 Tirira, J., Serruys, Y., and Trocellier, P. (1996). *Forward Recoil Spectrometry.* New York, London: Plenum Press.

46 Bethe, H. (1930). Zur Theorie des Durchgangs schneller Korpuskularstrahlen durch Materie. *Ann. Phys.* 5: 325. https://doi.org/10.1002/andp.19303970303.

47 Assmann, W., Davies, J.A., Dollinger, G. et al. (1996). ERDA with very heavy ion beams. *Nucl. Instrum. Methods Phys. Res., Sect. B* 118: 242. https://doi.org/10.1016/0168-583X(95)01183-8.

48 Ziegler, J.F., Biersack, J.P., and Littmark, U. (1996). *The Stopping and Range of Ions in Solids.* New York: Pergamon Press. https://doi.org/10.1007/978-3-642-68779-2_5.

49 Geissel, H., Weick, H., Scheidenberger, C. et al. (2002). Experimental studies of heavy-ion slowing down in matter. *Nucl. Instrum. Methods Phys. Res., Sect. B* 195: 3. https://doi.org/10.1016/S0168-583X(02)01311-3.

50 Greife, U., Bishop, S., Buchmann, L. et al. (2004). Energy loss around the stopping power maximum of Ne, Mg and Na ions in hydrogen gas. *Nucl. Instrum. Methods Phys. Res., Sect. B* 217: 1. https://doi.org/10.1016/j.nimb.2003.09.042.

51 Frey, C.M., Dollinger, G., Bergmaier, A. et al. (1995). Charge state dependence of the stopping power of 1 Mev/A ^{58}Ni-ions in thin carbon foils. *Nucl. Instrum. Methods Phys. Res., Sect. B* 99: 205. https://doi.org/10.1016/0168-583X(95)00218-9.

52 Assmann, W., Huber, H., Steinhausen, Ch. et al. (1994). Elastic recoil detection analysis with heavy ions. *Nucl. Instrum. Methods Phys. Res., Sect. B* 89: 131. https://doi.org/10.1016/0168-583X(94)95159-4.

53 Doolittle, L.R. (1985). Algorithms for the rapid simulation of Rutherford backscattering spectra. *Nucl. Instrum. Methods Phys. Res., Sect. B* 9: 344. https://doi.org/10.1016/0168-583X(85)90762-1.

54 Doolittle, L.R. (1986). A semiautomatic algorithm for Rutherford backscattering analysis. *Nucl. Instrum. Methods Phys. Res., Sect. B* 15: 227. https://doi.org/10.1016/0168-583X(86)90291-0.

55 Bergmaier, A., Dollinger, G., Frey, C.M., and Faestermann, T. (1995). Quantitative elastic recoil detection (ERD). *Fresenius J. Anal. Chem.* 353: 582. https://doi.org/10.1007/BF00321328.

56 Jeynes, C., Barradas, N.P., Marriott, P.K. et al. (2003). Elemental thin film depth profiles by ion beam analysis using simulated annealing - a new tool. *J. Phys. D: Appl. Phys.* 36: R97. https://doi.org/10.1088/0022-3727/36/7/201.

57 Jeynes, C., Bailey, M.J., Bright, N.J. et al. (2012). Total IBA – where are we? *Nucl. Instrum. Methods Phys. Res., Sect. B* 271: 107. https://doi.org/10.1016/j.nimb.2011.09.020.

58 Chanel, M., Hansen, J., Laurent, J.-M. et al. (2001). Experimental investigation of impact induced molecular desorption by 4.2-MeV/u Pb ions. *Proceedings of the 19th Particle Accelerator Conference*, Chicago, IL: IEEE [IEEE, Piscataway, NJ, 2001], p. 2165. https://doi.org/10.1109/PAC.2001.987311.

59 Bender, M., Kollmus, H., and Assmann, W. (2007). Desorption yields of differently treated copper samples characterized with ERDA. *Nucl. Instrum. Methods Phys. Res., Sect. B* 256: 387. https://doi.org/10.1016/j.nimb.2006.12.101.

60 Hedlund, E., Westerberg, L., Malyshev, O.B. et al. (2008). A new test stand for heavy ion induced gas desorption measurements at TSL. *Nucl. Instrum. Methods Phys. Res., Sect. A* 586: 377. https://doi.org/10.1016/j.nima.2007.12.020.

61 Meinerzhagen, F., Breuer, L., Bukowska, H. et al. (2016). A new setup for the investigation of swift heavy ion induced particle emission and surface modifications. *Rev. Sci. Instrum.* 87: 013903. https://doi.org/10.1063/1.4939899.

62 Baake, O., Seidl, T., Hossain, U.H. et al. (2011). An apparatus for in situ spectroscopy of radiation damage of polymers by bombardment with high-energy heavy ions. *Rev. Sci. Instrum.* 82: 045103. https://doi.org/10.1063/1.3571301.

63 Bozyk, L., Hoffmann, D.H.H., Kollmus, H., and Spiller, P. (2016). Development of a cryocatcher prototype and measurement of cold desorption. *Laser Part. Beams* 34: 394. https://doi.org/10.1017/S0263034616000240.

64 Bozyk, L., Spiller, P., and Kollmus, H. (2012). Development of a cryocatcher-system for SIS100. *Proceedings of IPAC 2012*, New Orleans, LA, USA [THEPPB004].

65 Maurer, Ch., Hoffmann, D.H.H., Bozyk, L. et al. (2014). Heavy ion induced desorption measurements on cryogenic targets. *Proceedings of IPAC 2014*, Dresden, Germany, [MOPRI105]. https://doi.org/10.18429/JACoW-IPAC2014-MOPRI105.

66 Maurer, Ch., Hoffmann, D.H.H., Bozyk, L. et al. (2015). Simulation and experimental investigation of heavy ion induced desorption from cryogenic targets. *Proceeding of IPAC 2015*, Richmond, VA, USA [THPF010]. https://doi.org/10.18429/JACoW-IPAC2015-THPF010.

67 Dong, Z.Q., Li, P., Yang, J.C. et al. (2017). Measurements on the gas desorption yield of the oxygen-free copper irradiated with low-energy Xe^{10+} and O^+. *Nucl. Instrum. Methods Phys. Res., Sect. A* 870: 73. https://doi.org/10.1016/j.nima.2017.07.025.

68 Bender, M., Kollmus, H., Krämer, A., and Mahner, E. *Ion-induced desorption yield measurements of bare and coated stainless steel, Nb, Mo, Ta, W, and Re samples irradiated with 1.4 MeV/u Zn^{10+} ions at GSI*, to be published.

69 Bender, M., Kollmus, H., Bellachioma, M.C., and Assmann, W. (2010). UHV-ERDA investigation of NEG coatings. *Nucl. Instrum. Methods Phys. Res., Sect. B* 268: 1986. https://doi.org/10.1016/j.nimb.2010.02.114.

70 Hedlund, E., Malyshev, O.B., Westerberg, L. et al. (2009). Heavy-ion induced desorption of a TiZrV coated vacuum chamber bombarded with 5 MeV/u Ar^{8+} beam at grazing incidence. *J. Vac. Sci. Technol., A* 27: 139. https://doi.org/10.1116/1.3032914.

71 CERN (2013). *Stainless steel forged blanks for ultra-high vacuum applications: 1.4429*. CERN Technical Specification No. 1001 - Ed. 5.

72 CERN (2013). *Stainless steel sheets/plates for vacuum applications: 1.4306*. CERN Technical Specification No. 1004 - Ed. 6.

73 Sasaki, Y.T. (2007). Reducing SS 304/316 hydrogen outgassing to 2×10^{-15} torr l / cm^2 s. *J. Vac. Sci. Technol., A* 25: 1309. https://doi.org/10.1116/1.2734151.

74 Bender, M., Kollmus, H., and Assmann, W. (2006). Understanding of ion induced desorption using the ERDA technique. *Proceedings of the 2006 European Particle Accelerator Conference*, Edinburgh [TUPCH173 (2006)].

75 Assmann, W. (2006). *Discussion of experimental results from ERDA measurements*. Private communication.

76 Ravi, R. and Paul, A. (2012). Diffusion and growth mechanism of phases in the Pd-Sn system. *J. Mater. Sci. - Mater. Electron.* 23. https://doi.org/10.1007/s10854-012-0832-4.

77 Mahner, E. (2004). *Sample coatings for low desorption yields*. Private communication.

78 Kollmus, H., Bellachioma, M.C., Bender, M. et al. (2006). Vacuum issues and challenges of SIS18 upgrade at GSI. *Proceedings of the 2006 European Particle Accelerator Conference*, Edinburgh, 2006 [TUPCH174 (2006)].

79 Molvik, A.W., Kollmus, H., Mahner, E. et al. (2007). Heavy-ion-induced electronic desorption of gas from metals. *Phys. Rev. Lett.* 98: 064801. https://doi.org/10.1103/PhysRevLett.98.064801.

80 Kollmus, H., Krämer, A., Bender, M. et al. (2009). Energy scaling of the ion-induced desorption yield for perpendicular collisions of Ar and U with stainless steel in the energy range of 5 and 100MeV/u. *J. Vac. Sci. Technol., A* 27: 245. https://doi.org/10.1116/1.3065979.

81 Thieberger, P., Fischer, W., Hseuh, H. et al. (2004). Estimates for secondary electron emission and desorption yields in grazing collisions of gold ions with beam pipes in the BNL relativistic heavy ion collider: proposed mitigation. *Phys. Rev. Spec. Top. Accel. Beams* 7: 093201. https://doi.org/10.1103/PhysRevSTAB.7.093201.

82 Warth, A., Bender, M., Bozyk, L. et al. (2016). *Ion-Induced Desorption and Cleaning Processes*. GSI Annual Report 2015. https://doi.org/10.15120/GR-2016-1.

83 Mahner, E., Bender, M., and Kollmus, H. (2005). A new cold-bore experiment for heavy-ion induced molecular desorption studies at low temperatures: first results obtained at 300 K, 77 K, and 15 K. *AIP Conference Proceedings*, Volume 773, p. 219. https://doi.org/10.1063/1.1949532.

84 Mahner, E., Evans, L., Küchler, D. et al. (2011). Heavy-ion induced desorption yields of cryogenic surfaces bombarded with 4.2 MeV/u lead ions. *Phys. Rev. Spec. Top. Accel. Beams* 14: 050102. https://doi.org/10.1103/PhysRevSTAB.14.050102.

85 Holzer, D.P., Mahner, E., Kollmus, H. et al. (2013). Heavy-ion-induced desorption yields of cryogenic surfaces bombarded with 1.4 MeV/u xenon ions. *Phys. Rev. Spec. Top. Accel. Beams* 16: 083201. https://doi.org/10.1103/PhysRevSTAB.16.083201.

86 Qu, W.W., Zhang, G.L., Zhang, H.Q., and Wolski, R. (2014). Comparative studies of Coulomb barrier heights for nuclear models applied to sub-barrier fusion. *Phys. Rev. C* 90: 064603. https://doi.org/10.1103/PhysRevC.90.064603.

87 Riviere, J.-R. (1995). Radiation induced point defects and diffusion. In: *Application of Particle and Laser Beams in Materials Technology* (ed. P. Misaelides). 53. Netherlands: Springer. ISBN: 978-94-015-8459-3, https://doi.org/10.1007/978-94-015-8459-3.

88 Heald, P.T. and Speight, M.V. (1975). Point defect behaviour in irradiated materials. *Acta Metall.* 23: 1389. https://doi.org/10.1016/0001-6160(75)90148-0.

89 Franklin, A.D. (2013). Statistical thermodynamics of point defects in crystals. In: *Point Defects in Solids: General and Ionic Crystal* (ed. J.H. Crawford and L.M. Slifkin), 1–101. Springer Science & Business Media. https://doi.org/10.1007/978-1-4684-2970-1.

90 Sigmund, P. (1969). Theory of sputtering. I. Sputtering yield of amorphous and polycrystalline targets. *Phys. Rev.* 184: 383. https://doi.org/10.1103/PhysRevdoi.184.383.

91 Emmoth, B. and Bergsaker, H. (1988). Sticking of sputtered particles to different surfaces. *Nucl. Instrum. Methods Phys. Res., Sect. B* 33: 435. https://doi.org/10.1016/0168-583X(88)90601-5.

92 Ahlen, S.P. (1980). Theoretical and experimental aspects of the energy loss of relativistic heavily ionizing particles. *Rev. Mod. Phys.* 52: 121. https://doi.org/10.1103/RevModPhys.52.121.

93 Wien, K. (1989). Fast heavy ion induced desorption. *Radiat. Eff. Defects Solids* 109: 137. https://doi.org/10.1080/10420158908220529.

94 Toulemonde, M., Dufour, C., and Paumier, E. (2006). The ion-matter interaction with swift heavy ions in the light of inelastic thermal spike model. *Acta Phys. Pol. A* 109: 311. https://doi.org/10.12693/APhysPolA.109.311.

95 Miotello, A. and Kelly, R. (1997). Revisiting the thermal-spike concept in ion-surface interactions. *Nucl. Instrum. Methods Phys. Res., Sect. B* 122: 458. https://doi.org/10.1016/S0168-583X(96)00665-9.

96 Wang, Z.G., Dufour, C., Paumier, E., and Toulemonde, M. (1994). The S_e sensitivity of metals under swift-heavy-ion irradiation: a transient thermal process. *J. Phys. Condens. Matter* 6: 6733. https://doi.org/10.1088/0953-8984/6/34/006.

97 Vineyard, G.H. (1976). Thermal spikes and activated processes. *Radiat. Eff.* 29: 245. https://doi.org/10.1080/00337577608233050.

98 Bender, M., Kollmus, H., Reich-Sprenger, H. et al. (2009). An inelastic thermal spike model to calculate ion induced desorption yields. *Nucl. Instrum. Methods Phys. Res., Sect. B* 267: 885. https://doi.org/10.1016/j.nimb.2009.02.039.

99 Spiller, P.J., Bozyk, L., Puppel, P., and Stadlmann, J. (2009). Synchrotron operation with intermediate charge state heavy ion beams. *Proceedings of the 23th Particle Accelerator Conference, Vancouver, BC, Canada*, [TU6PFP063], 1430.

Index

a

absorbed photons 69–72, 76, 143, 144
accelerator vacuum chambers 175, 265
  amount of ideal gas 217
  effective pumping speed 227
  gas close volume 216–218
  gas density and pressure 216–217
  gas density evolution, equations for 422–425
  gas dynamics modelling 265
  gas flow rate 217, 218, 220
  materials for
    aluminium alloys 83–84
    ceramics 85–86
    copper and copper alloys 84–85
    elastomer 87
    glass 82, 85–87
    stainless steel 82–83
    titanium and titanium alloys 85
  molar and molecular mass, of common gases 219
  NEG coating to 188
  operating requirements 81–82
  pumping characteristics 221–222
  purpose of 81
  surface treatments
    air bake 106
    bakeout method 105–106
    cleaning procedure 102–105
    surface coatings 108–109
    vacuum firing 106–107
  thermal outgassing 87–101
  total and partial pressure 218
  vacuum conductance
    of long tubes 224–225
    in molecular flow regime 223
    orifice 224
    series and parallel connection of tubes 226
    of short tubes 225–226
  vacuum system with pump 223
  velocity of gas molecules 218–219
accelerator vacuum system design 25, 62, 142, 215, 246, 265
activated charcoal 281, 324, 333–336, 338, 340–343
adsorption equilibrium 88–90
adsorption isotherm 89–91, 270, 272, 273, 275, 276, 279, 282–284, 286, 324, 331, 332, 334, 335, 337
alumina 83, 85, 86, 96, 375
aluminium based absorbers 338–340
amorphous carbon coating 337–338, 369
angular flux 32, 34, 50
antechambers 18–20, 69, 153–155, 175, 176, 181, 186, 187, 208, 227, 230, 232, 233, 240, 251, 252, 357, 359, 361, 363, 365, 408, 445

b

beam chamber size lower limit 17
beam chamber size upper limit 17
beam conditioning 131, 149, 205, 208, 371, 380, 384, 403
beam–gas interaction
  background noise in detectors 8
  beam particle loss 8
  Bremsstrahlung radiation 10
  lifetime 8

beam–gas interaction (*contd.*)
  particle accelerators associated with 8
  radiation damage of instruments 10–11
  residual gas ionisation 9
  risk to personnel safety 11
  sensitive surface, contamination of 9–10
beam-induced electron multipacting (BIEM)
  effects on vacuum 376
  electron cloud build-up, in LHC beam pipe 350
  electron sources 356
  negative impacts 350
  observation in CERN Large Hadron Collider (LHC) 380, 381
  observation tools 386
  synchrotrons 408, 409
  vacuum chamber wall properties 382–386
  vacuum pressure 381–382
  vacuum stability 405–408
beam-induced pressure instability 471
beam lattice 14–15, 19, 25
beam scrubbing 131, 140, 151, 481, 489, 496, 497
beam vacuum system design
  beam lattice 14–15
  check list for 14
  data extrapolations 23
  distributed pumping layout 19–20
  experimental data interpretation 23
  experimental errors 23
  gas dynamics model 24
  limiting factors 17–18
  lumped pumping layout 19
  magnet design 17
  mechanical engineering 17
  number and size of pumps 20
  required mechanical aperture 15–16
  sources of residual gas 20
  thermal outgassing 20
  total outgassing rate 20, 21

vacuum chamber cross sections 18–19
vacuum modelling and design considerations 21, 22
Boyle–Mariotte law 217
Bremsstrahlung radiation 8, 10, 11, 208, 279, 357
Brunauer, Emmett, Teller (BET) equation 275, 276
  isotherm 90, 275, 276
bunched-beam heavy ion accelerators 472, 473
  dynamic vacuum processes of 473

C
carbon based adsorbers 333–337
carbon fibre material 334, 335
charged particle accelerators 1, 2, 7, 29, 269
chemisorption 273, 274
Clausius–Clapeyron law 93, 286, 288
COLDEX
  beam screen calorimetric system 396
  BS with saw-tooth, PSD of 326–328
  Cu beam screen, PSD for 324
  experimental facility 321–324
  secondary PSD yields 325–326
  temperature oscillations 329–331
  vacuum transient effect 328
conductance-limited pumping 228, 242, 475, 478
copper based absorbers 340–341
critical current 421, 433
  beam–gas ionisation cross sections 465
  for cryogenic vacuum chamber 426–427
  for infinity long vacuum chamber 427–428
  ion impact energy 465
  ion-stimulated desorption yields 465–466
  pumping speed 466
  for short vacuum chamber 431–434
  total error in 466

cryogenic vacuum chambers
  adsorption isotherms   273–279
  cryosorbers   331–341
  cryotrapping   279–281
  desorbed gas, temperature of
    318–321
  equilibrium pressure   273
  hydrogen gas   276
  operating temperature, selection of
    286–289
  photon induced molecular cracking,
    of cryosorbed gas   312–318
  photon-stimulated desorption
    process
    discovery of secondary   301–306
    experimental facility for   301
    initial considerations   300
    primary PSD yields   308–310
    PSD yields calculation   306–308
    secondary PSD yields   310–312
  physisorption   281–282
  pressure and gas density   269–272
  synchrotron radiation   289
    infinitely long vacuum chamber
      291–294
    short vacuum chamber   294–300
  types   290
cryosorbers for beam screen
  aluminium based   338–340
  amorphous carbon coating   337–338
  carbon
    activated charcoal   333–334
    carbon fibre material   334, 335
  considerations   341
  copper   340–341
  criteria for   333
  distributed   342
  laser ablation surface engineering
    341
cryosorption   273, 298, 331, 338, 341,
  343
cryotrapping   279–281

d
Dalton law   218
desorption yield measurements
  481–482

Diamond Light Source (DLS)   11, 24,
  232, 258–259
differential pumping, vacuum system
  with   265, 266
DR equation   276
DRK equation   276, 278, 284
dynamic pressure and beam lifetime
  calculations   476–479
dynamic vacuum instabilities,
  mitigation of   504–505

e
effective pumping speed   88, 98,
  109–111, 156, 226–228, 251,
  258–260, 296, 301, 328, 407, 425,
  426, 476, 482, 486, 487
elastic recoil detection analysis (ERDA)
  195, 483
electron cloud (E-cloud)
  build-up in LHC beam pipe   350
  electron sources   356
  models   351–356
  observation in CERN Large Hadron
    Collider   379
  observation tools   386–390
  problem   349
  synchrotrons   408, 409
  vacuum chamber wall properties
    382–386
  vacuum effects   376
  vacuum pressure   381–382
electron energy estimation   376
electron stimulated desorption (ESD)
  bakeout effect   122–123
  copper surface preparation   125
  definition   109, 110
  for different materials as function of
    dose   112–113
  estimation   378
  as function of amount of desorbed gas
    113–114
  as function of electron energy
    119–122
  measurements   109–111
  vs photon stimulated desorption
    144
  pumping duration effect   114–119

electron stimulated desorption (ESD) (*contd.*)
  surface polishing   123–125
  surface treatment   125
  vacuum chamber temperature   125–128, 142
  vacuum firing   125
electronic stopping   495, 499, 500
equivalent free path, of gas molecules   220

f

fast ion instability   9, 14
forward scattering photon reflectivity   64, 66, 71
Freundlich equation   274, 276–278

g

Ga–As photocathode lifetime   9
gas close volume   216–218
gas dynamics model
  cryogenic vacuum chambers under SR   289–300
  infinitely long vacuum chamber   291–294
  short vacuum chamber   294–300
  of vacuum system   24–25
gas flow rate   215, 217, 218, 220–223, 231–233, 241, 246, 248, 249, 252, 255
gas injection, into tubular vacuum chamber   241
getters   108, 109, 176, 177, 179, 188, 193, 195, 230, 466, 472, 475, 492, 493, 505
Glidcop®   85, 112
grazing incidence   63, 64, 66, 142, 143, 308, 472, 474, 478, 487, 489–491, 493–496, 504

h

He and H_2 isotherms on Cu plated stainless steel   278
heavy ion-induced desorption   480
  angle dependence   495–496
  CERN set-up   486, 487
  cleaning methods   493–494
  conditioning methods   496–497
  copper   491–492
  cryogenic targets   488–489, 497–499
  desorption yield measurements   481–482
  energy loss scaling   494–495
  experimental results   489–499
  GSI set-up   486–488
  high-melting materials   492
  IMP set-up   489
  inelastic thermal spike model   500–504
  ions interaction with matter   499–500
  materials analysis   483–485
  noble getter surface coatings   493
  noble metal surface coatings   492–493
  stainless steel materials   490
  TiZrV-coated sample tubes   491
  TSL set-up   488
Henry's law   273, 274
He pressure wave measurements   280
H_2 isotherms on bare surface   283
H_2 isotherms on Cu plated stainless steel   276–278, 283, 285, 286
H_2 isotherms on electro-polished stainless steel   286
H_2 isotherms on stainless steel   285
H_2 isotherms, temperature dependence of   283
hydrogen adsorption isotherms   270, 272, 286, 331, 332

i

ideal volumetric pumping speed   222
impingement rates   221, 248, 260–263
incident electron flux estimation   376–377
inelastic thermal spike model   500–504
infinitely long vacuum chambers   425
  critical current   444–445
  gas density
    critical current   426, 427
    cryogenic vacuum chamber   426–427
    quasi-static case   427

room temperature vacuum
    chamber   425–426
  stability criteria   428
  vacuum chamber with beam screen
    426
  vacuum chamber with pumping
    slots   425
  vacuum chamber without beam
    screen   426
  vacuum chamber without pumping
    slots   425
ion stimulated desorption on gas
    density   441–443
synchrotron radiation   291
  with holes in beam screen
    292–294
  without beam screen   292
  two-gas system   443
insulating materials   364, 374–376
intersecting storage ring (ISR) vacuum
    system, 471
ion desorption stability   421, 465
ion energy in vacuum chamber
  dipole magnetic field   461
  quadrupole magnetic field   461–462
  solenoid magnetic field   462–464
  without magnetic field
    circular beams   455–458
    flat beams   458–460
ion-induced desorption   421, 423–426,
    428, 441, 449, 453–455, 472–474,
    480–504
ion induced pressure instability
  infinitely long vacuum chamber
    425–428
  multi-gas system   437–438
  short vacuum chamber   428–437
  two-gas system   438–440
  VASCO code   447–455
ion stimulated desorption (ISD)   155
  bakeout effect and argon discharge
    cleaning   161
  for condensed gases   163
  for different materials   160
  vs. electron stimulated desorption
    161
  function of ion dose   156–158

function of ion energy   158
function of ion mass   159–160
function of temperature   161–163
on gas density
  infinitely long vacuum chambers
    441
  single gas estimation   446
  two gases system   443, 446
  vacuum chambers with given
    pumping speed   441–442
ion trapping instability   9, 14

k
Knudsen number   220
Kovar®   86

l
Langmuir isotherm   89, 90, 274, 276
Large Hadron Collider (LHC)
  BIEM and E-cloud observations
    379–405
  cryogenic vacuum system   401–405
laser ablation surface engineering
    (LASE)   341, 360
Lorentz equation   32

m
M-branch   488
metallisation brazing   85
molecular beaming effect
  GaAs photocathode electron gun
    262–264
  heavy ion stimulated induced
    desorption   264
Molflow+   58, 59, 245, 246
Monte Carlo methods   2, 154, 190, 245

n
non-evaporable getter (NEG) coated
    vacuum chambers   20, 176
  ASTeC activation procedure   185
  beam path   187
  CERN–BINP activation procedure
    183–184
  manufacturing cost   176
  nature of getter materials   176
  NEG coatings   177–179

non-evaporable getter (NEG) coated vacuum chambers (contd.)
  operation conditions 176
  synchrotron radiation 196–199
  ultimate pressures in 195–196
  use of 208–209
non-evaporable getter (NEG) coatings
  activation temperature 178
  activation procedure 182–188
  benefits 207–208
  classification of sorption pumps 177
  cylindrical magnetron deposition configuration 180
  electron stimulated desorption yield 204
  film characterisation 181–182
  gas diffusion barrier 178
  high pumping speed 178
  lifetime 193–195
  at low temperature 207
  magnetron sputtering 179
  planar magnetron deposition 179, 180
  primary and secondary electron yields 204–206
  pumping optimisation at ASTeC 190
  pumping optimisation at CERN 188
  pure metal alloy 178
  reducing particle/electron stimulated desorption from 200
  sticking probabilities 200
  surface resistance 206–207
  Ti–Zr–V films 189, 190
  vacuum chamber cross sections 186, 187
non-evaporable getter (NEG) pumps 177

o

one-dimensional gas diffusion model 228
  accelerator vacuum chamber section 231–232
  advantage 231
  closed loop boundary conditions 233, 235, 237
  continuous flow fluid dynamics 229
  gas balance 229
  gas injection, into tubular vacuum chamber 241
  global and local coordinates 238–240
  molecular beaming effect 259–264
  open end boundary conditions 233, 237
  and TPMC model 257, 258
  uncertainties 240
  vacuum chambers
    with known pressures 244–245
    with known pumping speeds at the ends 241–244
OrAnge SYnchrotron Suite (OASYS) 56
OSCARS 59
oxide dispersion strengthened (ODS) copper 85
oxygen-free high conductivity copper 84, 491, 492

p

particle accelerators
  associated with beam gas-interaction background noise detectors 8
  beam particle loss 8
  Bremsstrahlung radiation 10
  radiation damage of instruments 10–11
  residual gas ionisation 9
  risk to personnel safety 11
  sensitive surface, contamination of 9–10
  synchrotron radiation 29
  vacuum design objectives 12
  vacuum specifications 6–13
photoelectron energy distribution curves
  of Cu co-laminated on stainless steel 74
  of gold 72, 73
  modification 75
photoelectron production
  incidence angle effect 76

photon energy effect 72–75
total photoelectron yield 69–72
photon induced molecular cracking, of
 cryosorbed gas 312–318
photon reflectivity 61, 64
 copper mirror 62, 63
 diffuse and forward scattered
  reflectivity 64
 experimental set-up 64, 65
 flat and sawtooth copper surface 67, 68
 flat LHC copper surface 69
photon stimulated desorption (PSD) 128
 bakeout effect 137–140
 critical photon energy 136–137
 for different materials 131–134
 vs. electron stimulated desorption 144
 as function of amount of desorbed gas 135
 as function of photon dose 131
 gauge method 129
 incident angle effect 142–144
 in-depth studies of 321
 measurements 129, 130
 vacuum chamber temperature 141
 yield data use
  distributed and lump SR absorbers 153–155
  lump SR absorber 151–153
  scaling the photon dose 145
  SR photon flux 148–151
  synchrotron radiation from dipole magnets 145–148
physisorption 273
 vs. condensation 273
 cryotrapping 279–281
 on gas condensates 281–282
pressure instabilities, in heavy ion
 accelerators 472, 480
 angle dependence 495–496
 CERN set-up 486, 487
 cleaning methods 493–494
 closed system (vessel) 476–478
 conditioning methods 496–497
 copper 491–492

cryogenic targets 488, 497
desorption yield measurements 481–482
dynamic pressure and beam lifetime calculations 476–479
energy loss-scaling 494–495
experimental results 489–499
getter surface coatings 492–493
GSI set-up 486–488
heavy ion-induced desorption 480
high-melting materials 492
IMP set-up 489
inelastic thermal spike model 500–504
ion-optical simulation 478
ions interaction with matter 499–500
longitudinal vacuum profile 478–479
materials analysis 483–485
noble metal surface coatings 492–493
stainless steel materials 490–491
TiZrV-coated sample tubes 491
TSL set-up 488
pressure rise method 481

r
radiated energy and power density 31–32
Redhead's equation 101
residual gas, in vacuum chambers 79–81
ring accelerator, stable operation of 479
Rutherford back scattering (RBS) spectrometry 483

s
secondary electron emission
 incidence angle 374
 insulating materials 374–376
 re-diffused electrons 371–374
 reflected electrons 371–374
 surface material effect 368–369
 surface roughness effect 369–371
 surface treatment effect 367

secondary electron emission (*contd.*)
  true secondary electrons   371–374
secondary electron yield (SEY)
  as function of incident electron energy   367
  incidence angle   374
  insulating materials   374
  measurement method   365–367
  re-diffused electrons   371–374
  reflected electrons   371–374
  surface material effect   368–369
  surface roughness effect   369–371
  surface treatment effect   367
  true secondary electrons   371–374
short vacuum chambers   428
  with gas density at the ends   428–431
  with pumping speed at the ends   431–434
  synchrotron radiation   294
    given pressure at the ends   296–298
    given pumping speed at ends   298–300
  two-gas system   439, 440
  without beam screens   434–437
spectra   56–59, 73, 74, 100, 101, 388, 389, 437, 492
sputter ion pump (SIP)   18, 20, 175, 177, 230, 466
sputtering   74, 179, 481, 492, 497, 500
sticking coefficient   222, 475
sticking probability of sorbing surface   221, 273, 325, 399
storage rings, good vacuum in   8
StrahlSim code   478, 479
superconducting magnets   11, 17, 269, 279, 282, 300, 320, 321, 331, 390, 472
synchrotron radiation (SR)   128
  angular flux   32
  charged particle emission, in magnetic field   29–32
  from dipoles
    emission duration and critical energy   33–34

    photon flux   34–37
    photon power   39–41
  from insertion devices   43–55
    angular aperture   51–52
    K-factor   47
    method for estimation absorbed power   54–55
    motion of charged particles   43–45
    power collected by simple geometry aperture   54
    power distribution in wiggler   52–54
    resonant wavelength   45–46
    undulator mode   46–51
    wiggler mode   46–51
  from quadrupoles   42–43
  radiated energy and power density   31–32
synchrotron radiation-vacuum chamber wall interaction   61
  photoelectron production   69–76
  photon reflectivity   61–69
Synchrotron Radiation Workshop (SRW)   56
SYNRAD+   58–59, 246

t

test particle Monte-Carlo (TPMC) method   245
  code input   246–248
  code output
    gas density and pressure   250
    gas flow rate   248–249
    mass of molecules   251
    pump effective capture coefficient   251
    temperature effect   251
    transmission probability   250–251
    vacuum conductance   250–251
  input parameters, defined set of   252–253
  molecular beaming effect   259–264
  one-dimensional gas diffusion model   257–264
  result accuracy   256–257
  vacuum chamber in   246

with variable parameters   253–256
virtual pumping surfaces   252
thermal desorption   22, 87, 93, 94, 100, 101, 112, 130, 156, 175, 185, 230, 252, 253, 289, 308, 337–339, 376, 407, 423, 424, 447, 449, 473–475, 480, 502
thermal outgassing   87
  equilibrium pressure   89–91
  measurement methods
    conductance modulation method   98
    gas accumulation method   99–100
    thermal desorption spectroscopy   100–101
    throughput method   97–98
    two-path method   98–99
  during pumping   88
  rate of materials   93–97
  vapour pressure   91–93
thermal transpiration effect   270–272
titanium sublimation pump (TSP)   20, 177, 466
total beam lifetime   8
transmitted photons   144
turbo-molecular pumps (TMP)   20, 83, 186, 265–267, 270, 323, 466

u

ultra-high vacuum/extreme high vacuum (UHV/XHV) vacuum system design   175

v

vacuum
  definition   5
  ranges/degrees of   5
vacuum chamber design
  active methods   363–365
  active vs. passive mitigation methods   365
  chamber wall properties   382
  DAΦNE dipole   364
  with given pumping speed   441, 442
  insulating materials   374–376
  KEK   359
  observations with beams   394–401
  passive mitigation methods   357
  pressure variation   381
  surface properties at cryogenic temperature   391–394
  wiggler   359
vacuum conductance
  of long tubes   224, 225
  in molecular flow regime   223
  orifice   224
  series and parallel connection of tubes   226
  of short tubes   225, 226
Vacuum Plumber's Formulas   223
vacuum specifications   11–12
  for particle accelerators   6–13
vacuum system designer   6, 13, 95
VASCO code
  basic equation   447–448
  cylindrical geometry   447
  finite elements   447–448
  multi-gas model
    in matrix form   448–451
    vs. single gas model   452–455
  second order linear differential equation, transformation of   450–451
  set of equations   451–452
  time invariant parameters   447

x

Xe adsorption isotherms   331
xenon adsorption isotherm   332
X-ray Oriented Programs (XOP)   56